Lecture Notes in Production Engineering

Lecture Notes in Production Engineering (LNPE) is a book series that reports the latest research and developments in Production Engineering, comprising:

- Biomanufacturing
- Control and Management of Processes
- Cutting and Forming
- Design
- Life Cycle Engineering
- Machines and Systems
- Optimization
- Precision Engineering and Metrology
- Surfaces

LNPE publishes authored conference proceedings, contributed volumes and authored monographs that present cutting-edge research information as well as new perspectives on classical fields, while maintaining Springer's high standards of excellence. Also considered for publication are lecture notes and other related material of exceptionally high quality and interest. The subject matter should be original and timely, reporting the latest research and developments in all areas of production engineering. The target audience of LNPE consists of advanced level students, researchers, as well as industry professionals working at the forefront of their fields. Much like Springer's other Lecture Notes series, LNPE will be distributed through Springer's print and electronic publishing channels.

Indexed in Scopus* *Indexed in EI Compendex To submit a proposal or request further information please contact Anthony Doyle, Executive Editor, Springer (anthony.doyle@springer.com).

Ludger Overmeyer · Bernd-Arno Behrens
Editors

Production at the Leading Edge of Technology

Proceedings of the 15th Congress of the German Academic Association for Production Technology (WGP), Hannover, September 2025

Editors
Ludger Overmeyer
Institute of Transport and Automation Technology
Leibniz University Hannover
Garbsen, Niedersachsen, Germany

Bernd-Arno Behrens
Institute of Forming Technology and Machines
Leibniz University Hannover
Garbsen, Niedersachsen, Germany

ISSN 2194-0525 ISSN 2194-0533 (electronic)
Lecture Notes in Production Engineering
ISBN 978-3-032-19523-4 ISBN 978-3-032-19524-1 (eBook)
https://doi.org/10.1007/978-3-032-19524-1

This work was supported by Leibniz University Hannover, Institute of Transport and Automation Technology, Garbsen, Germany and this publication received financial support from the publication fund NiedersachsenOPEN, funded by zukunft.niedersachsen.

This Springer imprint is published by the registered company Springer Nature Switzerland AG
The registered company address is: Gewerbestrasse 11, 6330 Cham, Switzerland

Preface

On behalf of the German Academic Association for Production Technology (WGP), we were pleased to welcome researchers and guests to the WGP Annual Congress 2025 in Hannover. The congress was held from September 30 to October 2 at the Production Technology Center Hannover (PZH), where the Leibniz University Hannover unites its production-oriented institutes of mechanical engineering.

The WGP Annual Congress served as a platform for researchers from both academia and industry to present their latest findings following a rigorous peer-review process, culminating in the publication of their contributions in this conference proceedings.

The 2025 congress, entitled "New Impulses for Production—Intelligent, Efficient, Digital," addressed key areas in modern production research. These included sustainable manufacturing processes that promote resource efficiency and circular economy practices, state-of-the-art manufacturing technologies enhancing precision and productivity, the integration of artificial intelligence for smarter automation and data-driven decision-making, and digital transformation through IoT, cloud computing, advanced analytics, and digital twins. The contributions provided an overview of current trends in production research and gave the readership an insight into ongoing research conducted by the German Academic Association for Production Technology. They showcased how production research continuously pushed the boundaries of what was feasible, advancing methodologies, efficiency, and interdisciplinary collaboration.

The congress also provided a forum for dialogue between generations, fostering connections between established experts and emerging researchers. We hope all participants found the congress stimulating, inspiring, and an opportunity for fruitful discussions.

The conference proceedings compile the contributions from production science and industrial research, offering readers a comprehensive overview of current trends in the field. They also provide insights into the ongoing research activities within the German Academic Association for Production Technology.

We warmly welcomed all to Hannover and trust that the WGP Annual Congress 2025 was engaging and insightful.

Garbsen, Germany
October 2025

Prof. Dr.-Ing Ludger Overmeyer
ITA
Prof. Dr.-Ing. Bernd-Arno Behrens

Organization

Leibniz University Hannover

Institute of Transport and Automation Technology
Ludger Overmeyer
Anna-Lena Fritze
Justus Lübbehusen

Institute of Forming Technology and Machines
Bernd-Arno Behrens
Kai Brunotte
Johanna Uhe
Jan-Ulrich Gellermann

Contents

Aritificial Intelligence in Production

Digital Transformation in Manufacturing

Sustainable Manufacturing Process

Integration of Ecological Sustainability into the Variant Management of Manufacturing Companies

Nikolai Kelbel[1](✉), Alexander Keuper[1], and Günther Schuh[1,2]

[1] Laboratory for Machine Tools and Production Engineering WZL, RWTH Aachen University, Aachen, Germany
n.kelbel@wzl.rwth-aachen.de

[2] Fraunhofer Institute for Production Technology IPT, Aachen, Germany

Abstract. In the course of the sustainability transition, integrating sustainability into the variant management is especially challenging for manufacturing companies with multi-variant product portfolios. As the established processes and frameworks of variant management are often revolving solely around financial metrics, this dimension is not yet systematically considered in decision-making. Existing approaches disregard target conflicts, rely on subjective metrics or lack the link to operative measures. This paper aims to address this by linking the sustainability controlling to the decision-making in variant management considering the ecological footprint of products from product development to production and usage. For this the metric of ecological target conformity is introduced. Special attention is paid to the weighting of different ecological targets and the subjectivity that such weighting may introduce into current practices and outcomes. Operative strategies to manage products with similar characteristics are derived based on existing tools of variant management.

Keywords: Product portfolio management · Ecology · Strategy · Decision-making

1 Introduction

With growing concerns about climate change, industries are under increasing pressure to adopt sustainable practices [1]. The manufacturing sector is pivotal in shaping environmental, social, and economic outcomes [2]. Encouraged by the sustainability goals and regulations of the European Union (EU), alongside customer-driven motivations, many companies are transforming their value creation processes. Some have committed to specific targets through initiatives like the Science Based Target (SBT) Initiative [3] or aim to qualify for sustainability accolades such as EcoVadis [4]. Although sustainability may offer a competitive edge, it often conflicts with short-term financial objectives, necessitating careful management and compromise [5]. This target conflict affects decision-making processes related to product modifications or discontinuations, highlighting the need for robust control mechanisms to guide decisions [6]. Companies with multi-variant product

L. Overmeyer and B.-A. Behrens (eds.), *Production at the Leading Edge of Technology*,
Lecture Notes in Production Engineering, https://doi.org/10.1007/978-3-032-19524-1_1

portfolios (PP) face particular challenges in achieving transparency. Product portfolio management (PPM), traditionally focused on variant management (VM) and market demand fulfillment [7], now also encompasses sustainability considerations. So far sustainability in PPM is often considered in isolation due to its historically low prioritization and the complexity of balancing multiple sustainability targets, resulting in tools that are impractical [6, 8–11]. This paper seeks to develop methodologies that assist manufacturing companies in effectively embedding sustainability into their VM processes. The aim is to enhance compliance with regulatory demands while fulfilling stakeholder expectations within sustainability criteria, primarily focusing on ecological aspects, and maintaining the profitability of the PP. In Sect. 2 the fundamentals and related research are elaborated. Section 3 presents the proposed methodology before Sect. 4 sums up the results of the paper.

2 Fundamentals and Related Research

In the following the fundamentals regarding sustainability targets of manufacturing companies and VM are explained. Subsequently, related work is presented.

2.1 Sustainability Targets in the Manufacturing Sector

In recent years, the EU has significantly advanced sustainability reporting through the introduction of the Corporate Sustainability Reporting Directive (CSRD). It mandates detailed disclosure of greenhouse gas emissions alongside other environmental metrics that were previously voluntary or less defined [12]. This directive marks a shift towards transparency and accountability in corporate practices, compelling many manufacturing companies to reassess their operations. As a result, companies are setting sustainability targets to align with regulatory demands and stakeholder expectations. The CSRD requires companies to disclose measurable, outcome-oriented targets related to material sustainability matters. These targets involve strategies such as reducing material consumption, switching to renewable resources, or phasing out environmentally detrimental products [12, 13]. However, no specific targets are mandated and targets are also not necessarily required if the emissions in question are not significant to the company [14]. Still, it is observed that companies have been engaging in sustainability reporting for many years [15, 16]. The motives of companies for setting sustainability targets can be of strategic or financial nature, focusing on future business opportunities, economic benefits or requirements by equity and debt capital providers, of risk management, focusing on long-term resilience, or even reputational, ethical or value-driven nature [17].

2.2 Product Portfolio Management and Variant Management

The difference between PPM and VM lies in the level at which the management is conducted. PPM operates at a higher level than VM. According to Cooper et al., PPM involves making decisions about the allocation of resources across various product groups, which are defined and justified based on long-term strategic considerations

[18]. Resource allocation and the associated strategic decisions are critical, as they significantly influence future competitiveness. Building on this, Cooper et al. identify four objectives for PPM: maximizing the value of the portfolio, achieving balance, ensuring strategic alignment, and maintaining an appropriate PP size [18]. Thus, PPM primarily operates at PP level or at product group level. The focus is on maintaining a balanced PP and ensuring alignment between the offered product groups and market demands, thereby meeting market expectations [18–21].

In contrast, "the task of managing the significant increase in product variants, primarily caused by the diversification of the enterprise, is categorically referred to as variant management" [22]. Therefore, VM is concerned with the operational implementation of the directives established by PPM, and focuses specifically on the individual product, also referred to as product variant (PV) in this paper. The focus shifts to how the concept of the PP can be translated into individual PVs, with the goal of optimally exploiting the market by targeting specific customer segments [22].

2.3 Related Work

Pinheiro et al. observed that studies integrating PPM and ecodesign in a unified approach are scarce, as these domains are often analyzed separately. Ecodesign involves the consideration of the ecological impact of a product across its life cycle to optimize the ecological sustainability of it. Therefore, they advocate for a holistic consideration of dimensions such as guidelines, methods, and tools, alongside organizational and strategic factors. While their theoretical framework proposes pathways to integrate sustainability aspects such as using checklists, ranking systems and different matrices to compare products or assembling interdisciplinary teams, it remains qualitative [20].

Silvius et al. seeks to establish an integrative connection of sustainability with corporate strategy, project portfolios, and projects. Given that a product within a PP can often be perceived as a project, particularly in the context of new product development, this framework aligns conceptually with the scope of this study but remains descriptive and generic. The interlinking practices outlined at the levels of organizational strategy, PPM, and project management emphasize fundamental aspects to be considered, but do not introduce a novel methodology or tools for evaluating individual PVs or projects in the context of ecological VM [8].

Medini et al. propose a novel customer-centric approach that captures preferences for specific sustainability factors through user inputs in a product configurator for subsequent processing. However, the exact nature of this subsequent processing remains unclear. While the customer-centric approach is conceptually valuable it lacks clarity on how data weighting and processing are to be conducted. Consequently, the paper does not provide further actionable insights or methodological guidance [21].

Building on the BCG Matrix [23], Moshrefi et al. introduce the Total Environmental Impact–Total Profit Matrix. It addresses a key criticism of the BCG Matrix by incorporating a profitability metric, thus shifting the focus from merely market share to a more nuanced consideration of market size and profit margin. Despite the advantages over other frameworks and methods, the approach proposed raises several critical concerns. The key among them are issues such as the "uncertain future, technology changes, environmental impacts of new technologies/products, and [the] life cycle of products," making it almost impossible to predict reliable environmental impacts. Moreover, it is virtually impossible to generate reliable sales forecasts over an extended period. While shorter timeframes are likely more feasible and could address this limitation, even these pose challenges. SBTs are not universally available across all industries or meaningful, since adherence to SBTs is not yet mandated and operational targets are often more short-term, calling into question both the practical applicability and relevance of this approach. Additionally, the method leaves unresolved how the Total Environmental Impact is to be calculated, particularly when multiple environmental factors must be considered simultaneously [24].

In conclusion, existing approaches for assessing the ecological performance of products are often overly qualitative, lack scalability, and fail to focus on the PV level while not aligning with specific corporate targets. These methods typically rely on generic national or global targets, which are impractical for operational use within a corporate context. Furthermore, current tools do not provide strategic foresight to anticipate regulatory changes or comprehensively address multiple environmental categories, often resulting in subjective evaluations presented on non-continuous scales. Consequently, there is no systematic approach available to resolve the trade-off between sustainability and profitability within VM effectively, nor a methodology that offers a clear pathway for managing PPs in alignment with comprehensive climate protection strategies. Thus, the aim of this paper is to complement established VM-strategies with the sustainability perspective by utilizing a data-based approach.

3 Methodology

The developed methodology consists of four steps that are described in the following. The first step can be interpreted as a prerequisite of the presented methodology. To manage the sustainability of a PP, PV-related sustainability data needs to be available. For this paper it is assumed that the ecological and quantitative impacts of all PVs within the management scope are available. Regarding a scalable provision of such data the authors refer to existing research leveraging machine learning [25, 26].

In the second step, an Ecological Target Conformity (ETC) is developed as a single metric to quantitatively evaluate the ecological performance of a PV in alignment with the sustainability targets of a company. The ETC applies specifically to individual PVs, and, depending on the level of aggregation, is categorized as either the Specific-ETC (S-ETC) or the Aggregated-ETC (A-ETC). The S-ETC is a dimensioned (€/quantity of the environmental category) and continuous metric, enabling the identification of the ecologic weaknesses of a product. Its calculation comprises the division of the revenue of the PV by the value of the specific category (e.g. an emission value). It is normalized by dividing through the target value of the category (see Formula 1).

In the third step, to capture the overall ecological performance of a PV in the context of corporate targets, the dimensionless and continuous A-ETC is introduced. This metric is calculated by forming a weighted sum of all S-ETC based on multi-criteria-decision-making theory and enables decision-making considering multiple targets. The weighting criteria are derived objectively and inherently from the available data, using a distance-to-target weighting approach (see Formula 2).

$$S\_ETC^{*}_{y,c,p} = \ln \frac{S\_ETC_{y,c,p}}{S\_ETC\ TV_{y,c}} = \ln \frac{\frac{Revenue_{y,p}}{EC\_Value_{y,c,p}}}{\frac{\sum_{p\in P} Revenue_{y,p}}{EC\_Target_{y,c}}} \quad (1)$$

$$A\_ETC_{y,p} = \sum_{c\in C} w_{y,c} \times S\_ETC^{*}_{y,c,p} \quad (2)$$

P	All PV under consideration
C	All environmental categories under consideration
$Revenue_{y,p}$	Revenue of PV p for year y
$w_{y,c}$	Weighting factor for year y and environmental category c
$S\_ETC_{y,c,p}$	S-ETC Score of PV p for year y and environmental category c
$EC\_Value_{y,c,p}$	Environmental value of PV p for year y and environmental category c
$S\_ETC\ TV_{y,c}$	S-ETC Target Value for year y and environmental category c
$EC\_Target_{y,c}$	Environmental target for year y and environmental category c
$S\_ETC^{*}_{y,c,p}$	Normalized S-ETC Score of PV p for year y and environmental category c
$A\_ETC_{y,p}$	A-ETC Score of PV p for year y

In the last step, to ensure that sustainability is incorporated into the VM process the ETC-Financial Performance Matrix is introduced (see Fig. 1). This method builds on the presented approach of Moshrefi et al. but incorporates ETCs to support effective PPM [24]. With that, each PV is analyzed individually to avoid the constraints of predefined product groupings. Furthermore, the temporal dimension, originating from the estimated impact of planned VM measures, is integrated into the matrix, facilitating strategic foresight and accommodating future dynamics.

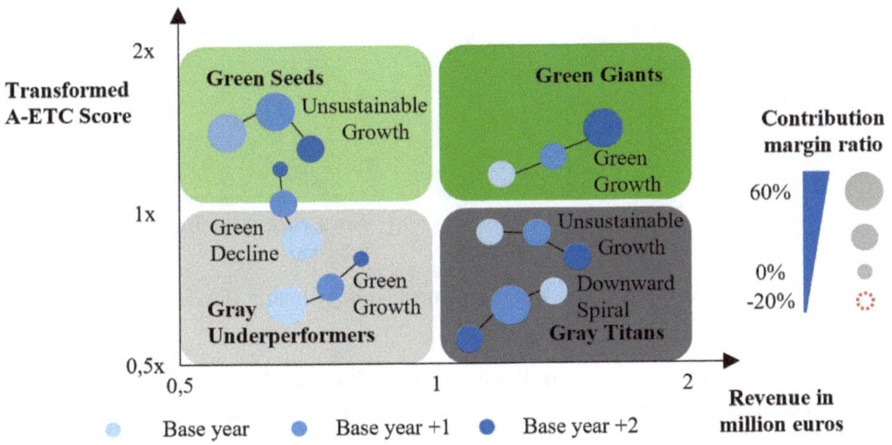

Fig. 1 ETC-financial performance matrix

Beyond simply categorizing PVs within a matrix, the Product Variant Management Strategy Matrix (PMSM) is developed to link products with measures of VM in dependance of their ecological and economic performance (see Fig. 2). The PMSM is structured in such a way that the y-axis indicates the severity of the current situation in relation to achieving sustainability targets, while the x-axis represents the problematic nature of trends affecting the ability of a company to meet these targets. Furthermore, the lower half of a quadrant represents unprofitable PVs, while the upper half depicts profitable ones. This graphical representation facilitates the identification of three primary strategic categories, *Limited Focus*, *Revenue Focus*, and *Sustainability Focus*, that can guide the development of targeted strategies.

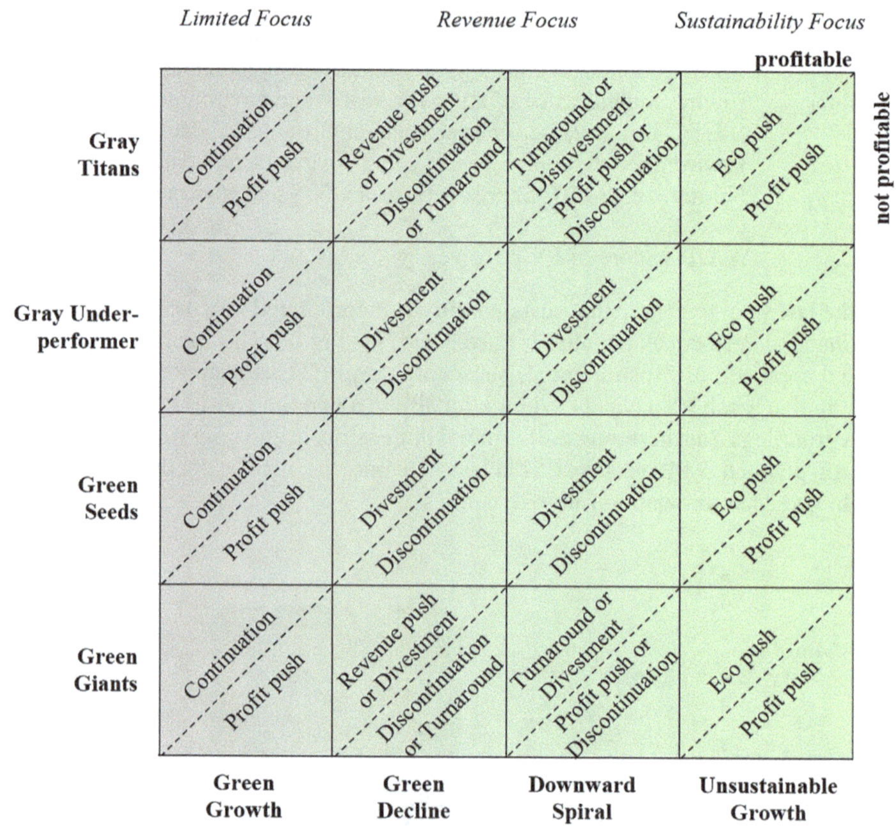

Fig. 2 Product variant management strategy matrix (PMSM)

Limited Focus strategies are applicable to PVs that are already experiencing *Green Growth.* Naturally, promoting growth through additional activities remains important, particularly if the growth rate is perceived as too slow or if improvements in the ETC are progressing inadequately. These PVs are not inherently critical under this strategy. In cases of unprofitability, countermeasures, called *Profit Pushes*, must be implemented.

Discontinuing (removing the PV from the portfolio) these PVs is generally not advisable, as they are projected to maintain *Green Growth*.

Revenue Focus strategies are relevant for PVs positioned within the cluster *Green Decline* or *Downward Spiral*. Here distinctions must be made regarding the profitability prospects of the PV. For profitable *Green Giants*, it is prudent to implement measures aimed at increasing revenue, while simultaneously counteracting potential declines in ETC. However, if revenue loss becomes unavoidable due to technological shocks or similar disruptions, the strategic objective should shift towards maximizing cash flow from the PV. This involves minimizing further investments (*Divestment*) and, once the PV transitions into a *Green Seed*, ultimately discussing a *Discontinuation* of the PV due to insufficient revenue generation. In the case of an unprofitable *Green Giant* experiencing a revenue decline, achieving a profitable turnaround becomes significantly more challenging. While the size of the PV makes it strategically important for the organization, it cannot be leveraged as a cash cow in the same way as its profitable counterpart. Consequently, the recommended course of action is either to discontinue the PV, after a thorough assessment of all potential consequences of such an action, or to undertake the considerable risk of making substantial investments.

Sustainability Focus strategies are particularly relevant for PVs exhibiting the trend of *Unsustainable Growth*. It is crucial to ensure that revenue growth does not escalate into an environmental issue, regardless of the static product clusters to which these PVs belong. However, this concern is especially critical for the clusters *Gray Titans* and *Green Giants*, as these already hold substantial importance for the company. Consequently, the overarching goal across all PVs must be to enhance the ecological performance (*Eco Push*) through measures in product design (e.g. reducing energy consumption), production processes, or supply chains. If *Unsustainable Growth* is observed in PVs with low or negative profitability, additional investments in corrective measures are justified. This is particularly the case when such PVs demonstrate significant growth potential or already account for substantial revenue, allowing the company to capitalize on this potential. A *Divestment* is not recommended in these situations.

The proposed methodology integrates sustainability in VM decision-making. It enables a systematic, transparent and efficient management of the sustainability of a PP while still considering traditional, economic dimensions and thus resolves potential target-conflicts between these decision dimensions.

4 Conclusions

Incorporating sustainability into decision-making within VM presents significant challenges, primarily due to potential target conflicts and insufficient transparency. This paper introduces a methodology for integrating sustainability considerations into VM focusing on objectivity and the link to operative management. This is done by modifying traditional VM tools and linking the sustainability targets of a company to an introduced metric for the sustainability performance of products. This aims for enabling VM to manage economic and ecologic targets systematically. The authors continue to work on a validation as well as the estimation of sustainability data via machine learning.

Acknowledgements. Funded by the Deutsche Forschungsgemeinschaft (DFG, German Research Foundation)—554929445.

Competing Interests. The author(s) has no competing interests to declare that are relevant to the content of this manuscript.

References

1. IPCC: Climate Change 2023. Synthesis Report. Summary for Policymakers (2023)
2. Stock, T., Seliger, G.: Opportunities of sustainable manufacturing in Industry 4.0. Procedia CIRP **40**, 536–541 (2016)
3. Science Based Targets Initiative. https://sciencebasedtargets.org/ (2025)
4. EcoVadis: https://ecovadis.com/de/ (2025)
5. Liedong, T.A., Taticchi, P., Rajwani, T., Pisani, N.: Gracious growth. How to manage the trade-off between corporate greening and corporate growth. Org. Dyn. **51**, 1–11 (2022)
6. Villamil, C., Schulte, J., Hallstedt, S.: Sustainability risk and portfolio management. Bus. Strat. Environ. **31**, 1042–1057 (2022)
7. Cooper, R.G., Edgett, S.J., Kleinschmidt, E.J.: New product portfolio management. J Prod. Innov. Manag. **16**, 333–351 (1999)
8. Silvius, G., Marnewick, C.: Interlinking sustainability in organizational strategy, project portfolio management and project management. Procedia Comput. Sci. **196**, 938–947 (2022)
9. Rossi, M., Germani, M., Zamagni, A.: Review of ecodesign methods and tools. J. Clean. Prod. **129**, 361–373 (2016)
10. Silvius, A. J. G., Kampinga, M., Paniagua, S., Mooi, H.: Considering sustainability in project management decision making. Int. J. Proj. Manag. **35**, 1133–1150 (2017)
11. Dekoninck, E.A., et al.: Defining the challenges for ecodesign implementation in companies. J. Clean. Prod. **135**, 410–425 (2016)
12. European Commission: Regulation (EU) 2023/2772. https://eur-lex.europa.eu/legal-content/EN/TXT/PDF/?uri=OJ:L_202302772 (2023)
13. Schuh, G., Keuper, A., Ruschitzka, C., Schümmelfeder, S., Prumbohm, F.: Managing Sustainable Innovations (2024)
14. European Parliament and the Council: Directive (EU) 2022/2464. https://eur-lex.europa.eu/legal-content/EN/TXT/?uri=CELEX%3A32022L2464 (2022)
15. Riesener, M., Keuper, A., Mertens, M., Recker, J., Schuh, G.: Development of a methodology for containment of solution spaces to identify eco-friendly product concepts. In: 35th CIRP Design 2025 (2025)
16. Greenhouse Gas Protocol: About Us | GHG Protocol. https://ghgprotocol.org/about-us (11/13/2024)
17. Richard, M.A.: Master thesis, RWTH Aachen University (2023)
18. Cooper, R.G., Edgett, S.J.: Portfolio Management for New Products, p. 382 (2001)
19. Cooper, R.G., Edgett, S.J., Kleinschmidt, E.J.: Portfolio management in new product development. Res. Technol. Manag. **40**, 16–28 (1997)
20. Pinheiro, M.A.P., Jugend, D., Filho, L.C.D., Armellini, F.: Framework proposal for ecodesign integration on product portfolio management. J. Clean. Prod. **185**, 176–186 (2018)
21. Medini, K., Wuest, T., Romero, D., Laforest, V.: Integrating sustainability considerations into product variety and portfolio management. Procedia CIRP **93**, 605–609 (2020)
22. Schuh, G., Riesener, M.: Produktkomplexität managen. Strategien—Methoden—Tools, 3rd edn. Hanser (2018)

23. BCG Global: What Is the Growth Share Matrix? https://www.bcg.com/about/overview/our-history/growth-share-matrix (2024)
24. Moshrefi, S., Abdoli, S., Kara, S., Hauschild, M.: Product portfolio analysis towards operationalising science-based targets. Procedia CIRP **90**, 377–382 (2020)
25. Kelbel, N., Yang, P., Riesener, M., Keuper, A., Schuh, G.: In: Drossel, W.-G., Ihlenfeldt, S., Dix, M. (eds.) Production at the Leading Edge of Technology, pp. 430–437. Springer, Cham
26. Sousa, I.: Approximate Life Cycle Assessment of Product Concepts Using Learning Systems (2002)

Solution Space Management for the Development of Ecologically Sustainable Products in the Early Innovation Phase

Matthias Sebastian Mertens[1(✉)], Alexander Keuper[1], Nikolai Kelbel[1], and Günther Schuh[1,2]

[1] Laboratory for Machine Tools and Production Engineering (WZL), RWTH Aachen University, Aachen, Germany
matthias.mertens@wzl.rwth-aachen.de
[2] Fraunhofer Institute for Production Technology IPT, Aachen, Germany

Abstract. For the successful realisation of ecologically sustainable products, it is crucial to be able to assess the ecological impact of products over their entire life cycle in the early innovation phase. Particularly when deciding between different possible technical solutions to fulfil the desired product functions, it is essential to correctly assess the impact on ecological sustainability of each solution at an early stage to choose the expected to be the most ecological sustainable one. This is particularly evident in the case of products that require for example energy-intensive production processes and technologies or tools with a high material input. However, this is a major challenge due to the uncertainties and lack of information about the environmental impact of technical product solutions and production processes associated with the early innovation phase. In this paper, the concept of a new methodology is derived, which is designed to narrow down technical solution alternatives for a planned product in the early innovation phase. The aim is to be able to estimate as early as possible which technologies, production processes and other product characteristics will later best fulfil the sustainability requirements of the product, with little available and uncertain information. By integrating approaches of environmental sustainability assessment, solution space management and multi-criteria decision making, the proposed methodology supports decision makers in selecting the most promising ecologically sustainable solutions under uncertainty. The method is based on a multi-criteria decision algorithm that is combined with approaches from fuzzy logic and scenario technique.

Keywords: Ecological sustainability · Early innovation phase · Solution space management · Multi-criteria-decision-analysis

1 Introduction

Innovations are crucial for a company's long-term market survival. At the same time, in today's globalised world, the need for environmental protection and sustainability is increasingly important. Companies acknowledge this need and are striving to make

L. Overmeyer and B.-A. Behrens (eds.), *Production at the Leading Edge of Technology*, Lecture Notes in Production Engineering, https://doi.org/10.1007/978-3-032-19524-1_2

their corporate culture and production processes more sustainable. Additionally, they are increasingly considering ecological sustainability in their new inventions [1].

Social pressure and an increasing number of laws and regulations from global organisations accelerate this development. The European Union intends to cut greenhouse-gas emissions by at least 55% by 2030 relative to 1990 [2]. Although the EU has delayed the Corporate Sustainability Reporting Directive's (CSRD) full rollout until 2028 and reduced its scope to about 20% of originally covered firms, it still obliges many companies to disclose their ecological performance, strengthening transparency and competitive pressure. Furthermore, the International Energy Agency reports that manufacture processes and processing of goods released roughly one-quarter of global energy system emissions in 2022 [3].

Product development's earliest phase carries a high responsibility for sustainability. Although the "fuzzy front end" of development generates few direct emissions, the decisions made at the concept stage have most of the impact on ecological sustainability [4]. Choosing unsuitable product and production technologies leads to costly corrections down the line.

However, generating sustainable innovations poses significant challenges. Decisions regarding technical solution principles for an idea are always uncertain due to a lack of information, resources, knowledge, and experience in the early phases. These decisions can have a significant impact on sustainability. A survey of 128 European manufacturers found that information and data deficits were among the three most critical obstacles to ensuring sustainable innovations and accurately meeting future corporate sustainability goals [5].

The aim of this paper is to develop and present an approach that supports decision-makers in steering early product concepts towards ecological sustainability by systematically narrowing down technical solution alternatives and based on limited and uncertain data, assessing which solutions are most likely to meet future sustainability requirements. After the introduction, Sect. 2 outlines the necessary fundamentals, reviews existing approaches from the scientific literature and describes the remaining research gap. Then the paper presents the four-step approach in detail, followed by a final section that summarizes the results and outlines possibilities for future research.

2 Fundamentals and Related Research

2.1 Fundamentals

The early phase of the innovation process, often termed the fuzzy front end (FFE), phase zero or pre-development, comprises every activity from the first vague idea to the money-gate decision that allocates resources for concept realisation. It is dynamic and loosely structured as teams explore diverse technical solutions while responsibilities, resources and documentation are still unclear [6].

The solution space, also known as the design space or as set-based engineering, denotes the totality of technically feasible solution alternatives to build up a new product for a given requirement set [7]. Instead of committing prematurely to a solution, development projects often first expand the space through systematic search and creativity and then, on an abstract level, successively evaluate and eliminate inferior or infeasible options until a single well-founded solution alternative remains that fitted best to the requirements [8].

Multi-criteria-decision-making (MCDM) is a decision methodology that addresses multi-criteria decision problems by structuring problems around at least two evaluation criteria and guiding decision makers through structured information processing and problem decomposition [9].

2.2 Related Work

When it comes to steering the solution space, Lüdtke shows that set-based design sets, an approach that maintains several design alternatives in parallel before progressively narrowing them, can increase development efficiency. His work also introduces initial concepts for describing a solution-space design and coping with requirement uncertainty, although the method remains complex to apply in the early phase [10].

With a stronger focus on sustainability assessment, Hallstedt et al. review 51 methods and introduce the "Qualitative Sustainability Fingerprint Tool". While well suited to whole-product assessment, it neither drills down to individual technical principles nor addresses technological uncertainty [11]. Omadara et al. illustrate a questionnaire-based path in which their "97-item PSAT" locates life-cycle sustainability hotspots but lacks by presenting mechanisms for comparing competing concepts or guiding subsequent decisions [12].

To turn Life Cyle Assessment (LCA) outputs into decisions, researchers employ Multi-Criteria-Decision-Analysis (MCDA). Haase et al. combine environmental, economic and social indicators with "Technique for Order Preference by Similarity to Ideal Solution" (TOPSIS) to rank automotive powertrains, but still require detailed data [13].

To ease this burden, Ghadimi et al. fuse fuzzy logic with MCDA, enabling nuanced decision-making under imprecise information. In a further step, the approach weights sustainability criteria against one another, delivering clear rankings of product alternatives even when assessment results fluctuate [14], an ability that makes their approach especially promising for the methodology proposed in this paper.

In 2022, the Helmholtz Working Group "MCDA for Sustainability Assessment" published a working paper summarising advances in multi-stakeholder criteria weighting and tool support, showcasing the promise of integrating sustainability assessments with MCDA [15], yet it still offers no remedy for the persistent data deficits that hinder decision-making in the early development phase.

The literature acknowledges the need to actively control the solution space for technical solution alternatives in the early innovation phase and to apply decision models, to guide sustainability-related trade-offs. However, no unified methodology or tool currently integrates at the same time qualitative and quantitative data to support decision-making under high uncertainty and information deficit. Equally challenging is the translation of a company's ecological sustainability goals into specific, actionable product criteria, which is essential for aligning solution alternative selections with the strategic objectives of companies, particularly in preparation for the Corporate Sustainability Reporting Directive (CSRD).

3 Methodology

The challenges underscore the need for a method enabling structured ecological decision-making despite limited clarity. Key requirements are the ability to address predictive and prescriptive questions, integrate qualitative and quantitative data, operate under scarcity and uncertainty, and provide decision value proportional to resources invested.

The method comprises four interlinked steps: (1) **structuring the description of technical solution alternatives in the FFE**, (2) **translating sustainability goals into actionable indicators to control and narrow the solution space**, (3) applying a **robust decision-making (RDM) approach with scenario techniques** in a quick-check to quantitatively reduce the large initial solution space, enabling efficient use of (4) a **multi-criteria decision analysis (MCDA) to rank the remaining alternatives**. The methodology focuses on the support of the development of breakthrough innovation, for which less knowledge already exists. However, it can also be applied to incremental innovations, whereby the effort involved must always be weighed against the benefits of the application if a wealth of experience already exists.

3.1 Description of Solution Spaces for a Technical Product in the Early Innovation Phase

A Framework was developed as the basis for the first step. This shows how a user of the methodology should describe a solution space of a technical product in the early innovation phase so that it can be controlled efficiently later. In order to be able to create a comprehensive and function-oriented description in the early idea phase, a function-oriented approach has been chosen. Due to the fact that high-level functions are realised through multiple technical solution principles (LP), each top-level function must be decomposed into subfunctions. This enables individual solution principles to be assigned to specific sub-functions that a new product idea must fulfil. An example of this would be the "energy storage" function in a vehicle, which could be achieved using the technical solution principle lithium-ion batterie or solid-state batterie.

A solution principle should be described, at least in the concept properties of physical effect, material and effective geometry in order to enable subsequent sustainability assessments to be carried out. Selecting a solution principle for each sub-function results in a solution alternative (Fig. 1). The combination of several solution principles leads to a solution alternative which describes the solution principles that could ultimately form the basis of the product.

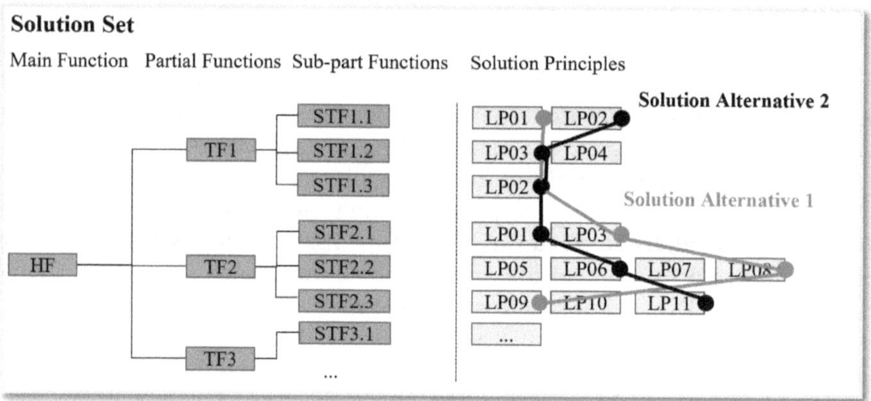

Fig. 1 Description of solution principles and the resulting possible solution alternatives

3.2 Identification of Indicators Controlling the Solution Space in the Context of a Sustainability Strategy

Control indicators are essential for effective solution-space management and the evaluation and selection of solution alternatives. To align decisions with corporate sustainability strategy while also considering production impacts, an ecological indicator catalogue was derived. Existing metrics from initiatives such as the Global Reporting Initiative were filtered and clustered via framework analysis, producing a list suitable for product-level solution-space control.

The task of the user of this methodological concept is to break down the strategic sustainability goals into individual performance indicators from the prepared list of indicators. This is the major difference to existing approaches, which focus on costs, for example, because in the context of ecological sustainability, global sustainability goals, such as portfolio indicators for CO_2 emissions, must be systematically broken down to individual product development projects and considered in the early innovation phase. At this point, approaches from strategic management can be applied. A systematic breakdown of strategic goals and target values into individual indicators and indicator values is helpful here. The appropriate method for this purpose is the Hoshin Kanri Matrix, which is frequently used in literature and practice [16]. This was adopted as part of the methodology development and supplemented by weighting the indicators so that, for example, CO_2 emissions and water consumption are weighted differently in the subsequent selection of alternative solutions. The weighting is done by using an Analytical Hierarchy Process (AHP).

3.3 Scenario-Dependent Preselection of Alternative Solutions

In order to do justice to the early innovation phase and the potentially very extensive solution space with many alternative solutions, step three provides a preliminary screening procedure to narrow down the solution space. This means that the more complex assessing of individual alternatives based on sustainability indicators within the multi-criteria decision-making approach only needs to be carried out on the basis of a promising selection of alternative solutions.

In this context, a methodological approach from the scientific field of robust decision-making is used. First, the individual solution principles and then the composite solution alternatives are qualitatively assessed on a scale of one to ten against the weighted sustainability indicators, based on the extent to which the solution alternative and its solution principles contribute positively or negatively to the fulfilment of the indicators. It is recommended to introduce a simple rating scale from one to ten that can be assigned by an interdisciplinary team. Initial validation shows that this allows a wide range of solutions to be narrowed down within the framework of a one-day workshop. A rating of one describes a highly negative deposit of the solution principle on the sustainability indicator in the scenario. A score of ten indicates that the solution principle makes a positive contribution to the sustainability indicator under the given circumstances. The average score from the qualitative assessment of the solution principles thus enables an initial assessment of the extent to which a solution alternative supports the achievement of the sustainability goals based on the indicators (Fig. 2). Since the ecological impact of product and production technologies can evolve markedly between initial implementation to series-production readiness, their impact must therefore be evaluated under multiple scenarios. For this reason, the assessment of the alternative solutions described must be carried out for each indicator within scenarios.

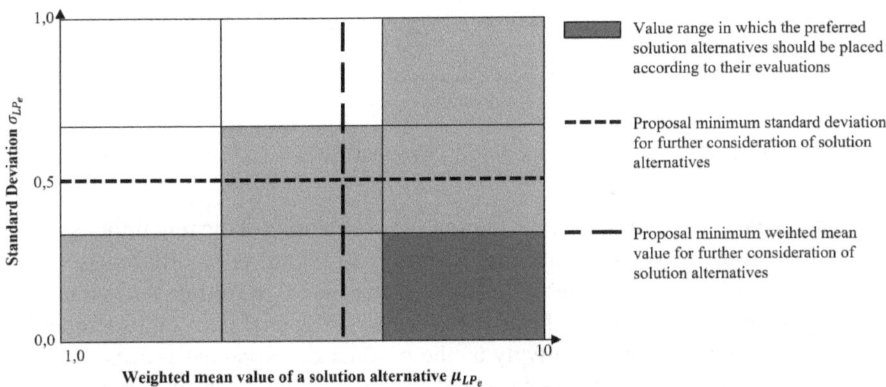

Fig. 2 Graphical visualization of the assessed solution space

The scenarios are derived from projections of key factors that may alter the ecological impact of products and production technologies, focusing on sustainability indicators from ideation to product phase-out. Methodological guidance is well-documented in the scientific literature [17]. In this context, both the weighted average and the standard deviation across all indicators and scenarios are calculated, the latter indicating an alternative's robustness under shifting futures. Step three produces a diagram plotting each alternative's weighted mean against its standard deviation. Initial validations recommend further consideration of alternatives with a mean >6 and standard deviation <0.5.

3.4 Limitations and Control of Sustainable Solution Alternatives

The surviving alternatives enter a mixed-method qualitative and quantitative multi-criteria-decision-analysis (MCDA). A consolidated evaluation matrix records, for every remaining alternative and each scenario, the corresponding sustainability indicator values. Due to the smaller number of solution alternatives, quantitative indicators such as LCA results can now also be included in the matrix. A comparison between qualitative and quantitative indicators is more complex but now only needs to be implemented for a reduced solution space. As only quantitative evaluations were used previously, normalization is required only in this final step. To do this, qualitative ratings must first be normalized and placed in context with the other quantitative values. Fuzzy logic models can help here if it is not possible to convert qualitative values into quantitative values [14]. The VIKOR algorithm ranks the alternatives based on a core scenario by calculating their distances to the ideal and anti-ideal reference solutions illustrated in Fig. 3.

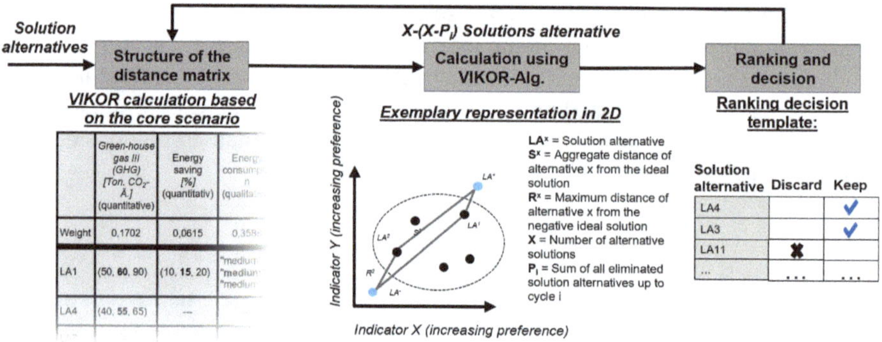

Fig. 3 Exemplary VIKOR distance plot with subsequent ranking and action recommendations

A sensitivity analysis based on the alternative evaluations in the remaining scenarios can be used to analyze how stable the determined ranking is in different scenarios. Generally weak scenarios and highly sensitive scenarios can be eliminated accordingly, depending on the risk tolerance of the decision-makers. The final step of the methodology, the MCDA, can be repeated iteratively as the product development process advances, as soon as new information has been added to the evaluation matrix. This allows the remaining solution space to be narrowed down quickly and efficiently.

4 Summary and Outlook

This paper presents a four-step methodology to help decision-makers find the most environmentally sustainable technical solution, even when not all detailed information about the impacts of possible product concepts is known.

The methodology forms a first concept based on solution-space management, coupled with scenario-based MCDA, that can embed ecological sustainability in product concepts before certain data are available, turning early uncertainty into a competitive advantage. A software application with an intuitive front-end will be developed to operationalise the methodology in practice. Validation will be conducted through industry pilots to test the usability of the method, measure decision-making efficiencies and assess practitioner acceptance in guiding sustainable concept selection.

Competing Interests. The author(s) has no competing interests to declare that are relevant to the content of this manuscript.

References

1. Hetherington, A., Borrion, A.L.: Use of LCA as a development tool within early research: challenges and issues across different sectors. Accessed 29 Feb 2024
2. European Comission: Mehr Ehrgeiz für das Klimaziel Europas bis 2030: In eine klimaneutrale Zukunft zum Wohl der Menschen investieren. Accessed 29 Feb 2024
3. IEA: Tracking Clean Energy Progress 2023. https://www.iea.org/reports/tracking-clean-energy-progress-2023. Accessed 16 Jun 2025
4. Herstatt & Verworn: The Fuzzy Front End of Innovation. Accessed 29 Feb 2024
5. Schuh, G., Keuper, A., Ruschitzka C., Schümmelfede, S., Prumbohm, F.: Managing Sustainble Innovations: Ergebnisbericht der Benchmarking Studie. https://complexity-academy.com/site/assets/files/7199/wzl-ls_ps-kbm-msi-whitepaper_ver_13_einzelseiten_druckqualitaet.pdf
6. Dornberger, U., Suvelza, A.: Managing the fuzzy front-end of innovation. International SEPT Program, the Leipzig University (2012)
7. Schuh, G.: InnoSpace - Risikobasierte Lösungsraum-Steuerung von Entwicklungsprojekten in der Arzneimittelbranche. Accessed 29 Feb 2024
8. Lenders, M.: Beschleunigung der Produktentwicklung durch Lösungsraum-Management. Accessed 29 Feb 2024
9. Labianca, C., de Gisi, S., Notarnicola, M.: Multi-criteria decision-making. In: Assessing Progress Towards Sustainability, pp. 219–243. Elsevier (2022)
10. Lüdtke, B.: Lösungsraum-Steuerung in der Produktentwicklung. Dissertation, Fraunhofer-Institut für Produktionstechnologie IPT; Rheinisch-Westfälische Technische Hochschule Aachen (2016)
11. Hallstedt, S.I., Villamil, C., Lövdahl, J., Nylander, J.W.: Sustainability fingerprint—guiding companies in anticipating the sustainability direction in early design. Sustain. Prod. Consum. **37**, 424–442 (2023). https://doi.org/10.1016/j.spc.2023.03.015
12. Omodara, L., Saavalainen, P., Pitkäaho, S., Pongrácz, E., Keiski, R.L.: Sustainability assessment of products—case study of wind turbine generator types. Environ. Impact Assess. Rev. **98**, 106943 (2023). https://doi.org/10.1016/j.eiar.2022.106943
13. Haase, M., et al.: Multi-criteria decision analysis for prospective sustainability assessment of alternative technologies and fuels for individual motorized transport. Clean Technol. Environ. Policy **24**(10), 3171–3197 (2022). https://doi.org/10.1007/s10098-022-02407-w
14. Ghadimi, P., Azadnia, A.H., Mohd Yusof, N., Mat Saman, M.Z.: A weighted fuzzy approach for product sustainability assessment: a case study in automotive industry. J. Clean. Prod. **33**, 10–21 (2012). https://doi.org/10.1016/j.jclepro.2012.05.010
15. Estrada, L.S.M., et al.: MCDA for sustainability assessment—insights to Helmholtz Association activities. (2022)

16. Kudernatsch, D.: Workbook Hoshin Kanri: Lean-Management-Strategien erfolgreich umsetzen, 1st edn. Schäffer-Poeschel; Imprint Schäffer-Poeschel, Stuttgart (2022). https://link.springer.com/book/10.34156/9783791053660
17. Gräßler, I., Thiele, H., Scholle, P.: Szenario-Technik. In: Vajna, S. (ed.) Integrated Design Engineering, pp. 689–717. Springer Berlin Heidelberg, Berlin, Heidelberg (2022)

Initiating the Corporate Sustainability Transformation: Internal Drivers for Emission Reduction

Niklas Bode(✉), Oskay Ozen, and Matthias Weigold

Institute for Production Management, Technology and Machine Tools (PTW), Technical University of Darmstadt, Darmstadt, Germany
n.bode@ptw.tu-darmstadt.de

Abstract. External regulatory requirements are increasing the pressure on companies to integrate sustainability measures. However, sustainability is not solely driven by external regulations but emerges from within the company. While the technological implementation of stated measures requires a systematic approach, the initiation of measures and the development of underlying ideas appear to follow an unstructured process. Thus, the purpose of this paper is to conduct a systematic literature analysis exploring initiation processes, connected decision-making processes as well as associated internal actors and drivers. This work assesses how internal corporate conditions influence sustainability efforts and how proactive transformations can be fostered. 41 peer-reviewed papers were identified and examined using qualitative content analysis. Results show that initiation involves top-down and bottom-up dynamics with strategic planning tools and emergent elements. Multi-criteria decision-making methods are beneficial, yet literature provides little information on their application in organizations. Top-management should ideally decide based on systematically structured information provided by all hierarchical levels. Successful sustainability integration depends on commitment, and organizational structure.

Keywords: Sustainability transformation · Initiation process · Decision-making

1 Introduction

A 2024 KfW Research study [1] of 1,805 companies found that over 50% consider sustainability important, primarily due to customer and societal expectations. Despite limited financial and personnel resources, these factors incentivize companies to introduce sustainability measures, especially regarding emission reduction [1]. Integrating sustainability measures requires internal transformation of structures, communication and innovation practices [2], which is driven by an internal initiation process that results in proposals for sustainability measures, followed by an organizational decision-making process that evaluates and approves them. Understanding how initiation and decision-making unfold within organizations provides valuable insights into associated internal drivers. Therefore, the purpose of this paper is to conduct a literature analysis examining

L. Overmeyer and B.-A. Behrens (eds.), *Production at the Leading Edge of Technology*, Lecture Notes in Production Engineering, https://doi.org/10.1007/978-3-032-19524-1_3

the internal initiation of sustainability measures, especially emission reduction, underlying decision-making processes, actors, and associated internal drivers. The paper proceeds with the theoretical background (Sect. 2), methodology (Sect. 3), findings (Sect. 4), discussion (Sect. 5), and concludes with implications for industrial application and future research (Sect. 6).

2 Theoretical Background

Regarding the integration of sustainability, Engert et al. [2] describe internal and external drivers as reasons or issues explaining why the integration of corporate sustainability is important and advantageous. The literature review conducted identified various drivers, among them legal compliance, competitive advantage, cost reduction, economic performance, and innovation [2]. Innovation usually takes place in an innovation ecosystem, which is "the evolving set of actors, activities, and artifacts, and the institutions and relations" [3], and emerges from initiative. Initiative refers to the specific behavior of a change agent resulting in a self-starting approach to cause change [4]. Initiation refers to the internal collection of processes that identify needs or opportunities, group change agents, develop proposals with justified objectives, benefits and investments, and prioritize proposals [5].

An organizational decision-making process follows in which proposals are selected, evaluated and approved or rejected. The levels of decision in an organization include strategic, tactical and operational decisions with different potential decision support methodologies [6]. Multiple-criteria decision-making (MCDM) support methodologies are used to structure and solve such decision problems. A common method is the Analytic Hierarchy Process (AHP), breaking down complex problems into a hierarchy, using pairwise comparison and calculating weighted priorities. Additionally, the Technique for Order Preference by Similarity to Ideal Solution (TOPSIS) ranks alternatives based on their geometric distance to the best- and worst-case solution [7]. In contrast to AHP, the Analytic Network Process (ANP) considers interdependencies, rather creating a network than a hierarchy [8]. TODIM (Portuguese: interactive multi-criteria decision-making) is based on prospect theory, accounting for perceived gains and losses for decisions under uncertainty [9]. Existing frameworks can be extended with the fuzzy set theory, which introduces linguistic terms and fuzzy numbers to manage vagueness and uncertainty in human judgment [7].

Both initiation and decision-making are shaped by internal conditions that influence how initiatives emerge, and how decisions are ultimately made. In this paper, these conditions are referred to as internal drivers. While previous studies have identified a range of factors for the integration of sustainability into organizations [2, 10], the specific link between the processes of initiation, decision-making and associated internal drivers remains insufficiently examined. This paper addresses this gap by examining internal initiation and decision-making processes and their associated internal drivers.

3 Literature Analysis Methodology

The search string "corporate" AND "sustainability" AND "initiation" AND "decision-making" was designed without OR operators and wildcards to maintain a focused search scope, as initiation and organizational decision-making processes are interconnected and well captured through their combination. The string was applied in the databases ScienceDirect and Web of Science, focusing on journal papers in a period from 2015 to 2025, yielding a total of 2,182 records. Systematic evaluation of titles, abstracts and full texts resulted in 31 reports. Reports focusing on organizational and technological measures regarding corporate sustainability, as well as reports focusing on initiation and decision-making in an organizational and technological dimension were included. An additional backward search, screening the references of the 31 identified core reports, and an explorative web search focusing on practitioner reports and grey literature yielded 10 additional reports resulting in the total of 41 reports [11]. The literature search is visualized in Fig. 1.

Deductive-qualitative content analysis, in which categories for coding are derived from previous theoretical considerations, is conducted to extract information from the identified papers [12]. Relevant categories, underlying sub-categories and the number of reports contributing to each sub-category is visualized in Table 1.

Table 1 Deductive-qualitative content analysis: coding system

Initiation				Decision-making				
Initiation process	Roles and actors	Connections and networks	Influence and barriers	Decision preparation	Organizational decision-making process	Decision-making support	Roles and actors	Influence and barriers
14	16	13	21	7	8	10	9	14

4 Findings of the Literature Analysis

4.1 Initiation

This section addresses the process of initiation and emergence of sustainability measures in organizations. Initiation occurs on strategic, tactical and operational levels. The strategic level includes scans of the environment, industry, resources, and includes the definition of vision and mission, whereas the tactical level focusses on the identification of stakeholders, requirements analysis, and preparations. On the operational level, the project charter is designed [13]. Borgert et al. refer to initiation as a checklist approach ensuring that relevant issues, holistically derived from sustainability standards, are included in initiatives. Prerequisite for this is a set of values aligned with stakeholders [14]. A case study conducted with a Swedish industrial company indicates the CEO's attendance of a presentation on global warming and competitors' actions towards sustainability as the initiation. Following this, senior management sets up a team of young employees to develop an approach for energy saving. After first results, a sustainability staff function is created, driving top-down initiatives, identifying overlapping ideas, and

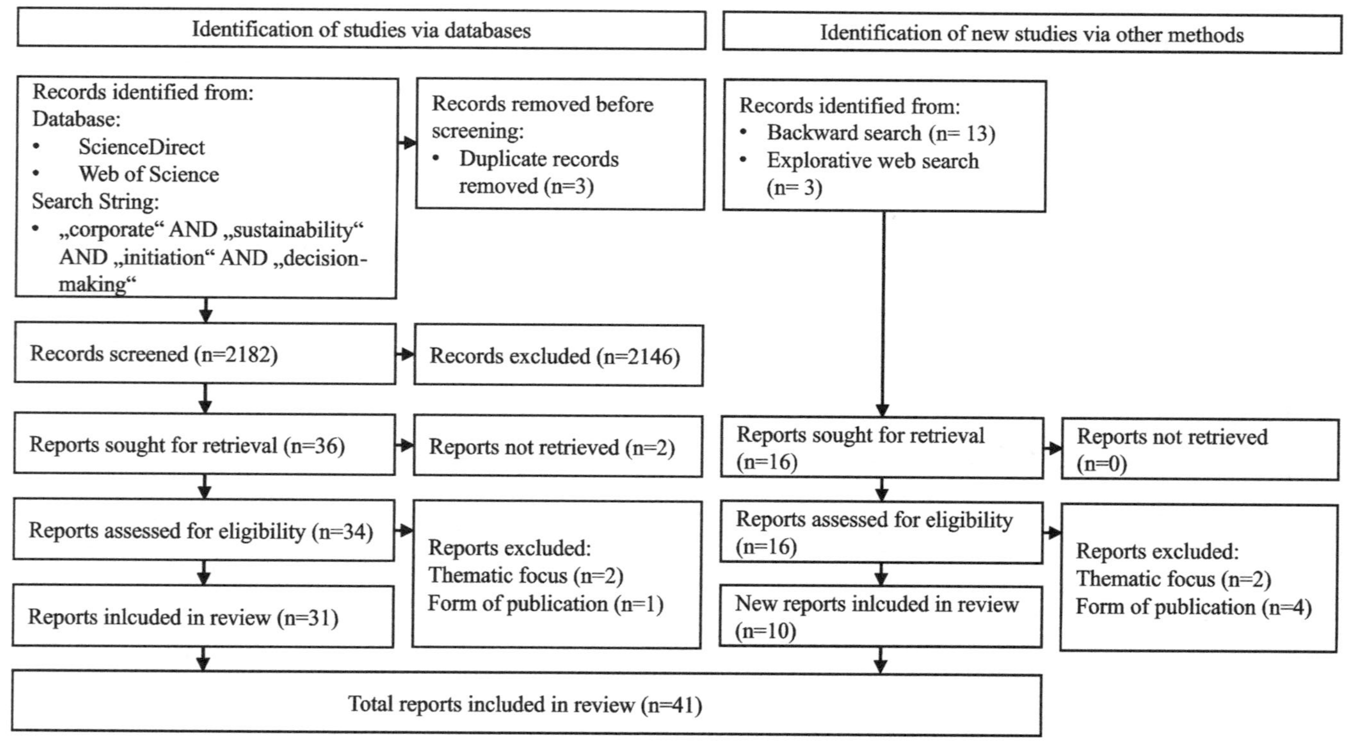

Fig. 1 Literature analysis: flow diagram of systematic literature search

creating networks [15]. In contrast to these top-down initiatives, other scholars emphasize the identification of a need through performance feedback or market opportunities, followed by idea generation with firm resources and skills as the initiation. Brainstorming sessions generate business ideas coupled with existing technologies. Engaging with diverse actors contributes to testing of value propositions. Bottom-up-movement of ideas is highlighted [16]. Vigneau promotes vertical and horizontal implementation to establish and diffuse sustainability measures. Commitment towards new measures is achieved vertically by adapting strategy and vision, meanwhile horizontally by developing leadership across different units. Including sustainability measures into operations requires communication on all levels [17].

The studies state different organizational actors as change agents. Since top-down and bottom-up movements are highlighted, all hierarchical levels seem to be involved in the initiation. Top management's responsibility to make necessary internal changes is emphasized [18, 19]. Managers come together in committees [20]. Others argue that middle managers and sustainability staff functions are more likely to play a key role, especially in small- and medium-sized enterprises (SMEs) [15]. Change agents collaborate across functions within a firm, especially with R&D, facilitating the generation of information, know-how and capabilities which is required for change [16]. Innovation ecosystems enable voluntary collaboration among startups, research institutions, suppliers, competitors, and the public sector to address external challenges by integrating resources, knowledge, and technology for sustainable development. Each actor's contribution to the ecosystem should be assessed to support a self-adjusting system [21]. Exchange and collaboration are emergent rather than organized [22].

Ameer and Khan identify factors influencing strategy orientation towards sustainable technologies on a personal, organizational and contextual level. On the personal level, personality traits, attitudes, abilities and emotions are relevant [19]. Essential characteristics of change agents are perceived responsibility and competence in the field [23]. In terms of initiation, individual acceptance of settings is relevant. An attitude toward measures emerges by weighing benefits, which leads to an associated intention to act [24]. Recognition provides change agents with capacity to act. Building this capacity requires autonomy at the individual level, caring practices at group level and justice at the institutional level. At the individual level, agency begins with self-imputation and is strengthened by imagination and value alignment. At the group level, mutual recognition and shared accountability, expressed through promises, build cooperation. Institutions support this capacity to act through stability, fairness, and procedural integrity [23]. On the organizational level, integration of sustainability and emission reduction into strategy is a prerequisite for the implementation of measures. Organizational structure, entrepreneurial culture, leadership support, management control, cross-hierarchical exchange, employee involvement, qualification, and resource buffers represent success factors for sustainability integration and change initiatives. The literature is divided on the influence of incentive systems [20]. On the contextual level, governmental interventions, stakeholders' involvement, and cross-organizational communication are success factors for change initiatives [19].

4.2 Decision-Making

In the organizational decision-making process, CEO integration plays a key role in identifying relevant issues, building knowledge, and sharing across the organization [25]. Interviews show that environmental capital expenditures require top-management approval before energy information is passed to the finance team [18]. This adds to Brigham et al. pointing out complex and multi-layered decision-making processes in organizations [26]. Kozioł-Nadolna and Beyer identify an organizational decision-making process, starting with problem and goal identification. Secondly, relevant information and resources are gathered, followed by the evaluation of alternatives. This evaluation is not specified. Decision-making follows the hierarchical structure to ensure clarity, especially in large companies. It can be stated that top-management makes decisions. Yet all hierarchical levels contribute to the decision by providing relevant information, since inclusive decision-making fosters commitment and accountability. Personal factors (commitment, values, experience, competence, age, stress), economic factors (resources, competition), and organizational factors (leadership style, organizational structure, organizational culture) influence sustainable decision-making [26, 27].

The lack of complete information regarding energy efficiency technologies poses a problem for SMEs with little capacities and expertise [28]. Gunarathne and Lee state that the use of information depends on the development stage of the organization. In the functional specialization stage, information is primarily used regarding efficiency strategies leading to emission reduction for setting performance targets, reporting, and meeting standards. In the internal integration stage, organizations also focus on consistency and sufficiency strategies, using information for budget preparing and product optimization. In the external integration phase, where focus lies on exploring competitive advantage, information supports make or buy and equipment replacement decisions. Especially within efficiency strategies, information is gathered in an initial audit conducted by an external party, often consultancies. Yet, identified potentials are not followed, indicating ad-hoc and unsystematic use of information within the decision-making process for sustainability measures, especially emission reductions [18].

Calabrese et al. systematize decision-making in an SME. Fuzzy AHP is applied for selecting sustainability issues with high relevance regarding the creation of shared value [25]. Other studies do not link the organizational decision-making process and decision-making support methodologies and maintain a technical focus. In sustainable energy systems, MCDM methods evaluate criteria such as efficiency, cost, and reliability. Hernández-Torres et al. combine AHP and TOPSIS, using AHP-derived weights in TOPSIS, and extend this with fuzzy AHP to handle linguistic uncertainty. The hybrid approach yields more robust outcomes [7]. Zhou et al. select locations for carbon capture and storage projects applying TODIM with criteria such as cost, climatic conditions, project safety, and public acceptance [9]. ANP is applied in the renewable energy sector to assess interrelated criteria with investment cost being most influential, followed by technical, environmental and social factors [8].

5 Discussion

Findings of the literature analysis demonstrate the relevance of structured initiative frameworks and emergent practices for initiating sustainability measures. On the one hand, planning with stakeholder identification, resource analysis and checklists are highlighted. On the other hand, idea generation from brainstorming, and collaboration are pointed out. Top-down and bottom-up movements are highlighted, indicating the inclusion of change agents in the initiation on all hierarchical levels. However, the presence of platforms for voluntary collaboration and the rather emergent than organized information exchange suggest that traditional, centralized models of leadership may not be well suited to develop conditions promoting sustainability initiatives, especially if top management commitment to sustainability lacks. Yet, concepts of collaboration, such as innovation ecosystems require qualified employees who feel responsible and develop agency. The conceptual understanding of agency and the organizational decision frame are under tension. It is unclear how reflective processes as recognition and acceptance can unfold within rigid hierarchies where individuals may feel little control. Literature suggests inclusive structures such as horizontal leadership, cross-functional and cross-hierarchical exchange to bridge this gap.

Top-management makes decisions based on personal values and intuition. It is stated that all hierarchical levels contribute to the decision by providing relevant information. Nevertheless, the use of information depends on the development stage of the organization. There is a lack of complete information in the early stage to support emission reduction by equipment replacement decisions. In this early stage, identified potentials are often not pursued, indicating ad-hoc and unsystematic use of information for decisions regarding long-term emission reduction. Relevant decisions may be overlooked in early development stages due to a lack of immediate economic incentive or internal capacity. This emphasizes the importance of building feedback loops, and integrating sustainability into decision-making to ensure that early insights lead to sustained action. While the literature acknowledges that all hierarchical levels contribute information, it offers limited detail on how the decision-making process is structured and which organizational actors are actively involved. Although identified MCDM methods offer structure in the objective, and criteria-based evaluation of complex trade-offs, their use is limited in industrial practice.

This paper is limited by the keyword-based search string, the focus on the English language, and the ten-year time frame, which may have excluded relevant literature.

6 Conclusion

The purpose of this paper was to conduct a literature analysis examining the internal initiation of sustainability measures, underlying decision-making processes, actors, and associated internal drivers, before the actual implementation of measures. The systematic literature analysis provides the findings of 41 papers. Initiation and decision-making for sustainability measures in organizations are multi-level processes influenced by personal, organizational, and contextual factors. Practical implications focus on the need of organizational structures to balance strategic direction with decentralized initiative-taking. Leaders play a key role by fostering commitment, recognition, and competence.

A shift from control to coordination, while enabling change agents through competence is necessary. Organizations must ensure that information is acted upon, and that decision-making tools are used meaningfully. Future research could investigate in detail how changes of organizational structure could promote initiatives, especially on the personal and organizational level. Furthermore, it should be examined how organizational decision-making processes are designed and how MCDM methods should be included to rationalize decision-making. Conducting a qualitative interview study could be expedient.

Competing Interests. The author(s) has no competing interests to declare that are relevant to the content of this manuscript.

References

1. Gerstenberger, J., Bauer, A.: Unternehmensbefragung (2024). https://www.kfw.de/%C3%9Cber-die-KfW/KfW-Research/Unternehmensbefragung.html. Accessed 21 July 2025
2. Engert, S., Rauter, R., Baumgartner, R.J.: Exploring the integration of corporate sustainability into strategic management. J. Clean. Prod. **112**(4), 2833–2850 (2016)
3. Granstrand, O., Holgersson, M: Innovation ecosystems. Technovation **90** (2020)
4. Frese, M., Kring, W., Soose, A., Zempel, J.: Personal initiative at work: differences between East and West Germany. AMJ **39**(1), 37–63 (1996)
5. ISO 21502:2020: Project, programme and portfolio management—guidance on project management (ISO 21502:2020), Berlin
6. Lara, C., Wassick, J.: The future of supply chain—a perspective from the process and online retail industries. Comput. Chem. Eng. **179**(4) (2023)
7. Hernández-Torres, J., Sánchez-Lozano, D., Sánchez-Herrera, R.: Integrated multi-criteria decision-making approach for power generation technology selection in sustainable energy systems. Renew. Energy **243** (2025)
8. Ozorhon, B., Batmaz, A., Caglayan, S.: Generating a framework to facilitate decision making in renewable energy investments. Renew. Sustain. Energy Rev. **95**, 217–226 (2018)
9. Zhou, J., Wu, Ṡ., Chen, Z.: Promoting the sustainable development of CCUS projects. Sustain. Cities Soc. **114** (2024)
10. Lozano, R.: A holistic perspective on corporate sustainability drivers. Corp. Soc. Respons. Environ. Manag. **22**(1) (2013)
11. Page, M., McKenzie, J., Bossuyt, P.: The PRISMA 2020 statement: an updated guideline for reporting systematic reviews. PLoS Med. **18**(3) (2021)
12. Mayring, P., Fenzl, T.: Qualitative Inhalts analyse. In: Baur, N., Blasius, J. (eds.) Handbuch Methoden der empirischen Sozialforschung, pp. 633–648. Springer VS, Wiesbaden (2019)
13. Gholamzadeh, A., Ariani, F., Jaromír, J.: Development of guidelines for the implementation of sustainable enterprise resource planning systems. J. Clean. Prod. **244** (2020)
14. Borgert, T., Donovan, J., Topple, C.: Initiating sustainability assessments: insights from practice on a procedural perspective. EIA Rev. **72**, 99–107 (2018)
15. Egels-Zandén, N., Rosén, M.: Sustainable strategy formation at a Swedish industrial company. J. Clean. Prod. **96**, 139–147 (2015)
16. Saka-Helmhout, A., Álamos-Concha, P., Polo-Barceló, C.: Corporate social innovation by multinationals: a framework for future research. IBR **33**(5) (2024)
17. Vigneau, L.: A micro-level perspective on the implementation of corporate social responsibility practices in multinational corporations. J. Int. Manag. **26**(4) (2020)

18. Gunarathne, N., Lee, K.: Environmental and managerial information for cleaner production strategies. J. Clean. Prod. **237** (2019)
19. Ameer, F., Khan, N.: Green entrepreneurial orientation and corporate environmental performance: a systematic literature review. EMJ **41**(5), 755–778 (2023)
20. Engert, S., Baumgartner, R.: Corporate sustainability strategy—bridging the gap between formulation and implementation. J. Clean. Prod. **113**, 822–834 (2016)
21. Tolstykh, T., Gamidullaeva, L., Shmeleva, N.: Elaboration of a mechanism for sustainable enterprise development in innovation ecosystems. JOItmC **6**(4) (2020)
22. Mirvis, P., Herrera, M., Googins, B.: Corporate social innovation: how firms learn to innovate for the greater good. J. Bus. Res. **69**(11), 5014–5021 (2016)
23. Niess, A., Duhamel, F.: The course of recognition and the emergence of change initiatives. JCM **31**(5), 1071–1083 (2017)
24. Griese, K., Franz, M., Busch, J.: Acceptance of climate adaptation measures for transport operations. Transp. Res. D: Transp. Environ. **101** (2021)
25. Calabrese, A., Costa, R., Levialdi, N.: Integrating sustainability into strategic decision-making: a fuzzy AHP method for the selection of relevant sustainability issues. Technol. Forecast. Soc. Change **139**, 155–168 (2019)
26. Brigham, M., Kiosse, P., Otley, D.: A structured framework to understand CSR decision-making: a case study of multiple rationales. EMJ **41**(3), 345–353 (2023)
27. Kozioł-Nadolna, K., Beyer, K.: Determinants of the decision-making process in organizations. Procedia Comput. Sci. **192**, 2375–2384 (2021)
28. Kounetas, K., Skuras, D., Tsekouras, K.: Promoting energy efficiency policies over the information barrier. IEP **23**(1), 72–84 (2011)

Consolidation Studies on Solid-State-Recycling by Forming Aluminum Chips

N. Gerke(✉), C. Glaubitz, J. Peddinghaus, K. Brunotte, and B.-A. Behrens

Institute of Forming Technology and Machines, Leibniz Universität Hannover, Hannover, Germany
gerke@ifum.uni-hannover.de

Abstract. In conventional recycling, where scrap is melted down, significant material losses occur due to oxidation and burn-off, especially with small elements like chips. In solid state recycling (SSR), chips are initially compacted into a green body before undergoing further heating and forming. The aim of this research is to determine whether the multiaxial forming of cold-pressed samples from forged aluminum alloy chips can produce a sufficient material bond for further processing. Cold pressing of chips out of EN AW-6082 and EN AW-7075 did not create sufficient material bonding to endure subsequent processes without damage. Therefore, Multi-Directional Forging (MDF) and the equal channel angular pressing (ECAP) process were employed. In order to evaluate the effects of the processes on the chip interfaces and the microstructure, micrographs and etchings were carried out. Metallographic analyses of the results showed that the quality of the interfacial bond before further forming is crucial in determining the final material properties. Furthermore, the MDF was able to eliminate larger defects in the form of pores.

Keywords: Solid-state-recycling · Wrought aluminum alloy · Chip consolidation

1 Introduction

The machining of high-quality aluminum alloys produces large quantities of chips, which, despite the associated material and energy consumption, have so far been insufficiently recycled [1, 2]. Conventional melting of these residues is not only energy-intensive, but also involves high material losses due to burning, oxidation, and slag formation when dealing with small chip fractions [3, 4]. In contrast, solid-state recycling (SSR) offers a promising way to consolidate aluminum chips without remelting, thereby achieving significant environmental and resource savings. The central goal of SSR processes is to obtain formable semi-finished products from aluminum chips without the losses associated with the remelting process [5].

The mechanisms on which SSR is based are the breaking up of oxide layers through shear stresses and high plastic deformation. High degrees of deformation are helpful in the creation of a material compound. These mechanisms also ensure grain refinement during deformation [6].

L. Overmeyer and B.-A. Behrens (eds.), *Production at the Leading Edge of Technology*, Lecture Notes in Production Engineering, https://doi.org/10.1007/978-3-032-19524-1_4

The representative AW6082 combines a Mg_2Si-hardened α-Al matrix with moderate strength, good toughness, and high corrosion resistance. The 7075 alloy, on the other hand, belongs to the high-strength 7xxx series and is used in a variety of load-bearing aircraft structures due to its high strength-to-weight ratio [7]. However, its increased Zn, Mg, and Cu content leads to segregation-promoting $MgZn_2$ and $CuAl_2$ formations at grain boundaries even in the undissolved initial state, which are described in the literature as a potential source of embrittlement and narrowed thermo-mechanical process windows [8, 9]. High-alloyed 7xxx chips (AW 7075), on the other hand, are significantly more sensitive. Above approximately 450 °C, the aforementioned $MgZn_2$ and $CuAl_2$ boundary films form, which act as microcrack nuclei and require narrow process windows and limited individual strains [8–11].

Previous work on the SSR of medium-alloyed 6xxx chips Haase and Tekkaya and Behrens et al. document full densification (>99%) and tensile strengths of up to 95% of the compact material after hot extrusion or hot backward extrusion [5, 12].

One SSR process approach is multidirectional forming (MDF). In this process, a square or rectangular sample is compressed in a die in alternating orientations. The effects of this process on the mechanical properties of different materials have already been demonstrated in several studies. Obara et al. [10] and Manjunath et al. [13] were able to show with samples of 7xxx aluminum alloys made from solid material, that a significant increase in tensile strength and hardness could be achieved after at least two forming processes at up to 400 °C.

It was also found that the influence of the forming speed on the force during MDF is only minor, but that the microstructure of the samples recovers more easily after slow forming [10].

In investigations into the processing of pre-compacted aluminum chips of a 6060 alloy, KOCH was able to produce a material composite using the Field assisted sintering process at 400 °C for 5 min and a press pressure of 80 MPa. Processing at higher temperatures reduces the defect sizes at the chip contact zones, but also promotes grain growth within the chips [14].

Another SSR approach is Equal Channel Angular Pressing (ECAP). In this process, a sample is pressed through a channel angle. The channel angle has a high influence on the homogeneity of the microstructure and the subsequent mechanical properties of the sample [15]. The plastic deformation and shearing of the sample material that occurs during this process results in a refinement of the microstructure. The pressing out of samples by subsequent material creates a counterpressure, which has a positive effect on the hardness and tensile strength of the samples [16]. Panigrahi et al. and Mckenzie et al. were able to demonstrate these effects using solid aluminum material [17, 18]. Lapovok et al. showed that directly filling the die with chips, compressing them in the die, and carrying out the process with counterpressure can produce a high material bond [19].

Both the ECAP and MDF processes have so far only been published on solid or previously consolidated material samples. This study aims to close this research gap. Therefore, this study investigates the direct processing of pre-compacted aluminum chips of alloys 6xxx and 7xxx. The comparison of these two alloys covers a broad spectrum from process-robust standard materials to process-sensitive high-performance materials and forms the basis for investigating the microstructural requirements for the solid-state recycling of aluminum chips. Cold-pressed aluminum chips are transformed by isothermal MDF and ECAP. The aim is to investigate whether the processes are suitable for consolidating samples under the same process parameters, with the goal of developing a parallel, efficient route for the further processing of SSR samples.

2 Material and Methods

First, samples were produced by cold pressing chips from EN AW-6082 and EN AW-7075 alloys. The samples were then subjected to the SSR processes MDF and ECAP with counterpressure. After one pass, the samples were analyzed for density and damage. Further passes were then carried out using the process that delivered the most promising results.

2.1 Material

For the tests, chips were produced from EN AW-6082 and EN AW-7075 alloys in the T6 condition by rotary machining. No cooling lubricant was used in order to avoid contamination of the samples. After machining, the alloys exhibited a chip volume ratio of 4.4 (EN AW-6082) and 3.6 (EN AW-7075).

2.2 Cold Compaction of Aluminum Chips

The chips of both alloys were cold-pressed into $21 \times 21 \times 100$ mm^3 briquettes. The pressing took place in one stroke. To ensure homogeneous compaction, the die was mounted on a floating bearing [12]. The green compacts were compressed to approximately 75% of their solid density. Samples measuring $21 \times 21 \times 14$ mm^3 were then taken from these briquettes through machining. The process parameters can be found in Table 1.

Table 1 Process parameters

	Die temperature in °C	Sample temperature in °C	Pressing speed in mm/s	Force limit in kN
Cold computations	Room temperature	Room temperature	10	–
ECAP	400	400	3	–
MDF	400	400	3	200

2.3 Forming Process

For multidirectional forming, dies and punches were manufactured made out of the hot-work steel 1.2367 (AISI H13) and used in the DYNSJ5590 servo-hydraulic forming simulator. The tool was designed that it could be heated to 400 °C in an electric chamber furnace together with the sample to ensure reproducible and isothermal conditions. This temperature was chosen to create a material compound while avoiding grain growth [14]. Once the tool had reached the target temperature, the sample was inserted and both components were heated together for another 10 min. The tool with the sample was then inserted into the press (Fig. 1b) and the forming operation was carried out at a press speed of 3 mm/s. The force was limited to 200 kN, which corresponds to a surface pressure of 453 MPa. The samples that had undergone the process were removed from the tool and reinserted into the die in a defined route, shown in Fig. 1a. The tool was then reheated together with the sample. Samples underwent the process up to three times, with the first stage representing warm compaction. The process parameters can be found in Table 1.

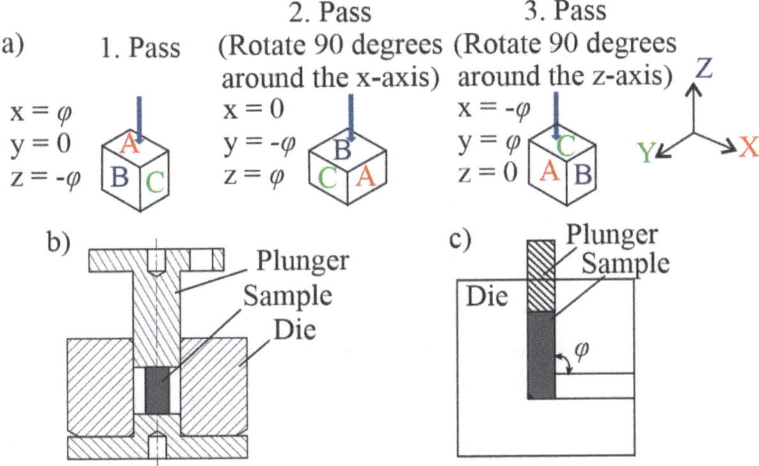

Fig. 1 MDF process route [13] (**a**), MDF tool (**b**), ECAP tool [16] (**c**)

For the ECAP process, the tool, with a channel angle of 90° (Fig. 1c), was also heated to 400 °C. After a heating phase, the sample bars (21 × 21 × 100 mm^3) made of EN AW-6082 alloy were placed in the tool and heated for 10 min. The tool and sample were then placed in the Schirmer und Plate hydraulic press, which has a maximum pressing force of 12,500 kN, and the process was carried out at a speed of 3 mm/s. The process parameters can also be found in Table 1. Samples were produced both without counterpressure and with counterpressure (from the prior sample still in the tool). After the samples were removed from the tools in both processes, they were cooled in air.

2.4 Analytical Methods

The metallographic analysis was carried out along the cross-section examination of the specimens. For this purpose, the MDF samples were cut in half. The ECAP samples were first cut longitudinally down the middle. Subsequently, an observation area was taken from the central area of the sample. The separation was performed using wet abrasive cutting.

To evaluate the change in grain size and reduction of brace boundaries, the samples were treated with the Barker etching method and evaluated under polarized light.

The overall density development was evaluated using immersion weighing based on the Archimedes principle.

3 Results and Discussion

When comparing the two processes after one run, compaction can be observed in both samples made from EN AW-6082. The MDF sample showed a density of 97%. The sample from the ECAP process with counter pressure had a density of 96%. A microscopic comparison of the two samples reveals clear differences. In the MDF sample, a large number of voids can be seen at the chip boundaries across the entire sample cross-section (Fig. 2a), which indicates insufficient compaction in the first process step.

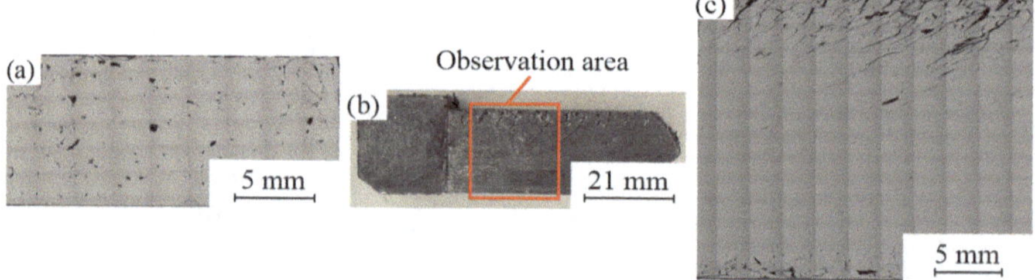

Fig. 2 EN AW-6082 after one pass MDF (**a**), ECAP viewing area (**b**), EN AW-6082 after one pass ECAP

For the examination of the ECAP sample out of EN AW-6082, a middle section was selected in order to examine an area from the continuous process (Fig. 2b). Here, clear damage can be seen in the upper and lower areas of the sample (Fig. 2c). This can be explained by shear stresses in the upper area and tensile stresses in the lower one [20, 21]. Insufficient material bonding between the chip material prior to the ECAP process led to inadequate consolidation and tearing of the sample in these areas.

Due to the cracking that occurred during the process, it can be assumed that the sample still in the outlet channel was unable to build up sufficient counterpressure to prevent cracking [17, 22].

When comparing the two processes, MDF and ECAP, a similar density was observed after one run. However, the MDF showed significantly fewer defects, as demonstrated here using the EN AW-6082 alloy. Subsequently, three forming processes were carried out using the MDF method.

During the MDF tests, it was found that the samples made of EN AW-7075 developed insufficient bonding. This inadequate bonding became apparent when the sample was removed and several areas separated. The metallographic image (Fig. 3) of a sample that had undergone the MDF-process three times shows clear cracks and defects. The cracks appeared at 400 °C, which is below the threshold of 450 °C for the formation of brittle phases mentioned in the literature [9–11]. Two factors could explain this: (i) local forming heat and friction peaks increase the temperature at the chip contacts by several tens of Kelvin for a short time [23]; (ii) zinc and magnesium enrichments on the chip surfaces lower the precipitation start temperature. Both effects shorten the incubation time of the η-/θ-boundary films and shift the effective process window downwards [8]. After three forming processes, it became apparent that the EN AW-7075 alloy could not be sufficiently strengthened with these process parameters.

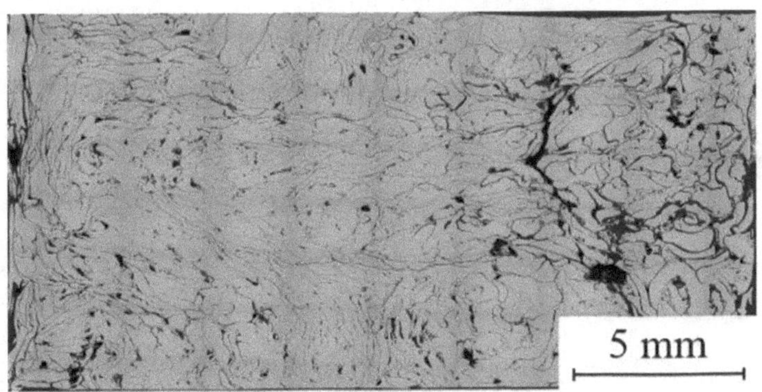

Fig. 3 EN AW-7075 after three passes MDF

The microstructural properties of the EN AW-6082 MDF samples were then analyzed in more detail.

However, after multiple passes through the MDF process, the EN AW-6082 samples showed a reduction in defects (Fig. 4a, b) and an increase in density from 97% to nearly 100%. The continuous increase in density and absence of cracks in AW 6082 due to kneading indicate successful oxide breakage and complete pore closure, as described by Haase and Tekkaya for 6xxx chips in similar shear deformation [5].

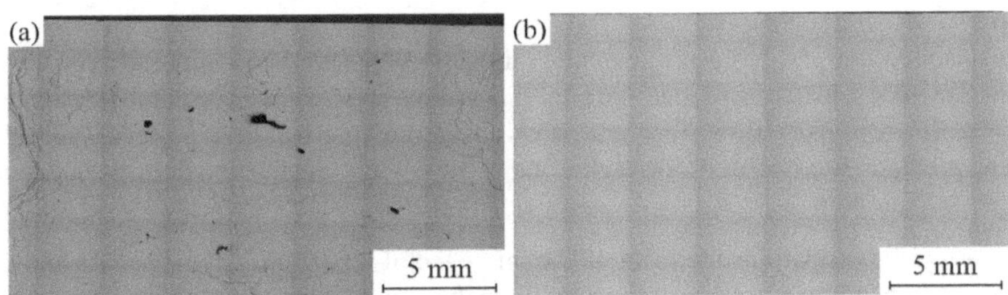

Fig. 4 EN AW-6082 after two passes MDF (**a**) and after three passes MDF (**b**)

When comparing the microstructure images from MDF EN AW 6082 after one and after three passes, a clear refinement can be seen. Kishchik et al. were able to attribute a comparable effect to continuous dynamic recrystallization for Al–Mg samples at 400 to 500 °C [23]. The previously clearly visible grain boundaries are no longer identifiable after three passes. However, Fig. 5b shows inhomogeneities in the grain size distribution, which can be explained by the varying degrees of deformation during upsetting [24]. These are process-related and can be made more uniform by increasing the number of forging strokes [25].

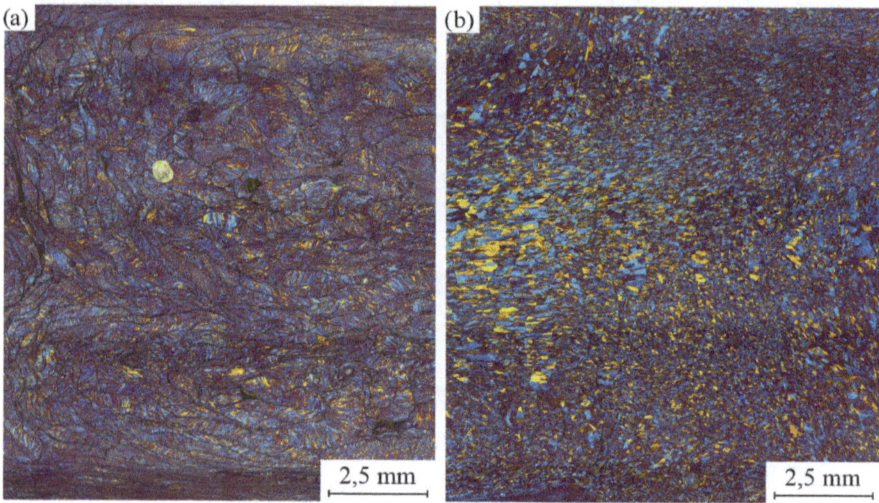

Fig. 5 EN AW-6082 etching after one pass MDF (**a**) and after three passes MDF (**b**)

4 Conclusion and Outlook

A clear difference was observed between the process parameters of the SSR methods MDF and ECAP and the samples used here from EN AW-6082 and EN AW-7075.

- Samples that already exhibit sufficient material bonding are required to perform the ECAP process, whereas the MDF process is suitable for processing cold pre-compacted chips.
- Consolidation and grain refinement of the EN AW-6082 alloy is possible with the parameters selected here, but more forging passes are required for homogenization.
- The EN AW-7075 alloy does not show any promising results for processing with the methods used here. A corresponding process window must be defined in further investigations.

Further investigations are to show the influence of the forming temperature and the number of kneading processes during MDF on the consolidation of both alloys, with the aim of finding a suitable process window for the direct processing of aluminum chips. In addition, the use of lubricants is to be investigated.

Acknowledgements. The results presented were developed while working on the project "Forming-based reconditioning of metal chips through sintering and forging", project number 215349600. The authors thank the German Research Foundation DFG (Deutsche Forschungsgemeinschaft) for financial support.

Competing Interests. The author(s) has no competing interests to declare that are relevant to the content of this manuscript.

References

1. Dhiman, S., Joshi, R.S., Singh, S., Gill, S.S., Singh, H., Kumar, R., Kumar, V.: A framework for effective and clean conversion of machining waste into metal powder feedstock for additive manufacturing. Clean. Eng. Technol. **4**, S. 100151 (2021)
2. Wang, J., Zhang, S., Lv, Z., Liu, B., Zhang, H., Du, S., Liu, J.: Microstructure evolution and properties comparation of industrial grade-maintained 7050–T7451 plate recycled from machining chips. J. Mater. Res. Technol. **25**, 6011–6026 (2023)
3. Duflou, J.R., Tekkaya, A.E., Haase, M., Welo, T., Vanmeensel, K., Kellens, K., et al.: Environmental assessment of solid state recycling routes for aluminium alloys: can solid state processes significantly reduce the environmental impact of aluminium recycling? CIRP Ann. **64**(1), S. 37–40 (2015)
4. Murray, J.W., Jin, X., Cleaver, Christopher J.; Azevedo, Jose M.C.; Liao, Zhirong; Zhou, Wenbin et al. (2024): A review of principles and options for the re-use of machining chips by solid, semi-solid or melt-based processing. J. Mater. Process. Technol. **331**, S. 118514
5. Haase, M., Tekkaya, A.E.: Cold extrusion of hot extruded aluminum chips. J. Mater. Process. Technol. **217**, 356–367 (2015)
6. Shamsudin, S., Lajis, M.A., Zhong, Z.W.: Solid-state recycling of light metals: a review. Adv. Mech. Eng. **8**(8), 1687814016661921 (2016)
7. Zhou, B., Liu, B., Zhang, S.: The advancement of 7XXX series aluminum alloys for aircraft structures: a review. Metals **11**(5), S. 718 (2021)
8. Birol, Y., Birol, F., Yuksel, B., Duygulu, O.: Corrosion behaviour of twin belt cast EN AW 7075 alloy. Mater. Corros. **64**(10), S. 881–889 (2013)
9. Abdollahi, A., Nganbe, M., Kabir, A.S.: On the elimination of solidification cracks in fusion welding of Al7075 by TiC-nanoparticle enhanced filler metal. J. Manuf. Process. **81**, 828–836 (2022)
10. Obara, C., Mwema, F.M., Jen, T.C.: Analysis of the multi-directional forging of aluminium alloy 7075 process parameters: numerical and experimental analysis. Key Eng. Mater. **924**, 61–72 (2022)
11. Ruhaizat, N.E., Yusuf, N.K., Lajis, M.A., Al-Alimi, S., Shamsudin, S., Tukiat, I.S.T., Zhou, W.: Effect of direct recycling hot press forging parameters on mechanical properties and surface integrity of AA7075 aluminum alloys. Metals **12**(10), S. 1555 (2022)
12. Behrens, B.-A., Frischkorn, C., Bonhage, M.: Reprocessing of AW2007, AW6082 and AW7075 aluminium chips by using sintering and forging operations. Prod. Eng. Res. Dev. **8**(4), S. 443–451 (2014)
13. Manjunath, G.A., Shivakumar, S., Avadhani, S.P., Sharath, P.C.: Investigation of mechanical properties and microstructural behavior of 7050 aluminium alloy by multi directional forging technique. Mater. Today Proc. **27**, 1147–1151 (2020)
14. Koch, A.: Verbindungsmechanismen und Leistungsfähigkeit von stranggepressten und feldunterstützt gesinterten Halbzeugen aus wiederverwerteten Aluminiumspänen. Springer (2024)

15. Ravikumar, K., Ganesan, S., Karthikeyan, S.: An overview on the influence of equal channel angular pressing parameters and its effect on materials: methods and applications. Adv. Mater. Process. Technol. **10**(3), 1814–1855 (2023)
16. Faraji, G., Kim, H.S., Kashi, H.T.: Severe Plastic Deformation: Methods, Processing and Properties. Elsevier (2018)
17. Panigrahi, A., Scheerbaum, N., Chekhonin, P., Scharnweber, J., Beausir, B., Hockauf, M., et al.: Effect of back pressure on material flow and texture in ECAP of aluminum. In: IOP Conference Series: Materials Science and Engineering, vol. 63, no. 1, p. 012153. IOP Publishing (2014)
18. Mckenzie, P.W.J., Lapovok, R.: ECAP with back pressure for optimum strength and ductility in aluminium alloy 6016. Part 2: Mechanical properties and texture. Acta Materialia **58**(9), 3212–3222 (2010)
19. Lapovok, R., Qi, Y., Ng, H.P., Maier, V., Estrin, Y.: Multicomponent materials from machining chips compacted by equal-channel angular pressing. J. Mater. Sci. **49**, 1193–1204 (2014)
20. Ghosh, A., Das, K., Eivani, A.R., Mohammadi, H., Vafaeenezhad, H., Murmu, U.K., et al.: Development of mechanical properties and microstructure for Al–Zn–Mg–Cu alloys through ECAP after optimizing the outer corner angles through FE modeling. Arch. Civ. Mech. Eng. **23**(2), 78 (2023)
21. Nagasekhar, A.V., Tick-Hon, Y.: Optimal tool angles for equal channel angular extrusion of strain hardening materials by finite element analysis. Comput. Mater. Sci. **30**(3–4), 489–495 (2004)
22. Lapovok, R., Tomus, D.: Production of Dense Compact Billet from Ti-Alloy Powder Using Equal Channel Angular Extrusion. ARC Centre of Excellence for Design in Light Metals, Dept. of Materials Engineering, Monash University, Clayton, Melbourne, Australia (2007)
23. Kishchik, M.S., Mikhaylovskaya, A.V., Kotov, A.D., Mosleh, A.O., AbuShanab, W.S., Portnoy, V.K.: Effect of multidirectional forging on the grain structure and mechanical properties of the Al–Mg–Mn alloy. Materials **11**(11), 2166 (2018)
24. Li, J., Dong, S., Zhao, C., Zeng, J., Jin, L., Wang, F., et al.: Multi-directional forging of large-scale Mg-9Gd-3Y-2Zn-0.5 Zr alloy guided by 3D processing maps and finite element analysis. Int. J. Adv. Manuf. Technol. **120**(9), 5985–5996 (2022)
25. Zhang, S., Zhang, G., Teng, D., Wang, C., Zhang, X., Guan, R.: Microstructural evolution and enhanced properties by multi-directional forging of 6201 aluminum alloy. Matéria (Rio de Janeiro) **29**(2), e20240020 (2024)

Evaluation of Measurement Techniques for Assessing Corroded Bolted Joints in Remanufacturing: A Reference Measurement for a Data-Driven Approach

Carolin Lange[1(✉)], Nehal Afifi[2], Sven Matthiesen[2], and Gisela Lanza[1]

[1] wbk Institute of Production Science, Karlsruhe Institute of Technology (KIT), Karlsruhe, Germany
carolin.lange@kit.edu

[2] IPEK Institute of Product Engineering, Karlsruhe Institute of Technology (KIT), Karlsruhe, Germany

Abstract. The circular economy emphasizes sustainability and resource efficiency, positioning remanufacturing as a key strategy. However, ensuring the quality and reliability of remanufactured components, particularly bolted joints, remains challenging due to corrosion-induced degradation. This study compares three measurement techniques: photographic evaluation, computed tomography (CT), and optical coordinate measurement with focus variation for assessing corrosion effects on bolts. The techniques are evaluated based on suitability for surface degradation detection and deformation assessment, limitations, measurement time, storage size and resolution. The results indicate that photography offers quick visual feedback and has the smallest storage size (4.6–22.6 MB), but has a medium resolution (10.7 μm/pixel) and lacks objectivity when evaluated manually. CT detects deformations and internal material defects well, but are less suitable for fine surface characterisation due to their medium resolution (19.83 μm/voxel), measurement time and storage limitations (45 min, 4.07 GB). Optical coordinate measurements allow precise analysis of surface roughness, efficiently capturing corrosion effects and deformations, while offering a balance between high resolution (0.5 μm vertically, 7.83 μm laterally) and medium measurement time (50 min) and moderate storage needs (80 MB), which can be significantly reduced if only the area of interest is scanned. These results serve as a reference for developing predictive, data-driven models to support quality assurance in remanufacturing within the circular economy. The study was limited by short corrosion durations and moderate preload differences, which reduced comparability across conditions. Future work should extend exposure times and load ranges to improve the method comparison.

Keywords: Corrosion measurement for bolts · Measurement comparison · Remanufacturing quality

L. Overmeyer and B.-A. Behrens (eds.), *Production at the Leading Edge of Technology*, Lecture Notes in Production Engineering, https://doi.org/10.1007/978-3-032-19524-1_5

1 Introduction

Under the increasing pressure of stringent resource scarcity, environmental regulations and the pursuit of economic efficiency, more companies are transitioning towards the circular economy model [1]. The circular economy introduces a systemic shift from a linear to a closed-loop model of production and consumption, aiming to maximize the value derived from materials by keeping them in use for as long as possible [2] and enable resource efficiency and sustainability.

Remanufacturing refers to the industrial process of disassembling, inspecting, cleaning, repairing, or replacing parts of end-of-life products to restore them to like-new or even superior functionality [2–4]. Products such as automotive parts, heavy machinery, and electronic equipment are commonly remanufactured, often backed by warranties comparable to new products [5]. The majority of remanufacturing activities focus on spare parts, especially in the automotive sector, driven by market demand and cost-saving potential [6]. However, ensuring the quality and reliability of remanufactured products remains a major challenge. A critical aspect of this challenge lies in bolted joints, which are essential for both product assembly and disassembly. Corrosion and mechanical degradation of these joints introduce significant uncertainty in the structural integrity and mechanical performance of reused parts.

The problem lies in the lack of a systematic comparison of the existing methods used to measure corrosion in bolted joints, which limits quantitative evaluation and the identification of the most suitable technique in remanufacturing contexts.

2 State of the Art

Bolted joints are essential mechanical elements widely used across various industrial products, enabling both assembly and disassembly—key requirements in remanufacturing processes [4]. Due to prolonged usage and exposure to environmental stressors, these joints are particularly susceptible to corrosion and mechanical wear, significantly affecting their reusability and reliability. Corrosion can impair the mechanical strength and complicate the disassembly of products. Common forms of surface degradation, such as pitting, alters the material condition and surface quality, introducing uncertainty in predicting joint performance [7, 8]. Deformation assessment focuses on measuring geometric deviations from the original shape, such as bending, flattening, or misalignment. Current industry practices predominantly rely on visual inspection by a worker to assess product condition. While fast and adaptable, they are subjective and not reliably reproducible. This subjectivity arises from varying lighting conditions, individual perception, fatigue, and differences in experience or training among inspectors, all of which can lead to inconsistent assessments of the same defect or surface feature.

According to the state of research, computed tomography (CT) was developed to capture internal and external geometries in high detail, and it is able to visualize volumetric deformations, but its high cost, time consumption [9]. Optical coordinate measurement with focus variation, on the other hand, was designed for precise surface topography, roughness and damage quantification, making them promising tools for objective evaluation [10].

Despite the efforts in research to develop different measurement techniques to assess the quality of returned products, there is still a research gap in having a comparative study or overview guiding which method to use under specific remanufacturing conditions to evaluate the reusability of bolted connections. The study would aim to enable data-driven modelling approaches to automate processes such as unscrewing and decisions regarding reuse strategies.

3 Contribution

This study aims to contribute to improved quality assurance in remanufacturing by evaluating and comparing three different measurement techniques for assessing surface degradation in corroded bolts. These methods include:

- photographic visual inspection,
- computed tomography (CT), and
- optical coordinate measurement with focus variation.

The study provides a comparative analysis of these techniques based on their ability to detect surface defects, deformations, and corrosion effects, while also considering their practicality in industrial contexts in terms of cost, measurement time, storage size and resolution. To assess the impact of corrosion on bolted joints and evaluate suitable measurement techniques, a structured experimental setup was developed. The focus lies on capturing corrosion-induced surface degradation and deformation using three different methods. Section 4 outlines the experimental setup, including bolt preparation, corrosion procedures, and handling. Section 5 presents and compares the applied measurement methods—photographic evaluation, computed tomography (CT) scanning, and optical coordinate measurement—along with selected results from each technique.

4 Experimental Setup

The experimental setup followed a 3×3 factorial design, with two controlled variables: preload force (12, 20 and 28 kN) and corrosion exposure time (2, 4, 8 h). Due to their common industrial application, standard M10 $\times$ 1.5 mm ISO 898, Class 8.8 bolts, nuts and square washers were used as test specimens (see Fig. 1).

First, all components were cleaned for two minutes in an ultrasonic bath to remove surface contaminants. They were then assembled and tightened using a Kistler INSPECTOR 5413–5103 electronic torque wrench to achieve one of three predefined preload forces (12, 20, 28 kN), simulating under-, standard-, and over-tightening conditions. Subsequently, the specimens were exposed to salt spray corrosion for 2, 4, or 8 h in a dedicated chamber using a 15% hydrochloric acid solution. Custom 3D-printed holders ensured uniform orientation and full exposure of the bolted joint to the corrosive mist. After corrosion, the bolts were dried and later disassembled.

Finally, three specimens from each combination of preload and corrosion time—resulting in $3 \times 3 \times 3 = 27$ samples—were examined using different measurement techniques: visual inspection using a camera, computed tomography (CT), and high-resolution optical 3D surface scanning. These methods were used to assess the extent and nature of corrosion-induced surface degradation and mechanical deformation.

Fig. 1 Experimental bolt assembly before corrosion: M10 × 1.5 mm ISO 898-1 Class 8.8 bolt, two DIN 436 square washers (11 × 30 × 3 mm), one DIN 125 round washer (10.5 × 20 × 2 mm), and a DIN 934 M10 nut

5 Comparison of Measurement Techniques

5.1 Photographic Evaluation

Multiple images were taken from different angles—frontal, top-down, and oblique—to capture various features of the corrosion damage. Among these, oblique views proved most effective for assessing the condition of the thread, as they provide better depth perception and visibility of the flanks (see Fig. 2).

Fig. 2 Oblique view of a bolt after 2 h of corrosion exposure. Minor deformations are visible on the thread crests, with corrosion primarily concentrated in the lower section of the bolt, outside the contact area with the nut

Corrosion on the bolt head could also be clearly identified in the top-down images. The lateral resolution was determined based on an image width covering a physical length, resulting in a lateral resolution of 5200 px / 55 mm = 10.7 μm per pixel. While photographs are quick and easy to produce with a small storage size (4.6–22.6 MB), their diagnostic value strongly depends on proper lighting conditions; without uniform and sufficient illumination, rust patterns and surface irregularities can be misrepresented or overlooked.

5.2 Computed Tomography

Computed tomography (CT) proved particularly effective for detecting deformations in the thread and head of the bolts. By overlaying a reference (non-corroded) bolt with the measured bolt, deviations in geometry such as flattening, bending, or loss of thread structure could be visualized and highlighted clearly. An example of such a comparative visualization is shown in Fig. 3. In the crest and flank areas of the engaged threads, deviations were predominantly observed in the positive direction, likely caused by material displacement during the tightening and loosening process. This suggests that mechanical interaction rather than corrosion alone contributed to changes in the thread geometry. While CT images also allow for the identification of heavily rusted areas on the bolt surface, the limited resolution makes it difficult to quantify surface roughness precisely. For the CT measurements the Zeiss Metrotom 1500 was used. Four bolts were scanned simultaneously in each run, resulting in a total scan time of 3 h. This corresponds to roughly 45 min per bolt, highlighting the trade-off between resolution and time efficiency.

Fig. 3 CT-based analysis of a corroded bolt with a non-corroded reference. *Left*: Deviation analysis showing geometric differences, particularly in the thread flanks and crests. *Right*: CT scanning setup with the custom holder used to scan multiple bolts simultaneously

5.3 Optical Coordinate Measurement with Focus Variation

In contrast, surface roughness was more accurately captured using optical coordinate measurement systems with focus variation performed with μCMM Alicona measurement system. The measurement was divided into two stages: first, the thread was scanned, followed by the underside of the bolt head. A custom 3D-printed adapter with a magnet was used to fix the bolts into the Alicona's clamping jaws, as shown in Fig. 4. The measurement captures 3D-surface structures and clearly reveals both corrosion patterns and localized deformations along the thread, highlighting the method's capability for detailed surface analysis. Each Alicona measurement took approximately 50 min. However, significant time savings are possible by limiting the scanning area to the most relevant functional surfaces instead of capturing the entire geometry. The optical coordinate measurement offers high precision with a vertical resolution of 0.5 μm and lateral resolution of 7.83 μm, producing detailed surface data stored in files of 80 MB.

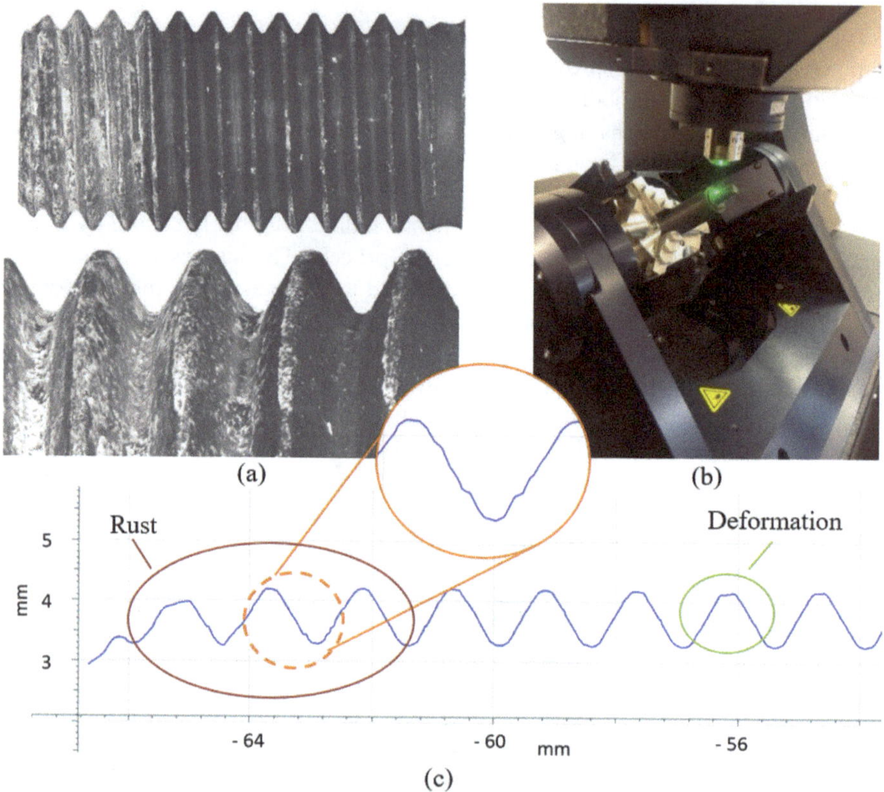

Fig. 4 Optical coordinate measurements of a corroded bolt. **a** 3D surface topography measurements showing corrosion and surface texture in high detail. **b** Measurement setup **c** Cross-sectional view through the thread, illustrating surface roughness and localized corrosion and deformation effects

5.4 Comparative Evaluation

In summary, photographic methods are useful for rapid visual documentation, CT scanning is useful for detecting internal deformation and optical coordinate measurements offers the most comprehensive, quantitative insight into surface degradation (see Table 1).

Table 1 Comparative evaluation of photographic, CT, and optical coordinate measurement methods based on advantages, limitations, suitability, measurement time, resolution and storage size

Method	Photographic	Computed tomography	Optical coordinate measurement
Advantages	Quick, low-cost, easy to perform	Precise deformation detection, internal structure analysis	High-resolution surface data, corrosion and deformation capture
Limitations	Qualitative, lighting-dependent, subjective	Limited surface roughness resolution, costly, time-consuming	Longer measurement time for detailed measurement of big parts
Suitability	Visual inspection, initial screening	Internal defects, cross-section, geometry check	Quantitative corrosion and roughness
Measurement time	1 min per bolt	~45 min per bolt	~50 min per bolt
Resolution of samples	10.7 μm/pixel	19.83 μm/voxel	Vertical resolution 0.5 μm Lateral resolution 7.83 μm
Storage size	RAW: 22.6 MB JPG: 4.6 MB	4.07 GB	80.03 MB

6 Discussion

The three measurement methods each offer unique advantages and limitations for assessing corroded bolted joints. Photographic evaluation is the fastest and most cost-effective, providing quick visual insights (1 min) into surface corrosion, but remains qualitative and lighting-dependent. CT scanning excels at detecting internal and external deformations and uniquely enables non-destructive cross-sectional analysis, though it is limited in surface detail resolution (19.83 μm/voxel) and requires high cost, long measurement times, and complex setups. Optical coordinate measurement delivers precise surface roughness data (0.5–7.8 μm) and captures corrosion patterns and deformations with small storage size (80 MB) and reproducibility. While slower than photography, it is possible to superimpose topographic data with camera images for enhanced spatial interpretation and comprehensive analysis.

A key limitation of this study is that the influence of different torque levels and corrosion durations could not be clearly distinguished, likely due to the relatively small variations in loading and exposure time. Additionally, considering the low unit value of typical components alongside the time and resource demands of these measurement methods, their use is most practical for reference measurements rather than for assessing every individual component. Future experiments should incorporate longer corrosion durations and more distinct preload levels to improve differentiation and reveal clearer trends. The data collected can then be used to develop data-driven models that automate decision-making processes in remanufacturing, enhancing efficiency and standardization. In this context, parameters such as storage size and resolution will remain crucial, as they determine both the feasibility of processing large datasets and the achievable level of detail in model training.

7 Conclusion

This study compared three methods for assessing surface degradation and deformation in corroded bolted joints: photographic evaluation, CT scanning, and high-resolution optical coordinate measurement. Photography enabled rapid visual documentation with 10.7 μm/pixel resolution, CT scans identified geometric deviations and internal defects, and optical coordinate measurement provided detailed surface data for quantitative analysis. Among them, optical coordinate measurement was most suitable for capturing both corrosion effects and mechanical deformation with 0.5 μm vertical resolution and 7.83 μm lateral resolution. Although none is scalable for high-throughput inspection, they are valuable as reference measurements to support data-driven models and decision logic in remanufacturing. Future work will link measurement data with mechanical performance to enable predictive, automated decision-making in circular economy applications.

Acknowledgements. This work was funded by the Deutsche Forschungsgemeinschaft (DFG, German Research Foundation)—"Functional modeling of bolted joints under uncertain product conditions for remanufacturing"—with the project number 525034540.

Competing Interests. The author(s) has no competing interests to declare that are relevant to the content of this manuscript.

References

1. Majumder, P., Groenevelt, H.: Competition in remanufacturing. Prod. Oper. Manag. **10**, 125–141 (2001). https://doi.org/10.1111/j.1937-5956.2001.tb00074.x
2. Jensen, J.P., Prendeville, S.M., Bocken, N.M.P., Peck, D.: Creating sustainable value through remanufacturing: three industry cases. J. Clean. Prod. **218**, 304–314 (2019)
3. Sitcharangsie, S., Ijomah, W., Wong, T.C.: Decision makings in key remanufacturing activities to optimise remanufacturing outcomes: a review. J. Clean. Prod. **232** (2019)
4. Kurilova-Palisaitiene, J., Sundin, E., Poksinska, B.: Remanufacturing challenges and possible lean improvements. J. Clean. Prod. **172**, 3225–3236 (2018)
5. Matsumoto, M., Yang, S., Martinsen, K., Kainuma, Y.: Trends and research challenges in remanufacturing. Int. J. Precis. Eng. Manuf. Green Technol. **3**, 129–142 (2016)
6. Charter, M., Gray, C.: Remanufacturing and product design. Int. J. Prod. Dev. **6** (2008). https://doi.org/10.1504/IJPD.2008.020406
7. Qiu, F., Qian, H., Wang, H., et al.: Study on the tensile performance degradation of bolt-ball joints. Structures **58**, 105627 (2023). https://doi.org/10.1016/j.istruc.2023.105627

8. Hoeppner, D., Arriscorreta, C.: Exfoliation corrosion and pitting corrosion and their role in fatigue predictive modeling: state-of-the-art review. Int. J. Aerosp. Eng. **2012** (2012). https://doi.org/10.1155/2012/191879
9. Withers, P.J., Bouman, C., Carmignato, S., et al.: X-ray computed tomography. Nat. Rev. Methods Primers **1**, 1–21 (2021). https://doi.org/10.1038/s43586-021-00015-4
10. Danzl, R., Helmli, F., Scherer, S.: Focus variation—a robust technology for high resolution optical 3D surface metrology. Strojniski Vestnik **2011**, 245–256 (2011). https://doi.org/10.5545/sv-jme.2010.175

Forecast-Based Assessment of Investment Requirements for Circular Business Models in Industrial Production

Luca Philipp[1](✉), Marcel Fischer[2], Ferenc Wolter[3], Joachim Metternich[2], Florian Stamer[3], and Matthias Schmidt[1]

[1] Institute of Production Systems and Logistics, Garbsen, Germany
philipp@ifa.uni-hannover.de
[2] Institute of Production Management, Technology and Machine Tools, Darmstadt, Germany
[3] Institute for Production Technology and Systems/Leuphana University Lüneburg, Lüneburg, Germany

Abstract. The transition to a circular economy is widely regarded as a promising approach to promoting sustainable economic practices. Within this context, business model innovation plays a pivotal role in the implementation of circular strategies. However, practical adoption remains limited due to challenges in translating conceptual frameworks into practices. A critical barrier is the lack of frameworks for quantitatively assessing the feasibility of circular business models during their development. In this regard, evaluating economic viability constitutes a crucial component. Important in this process is the ability to forecast product returns in terms of volume, timing, and condition, which must be taken into account in remanufacturing processes. This is particularly relevant for factory planning and design, as accurate product return forecasts are essential for appropriately dimensioning and optimizing production systems. Consequently, such forecasts support the quantification of investment costs arising from circular production systems in the context of circular business models. Therefore, a successful transition to a circular economy requires a comprehensive approach that integrates the business model perspective, product return forecasting, and factory planning. The interdependencies between these elements are discussed in the present paper, which also derives a method for quantifying factory-related investments.

Keywords: Circular business models · Product return forecasting · Factory planning

1 Introduction

According to a report by the Organization for Economic Co-operation and Development (OECD), global resource consumption is expected to double to 167 gigatons by 2060 unless political countermeasures are implemented [1]. A key driver of this development is the current linear economic and production model, which operates according to the 'produce–use–dispose' principle [2]. In this model, raw materials are primarily extracted

L. Overmeyer and B.-A. Behrens (eds.), *Production at the Leading Edge of Technology*, Lecture Notes in Production Engineering, https://doi.org/10.1007/978-3-032-19524-1_6

from the environment through energy-intensive processes and subsequently refined during production [3]. As a result, companies are increasingly confronted with resource scarcity and rising material costs.

To address these challenges at the European level, the EU Green Deal outlines concrete measures to counter these trends, including promoting a circular economy (CE) [4]. This paradigm shift from a linear to a CE aims to close material loops by ensuring that products and raw materials are reused, reconditioned or recycled after their initial use [5, 6]. Globally, however, the implementation of theoretical CE approaches into industrial practice remains at an early stage of development [7].

Business model innovation (BMI) is regarded as a key lever for implementing CE in industrial practice, owing to its holistic perspective [8]. Nevertheless, the complex interdependencies between product architecture, value creation structures, and the distribution of benefits pose substantial challenges for companies aiming to transform their existing business models.

In particular, quantifying the impact of different circular business models (CBMs) on company performance within specific business settings remains challenging [9]. The economic consequences for companies play a crucial role in this context. To address these circumstances, this paper examines the following research question: How can the required implementation investments for CBMs at the factory-level be quantified in the context of industrial production and BMI?

This paper, therefore, proposes a method for evaluating the investments required for factory-level implementation of CBMs. This assessment is based on specific CBM approaches and is heavily dependent on product return forecasts and their implications for factory planning. Thus, the transformation demands a holistic concept that integrates the CBMs, product return forecasting, and factory planning.

This paper is structured as follows. First, the theoretical background regarding CBMs and their various archetypes, the importance of product return forecasting, and their impact on factory planning as a major cost driver in the context of CBMs is outlined. Based on this foundation, a method for evaluating investments in different CBM approaches is introduced and illustrated through a use case. Finally, key insights and future research directions are discussed.

2 Theoretical Background

2.1 Circular Business Model

CE is defined as an economic system that replaces the traditional linear model and its associated ‘end-of-life’ concept with circular strategies across production, distribution, and consumption processes [10]. According to the Ellen MacArthur Foundation, CE is based on three core principles: eliminating waste and pollution, keeping products and materials in use, and regenerating natural systems [11]. Rather than representing a single, uniform strategy, CE functions as a broad conceptual framework that integrates a range of diverse and loosely connected approaches [12].

CBMs integrate the principles and objectives of CE into the business context. Four archetypes of CBMs can be distinguished [8]: (1) Cycling, which focuses on closing product life cycles through take-back systems and reconditioning; (2) Extending, which aims to prolong the use phase through durable product design and maintenance; (3) Intensifying, which seeks to increase product utilization through alternative usage models, such as sharing concepts; and (4) Dematerializing, which involves replacing physical hardware with software solutions and service-based models [8].

Given its comprehensive perspective, BMI is considered a central mechanism for translating CBMs into industrial practice [8]. Its methodical structure and supporting tools facilitate a systematic transformation or redesign of existing business models [13]. Figure 1 illustrates the generic process of BMI.

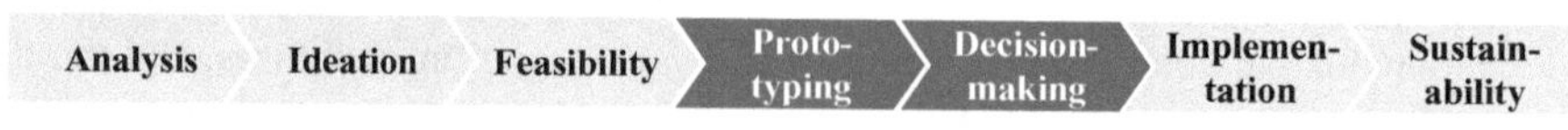

Fig. 1 Phases of business model innovation [13]

During the analysis phase, the current business context is assessed [13]. Building on this, the ideation phase generates preliminary business model alternatives, incorporating the four aforementioned circular archetypes, whose feasibility is subsequently evaluated [13]. Upon confirmation of feasibility, the prototyping phase plays a key role in shaping these alternatives through comprehensive and detailed design [13]. During the decision-making phase, the detailed alternatives are evaluated, and the approach to be implemented is determined [13]. To ensure a well-founded decision, the importance of a robust quantitative data basis must be emphasized. Finally, the sustainability phase safeguards the long-term viability of the business model [13].

2.2 Product Return Forecasting

A key factor for the implementation of CBMs is the design of efficient return processes, which are collectively referred to as reverse logistics. Reverse logistics is defined as the planning, implementation, and control of logistical material flows from the consumer back to the production site. The focal point of these return processes pertains to returned products. These returned products can vary greatly in quantity, timing, and quality and are significantly influenced by numerous factors, including product life and sales cycles, user behavior, warranty conditions, as well as marketing and service policies [14]. Product returns occur not only in cycling but also in extending and intensifying CBM types, because maintenance-driven returns and asset-sharing models likewise generate product returns to the facility. In the context of CBM implementation, an additional challenge is that the actual return flows often begin only after a considerable delay. The resulting uncertainty, stemming from both the variability and the delayed onset of product returns, impacts not only operational management but also poses a challenge for the strategic planning of circular value creation.

To address these challenges, various forecasting approaches are employed to enhance the management of uncertainties. Statistical methods, such as time series analysis and regression models, facilitate the forecasting of future product returns by leveraging historical data patterns and explanatory influencing variables [15, 16]. Simulation-based approaches, including Monte Carlo and agent-based simulations, enable the dynamic mapping of uncertainties and feedback effects in complex systems [17, 18]. Machine learning methods, such as tree-based algorithms and artificial neural networks, have been applied to achieve high forecasting accuracy, even when dealing with heterogeneous and non-linear data structures [19, 20].

2.3 Factory Planning

Product return forecasting is a critical instrument for factory planning, as it directly influences the dimensioning of capacities for remanufacturing processes, maintenance and servicing activities, as well as the allocation of inventory space. Insufficient forecast accuracy can result in inefficient capacity utilization or over-dimensioning, thereby affecting investment costs [17, 21].

Factory planning is crucial to the economic success of companies and is characterized by high investment [22]. Given a life-cycle of up to 40 years, decisions in the factory planning process are strategic and have a long-term impact on competitiveness and resource efficiency [23, 24]. Nevertheless, the factory remains a highly complex system whose planning requires the application of systematic principles [25].

While classic target variables, such as economic efficiency, logistics performance and changeability, remain relevant in the planning process, CE is becoming increasingly important [26]. This involves not only designing the factory building to be recyclable but also designing a factory system that supports circular value creation, both technically and organizationally [27]. Challenges such as the efficient reintegration of returned products into production processes and the handling of volatile return volumes must be considered at an early stage of the planning process.

3 Method for Assessing Factory-Level Implementation Investments for CBMs

To address the problem outlined in the introduction, the following section presents an initial decision-making method for the implementation of CBM approaches. This method provides a conceptual structure for systematically mapping the relevant elements and processes that must be considered when quantifying the required factory-level implementation investments for CBMs. To support this, Table 1 provides a tentative overview of the interrelations between types of CBMs, product return forecasting, factory planning, and their impact on potential production-related investments. These interrelations were identified based on key findings from the pertinent literature and the collective assessment of the authors [8, 21, 23, 27, 28].

Table 1 indicates that, in the context of closing product loops and reprocessing, the predictability of product returns is limited due to various uncertainties throughout the product life cycle. Implementing such business models requires substantial investments in production, such as restructuring value creation processes, adjusting material flows, or acquiring specific equipment. In contrast, business model approaches focused on extending the use phase include both durable product design and services such as repair or upgrades. While developing long-lasting products often requires higher research and

Table 1 Interrelation between CBMs, product return forecasting and factory planning

	Cycling	Extending	Intensifying	Dematerialization
Effect on product return forecasting	• Average quantity • Uncertain timing • Heterogeneous quality	• Low quantity • Timing depends on maintenance arrangements • Higher quality	• High quantity • Schedulable timing • Poor quality	• No returns
Impact on factory planning	• Structural and processual adjustments • Equipment for disassembly • Reintegration of returned products	• Minor to moderate planning adjustments depending on CBM specifics	• Minor differences in comparison to linear factory planning	• Decrease in required production capacities
Impact on the forecast of production investments	• Investments are primarily driven by production system adjustments	• Low investments are required in the context of production • Product development and the management of the usage phase are of greater relevance	• Low investments required in the context of production • Management of the usage phase of greater relevance	• Investments in factory deconstruction or integration of adaptability during factory planning to anticipate future capacity reductions

development investments and material changes, service-based extensions may also entail minor to moderate investments in infrastructure, diagnostics, and reverse logistics. By comparison, intensifying use, e.g., through sharing or leasing, typically requires fewer factory adjustments and mainly involves digital or organizational investments. Moreover, the predictability of product returns differs between the two approaches: life cycle extension increases uncertainty regarding the timing and quantity of product returns due to prolonged usage. Conversely, models that intensify product use through active management of the use phase offer greater transparency, thereby improving the predictability of product returns. In dematerialization concepts, where physical products are substituted by digital solutions or services, the relevance of product return forecasts diminishes significantly. From a factory planning perspective, the focus shifts toward the reduction of production capacities. This may require targeted investments or the divestment of existing assets, but can also be anticipated early on by incorporating changeability into factory planning and integrating it into investment decisions.

The findings indicate that the interdependencies between product return forecasting, factory planning, and expected factory-level implementation investments in manufacturing companies vary significantly depending on the type of CBM. Based on this, a decision-making process was developed that integrates CBM, product return forecasting, and factory planning, thereby enabling an estimation of the expected investments for implementing CBMs in manufacturing settings. The decision support process, as shown in Fig. 2, is divided into five consecutive phases, which are generally carried out sequentially but involve feedback on the previous steps. The presence of a CBM prototype initiates the process. In the first phase, company—and product-specific information, such as the number of units, the expected condition of the returned products, and the underlying service model, is compiled based on the CBM to be evaluated. This information forms the basis for product return forecasting, the second step in the process. As part of the forecast, statistical, simulation-based, or machine-learning methods are applied to determine the expected volume of product returns. In addition to the total number, possible seasonal or product-specific fluctuations are also analyzed, and conclusions about the reuse potential are drawn from the forecast regarding the quality of returned products.

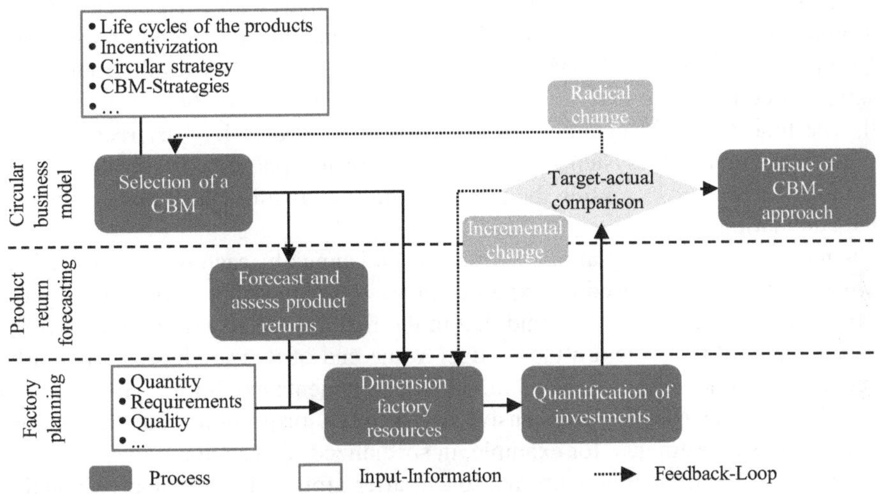

Fig. 2 Method for evaluating investments in the implementation of CBMs

The third phase of the method comprises factory planning. In this phase, the three central dimensions of space, production equipment and personnel are dimensioned based on information such as the number of units, product returns and quality.

Product returns are significant for dimensioning, as they influence the distribution between linear and circular value creation within the factory. This distribution is directly reflected in the capacities required for disassembly facilities or infrastructure for reprocessing, and potentially for maintenance activities, for example.

Once the dimensioning has been completed, the required factory-level implementation investments are quantified. These include, in particular, the investments needed to transform an existing linear factory into a circular factory. For the quantification key planning variables such as return quantities, product conditions, and space requirements must be estimated at an early stage of planning when there is still considerable uncertainty. These assumptions have a direct impact on the economic evaluation of the circular business model. To adequately account for the uncertainties associated with circular production systems, risk-adjusted capital cost rates should be employed.

The quantified factory-level implementation investments represent the central output of the third phase and form the basis for the target-actual comparison in the following phase. In this step, these investment requirements are compared with the previously defined budget for converting the business model. If the investment requirements exceed the budget, targeted feedback can be provided to address the issue. If incremental adjustments are required, the relevant information is fed back into the factory planning phase. If, on the other hand, a fundamental realignment of the business model approach is necessary, the process returns to the first phase. If the cost comparison is favorable, the CBM under consideration is pursued further.

4 Application of the Proposed Method

To illustrate the proposed method, the following section presents a use case from industrial production using dummy values. The company manufactures automation systems and seeks to evaluate two CBM approaches. These approaches result from the systematic procedure in the early phases of circular BMI, as described by Fischer et al. [29]. The first approach, cycling, aims to close the resource loop by recovering and refurbishing automation systems at the end of their use phase. The second approach, Extending, focuses on extending the use phase through regular maintenance and repair of the automation systems.

In general, 50,000 automation systems are sold annually, each with a service life of between 4 and 8 years. Based on expert estimates provided by the company regarding the circular approaches, it was found that in the first approach, approximately 50% of customers are willing to return the product at the end of its use phase, particularly in the context of new purchases. The individual components exhibit a high potential for refurbishment. However, comprehensive disassembly, inspection, and reconditioning require various investments, for example, in specialized workstations.

In the case of the maintenance approach, after around 4 years, approximately 75 percent of systems are inspected on a smaller scale and primarily wear parts are serviced. This maintenance enables an extension of the product life cycle to 10 to 14 years.

To determine the necessary investments for factory-level implementation in the context of factory planning, the use case includes a product return flow forecast. This forecast is used to quantify the necessary investments.

The resulting evaluation is presented in the following Table 2. The arrows in the table show how the required factory-level implementation investments compare in size to those for a linear production setup. A horizontal arrow roughly corresponds to the investment level required for linear production. For cycling, additional disassembly activities and machine operation are required. However, the personnel requirements are comparable to those in the linear case, as these activities are assumed to substitute for existing tasks rather than increase the overall staffing level.

Table 2 Exemplary use case illustrating factory-level implementation investments for the outlined CBM approaches

CBM-approach	Potential product returns	Dimensioning of the factory	Factory-level investments
Cycling	• 25.000±5.001 product returns that reach the end of life in a given year	• Space: Fluctuating demand for storage capacity • Equipment: Equipment for disassembly and storage of returns • Personnel: Trained personnel for disassembly activities	• Space: ↗ • Equipment: ↑ • Personnel: →
Extending	• 37.500±1.254 product returns that flow back for maintenance in the respective year	• Space: Plannable space requirement for product returns • Personnel: Trained personnel for maintenance activities	• Space: → • Equipment: → • Personnel: →

5 Conclusion

In this paper, the interrelations between product return forecasting, factory planning, and their influence on factory-level implementation investments were analyzed for the four different types of CBMs—Cycling, Extending, Intensifying, and Dematerializing. The findings indicate that these interrelations vary considerably depending on the specific type of CBM. To enable a systematic assessment of these dependencies in practice, this paper addressed the following research question: How can the required implementation investments for CBMs at the factory-level be quantified in the context of industrial production and BMI?

To answer this research question, an initial decision-making method is proposed to systematically capture the key elements and processes involved in quantifying the required factory-level implementation investments for CBMs. The five-phase process incorporates structured feedback loops, supporting both incremental adjustments and fundamental revisions to the CBM concept. Integrating CBM selection with product return forecasting and factory planning enables transparent estimation of the required investments. This quantification at an early stage is crucial for supporting well-informed decision-making in factory planning and for facilitating the successful implementation of CBMs in the transition to a circular economy.

Building on the presented method, further research should focus on a more detailed investigation of the interrelations between CBM selection, product return forecasting, and factory planning. A clear understanding of how these components interact on a process and data level is essential for enabling the practical application of corresponding decision support models in industrial contexts. In addition to the investment metrics presented here, further indicators such as life-cycle costs should be incorporated for a more comprehensive evaluation of transformation efforts. While established metrics from investment and production planning can be applied in part, certain indicators, such as net present value or payback periods, will need to be specifically developed or adapted to reflect the unique characteristics of circular business models and return-based production systems.

Competing Interests. The author(s) has no competing interests to declare that are relevant to the content of this manuscript.

References

1. OECD: Global Material Resources Outlook to 2060—Economic Drivers and Environmental Consequences. OECD Publishing, Paris (2019)
2. Wilts, H., Fink, P.: Deutschland auf dem Weg in die Kreislaufwirtschaft (2016)
3. Schaller, F., Randhan, A., Bösche, E., et al.: Kreislaufwirtschaft als Säule des EU Green Deal. In: Wittpahl, V. (ed.) Klima, pp. 233–251. Springer Vieweg, Berlin, Heidelberg (2020)
4. Ballester, N., Zaroffek, J.: Der Kreislaufwirtschaftsaktionsplan als wirtschaftliches Kernstück des Green Deals. In: Kurth, P., Oexle, A., Faulstich, M. (eds.) Praxishandbuch der Kreislauf— und Rohstoffwirtschaft, 2., überarbeitete und aktualisierte Auflage edn, pp. 379–400. Springer Vieweg, Wiesbaden, Heidelberg (2022)
5. Oluleye, B.I., Chan, D.W., Olawumi, T.O., et al.: Assessment of symmetries and asymmetries on barriers to circular economy adoption in the construction industry towards zero waste: a survey of international experts. Build. Environ. **228**, 1–13 (2023)
6. Wolf, S., Jordan, M.M., Seifert, I., et al.: Wie Industrieproduktion nachhaltig gestaltet werden kann. In: Wittpahl, V. (ed.) Klima, pp. 164–179. Springer Vieweg, Berlin, Heidelberg (2020)
7. Barreiro-Gen, M., Lozano, R.: How circular is the circular economy? analysing the implementation of circular economy in organisations. Bus. Strat. Env. **29**, 3484–3494 (2020)
8. Geissdoerfer, M., Pieroni, M.P., Pigosso, D.C, et al.: Circular business models: a review. J. Clean. Prod. 277 (2020)
9. Ferasso, M., Beliaeva, T., Kraus, S., et al.: Circular economy business models: the state of research and avenues ahead. Bus. Strat. Env. **29**, 3006–3024 (2020)
10. Kirchherr, J., Reike, D., Hekkert, M.: Conceptualizing the circular economy: an analysis of 114 definitions. Resour. Conserv. Recycl. **127**, 221–232 (2017)
11. Ellen MacArthur Foundation: Towards the Circular Economy—Vol. 1: an economic and business rationale for an accelerated transition (2013)
12. Blomsma, F., Brennan, G.: The emergence of circular economy: a new framing around prolonging resource productivity. J. Ind. Ecol. **21**, 603–614 (2017)
13. Wirtz, B., Daiser, P.: Business Model Innovation Processes: a Systematic Literature Review (2018)
14. Govindan, K., Soleimani, H., Kannan, D.: Reverse logistics and closed-loop supply chain: a comprehensive review to explore the future. Eur. J. Oper. Res. **240**, 603–626 (2015)
15. Canda, A., Yuan, X.M., Wang, F.Y.: Modeling and forecasting product returns: an industry case study. In: 2015 IEEE International Conference on Industrial Engineering and Engineering Management (IEEM), pp. 871–875. IEEE (2015)

16. Masui, K.: Calculation of amount of discarded end-of-life products by using multi-regression analysis. In: 2005 4th International Symposium on Environmentally Conscious Design and Inverse Manufacturing, pp. 624–625. IEEE (2005)
17. Tsiliyannis, C.A.: Prognosis of product take-back for enhanced remanufacturing. J. Remanuf. **10**, 15–42 (2020)
18. Mashhadi, A.R., Esmaeilian, B., Behdad, S.: Simulation modeling of consumers' participation in product take-back systems. J. Mech. Des. **138** (2016)
19. Heilig, L., Hofer, J., Lessmann, S, et al.: Data-driven product returns prediction: a cloud-based ensemble selection approach. In: Association for Information Systems: Proceedings of the 24th European Conference on Information Systems (ECIS 2016). AIS Electronic Library, Istanbul (2016)
20. Mishra, A., Dutta, P.: Return management in e-commerce firms: a machine learning approach to predict product returns and examine variables influencing returns. J. Clean. Prod. **477** (2024)
21. Krapp, M., Nebel, J., Sahamie, R.: Forecasting product returns in closed-loop supply chains. Int. J. Phys. Distrib. Logist. Manag. **43**, 614–637 (2013)
22. Grundig, C.G.: Fabrikplanung—Planungssystematik—Methoden—Anwendungen. 7, aktualisierte Auflage edn. Hanser, München (2021)
23. Wiendahl, H.H., Reichardt, J., Nyhuis, P.: Handbuch Fabrikplanung—Konzept, Gestaltung und Umsetzung wandlungsfähiger Produktionsstätten, 3. vollständig überarbeitete Auflage edn. Hanser Verlag, München (2024)
24. Dannapfel, H.M.: Ein systemisches Management-Modell für die Fabrikplanung. 1. Auflage edn (2019)
25. Kettner, H., Schmidt, J., Greim, H.R.: Leitfaden der systematischen Fabrikplanung—Mit zahlreichen Checklisten. Hanser; Books on Demand, München, Norderstedt (2010)
26. Lanza, G., Klenk, F., Martin, M., et al.: Sonderforschungsbereich 1574: Kreislauffabrik für das ewige innovative Produkt. Zeitschrift für wirtschaftlichen Fabrikbetrieb. **118**, 820–825 (2023)
27. Philipp, L., Schmidt, M.: Fabrik für die Kreislaufwirtschaft/factories for the circular economy. wt Werkstattstechnik online. **115**, 11–16 (2025)
28. Toktay, L.B., van der Laan, E.A., de Brito, M.P.: Managing product returns: the role of forecasting. In: Dekker, R., Fleischmann, M., Inderfurth, K., Wassenhove, L.N. (eds.) Reverse Logistics, pp. 45–64. Springer, Berlin, Heidelberg (2004)
29. Fischer, M., Lang, E., Sailer, S., et al.: Vorgehen bei der Entwicklung zirkulärer Geschäftsmodelle. Zeitschrift für wirtschaftlichen Fabrikbetrieb. **120**, 54–59 (2025)

Development of a Simulation-Based AI-Model for Optimizing Injection Molding Processes with Recyclate Content

Marcel Haab(✉), André Hürkamp, and Klaus Dröder

Technische Universität Braunschweig, Institute of Machine Tools and Production Technology, Braunschweig, Germany
marcel.haab@tu-braunschweig.de

Abstract. The incorporation of recyclates in plastic component production is an increasingly adopted approach to enabling a circular economy within the plastic industry. Combining recyclates with glass fiber reinforcement can help to maintain mechanical performance. However, using recycled plastics introduces variability in molding and product performance due to fluctuating recyclate quality, storage, and batch differences. Effective process control is needed to manage these variations. This work develops an artificial intelligence (AI) model based on a Multilayer Perceptron to predict material quality and enable adjustment of process parameters, aiming for real-time integration to optimize injection molding. To reduce resource-intensive data generation, simulations using Autodesk Moldflow were conducted to provide the training data. These simulations accounted for varying process settings and material properties such as viscosity. The simulation-based AI approach offers manufacturers valuable insights about the properties of used materials to improve product quality while supporting the increased use of recycled plastics in production.

Keywords: Injection molding · Circular production · AI

1 Introduction

The reduction in greenhouse gas emissions, such as CO_2, and the adoption of alternatives to fossil fuels have become increasingly important topics in recent years. This development particularly affects the plastics industry. For instance, the EU Directive [1] requires that all PET bottles contain at least 25% post-consumer recyclate (PCR) by 2025, increasing to 30% by 2030. The increasing demand for and use of PCR requires a detailed understanding of the material behavior during both pre- and postproduction stages. Noor et al. [2] and Schatz

L. Overmeyer and B.-A. Behrens (eds.), *Production at the Leading Edge of Technology*, Lecture Notes in Production Engineering, https://doi.org/10.1007/978-3-032-19524-1_7

[3] show that the mechanical properties of PCR not only differ from those of virgin materials but also vary significantly between batches and production cycles, which can challenge the reliability of component design. Tensile tests conducted by Aurrekoetxea et al. [4] reveal differences of 17% in the elastic modulus and 22% in the elongation at break between virgin polypropylene (PP) and its recycled variant. Ideally, such property variations can be identified during preproduction, for example by measuring the density of the PCR [5]. In addition to the varying property of the material itself, the production process can have different effects on the quality of the final part. Fernández et al. [6] present a simulation- and AI-assisted solution for predicting the part quality based on process settings and geometry during injection molding. In addition to pressure-based monitoring, Chang et al. [7] present a method that evaluates part quality by considering fluctuations in clamping force during injection molding, linking variations in recycled material properties to changes in the geometry of the final product. In Hürkamp et al. [8], a neural network has been trained to evaluate the product quality for overmolded thermoplastic composites during manufacturing. Xanthis et al. [9] present an approach where simulation-generated datasets are used to train machine learning models, demonstrating the potential of synthetic data as a substitute for real-world measurements.

This paper proposes an approach to predict PCR material quality using a simulation-based AI model during injection molding. Unlike traditional simulations, the AI model enables significantly faster predictions, making it suitable for real-time application. This enables a new integration of AI into the manufacturing process, aimed specifically at assessing material quality. Injection molding is a widely used manufacturing method for producing high volumes of identical parts by injecting molten plastic into a mold, where it subsequently solidifies [10]. The rheological properties will be reflected in the pressure differences occurring during the molding process. Javierre et al. [11] show that recycled content strongly affects viscosity and injection pressure, suggesting pressure as a quality indicator. Injection molding machines are typically equipped with sensors that measure the hydraulic pressure during mold filling at a constant flow rate, allowing detection of cycle-to-cycle variations caused by material differences. Combined with a simulation- and AI-based optimization system, this enables an efficient workflow for predicting postproduction quality.

2 Rheological Modeling of Recycled and Virgin Material

Most polymer melts exhibit non-Newtonian behavior, meaning their viscosity depends on temperature T, pressure p, and shear rate $\dot{\gamma}$. The published investigations within the work of Williams et al. [12] lead to the Cross-Williams-Landel-Ferry (Cross-WLF) viscosity model, which describes the relationship between these variables. The model describes the viscosity as

$$\eta(T,p,\dot{\gamma}) = \frac{\eta_0(T,p)}{1+\left(\frac{\dot{\gamma}\cdot\eta_0(T,p)}{\tau^*}\right)^{1-n}}, \tag{1}$$

where the zero shear viscosity

$$\eta_0(T,p) = D_1 \cdot \exp\left(-\frac{A_1(T-T^*)}{A_2+(T-T^*)}\right), \tag{2}$$

describes the viscosity the polymer melt approaches at very low shear rates. Here, the constant A_2 and the glass transition temperature T^* are given by

$$A_2 = \tilde{A}_2 + D_3 \cdot p \tag{3}$$

$$T^* = D_2 + D_3 \cdot p \tag{4}$$

The values of the power law index n, the critical stress level at the transition to shear thinning τ^* and the coefficients D_1, D_2, D_3, $\tilde{A}_2$ must be determined empirically. Once identified, these values allow the calculation of viscosity as a function of $\dot{\gamma}$, T, and p. The Cross-WLF viscosity model has been implemented in widely adopted simulation software for injection molding processes, such as Autodesk Moldflow [13]. Due to various factors such as alternating cross-sections of the mold, the flow velocity changes during the injection molding process, resulting in different shear rates [14]. To account for these variations, the Cross-WLF model is applied to simulate the filling behavior based on the locally varying shear rates across the part.

As shown in Fig. 1, the viscosity of a material described by the Cross-WLF model decreases with increasing melt temperature and shear rate. During data acquisition, simulations with varying rheological properties are conducted. In this contribution, the constant D_1, which is typically determined through data-fitting, is modified in order to account for changing viscosity with varying PCR share. Varying D_1 results in a change in viscosity: higher values lead to an increase in viscosity. This correlation is also shown in Fig. 1.

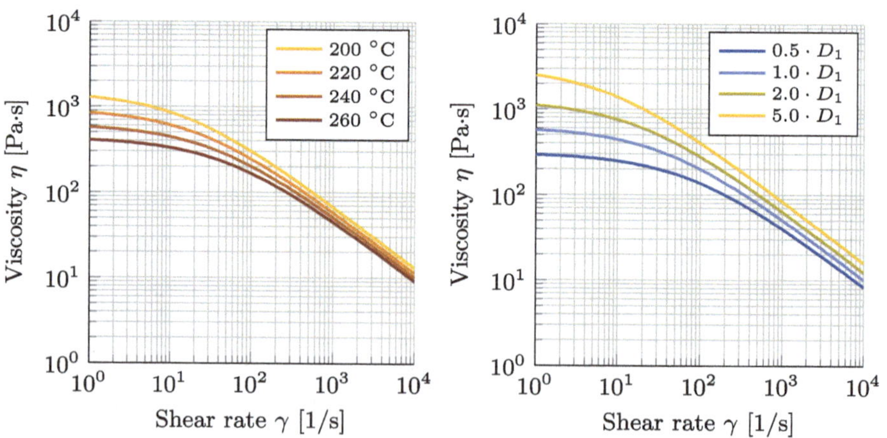

Fig. 1. Flow curves according to the Cross-WLF viscosity model

3 Data Acquisition and Application

3.1 Experimental Investigation

In this study, the material used is a long glass fiber reinforced PP. This material contains 30% long glass fibers. In addition to the virgin material, the manufacturer provided four other materials which contain a share of PCR. Two types of PCR were provided, labeled as low-quality (LQ) and high-quality (HQ). The LQ and HQ material types correspond to the worst- and best-sorted PCR materials provided by the manufacturer, respectively. Therefore, it can be assumed that HQ contains minimal contamination, while LQ exhibits a higher level of impurities. The composition percentages of these materials are listed in Table 1.

Table 1. Material composition for different compound variants

	V	HQ1	LQ1	HQ2	LQ2
LGF (%)	30	30	30	30	30
PP (Virgin) (%)	70	50	50	30	30
PCR (HQ) (%)	0	20	0	40	0
PCR (LQ) (%)	0	0	20	0	40

Differences in the materials V, HQ1, HQ2, LQ1 and LQ2 can occur in both their rheological and mechanical properties. Since this work aims to provide a solution for detecting differences during the injection molding process, the investigation focuses on the rheological properties of the polymer. In order to generate simulation data, these rheological differences must be quantified. In particular, the relationship between the PCR content and viscosity, as well as the influence of PCR quality on viscosity, must be examined experimentally. Using a capillary rheometer, rheological properties of the materials were determined. The measured viscosities are shown in Fig. 2. When using a capillary rheometer, systematic errors can occur, such as pressure losses at the entrance and exit of the die, temperature increases due to shear heating, and deviations caused by non-Newtonian behavior of polymer melts. As these effects are not accounted for in the viscosity calculation shown in Fig. 2, the measured values are better suited for relative comparisons than for determining absolute viscosity.

Analyzing the results, it can be observed that a higher PCR content leads to an increase in viscosity for the investigated materials. However, a decrease in PCR quality results in a reduction of viscosity. This suggests that present contaminants or aging effects reduce the viscosity of the material. As a result, rheological properties similar to those of the virgin material cannot necessarily be attributed to the material's composition alone. Nevertheless, if the PCR content is known, the rheological properties can be used to infer the quality of the material, and thereby estimate other characteristics, such as mechanical properties.

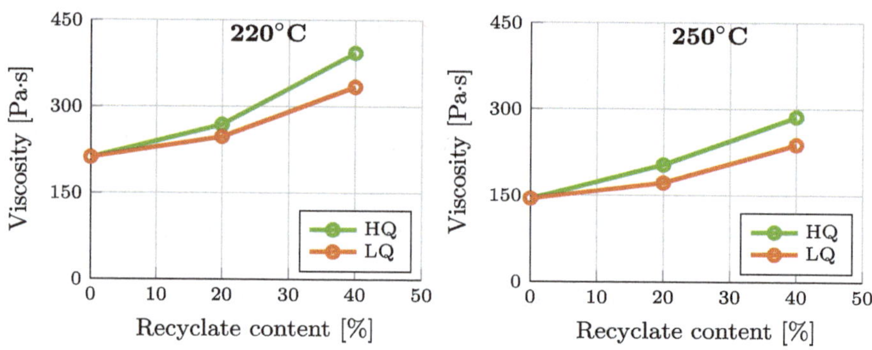

Fig. 2. Measured viscosity at $300\,\mathrm{s}^{-1}$ and different temperatures

3.2 Simulation Setup

To improve the accuracy of a simulation-based AI-model it is important to acquire a sufficiently large and meaningful dataset. For this purpose, relevant input and output parameters must be defined. Since the goal of the AI-model is to conclude from measured data to the quality of the material, the viscosity of the material is chosen as the quantity from which the quality should be deduced. To extract information about the melt's viscosity during the injection molding process, the pressure at the injection location can serve as an indirect indicator. A material with higher viscosity results in higher injection pressure compared to a material with lower viscosity.

To generate the required data, the constant D_1 from Eq. (2) was varied across different simulations by multiplying the value provided by the manufacturer with a value k_{D_1}. This variation affects the viscosity and thus leads to corresponding changes in injection pressure. In addition to D_1, mold temperature, melt temperature, and flow rate were also varied. These are common process parameters that influence the melt and the pressure at the injection location. In Table 2 are the values for the parameters used in the simulations listed.

Table 2. Ranges of the parameters used for the simulations

T_{mold} (°C)	T_{melt} (°C)	$\dot{V}$ (cm^3/s)	k_{D_1} (−)
20–80	200–280	1–10	0.1–10

The manufacturer provided values of the constants according to the Cross-WLF viscosity model for the virgin material as seen in Table 3. The corresponding values for the materials HQ1, LQ1, HQ2, and LQ2 are not available. In these cases, each material was assigned a k_{D_1} value reflecting the qualitative differences observed in the experimental results. Using the parameter ranges defined in Table 2 and the constants from Table 3, a total of 576 different simulations were performed in Autodesk Moldflow.

Table 3. Constant values for the virgin material

n (–)	τ^* (Pa)	D_1 (Pa s)	D_2 (K)	D_3 (K/Pa)	A_1 (–)	$\tilde{A}_2$ (K)
0.2683	22676.3	6.43156e+14	263.15	0	33.393	51.6

3.3 AI Surrogate Model

This work involves two distinct applications of AI using Multilayer Perceptrons (MLP). The first AI-model predicts the factor k_{D_1} based on the PCR content and the quality of the PCR material. The relation between these variables were obtained from the experimental data. To quantify material quality, the low-quality (LQ) material is defined as having 0% quality, and the high-quality (HQ) material as 100%. The quality range thus reflects the degree of sorting, from the least to the best sorted PCR material. For this work, a linear relation between quality and viscosity is assumed. However, since viscosity and D_1 are related in a non-linear manner within the Cross-WLF model, a numerical approach is required to calculate D_1 or k_{D_1} for a given PCR content and quality. Using this approach, a dataset was generated consisting of various PCR shares, corresponding quality levels, and the resulting k_{D_1} values. This dataset was then used to train the first AI-model, thereby reducing computation for the determination of k_{D_1} in future application. The second AI-model was trained using simulation data and is capable of predicting the injection pressure based on the mold temperature, melt temperature, flow rate, k_{D_1}, and time within the injection molding process. By combining both AI-models, pressure predictions can be made based on process parameters, PCR content, and its quality. This combined approach serves as a surrogate model, shown in Fig. 3, enabling significantly faster predictions compared to conventional simulations or measurements.

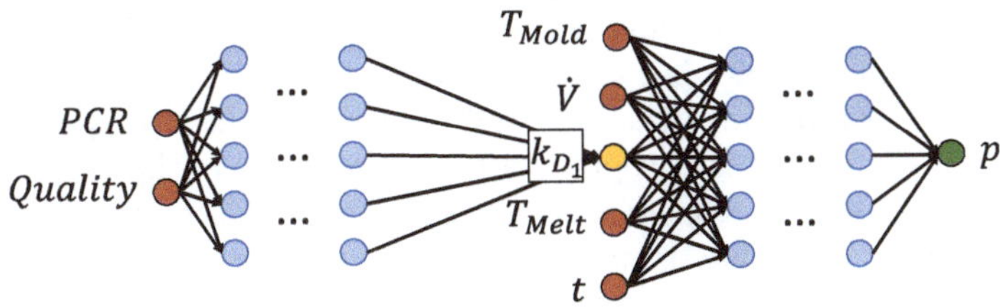

Fig. 3. Schematic depiction of the neural network of the surrogate model

4 Results

4.1 Surrogate Model Results

In Fig. 4, an exemplary comparison between the surrogate model and simulation results is presented. The AI model was evaluated using test data from 16 simulations featuring previously unseen parameter configurations, including some values outside the training range. The resulting R^2-value of 0.959 quantifies the close agreement between predicted and simulated pressure curves, indicating high predictive accuracy even under extrapolative conditions.

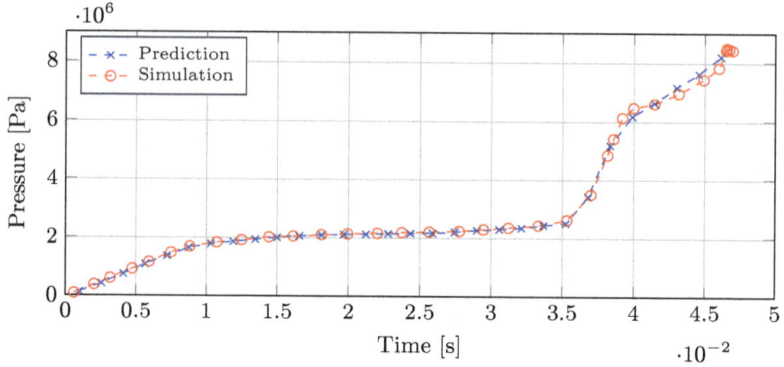

Fig. 4. Exemplary comparison between surrogate model and simulation

4.2 AI-Assisted Pressure Curve Optimization for PCR Quality Estimation

As illustrated schematically in Fig. 5, the user provides for the estimation of PCR quality input data such as process settings, the PCR content of the material, and the pressuretime curve measured during the injection molding process. Ideally, this information is obtained directly from the injection molding machine in real-time. Using the provided inputs, a developed Python code which functions as the optimization tool, applies the surrogate model to predict the corresponding pressuretime curve. Since the material quality is unknown, yet required for the prediction, the optimization process begins with an initial estimate of 50% quality. It then iteratively adjusts this value to minimize the mean squared error (MSE) between the predicted and measured pressure curves. After a defined number of iterations, the tool outputs the quality value that yields the best

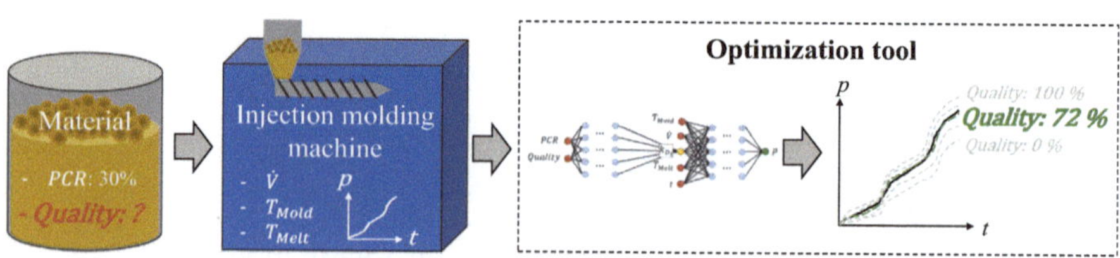

Fig. 5. Schematic workflow of the optimization tool

match. With this estimated material quality, the user can make informed decisions. For example, they may adjust the formulation by mixing in virgin material to achieve the desired performance.

5 Conclusion

In this paper a concept for optimizing the injection molding process in regard of PCR quality is presented. For that, rheological properties of different materials with different share of PCR and different quality of PCR were measured with a capillary rheometer. The experimental results were used for the Cross-WLF viscosity model in order to quantify the materials viscosity in regard of temperature and share rate. The model was used for conducting simulations in Autodesk Moldflow. There, data was acquired by gathering the pressure at the injection location for different process setting parameters. Such as temperature of the melt, the mold surface temperature, and the flow rate. The data was used to train an AI-model which can make corresponding predictions. An optimization tool was presented which uses the AI-model and can be used during the injection molding process to determine the materials quality and therefore allows manufactures to increase or decrease the share of PCR according to its quality. For future applications, real-time data from injection molding machines could be incorporated. In addition to predicting material quality, process settings also influence the final part quality and should be taken into account. A gradient-based approach, in which the part quality is maximized by systematically adjusting individual process parameters, could be a valuable method for further optimizing the injection molding process.

Acknowledgments. This research and development project is funded by the German Federal Ministry for Economic Affairs and Energy (BMWE) within the funding initiative "TTP LB Programm" and implemented by Projektträger Jülich (funding code: 03LB3025A). The authors are responsible for the content and gratefully acknowledge the assistance of Stefan Klare (IPI—Hochschule Kempten) in setting up the AI-assisted pressure curve optimization.

Competing Interests. The author(s) has no competing interests to declare that are relevant to the content of this manuscript.

References

1. European Union: Commission Implementing Decision (EU) 2023/2683 of 30 November 2023 (2023)
2. Noor, A.R., Ishak, A.: Mechanical properties of recycled plastics (2021)
3. Schatz, L.M.: Batch variations of post-consumer recyclates and their influence on material properties (2023)
4. Aurrekoetxea, J., Sarrionandia, M.A., Urrutibeascoa, I., Maspoch, M.Ll.: Fracture behaviour of virgin and recycled isotactic polypropylene (2001)

5. Gall, M., Freudenthaler, P.J., Fischer, J., Lang, R.W.: Characterization of composition and structure–property relationships of commercial post-consumer polyethylene and polypropylene recyclates (2021)
6. Fernández, A., Clavería, I., Pina, C., Elduque, D.: Predictive methodology for quality assessment in injection molding comparing linear regression and neural networks (2023)
7. Chang, H., Su, Z., Lu, S., Zhang, G.: Intelligent predicting of product quality of injection molding recycled materials based on tie-bar elongation (2022)
8. Hürkamp, A., Ossowski, T., Klaus, D.: In-mould assembly of functionally integrated structures: a surrogate model for fast quality assessment (2022)
9. Xanthis, C.G., Filos, D., Haris, K., Aletras, A.H.: Simulator-generated training datasets as an alternative to using patient data for machine learning: an example in myocardial segmentation with MRI (2021)
10. Koltzenburg, S., Maskos, M., Nuyken, O.: Polymere: synthese, eigenschaften und anwendungen (2024)
11. Javierre, A., Clavería, I., Ponz, L., Aísa, J., Fernández, A.: Influence of the recycled material percentage on the rheological behaviour of HDPE for injection moulding process (2006)
12. Williams, M.L., Landel, R.F., Ferry, J.D.: The temperature dependence of relaxation mechanisms in amorphous polymers and other glass-forming liquids (1955)
13. Autodesk Inc.: Cross-WLF viscosity model, 15 May 2025. https://help.autodesk.com/view/MFIA/2023/ENU/?guid=MoldflowInsight_CLC_Ref_Materials_sim_math_models_Cross_WLF_viscosity_model_html
14. Wu, W., Zhao, B., Mo, F., Li, B., Jiang, B.: In-line steady shear flow characteristics of polymer melt in rectangular slit cavities during thin-wall/micro injection molding (2022)

The Impact of CO_2 Pressure on Cryogenic Minimum Quantity Lubrication (cMQL) Performance in Deep Hole Drilling of Inconel 718

Martin Sicking(✉), Conrad Breer, and Dirk Biermann

Institute of Machining Technology, TU Dortmund University, Dortmund, Germany
martin.sicking@tu-dortmund.de

Abstract. Sustainability and the careful use of resources entail a demanding conglomerate of requirements for component properties to improve efficiency and service life. For this reason, the use of high-temperature, difficult-to-cut materials is increasingly becoming the focus of science and industry. Due to their properties, difficult-to-cut materials place high demands on the machining process; in the case of Inconel 718, heat dissipation is particularly difficult due to the low thermal conductivity. The machining process must also become more efficient and resource-saving, whereby conventional cooling lubricant systems often reach their limits in deep hole drilling with tools of the smallest diameters in materials that are difficult to cut, as insufficient cooling capacity severely impairs process reliability. The cryogenic minimum quantity lubrication (cMQL) can significantly improve performance in deep hole drilling, but is dependent on the interaction between lubricant and CO_2. Cryogenic MQL using CO_2 as a cryogenic medium offers an alternative, as very high cooling capacities can be achieved with low resource consumption. The aim of this work is to explore possible improvements for cryogenic minimum quantity lubrication in the deep hole drilling process with the smallest diameters, focusing on the interactions between the lubricant and liquid CO_2 pressure and their influence on the machining process. A higher CO_2 pressure has a positive effect, particularly with regard to the solubility of the lubricants in the liquid CO_2 phase, and the extent to which this influence also has a positive effect on the machining process will be determined as part of these investigations.

Keywords: Cryogenic MQL · Small size diameter · Deep hole drilling · Difficult to cut material

L. Overmeyer and B.-A. Behrens (eds.), *Production at the Leading Edge of Technology*, Lecture Notes in Production Engineering, https://doi.org/10.1007/978-3-032-19524-1_8

1 Introduction

Nickel-based alloys are regarded as essential materials in high-temperature technology, driven by continuous advancements in material development to meet the increasing demands of the aerospace, energy, and process industries [1]. Their out-standing mechanical and physical properties at elevated temperatures make them especially well-suited for highly stressed components, particularly in thermal engines. The use of nickel-based alloys enables higher combustion temperatures, thereby enhancing system efficiency [2].

However, machining these alloys poses a considerable technological challenge. Their strong adhesive tendencies, the presence of abrasive carbides, and low thermal conductivity lead to high thermomechanical stress on cutting tools, resulting in significant tool wear [3]. In this context, the thermal load at the cutting edge is a critical factor influencing tool life, process stability, and the quality of the machined surfaces. Oxidation and diffusion processes both of which intensify at high temperatures are among the primary causes of tool degradation [4].

Extensive scientific research has been dedicated to improving the machinability of nickel-based alloys, with a particular focus on Inconel 718. In addition to the development of efficient cooling and lubrication strategies, optimization at the tool level both in terms of macroscopic and microscopic design remains a central objective. Jaeger, for instance, introduced an innovative cryogenic minimum quantity lubrication (MQL) concept specifically for deep hole drilling applications [5]. Beer and Bücker demonstrated that process performance can be further improved by modifying the flank face geometry during drilling [6, 7].

A precise understanding of the thermomechanical conditions in the cutting zone is essential for the optimization of machining processes. While temperature is a key factor, the effectiveness of cryogenic minimum quantity lubrication (cMQL) is also strongly influenced by the composition of the CO_2-lubricant mixture. In particular, the pressure of the CO_2 plays a crucial role in determining the thermodynamic behavior and solubility of lubricating oils in CO_2. Recent studies have shown that an increase in CO_2 pressure leads to a significantly improved solubility behavior of ester-based lubricants in both liquid and supercritical CO_2 [8].

The present paper focuses on investigating the influence of CO_2 pressure varied between 57 and 100 bar on the performance of cMQL during deep hole drilling of nickel-based alloys. The experimental setup enables the systematic analysis of tool wear and process stability under defined pressure conditions. These findings offer valuable insights into the interaction between pressure-controlled lubricant delivery and thermomechanical tool loading.

2 Experimental Setup

2.1 Machining System

The experimental investigations were carried out using a modified three-axis special-purpose machine tool of the type *Körner CNC-607GK*. This CNC drilling and milling machine tool is equipped with a custom-designed programmable logic controller (PLC) of the type TC3.1 developed by Beckhoff and is modified specifically for cryogenic machining applications. Figure 1 illustrates the machine system along with the modifications implemented for this study.

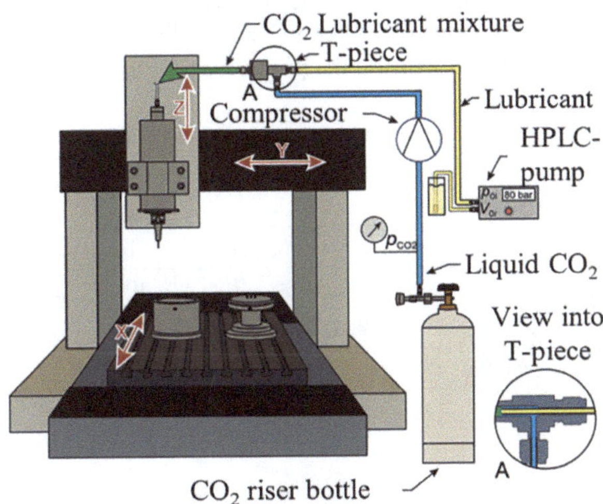

Fig. 1 Machine tool system for cMQL

Due to its modular architecture, the machine tool enables flexible integration of various cooling and lubrication strategies. The machining process can be performed using either conventional MQL or cMQL approach. For deep hole drilling operations with high length-to-diameter ratios (l/d), internal coolant supply is essential to ensure sufficient lubrication and thermal management in the contact zone. The cryogenic lubrication system employs liquid CO_2 as a carrier and cooling medium, to which a lubricant is added via a high-performance liquid chromatography (Latek HPLC P-402) pump. At the cooling channel outlet, the CO_2 phase is changed from liquid to gaseous. The lubricant flow rate can be precisely adjusted and is set on $\dot{V} = 30$ ml/h for all experiments, based on preliminary investigations [5]. In cMQL, maintaining single-phase conditions of CO_2 and lubricant before entering the spindle is essential to avoid phase separation due to centrifugal forces which leads to an inconsistent oil flow rate at the tool tip. Experimental investigations by Piche et al. showed that at pressures below approximately 60 bar, lubricants do not fully dissolve into the liquid CO_2 phase, resulting in a two-phase flow. This incomplete dissolution would lead to unstable lubrication and cooling conditions. Therefore, the CO_2 is supplied to the system through a CO_2-compressor, exceeding the typical cylinder pressure of p_{CO2} ~57 bar, to ensure complete dissolution of the lubricant in CO_2 prior to delivery to the cutting zone [8]. For the experiments a cutting speed of v_c = 31.25 m/min and a feed rate of f = 0.021 mm were selected, and all test series were repeated at least three times to ensure statistical reliability of the results.

2.2 Experimental Tools and Process Media

The primary drilling tool used in the tests is a solid carbide twist drill with a diameter of $d = 2$ mm, a length-to-diameter ratio (l/d) of 15, and a point angle of $\sigma = 135°$. The tool is manufactured by Gühring KG under the designation "ExclusiveLine" (type 6412). It features internal coolant channels for targeted lubrication of the cutting zone and is coated with TiAlN to improve wear resistance under thermal loads. As lubricants three ester-based oils were investigated. One is a polyester (POE) and two are monoester-based lubricants (V7 and S_{++}). Table 1 shows the properties of the oils [9–12].

Table 1 Properties of the lubricants

Characteristics	POE	V7	S_{++}	Unit
Base oil	Polyester	Monoester	Monoester	–
Viscosity, ν (40 °C)	68	21	40	g/mm^3
Density, φ (20 °C)	0.95	0.91	1.00	g/cm^3
Sulphur, S	0.05	–	16	%
Pourpoint	−27	−63	−6	°C

3 Results

3.1 Influence of CO_2 Pressure and Lubricant on the Flow Rate and Tool Life

The CO_2 pressure has a decisive influence on the solubility of different lubricants and, consequently, also indirectly affects the expected lubricant discharge. To experimentally capture the influence of pressure on solubility, discharge measurements are conducted, and their results are plotted alongside the pressure-dependent CO_2 flow rate and the pressure-dependent tool life in Fig. 2. The CO_2 flow rate was recorded directly at the CO_2-compressor using a flow meter CoriFlow M14 from Bronkhorst.

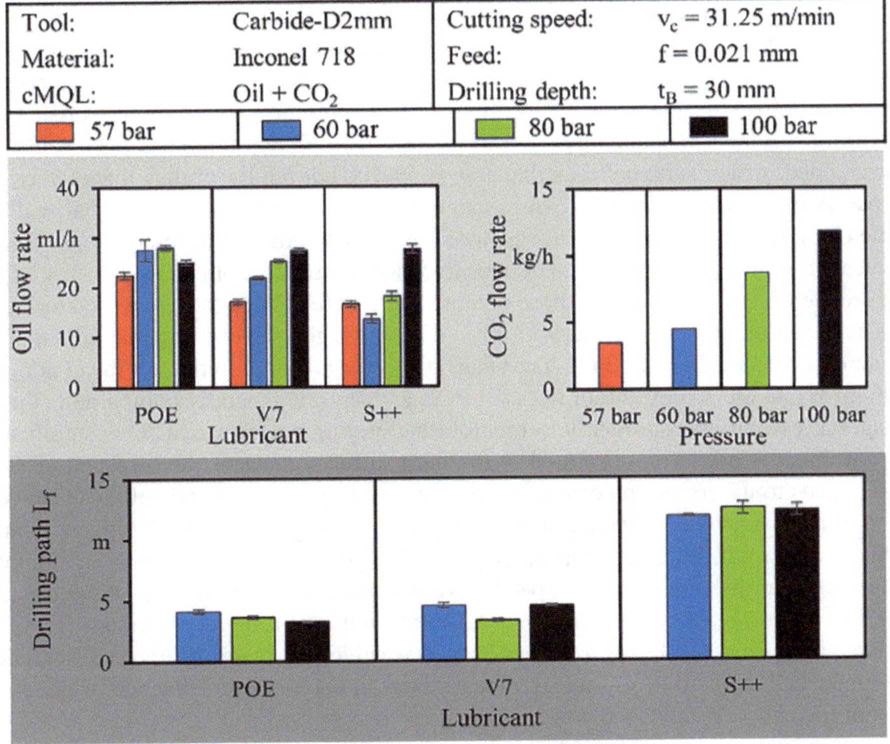

Fig. 2 Influence of CO_2 pressure on flow rate and drilling path length

For POE, the discharge quantity is already very high at a CO_2 pressure of $p_{CO2} =$ 60 bar, while at a CO_2 pressure of $p_{CO2} = 80$ bar, an almost identical discharge quantity can be observed, with the discharge quantity at $p_{CO2} = 100$ bar being slightly lower. In comparison, V7 exhibits somewhat lower solubility at $p_{CO2} = 60$ bar, with a deviation from the target value of $\Delta = 27\%$, indicating that some of the lubricant is precipitating in the spindle. Similarly to its higher solubility, a slight increase in lubricant discharge can be observed at $p_{CO2} = 80$ bar, with values falling within the range of POE's discharge quantities. For S_{++}, significantly lower discharge quantities are observed at $p_{CO2} =$ 60 bar compared to the other lubricants, with a deviation from the target value of $\Delta =$ 55%. The low discharge quantity can be attributed to S_{++}'s low solubility, especially at lower pressures. The CO_2 flow rate is directly dependent on pressure; while a mass flow rate of $\dot{m}_{CO2} < 5$ kg/h can be observed at a pressure of $p_{CO2} = 60$ bar, this value nearly doubles at $p_{CO2} = 80$ bar and reaches an even higher value of $\dot{m}_{CO2} \approx 12$ kg/h at p_{CO2} $= 100$ bar.

The tool life L_f, which is also shown in Fig. 1, is an important criterion for assessing the influence of the cooling and lubrication strategy on the machining process. Tool life refers to the distance that the tool has machined over its entire lifespan, with the end of tool life being determined by various criteria. One criterion is the maximum wear mark width VB, which is set at $VB = 300$ µm for the tools used. Additionally, a tool may reach the end of its lifespan due to cutting edge breakage or failure during the drilling process. First, the tool lives obtained with POE are compared as a function of CO_2 pressure, showing that they fall within a similar range of $L_f = 3.26...4.13$ m. Within this range, varying tool lives were achieved depending on CO_2 pressure, with the lowest recorded tool life at $p_{CO2} = 100$ bar being $L_f = 3.26$ m. At $p_{CO2} = 80$ bar, a tool life of $L_f = 3.62$ m was attained, which is higher than that at $p_{CO2} = 100$ bar but lower than that at $p_{CO2} =$ 60 bar. A tool life exceeding $L_f > 4$ m can only be achieved at $p_{CO2} = 60$ bar, with a value of $L_f = 4.13$ m; thus, the highest tool lives can be attained at reference pressure, albeit with minimal differences. The tool lives for V7 range from $L_f = 3.06...4.62$ m and are therefore comparable to those obtained with POE. The highest recorded tool life was achieved at $p_{CO2} = 100$ bar with $L_f = 4.62$ m, slightly exceeding that obtained at reference pressure ($L_f = 4.53$ m). The value at $p_{CO2} = 80$ bar is somewhat lower at $L_f =$ 3.06 m. While the achieved tool lives for POE and V7 are generally comparable, those using S_{++} exhibit approximately a threefold increase in performance. This significant improvement is primarily attributed to the high sulfur content of the lubricant (16%), which leads to the formation of a separation layer composed of metal sulfides between the tool and workpiece, thereby significantly enhancing lubrication effectiveness and reducing tool wear. When comparing different CO_2 pressures while using S_{++}, minimal differences can be observed. Comparable values were obtained at $p_{CO2} = 80$ bar (L_f $= 12.43$ m) and $p_{CO2} = 100$ bar ($L_f = 12.26$ m), while at reference pressure ($p_{CO2} =$ 60 bar), a drilling path length of $L_f = 12$ m was achieved. However, these differences are minor and fall within variability ranges. Thus, no significant influence of pressure on tool life can be established when using S_{++}.

In conclusion, increasing CO_2 pressure does not positively affect tool life across any of the lubricants examined. In the further evaluation of the experimental results, the focus will be laid on the two lubricants POE and S_{++}, as similar results were obtained with both POE and V7. Additionally, POE and S_{++} include a polyester and a monoester, as well as a low-additive and a high-additive lubricant.

3.2 Influence of CO_2 Pressure and Lubricant on Wear and Process Forces

Increasing wear leads to a deviation of the tool from its ideal geometry, which in turn results in an increase in cutting forces. Thus, cutting forces and tool wear are directly connected to each other thereby providing important information about the condition of the tool during the machining process. In this study, cutting forces were measured using a piezoelectrical dynamometer 9271A from Kistler, while tool wear was analyzed with a digital microscope VHX-5000 from Keyence. Figure 3 illustrates the wear development as a function of CO_2 pressure and lubricant used. When examining the width of wear marks, it becomes evident that all tools, except for the one used with POE at a CO_2 pressure of $p_{CO2} = 80$ bar, reach or nearly reach the maximum wear mark width VB_{max} $= 300$ µm. For POE, the development of wear mark width follows a similar trend for all pressure settings. Thus, no significant influence of CO_2 pressure on tool wear can be observed. Furthermore, when analyzing wear patterns, no substantial differences are noted among the various experimental series. In all tests, carbonization is evident on the flank face of the drilling tool, which can be attributed to high temperatures during the machining process. With S_{++}, the increase in wear mark width is significantly slowed compared to the other lubricant, which can primarily be attributed to the previously mentioned wear-reducing effect of the additive. Nonetheless, no influence of CO_2 pressure on the tool wear can be observed when using S_{++}.

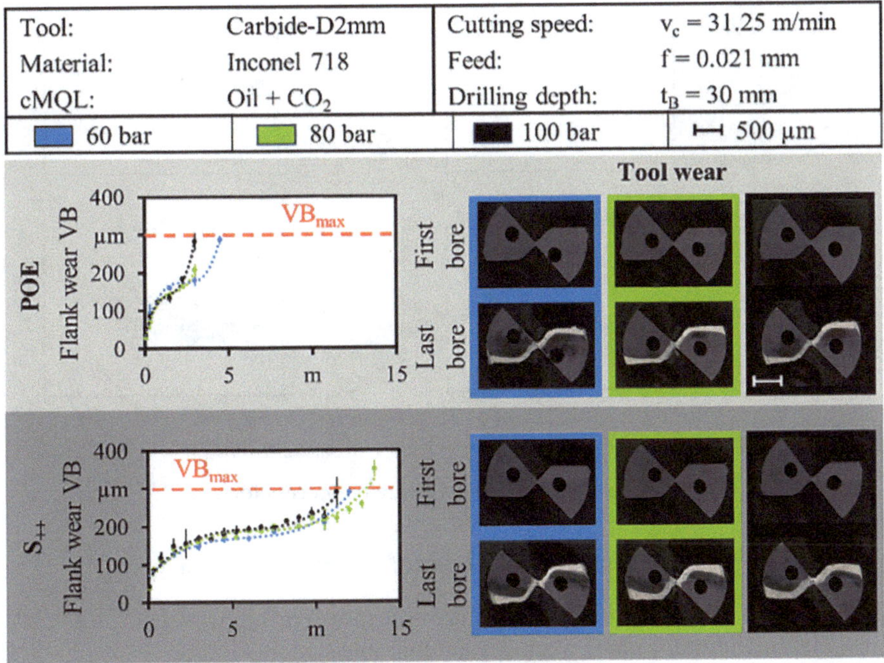

Fig. 3 Influence of CO_2 pressure on the tool wear

The feed forces shown in Fig. 4 follow a similar development over the tool life when using POE, such that, there are almost no noticeable differences between the various pressures. The drilling torque behaves similar to the feed forces across all experimental series; however, a significant variation in individual measurements occurs, which may be attributed to the low magnitude of the measured values. Overall, at higher pressure levels slightly increased feed forces and drilling moments are indicated, which may be related to insufficient expansion of CO_2 in the area of the cutting edge. However, the difference is minimal and has little impact on tool wear and life, leading to the conclusion that higher CO_2 pressure does not significantly affect tool wear. When using S_{++}, significantly reduced process forces are observed throughout most of the tool life compared to POE, which can primarily be attributed to the high lubrication effectiveness of this lubricant. Towards the end of tool life, process forces increase noticeably even when using S_{++}, although no influence of CO_2 pressure can be observed. In summary, a higher CO_2 pressure tends to result in slightly increased cutting forces; however, this influence remains minor.

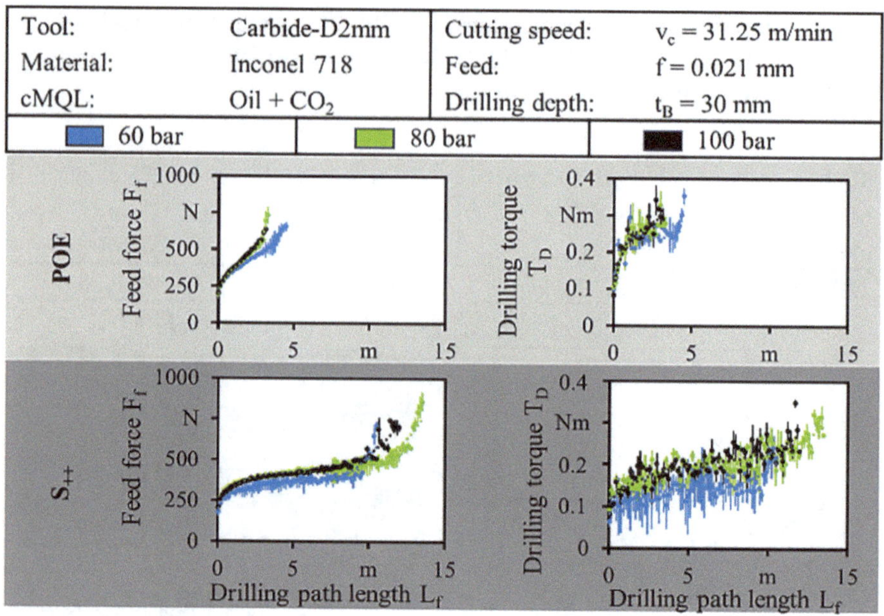

Fig. 4 Influence of CO_2 pressure on process forces

3.3 Influence of CO_2 Pressure and Lubricant on the Temperature

The influencing factors on cutting temperature in deep hole drilling processes with cMQL are diverse. In addition to tool wear, the cooling lubrication strategy has a significant impact. Therefore, the influence of CO_2 pressure on cutting temperatures, which was recorded by using a two-color pyrometer Fire 3 from en2Aix, is examined below. The fiber of the pyrometer is inserted directly into the workpiece and is machined alongside the material. Due to the angled insertion of the fiber, as drilling depth increases, the measurement point shifts from the center of the tool towards its edge. Thus, temperature progression along the cutting edge of the tool can be determined in a single measurement. The setup of the workpiece for temperature measurement and the different measuring areas along the cutting edge as well as the average temperatures along the cutting edge as a function of CO_2 pressure and lubricant are illustrated in Fig. 5.

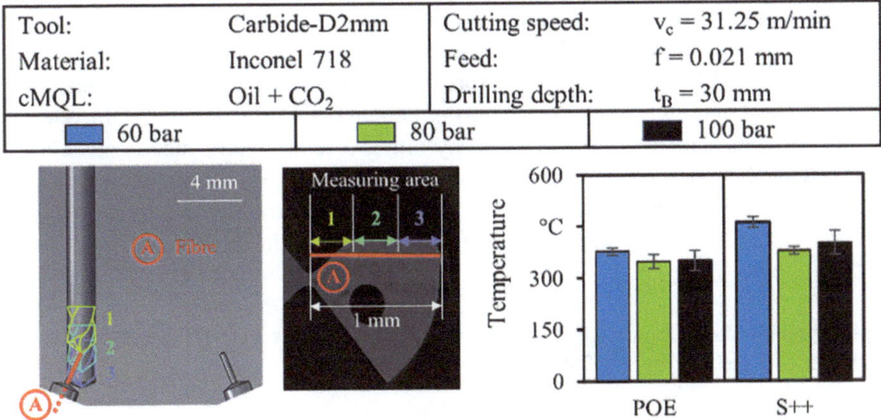

Fig. 5 Influence of the CO_2 pressure on the cutting temperature

It is evident that an increase in pressure from $p_{CO2} = 60$ bar to $p_{CO2} = 80$ bar results in a decrease in cutting temperature by $\Delta T \approx 30$ K to $\Delta T \approx 80$ K, depending on the lubricant used. Further increasing the pressure to $p_{CO2} = 100$ bar leads to an additional temperature reduction of approximately $\Delta T \approx 16$ K when using S_{++}. However, for POE, no further reduction in cutting temperature can be observed when increasing the pressure from $p_{CO2} = 80$ bar to $p_{CO2} = 100$ bar. This indicates that CO_2 may not fully vaporize in the cutting zone at a pressure level of $p_{CO2} = 100$ bar.

4 Conclusion

The objective of the conducted experiments was to explore potential improvements for the cMQL in helical deep hole drilling processes with minimal diameters. This paper focuses on the interactions between lubricant and CO_2 pressure and their influence on the machining process. A higher CO_2 pressure positively affects the solubility of lubricants in the liquid CO_2 phase; however, this work aimed to determine whether this influence also positively impacts the machining process. To achieve this, the actual quantities of lubricant and CO_2 at the tool were measured, allowing for conclusions to be drawn regarding the pressure-dependent behavior of the CO_2-lubricant solution. The lubricant discharge showed varying amounts depending on the lubricant used, with POE demonstrating discharge quantities that approached those delivered by the pump even at the lowest pressure level. In contrast, V7 and S_{++} exhibited a lower lubricant discharge, which significantly improved with increasing CO_2 pressure, resulting in very high discharge quantities for all lubricants at the highest pressure level. The CO_2 discharge is highly dependent on pressure, as an increase in pressure leads to a substantial rise in the CO_2 flow rate. The type of lubricant used and CO_2 pressure have differing effects on tool life, tool wear, and cutting forces. An increase in CO_2 pressure could not significantly improve tool life for any of the lubricants used. Regarding cutting forces, slight tendencies towards higher forces at elevated pressures were observed, which may indicate insufficient expansion of CO_2 in the cutting zone. In contrast to different pressure levels, a significant impact on tool lifespan was noted among different lubricants. Compared to POE, S_{++} achieved nearly a threefold increase in tool life. This can be attributed to its high content of EP-additives in the form of sulfur, which could offer considerable potential for enhancing tool longevity during helical deep hole drilling of Inconel 718. In addition to measuring cutting forces, pyrometric temperature measurements were conducted during machining. An influence of CO_2 pressure on cutting temperatures was observed; higher pressures tended to result in lower measured temperatures. However, no direct influence of lubricant on cutting temperature could be detected when using a new tool.

Acknowledgements. Supported by Deutsche Forschungsgemeinschaft (DFG, German Research Foundation)—Project number 452408713.

Competing Interests. The author(s) has no competing interests to declare that are relevant to the content of this manuscript.

References

1. Bargel, H.-J. (Hrsg.), Schulze, G. (Hrsg.): Werkstoffkunde. 12., bearbeitete Auflage, korrigierter Nachdruck. Springer-Vieweg, Berlin (2018)
2. Pollock, T.M., Tin, S.: Nickel-based superalloys for advanced turbine engines: chemistry, microstructure, and properties. J. Propuls. Power **22**(2), 361–374 (2006). https://doi.org/10.2514/1.18239
3. Wang, R., et al: Tool wear in nickel-based superalloy machining: an overview. Processes **10**(11), Artikel 2380 (2022). https://doi.org/10.3390/pr10112380
4. Longbottom, J.M., Lanham, J.D.: Cutting temperature measurement while machining—a review. Aircraft Eng. Aerosp. Technol. Int. J. **2**, 122–130 (2005)
5. Jaeger, J.: Wendeltiefbohren kleinster Durchmesser in schwer zerspanbare Werkstoffe mit kryogener Minimalmengenschmierung. Dissertation. Technische Universität Dortmund, Institut für Spanende Fertigung, Dortmund (2021)
6. Beer, N.: Systematische Untersuchung von Vollhartmetall-Wendelbohrern zum Bearbeiten von Inconel 718. Dissertation. Vulkan-Verlag GmbH, Dortmund (2015)
7. Bücker, M., Biermann, D.: Neue Methoden zur Entwicklung und Herstellung von Hochleistungswerkzeugen für die Bohrbearbeitung von Inconel 718. Dissertation. Technische Universität Dortmund, Institut für Spanende Fertigung, Dortmund (2021)
8. Piche, N., Pollak, S., Petermann, M.: Dissolution behavior of different lubricating oils in liquid and supercritical CO_2. J. Supercrit. Fluids **205** (2024)
9. Mortier, R.M. (Hrsg.), Fox, M.F., Orszulik, S.T.: Chemistry and Technology of Lubricants. Springer Netherlands, Dordrecht (2010)
10. Cavestri, R.C.: Potentially useful polyolester lubricant additives: an overview of antioxidants, antiwear and antiseize compounds. In: Lubrication Science (1996)
11. Pirro, D.M., Wessol, A.A., Dascher, E.: Lubrication Fundamentals, 3rd edn., Revised and Expanded. Noria Press, Houston, TX (2016)
12. Fahl, J.: Esteröle und ihre Eigenschaften. In: Tribologie-Fachtagung GfT (2000)

BY

Towards Automated and Robust Production Planning in Remanufacturing: Utilizing Semi Markov Decision Processes

Maurice Engels(✉), Martin Benfer, and Gisela Lanza

Wbk—Institute of Production Science, Karlsruhe Institute of Technology (KIT), Karlsruhe, Germany
maurice.engels@kit.edu

Abstract. Remanufacturing recovers critical components and materials from returned products, enhancing resource efficiency, supply chain resilience, and supporting circularity. To successfully implement remanufacturing operations, production planning must address significant challenges from uncertain return quality, fluctuating volumes, and variable process times, necessitating a stochastic planning approach. This paper introduces a Semi-Markov Decision Process framework that explicitly models stochastic return arrivals and dynamic processing conditions. Policy iteration guides lot-sizing and capacity-allocation decisions within a SimPy simulation. Benchmarking against an EOQ-based, bottleneck-first heuristic demonstrated increased cumulative output and marked the first promising results of the proposed methodology.

Keywords: Remanufacturing · Production planning

1 Production Planning in Remanufacturing Processes

In an era of dwindling resources, pollution and climate change, sustainable production paradigms such as remanufacturing are gaining importance. Remanufacturing is a process that preserves value by converting one or more used products into items that match or exceed the originals [1], thus maximizing the environmental and economic benefits of sustainable production. Implementing remanufacturing processes however is difficult, as return volumes and product quality are uncertain and disassembly process times vary. As a result, determining production batch sizes and assigning capacity is especially challenging when both inputs and outputs are uncertain. Effective production planning is therefore essential to increase profitability in remanufacturing and achieve environmental objectives, such as reducing resource consumption in production.

1.1 Challenges in Remanufacturing Production Planning

Uncertainty lies at the core of remanufacturing, affecting every stage of production planning and scheduling. Unlike traditional manufacturing, with forecastable raw-material supply and product demand, remanufacturing depends on

L. Overmeyer and B.-A. Behrens (eds.), *Production at the Leading Edge of Technology*, Lecture Notes in Production Engineering, https://doi.org/10.1007/978-3-032-19524-1_9

stochastic return streams whose timing and volume are often unpredictable, leading to either idle capacity or excess inventories [2]. Returned cores also vary widely in condition, therefore processes range from minor refurbishment to full disassembly, which induces large processing time variances and undermines fixed-batch strategies without costly pre-sorting [3]. As a result, planners must design schedules and capacity buffers that hedge against fluctuations in both supply of cores and market demand, effectively balancing service levels with cost efficiency [4].

Operational unpredictability also arises during the disassembly and reprocessing stages. Each returned unit may follow a unique routing through the facility, and variable failure rates can disrupt expected throughput. This transforms material requirements planning into a multi-stream inventory problem, as planners track and manage both recovered parts and new components—each with different quality grades, yields, and lead times [5].

Finally, economic, strategic, and regulatory considerations add another layer of complexity. Remanufacturing cost spanning inspection, disassembly, and reprocessing are highly variable and depend on return quality, necessitating integrated cost-accounting in planning models to maintain profitability across uncertain yield scenarios [6].

1.2 State of the Art in Remanufacturing Production Planning

Early remanufacturing planning extended linear models to include reverse logistics but often under estimated safety stocks by assuming known returns and fixed quality [7]. To address uncertainty, multi-stage stochastic methods (e.g., Stochastic Integer Programming (SIP) have been used to generate scenario trees for returns and demand in capacity allocation [8]. Furthermore, analytical cycle policies offer closed-form schedules based on return rates, capacities, and holding costs, refined via discrete-event simulations [9]. Markovian process-reliability models represent core degradation and rework as probabilistic state transitions, facilitating optimal costreliability trade-offs for remanufacturing cell design [6]. Additionally, Markovian Models have been used to optimize Order Dispatching in Remanufacturing Systems [10].

Despite these advances, remanufacturing models face practical limitations. Large scale stohastic methods such as SIP and SDDIP suffer from scenario explosion as uncertainties in product types, quality grades and lead time variations multiply, making exact solutions computationally infeasible. Stochastic models that treat uncertainties in isolation overlook their interactions, and separate decision support tools cannot reliably guide capacity or investment choices [8]. Furthermore, returns are often treated as independent and identically distributed, quality distributions are only represented as coarse fuzzy classifications, and correlations between the uncertainty factors are frequently not taken into account [11]. These simplifications potentially hide systemic risks, leading to schedule disruptions and higher costs when return patterns differ from expectations. An integrated modeling paradigm is thus required, combining scalable stochastic approaches and simultaneous management of interdependent uncertainties to create resilient and cost-effective remanufacturing systems. Therefore, this paper will present a novel framework that allows to effectively address uncertainties in remanufacturing production planning and accordingly improves the efficiency of remanufacturing.

2 Semi Markov Decision Processes

In this paper, a Semi-Markov Decision Process (SMDP) models and improves remanufacturing production planning. We first review Markov Decision Processes to expose the fixed time-step limitation, then show how SMDPs' yield greater planning flexibility.

2.1 From Markov to Semi Markov Decision Processes

A Markov Decision Process (MDP) is a tuple (S, A, P, R, γ) modeling sequential decisions under uncertainty [12]. Here, S and A are the state and action sets; $P(s' \mid s, a)$ the probability of moving to s' from s via a; $R(s, a, s')$ the immediate reward; and $\gamma \in (0, 1)$ the discount factor. The Markov property ensures that transitions depend only on the current stateaction pair. A stationary policy $\pi\colon S \to A$ maximizes the expected discounted return

$$V^{\pi}(s) = \mathbb{E}_{\pi}\big[\sum_{t=0}^{\infty} \gamma^{t} R(s_t, a_t, s_{t+1}) \mid s_0 = s\big]. \tag{1}$$

An SMDP generalizes the classical MDP by allowing transitions to occur at irregular, state-, and action-dependent time intervals rather than fixed steps. In an MDP, decisions and transitions occur at fixed intervals , without memory of elapsed time (Fig. 1a).

In contrast, an SMDP generalizes this by introducing random sojourn times, during which the system remains in a state-action pair before transitioning (Fig. 1b). This flexibility captures real-world timing variability essential to remanufacturing. MDP solution methods still apply to SMDP [13]. In this paper policy iteration will be used, which at iteration k computes V^{π_k} via the Bellman equations and then updates π_{k+1}, this process can be seen in Fig. 2. Through Policy Iteration the best possible solution can be approximated iteratively.

One advantage of the policy evaluation method shown in Fig. 2 is that if it used to approximate the policy rather than compute the exact policy, which is often infeasible, it can deal with large state spaces [14], making it suitable for complex systems such as remanufacturing production systems.

2.2 Advantages of SMDPs for Decisions in Remanufacturing

Unlike traditional stochastic programming, which requires discretization into a finite scenario tree that grows exponentially with the number of uncertainties and stages [15], SMDPs operate directly in continuous event time. This

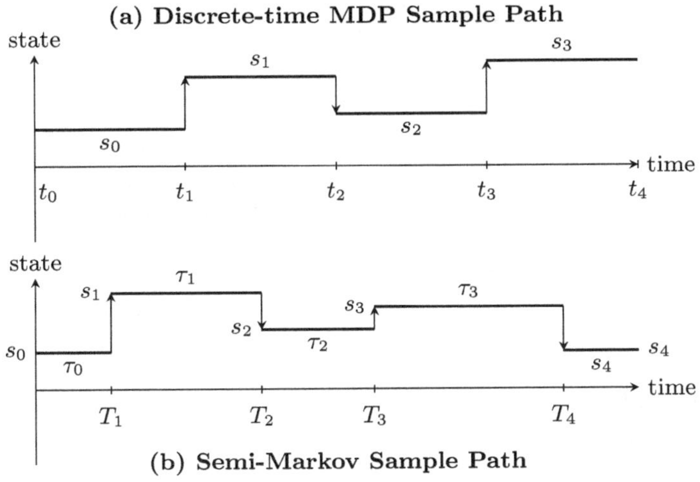

Fig. 1. **a** Discrete-time MDP; **b** semi-Markov decision process [13]

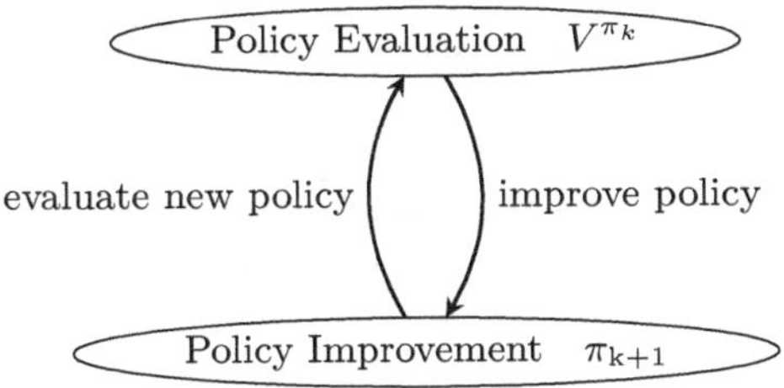

Fig. 2. Policy iteration alternates between evaluating the current policy and improving it by one-step lookahead [14]

allows compact, state-specific policies and continuous cost accrual based on real action durations [13]. This event-driven structure yields substantial planning benefits for remanufacturing. Capacity allocation becomes adaptive: resources can be reassigned immediately when delays or completions change availability. Lot-sizing improves as lots are released precisely when both cores and capacity align, minimizing setup waste and holding costs without introducing batching delays. Inventory is managed through continuous cost tracking, supporting finer-grained trade-offs between batch frequency and setup overhead. Finally, lead-time scheduling reflects the actual sequence of random durations, allowing urgent jobs to preempt ongoing ones and reducing both average and worst-case delays. These features collectively deliver lower inventory levels, increased throughput, and improved delivery reliability compared to discrete-time stochastic models.

3 Production Planning SMDP

After demonstrating the benefits of SMDPs, a simplified SMDP for remanufacturing that focuses on batch-sizing and short-term worker allocation is introduced, although the methodology can be extended to cover multiple planning activities. The remanufacturing production planning SMDP is defined with the following elements. The SMDP's state at any decision epoch is defined by the tuple $s = (I_c, I_f, W, L, d_j)$, which captures the current inventories of cores and finished goods (I_c, I_f), work-in-process levels (W), available labor (L), and the deadline of the next order (d_j). From each state s, the decision maker chooses an action $a = (e_1, \ldots, e_K, q)$, where e_k denotes the number of workers allocated to station k (for $k = 1, \ldots, K$) and q specifies the quantity of finished goods to release. Upon taking action a in state s, the process yields a reward.

$$r = R_{\text{base}} + \begin{cases} B, & t_{\text{delivery}} \leq d_j, \\ 0, & \text{otherwise}, \end{cases} - c_{\text{labor}} \sum_{k=1}^{K} e_k - h_c I_c - h_f I_f. \tag{2}$$

where R_{base} is the base reward from selling the finished goods, B is an incentive for finishing orders on time and h_c, h_f are holding costs for the cores and finished products respectively.

The Transition-Kernel $Q(s', \tau \mid s, a)$ is a probability distribution over the next state s' and sojourn time τ given the current state s and action a. It encapsulates both the dynamics of the system and the randomness of timing, in regards to core arrivals, new orders and completed orders. Decisions are made at event times T_n, with system dynamics driven by random durations between events—capturing variability in core arrivals, processing times, and resource availability. This SMDP is solved by using policy iteration (see Sect. 2.2).

4 Simulation-Based Production Planning Using an SMDP

The proposed SMDP for remanufacturing production planning is implemented using a simulation approach. This simulation driven approach is visualized in Fig. 3.

The remanufacturing workflow, including the processes inspection, disassembly, cleaning, assembly, and testing, is encoded as a discrete-event model, with hard machine and labor capacity constraints and minimum staffing to ensure feasibility. The Simulation was modeled after a real remanufacturing process of alternators. Core returns (timing, quantity, quality) are generated stochastically; to reduce state-space size, cores are grouped into three quality classes. A core's class determined defect probabilities during disassembly. The SimPy model allows dynamic allocation of workers and batch sizes based on policy decisions. The SMDP-derived production policy is deployed in SimPy to simulate operations and record discounted reward trajectories. These rewards were then fed back into the SMDP for iterative policy evaluation and improvement.

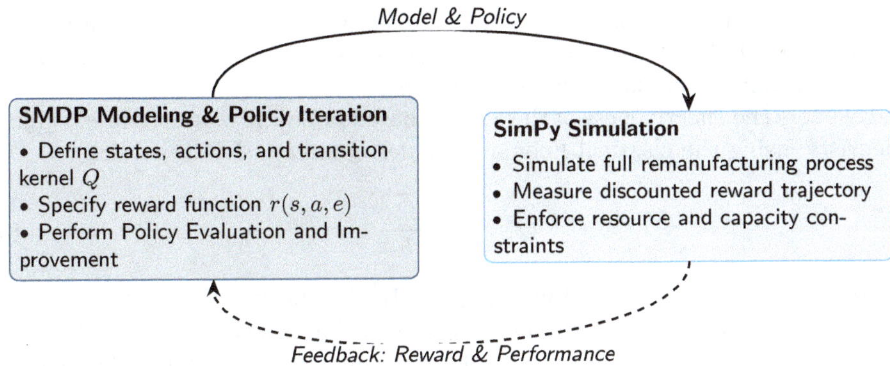

Fig. 3. Feedback loop between SMDP modeling (policy iteration) and SimPy validation for remanufacturing planning

Using the simulation framework illustrated in Fig. 3, production policies can be evaluated in a realistic environment before deployment in real-world settings. By leveraging a SimPy model, it is possible to closely replicates actual remanufacturing processes—capturing resource constraints, stochastic returns, and process dynamics—and thus enhances the validity and practical relevance of the performance metrics generated by the underlying SMDP.

4.1 Preliminary Results

The proposed and implemented SMDP model in the given model and parameters favours smaller lot sizes, which is apparent in Fig. 4, which depicts the lot-size decisions made by the SMDP policy.

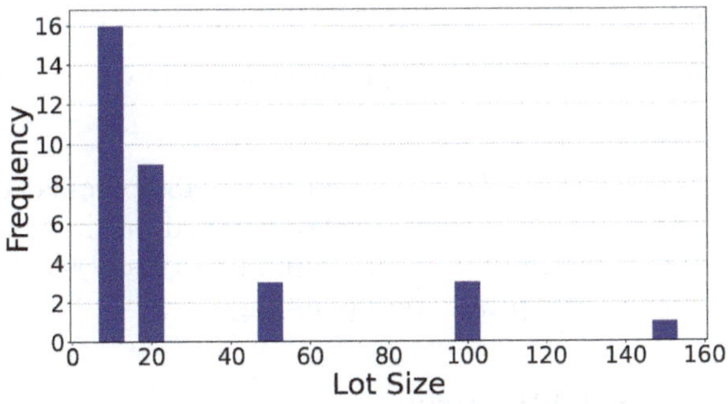

Fig. 4. Chosen lot sizes under the SMDP policy

The favouring of smaller lots is intuitive for remanufacturing, as smaller batches reduce exposure to potential component failure or unusable parts, lowering scrap and recovery costs. Additionally, smaller, more frequent batches accommodate random core returns more effectively, stabilizing inventory and enhancing flexibility. Furthermore, in the presented SMDP relatively high holding costs incentive the use of smaller batches.

To assess the effectiveness of the SMDP-based policy, it was compared against a heuristic using the classical Economic Order Quantity (EOQ) model,

$$Q^* = \sqrt{\frac{2\,D\,S}{h}} \tag{3}$$

where: D is the demand rate (units per period), S is the setup cost per batch, h is the holding cost per unit per period and Q is the lot-size calculated. This model minimizes total setup and holding costs. For capacity allocation, a bottleneck-first approach was applied: at fixed two-hour intervals, available extra capacity was preferentially assigned to the production process with the highest utilization. This combined heuristic balances responds to workload changes, providing a suitable benchmark for evaluating the SMDP policy. Figure 5 presents the results obtained from implementing this heuristic.

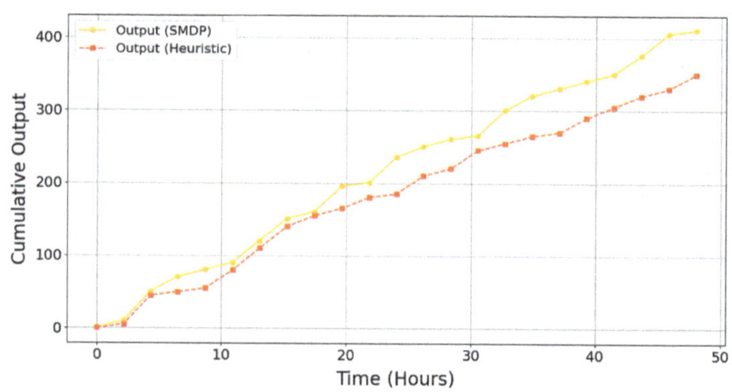

Fig. 5. Comparison finished products heuristic versus SMDP

Figure 5 illustrates the cumulative output over a 48-h simulation horizon. The SMDP policy clearly delivers a higher total of finished products than the benchmark heuristic, demonstrating its superior ability to navigate the complexities inherent in remanufacturing production planning.

5 Conclusion and Discussion

This paper introduces an SMDP framework for remanufacturing planning that integrates stochastic core returns, resource constraints, and process dynamics through policy iteration and a SimPy simulation. The resulting decision-support tool outperforms a classical heuristic in cumulative output while managing uncertainty effectively.

However, three main limitations remain. The model's finite-horizon, short term planning scope limits long-term planning and risks state-space explosion, though hierarchical or multi-timescale generalized formulations could mitigate this. Second, comprehensive benchmarking against industry-standard heuristics is needed to validate practical performance and uncover potential gaps. Third, incorporating finer quality grading, dynamic pricing, or correlated uncertainties makes explicit transition probabilities infeasible; sampling-based methods like Monte-Carlo Simulation offer scalable alternatives that bolster planning robustness. Addressing these areas will enhance the presented planning framework.

Acknowledgments. This research has received funding by the German Federal Ministry of Education and Research (BMBF)—funding number 02L19C250 and from the European Union's Horizon Europe research and innovation programme under Grant Agreement 101138930—Digitally-enhanced multi-level solution for smart human-centric remanufacturing (rEUman Project).

Competing Interests. The author(s) has no competing interests to declare that are relevant to the content of this manuscript.

References

1. DIN SPEC 91472:2023-06: Remanufacturing (Reman)—Qualitätsklassifizierung für Zirkuläre Prozesse (2023)
2. Lanza, G., Deml, B., Matthiesen, S., Martin, M., Brützel, O., Hörsting, R.: The vision of the circular factory for the perpetual innovative product. Automatisierungstechnik **72**(9), 774–788 (2024). https://doi.org/10.1515/auto-2024-0012
3. Zhang, W., Shi, J., Zhang, S., Chen, M.: Scenario-based robust remanufacturing scheduling problem using improved biogeography-based optimization algorithm. IEEE Trans. Syst. Man Cybernet. Syst. **53**, 3414–3427 (2023). https://doi.org/10.1109/TSMC.2022.3225443
4. Jing, Y., Wang, X., Li, W., Deng, L.: Robust production planning with remanufacturing and heterogeneous demands under return and demand uncertainty. Int. J. Appl. Math. Stat. **41**, 183–196 (2013). https://doi.org/10.1007/s13243-025-00149-8
5. Hoffmann, M., Krini, A., Mueller, A., Knorn, S.: Remanufacturing production planning and control: conceptual framework for requirement definition. J. Remanufact. (2025)
6. Stamer, F., Sauer, J.: Optimizing quality and cost in remanufacturing under uncertainty. Prod. Eng. (2024). https://doi.org/10.1007/s11740-024-01314-x
7. Guide, V.D., Jayaraman, V., Srivastava, R.: Production planning and control for remanufacturing: a state-of-the-art survey (1999). https://doi.org/10.46254/an12.20220224
8. Quezada, F., Gicquel, C., Kedad-Sidhoum, S.: A stochastic dual dynamic integer programming based approach for remanufacturing planning under uncertainty. Int. J. Prod. Res. **61**, 5992–6012 (2022). https://doi.org/10.1080/00207543.2022.2120924
9. Polotski, V., Kenné, J., Gharbi, A.: Set-up and production planning in hybrid manufacturing–remanufacturing systems with large returns. Int. J. Prod. Res. **55**, 3766–3787 (2017). https://doi.org/10.1080/00207543.2017.1293863

10. Wurster, M., Michel, M., May, M.C., et al.: Modelling and condition-based control of a flexible and hybrid disassembly system with manual and autonomous workstations using reinforcement learning. J Intell Manuf **33**, 575–591 (2022). https://doi.org/10.1007/s10845-021-01863-3
11. Ropi, N.M., Hishamuddin, H., Wahab, D.A., Saibani, N.: Optimisation models of remanufacturing uncertainties in closed loop supply chains—a review. IEEE Access 1 (2021). https://doi.org/10.1109/ACCESS.2021.3132096
12. Garcia, F. and Rachelson, E.: Markov decision processes. In: Sigaud, O., Buffet, O. (eds.) Markov Decision Processes in Artificial Intelligence (2013). https://doi.org/10.1002/9781118557426.ch1
13. Puterman, M.L.: Markov Decision Processes: Discrete Stochastic Dynamic Programming. Wiley (1994). https://doi.org/10.1002/9780470316887
14. Rachelson, E., Quesnel, G., Garcia, F., Fabiani, P.: A Simulation-Based Approach for Solving Generalized Semi-Markov Decision Processes, pp. 583–587 (2008). https://doi.org/10.3233/978-1-58603-891-5-583
15. Shapiro A., Dentcheva, D., Ruszczyński, A.: Lectures on Stochastic Programming: Modeling and Theory (2009)
16. Birge, J.R., Louveaux F.: Introduction to stochastic programming, Springer (1997)

Energy-Flexible Operation of a Cleaning Machine Through Mathematical Optimization

Lina Kramer(✉), Jonathan Magin, Jan Zangenberg, Marvin Zaun, and Matthias Weigold

Technical University of Darmstadt, Darmstadt, Germany
l.kramer@ptw.tu-darmstadt.de

Abstract. To promote renewable energy use and reduce the environmental impact of industrial facilities, we propose a novel demand response approach using aqueous parts cleaning machines. These machines offer high demand response potential due to their large water tanks, which can store thermal energy. A stochastic dynamic programming model is developed and integrated into a cyber-physical production system. It determines energy consumption for both the cleaning process and tank heating, using day-ahead electricity prices as a key decision variable. The model incorporates system constraints and controls tank temperature by switching heating elements on and off. An enhanced data model enables communication with the physical system. This work addresses the underexplored use of stochastic dynamic programming in systems with thermal storage. A case study shows that heating costs of the tanks were significantly reduced using day-ahead prices and the proposed approach, even resulting in net revenues due to negative electricity prices, while emissions decreased by 8.6%.

Keywords: Demand response · Optimal control · Stochastic dynamic programming · Single machine scheduling · Inherent energy storage · Cyber-physical production system

1 Introduction

More than 80 % of the world's energy consumption comes from fossil fuels like petroleum, natural gas, and coal [1]. Expanding the use of renewable energy is one approach to decreasing reliance on fossil fuels [4]. Global renewable energy generation almost doubled between 2011 and 2021, rising from 402 TW h to 763TW h [1]. Transitioning to renewable electricity generation requires adjustments to the electricity system [2].

Since renewable energy sources naturally fluctuate, consumers must adjust to changes in supply, for example, by using energy storage systems and applying demand response (DR) strategies as part of demand-side integration [17].

L. Overmeyer and B.-A. Behrens (eds.), *Production at the Leading Edge of Technology*, Lecture Notes in Production Engineering, https://doi.org/10.1007/978-3-032-19524-1_10

In previous work, we analysed the DR potential of an industrial aqueous parts cleaning machine (APCM) [7]. Building on this, we developed an automation data model designed to facilitate communication between DR services and machine automation, as demonstrated in [11]. To fully exploit this architecture, an optimization algorithm is necessary. In this paper, we propose a stochastic dynamic programming (SDP) approach for controlling the heating elements.

The paper is structured based on the Design Research Methodology (DRM) [3]. In Sect. 2, literature describing DR for APCM as well as modeling and optimization of inherent storages is analyzed as descriptive study I. The prescriptive study starts with the clarification of the research objective in Sect. 3 followed by the design (Sect. 4) and implementation (Sect. 5). The results are then analyzed and discussed in form of descriptive study II in Sect. 6. Finally, a summary and outlook is drawn.

2 Literature

This section reviews the literature addressing the identified research gap, as part of descriptive study I. For the literature review, we chose an approach based on [21]. First, the objective and scope are defined before moving on to the conceptualization phase, which involves formulating the central research questions. This is followed by the third phase, search and filtering, where scientific databases are queried using defined search terms. The final phase involves analyzing the remaining literature sources. The selection criteria are context, system, measures, and optimization. The literature review was conducted using the ScienceDirect and Web of Science database. The search results were filtered and evaluated through a multi-stage process.

Fuhrländer-Völker et al. [8] implemented a mixed-integer linear programming (MILP) optimization algorithm on an APCM. In a field test power changes of up to 49 % and energy shifts of up to 82 % were achieved, demonstrating the high DR potential of such machines.

Regarding storage capacities in production context, the following optimization approaches have been implemented.

Zangenberg et al. [24] present a data-driven method for system identification of inherent thermal energy storages. Component-level modeling is used to enable model predictive control (MPC) for DR in the overall system. Applied to a cooling system using MILP, the method achieved a 10.06 % cost reduction and 4.93 % lower carbon emissions by adapting cooling demand to electricity price variations.

Extending the scope beyond discrete manufacturing to include other sectors such as the energy industry or the food industry, the following sources can be identified. Liu et al. [13] investigated a continuous stirred-tank reactor operated with dynamic production levels. Production was increased during low electricity prices, storing excess output, and reduced during high prices using stored material to meet demand. A dynamic programming (DP) approach was employed to minimize operating costs under time-of-use electricity pricing.

Kelley et al. [10] examined the operation of an air separation unit (ASU). The study leverages the flexibility of the ASU by increasing production during periods of low electricity prices. The resulting overproduction is stored in dedicated tanks and can be retrieved later to meet demand. A linear programming (LP) approach was used to minimize operating costs. Wohlgenannt et al. [23] proposed a demand-side management strategy for an industrial food processing facility, combining the building as a passive thermal storage with an actively chilled water storage. The optimization problem was formulated as a LP. MPC was applied to manage forecast uncertainties, with simulation results showing significant reductions in energy consumption, electricity costs, and peak load. The literature analysis has shown that MILP is frequently used in combination with MPC algorithms to address optimization problems in the context of DR. In contrast, the use of SDP in this field remains largely unexplored, revealing a clear research gap that this work seeks to address.

3 Conceptualization

In this section, the prescriptive study is carried out by presenting the conceptualization. This paper aims to implement and evaluate an SDP framework tailored to a throughput cleaning machine (TPCM), with the intention of ensuring methodological transferability to a broader range of APCM systems. The approach exploits the inherent storage capacity of the washing and rinsing tanks of the TPCM by actively utilizing the permissible temperature fluctuations, thereby enabling their integration into DR strategies. An appropriate SDP-based optimization is developed and demonstrated through a representative use case, including system modeling, parameterization, and automation integration. In the evaluation and discussion, we present results from a case study. The assessment is based on a defined set of metrics, primarily comparing the optimized energy costs with an unoptimized reference process (see Sect. 5.3) as benchmark scenario.

4 Stochastic Dynamic Programming Approach

The SDP approach is particularly well-suited to our problem. At predefined time steps, a decision is made regarding the number of heating elements to activate. During operation (see Sect. 5.3), significant temperature fluctuations hindered accurate forecasting. The optimization problem is solved recursively using the *terminal cost function* V_n, which represents the expected cumulative costs from stage n to the end of the planning horizon. This yields the optimal policy δ^*. The recursion and its optimality properties have been thoroughly documented in the literature on dynamic programming (see, for example, [16]). A key advantage of the SDP method is that the optimal policy needs to be computed only once and can subsequently be applied throughout the process.

5 Implementation

The previous described approach is now demonstrated on an TPCM, which we consider a cyber-physical system due to its physical components and digital control [11].

5.1 System Description, Automation Structure and Data Model

This case study examines the cleaning of gearbox guide discs contaminated with cutting oil, using a TPCM from BvL Oberflächentechnik GmbH [6]. The process features aqueous detergent cleaning with continuous material flow, followed by convective drying. Cleaning and rinsing are carried out using an aqueous detergent stored in two 500 L tanks. Each tank is equipped with five individually controllable electric heating elements, each rated at 6 kW. These heaters account for the largest share of the machine's total average connected load of 96 kW, making them a key target for energy flexibility measures. Due to their thermal inertia, the heated detergent tanks offer significant DR potential as inherent thermal storage [19]. This enables the implementation of the DR measure *store energy (inherently)*, which is classified as an energy flexibility measure according to Association of German Engineers e.V. (VDI) 5207, under the assumption that the process is not adversely affected.

In contrast, other flexibility strategies, such as *interrupting the process* or *adjusting operational parameters*, are not applicable in this case, since the material flow is continuous and must remain uninterrupted, and the process parameters are fixed to ensure operational reliability [20]. This assumption was made in consultation with industrial cleaning experts, as it can be reasonably expected, based on the principles of Sinner's Circle, that a higher cleaning temperature is more likely to improve rather than impair the cleaning performance [18].

The automation architecture enables smooth operation through a control algorithm and is outlined below, with further details in [11]. The config structure is adopted from the eta-nexus package (version 0.1.1), which itself was developed further from the eta-utility package [9]. A standardized interface is essential for implementing energy flexibility on the TPCM. The central server uses day-ahead electricity prices from the ENTSO-E API to generate Energy Flexibility Data Model (EFDM) measures in JSON format, which are sent to machine-level servers to trigger flexibility actions [11]. The interface is essential for implementing energy flexibility measures on the TPCM.

5.2 Modeling

In the following section, the SDP approach is applied to our specific use case, and the relevant quantities are defined as follows [15,22].

(i) The *planning horizon* is defined by the total shift duration J and the time interval between decision stages $\Delta\tau$, yielding a total of N stages ($J = N \cdot \Delta\tau$). These stages are traversed sequentially in descending order: $N, N{-}1, \ldots, 1$.

(ii) The *state space* $\mathcal{S}$ comprises the feasible tank temperature levels, represented as a finite set of discretized values with step size $\lambda > 0$. It is divided into three subsets:

$$\mathcal{S} = \mathcal{S}_{\min}^{<} \cup \mathcal{S}_{\text{adm}} \cup \mathcal{S}_{\max}^{>},$$

where $\mathcal{S}_{\text{adm}} = \{T_{\min}, T_{\min} + \lambda, \dots, T_{\max}\}$ denotes the admissible range, with $T_{\min}$ and $T_{\max}$ representing the minimum and maximum operating temperatures intended to be maintained during operation. The set $\mathcal{S}_{\min}^{<} = \{T_{\text{LB}}, \dots, T_{\min} - \lambda\}$ defines the lower inadmissible region, and $\mathcal{S}_{\max}^{>} = \{T_{\max} + \lambda, \dots, T_{\text{UB}}\}$ defines the upper inadmissible region. The bounds T_{LB} and T_{UB} represent the extreme temperatures that can be reached in a single stage under maximum cooling starting at $T_{\min}$ or heating starting at $T_{\max}$, respectively.

(iii) The *action set* $\mathcal{A} = \{0, 1, \dots, 5\}$ defines all possible actions, including cooling ($a = 0$) and heating with a heating elements for $a > 0$. Not all of these actions are admissible in every state; admissibility depends on the current temperature level. In inadmissible regions, only specific actions are permitted, heating in the lower zone and cooling in the upper zone. The resulting state-dependent admissible action set $\mathcal{D}(T)$ is defined as follows:

$$\mathcal{D}(T) = \begin{cases} \{1, \dots, 5\}, & T \in \mathcal{S}_{\min}^{<} \\ \mathcal{A}, & T \in \mathcal{S}_{\text{adm}} \\ \{0\}, & T \in \mathcal{S}_{\max}^{>} \end{cases} \tag{1}$$

(iv) Cooling dynamics are modeled thermodynamically [12] and represented by a discrete random variable $\mathbf{W}$ with values $0 < w_1 < \cdots < w_l \leq 1$ and associated probabilities $\eta_i = P(\mathbf{W} = w_i)$. The values $w_i = e^{-\alpha_i \Delta_\mathcal{T}}$ denote the cooling factors derived from the thermodynamic model. The heating behavior is modeled using a piecewise linear approximation [12]. Temperature increase is described by a discrete random variable $\mathbf{X}$, which is supported on the finite set $\mathcal{X} = \{x_{\min}, \dots, x_{\max}\}$, where $x \in \mathcal{X}$ denotes the temperature increase in a single stage. The values $x_{\min}$ and $x_{\max}$ represent the minimal and maximal possible temperature increases, respectively. The corresponding probabilities are state- and action-dependent, given by $q_{T,a}(x) = P_{T,a}(\mathbf{X} = x)$ for $x \in \mathcal{X}$.

(v) The transition probabilities $p_{T,a}(T')$ from temperature T too T' between consecutive stages under action a are defined as follows. For cooling ($a = 0$), transition is given by

$$p_{T,0}(T') = \sum_{i=0}^{l} \eta_i \mathbf{1}_{\{T'\}}(\lfloor (1 - w_i)\vartheta_{HW} + w_i T \rfloor_\lambda), \tag{2}$$

where ϑ_{HW} denotes the temperature at the housing wall. Equation (2) sums the probabilities η_i of all cooling factors w_i that result in T'. For heating ($a > 0$), the transition probability is defined as:

$$p_{T,a}(T') = \begin{cases} q_{T,a}(T' - T), & \text{if } T' - T \in \mathcal{X} \\ 0, & \text{otherwise} \end{cases} \tag{3}$$

where the temperature difference $T'-T$ is assigned the probability $q_{T,a}(T'-T)$ if it lies within the feasible set $\mathcal{X}$.

(vi) Heating costs $r_{\mathrm{HC},n}(a) = L(a)\rho_n$ are incorporated, where n denotes the stage, ρ_n the known day-ahead price, and $L(a)$ the energy consumption associated with action a.

(vii) The penalty costs r_{PC} are imposed when the system leaves the admissible temperature range, calculated proportionally to the deviation from T_{min} or T_{max}, using coefficients $c^{<}_{\mathrm{min}}$ and $c^{>}_{\mathrm{max}}$, respectively. For cooling, they are given by

$$r_{\mathrm{PC}}(T,0) = c^{<}_{\mathrm{min}} \sum_{i=0}^{l} \eta_i \max\{0, T_{\mathrm{min}} - ((1-w_i)\vartheta_{HW} + w_i T)\} \tag{4}$$

and for heating by

$$\begin{aligned} r_{\mathrm{PC}}(T,a) &= c^{<}_{\mathrm{min}} \sum_{x\in\mathcal{X}} q_{T,a} \max\{0, T_{\mathrm{min}} - (T+x)\} \\ &\quad + c^{>}_{\mathrm{max}} \sum_{x\in\mathcal{X}} q_{T,a} \max\{0, (T+x) - T_{\mathrm{max}}\} \end{aligned} \tag{5}$$

.

(viii) Stage costs r_n at stage n combine penalty and heating costs:

$$r_n(T,a) = \begin{cases} r_{\mathrm{PC}}(T,0) & , a=0, \quad T \in \mathcal{S} \setminus \mathcal{S}^{<}_{\mathrm{min}} \\ r_{\mathrm{HC},n}(a) + r_{\mathrm{PC}}(T,a) & , a \in \{1,\ldots,5\}, \quad T \in \mathcal{S} \setminus \mathcal{S}^{>}_{\mathrm{max}} \end{cases} \tag{6}$$

(ix) The terminal cost function V_0 is zero. For stages $n = N,\ldots,1$, V_n is computed recursively by:

$$\begin{aligned} V_n(T_n) = \min \Bigg\{ & r_n(T_n,0) + \sum_{i=1}^{l} \eta_i \, V_{n-1}\left(\lfloor (1-w_i)\,\vartheta_{HW} + w_i\, T_n \rfloor_{\lambda}\right), \\ & \min_{a\in\mathcal{A}\setminus\{0\}} \left\{ r_n(T_n,a) + \sum_{x\in\mathcal{X}} q_{T_n,a}(x)\, V_{n-1}(T_n + x) \right\} \Bigg\}. \end{aligned} \tag{7}$$

5.3 Parameterization

A full shift consists of eight hours, but the shift duration for the experiment was set to four hours to reduce experimental effort. Control actions are executed at 3-min intervals to ensure system safety. This schedule yields a total of 80 decision stages. The admissible temperature range was defined between 48 and 70 °C. The lower limit reflects the two-point control of the reference process (48° to 52° around a 50 °C setpoint), while the upper limit corresponds to the machine's maximum allowable temperature specified by the manufacturer. A discretization parameter of $\lambda = 0.1\,^{\circ}\mathrm{C}$ was chosen to match the resolution of the temperature sensor.

To identify the heating and cooling parameters of the tank, cleaning experiments were conducted using the same parts and settings as in the reference process [14, S. 34].

The number of active heating elements was systematically varied. Heating was applied from $T_{\min}$ to $T_{\max}$, while cooling occurred passively in the reverse direction. The parameters were estimated using linear regression, and the associated uncertainty was quantified based on the regression residuals. A detailed description of the heating and cooling behavior is provided in the appendix in the GitHub repository [12].

6 Evaluation and Discussion

Within the scope of descriptive study II, the evaluation is carried out to demonstrate the functionality of the optimizer. To this end, the described SDP approach is implemented on the actual TPCM system to demonstrate its practical applicability.

For electricity pricing, we use European Power Exchange (EPEX) Spot day-ahead market prices from May 31, 2025, between 15:00 and 19:00 for the Germany-Luxembourg bidding zone [5]. The date was selected as a random day in May.

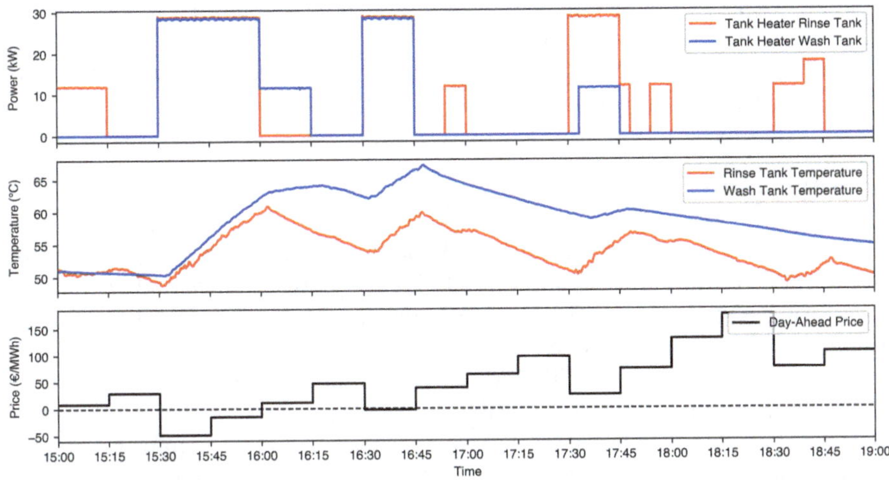

Fig. 1. Power of the tank heating elements (top), temperatures of tanks (center), and day-ahead electricity prices (bottom) for the optimized process

Fig. 1 presents the results of the field tests. Electricity prices were lower at the beginning of the shift and increased towards the end. Notably, prices were negative between 15:30 and 16:00 as well as between 16:30 and 16:45. During the early phase of the shift, heating was primarily performed when electricity prices were negative. In these periods, all heating elements were activated to take full advantage of the cost benefit. Towards the end of the shift, heating occurred only when necessary or during brief periods of favorable pricing. Compared to the rinse tank, the wash tank was heated more intensively at the beginning of the shift and only minimally towards the end. This strategy leverages its favorable thermal properties: the wash tank cools down more slowly and can be reheated more quickly. As a result, it acts as a thermal buffer, enabling the system to better bridge periods of high electricity prices later in the shift.

Compared to the reference process [14, S. 34], the heating elements exhibit longer activation periods and primarily operate when electricity prices are low. As a result, the tank temperature fluctuates in a wider temperature range compared to the reference process but consistently satisfies the specified temperature bounds throughout the experiment.

In the reference process, heating costs were 12.16€ with static electricity tariffs of 210€/MWh and 3.27€ with day-ahead prices. The proposed SDP approach reduced costs to –0.17€, which corresponds to a cost reduction of 105 % compared to the reference process with day-ahead prices and 101 % compared to static tariffs. CO_2 emissions decreased from 11.2 to 10.2 kg, a reduction of 8.60 %.

Due to the complexity of parameterizing the system, a linear approximation of the heating behavior is employed. Additionally, external disturbances are often correlated, complicating their accurate estimation. While the uncertainty associated with the tank system is represented relatively well, the overall computational complexity, though currently manageable, could become a limiting factor in larger-scale applications. Generalizing the approach to more extensive, integrated systems remains difficult due to increased complexity.

7 Conclusion and Outlook

In this work, we present a detailed SDP model for a DR service. The system uses a TPCM to implement energy flexibility measures. The model accounts for uncertainties, allowing precise adaptation and demonstrating significant savings compared to the reference process. We validate the model in a field test involving a thermally process-controlled TPCM and show that the DR measures effectively controls the machine in response to fluctuating electricity prices.

Furthermore, additional control variables, such as product quality, should be taken into account in order to avoid negative implications for the process. The implementation of a multi-objective optimization approach could also prove beneficial in systematically addressing potential trade-offs between economic and qualitative performance indicators. In addition, extended tests on selected representative days would be advisable to demonstrate the applicability of the optimizer for industrial use.

8 Acknowledgements and Appendix and Data Availability Statement

The authors gratefully acknowledge the financial support of the Kopernikus-Project "SynErgie" (Grant Number 03SFK3A0-3) by the Federal Ministry of Research, Technology and Space (BMFTR) and thank the Projektträger Jülich (PtJ) for the project supervision. General information on the TPCM is available at the TUdata lib of the Technical University of Darmstadt [6]. The developed optimization algorithm is available via GitHub [12]. The raw data and parameterization are provided under [6].

References

1. Abe, J.O., Popoola, A., Ajenifuja, E., Popoola, O.M.: Hydrogen energy, economy and storage: review and recommendation. Int. J. Hydrog. Energy **44**(29), 15,072–15,086 (2019). https://doi.org/10.1016/j.ijhydene.2019.04.068
2. Asadi Aghajari, H., Niknam, T., Shasadeghi, M., Sharifhosseini, S.M., Taabodi, M.H., Sheybani, E., Javidi, G., Pourbehzadi, M.: Analyzing complexities of integrating renewable energy sources into smart grid: a comprehensive review. Appl. Energy **383**, 125317 (2025). https://doi.org/10.1016/j.apenergy.2025.125317
3. Blessing, L.T., Chakrabarti, A.: DRM, a Design Research Methodology. Springer London, London (2009). https://doi.org/10.1007/978-1-84882-587-1
4. Da Silva, G.H.R., Nascimento, A., Baum, C.D., Mathias, M.H.: Renewable energy potentials and roadmap in Brazil, Austria, and Germany. Energies **17**(6), 1482 (2024). https://doi.org/10.3390/en17061482
5. ENTSO-E.: Transparency Platform–Day-Ahead Prices (2025). https://newtransparency.entsoe.eu/market/
6. Frank, M., Magin, J., TU Darmstadt: Throughput Cleaning Machine YUKON DAD-2 B (2024). https://doi.org/10.48328/tudatalib-1430
7. Fuhrländer-Völker, D., Magin, J., Weigold, M.: Demand Response on Aqueous Parts Cleaning Machines (2022). https://doi.org/10.13140/RG.2.2.14962.99529
8. Fuhrländer-Völker, D., Grosch, B., Weigold, M.: Modelling and Control of Aqueous Parts Cleaning Machines for Demand Response (2023). https://doi.org/10.15488/13498, https://www.repo.uni-hannover.de/handle/123456789/13608
9. Grosch, B., Ranzau, H., Dietrich, B., Kohne, T., Fuhrländer-Völker, D., Sossenheimer, J., Lindner, M., Weigold, M.: A framework for researching energy optimization of factory operations. Energy Inform. **5**, 15–16 (2022). https://doi.org/10.1186/s42162-022-00207-6
10. Kelley, M.T., Tsay, C., Cao, Y., Wang, Y., Flores-Cerrillo, J., Baldea, M.: A data-driven linear formulation of the optimal demand response scheduling problem for an industrial air separation unit. Chem. Eng. Sci. **252**, 117468, (2022). https://doi.org/10.1016/j.ces.2022.117468
11. Kramer, L., Fuhrländer-Völker, D., von Elling, M., Karnapp, S., Moser, M., Weigold, M.: OPC UA information model for energy-flexible aqueous parts cleaning machines. In: Kohl, H., Seliger, G., Dietrich, F., Vien, H.T. (eds.) Decarbonizing Value Chains. Lecture Notes in Mechanical Engineering, pp. 155–163. Springer Nature Switzerland, Cham (2025a). https://doi.org/10.1007/978-3-031-93891-7_18

12. Kramer, L., Magin, J., Zangenberg, J., Zaun, M., Weigold, M.: Energy-flexible operation of a cleaning machine through mathematical optimization (2025b). https://github.com/kramer-ptw/WGP2025-Energy-flexible-operation-of-a-cleaning-machine-through-mathematical-optimization
13. Liu, Y., Lavoie, D., El-Farra, N.H., Palazoglu, A.: A demand response strategy for continuous processes using dynamic programming approach. IFAC-PapersOnLine **51**(18), 180–184 (2018). https://doi.org/10.1016/j.ifacol.2018.09.296
14. LoTuS: LoTuS - Leistungsoptimierte Trocknung und Sauberkeit: Energetische Trockenprozessoptimierung und -vernetzung für eine energieeffiziente Reinigungsanlage (2023). https://www.enargus.de/pub/bscw.cgi/?op=enargus.eps2&q=lotus&m=2&v=10&s=8&y=1&id=1323683
15. Nickel, S., Rebennack, S., Stein, O., Waldmann, K.H.: Operations Research. Springer Berlin Heidelberg, Berlin, Heidelberg (2022). https://doi.org/10.1007/978-3-662-65346-3
16. Puterman, M.L.: Markov decision processes: discrete stochastic dynamic programming. Wiley series in probability and mathematical statistics. Applied Probability and Statistics Section, John Wiley & Sons, New York (2005). https://permalink.obvsg.at/
17. Robert, F.C., Sisodia, G.S., Gopalan, S.: A critical review on the utilization of storage and demand response for the implementation of renewable energy microgrids. Sustain. Cities Soc. **40**, 735–745 (2018). https://doi.org/10.1016/j.scs.2018.04.008
18. Rögner, F.H., Vohrer, U.: Die Erweiterung des Sinner'schen Kreises. JOT Journal für Oberflächentechnik **61**(S1), 8–10 (2021). https://doi.org/10.1007/s35144-021-1235-1
19. Strobel, N., Fuhrländer-Völker, D., Weigold, M., Abele, E.: Quantifying the demand response potential of inherent energy storages in production systems. Energies **13**(16), 4161 (2020). https://doi.org/10.3390/en13164161
20. VDI - Verein Deutscher Ingenieure.: VDI 5207 Energieflexible Fabrik. Blatt 1: Grundlagen (2020). https://www.vdi.de/richtlinien/details/vdi-5207-blatt-1-energieflexible-fabrik-grundlagen
21. Vom Brocke, J., Simons, A., Niehaves, B., Riemer, K., Plattfaut, R., Cleven, A.: Reconstructing the Giant: On the Importance of Rigour in Documenting the Literature Search Process (2009)
22. Waldmann, K.H., Stocker, U.M.: Stochastische Modelle: Eine anwendungsorientierte Einführung, zweite, überarbeitete und erweiterte auflage edn. EMIL@A-stat Medienreihe zur angewandten Statistik. Springer, Berlin and Heidelberg (2013). http://www.zentralblatt-math.org/zbmath/search/?q=an%3A1032.60001
23. Wohlgenannt, P., Huber, G., Rheinberger, K., Kolhe, M., Kepplinger, P.: Comparison of demand response strategies using active and passive thermal energy storage in a food processing plant. Energy Rep. **12**, 226–236 (2024). https://doi.org/10.1016/j.egyr.2024.06.022
24. Zangenberg, J., Wendt, J., Koch, T., Kapser, T., Weigold, M.: Data-driven identification and operational optimization of energy-flexible thermal supply systems. ACM SIGEnergy Energy Inform. Rev. **4**(4), 214–225 (2024). https://doi.org/10.1145/3717413.3717434

An Approach for the Reduction of Tungsten and Cobalt in Cutting Tools Using Additive Manufacturing

Julian Wisser(✉), Jacques Platz, and Jan C. Aurich

Institute for Manufacturing Technology and Production Systems, RPTU University Kaiserslautern-Landau, Kaiserslautern, Germany
julian.wisser@rptu.de

Abstract. Cutting tools made from cemented carbides have been the prevailing standard for industrial applications. Most commonly, they are made of fine tungsten carbide particles in a cobalt binder matrix. However, the extraction of the raw materials, cobalt and tungsten, is associated with significant shortcomings regarding social and ecological sustainability. Tungsten is categorized as a conflict mineral, the trade of which oftentimes directly finances armed conflicts. Approximately two-thirds of the worldwide cobalt demand is met by mines in the Democratic Republic of the Congo. A considerable proportion of these mines operate with minimal regulatory oversight, frequently engaging children in artisanal mining activities. The FairTools project outlined in this paper aims to address these issues by researching new high-performance materials that have the potential to substitute cemented carbides made from tungsten carbide and cobalt altogether. In case of technical necessities, a significant improvement in both recyclability and a reduction of tungsten and cobalt usage is pursued. The additive manufacturing technology of high-speed directed energy deposition (HS DED-LB) will be applied and improved with the aim of producing cutting tools with reduced tungsten and cobalt contents. This technology offers a promising approach to minimize defects and microstructural modifications prevalent in current additive manufacturing methods for cemented carbides, which complicate industrial applications. The potential to reduce the usage of cobalt and tungsten stems from the possibility to manufacture tool blanks that comprise a functional outer layer of cemented carbide, fused to a base body made from an alternative material.

Keywords: Sustainability · Cemented carbides · Additive manufacturing · Cutting tools

1 Introduction

Since their establishment in the 1920s, cemented carbides have been the preferred material for cutting tools in industrial machining [1]. The metal matrix composite is composed of a ceramic hard phase that is embedded in a ductile metallic binder phase. This results in a beneficial combination of hardness, impact resistance and high-temperature

L. Overmeyer and B.-A. Behrens (eds.), *Production at the Leading Edge of Technology*, Lecture Notes in Production Engineering, https://doi.org/10.1007/978-3-032-19524-1_11

strength. Consequently, tools made from cemented carbides demonstrate the capacity to operate at elevated machining speeds while exhibiting a prolonged service life when compared to other tooling materials, such as high-speed steel. The reduction in process times and workloads directly improves the feasibility of these tools, thereby making them the economic choice for most industrial applications.

The most prevalent cemented carbide is composed of tungsten carbide in a cobalt binder phase. This is due to the excellent wettability of tungsten carbide for molten cobalt, resulting in a remarkably strong bond that has yet to be matched by a different combination of hard phase and matrix [2].

Both tungsten and cobalt have been classified as critical resources by the European Union. This classification is due to their significance for modern technology and their uncertain security of supply [3]. Furthermore, both materials are subject to substantial criticism regarding their sustainability, in terms of ecological as well as social criteria. Tungsten, alongside tantalum, tin, and gold, is part of the so-called "3TG conflict minerals", the trade of which directly contributes to the perpetuation of armed conflict, violence and the abuse of human rights in regions, that are already characterized by weak development and political instability [4].

Most of the global cobalt demand is met by the Democratic Republic of the Congo, where it is extracted within the Katanga Copperbelt in the form of heterogenite. This mineral is found in thin and brittle layers, which are often located in close proximity to the surface. This has resulted in the emergence of numerous artisanal mines that operate in complete isolation from any regulatory oversight in hazardous conditions, frequently involving the employment of child labor. These artisanal mines have been found to cause significant long-term harm to public health and the surrounding ecosystem [5].

More than half of the world's natural cobalt reserves are located in the Democratic Republic of the Congo, which meets nearly 75% of the global demand. Similar proportions can be seen with tungsten, where the majority of the natural reserves and approximately 80% of current mining capacities are located in the People's Republic of China [6]. While the recycling of cemented carbides is not only a possibility but also a common practice, the secondary resources, however, are not yet sufficient to meet the global demand [7].

In order to address the adverse impacts associated with the extraction of tungsten and cobalt, it is necessary to minimize the extraction capacities in regions where social and environmental standards cannot be assured. Given the unequal distributions of these materials on the global market, it will be very challenging to meet the current demand solely from alternative sources. Therefore, it is imperative to reduce the general demand for these materials in the long term.

2 The FairTools Project

The FairTools Project, funded by the Carl-Zeiss-Stiftung, is an interdisciplinary research group of scientists from the University of Kaiserslautern-Landau and its associated institutes. The initiative is focused on an improvement of the current situation by pursuing three main aspects: The improved recycling, reduction and replacement of cemented carbides comprising tungsten and cobalt.

As part of the project, a detailed assessment of the present life cycle of cemented carbides, including various recycling methods, will allow the identification of promising approaches to enhance sustainability in a substantial manner. Concurrently, research will be conducted to identify and develop alternative high-performance materials to substitute for cemented carbides while matching the thermo-mechanical properties. This encompasses both modified steels and alternative hard and binder phases.

In circumstances where technical requirements necessitate the utilization of conventional tungsten carbide-cobalt cemented carbides, the objective is to achieve a minimum reduction of 50% cemented carbide usage. This will be pursued by the application and further development of the additive manufacturing technology of powder- and laser-based high-speed directed energy deposition (HS DED-LB). Using this technology, it is theoretically possible to produce tool blanks, that are composed of a functional outer layer made from cemented carbide fused to a base body made from a more sustainable material. Figure 1 illustrates this concept of layered tools. Subsequent sections will delve into a more thorough examination of the technology and its suitability for this particular application.

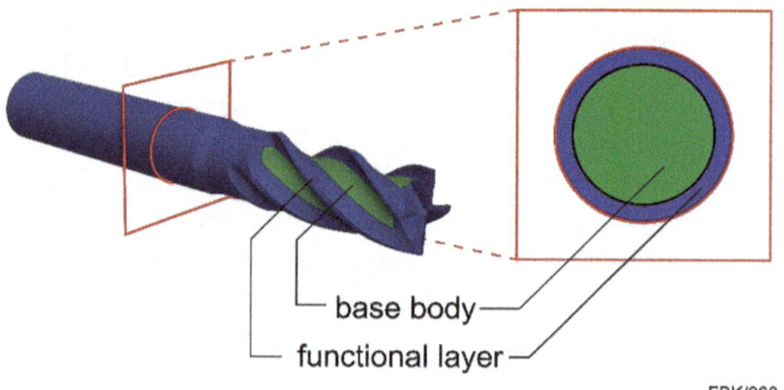

Fig. 1 The concept of end mills comprising a base body made from sustainable materials and an outer functional layer of cemented carbide that meets the requirements for steel machining applications

As a practical example, the FairTools project is working towards the manufacturing of sustainable end mills (Fig. 1), capable of machining the two industrially significant steels, 100Cr6 and 42CrMo4. To ensure the relevance of the conducted research and the developed tools, an advisory board has been established. This board is composed of representatives from various tooling and producing companies and supports the scientists by providing their own experiences as well as critically evaluating the results of the research.

3 State of the Art in Additive Manufacturing of Cemented Carbides and HS DED-LB

The production of cemented carbides is typically accomplished in an established liquid-phase sintering process, characterized by defined and controlled temperatures. The sintering process in an inert gas atmosphere has demonstrated to produce materials with a uniform tungsten carbide grain size distribution and a low prevalence of undesirable secondary carbides and porosities. The grain size as well as the cobalt content are critical factors in determining the high-temperature strength and toughness of the material. Consequently, these characteristics are considered to be the primary defining features [8].

3.1 Additive Manufacturing of Cemented Carbides

Additive manufacturing offers numerous advantages in the geometric design of cutting tools. Special tools with undercuts and integrated cooling channels are just two examples that are difficult to realize with conventional production technology [9]. Various additive manufacturing processes have already been evaluated for the fabrication of cemented carbides. The production of green compacts has been achieved by the implementation of binder jet 3D printing, 3D gel-printing and fused filament fabrication which are subsequently subjected to debinding and sintering processes. These workpieces are characterized by a highly uniform grain size distribution and minimal porosity. However, these benefits are counterbalanced by several factors. For instance, the process of sintering involves a significant contraction of the original workpiece, which might compromise the structural integrity of the material. Additionally, the distribution of the cobalt binder might be of poor uniformity as well as the overall complexity of the process which can contribute to the challenges in the application of these manufacturing technologies [10].

The direct additive manufacturing of components made of cemented carbide, without complex downstream processes, is possible using powder bed fusion or directed energy deposition [10]. Provided the process parameters are well configured, it is possible to produce cemented carbides with low porosity [11]. However, both processes exhibit an uneven distribution of heat within a layer, resulting in uneven growth of the tungsten carbide grains [12]. Additionally, the high energy densities, particularly in laser-based applications, have been observed to result in the formation of undesirable eta-phases (Co_3W_3C) [13] as well as the partial vaporization of the cobalt binder phase [14].

3.2 High-Speed Directed Energy Deposition (HS DED-LB) and Its Potential for the Processing of Cemented Carbides

The challenges associated with the direct additive manufacturing of cemented carbides are addressed by the previously mentioned application of HS DED-LB. In conventional powder- and laser-based directed energy deposition (DED-LB) as well as HS DED-LB, the powder material is focused by a coaxial nozzle onto a surface, where it is then fused to the substrate by a welding laser. The relative movement of the laser and powder nozzle against the workpiece surface leads to the formation of welding tracks, whose systematic alignment and layering results in the final, three-dimensional geometry.

In conventional DED-LB, the powder material is focused directly into the melt pool, where it is liquified in contact with the already molten substrate. In HS DED-LB, however, the powder material is focused above the substrate's surface, as illustrated in Fig. 2. Consequently, the material does not undergo liquefaction in the melt pool; rather, it reaches the workpiece in an already liquid state. This modification results in the formation of smaller melt pools, thereby reducing the amount of material that must be liquified. This facilitates the use of increased feed and build rates, consequently leading to diminished layer heights and heat-affected zones in comparison to conventional DED-LB [15].

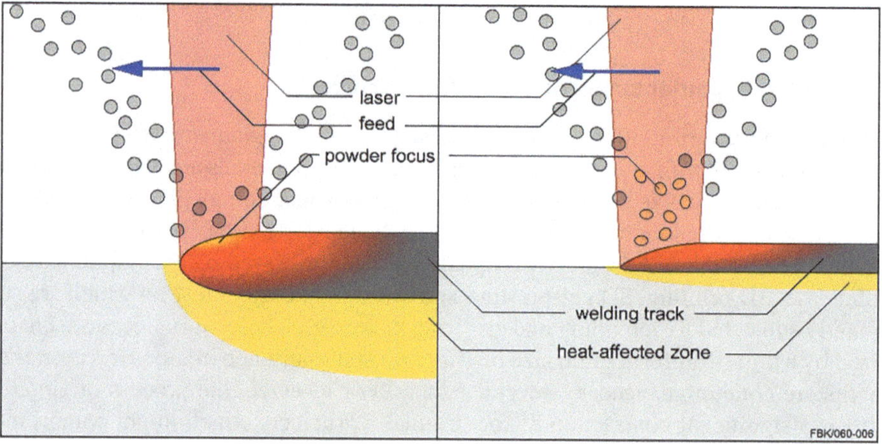

Fig. 2 Illustration of the functionality of conventional (left) and high-speed (right) powder- and laser-based directed energy deposition, highlighting the elevated powder focus as well as the reduced layer height and heat-affected zone

Due to these properties, HS DED-LB is a promising approach in addressing the challenges associated with the additive manufacturing of cemented carbides, as previously outlined. The high feed rates enable the realization of lower volumetric energy densities. In combination with reduced working times and the diminished heat-affected zone when applying additional layers, the tungsten carbide can be expected to exhibit diminished grain growth and reduced formation of undesirable tertiary phases. The reduced heat-affected zone also prevents a repeated growth of the tungsten carbide grains.

4 Feasibility Investigation

4.1 Methodology

IIn a preliminary series of tests, individual weld tracks made of cemented carbide via HS DED-LB were manufactured. For this purpose, a 5-axis pE3D-system from Ponticon GmbH*, is utilized. The laser system is an LDF 8000-6 high power diode laser with a maximum output of 8 kW by Laserline GmbH*. The powder material is fed by a Metco Twin 150 rotary powder feeder by OC Oerlikon* and focused on the substrate via a HighNo®—5.0 continuous coaxial nozzle by HD Sonderoptiken GmbH*.

The material used in this study is an agglomerated and sintered cemented carbide powder consisting of 88 wt% tungsten carbide and 12 wt% cobalt (WOKA 3102, OC Oerlikon*). The material has a particle size distribution of d10 = 24.9 μm and d90 = 62.1 μm, as determined by laser diffraction measurement (Bettersizer S3 Plus, 3P Instruments*). This particle size distribution has already proven effective for processing via HS DED-LB [16]. The substrate material is a cold-working tool steel without cobalt or tungsten (1.2842/AISI O2).

Initial trials featuring a wide-ranging variation of both laser power as well as powder flow rate and their respective visual inspection concluded a homogenous weld track at a powder flow rate of 50 g/min. To further evaluate the influence of the laser power on the processing of cemented carbides, the parameters in Table 1 were selected for a more thorough investigation.

Table 1 HS DED-LB process parameters

Laser power	2.0; 2.2; 2.4; 2.6; 2.8; 3.0 kW
Laser spot diameter	1.5 mm
Scan speed	20 m/min
Powder flow rate	50 g/min
Carrier gas flow rate	8 l/min
Shielding gas flow rate	14 l/min

The specimens were analyzed via a focus variation surface measurement using an InfiniteFocus G5 by Alicona Imaging GmbH*. This technology allows an optical as well as a topographic investigation of the specimen's surface. Based on this data, the track height was measured at 10 discrete positions for each set of parameters. The locations of the measuring points are marked in Fig. 3.

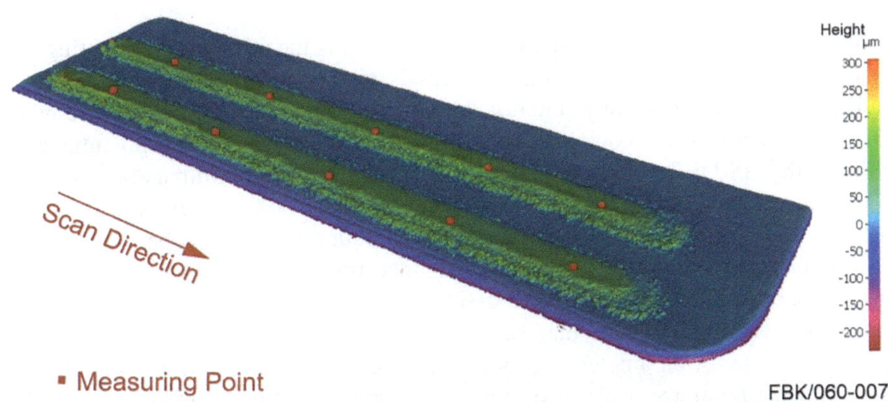

Fig. 3 Surface topography of weld tracks fabricated by HS DED-LB at a laser power of 3 kW with designated scan direction and measuring points

4.2 Results and Discussion

The visual inspection revealed the potential to produce uniform and coherent weld tracks. The presence of minor deposits of incompletely molten powder and slight tarnish has been observed on the majority of the samples examined. It has been demonstrated that cracks, which run orthogonally to the scan direction, begin to form when laser powers are elevated to a level exceeding 2.8 kW. This phenomenon is presumed to occur as a result of increased thermal stresses, that have been previously observed in the laser based processing of cemented carbides [17].

The topographical measurements obtained via focus variation, as illustrated in Fig. 3, yielded analogous results. A uniform weld track is observable alongside minor powder deposits. The heights of the individual weld tracks are consistently in the range of 140 to 180 μm. The mean height ($\overline{h}$) and standard deviation (σ_h) for each laser power are listed in Table 2. The increased laser power appears to result in a modest decline in track height, though this observation must be interpreted with caution. The relatively high σ_h values, in most cases exceeding 10% of $\overline{h}$, indicate substantial variation between tracks. This variation is partly caused by process-related instabilities, such as deviations in powder mass flow, but also by the reduced stability of the measurement itself due to powder depositions on the surface. Such unmelted powder particles, more pronounced at lower laser powers, can locally increase measured heights and therefore contribute to the scatter in the data.

Table 2 Average track heights at various laser power levels

P/kW	2.0	2.2	2.4	2.6	2.8	3.0
$\overline{h}$/μm	163.03	165.96	176.91	162.09	147.56	143.23
σ_h/μm	23.69	14.30	19.52	20.35	15.16	15.72

5 Conclusion and Outlook

One of the primary objectives of the FairTools Project is the fabrication of cutting tools with reduced cemented carbide content. This is to be achieved by the development of hybrid tools that consist of a functional, additively manufactured layer of cemented carbide surrounding a core composed of more sustainable materials. The potential applicability of the HS DED-LB process has been proven with a preliminary series of tests in which single weld tracks made from tungsten carbide-cobalt cemented carbide were successfully deposited on a tool steel substrate without tungsten or cobalt.

Subsequent metallographic investigations, incorporating optical and scanning electron microscopy, will be conducted to assess the microstructure of these samples and future specimens, with a particular focus on the tungsten carbide grain size. After a comprehensive set of parameters has been identified, the subsequent investigation will be supplemented with the analysis of multi-track and multi-layer samples.

In cooperation with the other participants of the research group, even more profound investigations of the achieved surface structure will be conducted, such as the evaluation of scanning electron microscope images on prepared specimens also featuring focused ion beam cuts at select locations, elemental mapping via energy-dispersive X-ray spectroscopy or crystallographic investigations via X-ray diffraction. Furthermore, the manufacturing of alternative materials will be evaluated.

The resulting potential for material savings, as a result of the aforementioned factors, contributes to an enhancement in the long-term sustainability for the environment and society.

Acknowledgements. This project is funded by the Carl-Zeiss-Stiftung as part of their framework "CZS Breakthroughs".

*The manufacturers are stated for reasons of completeness. This does not imply endorsement of or by the named companies nor that the specified products represent the best possible solution for the respective application.

Competing Interests. The author(s) has no competing interests to declare that are relevant to the content of this manuscript.

References

1. Ortner, H., Ettmayer, P., Kolaska, H.: The history of the technological progress of hardmetals. International Journal of Refractory Metals and Hardmetals 44, 148-159 (2014). https://doi.org/10.1016/j.ijrmhm.2013.07.014
2. Konyashin, I., Ries, B.: Cemented carbides. Elsevier, Amsterdam (2022).
3. Grohol, M., Veeh, C.: Study on the Critical Raw Materials for the EU 2023—Final Report. European Commission, Brussels (2023)
4. Public Law 111-203: Dodd-Frank Wall Street Reform and Consumer Protection Act. United States of America (2010)
5. Banza Lubaba Nkulu, C., Casas, L., Haufroid, V., de Putter, T., Saenen, N., Kayembe-Kitenge, T., Musa Obadia, P., Kyanika Wa Mukoma, D., Lunda Ilunga, J., Nawrot, T., Luboya Numbi, O., Smolders, E., Nemery, B.: Sustainability of artisanal mining of cobalt in DR Congo. Nat. Sustain. **1**(9), 495–504 (2018). https://doi.org/10.1038/s41893-018-0139-4
6. U.S. Geological Survey: Mineral Commodity Summaries 2025 (ver. 1.2, March 2025): U.S. Geological Survey, 212 p (2025). https://doi.org/10.3133/mcs2025
7. Freemantle, C., Sacks, N.: Recycling of cemented tungsten carbide mining tool scrap. J. S. Afr. Inst. Mining Metall. **115**(12), 1207–1213 (2015). https://doi.org/10.17159/2411-9717/2015/v115n12a9
8. García, J., Collado Ciprés, V., Blomqvist, A., Kaplan, B.: Cemented carbide microstructures: a review. International Journal of Refractory Metals and Hard Materials 80, 40-68 (2019). https://doi.org/10.1016/j.ijrmhm.2018.12.004
9. Ford, S., Despeisse, M.: Additive manufacturing and sustainability: an exploratory study of the advantages and challenges. J. Clean. Prod. **137**, 1573–1587 (2016). https://doi.org/10.1016/j.jclepro.2016.04.150
10. Yang, Y., Zhang, C., Wang, D., Nie, L., Wellman, D., Tian, Y.: Additive manufacturing of WC-Co hardmetals: a review. The International Journal of Advanced Manufacturing Technology 108(5-6), 1653-1673 (2020). https://doi.org/10.1007/s00170-020-05389-5

11. Uhlmann, E., Bergmann, A., Gridin, W.: Investigation on Additive Manufacturing of tungsten carbide-cobalt by Selective Laser Melting. Procedia CIRP 35, 8-15 (2015). https://doi.org/10.1016/j.procir.2015.08.060
12. Konyashin, I., Hinners, H., Ries, B., Kirchner, A., Klöden, B., Kieback, B., Nilen, R., Sidorenko, D.: Additive manufacturing of WC-13%Co by selective electron beam melting: Achievements and challenges. International Journal of Refractory Metals and Hard Materials 84, 105028 (2019). https://doi.org/10.1016/j.ijrmhm.2019.105028
13. Khmyrov, R., Safronov, V., Gusarov, A.: Obtaining Crack-free WC-Co Alloys by Selective Laser Melting. Physics Procedia 83, 874-881 (2016). https://doi.org/10.1016/j.phpro.2016.08.091
14. Fortunato, A., Valli G., Liverani, E., Ascari, A.: Additive Manufacturing of WC-Co Cutting Tools for Gear Production. Lasers in Manufacturing and Materials Processing 6(3), 247-262 (2019). https://doi.org/10.1007/s40516-019-00092-0
15. Schopphoven, T., Gasser, A., Backes, G.: EHLA: Extreme High-Speed Laser Material Deposition. Laser Technik Journal 14(4), 26-29 (2017). https://doi.org/10.1002/latj.201700020
16. Platz, J., Steiner-Stark, J., Kirsch, B., Aurich, J.: Investigation on different laser-powder alignments and their influence on part properties in high-speed Directed Energy Deposition. Procedia CIRP **124**, 190–193 (2024). https://doi.org/10.1016/j.procir.2024.08.097
17. 17.Yamashita, Y., Funada, Y., Kunimine, T., Sato, Y., Tsukamoto, M.: Formation of cemented tungsten carbide layer with compositional gradient processed by directed energy deposition. Mater. Sci. Forum 1016, 1676–1681 (2021). https://doi.org/10.4028/www.scientific.net/MSF.1016.1676

Design Concept for Circular Product Strategies

Sven Schümmelfeder[1](✉), Alexander Keuper[1], and Günther Schuh[1,2]

[1] Laboratory of Machine Tools and Production Engineering (WZL), RWTH Aachen University, Aachen, Germany
s.schuemmelfeder@wzl.rwth-aachen.de

[2] Fraunhofer Institute for Production Technology IPT, Aachen, Germany

Abstract. The increasing signs of climate change and the growing competition for scarce resources are putting more pressure on companies to develop new products that take into account sustainability aspects. At the same time sustainability embodies also a strategic opportunity to generate competitive advantage. Progressive, sustainable product and process innovations enable companies to differentiate themselves from the competition. To shape these innovations, concepts are needed that systematically link companies and their value-creation processes with their ecological and social environment. In this context, the principle of circular economy in the form of cross-life cycle management of products and the closing of material loops is an instrument for manufacturing companies to operationalize sustainability economically. So far, there are hardly any approaches that help companies to develop product strategies with the aim of achieving a circular economy. For this reason, this research approach develops a framework for the design of circular product strategies in the manufacturing industry, which enables companies to formulate strategic options for action and facilitate future research by systematizing the concept.

Keyword: Circular economy · Product strategy · Sustainability

1 Introduction

Sustainability has become increasingly critical for organizations to remain competitive. According to SCHOLZ, the development of sustainable products will be an essential prerequisite for a company to remain performant in the future [1]. The principle of circular economy (CE) in the form of closing material flows and the cross-life cycle management of products is an instrument for manufacturing companies to operationalize sustainability economically [2]. The transition to CE is in its early stages [3]. To push forward the transformation to CE towards operationalization, concepts are needed that connect companies and value creation with the economic and ecological environment in a systematic and success-oriented manner. In this context, a product strategy is generally used as a planning tool that helps a company to focus its value creation and to achieve strategic objectives. Although there are numerous approaches for describing product strategies,

L. Overmeyer and B.-A. Behrens (eds.), *Production at the Leading Edge of Technology*,
Lecture Notes in Production Engineering, https://doi.org/10.1007/978-3-032-19524-1_12

these do not adequately consider aspects of CE. Currently, there is insufficient knowledge in the manufacturing industry to plan corresponding strategic initiatives on product level. There are no suitable strategic frameworks that allow companies to holistically define and design appropriate circular product strategies.

To engage this issue, according to KUBICEK, the formulation of fundamental research questions can support the validation of the research process and the delimitation of the scope of work [4]. In response to calls for intensified research, the first goal is to understand how the principles of CE affect the targets of a product strategy (research question (RQ) 1: "What impact does CE have on the design of product strategies in the manufacturing industry?"). Since a sufficient systematization of the concept is needed for the individual formulation of circular product strategies, the second goal of the research approach is to deliver a comprehensive and systematic description of circular product strategies (RQ 2: "How can circular product strategies be systematically described?").

Thus, a concept is developed that supports companies in grasping and understanding the implications of CE for the design of product strategies. To answer the research questions, this approach first considers the theoretical background of CE and product strategy research. Building on this, the results of a systematic literature review are structured in a morphology of circular product strategies. This should enable practitioners to identify options for action for the design of circular product strategies and at the same time provide scholars as a systematization framework for future research.

2 Theoretical Background

2.1 CE in the Context of Manufacturing Industry

Sustainability is a reference frame for dealing with the interconnection of environmental, economic and social issues within a systems perspective characterized by complex interdependencies and long-term strategies. In view to the manufacturing sector this includes the pursuit of a CE [5]. According to ELLEN MACARTHUR FOUNDATION, CE is a system where materials never become waste [6]. Products and materials are kept in circulation through processes like maintenance, remanufacturing or recycling. The underlying cycles are distinguished in two segments. The biological cycle involves materials which are biodegradable. The technical cycle focuses on materials which are non-biodegradable. In the context of the application of CE in the manufacturing industry in particular the technical cycle is of relevance. It compels companies to make maximum use of limited natural resources and keep them at the peak of their utility for the longest time. To get this cycle right, it is essential that product manufactures focus on strategies for remanufacturing or refurbishing to rescue the product from early wear and tear.

A key concept of CE are circular strategies. These strategies are usually referred to as "R-strategies". R-strategies take a systematic view across company- and industry boundaries and aim at reducing the consumption of natural resources and minimizing negative environmental impacts. They can be considered as a basis for designing measures for slowing down, closing and narrowing material cycles [7]. According to KIRCHHERR ET AL., to be successful in the context of CE, companies must formulate and implement circular business strategies [5].

2.2 Business- and Product Strategies

The HARVARD BUSINESS SCHOOL defines business strategies as strategic initiatives a company follows to create value for the organisation, its stakeholders and to gain competitive advantage in the market [8]. Taking into account the principles of CE, a company has to formulate adequate circular business strategies to create, communicate and extract value based on circular principles. In the manufacturing industry, at the core of these business activities, the product of a company is located [9]. Accordingly, the product strategy is a long-term, comprehensive plan that determines the conception and overarching goal of a product. It captures the market, user needs, key performance and differentiating features and business objectives [10] and defines measures and resources to achieve certain goals. As an essential component of business planning, the product strategy is part of strategic product planning and offers statements on, e.g. the design of the product variety, the required number of variants, the required technologies, the product life cycle and the focus of value creation and production. An effective product strategy is characterized by its ability to react integratively and dynamically to changes in the market environment. In summary, PORTER describes a product strategy as a systematic approach to developing and positioning a product on the market to achieve competitive advantages and ensure the long-term success [11].

3 Research Design

To organize and guide entrepreneurial decisions, conceptual guidance and an appropriate ontology is required. In particular, the translation of CE objectives into product strategies for the manufacturing industry and the derivation of company-specific circular product strategies are not sufficiently considered [12, 13]. A concept is needed to support companies in understanding and consistently aligning product strategies with the objectives of the CE. The goal of this research approach is to provide a concept for an expanded, application-oriented framework for the clear description and delimitation of circular product strategies. Thus, this research approach focuses on the need to solve a real-world problem to enable the development of practical knowledge for designing new approaches. Figure 1 shows the underlying course of the investigation. Within this approach, a systematic literature research (SLR) according to VOM BROCKE ET AL. [14] allows to pre-structure the research area and to identify implications for the design of circular product strategies. Based on the approach of RIESENER ET AL. [15], a strategy for the search for existing literature was developed and a scheme for documentation and analysis of selected studies was designed. To analyse these contributions, the method of qualitative content analysis according to MAYRING [16] was applied. The aim was to identify dimensions and characteristics that represent overarching components of a circular product strategy. The results were examined in accordance to ZWICKY [17] and are presented in the form of a morphological box.

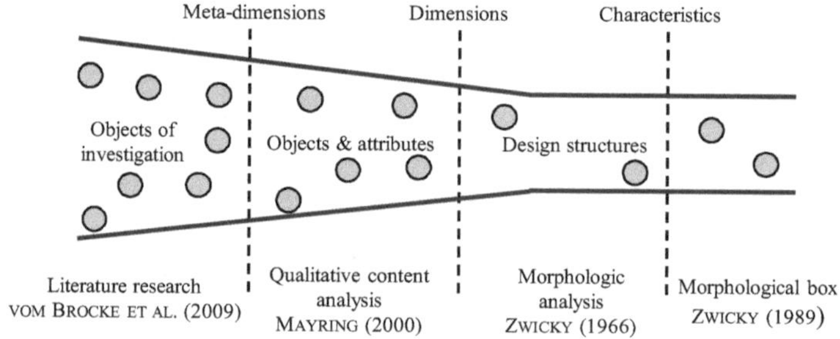

Fig. 1 Research design

4 Results

The following section presents the results of the research approach based on the methodology presented above. First the results of the SLR and the qualitative content analysis are introduced. Then, the dimensions and characteristics of a circular product strategy are presented in the form of a morphological box.

4.1 SLR and Qualitative Content Analysis

The SLR was conducted in the SCOPUS database, a widely used reference source for analyses in the field of manufacturing industry. The search focused on publications relating to product strategies in the context of CE. Relevant contributions were examined along the guidelines of VOM BROCKE ET AL. [14]. Components of the search string used on February 07, 2025 are shown below. The internal linking of terms was realized using an AND operator between the categories. Synonyms within the categories were considered using an OR operator (s. Fig. 2). The title and abstract were checked for relevance and agreement with the research objective. A total of 130 relevant articles were identified.

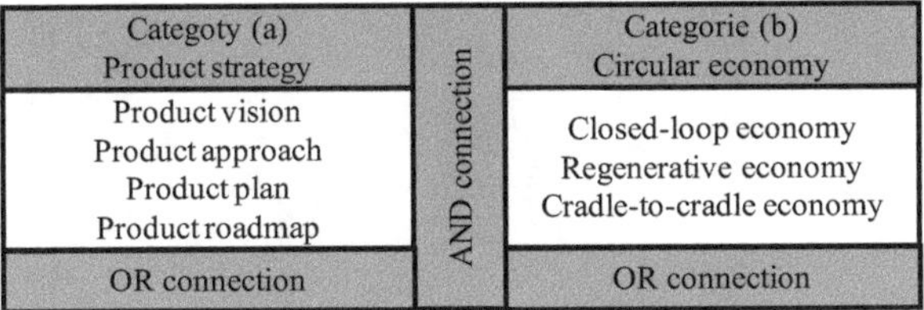

Fig. 2 Search string for the SLR

Within the first step of the methodology according to MAYRING [16], relevant topics were extracted from the identified contributions. To formulate these categories NICKERSON ET AL. suggest choosing meta-dimensions based on existing approaches [18]. Consecutive, the meta-dimensions of value proposition, value delivery, value creation and value skimming were selected in accordance with the research approaches by OSTERWALDER ET AL. [19] and GASSMANN ET AL. [20]. In addition the meta-dimension architecture [21] was added. In the second step, dimensions and characteristics were identified based on text passages that correspond to a specific meta-dimension.

4.2 Morphology for Circular Product Strategies

The resulting morphological box (s. Table 1) consists of five meta-dimensions (MD) which serve as structural elements. Between two and four dimensions (D) are assigned to each of these MDs. Finally, between three and nine characteristics (C) are allocated to each D, which are understood as possible manifestations.

Table 1 Morphological box for circular product strategies

Meta-dimensions	Dimensions	Characteristics								
Value proposition (MD1)	Ecological impact (D11)	Narrowing	Slowing	Closing						
	R-strategy (D12)	Reuse	Repair	Refurbish	Remanufacture	Recycle				
	Core functionality (D13)	Product transaction	Service transaction	Product and service transaction						
Value delivery (MD2)	Customer segment (D21)	Company-specific	Industry-specific	Cross-industry						
	Product-service-relation (D22)	Pure product	Pure service	Usage-oriented	Result-oriented					
	End-of-life scenario (D23)	Completely	Partially	Not reuse						
	Take-back system (D24)	Deposit	Reward	Exchange	Buyback	Leasing models	Collection service			
Architecture (MD3)	Modularity (D31)	Openness	Selection	Standardization	None					
	Perspective (D32)	Remanufacturing	Recycling	Lifetime	Scale effects	None				
Value creation (MD4)	Keyactivity (D41)	Production	Problem solving	Customer service	Matching	Research and Development	Others			
	Key resource (D42)	Tangible resources	Intangible resources	Multiple resource						
Value skimming (MD5)	Revenue source (D51)	Transaction	Utilization-oriented	Subscription	Fixed price	Licensing	Advertising	Additional services	Freemium	Others
	Revenue partner (D52)	Provider	Buyer	Third-party provider	Multiple					

The MD "Value Proposition" (MD1) is the "starting point" [1] for every circular product strategy. The value proposition represents the sum of all the benefits the customer derives from the product or service. The actual superior goal of the circular product strategy can be specified using the dimension of "Ecological impact" (D11). The derived characteristics "slowing", "narrowing" and "closing" address the underlying idea of closing material cycles and form an overarching framework along the product life cycle [22]. This is specified by the dimension "R-strategy" (D12) which provides a visualization and understanding of the different stages of resource use and waste management in CE. Espacially the strategies of "reuse", "repair", "refurbish", "remanufacture" and "recycle" are relevant for the manufacturing industry [15]. To further specify the generated value, it is necessary to describe the core functionalities (D13) of a circular product strategy. Consequently, strategies can be distinguished based on their transaction content. Pure "product transaction", pure "service-transactions" or "product and service transactions" can be identified.

The MD "Value delivery" (MD2) focuses on the market that is addressed. The "Customer segment" (D21) represents the target customers. Circular product strategies can be offered either "company-specific", with a focus on a single company, "industry-specific", with a focus on a single industry, or "cross-industry", with a focus across industry boundaries [1]. Within the "Product-service-relationship" dimension (D22), "pure product" transactions focus exclusively on the transfer of a physical product. "Pure service" is based on the transfer of an intangible service. The characteristic "usage-oriented" represents a system where the ownership of the tangible product is retained by the service provider, who sells the functions of the product. In "result-oriented" systems products are often replaced by services [23]. The dimension "End-of-life scenario" (D23) reflects how the circular product strategy addresses the end of the product's life cycle. It is possible to focus on "full reuse", where the product or all its components are reused at the end of the life cycle. "Partial reuse" describes the case in which a reuse strategy exists for only selected components. "No reuse" applies when no scenario is planed. The dimension "Take-back system" (MD24) describes how products can be returned to the manufacturer after use. "Deposit" reflects a financial incentive mechanism in which a deposit is collected when the product is purchased and is refunded when the product is returned. In the case of a "premium", customers receive a reward or other financial incentives for returning products. "Exchange" allows customers to exchange their used products for new or refurbished products. With a "buyback" program, a company offers its customers the opportunity to sell back used products. "Leasing models" allow customers to use products for a certain period of time and return them afterwards. Finally, a "collection service" describes a mechanism whereby the manufacturer collects used products from the customer.

The MD "Architecture" (MD3) describes how the underlying technical structure of the product enables circularity [2]. The dimension of "Modularity" (D31) describes the division of the overall system into modules. The characteristic of "openness" applies when actors are granted unrestricted access to the product or service and can add further value components. This design enables easy expansion and adaptation of modules. "Selection" refers to a architecture that provides a selection of specific modules that can be integrated into the product. Within "standardization" uniform standards as well as interfaces ensure easy integration of partial solutions from different actors. Also "no" modularity can be provided, so that there is no specific structuring of the product. The dimension "Perspective" (D32) describes the conceptualization of the architecture. "Remanufacturing" refers to a design for the overhaul and revaluation of a product at the end of its usage cycle. "Recycling" characterizes the design with the aim of recovering materials from used products in order to return them to the production cycle. "Lifetime" means maximizing the usage cycle of the product. "Scale effects" refer to the design of the product under the primary consideration of cost benefits. "None" means that the architecture is not based on any specific perspective for the definition and arrangement of modules.

The MD "Value creation" (MD4) describes how value is created. The "key activity" (D41) dimension refers to fundamental activities that are carried out to create value. "Production" describes the focus on creating a service by combining production factors in a transformation process. "Problem solving" represents identifying specific problems and deriving individual solutions. "Customer service" reflects the fulfillment of customer requests. The characteristic "matching" refers to the matching of supply and demand. "Research and development" embodies a continuous innovation orientation [24]. Also "other" key activities can be focused. Furthermore, "Key resources" (D42) can be distinguished between "tangible resources" or "intangible resources" or "tangible and intangible resources".

The MD "Value skimming" (MD5) describes the mechanism by which a company captures the created value in the form of profit or revenue. "Revenue source" (D51) embodies different models for generating income. "Transaction" describes a circular product strategy that charges a percentage fee or, alternatively, a predetermined amount for each successful transaction. Alternatively, revenue can be generated "utilization-oriented". An example of usage-based revenue generation is a fee charged per request for a specific service. The "subscription model" includes the contract-based use of the product at regular intervals in return for payment. Furthermore, alternative sources of revenue can be provided in the form of "fixed price", "licensing" or "advertising". Also, "additional services" can be sold. If individual groups of actors are granted free use of the value proposition in the form of a free basic product or service, as well as fee-based extensions, this is referred to as a "freemium" model [2]. "Revenue partner" (D52) refer to the actors involved in generating revenue. "Provider" is a company that sells products or services directly to the end customer. "Buyer" embodies the end users or businesses that purchases products or services. A "Third-party provider" is an external player that offers additional products or services that complement or extend the main provider's offering. "Multiple" means that different revenue partners are envisaged in the course of value extraction. Building on these results, Chap. 5 provides a brief outlook on the possible use of morphology in science and practice.

5 Conclusion and Derived Need for Further Research

To answer to the research questions about the implications of CE for product strategy research (RQ1) and a systematic description (RQ2) and on the basis of the application of the methods by vom Brocke et al. [14], Mayring [16] and Zwicky [17] a morphological representation of circular product strategies for the manufacturing industry was presented. This approach serves as a reference architecture that provides guidance for the description of circular product strategies. It includes consistent definitions of systems, design elements, as well as a common vocabulary to discuss specifications of implementations. This representation embodies a meta-theoretical framework to systematize the research area. For ease of understanding, the model contains a limited number of terms and characteristics and can consequently be described as concise and practicable.

From the results several implications for theory and practice can be drawn. Regarding scientific contributions, the research approach addresses the lack of systematic tools for the description of circular product strategies (e.g. [25]). This research approach confirms, existing product strategy frameworks take insufficient account of the complexity and specific characteristics of CE [15]. In this regard, "a good classification scheme [forms] the basis of theory development" [26]. Building on the high level of abstraction of generic product strategy definitions (e.g. [20]) this approach develops a contextual reference model in the form of a morphology. This morphology provides the research area with the required "common ground" for focusing further research initiatives. It facilitates the comparability and synthesis of existing and future research results and serves as an instrument for reducing complexity.

As for managerial contributions, the developed morphology addresses the lack of guidance for realizing product strategies in CE. As a simplified representation of reality, the model is intended to help minimize the cognitive demands on decision makers and overcome application difficulties regarding the missing structuring of the field of action. Morphologies are never perfect but exist to provide an appropriate solution in a given context. The approach is restricted by a few limitations that must be taken into account when interpreting the results. Due to its meta-theoretical orientation, the framework was developed through an inductive, qualitative analysis of the literature. As such, the identified dimensions and characteristics are not statistically representative but conceptually grounded. They may serve as a starting point for further deductive validation and refinement through empirical research such as structured case studies or quantitative content analysis. Such extensions could deepen the understanding of product strategies in CE, test the practical applicability of the framework, and identify gaps in research or implementation.

The SLR in conjunction with the development of a morphological box fundamentally strengthens the conceptual basis for further research on the realization of CE in the manufacturing industry. Research could focus on operationalizing the morphology as a support tool in strategic decision-making and product development processes. Classifications open the possibility of organizing abstract, complex concepts. Consequently, the research approach provides a solid and promising basis for conducting further qualitative and quantitative research initiatives.

Acknowledgements. Funded by the Deutsche Forschungsgemeinschaft (DFG, German Research Foundation) under Germany's Excellence Strategy—EXC—2023 Internet of Production—390621612.

Competing Interests. The author(s) has no competing interests to declare that are relevant to the content of this manuscript.

References

1. Scholz, U.: Opportunities for Sustainable Product Development. Springer, Heidelberg (2018)
2. Schuh, G., Keuper, A., Patzwald, M., Kuhn, M.: New Modularity and Technology Roadmapping. Conference on Empower Green Production (AWK) (2023)
3. Fontell, P., Heikkilä. P.: Model of Circular Business Ecosystem for Textiles. VTT Technology, Espoo
4. Kubicek, H.: Heuristische Bezugsrahmen und heuristisch angelegte Forschungsdesign als Elemente einer Konstruktionsstrategie empirischer Forschung. Institut für Unternehmungsführung im Fachbereich Wirtschaftswissenschaft der Freien Universität Berlin, Berlin
5. Kichherr, J., Hekkert, M.P., Relke, D.: Conceptualizing the circular economy: an analysis of 114 definitions. SSRN Electron. J. 221–232 (2017)
6. Ellen MacArthur Foundation: What is a circular economy. https://www.ellenmacarthurfoundation.org/topics/circulareconomy-introduction/overview (2013). Accessed 21 February 2025
7. Mast, J., Unruh, F. von, Irrek, W.: R-strategies as guidelines for the Circular Economy. https://prosperkolleg.ruhr/wp-content/uploads/2022/08/rethink_22-03_r-strategien_EN.pdf (2022). Accessed 24 February 2025
8. Harvard Business School: What is Business Strategy & Why is it important? https://online.hbs.edu/blog/post/what-is-business-strategy (2022). Accessed 24 February 2025
9. Riesener, M., Kuhn, M., Hellwig, F., Ays, J., Schuh, G.: Design for circularity—identification of fields of action for ecodesign for the circular economy. Procedia CIRP, 137–142
10. Pichler, R.: Strategisches Produktmanagement. Produktstrategien und -Roadmaps für digitale Produkte und agile Teams, 1st edn. dpunkt, Heidelberg (2023)
11. Porter, M.E.: Competitive Strategy: Techniques for Analyzing Industries and Competitors. Free Press, New York (1980)
12. Schmitt, R., Bodenbenner, M., Brings, H., Montavon, B.: Datenstrukturen für eine resiliente Life Cycle Sustainability. In: Conference on Empower Green Production (AWK) (2023)
13. Schuh, G., Keuper, A., Ruschitzka, C., Schümmelfeder, S., Prumbohm, Felix: Managing Sustainable Innovations - Ergebnisbericht der Benchmarking Studie. https://www.kbm-sustainable-innovation.de/de/ (2024). Accessed 21 February 2025
14. vom Brocke, J., Simons, A., Niehaves, B., Riemer, K., Plattfaut, R., Cleven, A.: Reconstructing the giant: on the importance of rigour in documenting the literature search process. In: 17th European Conference on Information Systems (ECIS)
15. Riesener, M., Keuper, A., Schümmelfeder, S., Schuh, G.: Identifying fields of action for the design of circular platform-based business models in the manufacturing industry based on phenotypes. J. Econ. Bus. Manage. **13**, 37–51 (2025)
16. Mayring, P.: Qualitative Content Analysis. Forum Qualitative Sozialforschung/Forum: Qualitative Social Research, vol. 1, no. 2 (2000); Qualitative Methods in Various Disciplines I: Psychology (2000). https://doi.org/10.17169/fqs-1.2.1089
17. Zwicky, F.: Entdecken - Erfinden - Forschen im morphologischen Weltbild. Knaur-Taschenbücher, vol. 264. Droemer Knaur, München, Zürich (1971)
18. Nickerson, R.C., Varshney, U., Muntermann, J.: A method for taxonomy development and its application in information systems. Eur. J. Inf. Syst. **22**, 336–359 (2013)
19. Osterwalder, A., Pigneur, Y., Tucci, C.L.: Clarifying business models: origins, present, and future of the concept. In: CAIS 16 (2005)
20. Gassmann, O., Frankenberger, K., Csik, M.: Geschäftsmodelle Entwickeln. Carl Hanser Verlag GmbH & Co. KG, München (2017)
21. Tiwana, A.: Platform Ecosystems. Morgan Kaufmann, Burlington (2014)

22. Mast, J., Unruh, F., von, Irrek, W.: R-strategies as guidelines for the Circular Economy. https://prosperkolleg.ruhr/wp-content/uploads/2022/08/rethink_22-03_r-strategien_EN.pdf. Accessed 24 February 2025
23. Tukker, A.: Eight types of product–service system: eight ways to sustainability? experiences from SusProNet. Bus. Strat. Env. **13**, 246–260
24. Riesener, M., Kuhn, M., Perau, S., Guenther, P., Schuh, G.: Characterization of continuous innovation for technical products in the usage phase. Procedia CIRP **119**, 987–992 (2023)
25. Kanda, W., Martin, G.: From circular business models to circular business ecosystems. Bus. Strateg. Environ. **30**, 2814–2829 (2021)
26. Lambert, S.C.: The importance of classification to business model research. J. Bus. Models **3**, 49–61 (2015)

Manufacturing Technologies

Towards Sustainable Hot Forging: An Experimental Assessment of Lubrication Parameters

Sascha Eckardt[1](✉), Mareile Kriwall[1], Malte Stonis[1], and Bernd-Arno Behrens[1,2]

[1] IPH—Institut für Integrierte Produktion Hannover gGmbH, Hannover, Germany
eckardt@iph-hannover.de

[2] Leibniz Universität Hannover, Garbsen, Germany

Abstract. Determining suitable lubrication settings is crucial for designing an effective hot forging process. Currently, this determination is predominantly carried out through trial and error at the beginning of production, with subsequent adjustments based on the operator's expertise. To reduce manual labor, costs, and especially resource consumption, it is advantageous to enable the selection of appropriate lubrication settings in advance. This supports the industry in achieving its carbon–neutral objectives while enhancing process stability. This paper presents simplified experimental methods to evaluate the tribological properties of hot forging processes; the ring compression test was chosen to determine the influence of the lubrication layer thickness on the tribological system. This study was structured in two stages: first, a design space exploration of spraying parameters; second, an evaluation of the friction factor for various lubrication layer thicknesses.

Keywords: Hot forging · Lubrication · Ring compression test

1 Introduction

Hot forging processes are complex, making their design a highly manual and iterative task. This involves not only designing the forged component itself but also developing the overall forging process. Engineers typically rely on prior experience when designing new processes for similar products. However, changing customer needs continuously reshape the portfolio of forged parts. For example, demand in the automotive sector is declining due to the shift toward electric mobility. Conversely, demand is growing in the energy and industrial machinery sectors. Many aspects of process design can be addressed prior to production through finite element analysis (FEA) and other digital tools. Extensive research has been conducted to optimize elements such as stage sequence planning, billet geometry, die design, and processing temperatures for specific requirements [1]. Nevertheless, discrepancies between simulation and real forging processes remain. One practical example is the optimization of lubrication methods (e.g., chipped wood, lubrication by flooding, or spraying) and the accompanying settings (e.g., amount/thickness

L. Overmeyer and B.-A. Behrens (eds.), *Production at the Leading Edge of Technology*,
Lecture Notes in Production Engineering, https://doi.org/10.1007/978-3-032-19524-1_13

of lubrication or spraying parameters). Initially, when producing a new component, the lubrication must be adjusted iteratively to achieve the intended forging results. Manual lubrication often yields inconsistent results due to operator-dependent variability [2]. Automated spraying systems improve repeatability, but the initial search for optimal parameter settings remains. To address this challenge, this paper introduces a method for evaluating different parameter settings and generating knowledge about the lubrication distribution and its effect on the tribological system. The method focuses particularly on how variations in lubrication layer thickness influence tribological behavior.

Over the years, numerous experiments have been conducted to study the influence of different lubrications and parameter settings on forming processes. In particular, the need for simplified experiments to understand the capabilities of different lubricants has led to a wide variety of studies. Basic experiments, such as pin-disc, ring-disc, and four-ball tests, are not considered here, as their high level of simplification limits their applicability to complex hot forging processes. Experiments that more closely resemble real hot forging processes and are widely used include the ring compression test (RCT), the warm hot upsetting sliding test (WHUST), the sliding compression test (SCT), and the double cup extrusion test (DCET). More experiments have also been introduced, including the container-less tapered plug penetration test [3], the open-die backward extrusion test (ODBET) [4], and different variations of existing experiments, such as the ring with inner boss compression test [5] and the triangular boss compression test [6].

The RCT is a method that uses the deformation of a ring-shaped sample between two parallel tools to evaluate the tribological system. Due to its high sensitivity to changes in friction, the change in the inner diameter corresponding to the height deformation is used to measure the friction coefficient (μ) or the friction factor (m). There are different approaches to derive the necessary nomograms to evaluate the coefficients for an executed experiment. One approach is to use basic correlations, such as those from Avitzur [7] or Male and Cockcroft [8]. Another approach is to use FEA. When basic correlations are used (based on theoretical approaches), a high degree of simplification is employed. In contrast, FEA is suitable for obtaining a more realistic representation of the process; however, these representations usually rely on a generalized friction law. Unlike other methods, this approach does not require the evaluation of process forces.

The WHUST is based on separating two stages. First, a contactor is pressed against a specimen to create a predefined deformation and contact pressure. Next, the contactor is moved with a constant velocity on the surface of the specimen. Separating the stages allows for individual adjustments of the process parameters. To determine the necessary penetration for a specific load case, an FEA is conducted. During the test, normal and tangential forces are recorded, which are necessary for evaluating the test results. In addition to the friction coefficient (μ), this method can also obtain other parameters, such as the critical length, which can be used to evaluate the lubrication efficiency [9]. The SCT uses the same division principle as the WHUST, but employs different specimen and tool geometries [10]. By adjusting the geometry, the normal stress and surface enlargement can be adjusted to the desired values.

The DCET is another simplified experiment used to determine the friction coefficient or friction factor during a forming process [11]. In the DCET, a cylindrical specimen is extruded into two cups: one around the stationary lower punch and one around the moving upper punch. The difference in extrusion height is used to measure the friction coefficient. Like the RCT, the DCET is an indirect measuring experiment that does not require forming force measurement; however, it has a more complicated tool geometry.

Due to its simplicity and broad applicability, this study employed the RCT to evaluate the tribological properties of different lubrication layer thicknesses. Furthermore, the goal was not to obtain knowledge about one specific constant tribological load, but rather to gain general knowledge about lubrication behavior. Thus, the RCT is suitable for this purpose, as its overall design most likely resembles industrial hot forging processes.

2 Methodology

This study consisted of two main stages. In the first stage, spraying parameters were explored to achieve a uniform distribution of lubrication across a wide range of layer thicknesses. To this end, Latin Hypercube Sampling (LHS) was employed to ensure that parameter combinations were evenly distributed within the operational limits of the spraying equipment.

The second stage examined the correlation between lubrication layer thickness and friction factor using RCT. For this purpose, FEA was applied to generate the required nomograms, while practical experiments were conducted to establish the relationship between lubrication layer thickness and tribological performance.

3 Case Study

3.1 Stage I—Exploration of Spraying Parameters

As outlined above, the first step was to investigate the spraying parameters. LHS generated randomized sets of spraying parameters to provide an overview of the selected range.

The experimental setup consisted of two heated flat plates serving as tools, placed side by side and sprayed with a robot-controlled spraying nozzle at three positions per plate. The tools were heated for 30 min (45 min for the first run of the day) using PID-controlled heating cartridges (1 kW maximum power per tool) to reach the desired temperature. After natural cooling to room temperature, the resulting lubrication distribution was examined using a Keyence VR-5200 digital profilometer. The applied lubrication layer was measured twice along a diagonal, and the mean value was calculated. Measurements from several repetitions per parameter set were compared, then averaged.

Since a maximum of 60 experiments could be carried out per day, and five repetitions per setting were preferable, 12 parameter sets were randomly tested per range. As shown in Table 1, the study used three different ranges.

Range 1 provided an initial estimate of the actual lubrication deposition rate. However, the initial time range of 0.5–2 s proved too short for measurable lubrication deposition, especially at high spraying distances and low lubrication pressures. Furthermore, the tests revealed that the spraying distance range had to be divided into two intervals. Additional spraying parameter tests beyond those generated by LHS led to divisions below and above 50 mm, with times of approximately 5 and 10 s, respectively.

Table 1 Parameter ranges for design space exploration

Parameter	Range 1	Range 2	Range 3
Lubrication pressure [bar]	1–4	1–3	1–4
Spray air pressure [bar]	5–6	4–6	4–6
Spraying distance [mm]	10–100	30–50	50–100
Spraying time [s]	0.5–2	3–7	5–15
Tool temperature [°C]	200 °C (surface measurement)		
Lubrication type	Graphite dispersion (Tribo-Chemie Graphitex® 289)		
Lubrication/water ratio	1/10		

Regardless of the parameter combinations, experiments in range 2 showed that the lubrication deposition cross-section could be approximated by a quartic polynomial, as shown in Fig. 1 for two averaged exemplary parameter sets (Set 18 and 24).

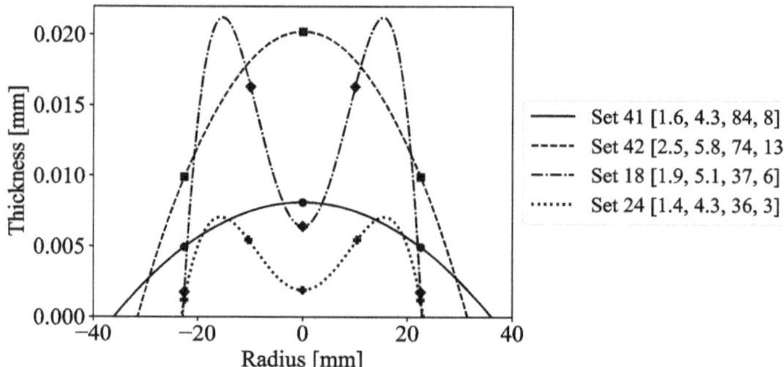

Fig. 1 Examples of lubrication deposition cross-sections in ranges 2 (Set 18 and 24) and 3 (Set 41 and 42). Spraying parameter sets are named as 'set number [lubrication pressure, spray air pressure, spraying distance, spraying time]'

The individual measures (maximum layer thickness, diameter, and local maxima) varied among parameter sets, and significant temporal variations were also observed within a single set over time.

Experiments in range 3 demonstrated that the cross-section of lubrication deposition significantly changed according to the spraying distance. At a spraying distance of about 60 mm, the lubrication deposition profile shifted toward a quadratic polynomial, as shown in Fig. 1 for two averaged exemplary parameter sets (Set 41 and 42). Even within range 3, variations in lubrication deposition occurred over time, but results from some parameter sets deviated more than others. This is exemplified in Fig. 2 for the maximum layer thickness of parameter sets with a quadratic deposition cross-section.

Greater spraying distances combined with lower lubrication pressure yielded more consistent results. For the subsequent RCT, parameter sets 41 and 42 were chosen because

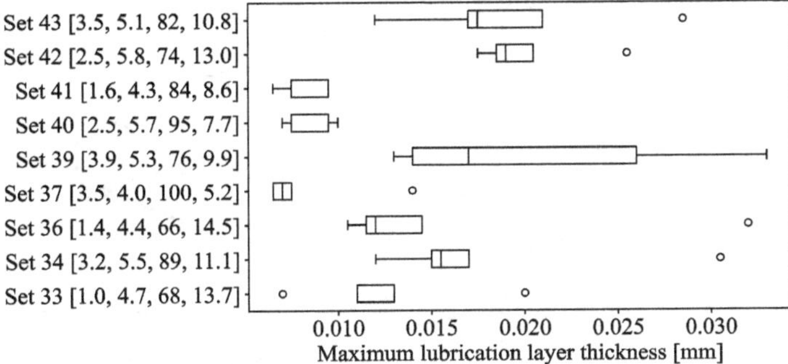

Fig. 2 The distribution of maximum lubrication layer thickness for exemplary parameter sets with a quadratic deposition cross-section

they had higher uniformity in lubrication layer thickness and lower standard deviations relative to the mean, thereby ensuring higher reproducibility.

3.2 Stage II—Identification of the Correlation Between Lubrication Layer Thickness and Friction Factor

The experimental phase of the RCT was carried out on a Weingarten PSH 4.265f press with a maximum energy of 100 kJ, of which 5% was used. The applied parameters are summarized in Table 2.

Table 2 Experimental parameters for the RCT

Parameter	Setting(s)
Lubrication settings	Set 41 and 42 with varying spraying time (between 0 and 60 s)
Lubrication type	Graphite dispersion (Tribo-Chemie Graphitex® 289)
Billet	Ring (30 mm outer diameter, 15 mm inner diameter, 10 mm height) made of 42CrMo4 heated up to 1250 °C (with an induction furnace)
Tool	55NiCrMoV7, hardened and annealed 46 HRC, ground to Ra 0.4, heated up to 200 °C (surface temperature)

During the forming process (50% compression), temperatures and forming velocity were measured to enable the creation of suitable FEAs. Figure 3 shows a schematic of the tool setup used.

The inner diameter and height were measured before and after deformation using a Helios Preiser DIGI-MET digital caliper. The inner diameter was measured at two positions and the height at four points along the mid-circumference of the ring.

At the start of each forging sequence, three positions on each tool insert were lubricated per setting. Two of these positions were used for forging, while one served as a reference spray. After two consecutive forgings, the tool cooled, and the reference spray

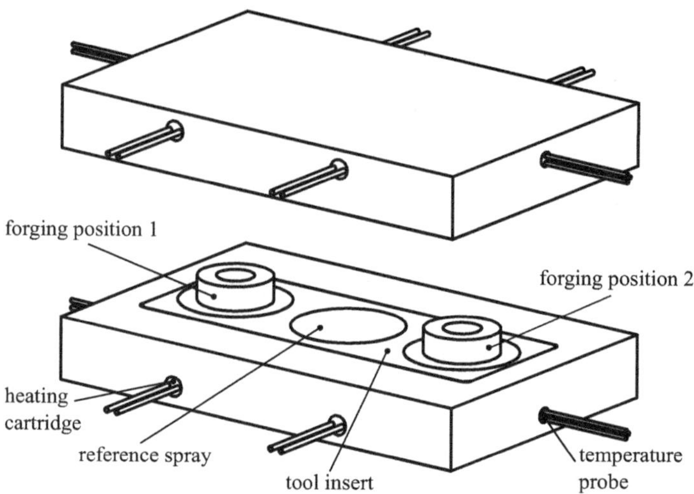

Fig. 3 Tool schematic for the RCT

was measured with the profilometer mentioned above to determine the actual thickness of the deposited lubrication layer. For each parameter set, a new tool insert was used to prevent changes in surface topology and lubrication residues influencing the tribological system.

The required nomogram was created using Simufact Forming 2024.2 and a Python post-processing script. A 2D axisymmetric simulation with quadratic elements (Quads (10)) was employed. The nomogram was then supplemented with the mean experimental results for each parameter combination to determine the friction factor corresponding to a given lubrication layer thickness, as illustrated in Fig. 4.

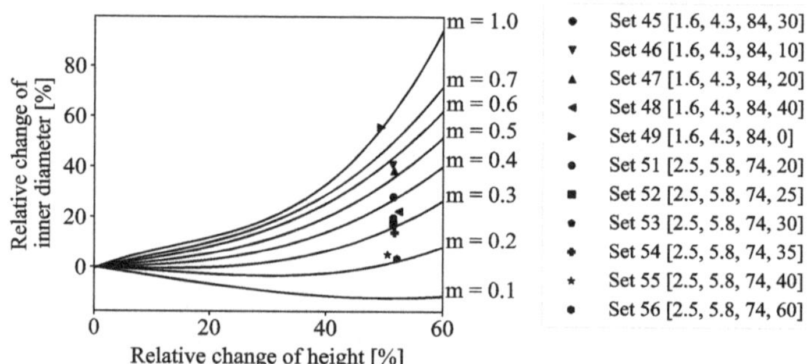

Fig. 4 Nomogram with marked experimental results

An increase in lubrication layer thickness clearly reduced the friction factor, as shown in Fig. 5. While dry forging yielded a friction factor of about 1, a lubrication layer thickness of 59 μm (the mean of the maximum layer height and the height at 45 mm diameter) decreased it to 0.2.

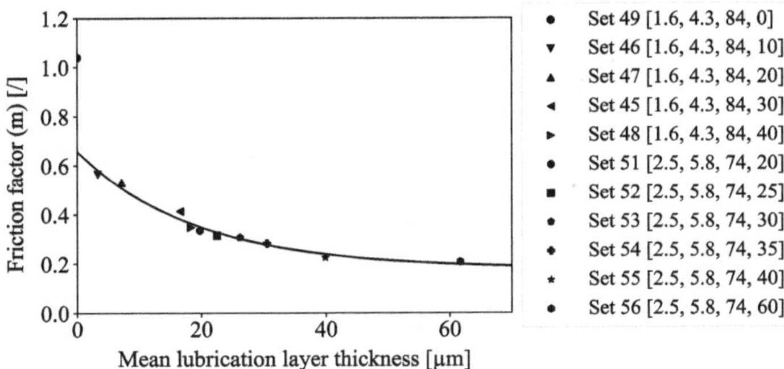

Fig. 5 Friction factor depending on the lubrication layer thickness

As shown in Fig. 5, the friction factor was approximated by an exponential decay across the observed thickness range. To achieve a friction factor of 0.3, commonly used in FEA, an average lubrication layer thickness of around 26 µm was required.

4 Summary and Outlook

This study demonstrates that LHS is effective for exploring the design space of spraying parameters. However, results strongly depend on the manually defined parameter ranges. The experiments showed further that increasing the lubrication layer thickness significantly reduces the friction factor during hot forging. Whether a threshold exists beyond which the friction factor rises again, as suggested by the Stribeck curve, remains uncertain and requires further investigation. Additionally, the relationship between lubrication layer thickness and friction factor may vary for other load cases, which warrants further study. Lubrication deposition is strongly influenced by spraying distance. Two characteristic deposition regimes were identified: quartic and quadratic cross-sections. In the transition region between these regimes, certain settings produced nearly uniform layer thicknesses (at a 45 mm spraying diameter), which may be advantageous in forging. However, this transition region is difficult to reproduce consistently, as is quartic spraying.

Overall, the results highlight the potential of tailoring lubrication layer thickness to optimize forging processes. Nevertheless, further research is needed to fully understand the influence of thickness variation under varying process conditions.

Acknowledgements. The project "Entwicklung eines Modells zur automatisierten Ermittlung der idealen Schmiermittelschichtdicke bei der Warmmassivumformung" (funding code 524639810) receives financial support from the German Research Foundation (DFG).

Competing Interests. The author(s) has no competing interests to declare that are relevant to the content of this manuscript.

References

1. Kitayama, S.: Technical review on design optimization in forging. Int. J. Adv. Manuf. Technol. **132**, 4161–4189 (2024). https://doi.org/10.1007/s00170-024-13593-w

2. Hawryluk, M., Dudkiewicz, Ł, Marzec, J., Rychlik, M., Tkocz, R.: Selected aspects of lubrication in die forging processes at elevated temperatures—a review. Lubricants **11**, 206 (2023). https://doi.org/10.3390/lubricants11050206
3. Asai, K., Kitamura, K.: Estimation of frictional coefficient between tool and specimen in container-less tapered plug penetration test. IOP Conf. Ser. Mater. Sci. Eng. **1270**, 12101 (2022). https://doi.org/10.1088/1757-899X/1270/1/012101.
4. Sofuoglu, H., Gedikli, H.: Determination of friction coefficient encountered in large deformation processes. Tribol. Int. **35**, 27–34 (2002). https://doi.org/10.1016/S0301-679X(01)00076-7
5. Hu, C., Yin, Q., Zhao, Z., Ou, H.: A new measuring method for friction factor by using ring with inner boss compression test. Int. J. Mech. Sci. **123**, 133–140 (2017). https://doi.org/10.1016/j.ijmecsci.2017.01.042
6. Dwivedi, R., Choudhary, M., Sahgal, K., Purohit, R.: Analysis of friction factor in cold forging by using ring with triangular boss compression test. Adv. Mater. Process. Technol. **8**, 1422–1440 (2022). https://doi.org/10.1080/2374068X.2021.1945265
7. Avitzur, B.: Forging of hollow discs, Israel. J. Technol. **2**, 295–304 (1964)
8. Male, A.T., Cockcroft, M.G.: A Method for the Determination of the Coefficient of Friction of Metals under Conditions of Bulk Plastic Deformation. Journal of the Institute of Metals 38–46 (1964)
9. Dubois, A., Dubar, M., Dubar, L.: Warm and hot upsetting sliding test: tribology of metal processes at high temperature. Procedia Eng. **81**, 1964–1969 (2014). https://doi.org/10.1016/j.proeng.2014.10.265
10. Groche, P., Müller, C., Stahlmann, J., Zang, S.: Mechanical conditions in bulk metal forming tribometers—part one. Tribol. Int. **62**, 223–231 (2013). https://doi.org/10.1016/j.triboint.2012.12.008
11. Buschhausen, A., Weinmann, K., Lee, J.Y., Altan, T.: Evaluation of lubrication and friction in cold forging using a double backward-extrusion process. J. Mater. Process. Technol. **33**, 95–108 (1992). https://doi.org/10.1016/0924-0136(92)90313-H

Enhancing Building-Integrated Photovoltaics: A Gripper System for Precise Layup of Solar Cell Strings in PV Module Production

Jessica Schönburg[1](✉), Jens Eilrich[2], Anja Mercker[2], Henning Schulte-Huxel[2], and Annika Raatz[1]

[1] Institute of Assembly Technology and Robotics, Leibniz University Hannover, Garbsen, Germany
schoenburg@match.uni-hannover.de
[2] Institute for Solar Energy Research, Hamelin, Germany

Abstract. Photovoltaic systems have significant potential for generating and utilizing of renewable energy, and research is being conducted to expand the range of applications for photovoltaic modules. One promising solution for activating previously unutilized potential is to extend the yield areas of roofs or open spaces to building facades. However, achieving high utilization of this area requires a high degree of geometric flexibility and a wide variety of PV module shapes, which are currently very limited due to rigid manufacturing processes. To address these challenges, it is essential to enhance the flexibility and optimize the production process. A crucial aspect of this is the safe handling of cell formats of different sizes with consistently high process accuracy. In this article, we present an approach for a modular vacuum gripper that can safely handle solar strings of variable cell size and length. A detailed analysis of the geometric cell parameters was used to precisely adapt the gripper design to the respective specifications. Tests regarding gripper quality validate the safe handling of different formats without cell damage. Investigations into positioning accuracy, on the other hand, reveal unstable and complex string behavior, which currently prevents the accuracy requirements from being met and requires further investigation.

Keywords: Gripper · Building-integrated photovoltaics · Flexible handling

1 Introduction

In order to counteract climate change, which has been greatly accelerated by human influences and rising energy demands, science is researching sustainable alternatives for the long-term replacement of fossil fuels [1]. One key solution to this problem is the increased use of renewable energies, in particular photovoltaic (PV) modules. [2]. With the Federal Government's new Climate Protection Act, Germany is aiming to achieve greenhouse neutrality by 2045 by continuously reducing CO_2 emissions [3]. One approach is to anchor the production of solar modules more locally again and to further research the possible applications, such as building-integrated photovoltaics (BIPV) [4] as it combines the country-specific culture of construction with PV technology.

L. Overmeyer and B.-A. Behrens (eds.), *Production at the Leading Edge of Technology*, Lecture Notes in Production Engineering, https://doi.org/10.1007/978-3-032-19524-1_14

Initial studies on the quantification of urban solar potential in cities have shown that an extension of previously used roof and open spaces to building facades offers a promising approach to significantly increase the ecological energy yield in a space-saving manner [5]. However, a high degree of geometric flexibility and a wide variety of PV module shapes and designs are required in order to make efficient use of these additional areas [6]. The development of customized and aesthetically convincing solutions that can be harmoniously integrated into various facade structures and visually adapted to different architectural conditions should also contribute to an increase in social acceptance and at the same time create additional incentives for the expansion of renewable energy generation [7]. These requirements for building-integrated power generation are currently in conflict with the rigid manufacturing processes of conventionally modules. Current production facilities are optimized for the series production of identical modules and have a high degree of automation [8]. In particular when handling individual module components, customized gripping systems are used which are designed exclusively for their application of a specific string format. Adapting these existing, rigid production lines to frequently changing module formats is only possible with increased effort and affects the BIPV approach.

In order to overcome the challenges described above, a consistent further development of conventional production processes is required, which allows both the production of individual module geometries and the precise handling of varying cell formats. An essential component of this optimization is the flexibilization of handling technologies for solar cells. Due to their material-specific and design properties, solar cells are considered to be extremely fragile and susceptible to surface damage [9]. The requirements for an automated gripping system that ensures gentle, adaptive handling of the fragile components are correspondingly high. At the same time, consistently high repeatability must be ensured throughout the entire process. In order to meet these requirements, innovative gripper concepts are needed that can be adapted to different cell sizes and geometries without introducing mechanical or thermal loads that could jeopardize the structural integrity of the cells. Building on the work of Blankemeyer et al. [10], who investigated the flexible handling of solar cells with simply curved surfaces, this article presents an approach for a modular vacuum gripper that enables the handling of solar strings of different cell formats and lengths. To this end, the design and manufacture of the currently most common cell sizes were analyzed and the possibility of combining and standardizing the handling process for the various cell formats was investigated.

2 Structure of a Solar Cell and Production of a Solar Module

The main component of a solar module is the solar cell. It consists mainly of the semiconductor material crystalline silicon and is manufactured from so-called Si wafers with layer thicknesses between 100 and 300 μm. In addition to conventional full-cell designs, technologically advanced half-cut-cell formats that achieve an even higher level of efficiency are becoming increasingly established [11].

Once the individual solar cells have been manufactured, they can be further processed for the production of PV modules. The process comprises several precisely coordinated steps and begins with the interconnection of the individual cells. They are coupled together—in series to increase the voltage. In a system, the individual cells are linked together forming a string of solar cells and, in our application, prepared into strings of up to 12 cells. For this purpose, the positive pole of one cell is connected to the negative pole of another cell via fine copper wires, which are applied to the cell contacts using soldering technology. The cells can be soldered a short distance apart, adjacent or shingled, with a slight overlap. To set the desired electrical properties, these strings of solar cells are then connected further in series or in parallel. The latter increase the current. The strings are then assembled together with other components, such as the glass panel and encapsulant, to form a complete module. Figure 1 shows the individual layers and structural design of the components that are installed in a module.

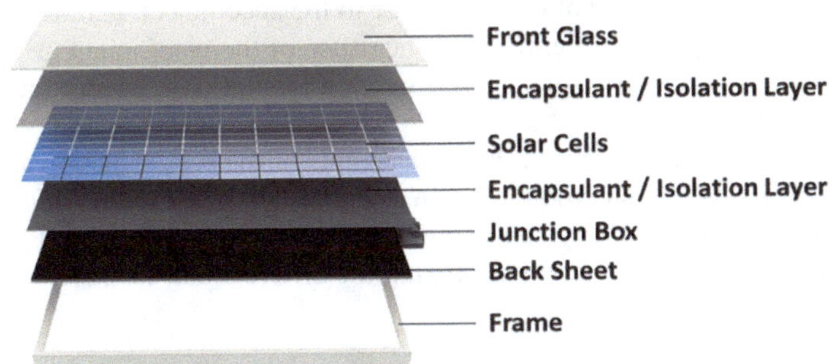

Fig. 1 Schematic representation of layers in the construction of a solar module

While the production process of conventional PV modules is already largely automated, the production of building-integrated photovoltaic modules requires the existing processes to be adapted and expanded. Due to the high variability in terms of module geometries and design specifications for the production of BIPV, a more flexible and optimized process chain is required, in particular its handling methods.

3 Handling System for Modular Cell Formats

As part of the research for the realization of BIPV, an automated process chain is currently being developed at the Institute for Solar Energy Research in Hamelin (ISFH), which enables the flexible processing of different cell and module formats within a continuous production process. A key process step in this system is the fully automated positioning of the solar strings produced by the stringer machine. In order to reliably implement this processing step, a gripper system is required that fits seamlessly into the automated line and picks up different string formats.

The following section outlines an approach for a gripping system that can securely grip and handle strings of variable length and size. This creates the basis for being able to place them at varying deposit positions with consistently high positioning repeatability. Due to their geometry, sensitivity and brittle material properties, gripping mechanisms based on a pneumatic principle of action, such as Bernoulli grippers, vacuum grippers or beam/surface suction pads, have mainly been established for handling solar cells

[12]. These technologies are compared in advance with the requirements for a flexible gripping system for the BIPV production line.

3.1 Analysis of the Gripper Technology and Preliminary Tests

To evaluate the gripping principles, preliminary tests are carried out on solar cells with different cell formats. The Bernoulli principle is already ruled out in advance, as the expected transverse forces on the string prevent reliable handling at high robot trajectory speeds. Preliminary tests with surface suction pads, on the other hand, did not result in satisfactory, non-destructive gripping quality: longer strings could not be gripped completely with a moderate vacuum, while an increase in negative pressure already lead to cell damage. High leakage volume flows due to the specific cell geometry and shape instability of the strings were identified as the cause of the insufficient gripping force, which prevented a sufficient vacuum level. In contrast, flat or bellows vacuum suction pads with a round shape showed a promising approach. Under the condition that the suction cups can be placed between the solder joints of the cell, a stable vacuum could be generated and solar cells could be handled reliably. In order to use this concept for the development of a modular gripping system for processing flexible BIPV strings, universal applicability was tested by analyzing the different cell formats.

3.2 Analysis of the Cell Formats

As part of the preparation for the automated handling of BIPV, the structural design of three established cell formats M2, M6 and M12, as they are currently used on the BIPV system, was first systematically analyzed. The aim of this preliminary investigation was to identify commonalities in terms of shape and dimensions that would enable a uniform gripper base in the process. Figure 2 shows the different cell formats in relation to each other and illustrates the non-uniform arrangement of the copper wire spacing.

By superimposing the cell formats, two potential areas at a distance of 90 mm could be determined in Fig. 2, which are suitable for a point of application of the vacuum nozzles on the cells and are decisive for the subsequent design of the gripping system.

4 Development of the Gripper

Based on the findings from 3.1 and 3.2, the specific functional requirements and design conditions for a flexible gripping system that can handle heterogeneous cell strings safely and gently were derived. A pair of vacuum nozzles was designed to pick up the cells, which grip the cell surfaces that had been identified as suitable at a distance of 90 mm. The double contact of the cell ensures improved stability and reduces the risk of tilting during handling. Due to the varying cell formats and differing interconnection distances of the cells, it was necessary to equip the design of the gripping system with horizontal adjustability of the nozzles. This was achieved by mounting the pairs of nozzles on a profile rail. The Fig. 3 visualizes the structural implementation and the spatial arrangement of the components. As part of a future development, it is planned to automate the mechanical adjustment after successful concept validation in order to enable

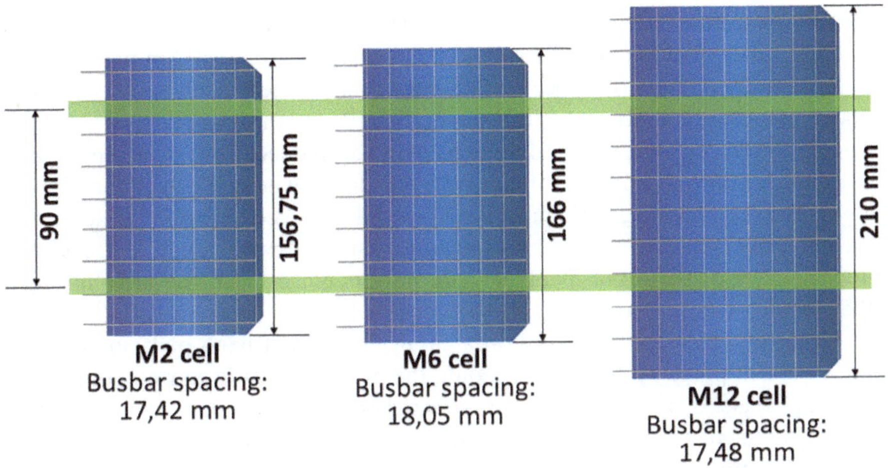

Fig. 2 Illustration of the three analyzed cell formats and the potential gripping surface

process-integrated, adaptive adjustment to different handling situations. In addition, each gripper unit is equipped with a pneumatic valve, enabling selective activation of individual vacuum nozzles. This design allows flexible gripping of string lengths with 1 up to 12 cells without any significant volume flow losses. Assuming a structurally intact cell connection, this concept should ensure a reliably maintainable vacuum level.

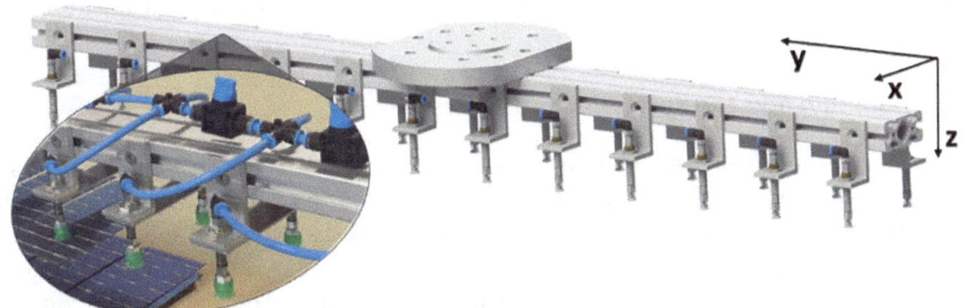

Fig. 3 CAD drawing of the vacuum stringer gripper

4.1 Investigations of Gripping Quality and Repeatability

In order to validate the functionality of the developed vacuum gripper, the safe handling of the different cell formats was examined and trials were carried out on the repeatability of the string placement. To evaluate the gripping quality, strings of all three cell formats were picked up with a vacuum gripper and subjected to a defined robot trajectory. The aim of the tests was to validate the gripping force to be applied while simultaneously observing any cell deflections. For this purpose, a two-stage movement was carried out—a vertical lifting movement followed by a horizontal traverse—whereby the acceleration was set to 1 m/s^2 and the path speed to 1 m/s. The choice of these parameters is based on the cycle times of about 10 s/String and travel distance of about 3 m to be expected in the subsequent production process. The evaluation of the repeatability was carried out under the same conditions. A tolerance of ± 100 μm is targeted for the precise positioning of the cell strings in order to ensure a gap-reduced row arrangement in the module and to achieve optimum efficiency. Even slight twisting or overlapping of the strings leads to cell breakage and must be avoided. For the cell handling process, three key factors influencing the overall positioning repeatability were identified in advance. The repeatability of the robot together with the vacuum gripper, the pick-up process and the placing process. To quantify these effects, a series of tests were carried out with an M6 string in which the robot performs 30 cycles from a fixed pick-up position to a defined deposit point. At the deposit point, the measurement is carried out along the y-axis using a laser displacement sensor mounted on a linear guide (line laser with 800 measuring points, resolution 50 μm, repeatability 10 μm), as shown in Fig. 5. If there is a change in the measured height profile, the edges of the string can be detected and their position determined. The data obtained in this way forms the basis for the first process validation. In total, the following series of tests were carried out.

- Determining the system accuracy: To determine the isolated system accuracy, the pick-up and deposit processes were specifically excluded in the first test. A cell string was moved repetitively (30 repetitions) from the starting point to the deposition position and returned after a measurement without deposition.
- Determining the positioning repeatability: In this series of tests, the existing procedure was extended to include the picking and depositing process. The cell string was gripped cyclically at the pick-up position, transported to the deposit position and then deposited and measured.

4.2 Evaluation and Interpretation of the Results

The results of the tests on gripping quality show that all tested strings can be handled safely, regardless of cell format and length. The deflection of the solar cells observed during the movement remained consistently far below the damage limit and can therefore be classified as negligible. Figure 4a documents an example of a handled M6 string. A subsequent electroluminescence (EL) test shown in Fig. 4b did not reveal any evidence of cell damage caused by gripping contacts.

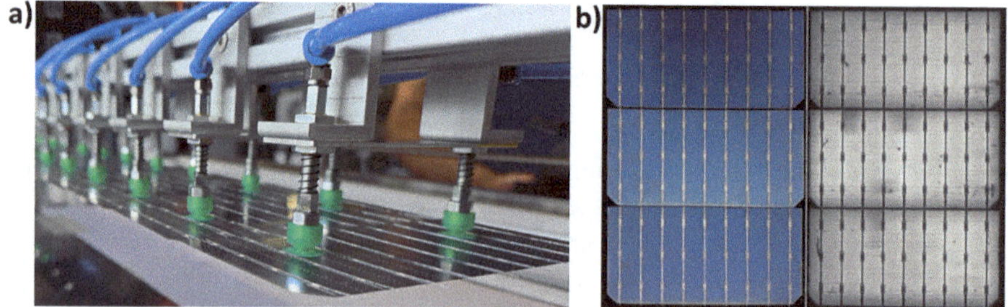

Fig. 4 **a** Gripped M6 string, **b** EL analysis of the M6 string after gripping

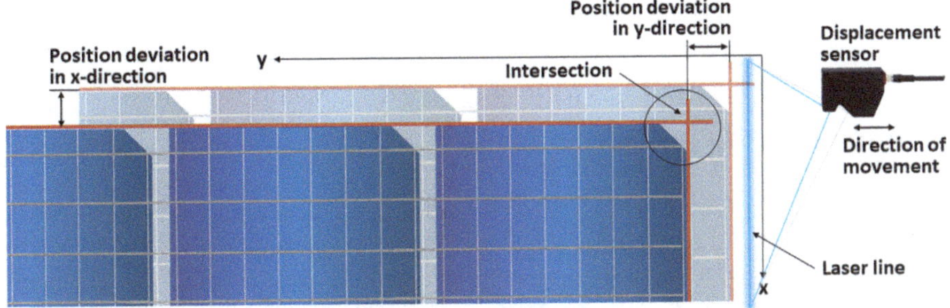

Fig. 5 Measured value evaluation of the relative repeatability

These test results can be applied to other module formats. Initial tests with the M2 and M12 cell formats showed that adjusting the gripping distances to different cell spacing works reliably. All cell formats whose busbar spacing allows a gripping surface between two wires can also be handled.

To analyze the relative positioning repeatability of the cell strings, the evaluation was carried out using MATLAB. Based on the measured values recorded in the x and y directions of the string, an edge detection was carried out and two regression lines were determined. Figure 5 shows an example of an edge detection with associated regression analysis. The intersection of the two lines serves as a normalized reference point for quantitative comparability. The coordinates obtained in this way make it possible to calculate the position deviations in both spatial directions. This procedure is applied uniformly in both test series. The potential influence of a possible twisting of the strings is not yet considered at this stage of the research.

The results for determining the system accuracy showed a deviation in the x-direction of $\pm 91.55\ \mu m$ and in the y-direction of $\pm 70.98\ \mu m$. The values are therefore slightly above the specified repeatability of the robot (60 μm), which is in line with expectations and can be attributed to the flexibility of the handling system. In particular, the deliberately implemented flexibility of the spring plungers in the z-direction, required for damage-free cell pick-up, leads to unavoidable bearing play in the horizontal plane. In addition, the elastic properties of the polysiloxane suction cups must also be taken into account as a possible cause of the measured position deviations.

The analysis of the test series for pick-up and placement repeatability revealed a relative positioning deviation of $\pm 486.53\ \mu m$ in the x-direction and $\pm 77.73\ \mu m$ in the y-direction. These results illustrate a significant deterioration in the repeatability along the x-axis and clearly exceed the target limit value of $\pm 100\ \mu m$, while the repeatability in the y-direction remains constant within the tolerance range. The observed positional deviations can most likely be attributed to process-induced tension in the strings during the pick-up process, which relaxes as a result of the vacuum loss during deposition and leads to uncontrolled springback. This is due to the fact that the wires interconnecting two neighboring cells in the string have a lower stiffness in the x-direction than in the y-direction. This string-dependent behavior is inherently complex and is difficult to eliminate by design measures alone. It is therefore necessary to optimize the process by using additional sensors for position detection and trajectory correction in order to detect such shifts during handling at an early stage and actively compensate for them.

5 Summary and Outlook

A key aspect for the production of BIPV modules is the development of form-flexible handling systems that enable the reliable handling of solar strings with different geometric properties. The modular vacuum gripper presented enables break-free and precise handling of solar strings of varying formats and lengths. Electroluminescence measurements prove that the contact points of the gripper do not impair the subsequent cell functionality. In addition, tests on repeatability show that the system-related deviations are within the permissible tolerance of $\pm 100\ \mu m$, which validates the basic functionality of the concept. Nevertheless, the results of the pick-up and deposit process revealed an insufficient reproducibility of the positioning in the x-direction, and thus indicate a need for optimization. The integration of sensors in the handling unit is proposed as a possible solution. This should enable position detection of the strings picked up and ensure more precise alignment to the deposit position via real-time correction of the robot trajectory. In addition, the integrated sensor can be used to check the quality of the depositing process.

Acknowledgements. The authors gratefully acknowledge the funding by the Bundesministerium für Wirtschaft und Energie (BMWE Federal Ministry for Economic Affairs and Energy)—Project no. 03EE1187D and 03EE1187E. The authors would like to thank the BMWE for the support within the project "Digitale Planung und automatisierte Produktion von gebäudeintegrierter Photovoltaik"

Competing Interests. The author(s) has no competing interests to declare that are relevant to the content of this manuscript.

References

1. Brasseur, G.P., Jacob, D., Schuck-Zöller, S.: Klimawandel in Deutschland. Springer, Heidelberg (2017)
2. Urbansky, F.: Schafft die Welt die Wende hin zu erneuerbaren Energien. Zeitschrift für Energiewirtschaft, Band **47**, 8–13. Springer Professional (2013)
3. Klimaschutzprogramm der Bundesregierung, bundesregierung.de. Last accessed 13 June 2025
4. Bett, A.W., Rentsch, J., Preu, R.: PV-Produktion in Europa—Aktueller Status. Fraunhofer Institut für Solare Energiesysteme (2023)
5. Bredemeier, D., Schinke, C., et al.: Fast evaluation of rooftop and façade PV potentials using backward raytracing and machine learning. In: 48th PVSC, pp. 294–299 (2021)
6. Chandrasekar, M.: Building-integrated solar photovoltaic thermal (BIPVT) technology. In: Advances in Building Energy Research, vol. 17, p. 234. TaylorFrancis Group (2023)
7. Brink, G.: Energiewende 2.0, pp. 94–106. Springer, Wiesbaden (2025)
8. Keiber, M.: Wie Solarzellen Hergestellt Werden. https://www.springerprofessional.de. Last accessed 28 June 2025

9. Blankemeyer, S., Schulte-Huxel, H., et al.: Assembly cell for the manufacturing of flexible solar modules in bipv. In: 56th CIRP Conference, vol. 120, pp. 952–957. Elsevier (2023)
10. Zahoransky, R., Fichter, C.; Energietechnik.10 Auflage, p. 216. Springer, Wiesbaden (2025)
11. Witteck, R., Hinken, D., et al.: Optimized interconnection of passivated emitter and rear cells by experimentally verified modeling. In: IEEE J. PVs **6**(2), S. 432–439 (2016)
12. Fischmann, C.: Verfahren zur Bewertung von Greifern für Photovoltaik-Wafer, In: Stuttgarter Beiträge zur Produktionsforschung Band 26, p. 46. Fraunhofer (2013)

Observer-Based Disturbance Compensation for Force Plates in Dynamic Environments

M. Richter[1(✉)], A. G. de Mendoza[2], K. Güzel[1], and H.-C. Möhring[1]

[1] Institute for Machine Tools (IfW), University of Stuttgart, Stuttgart, Germany
max.richter@ifw.uni-stuttgart.de
[2] Nuton GmbH, Berlin, Germany
mail@nuton.de https://www.nuton.de

Abstract. In machine tools, forces are measured in many different scenarios. In certain cases, such as when probing thin-walled workpieces or monitoring micro-cutting processes, these forces can be less than 1 N. If a force plate is used on a moving machine table, inertial forces are superimposed on the forces to be measured and therefore falsify the measurement. This paper shows a way of compensating for the effects of such disturbances with the aid of a disturbance sensor. First, the dynamics of the force plate and the acceleration sensor are modeled. The model is then extended by the estimated dynamics of the disturbance and used within a signal observer to compensate for the disturbances. The disturbance compensation provides appropriate force measurements, even if the amplitudes of the disturbances occurring are several times larger than the forces to be measured. The procedure described above can be transferred and applied analogously to many different force measurement scenarios.

Keywords: Force measurement · Disturbance · Observer

1 Introduction

Force plates are widely used in research and industry to measure forces by sensing the relative motion between top and base plates with strain gauges on the connecting legs. After calibration—i.e., establishing proportionality between applied forces and changes in gauge resistance—the assembly behaves conceptually like a multidimensional spring system.

This approach is accurate for static or quasi-static loads, but at higher excitation frequencies near structural modes the signal is distorted: an impulse on the top plate produces prolonged ring-down because the plate reports interaction forces between base and top, not the true toolworkpiece force. When the plate is mounted on a moving or vibrating machine table, additional inertial loads from accelerating the top-plate/vice/workpiece stack are superimposed on

L. Overmeyer and B.-A. Behrens (eds.), *Production at the Leading Edge of Technology*, Lecture Notes in Production Engineering, https://doi.org/10.1007/978-3-032-19524-1_15

the signal, and orientation-dependent gravity can bias readings. Together, these effects impede a clean separation of process and structure-induced contributions.

In previous work [1], machine learning was used to calibrate the same strain-gauge force unit, addressing nonlinearities and crosstalk under static conditions. The present study extends to dynamic operation on moving tables and presents an observer-based method that compensates motion-induced disturbances.

2 State of the Art

Recent surveys underline that on-machine force signals are confounded by structural dynamics and noise, complicating the separation of process from non-process effects [2]. On multi-axis machines, orientation-dependent gravity and motion-induced inertial loads further corrupt measurements; Klocke et al. formulated this disturbance taxonomy and subtracted gravity/rotation terms using machine axis signals [3].

A complementary line identifies the machinedynamometer transmissibility and applies (often MIMO) inverse filtering to recover forces-improving accuracy and bandwidth but requiring prior dynamic characterization/regularization and typically assuming a fixed base [4–6]. Model-embedded estimators have also been proposed directly on dynamometer outputs (regularized deconvolution and sliding-mode observers) for stationary platforms [7], with related two-stage observer ideas in tool-force sensing [8]. Sensorless estimators with machine-in-the-loop pre-compensation have been reported [9], and output-only, in-process dynamics identification tracks mode shifts for stability rather than force reconstruction [10]. Closest to the present disturbance-sensing approach, Totis et al. augment a piezo dynamometer with base accelerometers and reconstruct forces via a universal inverse filter learned from base excitations [11].

Prior schemes either (i) subtract encoder/kinematics terms without modeling sensor dynamics, (ii) invert identified transmissibilities requiring re-identification under boundary changes, or (iii) embed observers but assume a fixed base. In contrast, the proposed method fuses *encoder-derived acceleration* with a *base-mounted strain-gauge acceleration sensor* inside a *two-stage state-space observer* that explicitly estimates base-motion states *and* compensates force-plate dynamics, enabling *sub-Newton* force reconstruction on moving tables without an exogenous FRF training phase.

3 Modeling and Parameter Identification

Before an observer can be implemented, a physical model of both the top plate and the acceleration sensor must be developed. The force plate used is an experimental, strain-gauge-equipped plate that has been developed in a collaboration between the IfW and the Nuton GmbH. A detailed description of the mechanical construction of the force plate, including photographs and schematics, is provided in [12]. In brief, the strain-gauge-equipped sensor legs connect the top and base plates, while the acceleration sensor is mounted on the base plate with its bronze mass supported by a single sensor leg. This arrangement isolates the sensor mass from process forces while making it sensitive to inertial disturbances.

In the context of this work, the dominant modeling errors originate from the first two natural modes of the force plate, which correspond to in-plane translational vibrations of the top plate in the X and Y directions. These modes are modeled as second-order mass - spring - damper systems, resulting in the PT2 (second-order low-pass) transfer behavior

$$\begin{aligned} m_{\text{top}}\, \ddot{x}_{\text{top}}(t) + d_{\text{top}}\, \dot{\Delta x}_{\text{top}}(t) + k_{\text{top}}\, \Delta x_{\text{top}}(t) &= F_{\text{proc}}(t), \\ m_{\text{sen}}\, \ddot{x}_{\text{sen}}(t) + d_{\text{sen}}\, \dot{\Delta x}_{\text{sen}}(t) + k_{\text{sen}}\, \Delta x_{\text{sen}}(t) &= 0, \end{aligned} \tag{1}$$

where

$$\begin{aligned} \Delta x_{\text{top}}(t) &= x_{\text{top}}(t) - x_{\text{base}}(t) \\ \Delta x_{\text{sen}}(t) &= x_{\text{sen}}(t) - x_{\text{base}}(t). \end{aligned} \tag{2}$$

Here, $x_{\text{top}}(t)$, $x_{\text{sen}}(t)$ and $x_{\text{base}}(t)$ are absolute positions; $F_{\text{proc}}(t)$ is the external process force. The inertial terms use absolute acceleration, whereas damping/stiffness terms use displacements relative to the base. The goal is to reconstruct $F_{\text{proc}}(t)$ from the available $\Delta x_{\text{top}}(t)$ and $\Delta x_{\text{sen}}(t)$. Assuming geometric symmetry (identical moving masses, stiffnesses and damping in X/Y), the parameters are summarized in Table 1.

Table 1. Physical parameters used in the dynamic model and their identification methods

Parameter	Description	Value	Unit	Identification method
m_{top}	Mass of top plate	36.13	kg	Weighed directly
k_{top}	Plate stiffness (X/Y)	670×10^6	N/m	Force/displacement test
d_{top}	Damping coefficient	3.35×10^3	N s/m	Logarithmic decrement
m_{sen}	Mass of sensor element	1.602	kg	Weighed directly
k_{sen}	Sensor stiffness	55.8×10^6	N/m	Force/displacement test
d_{sen}	Sensor damping	279	N s/m	Logarithmic decrement

4 Observer Structure

The dynamic model of the system has been established. This section describes how it is embedded into a layered observer structure to estimate and compensate disturbances acting on the force plate. These disturbances primarily result from movements of the machine table, which include both controlled axis motion and additional vibrations. Since only part of the table motion is directly observable via the machine's linear encoders, a combination of signal processing and physical modeling is used.

Acceleration Estimation from Position Encoders

The position of the machine table in X-direction is measured in real time by passively splitting the $1\,\mathrm{V}_{pp}$ signal from the linear encoder to the machine control. To estimate the second derivative $\ddot{x}_{\mathrm{enc}}(t)$ of the position signal $x_{\mathrm{enc}}(t)$, a third-order low-pass differentiator is used. Its transfer function

$$G(s) = \frac{\gamma^3\, s^2}{(s+\gamma)^3} \tag{3}$$

provides an approximation of $\ddot{x}_{\mathrm{enc}}(t)$. The parameter γ (in rad/s) determines the location of all three poles and thus governs the trade-off between responsiveness and noise suppression. Ideally, γ is set higher than the dominant natural frequency of the structure to ensure that the relevant dynamic content is preserved in the acceleration estimate. In practice, the optimal value also depends on the measurement noise spectrum: increasing γ improves responsiveness up to a point, beyond which noise amplification offsets any gain in dynamic accuracy.

Disturbance Observer for Table Vibrations

The actual acceleration of the sensor is

$$\ddot{x}_{\mathrm{base}}(t) = \ddot{x}_{\mathrm{enc}}(t) + a_{\mathrm{corr}}(t), \tag{4}$$

where $a_{\mathrm{corr}}(t)$ is a correction term accounting for structural dynamics between the encoder and the base plate, that cannot be measured by the encoder. This correction typically comes with small amplitudes but can significantly influence force measurements during dynamic excitation. The disturbance itself is modeled as a second-order system representing the dominant mode of the table:

$$\ddot{a}_{\mathrm{corr}}(t) + \omega_{\mathrm{corr}}^2\, a_{\mathrm{corr}}(t) = 0. \tag{5}$$

Defining the state vector

$$\boldsymbol{z}_{\mathrm{sen}}(t) = \left[\Delta x_{\mathrm{sen}}(t)\ \dot{\Delta x}_{\mathrm{sen}}(t)\ a_{\mathrm{corr}}(t)\ \dot{a}_{\mathrm{corr}}(t)\right]^{\mathrm{T}} \tag{6}$$

leads to the state-space system

$$\dot{\boldsymbol{z}}_{\mathrm{sen}}(t) = \boldsymbol{A}\boldsymbol{z}_{\mathrm{sen}}(t) + \boldsymbol{b}\,\ddot{x}_{\mathrm{enc}}(t), \quad \Delta x_{\mathrm{sen}}(t) = \boldsymbol{c}\boldsymbol{z}_{\mathrm{sen}}(t) \tag{7}$$

with matrices:

$$\boldsymbol{A} = \begin{bmatrix} 0 & 1 & 0 & 0 \\ -\frac{k_{\mathrm{sen}}}{m_{\mathrm{sen}}} & -\frac{d_{\mathrm{sen}}}{m_{\mathrm{sen}}} & -1 & 0 \\ 0 & 0 & 0 & 1 \\ 0 & 0 & -\omega_{\mathrm{corr}}^2 & 0 \end{bmatrix}, \quad \boldsymbol{b} = \begin{bmatrix} 0 \\ -1 \\ 0 \\ 0 \end{bmatrix}, \quad \boldsymbol{c}^{\mathrm{T}} = \begin{bmatrix} 1 \\ 0 \\ 0 \\ 0 \end{bmatrix}. \tag{8}$$

The state observer is

$$\dot{\hat{\boldsymbol{z}}}_{\mathrm{sen}}(t) = \boldsymbol{A}\hat{\boldsymbol{z}}_{\mathrm{sen}}(t) + \boldsymbol{b}\,\ddot{x}_{\mathrm{enc}}(t) + L\big(\Delta x_{\mathrm{sen}}(t) - \boldsymbol{c}\hat{\boldsymbol{z}}_{\mathrm{sen}}(t)\big). \tag{9}$$

designed via pole placement. Placing the poles further into the left open half-plane improves the observer's tracking behavior, but also results in a higher amplification of measuring noise. Estimated variables are marked with a hat.

The estimated correction $\hat{a}_{\text{corr}}(t)$ is taken from the third component of $\hat{\boldsymbol{z}}_{\text{sen}}(t)$ and, together with $\ddot{x}_{\text{enc}}(t)$, forms the estimated base acceleration $\hat{\ddot{x}}_{\text{base}}(t) = \ddot{x}_{\text{enc}}(t) + \hat{a}_{\text{corr}}(t)$ used in the final compensation stage.

Process Force Observer for the Top Plate

A second observer is implemented to estimate the unknown process force $F_{\text{proc}}(t)$ acting on the top plate. The model receives three inputs: the axis acceleration $\ddot{x}_{\text{enc}}(t)$, the previously estimated base disturbance $\hat{\ddot{x}}_{\text{base}}(t)$, and the measured relative displacement $\Delta x_{\text{top}}(t)$ between top plate and machine base.

The system state vector is defined as

$$\boldsymbol{z}_{\text{top}}(t) = \left[\Delta x_{\text{top}}(t)\ \Delta\dot{x}_{\text{top}}(t)\ F_{\text{proc}}(t)\ \dot{F}_{\text{proc}}(t)\right]^{\text{T}} \tag{10}$$

and the dynamics are given by:

$$\dot{\boldsymbol{z}}_{\text{top}}(t) = \boldsymbol{A}\,\boldsymbol{z}_{\text{top}}(t) + \boldsymbol{b}\,\ddot{x}_{\text{base}}(t), \quad \Delta x_{\text{top}}(t) = \boldsymbol{c}\boldsymbol{z}_{\text{top}}(t) \tag{11}$$

with matrices

$$\boldsymbol{A} = \begin{bmatrix} 0 & 1 & 0 & 0 \\ -\frac{k_{\text{top}}}{m_{\text{top}}} & -\frac{d_{\text{top}}}{m_{\text{top}}} & \frac{1}{m_{\text{top}}} & 0 \\ 0 & 0 & 0 & 1 \\ 0 & 0 & 0 & 0 \end{bmatrix}, \quad \boldsymbol{b} = \begin{bmatrix} 0 \\ -1 \\ 0 \\ 0 \end{bmatrix} \quad \text{and} \quad \boldsymbol{c}^{\text{T}} = \begin{bmatrix} 1 \\ 0 \\ 0 \\ 0 \end{bmatrix}. \tag{12}$$

The observer is again designed by pole placement. The estimated process force $\hat{F}_{\text{proc}}(t)$ is extracted from the third state component. This final stage allows the system to distinguish between inertial effects and actual process forces acting on the top plate.

5 Validation

Once the observer has been designed, its effectiveness must be validated through experiments. The observer has three primary objectives:

1. Ensuring that actual process forces acting on the top plate are faithfully represented in the output signal.
2. The suppression of inertial forces resulting from table movements and vibrations.
3. Compensating for the dynamics of force plate and acceleration sensor.

Experimental validation must be structured to evaluate these three objectives and to provide a known reference for the actual process forces. Two experiments carried out on a Hermle UWF 1202H are described below.

The first experiment involves striking the machine table with an impulse hammer to generate a broadband vibrational disturbance. This excitation produces inertial forces that the observer should suppress. Figure 1 compares the estimated force signals with and without observer-based compensation. Without the observer, the signal includes oscillations exceeding 300 N due to inertial effects. With the observer, the residual signal remains below 1 N. The small remaining error is likely due to modeling inaccuracies. This experiment confirms that the observer effectively suppresses table-induced inertial forces.

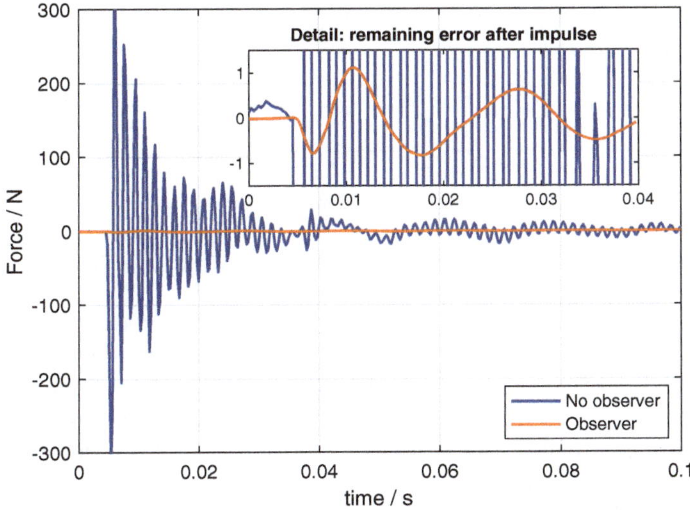

Fig. 1. Impulse hammer test: inertial peaks reduced from 300 to < 1 N with observer

The second experiment demonstrates that the observer correctly preserves genuine process forces, rather than merely attenuating all dynamics like a low-pass filter. A titanium rod with known stiffness is mounted in the spindle. The machine table moves the force plate against the rod tip and back, generating an elastic deformation force of approximately 1 N. Figure 2 shows the comparison, analogous to the previous setup. As expected, the observer successfully reconstructs the step in force, while suppressing the inertial contributions.

6 Conclusion

This paper presented an observer-based approach for compensating inertial disturbances in force plates mounted on moving machine tables. By modeling the dynamics of the force plate and integrating measurements from a strain-gauge-based acceleration sensor, the method enables accurate reconstruction of process forces, even when these are significantly smaller than the superimposed inertial

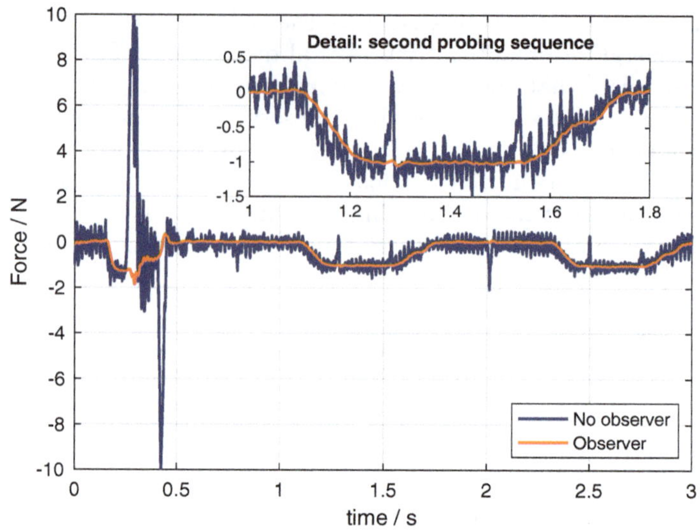

Fig. 2. Probing a titanium rod (1 N): step recovered; inertia suppressed

forces. Validation experiments confirmed that the observer effectively suppresses vibrational disturbances and accurately tracks small process forces.

While the presented method was developed for a specific force plate setup, the underlying approach is broadly applicable to other measurement systems affected by structural dynamics and inertial disturbances. Sensor configurations that can be modeled as coupled massspringdamper systems—such as torque sensors, load cells in dynamic environments, or multi-axis dynamometers—can benefit from the integration of physical modeling and observer-based disturbance compensation.

The present work focuses on the dynamic modeling and observer design; a detailed quantification of the resulting measurement uncertainty is beyond the scope of this paper. Since uncertainty evaluation is essential when assessing overall measurement accuracy, this aspect will be addressed in a forthcoming publication, which will also consider how disturbance compensation interacts with sensor calibration and error propagation.

Several limitations of the method must be acknowledged. First, the observer relies on a physical model whose accuracy depends on known system parameters. Changes in the mass of the top plate assembly—e.g., due to different workpieces or material removal—can lead to estimation errors. Second, only the dominant translational modes of the system are modeled; if higher modes are excited, the observer may fail to compensate for their effects. While additional modes could be modeled, this would increase complexity and sensitivity to noise. Third, the assumption that the process force varies smoothly over time, while useful for robustness, introduces modeling error when abrupt or non-smooth force profiles occur.

Nonetheless, the method offers a promising basis for more advanced applications. Future work could extend the model to multiple degrees of freedom and explore adaptive or gain-scheduled observers that cope with parameter changes. Furthermore, since the dynamic behavior of the acceleration sensor and the observer jointly determine signal quality and usable bandwidth, an optimization framework could co-tune physical sensor parameters and observer gains to maximize overall system performance.

Competing Interests. The author(s) has no competing interests to declare that are relevant to the content of this manuscript.

References

1. Richter, M., et al.: Calibration of a strain gauge-equipped force measuring unit using machine learning algorithms. Proc. CIRP **118**, 181–186 (2023). https://doi.org/10.1016/j.procir.2023.06.032
2. Shokrani, A., et al.: Sensors for in-process and on-machine monitoring of machining operations. CIRP J. Manuf. Sci. Technol. **51**, 263–292 (2024). https://doi.org/10.1016/j.cirpj.2024.05.001
3. Klocke, F., Blattner, M., Adams, O., Brockmann, M., Veselovac, D.: Compensation of disturbances on force signals for five-axis milling processes. Proc. CIRP **14**, 472–477 (2014). https://doi.org/10.1016/j.procir.2014.03.098
4. Castro, L.R., Viéville, P., Lipinski, P.: Correction of dynamic effects on force measurements made with piezoelectric dynamometers. Int. J. Mach. Tools Manuf **46**(14), 1707–1715 (2006). https://doi.org/10.1016/j.ijmachtools.2005.12.006
5. Girardin, F., Rémond, D., Rigal, J.-F.: High frequency correction of dynamometer for cutting force observation in milling. J. Manufact. Sci. Eng. **132**(3), 031002 (2010). https://doi.org/10.1115/1.4001538
6. Korkmaz, E., Gozen, B.A., Bediz, B., Ozdoganlar, O.B.: Accurate measurement of micromachining forces through dynamic compensation of dynamometers. Precis. Eng. **49**, 365–376 (2017). https://doi.org/10.1016/j.precisioneng.2017.03.006
7. Jullien-Corrigan, A., Ahmadi, K.: Measurement of high-frequency milling forces using piezoelectric dynamometers with dynamic compensation. Precis. Eng. **66**, 1–9 (2020). https://doi.org/10.1016/j.precisioneng.2020.07.001
8. Knechtelsdorfer, U., Saxinger, M., Schwegel, M., Steinboeck, A., Kugi, A.: A two-stage observer for the compensation of actuator-induced disturbances in tool-force sensors. Mech. Syst. Sig. Process. **146**, 106989 (2021). https://doi.org/10.1016/j.ymssp.2020.106989
9. Yamato, S., Kakinuma, Y.: Precompensation of machine dynamics for cutting force estimation based on disturbance observer. CIRP Ann. **69**(1), 333–336 (2020). https://doi.org/10.1016/j.cirp.2020.04.068
10. Liu, Y.-P., Altintas, Y.: In-process identification of machine tool dynamics. CIRP J. Manuf. Sci. Technol. **32**, 322–337 (2021). https://doi.org/10.1016/j.cirpj.2021.01.007

11. Totis, G., Bortoluzzi, D., Sortino, M.: Development of a universal, machine tool independent dynamometer for accurate cutting force estimation in milling. Int. J. Mach. Tools Manuf **198**, 104151 (2024). https://doi.org/10.1016/j.ijmachtools.2024.104151
12. Palalic, M., et al.: Multiaxial force platform with disturbance compensation for machine tools. J. Mach. Eng. **20**(3), 5–16 (2020). https://doi.org/10.36897/jme/127105

Numerical Investigation of Bending-Critical Eigenmodes and Stable Operating Conditions in the Utilization of Slim Tool Extensions—The Influence of Resonance and Nutation Phenomena

Thomas Pache[1(✉)], Eckart Uhlmann[1,2], Mitchel Polte[1,2], and Nils Bergström[1]

[1] Institute for Machine Tools and Factory Management, Technische Universität Berlin, Berlin, Germany
pache@iwf.tu-berlin.de

[2] Fraunhofer Institute for Production Systems and Design Technology IPK, Berlin, Germany

Abstract. State-of-the-art machining of complex integral workpieces requires the utilization of slim tool extensions in machine tools. Previous investigations indicate that resonant excitation of slim tool extensions at their bending-critical eigenfrequency can result in complex plastic deformation and subsequent failure. Significantly increased eccentricity of slim tool extensions' mass causes an abrupt increase in the accumulated kinetic energy of potential fragments released in the event of a resonance catastrophe. This exceeds the retention capacity of standard safety guards by orders of magnitude. A novel approach for resolving this issue is to induce defined failure in slim tool extensions by means of design measures. Therefore, preliminary knowledge of the excitation conditions is essential. This paper presents findings from numerical analyses with specific focus on the interrelationship between resonance and the rotodynamic phenomenon of synchronous and asynchronous nutation. Modal and frequency response analyses on finite element models are conducted to determine bending-critical eigenfrequencies of slim tool extensions and tool holders along with their respective eigenmodes. The insights gained throughout these analyses facilitate a more profound comprehension of slim tool extensions' behavior in the range of bending-critical eigenmodes and the associated limits of safe operation.The results obtained provide substantial evidence that, beyond the mere phenomenon of resonance, failure of slim tool extensions due to bending-critical excitation is caused by a complex resulting state of stress that cannot be attributed to mere bending. Furthermore, exemplarily conducted transient run-up simulations could demonstrate that a distinct relationship between invoked resonance and stress states can be reproduced in a non-linear time domain.

Keywords: Safety · Machine tools · Slim tool extensions · Rotational dynamics · Failure

L. Overmeyer and B.-A. Behrens (eds.), *Production at the Leading Edge of Technology*, Lecture Notes in Production Engineering, https://doi.org/10.1007/978-3-032-19524-1_16

Abbreviations

BCE	Bending critical eigenfrequency
BW	Backward whirl
FW	Forward whirl
STE/extension	Slim tool extension

1 Introduction

In the field of machining, utilization of slim tool extensions, hereafter referred to as extensions, has gained significant prevalence in the fabrication of integral components characterized by large cavities [1, 8]. A broad variety of extensions is currently available and in use on the global market [10]. Concurrently, the utilization of such elongated, slim components in machine tools entails the potential for substantial deflections at the tool center point and intense vibrations. This can lead to diminished surface quality of the workpiece and potential risks to machine safety [5, 9, 11].

From a rotor dynamics perspective, extensions and tool holder can be modeled as cantilevered, vibration-capable shaft that undergoes rotational movement around its longitudinal Z-axis forced by the machine tool spindle's torque T, see Fig. 1 [6].

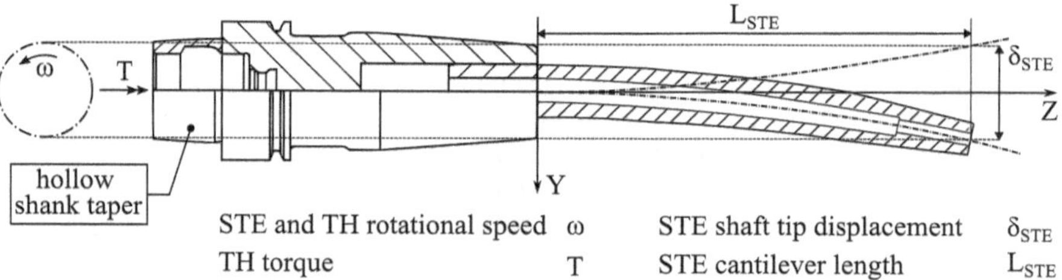

Fig. 1 Schematic representation of a rotating slim tool extension (STE), including shrink chuck tool holder (TH), with STE's cantilever length L_{STE} and STE shaft tip displacement δ_{STE} in the first non-rigid body mode

The manner in which rotation is induced, represents stability-critical excitation type, even in instances of otherwise load-free operation of extensions. Notably, this occurs in situations wherein external excitation corresponds to the eigenfrequencies f_i of extensions, thereby inducing resonance [4]. In the mechanical schematic depicted in Fig. 1, the shape of the first non-rigid body mode typically corresponds to a knot-free bending with respect to its clamping, i.e. the hollow shank taper (HSK). The eigenfrequencies f_1 and f_2 corresponding to these rotationally symmetric eigenmodes are therefore also referred to as bending-critical eigenfrequency (BCE) f_b, and the corresponding extension shaft tip displacement is denoted by δ_{STE}.

As previous research has repeatedly demonstrated, the utilization of extensions in machine tool-based machining processes is associated with significant safety concerns. The machine tool's speed control enforces constant rotational speed ω of the extension.

In conjunction with the invoked tilting movements of the extension induced at resonance and the resulting variation in its polar moment of inertia $J_{p,STE}$, this issues a periodically alternating torque Uhlmann et al. [10] have found that in the event of a resonance catastrophe and the resulting component failure, high kinetic energies E_{kin} are accumulated by the extension and potential fragments. These exceed impact resistance Y of standard machine tool safety guards by up to several orders of magnitude. In those previous considerations, a limited parameter space was assumed, which took into account extensions with a slenderness ratio $g_S := L_{STE}/d_{STE} \geq 6$. In addition, the tool holder itself was not subjected to those investigations. Therefore, resulting limitations lead to the assumption that, within the selected model boundaries, gyroscopic effects could be neglected[1] when determining the BCE f_b [2, 4].

Furthermore, secondary effects resulting from the aforementioned speed control were not taken into account. These include, for example, time-varying behavior of acting load torque L and torque T, what in turn can constitute an indirect external excitation. When torque T is dynamically modulated by the speed controller at a frequency $f_T \approx f_b$, resonant excitation may occur.

The findings presented in this paper demonstrate that the assumption of negligible gyroscopic effects is no longer valid when taking into account tool holders and extensions exhibiting a slenderness ratio of $g_S < 6$. In particular, numerical investigations indicate that gyroscopic effects cause the BCE f_b to depend on the rotational speed ω. Conversely, the influence of gyroscopic effects provides a potential explanation for the empirically observed damage patterns on extensions in the context of resonant failure, depicted in Fig. 2.

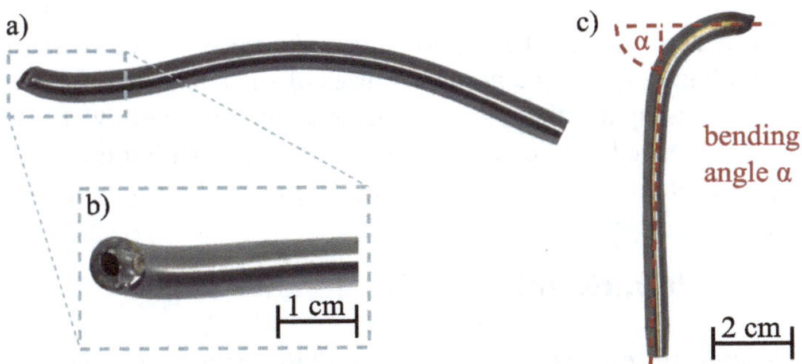

Fig. 2 Experimentally observed damage patterns on slim tool extensions (STE) in case of resonance catastrophe; **a**) severe bending and torsional deformation; **b**) constriction at the fracture surface indicating ductile failure resulting from additional normal stress loading; **c**) bending angle of $\alpha \approx 90°$ prior to resonant failure of the STE

These observations defy explanation in terms of a scalar and constant fracture strain A and mere bending deformation, as evidenced by the extreme bending angle $\alpha \approx 90°$, multiple torsional deformation, and the constriction on the fracture surface characteristic of normal stress-induced failure. It can be hypothesized that these occasionally extreme

[1] The ratio of polar and axial mass moments of inertia satisfies the requirement $J_p/J_a \ll 1$.

deformations are caused by complex multi-axis stress states. Conventional assumptions that bending-critical excitation of extensions necessarily results in pure bending deformation are not in alignment with empirical observations and the results presented in this paper.

The objective of the conducted numerical investigations is twofold: first, to examine the theoretically anticipated bifurcation of the BCE f_b into a backward whirl (BW) and forward whirl (FW) branch depending on rotational speed ω; and second, to quantify the extent to which secondary excitation due to various load types with the determined eigenfrequencies f_i, particularly the BCE f_b, may also induce resonance. These findings, especially the quantified determination of excitation magnitude V, establish a foundation for subsequent investigations of design measures aimed at limiting the deformation of extensions and thereby ensuring their operation within safe limits.

In standstill, the extension's constant BCE f_b corresponds to that of a cantilevered beam clamped at one end. When subjected to forced rotation, the structure's mass eccentricity and induced tilting moments result in the nutation of the extension's longitudinal axis with respect to the axis of rotation. Friswell et al. [3] found that these compensatory movements cause oscillating increases and decreases in the stiffness of the system and, consequently, in the BCE f_b, depending on rotational speed ω.

External excitation of rotating extensions is primarily composed of singular or harmonic shocks, arising from geometric or material-related imperfections, seismic excitation, or a time-varying rotational acceleration η, resulting from variable load torque L(t) and chatter. The more critical case of BW nutation, i. e. the excitation of the BCE $f_{b,bw}$, can only be caused by a spatially aligned, harmonically variable disturbance force F_D ~ $\cos(2\pi f_b t)$ under the valid assumption of rigid and isotropic clamping for the HSK under consideration [4, 7]. Deviations in the concentricity of the extension, microscopic material anisotropy, and asymmetries, particularly those associated with the used tool, in conjunction with gravity, are the primary causes of such disturbance forces F_D. Consequently, it can be anticipated that the dynamic behavior and stability of extensions will be considerably influenced by the degressive BCE $f_{b,bw}$. This, in turn, also determines the extension's-specific limit of safe operation in high-speed machining processes.

2 Modelling Setup, Results and Discussion

In order to determine the effects of resonant excitation, various parameter combinations of the structure shown in Fig. 1, consisting of extension and tool holder, were subjected to structural dynamic analyses using the finite element software LS-Dyna from LIVERMORE SOFTWARE TECHNOLOGY CORPORATION, Livermore, USA. In the following sections, the methodology and technical specifics of these analyses are explained. The parameter combinations that were investigated are presented in Table 1. Notably, the tool holder type employed in each sample is not a free parameter, as it is required to be selected based on the extension's outer diameter d_{STE}. The motor spindle, as source of torque T, was modeled based on the MFW 1230/42/4 type from FISCHER AG, Herzogenbuchsee, Switzerland, in the form of variable boundary conditions at the tool holder's HSK-interface. For the sake of comparability, material properties of extension and tool holder were assumed according to hot-work steel X38CrMoV5-1 as proposed by Uhlmann et al. [10].

Table 1 Parameter combinations of slim tool extension (STE) and tool holder (TH) investigated in modal and frequency response analyses; TH types: I—shrinking chuck, II—collet chuck; backward whirl (BW) bending critical eigenfrequencies (BCE) $f_{b,bw}$, according to results presented in Sect. 2.1, are provided

Parameter	Symbol	Sample #				
		1	2	3	4	5
STE outer diameter	d_{STE}	12 mm	16 mm	20 mm	16 mm	20 mm
STE cantilever length	L_{STE}	140 mm	140 mm	140 mm	280 mm	280 mm
STE inner bore diameter	d_B	4 mm	6 mm	8 mm	6 mm	8 mm
TH type	–	I	I	II	I	II
BW BCE	$f_{b,bw}$	244 Hz	337 Hz	377 Hz	93 Hz	110 Hz

2.1 Modal Analysis

To obtain the extensions' specific BCE f_b, the model was initially subjected to a numerical modal analysis. The speed dependence described in Chap. 1 was duly considered in this context. The sole boundary condition that was imposed was a rigid clamping at the HSK, see Fig. 1, whereby the rotation around the Z-axis was unrestricted. In Fig. 3 results are presented by means of a CAMPBELL diagram.

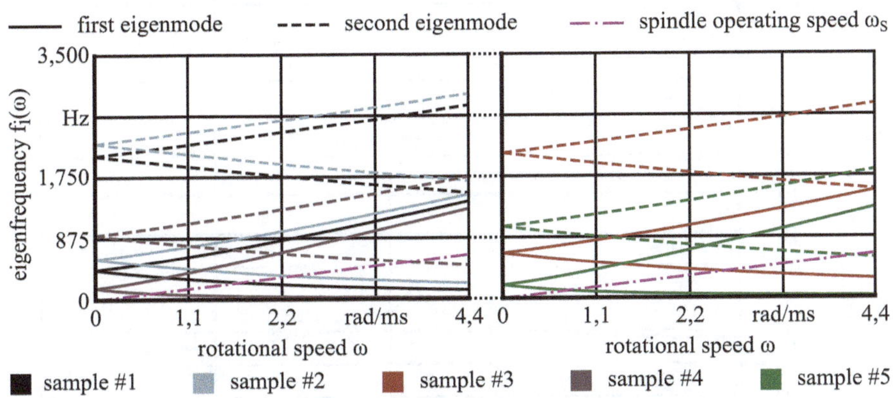

Fig. 3 CAMPBELL diagrams of determined first and second eigenmodes and their respective ascending forward whirl and descending backward whirl branches

The BCE of FW $f_{b,fw}$ and BW $f_{b,bw}$ of the first and second eigenmodes were obtained in the range of rotational speed $0 \leq \omega \leq 4{,}400$ rad/s, which corresponds to range of the modelled spindle type. The bifurcation of the first and second eigenmode into a FW and BW branch with increasing rotational speed ω is distinctly evident. The intersections with spindle operating speed mark the respective BCE $f_{b,bw}$, see Table 1.

2.2 Frequency Response Analysis

Based on the modal analysis' results, a frequency response analysis was then performed on model samples 1, 2 and 4 using torque T, rotational speed ω and base excitation by force F_D as excitation type. With the exception of the latter all excitation types acted on the tool holder in Z-rotation. The spectrum in which BCE f_b occurred, as determined by modal analyses, was utilized for frequency variation up to a maximum excitation frequency $f_{e,max} \leq \omega_{max}/(2\pi)$. Nodes at which system response was evaluated were defined at the tip of each extension sample under consideration according to Table 1. The results of frequency response analyses are displayed in Fig. 4. This analytical step entailed the quantification of excitation magnitude V of excited eigenmodes that were determined in modal analyses.

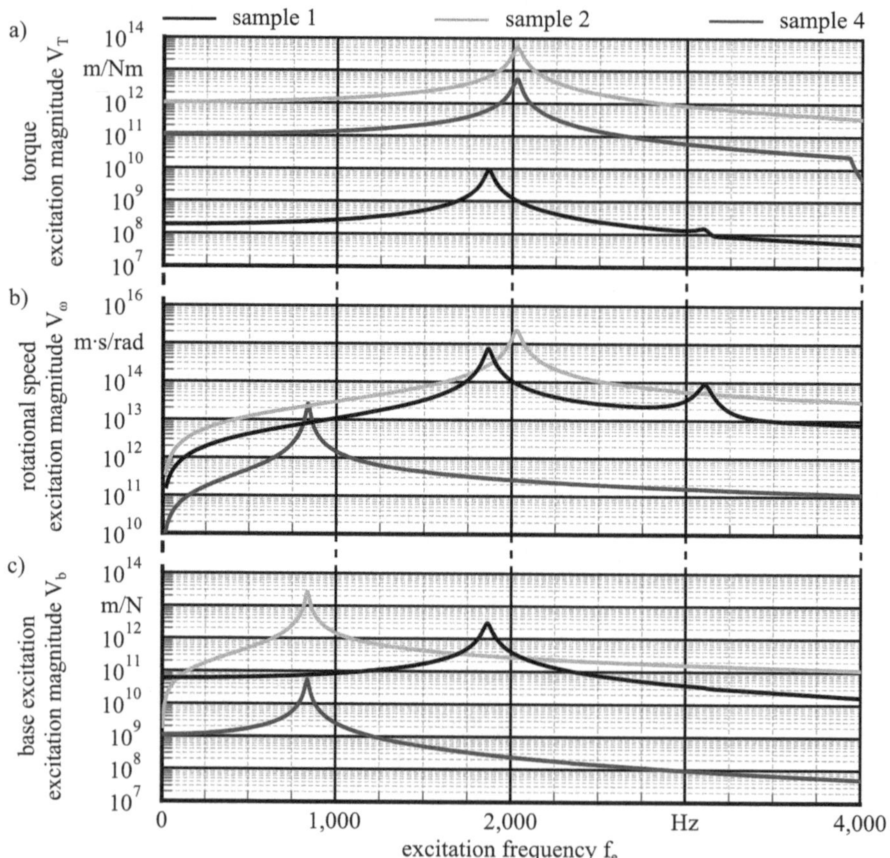

Fig. 4 Excitation magnitudes V at the tip of slim tool extensions (STE); **a**) torque T with respect to Z-axis; **b**) rotational speed ω with respect to Z-axis; **c**) base excitation by force F_D in direction of gravity (Y-axis)

The amplitude frequency responses of the samples under consideration exhibit distinct excitation magnitudes V at the respective second eigenmode, i.e. one knot bending with respect to the extension's clamping in the tool holder. It is evident that the BCE $f_{b,bw}$ is not excited by any of the three excitation types. This finding indicates that alternative excitation types are likely to be the predominant cause of resonant failure during the operation of extensions in motor spindles. While resonant failure may be indirectly attributable to the rotational movement, no direct relationship was identified based on the obtained results.

2.3 Transient Run-Up Simulation

In a transient, explicit simulation, the run-up of extension and tool holder was simulated for sample 1, see Table 1. The objective of this study was first to simulate the pass-through of the BCE $f_{b,bw}$ under speed control. Secondly, the objective was to verify the findings obtained in the frequency domain in the time domain. The speed control used in this model was implemented in LS-Dyna, employing a virtual PID architecture analogous to the modeled motor spindle. The disturbance force F_D, as described in Chap. 1, was introduced through a geometric imbalance U = 80 gmm at the extension's tip in conjunction with gravity in the Y-direction. Figure 5 illustrates the simulated deformation state at the time of resonant failure at rotational speed $\omega_{res} \approx 2{,}160$ rad/s $\widehat{=}$ 344 Hz.

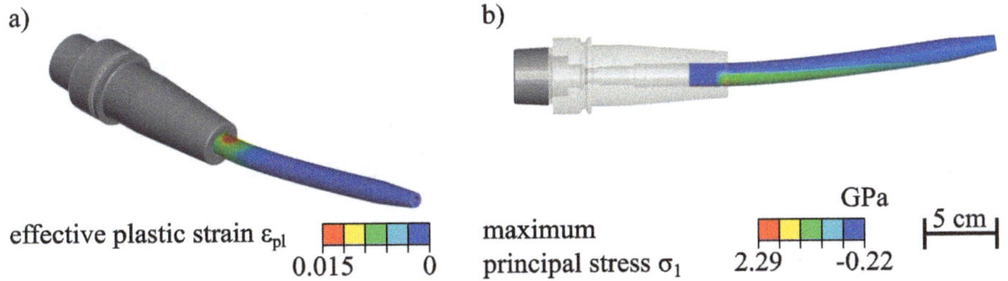

Fig. 5 Simulated resonant failure in run-up simulation; **a**) effective plastic strain ε_{pl} at initiation of resonance failure; **b**) corresponding maximum principal stress σ_1

Hence, significant plastic deformation occurred only at approximately 41% higher rotational speed ω_{res} than expected from obtained BCE $f_{b,bw}$. However, rotational speed at resonant failure ω_{res} is also about 21% lower than BCE $f_{b,0}$ at standstill. The stress state at the extension's clamping in the tool holder at the moment of resonant failure indicates a bending load superimposed by normal load.

3 Conclusion

In this paper, finite element simulations in the frequency and time domain were conducted to investigate on rotor dynamic behavior of slim tool extensions and tool holders with regard to bending-critical excitation. The modal analyses yielded anticipated bifurcation of the BCE f_b into an FW and BW branch, as predicted by theory. The subsequent frequency response analyses were performed to investigate the distinct load type that causes the excitation of slim tool extensions and tool holders with BCE f_b. The investigation revealed that none of the excitation types examined were capable of inducing resonant behavior at BCE f_b. Given the absence of damping implementation in model, the possibility of supercritical damping ratios is not a plausible explanation. In the transient simulation, the objective was to verify the failure of slim tool extensions under resonant excitation with BCE f_b in the extended parameter range. Evidence of this had been demonstrated in experimental tests conducted in previous studies [5, 10]. However, the exemplary run-up simulations demonstrated that the onset of plastic deformation occurs at rotational speeds within the range of $f_{b,bw} < \omega / (2\pi) < f_{b,0}$, i. e. between BW and standstill. Here, as well, any influence from system damping can be disregarded. Given the fact that explicit nonlinear effects were also included in the transient analysis, such as plasticity and dynamic control of torque T, where modal and frequency response analyses consider only a linearized representation of the system, the observed deviations may also be attributable to this aspect. However, the stress distribution at the slim tool extension's bending point, see Fig. 5, supports the hypothesis that bending-critical excitation can also lead to stress states not associated with mere bending. This finding provides a possible explanation for the observed damage patterns depicted in Fig. 2.

Future research, particularly in comparison with experimental validation throughout run-up tests and impulse hammer modal analyses, should elucidate the origins of discrepancies between the different structural dynamic investigations and identify which results correspond to actual physical behavior. This is mandatory for the prospective design of a predetermined breaking point to limit slim tool extensions' deformation.

Acknowledgements. This research was funded by the Federal Ministry for Economic Affairs and Climate Action (BMWK) based on a resolution of the German Bundestag.

Competing Interests. The author(s) has no competing interests to declare that are relevant to the content of this manuscript.

References

1. Dispan, J.: Entwicklungstrends im Werkzeugmaschinenbau. In: Working Paper Forschungsförderung. Hans-Böckler-Stiftung, Düsseldorf (2017)
2. Dresig, H.; Holzweißig, F.: Maschinendynamik. Springer Berlin Heidelberg, Berlin, Heidelberg (2016)
3. Friswell, M.I., Penny, J.E.T., Garvey, S.D., Lees, A.W.: Dynamics of Rotating Machines. Cambridge University Press, Cambridge (2010)
4. Gasch, R., Nordmann, R., Pfützner, H.: Rotordynamik. Berlin Heidelberg. Springer (2006)
5. Landi, L., Uhlmann, E., Hörl, R., Thom, S., Gigliotti, G., Stecconi, A.: Evaluation of testing uncertainties for the impact resistance of machine guards. ASCE-ASME J. Risk Uncertainty Eng. Syst. Part B: Mech. Eng. **2**(8), S. 021001-1 (2022)
6. Mang, H.A., Hofstetter, G.: Festigkeitslehre. Springer, Berlin, Heidelberg (2018)
7. Park, J.-H.; Kim, J.-S.; Ku, M.-S.; Kang, I.-S.; Kim, K.-T.: An analysis of static and dynamic behavior of the HSK tooling system according to bearing characteristics. J. Korean Soc. Mach. Tool Eng. **19** (2010)
8. Schneider, M., Michelberger, M.: Schlanke Spannlösungen erleichtern die Zugänglichkeit. VDI-Z Special Werkzeuge + Fertigungstechnik **1**, S. 80–81 (2017)

9. Sørby, K.: Development and optimization of vibration-damped tool holders for high length-to-diameter boring operations. High Speed Mach. **1**, 2 (2016)
10. Uhlmann, E.; Thom, S.; Barth, E.; Pache, T.; Prasol, L.: Safety of slim tool extensions for milling operations. In: Proceedings of the ESREL (2019)
11. Yadav, A., Talaviya, D., Bansal, A., Law, M.: Design of chatter-resistant damped boring bars using a receptance coupling approach. JMMP **4**(2), S. 53 (2020)

Influence of the Dressing Process on the Ultrasonic-Assisted Grinding of Technical Ceramics with Different Bonding Systems

Lukas Hagenbach[1(✉)], Bernhard Gülzow[1], and Eckart Uhlmann[1,2]

[1] Institute for Machine Tools and Factory Management (IWF), Technische Universität Berlin, Berlin, Germany
lukas.hagenbach@tu-berlin.de

[2] Fraunhofer Institute for Production Systems and Design Technology IPK, Berlin, Germany

Abstract. The application of ultrasonic superposition is utilized in industrial manufacturing to improve chip formation mechanisms, particularly when machining hard-to-machine materials. The Institute for Machine Tools and Factory Management (IWF) systematically investigates the interactions between tool and workpiece during ultrasonic-assisted grinding of high-performance ceramics. To ensure a holistic view, three tool concepts with different bonding systems were compared. In particular, the influence of the truing and dressing process on the resulting macro- and micro topography of the grinding tools was analyzed. In order to create varying topographies, the grinding tools were conditioned using different dressing methods and parameters adapted to the respective bonding systems. The vitrified bonded grinding tools were dressed using a diamond roller. In contrast, the metal-bonded grinding tools were sharpened with a corundum stone after they were profiled with a silicon carbide (SiC) profile roller. To compare the differences in one bonding system, the resin-bonded grinding tools were dressed in both of these described methods. All dressing processes were carried out with and without ultrasonic superposition to investigate the influence on the dressing process and the resulting topographies. After truing and dressing of the grinding tools, the resulting topographies were measured. Subsequently a reference circumferential longitudinal ultrasonic-assisted grinding process was performed to evaluate the effect of different truing and dressing methods in relation to process forces and overall work results. The findings of these experimental investigations enabled a detailed assessment of the influence of different dressing methods and the resulting topographies on the ultrasonic-assisted grinding process.

Keywords: Grinding · Ceramic · Ultrasonic · Dressing

1 Introduction

Advances in the field of technical ceramics are expanding their applications through improved toughness while maintaining beneficial properties like high temperature resistance, minimal thermal expansion, and outstanding hardness. These materials enable

L. Overmeyer and B.-A. Behrens (eds.), *Production at the Leading Edge of Technology*, Lecture Notes in Production Engineering, https://doi.org/10.1007/978-3-032-19524-1_17

more complex components requiring stringent shape and dimensional accuracy, with grinding offering efficient and economical processing. Manufacturing process optimization and adapted machine technologies enhance both cost-effectiveness and quality. However, the high hardness of advanced ceramics and new ceramic matrix composites (CMCs) increases machining forces and tool loads, presenting new challenges. This results in higher process temperatures and accelerated tool wear which can cause fluctuations in process forces and surface quality [1]. By employing appropriate process parameters, optimized tooling and advanced process kinematics, it is possible to mitigate machining challenges by reducing process forces, improving surface quality results and minimizing tool wear [2].

The material-specific critical chip thickness $h_{cu,crit}$, as defined by Bifano et al. [3], serves as a key indicator for the chip formation mechanisms during machining. This critical value marks the transition between ductile and brittle chip formation in brittle-hard materials. Brittle chip formation typically results in lower process forces. This is due to reduced material breakouts. However, increased material toughness raises the critical chip thickness. As a result, material removal becomes more demanding.

By combining conventional grinding kinematics with oscillating ultrasonic (US) movements, it is possible to achieve reduced machining forces [4], higher material removal rates [5], and improved surface quality [6]. Studies on ultrasonic-assisted grinding of CMC confirm a significant decrease in process forces and tool wear compared to conventional grinding methods [7]. This improvement is attributed to the changing engagement directions of the abrasive grains which creates a self-sharpening effect, and the optimized chip formation mechanisms described above [8]. Excitation is accomplished using piezo-ceramic high-performance transducers which convert electrical energy into ultrasonic vibrations. Periodic elastic deformations within the oscillating system cause the generated longitudinal waves. These resulting deflections, typically measuring just a few micrometers, occur at frequencies of $f > 16$ kHz [9]. The benefits of ultrasonic assistance during the machining of high-performance ceramics and CMCs, such as higher material removal rates, lower grinding forces and reduced tool wear have been demonstrated in numerous scientific papers [7, 8, 10–14].

2 Experimental Setup

The experiments are performed using the Sauer ULTRASONIC 260 Composite 5-axis machine, manufactured by DMG Mori AG in Stipshausen, Germany. For this investigation, cylindrical tools equipped with diamond grains from GÜNTER EFFGEN GMBH, Herrstein, Germany, are used. The tools differ in their bonding systems whereas vitrified, resin and metallic bonds are investigated. The tool utilizes a grain size of $D_g = 64$ µm. The workpiece material is sintered silicon carbide (SSiC) from SITUS TECHNICALS GMBH, Wuppertal, Germany. For dressing, a diamond dressing wheel with a dressing grain size $D_{gd} = 356$ µm and a dressing spindle from DR. KAISER DIAMANTWERKZEUGE GMBH & CO. KG, Celle, Germany is used. For profiling, silicon carbide (SiC) profile roller SCG 60 I10 V40G from HERMES SCHLEIFMITTEL GMBH, Hamburg, Germany, are employed. Additionally, corundum stone Nr. 5 from SAINT-GOBAIN DIAMANTWERKZEUGE GMBH, Norderstedt, Germany is used for sharpening. Process forces are recorded using a multi-component dynamometer from KISTLER INSTRUMENTE GMBH, Sindelfingen, Germany. The generated tool topography is analyzed after dressing. For this purpose, the tool surfaces are recorded using focus variation imaging with the InfiniteFocus SL from ALICONA IMAGING GMBH. Raaba/Graz, Austria. The raw data then is analyzed by an internally developed tool for surface evaluation in the Matlab software environment from THE MATHWORKS INC., Massachusetts, USA.

3 Technological Investigations

To investigate the influence of ultrasonic excitation during the dressing process, three different bonding systems were investigated. Experiments were conducted with and without ultrasonic excitation during dressing.

After dressing, sintered SSiC components were machined by circumferential longitudinal grinding, shown in Fig. 1a, with a depth of cut $a_e = 0.2$ mm for 5 repetitions, cutting speed $v_c = 15$ m/s and feed velocity $v_f = 200$ mm/min, all with ultrasonic assistance to evaluate the effects of the different dressing conditions. The ultrasonic amplitude was set to $A_{US} = 1$ µm due to tool clamping constraints. The natural frequency of the corresponding tool was determined on the machine prior to each test and was consistently in the range of f = 25 kHz for all experiments. This approach allowed for a systematic comparison of the impact of ultrasonic excitation on the resulting tool topography and subsequent grinding performance. In particular, the influence of the truing and dressing process on the resulting macro- and micro topography of the grinding tools were analyzed. To create varying topographies, dressing methods and parameters were adapted to the respective bonding systems. Vitrified bonded tools were dressed using a diamond roller, shown in Fig. 1b. In contrast, the metal-bonded tools were profiled with a SiC profile roller similar to Fig. 1b and then sharpened with a corundum stone. To compare the differences in one bonding system, the resin-bonded tools were dressed in both described methods. For dressing with diamond form roller, both vitrified and resin bond systems were investigated. Various velocity ratios were examined during dressing while maintaining constant overlapping rate in dressing of $U_d = 4$, grinding wheel circumferential speed during dressing $v_{sd} = 15$ m/s similar to the grinding process and depth of dressing cut $a_{ed} = 5$ µm for 2 repetitions. The results shown in Table 1 demonstrate that vitrified bonded tools exhibited significant higher core roughness S_k and mean grain protrusion k values compared to resin bonded wheels.

However, the static cutting edge density was higher for resin bonded tools compared to vitrified bonded systems. The application of ultrasonic vibration during dressing tends to promote grain fracture and cutting edge breakage for vitrified bonding. A similar behavior was observed for resin bonded tools, confirming the ultrasonic effects across different bond systems. The enhanced grain fracture mechanism under ultrasonic conditions can be attributed to the high-frequency mechanical vibrations that create additional stress concentrations at grain-bond interfaces, leading to preferential fracture along crystallographic planes. The calculation of the average bonding level was consistently maintained. The observed greater mean grain protrusion k in ceramic-bonded tools can be attributed to the higher porosity which leads to a comparatively lower average bonding level relative to the other investigated bonding systems.

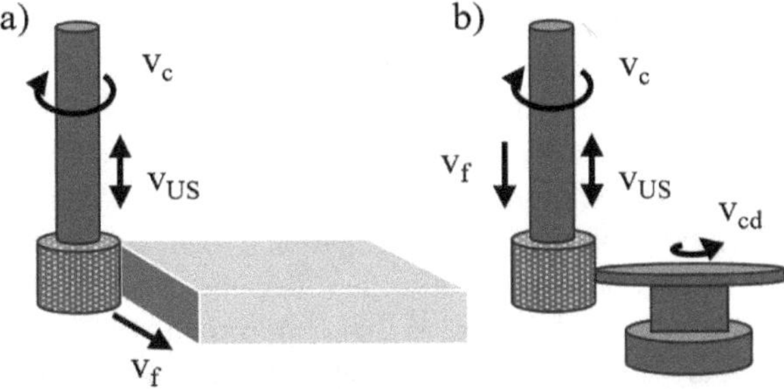

Fig. 1 Schematic representation of **a** the process of circumferential longitudinal grinding and **b** the dressing process with a diamond dressing wheel

Table 1 Topography data for dressing with diamond form roller

	Ratio of dressing speeds	Supersonic amplitude	Core roughness	Static number of active cutting edges	Mean grain protrusion
Symb	q_d	A_{US}	S_k	N_{stat}	k
Unit	[/]	[µm]	[µm]	[/]	[µm]
Vitrified bonding	−0.7	0	59.90	44.09	66.99
		1	52.69	38.79	58.74
	−0.4	0	60.41	38.79	67.89
		1	63.76	36.38	70.64
	0.7	0	62.64	41.44	69.32
		1	61.55	32.28	67.49
Resin bonding	−0.7	0	25.49	58.54	33.81
		1	19.63	42.64	22.53
	−0.4	0	23.02	56.85	31.68
		1	24.98	58.54	32.99
	0.7	0	23.37	55.89	32.49
		1	30.73	48.97	41.25

To evaluate dressing operations using SiC profile roller and corundum sharpening stones, both resin and metallic bonded tool systems were investigated. For this dressing process, constant depth of dressing cut $a_{ed} = 5$ µm for 2 repetitions and ratio of dressing speeds for profiling $q_{dp} = -0.7$ was maintained during profiling. Simultaneously, the specific material removal in sharpening V'_{Sb} was varied systematically. The results indicate that both bonding systems produce comparable surface topographies under conventional dressing conditions, shown in Table 2. N_{stat} increased with increasing specific material removal in sharpening for resin bonded tools, as the bond material was progressively receded, exposing more active cutting edges. For metallic bonded systems, the specific material removal in sharpening showed no significant influence on static cutting edge density. However, an increase in the specific material removal in sharpening could potentially lead to greater bond recession, thereby increasing N_{stat}. The different behavior between resin and metallic bonds reflects their distinct material properties and wear mechanisms under mechanical and ultrasonic loading. To enable direct comparison between the two dressing methods, resin bonded tools were dressed using both diamond form roller and SiC profile roller with corundum sharpening stone. The topographical evaluation results reveal that core roughness S_k and mean grain protrusion k are substantially higher when dressing with diamond form roller. Conversely, the static cutting edge density achieved through SiC/corundum dressing. Particularly under ultrasonic excitation during sharpening with corundum stones, the static cutting edge density was significantly elevated. This suggests that dressing with diamond form roller results in increased grain fracture and breakage.

Table 2 Topography data for dressing with SiC profile roller and corundum sharpening stone

	Specific material removal in sharpening	Supersonic amplitude	Core roughness	Static number of active cutting edges	Mean grain protrusion
Symb	V'_{Sb}	A_{US}	S_k	N_{stat}	k
Unit	[mm^3/mm]	[μm]	[μm]	[/]	[μm]
Resin bonding	50	0	17.76	54.93	23.69
		1	18.10	60.95	24.04
	150	0	20.07	53.96	26.78
		1	13.35	64.56	18.17
	250	0	15.03	63.84	19.06
		1	15.26	61.19	21.34
Metal bonding	50	0	13.37	60.23	20.04
		1	10.08	59.26	14.41
	150	0	16.82	61.67	23.36
		1	18.10	59.99	26.51
	250	0	11.14	66.36	16.68
		1	8.99	62.51	12.81

The enhanced cutting edge density achieved through ultrasonic-assisted corundum stone sharpening can be attributed to the combined effects of ultrasonic cavitation and mechanical impact. The high-frequency vibrations create localized stress concentrations that promote controlled grain fracture and expose fresh cutting surfaces while simultaneously removing dulled grain surfaces and bond material. This mechanism results in a more uniform distribution of active cutting edges with optimized sharpness characteristics for improved grinding performance. Vitrified-bonded tools showed increased mean normal grinding forces F_n after ultrasonic dressing, shown in Fig. 2. This increase results from altered topography due to enhanced grain fracture during ultrasonic dressing, creating more aggressive cutting edges but potentially reducing cutting efficiency through increased grain-workpiece interference. Metallic-bonded tools exhibit a slight increase in F_n with increasing specific material removal during ultrasonic-assisted sharpening, shown in Fig. 3. Mean tangential forces F_t decreased with increasing removal volume under ultrasonic excitation. Thus, suggesting ultrasonic dressing to induce a favorable cutting geometry that reduces material removal energy requirements. Resin-bonded tools dressed with SiC/corundum yielded decreased forces F_n and F_t compared to diamond-dressed tools. This reflects distinct topographical characteristics, where SiC/corundum creates more efficient cutting surfaces. Under ultrasonic-assisted dressing F_n progressively declined with increasing V'_{Sb}.

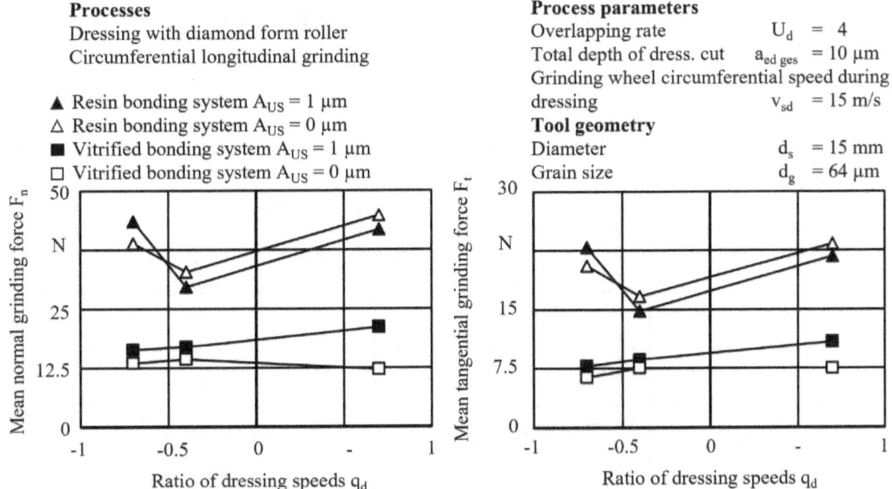

Fig. 2 Grinding forces F_n and F_t for dressing with diamond form roller

To assess post-grinding workpiece quality, the arithmetic mean surface height Ra of the workpiece surfaces were measured. Predominantly, ultrasonic-assisted dressing resulted in lower Ra values, indicating improved surface finish, shown in Fig. 4.

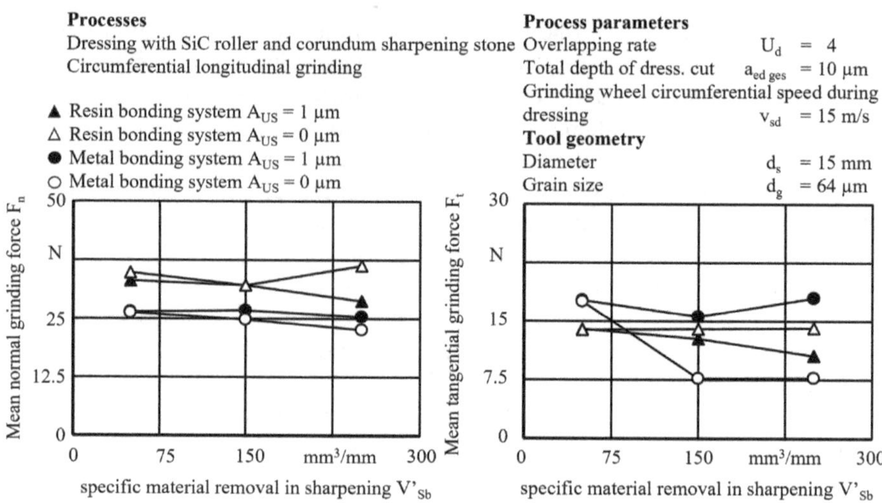

Fig. 3 Grinding forces F_n and F_t for dressing with SiC roller and corundum stone

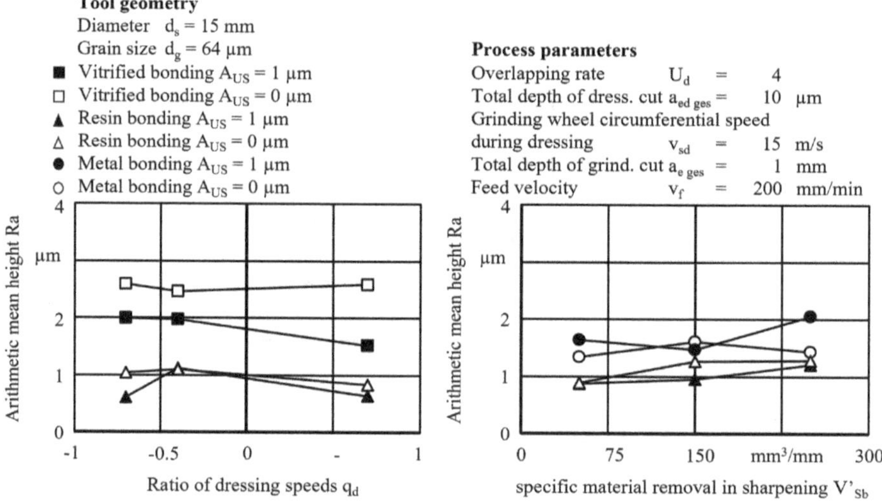

Fig. 4 Arithmetic mean heights Ra of the workpieces after grinding for different dressed tools

4 Conclusion and Outlook

Ultrasonic-assisted dressing measurably enhances the performance of grinding tools across various bond types when machining sintered silicon carbide, primarily by promoting controlled grain fracture and optimal abrasive exposure. This leads to reduced grinding forces and improved surface finishes due to enhanced self-sharpening and cutting efficiency. The effectiveness depends on both the bond material and dressing method. Future work focuses on validating ultrasonic amplitude stability under process loads and investigating long-term tool wear to ensure industrial viability.

Acknowledgements. This publication is based on the results of the research project "Tool-workpiece interaction in ultrasonic-assisted grinding" (project number 326610871), which was funded by the German Research Foundation (DFG).

Competing Interests. The author(s) has no competing interests to declare that are relevant to the content of this manuscript.

References

1. Uhlmann, E., Bruckhoff, J.: Interaction of tool and workpiece in ultrasonic-assisted grinding of high performance ceramics. In: 16th Global Conference on Sustainable Manufacturing—Sustainable Manufacturing for Global Circular Economy (2019)
2. Tawakoli, T., Azarhoushang, B.: Entwicklungen im ultraschallunterstützten Schleifen für Keramik- und Metallwerkstoffe. In: Conference: 9. Seminar Moderne Schleiftechnologie und Feinstbearbeitung, Stuttgart, Germany (2012)
3. Bifano, T.G., Dow, T., Scattergood, R.: Ductile-regime grinding of brittle materials. In: Procedia of the UME, Aachen, Germany, pp. 22–40 (1988)
4. Liang, Z., Wu, Y., Wang, X., Zhao, W.: A new two-dimensional ultrasonic assisted grinding (2D UAG) method and its fundamental performance in monocrystal silicon machining. J. Mach. Tools Manuf. **50**(8), 728–736 (2010)
5. Tawakoli, T., Azarhoushang, B.: Influence of ultrasonic vibrations on dry grinding of soft steel. Int. J. Mach. Tools Manuf. **48**(14), 1585–1591 (2008)
6. Denkena, B., Friemuth, T., Reichstein, M.: Potentials of different process kinematics in micro grinding. CIRP Ann. Gen. Assembly CIRP **52**(1), 463–466 (2003)
7. Huda A.H.N.F., Ascroft, H., Barnes, S.: Machinability study of ultrasonic assisted machining (UAM) of carbon fibre reinforced plastic (CFRP) with multifaceted tool. In: Procedia CIRP, vol. 46, p. 488–491 (2016)
8. Uhlmann, E., Hübert, C.: Ultrasonic assisted grinding of advanced ceramics. In: ASPE Proceedings of Spring Topical Meeting on Vibration Assisted Machining Technology, Chapel Hill, U.S.A (2007)
9. DIN 1320 (2009) Akustik. Grundbegriffe. Beuth, Berlin
10. Tawakoli, T., Westkämper, E., Azarhoushang, B.: Effects of vibration-assisted grinding on wear behaviour of vitrified bond Al_2O_3 wheel. Adv. Mater. Res. 76–78, p. 21–26 (2009)
11. Liang, Z., Wu, Y., Wang, X., Zhao, W.: A new two-dimensional ultrasonic assisted grinding (2D-UAG) method and its fundamental performance in monocrystal silicon machining. J. Mach. Tools Manuf. **50**(8), 728–736 (2010)
12. Zaitsev, G.N., Nikitkov, N.V., Buturovich, I.K.: Effectiveness of surface ultrasonic grinding of ceramics with diamond disks. Glass Ceram. **37**(3–4), 185–196 (1980)
13. Zapp, M.: Ultraschallunterstütztes Schleifen von Hochleistungskeramik—Ein Beitrag zur gezielten Beeinflussung der Bauteileigenschaften durch eine ganzheitliche Prozeßkettenbetrachtung. Dissertation, Universität Kaiserslautern. Kaiserslautern, Germany (1998)
14. Ding, K., Fu, Y., Su, H., Cui, F., Li, Q., Lei, W., Xu, H.: Study on surface/subsurface breakage in ultrasonic assisted grinding of C/SiC composites. Int. J. Adv. Manuf. Technol. **91**(9), 3095–3105 (2017)

Modelling the Wear Effects During the Forming of Steel and Aluminium for Tailored Forming

Simon Peddinghaus(✉), Armin Piwek, Hendrik Wester, Julius Peddinghaus, Johanna Uhe, and Bernd-Arno Behrens

Institute of Forming Technology and Machines, Garbsen, Germany
s.peddinghaus@ifum.uni-hannover.de

Abstract. In industrial bulk metal forming processes the die wear has a major influence on the process stability and the overall economical profitability. In Tailored Forming processes, the semi-finished products consist of different materials, such as aluminium and steel. These hybrid semi-finished products are formed together. The predominant wear mechanism in steel forming processes is abrasive wear, whereas in aluminium forming processes mainly adhesive wear occurs. Due to the alternating material contact, the dies are exposed to alternating wear mechanisms. Although both wear mechanisms can separately be predicted numerically using models such as the approach of Archard, interactions between the mechanisms are not taken into account. The present work is therefore dedicated to investigating the wear effects during a hybrid bulk forming process. Therefore, first upsetting experiments with mono-materials were carried out. The experiments were then repeated whilst alternating the workpiece materials each stroke and then compared to the first trial. Finite element (FE) simulations were used to recreate the upsetting processes numerically. The wear depth was determined numerically through an adapted simulation model to take the alternating wear mechanisms into account. The numerical results were compared to the experiments to analyse the interactions of the wear mechanisms and derive suitable wear coefficients. The findings allow the development of an improved numerical approach for the prediction of the alternating die wear effects in Tailored Forming processes in the future.

Keywords: Tailored forming · Wear · Adhesion · Abrasion · FEM · Modelling

1 Introduction

Through Tailored Forming, components can be produced by combining different materials to reduce the weight whilst increasing or maintaining their performance. Thereby the properties of the different materials, for example steel and aluminium, are combined and formed together to reach specific properties at different areas of the component adjusted to loads or requirements. Tailored Forming is characterised by the forming of previously joined semi-finished parts [1]. In hot bulk forming processes the dies are subject to high tribological loads such as contact pressures or temperature gradients from the workpiece to the die [2]. As a result, die wear occurs and influences not only the individual

L. Overmeyer and B.-A. Behrens (eds.), *Production at the Leading Edge of Technology*,
Lecture Notes in Production Engineering, https://doi.org/10.1007/978-3-032-19524-1_18

workpiece geometry, but also the economical profitability of the entire process [3]. In industrial processes, in addition to the major cost factor of manufacturing the dies, die failure through wear causes forced interruptions of the production and can lead to delays and additional costs [4]. Thus, the effects of die wear and methods to prevent or predict the wear behaviour are being intensively researched [5].

As the term 'wear' itself describes the progressing material loss caused through the surface interactions of two bodies in contact, it consists of multiple overlapping mechanisms such as abrasion, plastic deformation or adhesion [6]. In the context of hot bulk metal forming, studies have shown that for steel the predominant wear mechanism is abrasion. Thereby the softer body is carved out through particles on the contact surface [7]. The most well-known numerical approach for abrasive wear was developed by Archard. In its initial form, as shown in Eq. 1, the wear depth w is calculated through the die hardness H, the normal stress σ_N, the sliding distance L, which are weighted through a process specific wear coefficient k [8].

$$w = k \cdot \frac{\sigma_N \cdot L}{H} \tag{1}$$

For hot forming processes with aluminium, adhesion has been identified as most relevant wear mechanism because of its strong sticking tendency. It occurs when sticking bonds at the interacting surfaces are formed which exceed the material strength of the weaker body in contact. In forming processes, this leads to build-ups on the die surface which then tear off once they reach a critical size and remove particles of the die material [9].

To describe adhesive wear numerically, several works have developed alternatives or modifications to the model of Archard for the asperity level, as compared in [10]. Such approaches, as for example the modification of Archard by Brink et al., describe abrasion processes at a microscopic level. Although they are suitable to describe the build-up and tear-off behaviour of adhesive wear in details, they are not yet applicable to macroscopic processes such as forming operations [11]. For this, a suitable approach is the modification of Archard's model by Painter et al. Thereby the wear is predicted according to the model of Archard, but the hardness of the workpiece is used instead of the die, as it is the softer body in contact. However, this approach only allows the prediction of adhesive build-ups but not the tear-off behaviour [12].

The prediction and prevention of die wear in the context of Tailored Forming presents a new challenge. As the hybrid semi-finished products are formed together, local die surfaces come into contact with both aluminium and steel during each forming stroke. Due to this alternating material contact, these surfaces are exposed to alternating adhesive and abrasive wear, instead of mostly consistent wear mechanisms as in mono-material forming processes. In a preliminary study, Piwek et al. investigated an upsetting process, where the material of the semi-finished products was alternated between aluminium and steel for each stroke. The wear was compared to experiments where the mono-materials were formed separately. This comparison indicated interactions between adhesive and abrasive wear, whilst the adhesive wear was predominant in the alternating experiments [13]. However, only qualitative observations could be made, as the locally concentrated adhesions were not quantified. To understand the interactions of adhesive and abrasive wear, and to numerically consider them in simulations of Tailored Forming processes, deeper insights as well as a quantitive analysis are required [13].

Therefore, the aim of this paper is to utilise the results of the studies of Piwek et al. to gain a deeper understanding of the interactions of adhesive and abrasive wear and to investigate the capabilities of current numerical approaches to describe these interactions. By analysing the worn die locally, it was possible to determine the local wear distribution. As a result, local differences between monolithic and alternating wear could be determined. The wear coefficients for aluminum and steel were determined iteratively by comparing the local wear depth. Finally, an alternating cycle was simulated in which the Archard model for abrasive wear was used for steel and the modified model for adhesion according to Painter et al. [8] was used for aluminium. The calculated wear depths were then extrapolated to the total number of strokes of the experimental series, compared with the experimentally determined wear depths and discussed.

2 Material and Methods

The analysed forming process is an upsetting process as described in the work by Piwek et al. The forming experiments were carried out whilst alternating between the aluminium alloy AA-6082 and the steel alloy AISI5120H (20MnCr5) as material for the semi-finished products with a diameter of 30 mm and a height of 40 mm. The workpieces were formed on a LASCO SPR 500 screw press to a thickness of 8 mm [13]. To achieve comparable yield stresses as desired for Tailored Forming processes, the workpieces were heated inductively to the temperatures of 900 °C for steel and 200 °C for aluminium. The process parameters as f. e. the lubrication and ram speed are described in the work of Piwek et al. [13]. According to the model of Archard and modifications of it, the die wear is strongly dependent of the sliding distance. Therefore, the dies were measured optically along specific radii from the centre of the dies using a Keyence VR3200 3D-profilometer as shown schematically for three exemplary radii in Fig. 1.

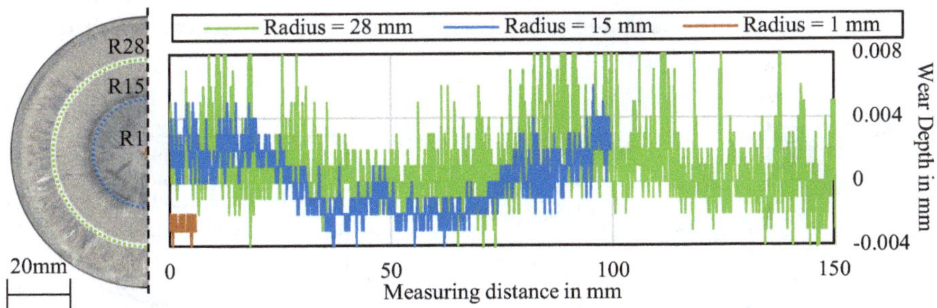

Fig. 1 Wear measurement along specific radii of the die after experiments

The datapoints of each specific radii were then averaged as they showed strong deviations through the locally concentrated effects of adhesion. The reference plane was defined at the outside of the die surface, which does not come in contact with the workpiece during the forming process and therefore can be assumed as free of wear. In relation to this plane, adhesion was defined as positive wear depth and abrasion as negative. The averaged wear depths were then summarised in wear curves for each experiment in dependence of the specific radius. As the work of Piwek et al. consistently showed stronger wear phenomena on the upper dies than on the bottom dies, only the upper dies are investigated in this study [13]. To investigate the fundamental wear effect experiments were carried out with quenched and tempered dies of 1.2343 with a hardness of 48 HRC. Multiple series with separate sets of dies were investigated: first the dies were analysed after 50 parts were formed with both of the workpiece materials individually. These were compared to a separate set of dies with which a total of 100 parts were formed with alternating workpiece materials as these included 50 forming operations of each material. The wear curves of the dies were compared to each other and to the results of the simulations.

To recreate the experiments, a 2D axially symmetrical simulation model was setup in the software Simufact Forming v16.0 as schematically shown in Fig. 2a. Thereby, the approach of [12], based on the Archard model, was applied for the forming process with an aluminium workpiece and the hardness of the workpiece was taken into consideration with 110 HV according to [14]. The material properties of the workpieces, such as flow curves, were taken from previous work [15]. For modelling the friction, a combined friction model consisting of a friction coefficient of $\mu = 0.1$ to describe the friction according to Coulomb and a friction factor of $m = 0.3$ with the approach of Tresca, was selected based on the work of [16]. Through a preliminary mesh study, a quadrilateral mesh with an element size of 0.6 mm and a remeshing criterion for strain changes of more than 0.4 occur, was chosen as it showed a good prediction quality with an acceptable computational effort. The upper die was defined as elastic deformable with material properties from the simufact database for 1.2343. As the focus of the experimental analysis is on the upper dies, the lower die was modelled as rigid die in order reduce the computation time. For both dies a temperature of 150 °C was selected [13]. As presented in Fig. 2b, the force–stroke curves of the simulation and the mono-material experiments are both in good agreement which indicates that the simulation model is suitable. The similar force requirements for aluminium and steel confirm the choice of forming temperatures with the aim of achieving comparable flow stress for both materials.

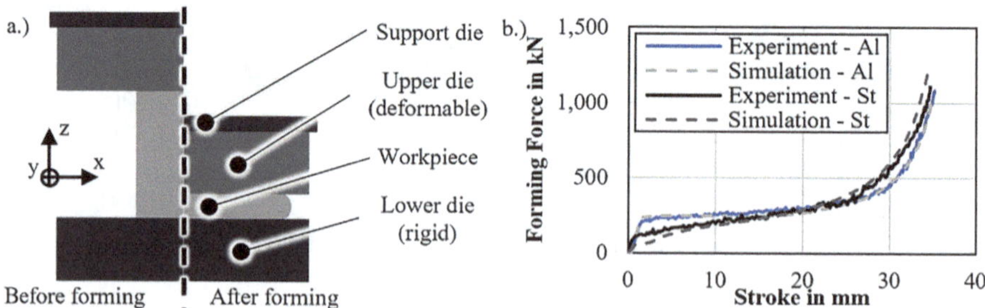

Fig. 2 **a** Investigated forming process **b** Force–stroke curves for both workpiece materials

The work of Piwek et al. showed that abrasive and adhesive wear both occur in the forming with alternating workpiece materials and also indicates that they are overlayed through interactions of the wear mechanisms [13]. As currently no wear modelling approach exists to specifically describe these interactions, a simplified approach was chosen. First, the wear curves of the mono-material experiments were used to determine wear coefficients for both materials separately. For this, in the simulation model the wear coefficients were varied iteratively until the amount of the maximum wear depth was reached. Thereby, the numerically determined wear of a single stroke was multiplied with the number of strokes of the series. The maximum wear depth was chosen as reference, as it represents a critical value for the service life of the dies. For steel the conventional model of Archard was used and for aluminium the approach of Painter et al. [12]. The wear coefficients were then implemented into a combined simulation model consisting of linked simulations of the aluminium and steel upsetting processes, each with separate wear models. Thereby the amount of wear was transferred from the results of the aluminium process and taken into account for the simulation of the steel process. This allowed to sum up the amounts of adhesive an abrasive wear for single alternation of the workpiece material. As in the mono-material simulations, this wear was then multiplied with the total amount of 50 alternations. The results were compared to the experiments.

3 Results and Discussion

As compared in Fig. 3a, the dies exhibited varying wear behaviour in dependence of the workpiece material. In the mono-material experiments the dies for steel showed a low amount of abrasive wear, which is represented as negative wear depth. For aluminium, a considerable amount of adhesive wear could be noticed, which is indicated through positive wear depth values. For the experiments with steel and with alternating workpiece materials an increase of the wear depth towards the middle of the die occurs.

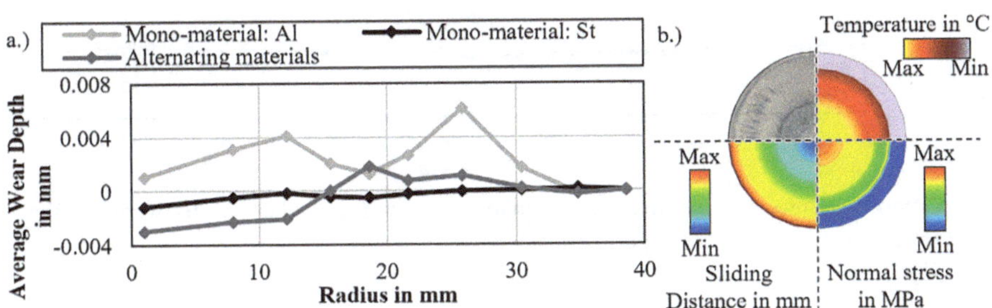

Fig. 3 **a** Experimental wear curves of mono-material compared to alternating workpieces **b** numerically determined tribological results for a steel workpiece

The numerical results presented in Fig. 3b show that in the middle of the die the lowest sliding distances occur. With regards to the model of Archard this indicates that the measured wear in the middle of the die is neither abrasive nor adhesive. As the highest temperatures and normal stresses develop in the middle of the die, plastic deformation could have occurred in the middle of the dies, particularly since the forming temperatures were relatively low in the context of hot forming leading to comparably high flow stresses. The wear behaviour at the middle of the die cannot be observed for the mono-material aluminium experiments. As the flow stresses of both workpiece materials are comparable, this indicates that the suspected plastic deformation could have been superimposed with thermal softening of the dies through the contact with the hot steel workpieces.

With alternating workpiece materials, the dies showed a higher degree of wear in the middle of the die than in the mono-material steel experiments. As the amount of steel workpieces formed and therefore the inflicted thermal softening is the same for these two trials, it can be assumed that the plastic deformation was the predominant mechanism is the middle of the dies. The dies with alternating workpiece materials showed a lower amount of adhesive wear after 50 alternations (consisting of 50 aluminium and 50 steel parts) than the dies of the mono-material aluminium experiments after 50 strokes. This leads to the assumption that through the interactions between the two wear mechanisms the adhesive wear is reduced.

For the simulations, the wear coefficients were determined from the amount of the maximum wear depths of the experimental results whilst ignoring the wear area in the center of the dies. The experimental wear curves are compared to the results of simulations with wear coefficients from the mono-material experiments in Fig. 4. It can be observed, that on the one hand the wear distributions are only recreated to limited degree.

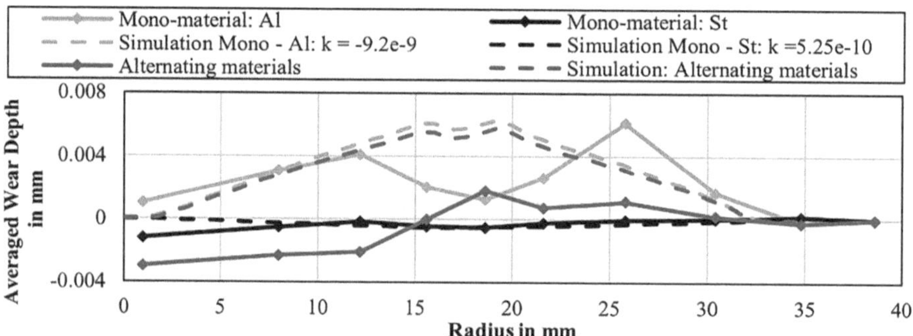

Fig. 4 Experimental wear curves in comparison to the simulated wear curves for wear coefficients derived from the mono-material experiments

For the mono-material aluminium experiments, the qualitative form of the wear curve, consisting of two peaks with a reduction in between, is vaguely recreated through the simulation. The simulation shows a qualitative agreement for the areas outside of the peaks of the experiment. The numerical peak is predicted in the area of the reduction of the experimental wear curve. A possible explanation for this discrepancy is, that a critical amount of adhesion could have been reached at the radius of the numerically predicted peak. Through this tear-offs might have occurred, which result in a lower adhesive build-up. This phenomenon cannot be replicated through the chosen wear modelling approach.

On the other hand, the wear curve of the combined simulation neither describes the abrasive nor the adhesive wear of the alternating process correctly. It rather shows a resemblance to the simulation of the aluminium mono-material process and only consists of adhesive wear. The characteristic of the curve indicates that the abrasive wear was assumed to low and the adhesive wear to high. This demonstrates, that with regard to the interactions between the wear mechanisms, a simple addition of the individual wear behaviour is not a sufficient approach. This also shows that adapted wear coefficients are required to describe the alternating wear mechanisms and wear coefficients of mono-material processes cannot be applied directly.

For the experiments with mono-material steel and alternating workpiece materials, the wear increase towards the center of the die is not reproduced through the simulations. This might be caused through plastic deformation, which also cannot be predicted through the chosen wear modelling approach. The insufficient agreement between the numerically determined wear curves and the experimental curves leads to the conclusion, that an alternative numerical method is required for an improved wear calculation in the context of Tailored Forming.

Also, as Piwek et al. [13] suggested that over the course of a serial process, global adhesive build ups followed by tear-offs occur, it can be assumed that the adhesive wear fluctuates considerably throughout the process. In order to exclude the possible influence of such fluctuations, additional investigations at different points of a process series are required.

4 Conclusion and Outlook

In this study, mono-material upsetting experiments with aluminium and steel were carried out. The experiments were then repeated whilst alternating the workpiece materials each stroke, investigated in detail and connected with numerical simulations. The alternating process was separately simulated with an adapted model with wear coefficients which determined from the mono-material experiments. Although the simulations showed a qualitative agreement with the mono-material experiments, for the alternating experiments only a limited agreement was reached. The results demonstrate, that an extended numerical approach is required for the modelling of wear in the context of Tailored Forming. In order to evaluate the influences of unknown aspects in this study, such as the fluctuations of the adhesive wear or the plastic deformation of the dies, further investigations are needed. Future work will also focus on developing an adapted model for Tailored Forming processes.

Acknowledgements. Funded by the Deutsche Forschungsgemeinschaft (DFG, German Research Foundation)—SFB 1153—252662854 subprojects B03, C01.

Competing Interests. The author(s) has no competing interests to declare that are relevant to the content of this manuscript.

References

1. Behrens, B.-A., Uhe, J.: Introduction to tailored forming. Prod. Eng. Res. Devel. **15**(2), 133–136 (2021). https://doi.org/10.1007/s11740-021-01022-w

2. Shirgaokar, M.: Technology to Improve Competitiveness in Warm and Hot Forging: Increasing Die Life and Material Utilization
3. Ficak, G., Łukaszek-Sołek, A., Hawryluk, M.: Durability of forging tools used in the hot closed die forging process—a review. Materials **17**(22) (2024). https://doi.org/10.3390/ma17225407
4. Lange, K., Cser, L., Geiger, M., Kals, J.: Tool life and tool quality in bulk metal forming. CIRP Ann. **41**(2), 667–675 (1992). https://doi.org/10.1016/S0007-8506(07)63253-3
5. Felder, E., Montagut, J.L.: Friction and wear during the hot forging of steels. Tribol. Int. 61–68 (1980)
6. Rabinowicz, E.: Friction and Wear of Materials, 2nd edn. Wiley, New York, NY (1995)
7. Altan, T., Ngaile, G., Shen, G.: Cold and hot forging. Fundamentals and applications. ASM Int. Mater. Park, OH (2010)
8. Archard, J.F., Hirst, W.: The Wear of Metals under Unlubricated Conditions
9. Podgornik, B., Leskovšek, V.: Wear mechanisms and surface engineering of forming tools. Mater. Tehnol. **49**(3), 313–324 (2015). https://doi.org/10.17222/mit.2015.005
10. Zhang, H., Goltsberg, R., Etsion, I.: Modeling adhesive wear in asperity and rough surface contacts: a review. Materials **15**(19) (2022). https://doi.org/10.3390/ma15196855
11. Brink, T., Frérot, L., Molinari, J.-F.: A parameter-free mechanistic model of the adhesive wear process of rough surfaces in sliding contact. J. Mech. Phys. Solids **147**, 104238 (2021). https://doi.org/10.1016/j.jmps.2020.104238
12. Painter, B., Shivpuri, R., Altan, T.: Prediction of die wear during hot-extrusion of engine valves (n.d.). https://doi.org/10.1016/0924-0136(96)02294-7
13. Piwek, A., Peddinghaus, J., Uhe, J., Brunotte, K.: Wear Effects on Coated Tools during the Alternating Forming of Case-Hardening Steel and Aluminium Wrought Alloy
14. Ostermann, F.: Anwendungstechnologie Aluminium. Springer, Berlin, Heidelberg (2014)
15. Behrens, B.-A., Uhe, J., Ross, I., Peddinghaus, J., Ursinus, J., Matthias, T., Bährisch, S.: Tailored forming of hybrid bulk metal components. Int. J. Mater. Form. **15**, 42 (2022). https://doi.org/10.1007/s12289-022-01681-9
16. Wester, H., Stockburger, E., Peddinghaus, S., Uhe, J., Behrens, B.-A.: Modelling failure of joining zones during forming of hybrid parts. In: Materials Research Forum LLC, pp. 717–726 (2023)

Numerical Process Design for Improved Surface Residual Stresses in Hot Formed Shafts

Matthias Hammes(✉), Hendrik Wester, Kai Brunotte, and Bernd-Arno Behrens

Institute of Forming Technology and Machines (IFUM), Leibniz Universität Hannover, Garbsen, Germany

hammes@ifum.uni-hannover.de

Abstract. Due to the achievable strength, reliability and cost effectiveness, hot forging is an established method for the manufacturing of highly stressed components like shafts, gears or bearings. Premature failure of those components is often attributed to tensile residual stresses remaining in the surface which are induced during production. However, it is known that residual stresses can also have a positive effect on the component properties. For example, compressive residual stresses close to the surface can increase the service life and wear resistance. In order to prolong service life, the aim of this study is to achieve a tailored residual stress distribution in a shaft after impact extrusion. Therefore, compressive stresses are induced after forming by means of adapted spray coo-ling from the forming heat. Since a subsequent heat treatment would revert the induced residual stresses, a bainitic microstructure of the utilised 42CrMo4 (AISI 4140) steel is also to be set during spray cooling. Different cooling strategies, e.g. by air, water or spray cooling, are tested in numerical studies with Simufact Forming regarding the evolution of the near surface residual stresses in axial and tangential direction. To predict residual stress development, a coupled mechanical, thermal and metallurgical material model is used. Both the magnitude and the reachable depths of the residual stresses are analysed, since those criteria affect crack initiation as well as propagation and are therefore crucial for the attainable fatigue durability. As a result, promising process routes for a later experimental validation are identified.

Keywords: Hot forming · FE-based process design · Residual stresses

1 Introduction

Hot forming is an established method for the manufacturing of highly stressed components in industrial mechanical engineering [1]. Subsequent heat treatments are used to form the desired microstructure and therefore set the desired mechanical properties. In the end, tensile residual stresses often remain at the surface which negatively affect service life. Residual stresses form due to plastic deformation, different local temperature gradients or microstructure changes and are present while no external loads are applied. Research has shown that the remaining residual stresses greatly influence crack initiation and propagation which in turn affect component failure and therefore service life [2]. Further studies revealed that compressive residual stresses have a positive effect

L. Overmeyer and B.-A. Behrens (eds.), *Production at the Leading Edge of Technology*, Lecture Notes in Production Engineering, https://doi.org/10.1007/978-3-032-19524-1_19

on service life. Treatments for inducing these stresses are for example shot-peening and induction hardening. Shot-peened parts after forging exhibited increased fatigue strength properties in cyclic bending tests [3], while the crack initiation location in induction hardened shafts was shifted from the surface more towards the core which led to longer service life [4]. Figure 1 shows the results of fatigue tests with shot and cavitation peened specimens compared to non-peened ones, where the peened specimens sustained significantly higher stresses with the same number of cycles before failure due to the induced compressive residual stresses at the surface [5].

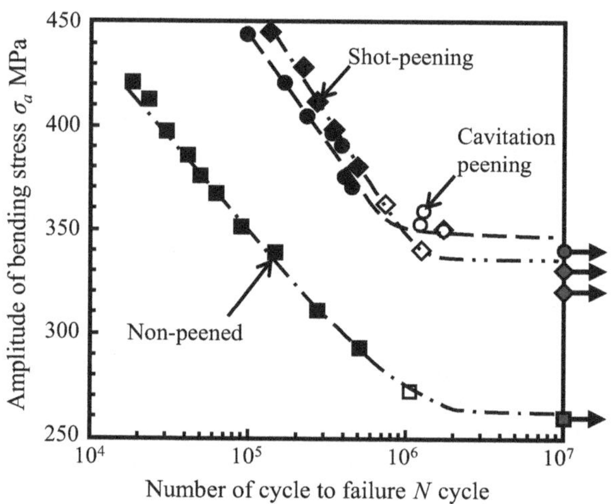

Fig. 1 Stress-life curves of non-peened specimens compared to specimens treated with shot and cavitation peening. The open and filled symbols indicate different crack initiation locations [5]

All the aforementioned treatments are characterised by the drawback that another subsequent step in production is necessary to induce the favourable compressive stresses. Previous work on cylindrical test specimens has shown that compressive surface residual stresses can be achieved during adapted cooling from the forming heat after upsetting, thus no additional treatment step is necessary [6]. Different microstructures, martensite and pearlite, were obtained in the steel 100Cr6 (AISI 52100) through adapted spray cooling strategies. This research however focuses on 42CrMo4 (AISI 4140), which is a common steel used for highly stressed components such as drive shafts where alternating loads are present. Since no subsequent heat treatment is to be performed, which would relax the residual stresses induced during cooling, a bainitic microstructure is desired which offers a good trade-off between strength and toughness.

This paper is focused on the achievable local compressive residual stresses in component-like shafts manufactured by hot impact extrusion. Through tailored cooling from the forming heat as one integrated step in production, time and cost can be saved compared to the conventional route consisting of separate heat and surface treatment. By comparing the average cost distribution with [7] and without secondary operations costs [8], where heat and surface treatments like shot peening are the main factors, around 10% of the overall component costs can be saved. The cost of cooling still needs to be considered, but should be small compared to an additional process step. In order to assess viable cooling strategies, finite element (FE) simulations are carried out in Simufact Forming v16 and the resulting residual stress distributions are compared. Therefore, the optimal cooling route for inducing compressive residual stresses is identified for an improved service life.

2 Materials and Methods

The semi-finished products used in this study were cylindric steel bars made of 42CrMo4 with Ø30 mm and a length of 130 mm. The process route is illustrated in Fig. 2 and consists of three steps. Homogenous austenitisation of the semi-finished product at 1150 °C, followed by a hot impact extrusion operation and a subsequent adapted cooling from the forming heat.

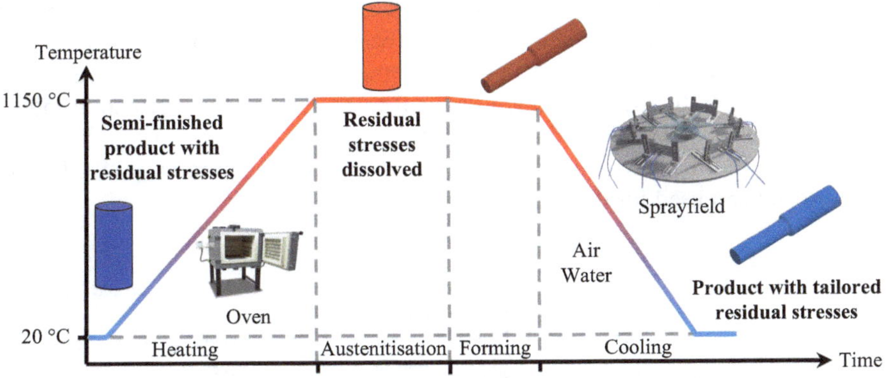

Fig. 2 Process route of the hot formed shafts with subsequent cooling

For the simulations, both punch and die for the hot impact extrusion are modelled as rigid heat-conducting tools made of high-speed steel HS 6-5-2 C (AISI M2). The corresponding specific heat capacity of 510 J/(kg K) and the thermal conductivity of 27.6 W/(m K) were taken from the Simufact Forming material library. For contact modelling, the combined Coulomb-Tresca model with a friction coefficient of 0.3 [9] and a friction shear factor of 0.4 [10] were used. The flow curves for 42CrMo4 are calculated with the GMT model (Gesellschaft für Maschinentechnik mbH) given in Eq. 1, which was calibrated with experimental data in [11].

$$k_f = c_1 \cdot e^{c_2 \cdot T} \cdot \varepsilon^{n_1 \cdot T + n_2} \cdot e^{\frac{l_1 \cdot T + l_2}{\varphi}} \cdot \dot{\varepsilon}^{m_1 \cdot T + m_2} \tag{1}$$

The yield stress k_f is calculated as a function of the true strain ε, the true strain rate $\dot{\varepsilon}$ and temperature T with the material specific coefficients shown in Table 1.

Since phase changes and the final phase distribution play an important role in the evolution of residual stresses, material data in the form of time–temperature-transformation

Table 1 Fitted GMT model parameters

c_1 in MPa s	c_2 in $\frac{1}{°C}$	n_1 in $\frac{1}{10^5 \cdot °C}$	n_2 in –	l_1 in $\frac{1}{10^5 \cdot °C}$	l_2 in –	m_1 in $\frac{1}{10^5 \cdot °C}$	m_2 in –
6054.21	−0.0039	−3.0846	0.3000	1.8926	−0.03548	9.4422	0.2484

diagrams are implemented which were experimentally determined in [12]. The transformation induced plasticity (TRIP) effect is also taken into account by implementing phase-specific TRIP-coefficients, evaluated with an experimental–numerical approach in [12]. In order to determine the remaining material parameters for each phase, the thermodynamic calculation software JMatPro was used [13]. Based on the chemical composition of the alloy and empirical equations, temperature-dependent and phase specific values were calculated for the specific heat capacity, thermal conductivity, thermal expansion coefficient, density, Young's modulus, Poisson's ratio, latent heat and hardness of 42CrMo4. For the simulations, the additive strain decomposition method in the implicit MSC.Marc solver was used [14], where the total strain is calculated as the sum of the elastic and plastic strain increments as well as the thermal, transformation-induced and transformation-plasticity strain increments.

Due to the rotational symmetry of the forming and cooling process, a 2D FE model was developed and used for the simulations (see Fig. 3).

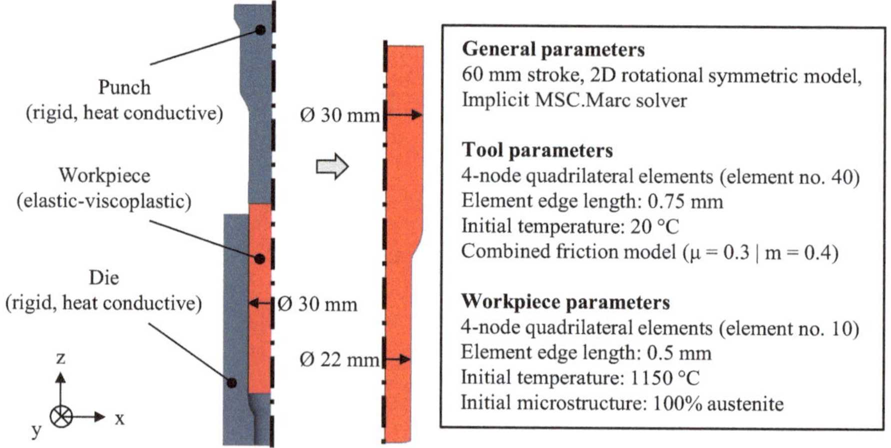

Fig. 3 FE model of hot impact extrusion

A mesh sensitivity study revealed that an element edge length of 0.5 mm for the workpiece offered a good trade-off between computation time and result accuracy. The simulation starts with a numerically calculated thermally expanded round billet at 1150 °C and a 100% austenitic microstructure. Two different impact velocities of 300 and 600 mm/s, representing typical forming speeds for hot impact extrusion, were investigated in the forming simulation. The workpiece is extruded to a length of approximately 160 mm with a stroke of 60 mm, resulting in a reduced diameter of 22 mm (see Fig. 3). Afterwards, the geometry and the stress and temperature field of the workpiece were imported in a subsequent cooling simulation. For the cooling simulations, the effect of different media as well as the starting time for cooling is investigated.

The different cooling characteristics of air, water and sprayfield were modelled through temperature-dependent surface heat transfer coefficients (HTC). The HTC were determined with an experimental–numerical approach, as detailed in [15], while the experimental tests concerning the spray cooling behaviour were carried out with 100Cr6 [16]. Both air and water had a pressure of 0.04 MPa when mixing in the spray nozzle. Since the cooling characteristic of the sprayfield is mainly influenced by droplet size and surface condition and less by the material itself [17], the spray HTC curve of 100Cr6 was also used in this study for the spray cooling simulations with 42CrMo4. Figure 4 gives an overview of the implemented HTC curves.

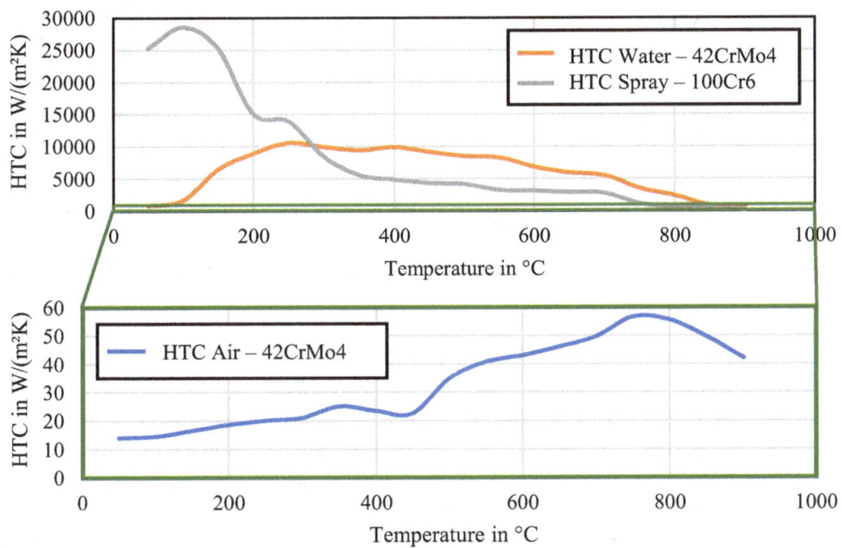

Fig. 4 HTC curves between air, water or spray and steel [16]

In order to precisely assign the specific HTC for spray cooling to a defined surface, the near-field contact function of Simufact Forming was used. For the remaining surfaces not exposed to the spray, the HTC between steel and air was assigned. The same was done in case of water cooling, with the difference that the HTC for water was assigned to the entire workpiece surface during contact with water.

3 Results and Discussion

Since the aim is to improve fatigue life by inducing compressive residual stresses, they have to be present at the location of the highest stresses due to the applied loads during operation, which is prone to crack initiation. Typical loads for shafts are bending and torsion, and the resulting stresses are higher in the zone with the smaller cross-section area. The transition to the shoulder area leads to a local excessive stress state due to the geometric discontinuity, which is why this area is focused on.

Figure 5 shows that the punch speed has no significant effect on the temperature and tangential residual stress distribution after impact extrusion.

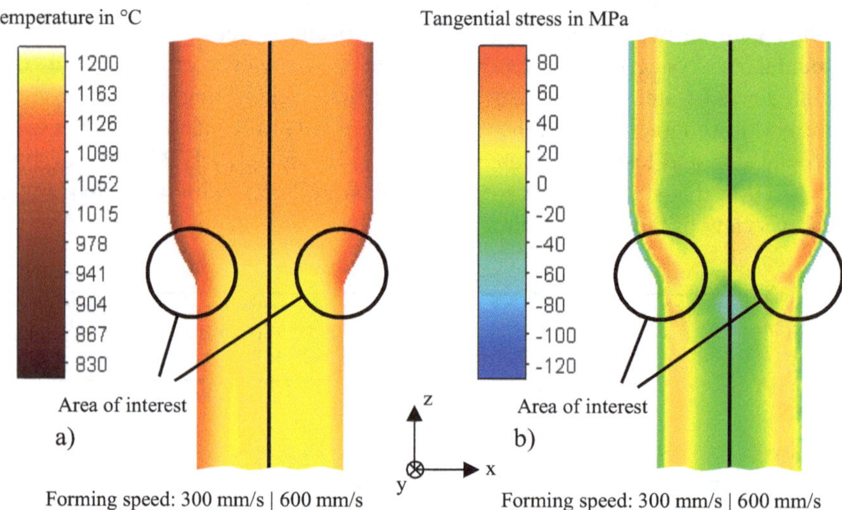

Fig. 5 **a** Temperature and **b** tangential stress distribution after forming simulation

In the shoulder area with transition to the smaller diameter, the surface temperature is in the range between 1100 and 1150 °C after forming. Only small compressive surface residual stresses are to be found of around −20 to −30 MPa after tool ejection, independent of the forming speed. The results indicate that the forming speed has no significant impact on the properties after impact extrusion. Consequently, the following results only consider the higher speed of 600 mm/s.

After the forming process, cooling simulations were performed. At first, in order to assess a possible cooling route leading to a bainitic microstructure, the workpiece was air cooled to room temperature to avoid high cooling rates resulting in martensite. A complete bainitic microstructure was achieved since the critical cooling rate of about 10 K/s was not exceeded, but also no compressive residual stresses were present. Due to the requirement of a final bainitic microstructure, higher cooling rates can only be applied after phase formation, which is finished after approximately 750 s of air cooling with an average work piece temperature of around 470 °C. Figure 6 shows the results when stronger cooling is applied by means of water or spray cooling after the complete bainite formation.

In both cases the surface temperature drops faster than in the core, resulting in a stronger thermal contraction on the surface than inside. Therefore, tensile stresses are evolved at the surface and opposing compressive stresses in the core when higher cooling rates are applied (see Fig. 6a). Only when the cooling rate in the core is higher than on

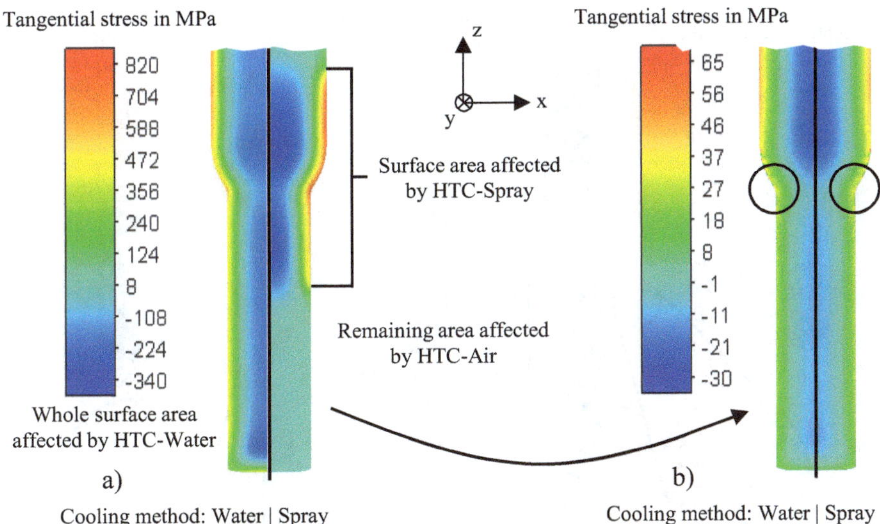

Fig. 6 **a** Tangential stress distribution in the shaft directly at the start of water and spray. Cooling and **b** final distribution at room temperature

the surface, the stress distribution begins to shift. However, the remaining temperature level after bainite formation is low, and the shift of higher cooling rates to the core starts too late with both cooling strategies. Therefore, minimal final tensile residual stresses on the surface of around 20 MPa are formed in the area of interest (see Fig. 6b).

When the formation of a small amount of martensite is allowed, high compressive residual stresses can be achieved at the surface. Since the spray offers the possibility of targeting only a small surface area, the formation of martensite can be kept at a minimum. The phase transformation has the positive effect of offsetting the stress distribution at the beginning of cooling. Due to the rapid cooling at the surface, martensite forms, resulting in a volume increase and an additional compressive stress portion and therefore reducing the high tensile stresses due to thermal shrinking. This means the compressive stresses near the core are also lowered. In order to keep the transformation of martensite to a minimum, the spray-affected area was shifted more towards the smaller diameter and the cooling times were adapted. Spray cooling starts after 690 s for 3 s and is paused again for about 110 s. This pause allows for the rest of the austenite in the core to transform to bainite. After all transformations are completed, spray cooling is continued in order to increase the final compressive residual stresses at the surface due to thermal shrinkage. While at first the surface develops additional tensile stresses, during further cooling the temperature in the core starts to drop faster than at the surface, so thermal contraction starts at the core and tensile stresses are developed. Since the residual stresses are balanced in the whole part, further compressive stresses build up at the surface. In the end, minimal residual stresses of about −250 MPa are achieved at the surface while compressive stresses are still present at a depth of approximately 1.7 mm (see Fig. 7a).

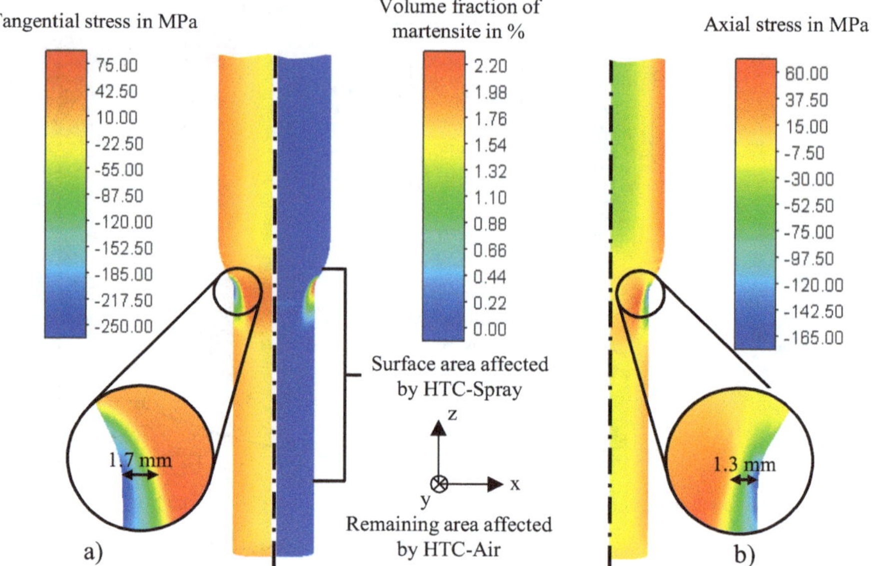

Fig. 7 **a** Final tangential stresses and volume fraction of martensite and **b** final axial stresses after modified spray cooling

Focusing on the results for the axial residual stress distribution shown in Fig. 7b, relevant compressive stresses are induced of about −165 MPa near the surface, while a depth of around 1.3 mm is reached.

4 Conclusion and Outlook

The presented results show that it is challenging to induce compressive residual stresses after forming through adapted cooling for the investigated shaft while ensuring a bainitic microstructure. Only by allowing a small amount of martensite formation, final compressive residual stresses can be achieved with the superimposed thermal stresses. The martensite formation leads to compression at the surface due to the increase in volume, while the shift of the higher cooling rate from the surface to the core leads to tension in the core and further compression at the surface.

So far, the spray cooling parameters were kept in one pressure setting for mixing. In addition to a localised cooling, the spray also offers the advantage of different cooling characteristics based on different pressures of air and water. Further testing of different spray cooling settings has to be done and the corresponding HTC curves determined. Finally, the resulting residual stresses are to be experimentally validated.

Acknowledgements. This research was funded by the Deutsche Forschungsgemeinschaft (DFG, German Research Foundation)—project number 530125423 ("Experimental and numerical modelling and analysis of microstructural residual stresses of hot formed components with adapted cooling").

Competing Interests. The author(s) has no competing interests to declare that are relevant to the content of this manuscript.

References

1. Semiatin, S.L.: ASM Metals Handbook Volume 14—Forming and Forging. ASM International, Novelty, Ohio (1998)
2. Wildeis, A., Christ, H.-J., Brandt, R.: Influence of residual stresses on the crack initiation and short crack propagation in a martensitic spring steel. Metals **12**(7), 1085 (2022). https://doi.org/10.3390/met12071085
3. Gerin, B., Pessard, E., Morel, F., Verdu, C.: Competition between surface defects and residual stresses on fatigue behaviour of shot peened forged components. Procedia Struct. Int. **2**, 3226–3232 (2016). https://doi.org/10.1016/j.prostr.2016.06.402.
4. Zhang, H.-Y., Stephens, R.I., Glinka, G.: Subsurface fatigue crack initiation and propagation behavior of induction-hardened shafts under the effect of residual and applied bending stresses. In: Jerina K.L., Paris P.C. (eds.) Fatigue and Fracture Mechanics, vol. 30, pp. 240–260. ASTM International, West Conshohocken, Pennsylvania (2000)
5. Soyama, H., Chighizola, C.R., Hill, M.R.: Effect of compressive residual stress introduced by cavitation peening and shot peening on the improvement of fatigue strength of stainless steel. J. Mater. Process. Technol. **288**(116877) (2021). https://doi.org/10.1016/j.jmatprotec.2020.116877
6. Behrens, B.-A., Brunotte, K., Wester, H., Kock, C.: Targeted adjustment of residual stresses in hot-formed components by means of process design based on finite element simulation. Arch. Appl. Mech. **91**(8), 3579–3602 (2021). https://doi.org/10.1007/s00419-021-01928-y
7. Raj, M.P., Kumar, M., Pramanick, A.K.: Yield improvement in hot forging of differential spider. Mater. Today Proceed. **26**, 3107–3115 (2020). https://doi.org/10.1016/j.matpr.2020.02.642.
8. Knight, W.A.: Simplified early cost estimating for hot-forged parts. Int. J. Adv. Manuf. Technol. **7**(3), 159–167 (1992). https://doi.org/10.1007/BF02601619.
9. Serebriakov, I., Puchi-Cabrera, E.S., Dubar, L., Moreau, P., Meresse, D., La Barbera-Sosa, J.G.: Friction analysis during deformation of steels under hot-working conditions. Tribol. Int. **158**, 106928 (2021). https://doi.org/10.1016/j.triboint.2021.106928
10. Jeong, D.J., Kim, D.J., Kim, J.H., Kim, B.M., Dean, T.A.: Effects of surface treatments and lubricants for warm forging die life. J. Mater. Process. Techno. **113**(1–3), 544–550 (2001). https://doi.org/10.1016/S0924-0136(01)00693-8
11. Behrens, B.-A., Schröder, J., Wester, H., Brands, D., Uebing, S., Kock, C.: Experimental and numerical investigations on the development and stability of residual stresses arising from hot forming processes. In: Daehn, G., Cao, J., Kinsey, B., Tekkaya, T., Vivek, A., Yoshida, Y. (eds.) 13th International Conference on the Technology of Plasticity (ICTP 2020), pp. 2289–2301. Springer International Publishing (The Minerals, Metals & Materials Series), Basel, Switzerland (2021)
12. Behrens, B.-A., Chugreev, A., Kock, C.: Macroscopic FE-simulation of residual stresses in thermo-mechanically processed steels considering phase transformation effects. In: Oñate, E., Owen, D.R.J., Peric, D., Chiumenti, M., De Souza Neto, E. (eds.) XIV International Conference on Computational Plasticity: Fundamentals and Applications (COMPLAS 2019). Barcelona, Spain (2019)
13. Saunders, N.; Guo, U. K. Z.; Li, X.; Miodownik, A. P.; Schillé, J.-Ph.: Using JMatPro to model materials properties and behavior. J. Miner. Metals Mater. Soc. **55**(12), 60–65 (2003). https://doi.org/10.1007/s11837-003-0013-2

14. MSC.Marc® User's Handbook, Volume A: Theory and User Information. MSC Software GmbH, München, Germany (2018).
15. Behrens, B.-A., Schröder, J., Brands, D., Scheunemann, L., Niekamp, R., Chugreev, A., Sarhil, M., Uebing, S., Kock, C.: Experimental and numerical investigations of the development of residual stresses in thermo-mechanically processed Cr-alloyed steel 1.3505. Metals **9**(4), 480 (2019). https://doi.org/10.3390/met9040480
16. Behrens, B.-A., Gibmeier, J., Brunotte, K., Wester, H., Simon, N., Kock, C.: Investigations on residual stresses within hot-bulk-formed components using process simulation and the contour method. Metals **11**(4), 566 (2021). https://doi.org/10.3390/met11040566
17. Chabičovský, M., Horský, J.: Factors influencing spray cooling of hot steel surfaces. In: 26th International Conference on Metallurgy and Materials (METAL 2017). Tanger Ltd., Brno, Czech Republic (2017)

Grinding Burn Prediction by Sensor Data in Generating Gear Grinding

Mariana Mendes Wilfinger[1,2(✉)], Mareike Davidovic[1,2], Christian Westphal[1,2], and Thomas Bergs[1,3]

[1] MTI of RWTH Aachen University, Aachen, Germany
m.mendes_wilfinger@wzl.rwth-aachen.de
[2] WZL of RWTH Aachen University, Aachen, Germany
[3] Fraunhofer IPT, Aachen, Germany

Abstract. Continuous generating grinding is a highly productive process for hard finishing gears to meet quality and surface integrity standards. Due to its abrasive characteristics and the resulting significant material deformation, the process requires substantial energy per volume of material removed. This energy results in thermal and mechanical loads into the workpiece. Thermal loads account for the largest share, resulting in a risk to surface integrity through grinding burn. To ensure sufficient surface integrity while enhancing productivity, the current approach involves iterative adjustment of process parameters along with monitoring for the presence of grinding burn. Previous research has shown a correlation between the grinding machine tool spindle power signals and analytically calculated thermal energy, indicating potential for in-situ assessment of surface integrity. Therefore, the objective of this paper is to systematically investigate the correlations between machine signals and the presence of thermal damage. To verify the effects in the high frequency range, where micro-cutting components are dominant, acoustic emission and acceleration sensors were employed to record the energy responses during the process. To determine which sensors effectively acquire process-relevant information correlated to grinding burn, the sensors were installed at different locations in the grinding machine. Generating gear grinding trials were conducted, varying the feed and infeed to generate different levels of grinding burn. The sensor data were processed and evaluated, and the observed patterns were correlated with the presence of grinding burn in the ground gears.

Keywords: Generating gear grinding · Machine signals · Sensors · Grinding burn

1 Motivation and Approach

Generating gear grinding is a process commonly applied for the hard finishing of gears with high quality requirements. During the process the tool and workpiece are in constant engagement, and multiple teeth are ground simultaneously [1]. As a manufacturing process with an undefined cutting edge, generating gear grinding inherently involves the

L. Overmeyer and B.-A. Behrens (eds.), *Production at the Leading Edge of Technology*, Lecture Notes in Production Engineering, https://doi.org/10.1007/978-3-032-19524-1_20

introduction of process-induced thermal energy into the workpiece [2]. During the tool-workpiece engagement, a portion of this heat is dissipated into the surroundings through the grinding wheel, chips and coolant. However, most of the heat is transferred into the workpiece due to the high thermal conductivity of steel, leading to a localized temperature increase within the grinding zone [3]. The magnitude of this temperature increase, and the duration of thermal exposure can critically influence the metallic microstructure within the workpiece subsurface zone [4]. Such alterations of the workpiece microstructure are referred to as thermal damage, or commonly as grinding burn.

To detect the presence of grinding burn in the surface layer of the workpiece, both destructive and non-destructive methods have been established. As destructive methods, metallography analysis, hardness and residual stress measurements can be applied. Such methods are usually not performed directly on the manufactured part itself, but rather on a sample cut from it. Moreover, some of these techniques inherently destroy the sample during application [5]. This requirement for component destruction therefore makes the grinding burn assessment a time-consuming process. For that reason, non-destructive methods are commonly used in the industry. Nital etching is established as one such method, based on the resistance of the workpiece material to nitric acid [5]. The application of nital etching results in the surface regions which are affected by grinding burn appearing darker. Besides nital etching, the procedure of Barkhausen noise measurement is commonly applied, based on the difference of magnetic properties between thermally damaged and undamaged surfaces. As a result, the procedure requires calibration with a component of known properties [5].

For both destructive and non-destructive methods, the gears must be analyzed after completion of the grinding process. Therefore, the manufacturing process cannot continue until the grinding burn assessment is completed. In that regard, as an alternative to the current iterative procedure of grinding burn detection, on-line process monitoring methods offer the advantage of immediate assessment during the process. However, real-time monitoring of grinding processes is challenging due to the complex, highly dynamic tool-workpiece interactions, coupled with high speeds, temperatures, and machine stiffness [3]. Nevertheless, acceleration and acoustic emission (AE) sensors have been proven suitable to provide relevant on-line information on the grinding process [6]. Acceleration sensors are able to capture vibrations arising from tool-workpiece interaction and machine dynamics, characterizing the overall kinematic and structural behavior during grinding. Meanwhile, AE sensors detect high-frequency stress waves from dynamic events at the grinding interface, caused by micro-cutting components [6]. The inhomogeneities and phase transformations associated to grinding burn contribute significantly to the sudden release of energy which generates acoustic emissions [8]. Hence, the objective of this paper is to investigate the detectability of grinding burn during generating gear grinding based on acceleration and acoustic emission sensor signals.

To achieve the objective, the approach is divided into four phases. In phase 1, experimental trials were conducted for generating gear grinding of case-hardened gears. During the trials, spindle power as well as acceleration and acoustic emission signals of the process were recorded. In phase 2, the ground gears were characterized in terms of grinding burn presence, by means of nital etching and Barkhausen noise measurements. In phase 3, the signals recorded during the trials were treated and evaluated through joint time frequency analysis. Finally, in phase 3, the patterns observed on the sensor data were evaluated comparatively to the presence of grinding burn on each gear.

2 Experimental Methodology

2.1 Experimental Setup

The generating gear grinding trials were performed on a Klingelnberg Viper 500 KW grinding machine, with the setup shown on the left side of Fig. 1. The tool and workpiece properties are shown on the top right side of the figure. The workpiece was centered and clamped hydraulically.

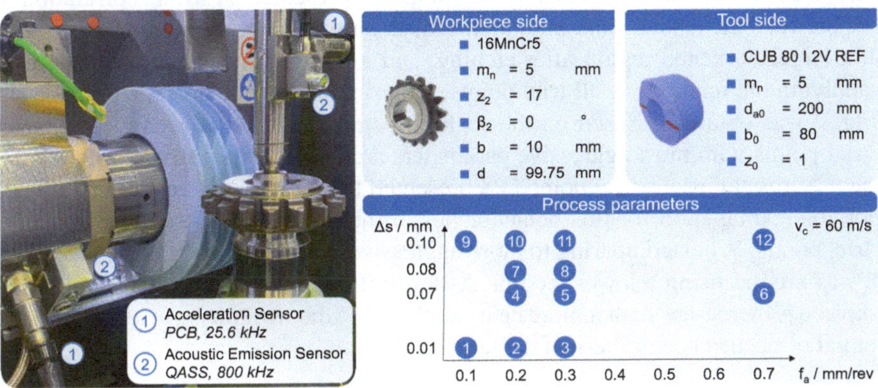

Fig. 1 Experimental setup

To record different signals from the process, acoustic emission and acceleration sensors were installed in the machine. Such sensors should ideally be installed as close to the contact region as possible. To identify the most appropriate placement, one of each sensor was installed on both the tool side and the workpiece side. On the tool side, the sensors were mounted in the stationary section of the tool spindle, while on the workpiece side they were installed on the counterholder. The counterholder had no contribution in clamping the workpiece and was only mounted for placement of the sensors. While the acceleration sensors were glued directly onto the surface of the spindle/counterholder, the acoustic emission sensors were mechanically fixed onto an aluminum plate, which was then glued onto the machine surfaces. The properties and positions of the sensors are detailed on the left side of Fig. 1.

The process parameters were varied to achieve different levels of grinding burn. Therefore, the axial infeed Δs and axial feed f_a were varied, given their influence on the changes in the workpiece surface integrity [7]. The axial infeed Δs was varied between 0.01 and 0.10 mm, while the axial feed f_a was varied between 0.1 and 0.7 mm/rev, as seen in the bottom right of Fig. 1. The cutting speed was kept constant at $v_c = 60$ m/s for all trials. The number of strokes was varied for each axial infeed Δs to obtain the designed tooth thickness. To ensure correct material stock removal, the gear geometry was measured after every trial. For the analysis, the sensor signals were recorded for the last stroke of each trial point.

2.2 Procedure for Grinding Burn Detection

On the left side of Fig. 2, the first column of pictures illustrates the gear flanks of 4 trial points directly after the grinding process. It is visually not possible to identify the presence of grinding burn. Therefore, to characterize the ground gears in terms of grinding burn, the procedures of nital etching and Barkhausen noise measurement were applied. In the second column of pictures in Fig. 2, the gear flanks of the same 4 trial points after nital etching are displayed. The Barkhausen noise values were measured for each gear and normalized based on the reference trial TP1. In the left corner of each picture, the normalized Barkhausen noise values are presented. A strong correlation was found between the results from both methods, as flanks which are more burned present both a larger darkened region after etching, and a higher value of Barkhausen noise. By applying the methods to all trial points, the grinding burn scale shown on the right of Fig. 2 was obtained. Based on the scale in comparison to the process parameters, the trial points with most aggressive parameters are generally not necessarily the most burned. Across all trials, trial point TP8 presented the highest degree of grinding burn. A possible explanation for this behavior lies in the difference of the feed fa between the trial points. When comparing to more aggressive trial points such as TP6 and TP12, TP8 was ground using a lower feed fa. As a result, the contact time between tool and workpiece is increased, so that more heat can flow into the workpiece, leading to a higher potential of occurrence of thermal damage [3].

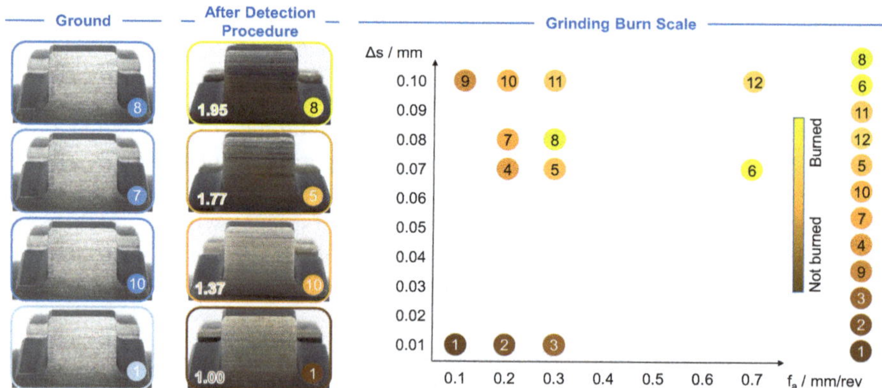

Fig. 2 Grinding burn scale

2.3 Preparation of Sensor Data

The mechanisms leading to the occurrence of grinding burn involve material phase transformations and contact between grinding worm grains and the workpiece. These occurrences manifest at high frequencies, and their magnitude changes continuously throughout the process. For that reason, both the time and frequency spectrum of the acceleration and acoustic emission signals are relevant for the analysis. Therefore, to evaluate each trial point, a joint time frequency analysis (JTFA) of the signal was applied. For this purpose, the signal in the time domain was segmented into multiple discrete time windows. For each time window, a Fast Fourier Transform (FFT) of the signal was computed. The FFT of each window was then aligned sequentially. By reorienting the visualization for a two-dimensional representation, a waterfall diagram was generated. In the diagram, the time in seconds and frequency in Herz are depicted on the X- and Y-axes, respectively, while the signal amplitude is represented by the color scale. To evaluate a comparable region across all trial points, the last instant of the last grinding stroke was analyzed for all trials.

3 Analysis of the Influence of the Presence of Grinding Burn on the Sensor Data

As mentioned in Chap. 2, the sensors were installed on both the tool and on the workpiece side. Figure 3 shows the acceleration (top row) and acoustic emission (bottom row) JTFA diagrams for the tool side. The results are shown for trial points TP1 (reference), TP10 (partially burned) and TP8 (most burned). When comparing the diagrams for each trial point, no visible change is noticed, despite the different levels of grinding burn and aggressiveness of process parameters. This observation suggests that the tool side is not suitable for sensor installation. The robustness and stiffness of the tool spindle restrict the accurate transmission of subtle behaviors from the contact zone to the sensors.

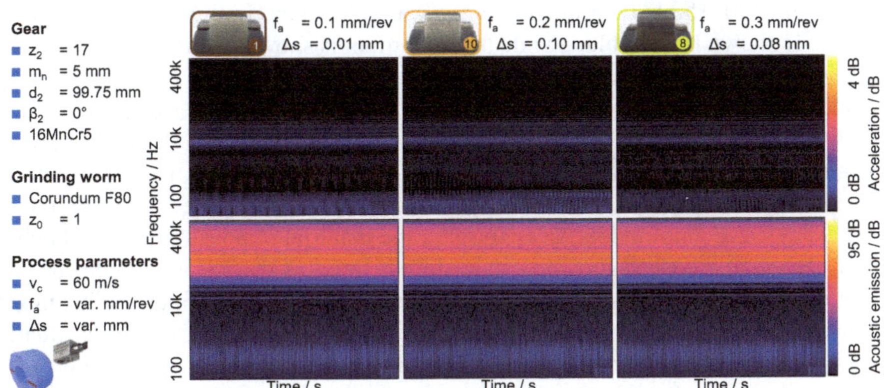

Fig. 3 Joint time frequency diagrams of acceleration and AE signals in the tool side

For the sensors installed on the workpiece side (counterholder of the gear), a visible change is noticeable across trial points TP1, TP10 and TP8, see Fig. 4. On the acceleration signals (top row), the most significant change happens at the maximum frequency of 15 kHz. At this frequency, most of the captured behavior is characteristic of the grinding process dynamics, such as vibration of mechanical components and increase of grinding forces due to more aggressive process parameters. However, on the acoustic emission signals (bottom row), a clear change can be seen when comparing the trial points in the high frequency range. As the level of grinding burn gets higher, an overall increase in the AE amplitude, as well as excitation of larger frequency bands, are identified. For further comparison of all trial points, a section of the signal between the frequencies of 100 and 150 kHz was selected, as shown by the yellow box in Fig. 4. Previous research has shown a correlation between an increase of the AE amplitude in this frequency range, and the presence of grinding burn [8].

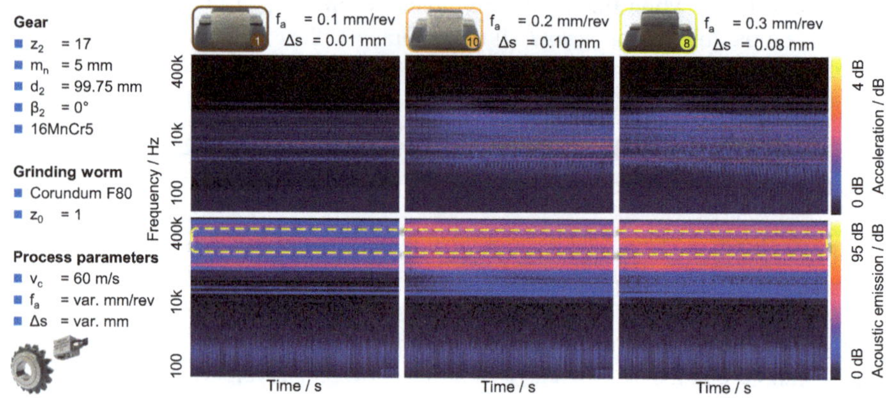

Fig. 4 Joint time frequency diagrams of acceleration and AE signals in the workpiece side

For all trial points, the average and maximum amplitude of the acoustic emission signals in the 100–150 kHz frequency range was calculated. In Fig. 5, the average AE amplitude is shown in light blue, while the maximum is shown in dark blue. The values were normalized to the maximum across all trial points and are shown as a percentage for comparison. On the X-axis, the Barkhausen noise measurement is displayed, corresponding to each trial point. Initially, a positive correlation between AE amplitude and Barkhausen noise is evident, as shown by the R2 values shown in the figure with the corresponding colors (R2 = 0.7107 for the maximum AE amplitude, and R2 = 0.8314 for the average AE amplitude). An analysis of the material removal rate Q'_W, shown in lightest blue with a dashed line, reveals a similar trend as the AE amplitude for most trial points. The same is true for the average tool spindle power, which can be mostly correlated to the grinding forces during the process. For that reason, from this evaluation it is not clear whether the increase in AE amplitude is a consequence of the presence of grinding burn, or the change of process parameters. Nevertheless, an exception to that behavior is the most burned trial point, TP8. For this trial point, there is a decrease in the values of material removal rate Q'_W and tool spindle power, in comparison to an increase of the Barkhausen noise measurement. However, an increase in the AE amplitude is observed. This behavior suggests that, although the process parameters and grinding forces are less aggressive for TP8, the occurrence of thermal damage (and with that the increase of the Barkhausen noise measurement), could have been the source of emissions which had an influence in the AE amplitude.

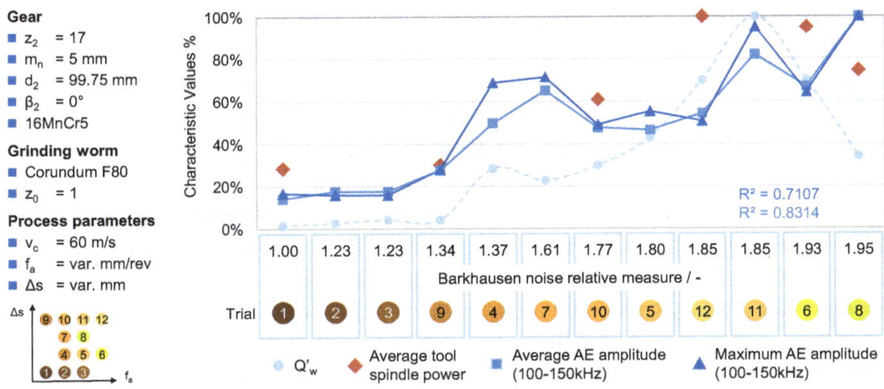

Fig. 5 Comparison of AE amplitude in the 100–150 kHz frequency range for all trial points

To analyze whether AE amplitude changes are caused by the change of process parameters or presence of grinding burn, results from repetition trials for trial point TP7 (slightly burned) are presented in Fig. 6. For this trial point, three gears were ground with the same process parameters (TP7-1, TP7-2 and TP7-3). However, both Barkhausen and nital etching results showed that only the gear for TP7-3 was thermally damaged, see left of Fig. 6. An evaluation of the AE amplitude revealed an increase in the signal amplitude for the thermally damaged gear compared to the others, despite the constant process parameters (see right side of Fig. 6). According to the gear measurements done after the trials, a comparable material stock has been removed in all trials. This effect supports the hypothesis that the amplitude increase may be a consequence of the presence of grinding burn, and not the change of process parameters.

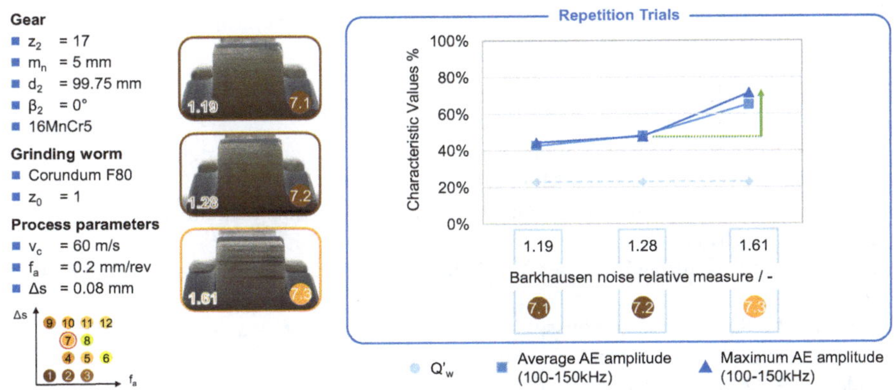

Fig. 6 Comparison of AE amplitude in the 100–150 kHz frequency range for repetition trials

4 Summary and Outlook

The objective of this research was to investigate the presence of grinding burn during generating gear grinding using sensor data. To achieve this, experimental trials were conducted with acoustic emission and acceleration sensors installed at different positions on the grinding machine. The process parameters were systematically varied to achieve different levels of grinding burn. The ground gears were characterized for grinding burn using nital etching and Barkhausen noise measurements. The signals acquired from the sensors were then analyzed through joint time frequency analysis (JTFA). Evaluation of the signals showed no variation on the signals from the tool side sensors, and no correlation between workpiece side acceleration signals and grinding burn. However, trials with grinding burn showed an increased acoustic emission amplitude on the workpiece side. This increase occurred specifically within the 100–150 kHz frequency range. Repetition trials further validated that a 100–150 kHz AE amplitude increase is observed when grinding burn is detected, even when there is no change in process parameters.

In the next steps, further experiments, including repetition trials, should be conducted to investigate more severe cases of grinding burn and a broader range of process parameters. In the trials, installing additional sensors may aid in understanding the isolated effect of grinding burn on the AE signal. With a clearer definition of the influence of the presence of grinding burn to AE signals, it will then be possible to develop a model for definition of a limit value of AE amplitude for detection of grinding burn. Additionally, with different gear geometries must be carried out to validate the frequency range where grinding burn is identified. With those steps, the in-process detection of grinding burn will be possible, reducing the need for manual inspection or laboratory analysis through nital etching and Barkhausen noise measurements.

Competing Interests. The author(s) has no competing interests to declare that are relevant to the content of this manuscript.

References

1. Klocke, F., Brecher, C.: Zahnrad- und Getriebetechnik: Auslegung-Herstellung-Untersuchung-Simulation, 1st edn. Carl Hanser, München (2017)
2. Rowe, B.: Principles of Modern Grinding Technology, 2nd edn. William Andrew Publishing, Norwich, NY (2014)
3. Klocke, F.: Manufacturing Processes 2: Grinding, Honing, Lapping. Bd. Nr. 2. Springer, Berlin (2009)
4. Reimann, J.: Randzonenbeeinflussung beim kontinuierlichen Wälzschleifen von Stirnrad-verzahnungen. RWTH Aachen University, PhD-thesis (2014)
5. Seidel, M.W., Zösch, A., Härtel, K.: Grinding Burn and its Testing, 1st edn. Hanser Fachverlag, Munich (2020)
6. Pandiyan, V., Shevchik, S., Wasmer, K., Castagne, S., Tjahjowidodo, T.: Modelling and monitoring of abrasive finishing processes using artificial intelligence techniques: a review. J. Manuf. Process. **57**, 114–135 (2020)

7. Karpuschewski, B., Beutner, M., Eckebrecht, J., Heinzel, J., Hüsemann, T.: Surface integrity aspects in gear manufacturing. In: Procedia CIRP, vol. 87, pp. 3–12 (2020)
8. Saxler, W.: Erkennung von Schleifbrand durch Schallemissionsanalyse. RWTH Aachen University, PhD-thesis (1997)

The Influence of the Stand-Off Distance on Surface Topography and Related Roughness Values for Titanium Implant Processing by Pure Water Jet

Zhuxi Lang[1](✉), Frank Pude[2,3], Joachim Regel[1], Florian Morczinek[1], and Martin Dix[1,4]

[1] Professorship Production Systems and Processes, Chemnitz University of Technology, Chemnitz, Germany
zhuxi.lang@mb.tu-chemnitz.de
[2] Inspire AG, Zurich, Switzerland
[3] Steinbeis Consulting Center High-Pressure Waterjet Technology, Horgau, Germany
[4] Fraunhofer Institute for Machine Tools and Forming Technology IWU, Chemnitz, Germany

Abstract. In cementless arthroplasty, the prosthetic surface is intentionally roughened to enhance the initial mechanical stability and to facilitate long-term osseointegration—the process by which bone gradually grows onto and integrates with the implant over time. For this goal this study investigates the processing of titanium implant surfaces, focusing on the achievable surface topography and roughness variations resulting from changes in stand-off distance (jet length) when using a pure water jet (without abrasive). This particle-free processing approach serves as an alternative to the conventional method of sandblasting followed by chemical etching. By eliminating the use of abrasive particles, this technique entirely mitigates the risk of implant failure caused by undetected particle contamination on or beneath the surface. The ability to achieve the desired surface topography depends on the jet profile as it impacts the titanium surface, as well as the disintegration of the water jet, which exits the nozzle as a compact stream but subsequently breaks into discrete fluid packets or, depending on the nozzle type, droplets. The dynamics of this jet profile and impact behavior evolve with increasing stand-off distance. In this laboratory study, the stand-off distance was varied as a processing parameter and its influence on the resulting surface characteristics was quantitatively assessed. The ultimate goal of optimizing the implant surface topography is to enable the use of titanium components as cement-free implants, ensuring direct contact between the implant surface and bone.

Keywords: Processing of titanium implant · Surface topography · Water jet technology · Stand-off distance

1 Introduction

With the trend of an aging population and rising obesity rates, the number of people undergoing joint replacement surgery is increasing every year [1]. A total of 378,812 hip and knee prosthesis implantation procedures were recorded in the Endoprosthesis

L. Overmeyer and B.-A. Behrens (eds.), *Production at the Leading Edge of Technology*, Lecture Notes in Production Engineering, https://doi.org/10.1007/978-3-032-19524-1_21

Register of Germany in 2023, representing an increase of almost 7% compared to 2022. About 10% of these are subsequent repair surgeries where the initial implant failed. The two most common reasons for revision surgery are prosthesis loosening and infection [2]. Compared to initial implantation, revision surgery carries a higher risk of infection and complications, and causes pain to the patient. Therefore, the improvement of the success rate of the initial surgery and prolong the life of the prosthesis has become the key to joint replacement surgery [3–5].

Arthroplasty can be categorized as cemented or cementless depending on how the built-in prosthesis is anchored in the bone cavity. In cementless arthroplasty, the surface of the implant is in direct contact with the bone, and therefore the success of the procedure largely depends on the quality and function of the implant-bone interface. In this type of surgery, the surface of the prosthetic implant is purposely roughened to enhance initial mechanical locking and to promote long-term biological fixation through osseointegration [6, 7]. The dominant method for surface preparation of cementless prostheses on the market today is sandblasting followed by chemical etching. However, this method inevitably leaves abrasive particles on and under the surface of the prosthesis. These residues may trigger inflammatory reactions, impair biocompatibility and affect the long-term stability of the implant [8, 9].

Arola et al. [10] first used abrasive water jet (AWJ) to treat the surface of Ti6Al4V implants. Although the desired surface texture and roughness were obtained, contamination with abrasive particles was still detected on the implant surface after processing. Later Arola et al. [11] tried to roughen the implant surface using pure water jet (PWJ). While this avoided particle contamination, the machining efficiency was too low, with no significant change in surface roughness after processing. Pulsating water jets, which are generated by introducing periodic pressure variations into a pure water jet, greatly improve processing efficiency and reduce the required working pressure [12–14]. This is because the water hammer effect produced by intermittent water impacts greatly increases the erosion efficiency on the target material. Surface treatment with pulsating water jet achieves both the desired roughness and avoids contamination by abrasive particles. However, this innovative approach requires an additional device to modulate the continuous water jet, such as an ultrasonic generator, which increases the complexity and costs.

When a high-speed water jet passes through the air, the resulting of air resistance causes it to oscillate. As the jet increases in length, the air disturbances gradually intensify, eventually causing the jet to break down. This phenomenon is known as the Plateau-Rayleigh instability [15]. High-speed camera images record the changes in the morphology of water jet as the distance increases [16, 17]. The jet transitions from a continuous state to a discontinuous state, forming droplets that then disintegrate and dissipate into the air. Although the kinetic energy of the water jet is reduced by air resistance, the impact force of the droplets is much higher than that of the continuous fluid. This study aims to use droplets by increasing the stand-off distance Z_{st}, i.e. the length of the pure water jet, with the aim of improving the erosion efficiency on titanium implant and evaluating the influence of Z_{st} on the final surface topography.

2 Material and Experimental Setup

Ti6Al4V is one of the most common implant material on the market due to its excellent mechanical properties and biocompatibility [18]. In this experiment, 1.5 mm thick Ti6Al4V plates were used as samples for erosion testing with pure water jets at Z_{st} (nozzle-to-sample distance) of 2 mm, 10 mm, 20 mm, 50 mm and 100 mm, respectively. The experimental setup is shown in Fig. 1a. The X- and Y-axis control the erosion path, while the Z-axis controls the Z_{st}. The distance from the nozzle to the machine table is limited to the travel range of the Z-axis, which is between 40 and 110 mm. Therefore, to achieve a Z_{st} less than 40 mm, a fixture with a height of 50 mm was required to lift the sample. For Z_{st} of 50 and 100 mm, the samples were fixed directly to the machine table. The nozzle diameter D used was 0.25 mm, the supply water pressure p was 3000 bar, and the water jet processing traverse speed v_p was held constant at 600 mm/min.

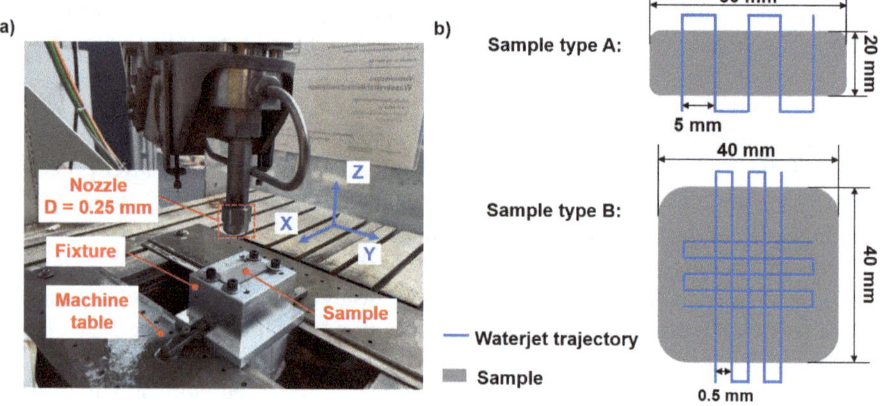

Fig. 1 **a** Construction of the experimental setup, **b** geometry of the samples and water jet processing trajectories

Figure 1b shows the geometry of the two sample types used in the tests and the processing trajectory of the water jet. First, five parallel tracks were made on sample A for the same Z_{st} with a spacing of 5 mm to evaluate the material removal rate ε_r, which is an indicator of productivity. With v_p known, ε_r can be calculated by Eq. (1) with the volume of removed material V_r within a certain sampling length L_s. Five measurements were taken for each sample, corresponding to five grooves, to minimize measurement errors.

$$\varepsilon_r = \frac{V_r \cdot v_p}{L_s} \tag{1}$$

To create the surface topography, perpendicular cross-tracks were made on sample B with a spacing of 0.5 mm. The area roughness arithmetical mean height S_a, maximum height S_z, and kurtosis S_{ku} were then measured in a 1 mm^2 area at the center of the sample, where the trajectories overlap. In order to compare the surface topography after different treatment methods, an additional sample B was processed with AWJ. In order to obtain a comparable surface roughness to PWJ, the process parameters for AWJ were set to $p = 1500$ bar, $v_p = 1200$ mm/min, $Z_{st} = 10$ mm. The experimental parameters are summarized in Table 1.

Table 1 Process parameters for surface treatment experiments

Method	D [mm]	p [bar]	v_p [mm/min]	Z_{st} [mm]
PWJ	0.25	3000	600	2/10/20/50/100
AWJ	0.25	1500	1200	10

As demonstrated in Fig. 2, the samples were grinded before the jetting, leaving scratches of 2–10 μm on the surface. However, due to the shallow nature of the scratches, there was no influence on subsequent experimental measurements and evaluations. The roughness parameters of the samples surface before water jet processing was summarized in Table 2. The measurements were taken with a KEYENCE® VK9700 laser scanning microscope with a scanning height of 0.1 μm.

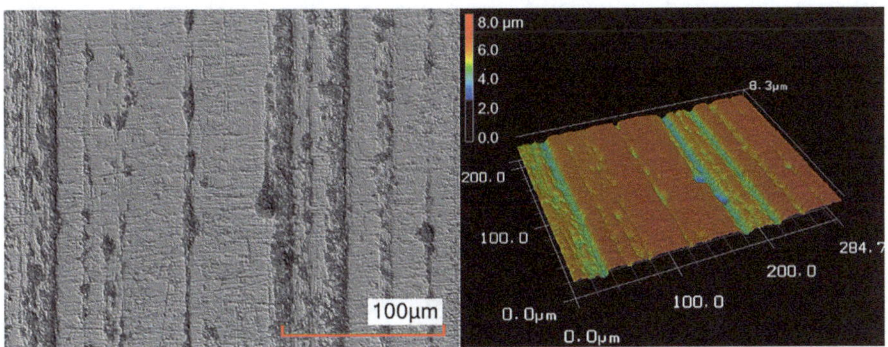

Fig. 2 An example of the initial surface topography of samples before water jet processing

Table 2 Material removal rates and surface roughness under different treatment conditions

Treatment conditions	ε_r [mm^3/min]	S_a [μm]	S_z [μm]	S_{ku}
Initial surface	–	2.9	53.6	2.9
PWJ at Z_{st} [mm]				
2	–	2.7	52.0	2.8
10	–	3.8	63.3	4.1
20	0.898	5.3	76.1	3.1
50	3.053	5.7	81.5	2.9
100	6.130	10.1	108.4	2.5

(*continued*)

Table 2 (*continued*)

Treatment conditions	ε_r [mm^3/min]	S_a [µm]	S_z [µm]	S_{ku}
AWJ ($Z_{st} = 10$ mm)	–	6.3	130.5	3.7

3 Results and Discussion

In order to visualize the effect of the water jet profile on the erosion (see Fig. 3), an image was used to record the dynamic characteristics of the water jet for this experimental condition. As can be seen in Fig. 3a, the jet takes on a regular columnar shape in the region close to the nozzle ($Z_{st} < 20$ mm). As the length of the jet increases, gradually water droplets peel off from the central region and disintegrate. When Z_{st} reaches 50 mm, corresponding to 200 D in Ref. [7], the stream has spread out significantly and become discontinuous. At $Z_{st} = 100$ mm (400 D), the center of the stream is almost indistinguishable, and a large amount of the water is dispersed into the air. Figure 3b shows the erosive effect of the water jet on sample A at various Z_{st}. . At shorter distances ($Z_{st} < 10$ mm), the water jet leaves almost no trace on the Ti6Al4V sample. This is because the droplets have not formed yet, and the impact of continuous water jet is insufficient to remove the material. As the Z_{st} increases, the water hammer effect caused by the droplets in the water stream makes it much more efficient at eroding material. Clear grooves were left on the surface of the samples when Z_{st} was greater than 50 mm.

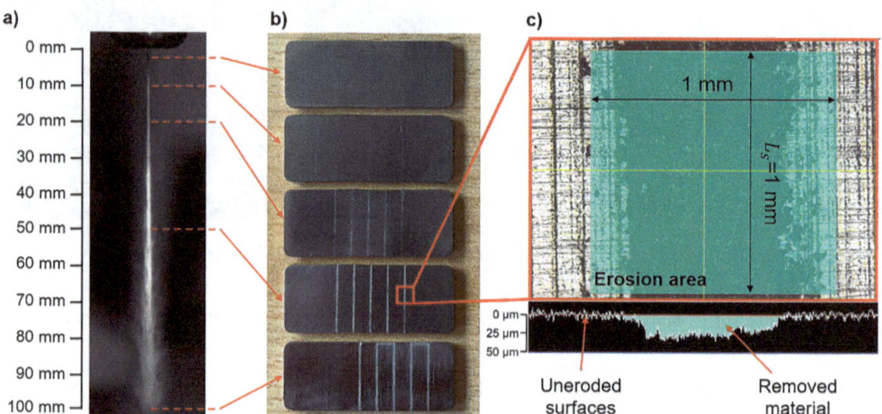

Fig. 3 **a** Profile of water jet and **b** erosion effects at different stand-off distances, as well as **c** the microscopic measurements of material removal

KEYENCE®'s analysis software can calculate the volume of removed material by integrating the distance between the upper and lower surfaces. As shown in Fig. 3c, an erosion area is randomly selected on each groove. By defining the height of the uneroded surface, the volume of removed material V_r can be obtained for the calculation of ε_r. The average ε_r for each Z_{st} are summarized in Table 2. It is not possible to take a measurement when Z_{st} less than 10 mm because there is no optical difference between the initial state of the surface and the grooves produced. ε_r increases gradually with rising Z_{st} and reaches 6.13 mm^3/min at $Z_{st} = 100$ mm. Due to the limitation of the Z-axis, it is not possible to increase Z_{st} further in this work. However, it can be predicted that a larger Z_{st} will slow down the jet because of the air friction, thereby reducing the kinetic energy of the water droplets. Consequently, an excessively large Z_{st} will instead lead to a decrease in ε_r.

The results after cross-tracks pure water jetting are shown in Fig. 4. The microscope images (Fig. 4b) represent the area where the trajectories overlap in the center of sample. The surface roughness parameters S_a, S_z, and S_{ku} under different treatment conditions are summarized in Table 2. It can be concluded that S_a and S_z both increase incrementally with Z_{st}. Although the kinetic energy of the water jet is greatest at $Z_{st} = 2$ mm, it has almost no effect on the surface roughness. There are different theories regarding the optimal surface roughness for osseointegration [19]. The results of the present experiments are within the range of S_a values for the most widely used cementless implants, i.e., 2–10 μm. S_{ku} describes the sharpness of the surface height distribution. A higher S_{ku} value indicates a surface with more extreme profiles, while a lower S_{ku} suggests a smoother and more uniform surface texture. When $Z_{st} = 10$ mm, a grid-like pattern was observed. This occurred because only a small amount of material along the path direction was removed, creating distinct peaks and valleys in the microstructure, and thus the largest S_{ku} values. As Z_{st} continues to increase ($Z_{st} \geq 20$ mm), the eroded area becomes more uniform and the S_{ku} value decreases.

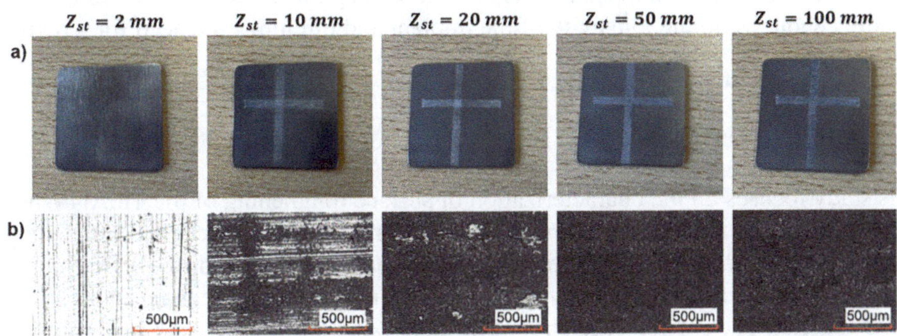

Fig. 4 **a** Surface topography of the samples after cross-track processing using pure water jet at different stand-off distances and **b** microscope images of the overlapping area of the processing tracks

Surface treatment with AWJ is analogous to sandblasting, as both methods rely on high-velocity abrasive particles carried by a fluid to erode the surface material. Figure 5 presents a comparison of surface topographies resulting from PWJ and AWJ treatments.

The corresponding surface roughness parameters for AWJ are provided in Table 2. Similar to sandblasting, residual abrasive particles were detected on the surface machined by AWJ. In contrast, the surface processed by PWJ with $Z_{st} = 100$ mm exhibits a more uniform texture and a lower S_{ku} value. This difference can be attributed to the cutting mechanism of AWJ, where abrasive particles cut the material at the microscopic level, leaving sharp edges on the surface. Such sharp edges may cause stress concentrations, leading to bone resorption and affecting the long-term stability of the implant [20, 21].

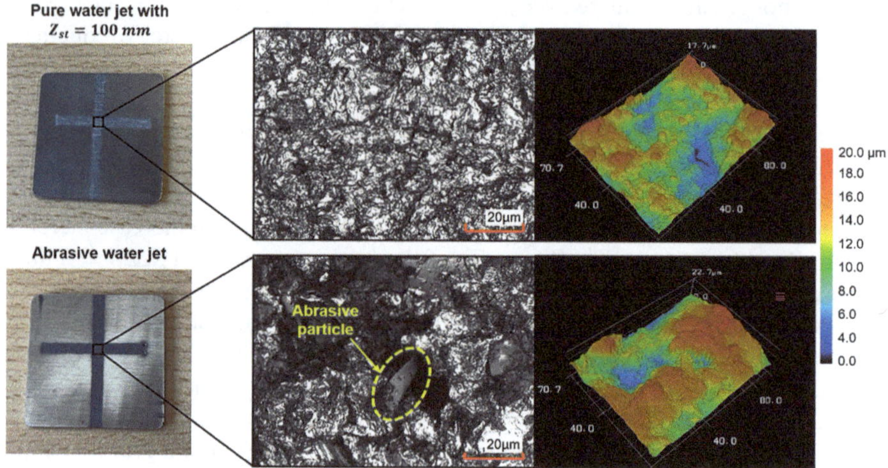

Fig. 5 Comparison of surface topography after surface processing using PWJ (at $Z_{st} = 100$ mm) and AWJ

The results show that surface treatment with PWJ not only does not leave particles on the implant surface, but also results in a more uniform surface texture, all of which is more conducive to the longevity of the implant and minimize the need for a revision surgery.

4 Conclusions

This study focuses on the characterization of surface roughening of the Ti6Al4V using pure water jets. The experimental results demonstrate the limited erosion efficiency of pure water jets on the material at a short Z_{st}. As the Z_{st} increased, the water hammer effect induced by dispersed droplets in the jet stream significantly enhanced the erosion of the material. The erosion efficiency of the pure water jet reached 6.13 mm^3/min using a 3000 bar pressure, 0.25 mm nozzle at Z_{st} of 100 mm. The surface roughness of the treated surfaces using pure water jet can be as high as $S_a = 10.1$ μm and $S_z = 108.4$ μm, with a low S_{ku} of 2.5. Higher S_a and S_z values promote initial mechanical stability and long-term osseointegration. Lower S_{ku} values prevent bone resorption caused by stress concentration. Compared to traditional surface treatments such as sandblasting and abrasive water jet, pure water jet processed surface not only leave no particulate contaminants, but also have a more uniform surface texture.

Acknowledgements. The authors would like to thank the Federal Ministry of Research, Technology and Space (BMFTR) for their support in the project "OrthoJet" (Project no. 13GW0586F), in the Action Field "Gesundheitswirtschaft im Rahmenprogramm Gesundheitsforschung" via the "KMU-Innovativ: Medizintechnik".

Competing Interests The author(s) has no competing interests to declare that are relevant to the content of this manuscript.

References

1. OECD Indicators: Health at a Glance (2023), https://www.oecd.org/en/publications/2023/11/health-at-a-glance-2023_e04f8239/full-report/hip-and-knee-replacement_687dec46.html#WHO14_e2151bdeee.
2. EPRD Jahresbericht (2024), https://www.eprd.de/fileadmin/user_upload/Dateien/Publikationen/Berichte/Jahresbericht2024-Status5_2024-10-22_F.pdf.
3. Pellicci, M.P., Wilson, P.D., Sledge, C.B., et al.: Revision total hip arthroplasty. Clin. Orthop. Relat. Res. **170**, 34–41 (1982)
4. Amstutz, H.C., Ma, S.M., Jinnah, R.H., et al.: Revision of aseptic loose total hip arthroplasties. Clin. Orthop. Relat. Res. **170**, 21–33 (1982)
5. Bryan, R.S., Rand, J.A.: Revision total knee arthroplasty. Clin. Orthop. Relat. Res. **170**, 116–122 (1982)
6. Morshed, S., Bozic, K.J., Ries, M.D., et al.: Comparison of cemented and uncemented fixation in total hip replacement. Acta Orthop. **78**(3), 315–326 (2009). https://doi.org/10.1080/17453670710013861
7. Goldberg, V.M., Stevenson, S., Feighan, J., Davy, D.: Biology of gritblasted titanium alloy implants. Clin. Orthop. **319**, 122–129 (1995)
8. Böhler, M., Kanz, F., Schwarz, B. et al.: Adverse tissue reactions to wear particles from Co-alloy articulations, increased by alumina-blasting particle contamination from cementless Ti-based total hip implants. J. Bone Joint Surg. (Br) **84**-B(1), 128–136 (2002)
9. Schuh, A., Uter, W., Kachler, W. et al.: Comparative surface examinations on corund blasted titanium implants and explants in total hip arthroplasty. Arch Orthop Trauma Surg. **125**, 676–682, Springer (2005)
10. Arola, D.D., McCain, M.L.: Abrasive waterjet peening: a new method of surface preparation for metal orthopedic implants. J. Biomed. Mater. Res. **53**(5), 536–546 (2000)
11. Arola, D.D., McCain, M.L., Kunaporn, S., et al.: Waterjet and abrasive waterjet surface treatment of titanium: a comparison of surface texture and residual stress. Wear **249**(10–11), 943–950 (2001)
12. Steeves, A., Variola, F., Vijay, M.: In vitro activity of forced pulsed WATERJET (FPWJ) modified titanium, WJTA-IMCA Conference and Expo (2017)
13. Stolárik, G., Svobodová, J., Klichová, D., et al.: Titanium surface roughening with ultrasonic pulsating water jet. J. Manuf. Process. **90**, 341–356 (2023)
14. Klichová, D., Stolárik, G., Nag, A. et al.: Surface modification of titanium alloy Ti6Al7Nb by high-frequency droplet impingement. Wear vol. **570** (2025)
15. Lord Rayleigh, F.R.S.: On the Instability of Jets. Proceedings of the London Mathematical Society Vol. s1–10, pp. 4–13 (1878), https://doi.org/10.1112/plms/s1-10.1.4
16. Zelenak, M., Riha, Z., Votavova, H. et al.: Methods for the behaviour analysis of continuous flat water jet structures. Measurement **202** (2022)
17. Jiang, M., Wang, F., Yuan, M., et al.: Analysis of the whole main structure morphological evolution and velocity distribution characteristics of a high-pressure water jet by an imaging experiment. Measurement **214**, 15 (2023)

18. Brunette, D.M., Tengvall, P., Textor, M., Thomsen, P.: Titanium in Medicine, 1st edn. Springer, Berlin (2001)
19. Wennerberg, A., Albrektsson, T.: Effects of titanium surface topography on bone integration: a systematic review. Clin. Oral Impl. Res. **20**(Suppl. 4), 172–184 (2009)
20. Kitamura, E., Stegaroiu, R., Nomura, S., et al.: Influence of marginal bone resorption on stress around an implant—a three-dimensional finite element analysis. J. Oral Rehabil. **32**, 279–286 (2005)
21. Hansson, K.N., Hansson, S.: Skewness and Kurtosis: important parameters in the characterization of dental implant surface roughness—a computer simulation. International Scholarly Research Notices vol. 2011 (2011), https://doi.org/10.5402/2011/305312

Influence of Relative Transducer Positioning on Cavitation Performance in Ultrasonic Cleaning Baths

Felix Weiß[1(✉)] and Welf-Guntram Drossel[1,2]

[1] Professorship Adaptronics and Lightweight Design, Chemnitz University of Technology, Chemnitz, Germany
felix.weiss@mb.tu-chemnitz.de

[2] Fraunhofer Institute for Machine Tools and Forming Technology IWU, Chemnitz, Germany

Abstract. Cleaning of parts using ultrasonic cleaning baths (UCBs) is common in industrial manufacturing, especially for fields with high demands regarding the cleanliness of parts and products. However, so far, the design principles for the ultrasonic transducers and UCBs mostly remain as institutional knowledge of the manufacturers with limited publicly available resources. Therefore, the aim of this study is to identify design principles for improving the cleaning performance while reducing the power consumption and required number of transducers. A promising approach to improve these parameters is to optimize the wave interference within the UCB. Thus, aluminum foil tests were performed to compare the cavitation performance with a distance of 85 mm and 134 mm between the transducers. Further tests were performed with a movable back plate to simulate different lengths of the UCB and investigate the effects thereof on cavitation performance. Experiments using transducers with an operating frequency of 25 kHz showed that increasing the distance between transducers from 85 mm to 134 mm improved the cavitation performance, determined by total hole area, more than 6-fold despite a reduction in the amount of transducers from 35 to 23. Furthermore, positioning the movable plate at distances of integer-multiples of the wavelength increased the cavitation performance by about 80% over distances offset by half a wavelength relative to integer multiples. Overall, the results suggest that optimal transducer positioning depends on the wavelength of the ultrasound at the operating frequency and should be chosen accordingly to optimize cleaning performance and efficiency.

Keywords: Ultrasonic cleaning · Design principles · Energy efficiency

https://www.tu-chemnitz.de/mb/adaptronik/.

L. Overmeyer and B.-A. Behrens (eds.), *Production at the Leading Edge of Technology*, Lecture Notes in Production Engineering, https://doi.org/10.1007/978-3-032-19524-1_22

1 Introduction

Gentle cleaning of fragile items or parts with complex shapes is a challenging but essential task in a variety of industries such as life sciences, aerospace engineering and microelectronics [1,2]. Ultrasonic cleaning is well-suited for this task because it can reach difficult-to-access areas of parts. It can be adjusted to different types of contaminants and is typically less damaging than alternative cleaning methods, which has lead to its widespread adoption [3,4]. Due to this prevalence of ultrasonic cleaning in industrial processes, improvements in energy efficiency could help reduce energy consumption and production costs.

Ultrasonic cleaning baths (UCBs) typically consist of one or more arrays of ultrasonic transducers affixed to the walls of a liquid-filled tank. Excitation of the transducers with a signal generator then utilizes the walls as diaphragms to vibrate the liquid in which the soiled parts are placed [4,5]. The cleaning effect is primarily achieved through a sequence of repeated formation and subsequent implosion of microbubbles in the cleaning liquid in a process known as cavitation [4,6]. The cleaning performance and thus efficiency of ultrasonic cleaning at a given level of power is influenced by a variety of different factors, such as the type and temperature of the cleaning medium [7,8], the design and material composition of the ultrasonic transducers [9,10], the frequency and amount of power provided to the transducers by the signal generator [11], and the overall design and material composition of the UCB [9].

However, because of the complexity of the subject and a lack of available data, there remains potential for improvement [4]. Since ultrasonic waves are affected by wave superposition [12], and the size and amount of microbubbles and the intensity of their collapse is frequency and amplitude dependent [13], one possible concept for efficiency improvements is the positioning of the transducers.

2 Materials and Methods

This research aims to identify design principles for better cleaning performance and increased energy efficiency of UCBs. To achieve this, two experiments were conducted to determine how the positioning of ultrasonic transducers, relative to each other and the opposing wall of the tank, affects the cavitation performance.

The influence of transducer spacing is investigated by analyzing the cavitation performance of the UCB using diaphragms with different transducer spacings. Additionally, the impact of the distance between the diaphragm and the opposing wall is examined by analyzing the cavitation performance for various tank widths using a backplate in the tank to simulate different tank widths.

All experiments were performed using an industrial UCB equipped with three exchangeable diaphragms positioned in the bottom, left, and right walls of the 350-liter tank. The bottom of the tank was slightly sloped at a 3° incline from front to back to facilitate drainage. The UCB was also equipped with a heating function for temperature control of the cleaning medium, as well as a variable frequency signal generator with a function for automatic identification of the resonant frequencies of the installed transducer arrays.

A fixed frequency was chosen for these experiments to eliminate confounding variables related to the frequency sweep. However, any insights gained should also apply to a swept frequency, since a frequency sweep only covers a narrow symmetric band centered around the resonant frequency of the combined transducer array. Therefore, it should show a similar degree of wave superposition. To further reduce the number of confounding variables all experiments were performed using tap water without additives as the cleaning medium to eliminate any kind of chemical cleaning effects.

The optimal spacing between transducers is assumed to be the distance of one transverse wavelength in the diaphragm material between the centers of the transducers, as this should maximize constructive wave interference. The transverse wavelength λ_t was calculated as

$$\lambda_{t,d} = \frac{c_{t,d}}{f} \tag{1}$$

from target frequency f and the transverse speed of sound in the diaphragm material $c_{t,d}$ which was calculated as

$$c_{t,d} = \sqrt{\frac{E}{2\rho(1+\nu)}}. \tag{2}$$

For the given diaphragm material (stainless steel) with Young's modulus $E = 200\,\mathrm{GPa}$, density $\rho = 8\,000\,\mathrm{kg/m^3}$, and Poisson's ratio $\nu = 0.3$, this results in a transverse wavelength of 124.0 mm at 25.0 kHz and 104.8 mm at 29.6 kHz.

For the experiments regarding the optimal UCB width the wavelength λ_m in the cleaning medium (water at 40°C) was calculated analogously to (1) as

$$\lambda_m = \frac{c_m}{f} \tag{3}$$

from the frequency f and speed of sound in the cleaning medium c_m.

The three diaphragms used for both experiments are shown in Table 1. The two transducer types are mostly identical but differ in their resonant frequencies. The positioning of the transducers on the diaphragms is shown in Fig. 1.

Table 1. Diaphragms used in experiments

Transducer type	Resonant frequency (kHz)	Spacing (mm)	No. Transducers
Design A	29.6	85	35
Design A	29.6	134	23
Design B	25.0	134	23

The aluminum foil test was chosen to evaluate the cavitation performance of the UCB for each experimental setup. This is a commonly employed method for testing the cleaning performance of UCBs [4]. A thin sheet of aluminum foil is used to visualize the distribution and intensity of cavitation activity. The foil is placed in the cleaning solution and sonicated. Due to the foil's fragility, cavitation causes erosion of the foil and the formation of holes, with the size of the holes correlating to the cavitation performance. That being the amount and intensity of cavitation produced under a given set of experimental settings.

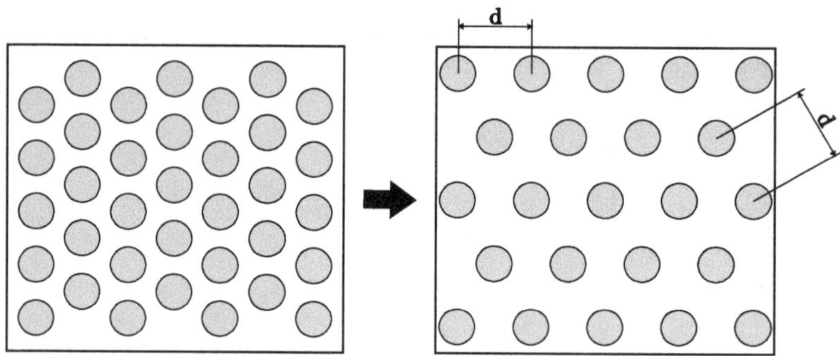

Fig. 1. Current (left) and wavelength optimized transducer spacing (right)

All of the aluminum foil tests conducted for this research followed the same general procedure. After filling the tank, the water was first degassed for several hours until no more formation of bubbles was visible. On each day of the experiments, the water was then degassed again for at least 1 h while it was heated up to 30°C for the transducer spacing experiments or 40°C for the UCB width experiments. For each experiment a sheet of industrial aluminum foil (thickness 0.03 mm) was placed in between two wire racks which were fastened together with screws through the foil at each corner. For the transducer spacing experiments, the foil was then placed in the middle of the tank in parallel to the diaphragm. For the UCB width experiments, only the diaphragms with 134 mm spacing were used. Furthermore a backplate was placed in the UCB at a distance to the diaphragm corresponding to either an integer multiple of the wavelength in water or an integer multiple offset by half a wavelength, to simulate a UCB of a given width (see Table 3 in Results). In accordance with the setup for foil tests using the whole width of the UCB, the foil was then placed in the middle between the diaphragm and the backplate (see Fig. 2). Afterwards the foil was treated with ultrasound at the resonant frequency of the transducers for 40 min. Finally, the results were photographed, quantified, and analyzed. For the transducer spacing experiments, this procedure was repeated twice per diaphragm for a total of three measurements, while only two measurements were performed for each simulated UCB width.

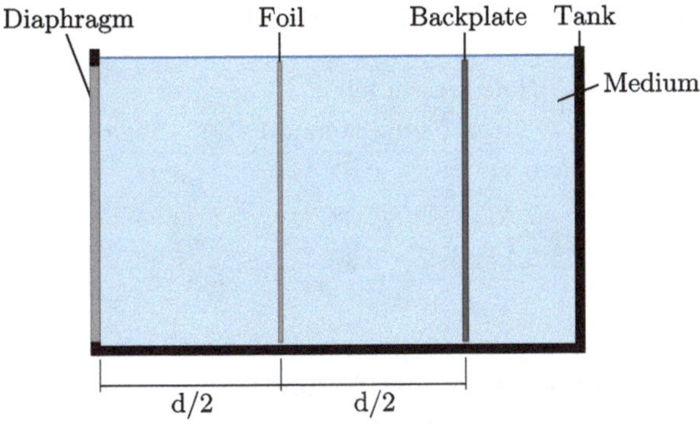

Fig. 2. Experimental setup for UCB width experiments

To make the results of the foil tests comparable, a quantification procedure was used (see Fig. 3). First, the eroded foil was removed from the wire cage, placed on a mono-colored surface and photographed. Then, the image was cropped to the area between the screw holes (Fig. 3a). Next, the colors of the image were filtered (Fig. 3b) and posterized (Fig. 3c) to reduce the number of colors to two and to create clear contrast between the foil in black and holes in red. Finally, the percentage of hole area in the foil was then determined by automated counting of pixels by color and calculation of the percentage of red pixels in the image (Fig. 3d).

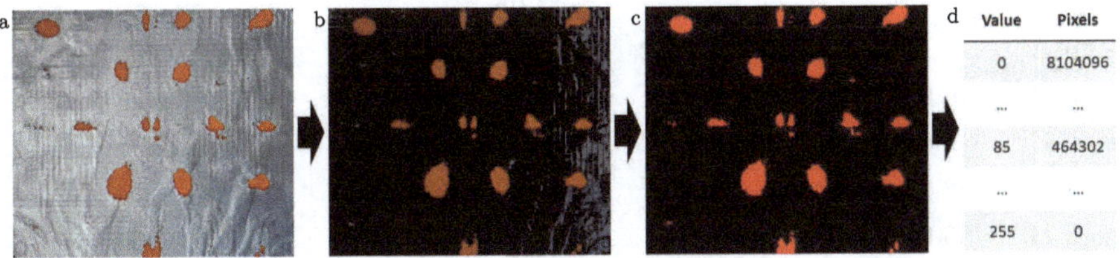

Fig. 3. Procedure for quantification of foil test results

3 Results

The transducer spacing experiments using transducer design A show on average a more than 6-fold larger hole area in the foil after ultrasonic treatment if the diaphragm with 134 mm spacing was used instead of the diaphragm with 85 mm spacing (see Table 2). The diaphragm using transducer design B shows an even larger hole area, with an average hole area about seven times larger. Overall, the results were highly variable, with the hole area increasing with each measurement in all three setups. However, this trend is not visible across setups.

Table 2. Results for transducer spacing experiments

Setup		Hole area in foil			
Transducers	Spacing (mm)	Mean (%)	Sample 1 (%)	Sample 2 (%)	Sample 3 (%)
Design A	85	0.34	0.02	0.24	0.77
Design B	134	2.46	1.28	2.55	3.55
Design A	134	2.19	0.30	1.96	4.30

The UCB width experiments showed an 80% (transducer design A) and 85% (transducer design B) larger hole area in the foil after ultrasonic treatment if the backplate was positioned at a distance of an integer multiple of the wavelength ($n\,\lambda$) rather than offset by half a wavelength ($n\,\lambda + \lambda/2$) as is shown in Table 3.

Table 3. Results for UCB width experiments

Setup			Hole area in foil		
Transducers	Width (cm)	Wavelengths	Mean (%)	Sample 1 (%)	Sample 2 (%)
Design A	31.0	$6.0\,\lambda$	2.50	3.13	1.88
Design A	41.5	$8.0\,\lambda$	3.88	2.94	4.82
Design A	44.0	$8.5\,\lambda$	2.07	1.24	2.89
Design A	49.0	$9.5\,\lambda$	1.32	1.25	1.08
Design A	51.5	$10.0\,\lambda$	2.76	2.79	2.73
Design A		$n\,\lambda$	3.05		
Design A		$n\,\lambda + \lambda/2$	1.69		
Design B	46.0	$7.5\,\lambda$	1.17	1.10	1.23
Design B	49.0	$8.0\,\lambda$	2.68	2.39	2.98
Design B	52.0	$8.5\,\lambda$	1.09	1.03	1.15
Design B	55.0	$9.0\,\lambda$	1.30	1.35	1.24
Design B	61.0	$10.0\,\lambda$	2.29	1.84	2.74
Design B		$n\,\lambda$	2.09		
Design B		$n\,\lambda + \lambda/2$	1.13		

4 Discussion

The aim of this research was to identify design principles for improved cleaning performance and energy efficiency of UCBs. The results of the transducer spacing experiment show more than 6-fold cavitation performance, calculated as the average total hole area in the foil after the aluminum foil test, with 134 mm than with 85 mm spacing, despite a reduction in the number of transducers from

35 to 23 (−34%). The offset from the assumed optimal spacing of 104.8 mm for transducer design A is greater at 134 mm (+27.9%) than at 85 mm (−18.9%). Therefore, 85 mm spacing should show greater cavitation performance. However, the opposite is true. These results seem to refute the hypothesis that cavitation performance is optimal when transducers are spaced one transverse wavelength apart. One possible explanation for the increased cavitation performance with 134 mm spacing might be a combination of reduced mass and stiffness of the diaphragm due to the reduced number of transducers allowing for easier movement of the diaphragm. In this case, wave superposition of transverse waves might still affect cavitation performance, although to a lesser degree than the number of transducers. The hole area also increases with each measurement, possibly due to insufficient degassing of the cleaning medium. However, since the cleaning medium was degassed for several hours before first use and then again for at least 1 h directly before the experiments, this seems unlikely. Another possible explanation is that the general random variability in the foil test results aligns to appear an upward trend due to the low number of measurements (three) per variant. Overall, the results suggest that spacing transducers farther apart can improve performance, either despite or potentially because of the reduced number of transducers. However, more data is needed to confirm and explain this behavior. Therefore, additional experiments must be conducted examining a greater number of transducer types and spacings and with a greater number of measurements per diaphragm.

The results of the UCB width experiment show an increase in cavitation performance by about 80% for both transducer designs. These results seem to support the hypothesis that cavitation performance is optimal when the distance between the diaphragm and the opposing wall of the tank is an integer multiple of the wavelength in the cleaning medium. It also fits the formation of standing waves creating dead zones and zones of high cavitational activity within the tank. However, the results indicate that, although the distribution of the cavitation in the tank is less even, an overall increase in cavitation performance can be observed. The unevenness of distribution could be mitigated by moving the soiled parts within the tank or using a swept frequency, although further experiments are necessary since this might also negate the increase in cavitation performance.

The procedure for quantification of aluminum foil test results worked well overall, though improvements like a custom recording fixture with a backlight and protection from ambient light would reduce the necessary amount of processing steps, thus reducing potential for error and improving accuracy. Since the distribution of the cavitation activity in the cleaning medium is an important factor for even cleaning of parts, implementation of a metric for the uniformity of the distribution would also be beneficial for improved analysis of the results. Optimally the experiments should also be repeated with a representative soiled item to verify that the cleaning performance actually correlates to the cavitation performance and that uniform cleaning occurs. This is especially relevant for the UCB width experiments since the presence of the soiled item affects the spread of the vibration in the medium.

5 Conclusion

In this study experiments regarding two factors were conducted: the spacing of transducers and the distance of the diaphragm from the opposing wall of the tank. The results suggest that both the transducer spacing and the UCB width have a noticeable impact on cavitation performance and should be considered where possible to optimize cavitation performance. However only a limited variety of diaphragms were tested. Therefore further research is needed to verify the results under different conditions and for different UCB designs. Since aluminum foil tests can only assess cavitation performance, the actual impact on cleaning performance could not be verified and has to be assessed separately as well.

Acknowledgments. This research was funded by the Federal Ministry for Economic Affairs and Energy as part of the project "OptiSonic" (16KN087169) within the ZIM (Central Innovation Programme for SMEs) funding program.

Competing Interests. The author(s) has no competing interests to declare that are relevant to the content of this manuscript.

References

1. Zhou, W., Sarpong, F., Zhou, C.: Use of ultrasonic cleaning technology in the whole process of fruit and vegetable processing. Foods (Basel, Switzerland) **11**(18), 2874 (2022). https://doi.org/10.3390/foods11182874
2. Nguyen, D.D., Ngo, H.H., Yoon, Y.S., Chang, S.W., Bui, H.H.: A new approach involving a multi transducer ultrasonic system for cleaning turbine engines' oil filters under practical conditions. Ultrasonics **71**, 256–263 (2016). https://doi.org/10.1016/j.ultras.2016.07.001
3. Mason, T., Peters, D.: An introduction to the uses of power ultrasound in chemistry. In: Practical Sonochemistry, pp. 1–48. Woodhead Publishing, Sawston (2002). https://doi.org/10.1533/9781782420620.1
4. Fuchs, F.J.: 19 - Ultrasonic cleaning and washing of surfaces. In: Gallego-Juárez, J.A., Graff, K.F. (eds.) Power Ultrasonics, pp. 577–609. Woodhead Publishing, Sawston (2015). https://doi.org/10.1016/B978-1-78242-028-6.00019-3
5. Phophayu, S., Kliangklom, K., Thongsri, J.: Harmonic response analysis of tank design effect on ultrasonic cleaning process. Fluids **7**(3), 99 (2022). https://doi.org/10.3390/fluids7030099
6. Tangsopa, W., Thongsri, J.: Development of an industrial ultrasonic cleaning tank based on harmonic response analysis. Ultrasonics **91**, 68–76 (2019). https://doi.org/10.1016/j.ultras.2018.07.013
7. Niemczewski, B.: A comparison of ultrasonic cavitation intensity in liquids. Ultrasonics **18**(3), 107–110 (1980). https://doi.org/10.1016/0041-624X(80)90021-9
8. Niemczewski, B.: Observations of water cavitation intensity under practical ultrasonic cleaning conditions. Ultrasonics Sonochemistry **14**(1), 13–18 (2007). https://doi.org/10.1016/j.ultsonch.2005.11.009
9. Duran, F., Teke, M.: Design and implementation of an intelligent ultrasonic cleaning device. Intell. Autom. Soft Comput. **25**(3), 1–10 (2019). https://doi.org/10.31209/2018.11006161

10. Tangsopha, W., Thongsri, J., Busayaporn, W.: Simulation of ultrasonic cleaning and ways to improve the efficiency. In: 2017 International Electrical Engineering Congress (iEECON), pp. 1–4. IEEE, Pattaya (2017). https://doi.org/10.1109/IEECON.2017.8075747
11. Mason, T.J.: Ultrasonic cleaning: an historical perspective. Ultrason. Sonochemistry **29**, 519–523 (2016). https://doi.org/10.1016/jultsonch.2015.05.004
12. Ni, P., Lee, H.-N.: High-resolution ultrasound imaging using random interference. IEEE Trans. Ultrason., Ferroelectr., Freq. Control. **67**(9), 1785–1799 (2020). https://doi.org/10.1109/TUFFC.2020.2986588
13. Lee, Y., Kim, J., Ha, T., Kang, B., Kim, Y.: Effect of stable and transient cavitation on ultrasonic degassing of Al alloy. Metals **14**(12), 1372 (2024). https://doi.org/10.3390/met14121372

Model-Based Approach for a Control System for Ultrasonic Systems Using Mechanical Sensors

Jonas M. Werner[1(✉)], Marcus Gerber[1], and Welf-Guntram Drossel[1,2]

[1] Professorship Adaptronics and Lightweight Design, Chemnitz University of Technology, Chemnitz, Germany
jonas-maximilian.werner@mb.tu-chemnitz.de

[2] Fraunhofer Institute for Machine Tools and Forming Technology IWU, Chemnitz, Germany

Abstract. Ultrasonic vibration assistance can significantly enhance the performance of manufacturing processes and reduce process limitations. Changing external loads during manufacturing, e.g., process forces or heat flow, require a continuous adjustment of the system's operating frequency. Conventional approaches based on the readily available electrical signals of the ultrasonic transducer, like a phase-locked-loop approach, are effective but not ideal as the electrical signals are gathered far from the cutting edge. Since the resonance frequencies of the electrical and mechanical resonant circuit do not necessarily match, using signals based on mechanical quantities close to the cutting edge, like acceleration, might offer benefits regarding the current vibration behavior of the system. Therefore, a better control of the ultrasonic system during operation can be provided. In this work a control model for an ultrasonic system using signals based on the acceleration at the cutting edge is designed and evaluated. The data for the modeling of the ultrasonic system is obtained from experiments using an existing system. Two control algorithms, phase-locked-loop and autoresonant control, are modeled and adapted for using mechanical quantities as the input for the control model. The results show the differences between the resonance frequencies of the electrical and the mechanical resonant circuit. The findings support the benefits of using a control algorithm based on mechanical quantities to better control the process. Furthermore, larger displacements at the tool could be achieved using acceleration in the proposed control algorithms.

Keywords: Ultrasonic systems · Control · Modeling

L. Overmeyer and B.-A. Behrens (eds.), *Production at the Leading Edge of Technology*, Lecture Notes in Production Engineering, https://doi.org/10.1007/978-3-032-19524-1_23

1 Introduction

Applying ultrasonic vibrations to machining processes can be linked with achieving different benefits depending on the process. In cutting, ultrasonic vibration assisted processes can reduce friction [1] and cutting forces while also generating a better surface regarding its uniformity and roughness [2]. Additional effects include a lower formation of burr, better chip breakage [3,4] as well as a longer tool life [5]. An application of an improved surface finish is the microstructuring of functional surfaces. Here, ultrasonic vibration assistance is used to generate a defined microstructure by using the vibration of the tool [6–8]. To achieve these advantages a control system is necessary to compensate for the influence of internal and external loads and their effect on the dynamic behavior of the system. These include, but are not limited to, changing loads from the manufacturing process, thermal influences from the process or from the ultrasonic transducer itself. All of these have an effect on the system's resonance frequency [9]. The control system needs to constantly match the operating frequency to the varying resonance frequency of the system during operation. A mismatch leads to lower displacements at the tool as well as a lower effiency of the ultrasonic system. Widely used control concepts use the readily available electrical signals of the transducer. Most commonly the phase shift between the supplied voltage and current is used to determine the resonance frequency. These control concepts, e.g. phase-locked-loop (PLL) [10] offer high precision when matching the operating frequency to the resonance frequency of the electrical resonant circuit, but resulting from the non-linear behavior of ultrasonic systems, the electrical resonance frequency does not necessarily match the resonance frequency of the mechanical resonant circuit [11]. Therefore, an approach using mechanical signals, ideally close to the cutting edge, seems advantageous to better control the ultrasonic system. Firstly, it might offer a better match of the operating frequency to the mechanical resonance frequency. Secondly, sensors measuring mechanical dimensions like a displacement or acceleration close to the tool might offer more information regarding the current vibration behavior during operation. This approach was examined for an autoresonant control concept [12,13]. It is stated, that using acceleration for controlling the system offers the benefit of only exhibiting one mode where the phase shift is $0°$ as opposed to two modes using electrical dimensions[12]. Several works have already investigated the difference between the mechanical and electrical resonance frequencies of ultrasonic systems [14,15], but the research regarding control concepts is focused on control systems using electrical signals [10]. Therefore, in this paper a control model designed in MATLAB Simulink to examine two different control concepts using mechanical instead of electrical dimensions is presented. For this, a mathematical model based on an existing ultrasonic transducer and sonotrode is build using experimental data. The two already existing control concepts, phase-locked-loop and autoresonant control, are implemented and adapted for using mechanical dimensions in the control loop. Using these control systems simulations are carried out to compare their performance using varying parameters.

2 Ultrasonic System

The ultrasonic system used as a basis for the considered model consists of a piezoelectric transducer and a sonotrode. The transducer consists of two piezoelectric disks (PIC181, PI Ceramic, thickness $t_{\text{disk}} = 5\,\text{mm}$, piezoelectric charge coefficient $d_{33} = 265 \cdot 10^{-12}\,\text{C/N}$, area of the disk $A_0 = 999\,\text{mm}^2$, mechanical compliance $s_{33} = 14.2 \cdot 10^{-12}\text{m}^2/\text{N}$) sandwiched between three electrodes made of copper. Two masses made of Ti6Al4V (Young's modulus $E = 118\,\text{GPa}$) are used to tune the resonance frequency of the transducer to 19,300 Hz. The sonotrode is also made of Ti6Al4V and designed to match the resonance frequency of the transducer. The diameter of all parts is 38 mm, therefore there is no amplification of the vibration amplitude. An indexable insert is fixed with a screw at the tip of the sonotrode enabling ultrasonic superimposed turning. The design of the ultrasonic system is shown in Fig. 1.

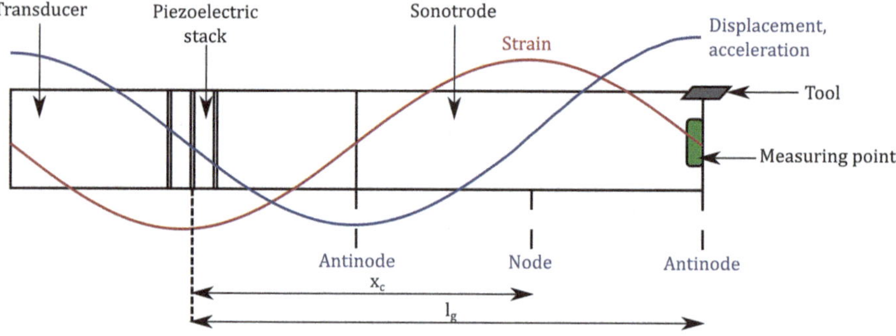

Fig. 1. Design of the ultrasonic system showing the displacement, acceleration and strain along the length of the system

3 Methodology

3.1 Modeling of the Ultrasonic System

Numerous approaches for modeling the behavior of ultrasonic systems exist, but most of them use three-dimensional finite element models [13], while few of them focus on one-dimensional numerical models, which would be beneficial regarding their use to develop control concepts as they offer faster processing than FE models. Voronina et al. proposed an approach to model an ultrasonic system consisting of a transducer and a sonotrode as a two degree-of-freedom model [13]. As this approach proved that the model can be used to examine a control system, this approach will be adapted. For the modeling of the system several simplifications can be used to reduce the complexity of the model while still achieving a high consistency of the model behavior with the behavior of the real system. Firstly, the indexable insert as well as the mount of the sonotrode are not taken into account in the model. Instead, the sonotrode is modeled as a cylinder with a constant cross-sectional area for the whole length. Furthermore, because of the symmetry of the transducer, the back end as well as one piezoelectric disk can be neglected for the model. As there is no amplification caused by the geometry of the sonotrode, the displacement amplitudes at the end of the sonotrode and the first antinode can be assumed to be equal. The node of the system $x_c = 127\,\text{mm}$ is placed at $2/3\,l_g$ with $l_g = 185\,\text{mm}$.

The model of the front end of the transducer and the sonotrode is assumed to be two cylinders with circular cross-sections A_1 and A_2 (both $= 1134\,\text{mm}^2$). Both cylinders are connected at x_c which corresponds to the mount of the system. For the considered system both cross-sections are equal. Taking different cross-sections into account enables an easier adaption of the model for different geometries in future works. To calculate the parameters for the model the approach proposed by [13] is used. Using the equations given in therein the modes of oscillation can be calculated. The masses and stiffness of the 2-DOF system are then chosen so that they match the eigenvalues, the experimentally determined eigenvectors and the energy of the system during operation. The eigenvectors are gathered from the measured velocities at the front end, see Fig. 1, of the sonotrode during operation using a Laser Vibrometer CLV-2534 (Polytec Gmbh, Germany):

- $u = 100$ V: velocity = 130 mm/s, displacement: 1.07 µm
- $u = 200$ V: velocity = 250 mm/s, displacement: 2.06 µm
- $u = 300$ V: velocity = 370 mm/s, displacement: 3.05 µm
- $u = 570$ V: velocity = 730 mm/s, displacement: 6.02 µm

The model of the piezoelectric transducer has to describe the correlation between the mechanical and electrical properties and has to take the supplied voltage, the load from the sonotrode and the displacement of the transducer itself into account. The coupled equations of the piezoelectric properties then enable the calculation of the displacement. With the model of the piezoelectric transducer as well as the model of the front mass and the sonotrode a complete model of the ultrasonic system can then be derived. The damping coefficients d_1 and d_2 are calculated to adjust the vibration amplitudes of the model to those that were determined experimentally. Finally, based on this the spring stiffnesses of the system can be calculated.

An external load resulting from a manufacturing process is modeled as an additional spring k_L and dampler d_L, located at a distance of δ in relation to the front end. Increasing k_L increases the load on the ultrasonic system. Additionally, the contact time of the tool and workpiece as well as the maximum forces can be increased by lowering δ. As a result of this model only the passive force of the turning process is modeled. Additional force components like the cutting force and the feed force are not considered as only the passive force has an influence in the one-dimensional model. In this work two control algorithms are modeled and their performance using electrical and mechanical signals is compared. The concept most commonly used is the phase-locked-loop concept

(PLL) [10]. The goal of this control concept is to keep the phase shift φ between the voltage u and the current i at 0°, which corresponds to the electrical resonance frequency. As the mechanical speed of the first mass is in phase with the current, both can in theory be used for controlling the frequency of the system and ensure an operation in resonance. Depending on this frequency shift the operating frequency is adjusted accordingly. As a gain factor K_{pll} is used in the model. The model calculates the error between the phase shift φ from the ideal value of 0°. This error is integrated and then multiplied with the gain factor K_{pll} to determine the supplied voltage.

Higher K_{pll} should enable a faster response time, but the control might overshoot. Lower K_{pll} should result in a slower response to deviations of the phase shift. Another similar concept is the autoresonant control. Here the root-mean-square (RMS) value of a dimension is constantly calculated for a small time step [16]. This value is then compared to the value of the previous step. The operating frequency is shifted by a given $d\varphi$ each step. If the RMS value of the current step is smaller than in the previous step, the sign of the frequency shift $d\varphi$ is reversed. Therefore the phase difference between the two considered values is constantly fluctuating around 0°. For the autoresonant control the considered dimension itself is used to generate the signal for the excitation voltage for the system. For this, the measurand is multiplied with a gain factor K_{ar}. This is necessary adjust the signal for the correct dimensions. If K_{ar} is too low, the supplied voltage should get a smaller with each iteration, which may cause the vibration to stop. If it is too high, the supplied voltage may increase too much and therefore causes damages of the piezoelectric elements. To prevent this, a limiter needs to be used to avoid unnecessarily high voltages. Schematics of the models are given in Fig. 2.

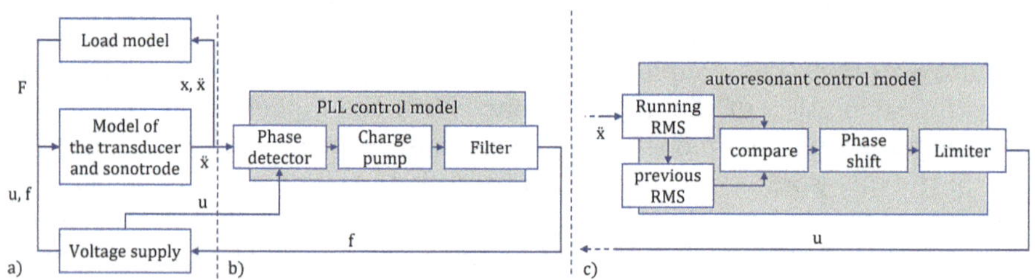

Fig. 2. Block diagram of **a** the load model and the model of the ultrasonic system, **b** the PLL control model and **c** the autoresonant control model

3.2 Design of Experiments

For the evaluation the parameters in Table 1 are varied. $\hat{u}$ is the maximum voltage supplied to the transducer and K_{pll} and K_{ar} are the gain factors for the phase-locked-loop and autoresonant control, respectively. Furthermore, the initial operating frequency f_0 is varied to analyze the influence of larger deviations to the resonance frequency. Also, the influence of noise in the measured signal on the functionality of the control algorithm is evaluated. All parameters are examined for both control algorithms using conventional electrical as well as mechanical dimensions, in this case the acceleration $\ddot{x_2}$ at the front end of the sonotrode. Additionally, the force which is applied from the manufacturing process is varied with the spring stiffness k_L and the initial distance of the workpiece to the ultrasonic system δ. As a reference point a simulation without any control system is carried out. These parameters allow comparing both control algorithms, the use of a mechanical dimensions as a measurand as well as the influence of the various parameters on the stability of the control system. For each iteration of the simulation, the following process is carried out: At the beginning the system operates in free oscillation at the starting frequency f_0. At 0.0005 s the control system is activated. Then at 0.015 s the load is applied and the control system has to adjust the operating frequency accordingly. The total length of the simulation is 0.5 s.

Table 1. Model parameters, initial values given in bold

Parameter	Parameter settings	Unit
Maximum voltage $\hat{u}$	100; **300**; 500	V
Gain factor K_{pll}	-5; -10; -50; -100; **-500**; -5,000	e4
Gain factor K_{ar}	1; 100; **1,000**; 10,000; 100,000	e3
Starting frequency f_0	10,000; 15,000; 18,000; **19,300**; 20,000; 25,000	Hz
Noise	**0**; 100; 300; 500; 1,000; 10,000	m/s^2
Distance workpiece δ	0.1; **0.5**; 1	μm
Spring stiffness k_L	1; **3**; 7; 10	e9 N/m

4 Results

Results show that both control systems can keep the vibration amplitude at the tool constant. The outcome of increasing the maximum supplied voltage are higher accelerations and displacements, as was to be expected. An increase of the amplification factor K_{pll} enables a faster match of the operating frequency to the resonance frequency during operation, but should be chosen carefully. If the amplification factor is too high, the control algorithm might overshoot and cause the ultrasonic system to stop, which is the case for $K_{pll} = -5,000$ e4. For the autoresonant control smaller amplifaction factors K_{ar} seem advantageous, despite a slower reaction to changing loads. Higher K_{ar} offer a faster adjustment of the frequency and therefore higher vibration amplitudes with lower fluctuations. But it may also cause the operating frequency to rapidly change in a range of 15 Hz, which might make the system more unstable. For the PLL concept the initial operating frequency can be chosen when setting the control parameters. Using the acceleration as the measurand enables finding the correct operating frequency even with large initial mismatch between the initial operating and the resonance frequency, see Fig. 3a. Contrary, using the conventional approach using the phase shift between the voltage and current fails to do so. This is in line with the findings in [13]. However, the assumption for the adapted model is that it is only applicable near the resonance frequency. Therefore the limitations of changing the starting frequency in this adapted model are questionable. As the spring stiffness of the load increases, so does the resulting resonance frequency of the system during operation. Also, it is important to note, that the displacement at the tool during operation exhibits a shift for higher forces resulting in lower penetration depth at the workpiece. This was to be expected from the higher stiffness. The same behavior can be observed for the adapted autoresonant control. The results also show that after the contact between the sonotrode and the workpiece ends, the system returns to the same steady state as before. The distance to the workpiece δ has the same influence as an increase of the spring stiffness. In this case the tool also exhibits a higher penetration depth for lower δ resulting in higher forces affecting the system and therefore causing a higher resonance frequency of the system. Both control models can correctly adjust the operating frequency to this change and keep the vibration amplitudes at the tool constant.

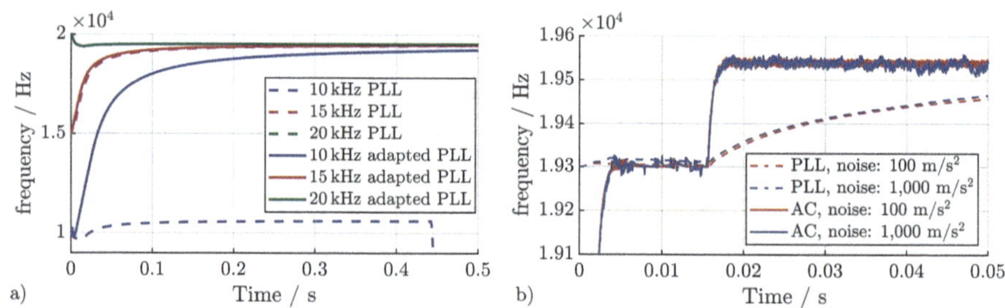

Fig. 3. Effect of the **a** starting frequency using PLL control and **b** the noise level on the controllability using PLL and autoresonant control (AC)

As the model stands, it calculates every dimension ideally, i.e. as an ideal sine wave. Real measurements are always afflicted with random noise. To examine the influence of noise on the controllability simulations are carried out with increasing amplitudes of random noise. During ideal operation the ultrasonic system achieves displacements of approx. 4 μm, which corresponds to maximum accelerations of approx. 58,800 m/s^2. Random noise with amplitudes up to 1,000 m/s^2 have a small effect on the controllability using the adapted autoresonant control as it can keep the vibration amplitude constant, though the operating frequency experiences higher fluctuations, see Fig. 3b. For higher noise the adapted autoresonant control fails to find the correct resonance frequency and the displacement at the tool drops off completely. The adapted PLL model already starts to fail for lower levels of noise. Starting from noise amplitudes of 500 m/s^2 the operating frequency is increased more than necessary. This results in a 15 % lower displacement at the tool. For noise amplitudes of 10,000 m/s^2 the adapted PLL control model fails to keep the operating frequency constant.

5 Conclusions

In this work a phase-locked-loop and an autoresonant control model is adapted to use the acceleration at the front end of a sonotrode as the input value for the control systems instead of electrical parameters. Both are compared with one another as well as with the conventional version of both control models. Results show that both control models enable a stable control of the considered ultrasonic system when choosing the control settings carefully. Using acceleration as the input value for the control system comes with an increased amount of information regarding the current vibration behavior as the vibration amplitude can be calculated at each moment, which is not possible using electrical signals. Furthermore, as the relation of the phase shift between the acceleration and the supplied voltage is biunique contrary to the phase shift between the current and the supplied voltage, the adapted systems therefore enables the identification of the correct operating frequency [13].

The proposed model enables the design of control models for ultrasonic systems based on mechanical dimensions. While functional there are some limitations of the model that need to be accounted for in the future. Several simplifications have been made to reduce the complexity of the model. It needs to be researched how this affects the match between the model and the real system. Furthermore, the model is only one-dimensional and exludes the cutting and feed force of the manufacturing process. These may cause a bending of the front end of the sonotrode, which may affect the vibration shape.

Although the presented approach seems beneficial there are some challenges applying this to real world systems. Currently there is no feasible solution for measuring the acceleration of ultrasonic systems during operation as the sensor choices are limited regarding their mass, frequency range and maximum accelerations. The highest amount of information regarding the process behavior would be achievable using accelerometers, but vibrations at a frequency of 25,000 Hz and displacements of 10 µm result in maximum accelerations of approx. 20,000 g. Combined with a high frequency range and ideally a low mass of the sensor only few accelerometers, like a PCB 352A92 (PCB Synotech) are suited. Strain gauges might also be used to monitor the system, as they are relatively cheap, low weight and offer high frequency ranges. For all sensors their mounting on the ultrasonic system is challenging. For accelerometers a connection with a bolt would be ideal, but this comes with additional weight and might not be suitable for small sensors. Adhesive bonding is possible but results in additional damping and might affect the signal quality. Laser vibrometers offer a precise and contactless measurement, but the optical accessibility is in most cases not given during manufacturing processes. In the future, suitable sensors need to be evaluated regarding their sampling rates, occuring noise and the quality of the generated data to implement and apply the control models to a real system.

Acknowledgments. This research was funded by the Deutsche Forschungsgemeinschaft (DFG, German Research Foundation)–grant number 510749881.

References

1. Amini, S., Khosrojerdi, M.R., Nosouhi, R.: Elliptical ultrasonic-assisted turning tool with longitudinal and bending vibration mode. Proc. Inst. Mech. Eng. **231**, 1389–1395 (2017)
2. Hahn, M., Cho, Y., Jang, G., Kim, B.: Optimal design and experimental verification of ultrasonic cutting horn for ceramic composite material. Appl. Sci. **11**(4), 1954 (2021)
3. Brehl, D.E., Dow, T.A.: Review of vibration-assisted machining. Precis. Eng. **32**, 153–172 (2008)
4. Kandi, R., Kumar Sahoo, S., Kumar Sahoo, A.: Ultrasonic vibration-assisted turning of Titanium alloy Ti-6Al-4V: numerical and experimental investigations. J. Braz. Soc. Mech. Sci. Eng. **42**, 399 (2020)
5. Bulla, B., Klocke, F., Dambon, O., Hünten, M.: Ultrasonic assisted diamond turning of hardened steel for mould manufacturing. Key Eng. Mater. **516**, 437–442 (2012)
6. Zhang, R., Steinert, P., Schubert, A.: Microstructuring of surface by two-stage vibration assisted turning. Procedia CIRP **14**, 136–141 (2014)
7. Schubert, A., Nestler, A., Pinternagel, S., Zeidler, H.: Influence of ultrasonic vibration assistance on the surface integrity in turning of the aluminium alloy AA2017. Materialwissenschaft und Werkstofftechnik **42**, 658–665 (2011)
8. Xu, S., Kuriyagwa, T., Shimada, K., Mizutani, M.: Recent advances in ultrasonic-assisted machining for the fabrication of micro/nano-textured surfaces. Front. Mech. Eng. **12**, 33–45 (2017)
9. Hreha, P., Radvanská, A., Hloch, S., Perzel, V., Krolczyk, G., Monkova, K.: Determination of vibration frequency depending on abrasive mass flow rate during abrasive water jet cutting. Int. J. Adv. Manuf. Technol. **77**, 763–774 (2015)
10. Ille, I., Twiefel, J.: Model-based feedback control of an ultrasonic transducer for ultrasonic assisted turning using a novel digital controller. Phys. Procedia **70**, 63–67 (2015)
11. Yokozawa, H., Twiefel, J., Weinstein, M., Morita, T.: Dynamic resonant frequency control of ultrasonic transducer for stabilizing resonant state in wide frequency band. Jpn. J. Appl. Phys. **20**, 07JE08 (2017)
12. Babitsky, V., Astashev, V.: Nonlinear dynamics and control of ultrasonically assisted machining. J. Vib. Control. **13**, 441–460 (2007)
13. Voronina, S., Babitsky, V.: Autoresonant control strategies of loaded ultrasonic transducer for machining applications. J. Sound Vib. **313**, 395–417 (2008)
14. Ji, H., Lin, L., Zou, H., Hu, X.: Study on the influence of force load on output amplitude in ultrasonic vibration system. J. Mech. Sci. Technol. **38**, 6287–6296 (2024)

15. Puga, H., Grilo, J., Oliveira, F.J., Silva, R.F., Girao, A.V.: Influence of external loading on the resonant frequency shift of ultrasonic assisted turning: numerical and experimental analysis. Int. J. Adv. Manuf. Technol. **101**, 248–2496 (2019)
16. Voronina, S., Babitsky, V., Meadows, A.: Modelling of autoresonant control of ultrasonic transducer for machining applications. Proc. Inst. Mech. Eng., Part C: J. Mech. Eng. Sci. **10**, 1957–1974 (2008)

Analysis of the Surface Integrity in Grinding of Cemented Carbide at High Cutting Speeds

Ulf Hensler(✉), Monika Kipp, and Dirk Biermann

Institute of Machining Technology, TU Dortmund University, Dortmund, Germany
ulf.hensler@tu-dortmund.de

Abstract. Regarding the necessity for increases in productivity as well as a more intensive focus on sustainable manufacturing processes, resource-optimized processes are an aspect of current research. This also applies to the field of the grinding of cutting tools. In grinding of cemented carbide shaft tools, a significant proportion of the production time is taken up by the grinding of the flutes. One approach to enhance productivity is to increase the material removal rate through creep feed grinding with significantly higher cutting speeds and increased feed speeds at the same time. Increased cutting speeds go along with the need to adapt the process conditions and process control to the changed process characteristics. The increased cutting speeds have a variety of effects on the process behavior. In addition to a change in the wear behavior of the grinding wheels, the effects occurring on the cemented carbide specimens are diverse. Changing qualitative optical effects can be seen on the specimens with varying cutting speeds and process conditions. Furthermore, the surface roughness of the ground surfaces decreases to a critical point at higher cutting speeds. However, residual stress measurements as a means of assessing the surface integrity of the ground carbide specimens show a correlation between cutting speed and a change in the residual stress state. With increasing cutting speed, the residual compressive stresses introduced by the grinding process and the Full-Width Half Maximum (FWHM) values decrease. Even at significantly higher cutting speeds, the stress level does not reach the initial level.

Keywords: Grinding at high cutting speeds · Carbide · Surface integrity

1 Introduction and Motivation

A significant proportion of the manufacturing costs of cemented carbide shaft tools, such as twist drills or end mills, is attributed to grinding processes. With regard to grinding these shaft tools, most of the production time is spent on manufacturing the flutes. For this reason, there is high potential to reduce manufacturing costs by increasing the productivity of the flute grinding process [1, 2]. One approach to enhance the productivity of the flute grinding process is to increase the material removal rate. Therefore, it is the aim to increase both the circumferential speed of the grinding wheel and the feed speed simultaneously [3]. However, it is essential to ensure that the surface quality of the ground

L. Overmeyer and B.-A. Behrens (eds.), *Production at the Leading Edge of Technology*,
Lecture Notes in Production Engineering, https://doi.org/10.1007/978-3-032-19524-1_24

workpieces is at least equivalent to that achieved with conventional grinding parameters. Furthermore, the surface integrity of the ground carbide tools must be appropriate to guarantee their suitability for use as cutting tools.

The variation of the process parameters, their interdependencies as well as the process conditions directly influence the output quantities of the grinding process, such as process forces, surface roughness or surface integrity. With respect to the grinding of cutting tool materials, it could be shown that increasing the cutting speed results in a decrease of compressive residual stresses. This is due to higher thermal impact during grinding [4]. Hence, the grinding conditions also affect the wear behavior of the cutting tool [4]. In addition to the process parameters, the grinding wheel condition also affects the grinding process. The conditioning process, and consequently the grinding wheel topography, show an impact on grinding forces, the surface roughness and the residual stresses of the ground workpiece [5, 6]. Therefore, several parameters have to be taken into account regarding the design, understanding and enhancement of the grinding process.

The objective of the investigations presented in this paper is to enhance the productivity of the creep feed grinding process of cemented carbides on the basis of an increase in cutting and feed speed. As a basis, there is the need for the fundamental analysis and understanding of the effects of different process conditions and adjusted process parameters on the mechanical load and the surface integrity of the ground cemented carbide specimens during grinding at higher cutting speeds.

2 Process Conditions for Creep Feed Grinding at High Cutting Speeds

To achieve considerably higher cutting speeds during creep feed grinding of cemented carbide, it is necessary to analyze the process boundary conditions and make targeted adjustments to reduce the thermal load on the tool and workpiece. Important influencing factors include the process design, the grinding wheel conditioning, and the coolant supply [7].

2.1 Experimental Setup

Regarding the process design, down grinding with a constant equivalent chip thickness has proven to be an effective approach. As a basis for increasing the cutting and feed speed, conventional grinding parameters of v_c ($v_c = 18$ m/s) and v_f ($v_f = 54$ mm/min) were chosen. Based on this, the equivalent chip thickness h_{eq} was determined using Eq. 1 and both the cutting speed and feed speed were increased stepwise at a constant equivalent chip thickness of $h_{eq} = 0.1$ µm.

$$h_{eq} = \frac{1}{60} \cdot \frac{a_e \cdot v_f}{v_c} \hat{=} v_f = 60 \cdot \frac{h_{eq}}{a_e} \cdot v_c \quad (1)$$

To analyze the influence of different bonding systems and grain sizes in the context of investigations into creep feed grinding at high cutting speeds, grinding wheels with different grain sizes were examined. Different bonding systems with increased porosity were also taken into account. Oil was used as a lubricant in all grinding processes. For the investigations presented in the first part of this paper, a hybrid-bonded grinding wheel with a grain size of D126 and a diameter of d = 175 mm a range of cutting speed of $v_c = 40\ldots80$ m/s was used. In addition to conventional grinding wheel truing using silicon carbide tools, followed by a sharpening process, the influence of a wheel topography realized by electric discharge machining (EDM) on the grinding behavior was analyzed as well [8]. The grinding wheel conditioning aimed to generate a surface topography characterized by a high grain protrusion height, in order to ensure effective chip formation. In this context, previous investigations into grinding wheel conditioning revealed that a low ratio of dressing speeds results in a rough surface topography.

A cylindrical grinding wheel was selected for the basic analysis and to grind straight flutes. The depth of cut was set to $a_e = 2$ mm and the width of cut to $a_p = 4$ mm. Figure 1 summarizes the most important information about the tool and the workpiece material.

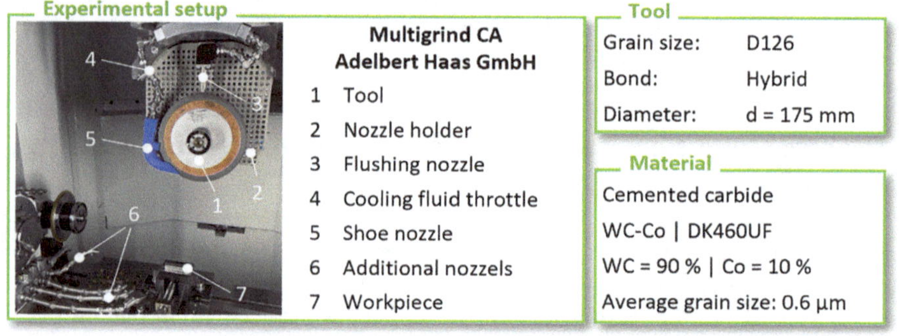

Fig. 1 Process conditions for creep grinding with high cutting speeds

In addition to targeted grinding wheel conditioning, a suitable conceptual design of the coolant supply system is required for the grinding process at high depth of cut and high cutting speeds. The setup consists of a shoe nozzle, arranged close to the grind zone, and a flushing nozzle. Both nozzles are fixed on a nozzle holder. This ensures precise and reproducible positioning of the coolant nozzles, even under high-pressure conditions. At the same time, it enables a flexible adjustment of the nozzle configuration and alignment to changing process conditions.

2.2 Process Analysis with Respect to Surface Roughness, Process Forces and Tool Wear

The effects of increased cutting speeds on surface roughness, process forces and the operating characteristics of the tool are diverse. The resulting surface roughness values and measured process forces are shown in Fig. 2.

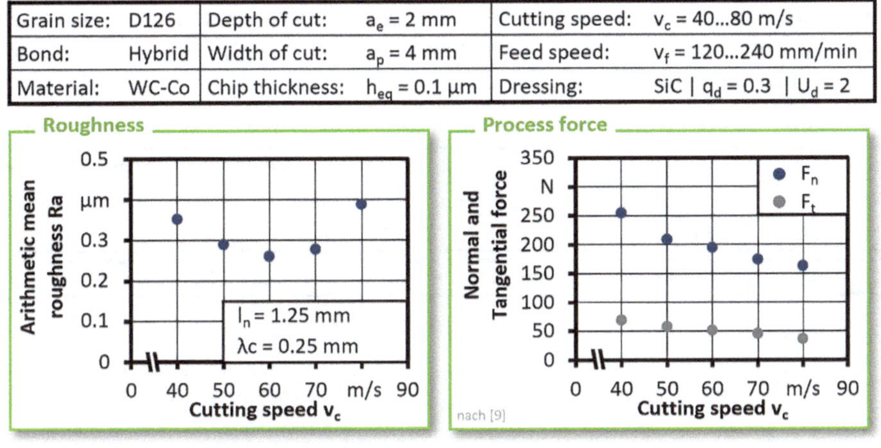

Grain size: D126	Depth of cut: $a_e = 2$ mm	Cutting speed: $v_c = 40...80$ m/s
Bond: Hybrid	Width of cut: $a_p = 4$ mm	Feed speed: $v_f = 120...240$ mm/min
Material: WC-Co	Chip thickness: $h_{eq} = 0.1$ µm	Dressing: SiC | $q_d = 0.3$ | $U_d = 2$

Fig. 2 Influence of increased cutting speed on process characteristics

The force measured during the application tests using piezoelectric sensors installed in the headstock of the grinding machine reveals that both normal and tangential force decrease with increasing cutting speed and seem to approach a lower limit. Surface roughness measurements of the specimens show that with increasing cutting speed the surface roughness decreases up to a critical point. With further increase in cutting speed, the surface roughness increases as well. It can be assumed that the decrease in surface roughness at increased cutting speeds can be attributed to the higher overlap of grinding paths. Vibrations occurring at higher cutting speeds and increasing changes in the grinding wheel topography indicate changes in the process conditions that appear to have a negative effect on surface roughness.

2.3 Analysis of the Residual Stresses Depending on the Process Conditions

To further assess the material condition, residual stress measurements were carried out on selected test specimens. In addition to the residual stress, the FHWM was also considered as a parameter. This allows for the derivation of conclusions about the crystallite sizes and microstrains in the material, which can be influenced by mechanical processing. For this purpose, the Xstress DR45 diffractometer at Stresstech GmbH was used. The analysis was performed using two-dimensional X-ray diffraction with the $\sin^2\psi$-method [10]. In addition to a test series conducted with a conventionally dressed grinding wheel, comparative investigations were carried out using an EDM-conditioned grinding wheel. Due to the use of an inaccurate X-ray elasticity constant, the measurements are relatively comparable to each other. It becomes clear that both the compressive residual stresses and the FWHM decrease with increasing cutting speed. Furthermore, it becomes evident that the initial stress level of the cemented carbide specimen is not exceeded within the considered cutting speed range. A clear relationship between cutting speed and residual stresses can be identified. The compressive residual stresses introduced by the grinding process are counteracted by higher cutting speeds. It can be assumed that this is a result of increasing temperature with higher cutting speed and material removal rate.

Furthermore, the grinding wheel conditioning process does not appear to reasonably influence the residual stress state of the cemented carbide sample. The residual stress in both cases is comparable, Fig. 3.

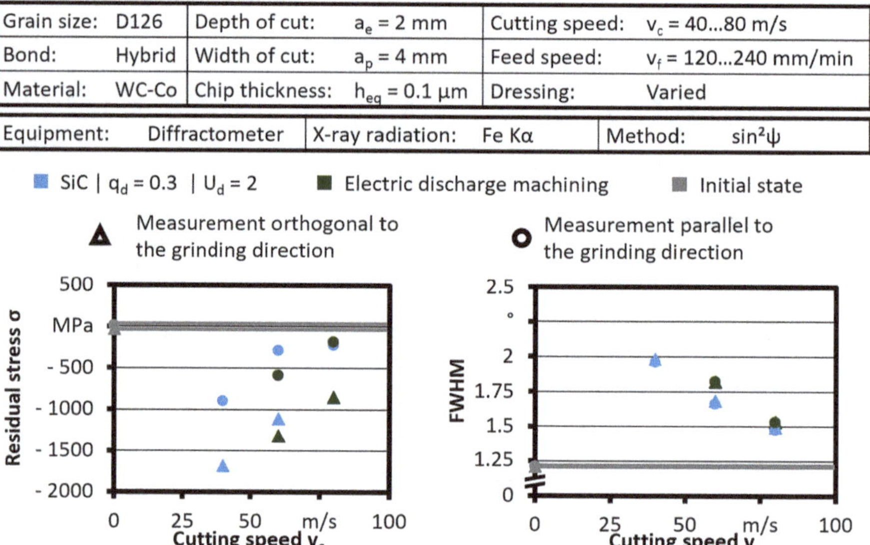

Fig. 3 Development of the residual stresses when grinding cemented carbide with higher cutting speeds and hybrid bond grinding wheel, depending on the grinding wheel conditioning

3 Additional Investigations on Surface Integrity

Residual stresses are a criterion for assessing the condition of a workpiece and evaluating its suitability for later application. Additional specimens were analyzed to examine, in particular, a correlation between residual stress and optical appearance of the cemented carbide samples.

3.1 Experimental Setup

In these tests, the parameters of the grinding wheel truing, the bonding system (metallic-porous) and the grain sizes (D54/D126) of the diamond grinding wheels were all varied. In contrast to the investigations presented in Chapter 2, the investigations were performed by using the 305micro tool grinding machine of Alfred H. Schütte GmbH & Co. KG. The sample design and the process conditions with regard to the process parameters were similar to those of the previously presented test series.

3.2 Analysis of the Correlation Between Optical Effects and Residual Stress

During the grinding process, optical changes on the cemented carbide specimens as well as smoke formation were observed in individual tests of this series. These effects indicate a high thermal load on both the tool and the specimen. It becomes evident that the areas of optical change, which were analyzed using digital light microscopy, increase with rising cutting speeds (Fig. 4).

Grain size:	Varied	Depth of cut:	a_e = 2 mm	Cutting speed:	v_c = 18...80 m/s
Dressing:	Varied	Width of cut:	a_p = 4 mm	Feed speed:	v_f = 54...240 mm/min
Material:	WC-Co	Chip thickness:	h_{eq} = 0.1 µm	Bond:	Metallic-porous

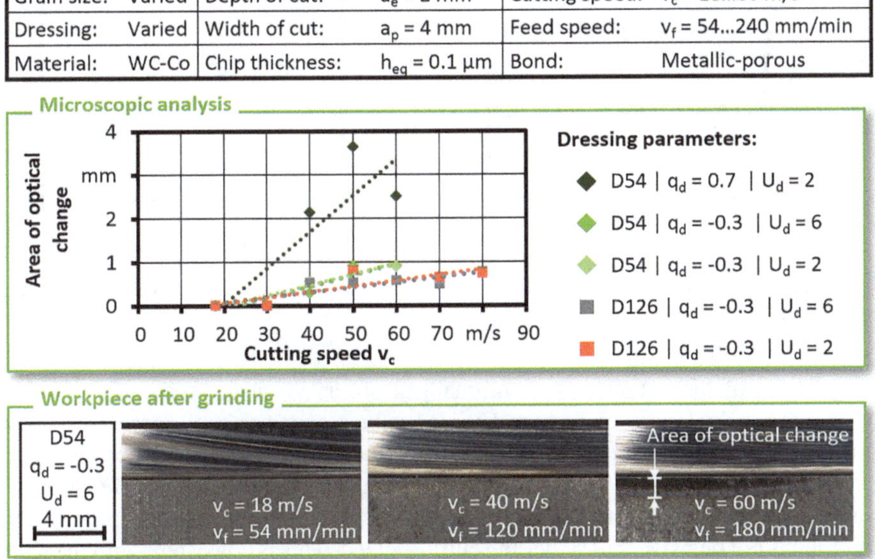

Fig. 4 Optical sample change depending on cutting speed, grain size and dressing of the tool

Furthermore, differences in the development of the areas of optical change can be observed depending on the process boundary conditions. These differences can be attributed to the topography of the grinding wheel used in the process. This is adjusted by the bonding system and the grain size specification, as well as by the targeted conditioning of the grinding wheel. The results indicate that a rougher grinding wheel topography causes fewer optical changes to the specimens when grinding with increased cutting speeds. It can be assumed that this is due to lower thermal stress.

Comparative residual stress measurements conducted on selected test specimens indicate an influence of the cutting speed on the resulting stress state, as is shown in Fig. 5.

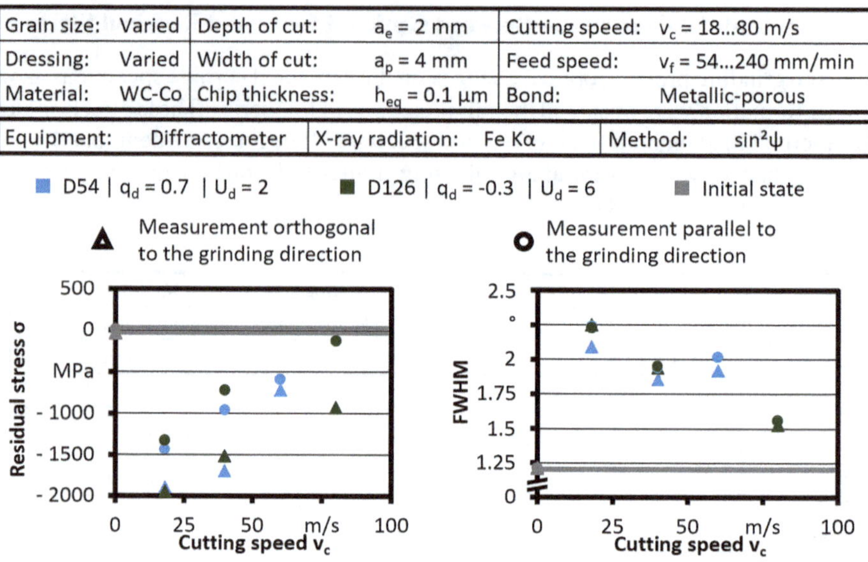

Fig. 5 Development of the residual stresses when grinding with metallic-porous bond grinding wheels and higher cutting speeds

Regardless of the extent of optical changes observed on the specimens, both the compressive residual stresses and the FWHM values decrease with increasing cutting speed. However, within the considered cutting speed range, these values do not fall below the residual stress levels of the initial cemented carbide specimens. Furthermore, it can be seen that, unlike in the optical analysis, the influence of the grinding wheel topography is not decisive, but rather the cutting speed. This is consistent with the results of previously presented test series.

4 Summary and Conclusion

The investigations confirm the influence of process boundary conditions on the grinding process and demonstrate that targeted adjustments of these boundary conditions are essential for grinding cemented carbide at increased cutting speeds. The conditioning of the grinding wheel and the resulting effective topography, the coolant supply, and the process design are of particular importance. Considering these factors and their targeted adjustment, an increase in cutting speed is realizable in creep feed grinding of cemented carbide.

In addition to an increase in productivity, simultaneously rising cutting and feed speed leads to reduced process forces and, up to a critical point, to a decrease in surface roughness. Furthermore, residual stress measurements reveal a correlation between cutting speed and the residual stress state of the specimens. While conventional grinding parameters introduce compressive residual stresses into the specimen, these stresses, as well as the FWHM values, decrease within the applied cutting speed range. At the same time, the investigations indicate that a purely optical assessment of the test specimens provides only limited information regarding the actual residual stress rate.

Acknowledgements. The IGF-project 01IF21687N "Griding of cemented carbide with high cutting speeds" was supported by the Federal Ministry for Economic Affairs and Energy on the basis of a decision by the German Bundestag.

As part of industrial joint research, the project was supported by a project advisory committee. We would like to thank the members of this committee for their support in this research project and, especially, Stresstech GmbH for conducting the residual stress measurements. In addition to this, also a big thanks to Mitsubishi Electric performing the grinding wheel conditioning by EDM.

Competing Interests The author(s) has no competing interests to declare that are relevant to the content of this manuscript.

References

1. Friemuth, T.: Herstellung spanender Werkzeuge, Dissertation, Universität Hannover (2002)
2. Byrne, G., Dornfeld, D., Denkena, B.: Advancing cutting technology. CIRP Ann. **52**(2), 483–507 (2003)
3. Klocke, F., Brinksmeier, E., Evans, C., Howes, T., Inasaki, I., Minke, E., Tönshoff, H.K., Webster, J.A., Stuff, D.: High-speed grinding—fundamentals and state of the art in Europe, Japan, and the USA. CIRP Ann. **46**(2), 715–724 (1997)
4. Weinert, K., Johlen, G., Schneider, M.: Schleifen von Hartmetall- und Cermet-Werkzeugen. VDI-Z Special Werkzeuge Mai **98**, 46–49 (1998)
5. Wegener K., Hoffmeister H.-W., Karpuschewski, B., Kuster, F., Hahmann, W.-C., Rabiey, M.: Conditioning and monitoring of grinding wheels. CIRP Ann. Manuf. Technol. **60**, 757–777 (2010)
6. Lierse, T.: Schleif- und Abrichttechnik. Carl Hanser Verlag, München (2020)
7. Hensler, U., Kipp, M., Biermann, D.: Schleifen von Hartmetall mit deutlich erhöhten Schnittgeschwindigkeiten. In: Broeckmann, C., Danninger, J., Hein, S. (eds.), Pulvermetallurgie in stürmischen Zeiten, Band 39, 349–352. Heimdall Verlag, Rheine (2024)
8. Wegener, K., Weingärtner, E., Walter, C., Dold, C., Stirnimann, J.: Konditionieren von Schleifscheiben mit Licht und Strom, ETH Zürich (2012)
9. Hensler, U., Kipp, M., Biermann, D.: Schleifen von Hartmetall mit deutlich höheren Schnittgeschwindigkeiten. VDW Branchenreport Februar **2025**, 8–10 (2025)
10. Send, S., Dapprich, D., Palosaari, M.: Fast residual stress determination by means of two-dimensional x-ray diffraction using the $\sin^2\psi$-method. ICRS11-The11th International Conference of Residual Stresses, SF2M; IJL, Mar2022, Nancy, France (2023)

3D Printing Technology for High Viscous Rubber Materials

Process Challenges and Technical Equipment in Additive Manufacturing of Rubber

Hans Helge Schwieger[1(✉)], Shubham G. Kirve[2], Sebastian Leineweber[1], Benjamin Klie[2], Ulrich Giese[2], and Ludger Overmeyer[1]

[1] Institute of Transport and Automation Technology (ITA), Leibniz University Hannover, Garbsen, Germany
helge.schwieger@ita.uni-hannover.de
[2] Deutsches Institut für Kautschuktechnologie e.V. (DIK), Hannover, Germany

Abstract. Additive manufacturing (AM) technologies have grown significantly in recent years. They offer advantages such as an extensive freedom in design, easy spare parts reproduction, product customization, and resource-efficient material usage. Even though a variety of materials is usable, challenges still remain in processing elastomers and especially with often used high viscosity rubbers. To address this, the German Institute of Rubber Technology (DIK) and the Institute of Transport- and Automation Technology (ITA) developed the Additive Manufacturing of Elastomers (AME) process. It utilizes layered deposition of raw rubber mixtures followed by a vulcanization step. The process was demonstrated with a prototype 3D printer, but improvements in part quality and process reproducibility are required for industrial implementation. To overcome these issues, this paper describes challenges that arise in 3D printing of highly viscous rubber materials and presents technical concepts to address them. Many challenges are directly linked to different steps in the printing workflow. For this reason, the process is analyzed methodically from the filament feeding to the final cured part. Quality relevant steps are identified and technical opportunities to influence them are given. Finally, a new concept of a 3D printer is derived, which forms the basis for further research in rubber 3D printing.

Keywords: Rubber 3D printing · Additive manufacturing rubber · Additive manufacturing elastomers · 3D printing technology · High viscosity printing

1 Theoretical Background

Over the last two decades, additive manufacturing has undergone rapid development. Expiry of restrictive patents at the turn of the millennium, in combination with simultaneous progress in computer and software technologies, led to an increasing popularity of the technologies [1, 2]. A wide range of materials such as ceramics, plastics, or metals can be processed, allowing individualized production on demand, in variable quantities

L. Overmeyer and B.-A. Behrens (eds.), *Production at the Leading Edge of Technology*, Lecture Notes in Production Engineering, https://doi.org/10.1007/978-3-032-19524-1_25

[3]. One subgroup that is still lagging behind this revolution is the processing of elastomer materials [4] and especially in additive manufacturing of vulcanized rubber parts [5, 6], as they are used in seals, elastic brackets, dampers or tires [7]. These components are created from raw rubber mixtures, which are first mixed, then shaped by a processing method and thermally cured in a vulcanization step afterwards. The ingredients and their proportions within the mixtures are strongly adapted to application dependent requirements and the resulting material properties [8].

There is already a variety of 3D printing processes for highly elastic materials. In general, these technologies are based on specially adapted polymers, such as thermoplastic elastomers, fast-curing reactive resins or photopolymerizable liquids [9]. There are also several studies on additive manufacturing of vulcanized rubber particles in composite structures using a bonding agent [10–12]. The chemical, thermal or UV resistance of these materials is often limited, compared to vulcanized rubber and these substances are often associated with high costs.

With this background, several concepts have been generated for processing highly viscous rubber mixtures using 3D printing techniques [13–15]. The first exemplary tests followed in studies at the University of Chemnitz, with a modified standard 3D printer, but the technology failed to extrude mixtures with higher viscosities [6].

The first practical implementation of a rubber 3D printing process for complex geometries and highly viscous compounds was achieved by the research of the ITA and DIK. It resulted in the "Additive Manufacturing of Elastomers" (AME) process [16]. The technology is based on commonly used fused deposition modelling (FDM) and is described schematically in Fig. 1. An axis system moves a rubber extruder relative to an underlying printing bed. In contrast to FDM with thermoplastic materials, the rubber mixture does not harden after extrusion and remains in a viscous state. To prevent the material from flowing apart, an enveloping structure is printed around the component using a second thermoplastic print head. Water-soluble polyvinyl alcohol (PVA) is mainly used for this purpose. After printing, the body undergoes a vulcanization reaction in an autoclave or an oven. The thermoplastic shell survives this process and can be removed mechanically or by dissolving in water in a final step [17].

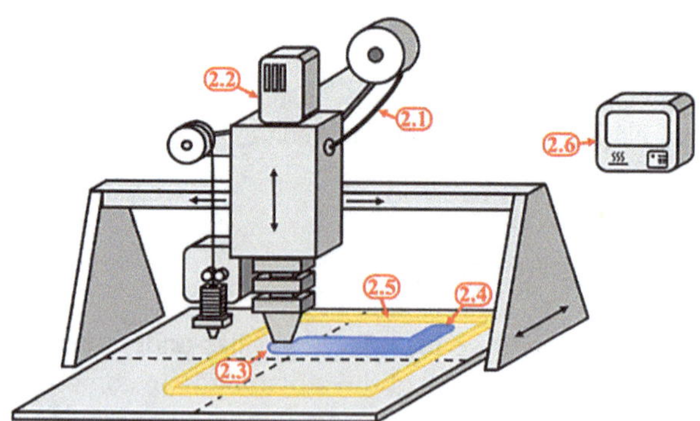

Fig. 1 Schematic structure of the AME printing system

2 Process Analysis

The dual extruder combination and the unique properties of highly viscous rubber mixtures create specific challenges for the AME process. Since these challenges have not yet been explicitly identified, a systematic process analysis will be caried out in this section. Difficulties in the printing process are clarified and shown on conducted test prints. Possible ways of overcoming them are given. The analysis follows an exemplary run through the individual production steps of the current technology. They are marked with red circles in Fig. 1, which correspond to detailed descriptions in the following subchapters. Additionally, differences to FDM elastomer printing are highlighted in order to emphasize the special characteristics of the AME process.

2.1 Filament Tearing at the Extruder Inlet

The first issue in the AME process arises at the rubber extruder's filament feed. The raw filament, produced in a prior extrusion process and stored on a roll, is pulled in by the twin screw extruder. It must transmit low tensile forces to overcome storage roll friction. Unlike FDM, the filament mixtures generally have a significantly lower tensile strength, leading to frequent tearing of the filament at the extruder inlet and thus interruptions during the printing process. Therefore, the roller is currently pushed forward manually by the machine operator, which does not correspond to an automated process.

As a remedy, the tensile stresses in the filament must be reduced. While an alternative feed technology might be considered, the storage roller concept has proven itself in handling, especially during prior filament production steps.

A simple solution would be an electronic drive of the roller. Various options of controlling this drive exist. Since the flowrates differ by filament type and nozzle diameter, it is not useful to control the drives speed by the rotational speed of the screw extruder. By that a simple solution could be a mechanism which detects the filament's tension and emits a signal for further unwinding. This could be implemented with a light barrier that gets interrupted by the stretched course of the strand. The practical implementation of this system has already shown promising results in own initial test trials, as no manual interventions are necessary with such a mechanism.

2.2 Setting the Extrusion Flow Rate

A major issue in AME process is extrusion rate control. Unlike conventional FDM gear extruders, where volume flow is mainly controlled by drive gear speed, there are more influencing factors on screw extruders. The screw's pitch creates a flow field in the barrel in forward, cross, or reverse direction, depending on the upstream pressure conditions near the nozzle [18]. Due to this behavior, viscosity of the material, as well as the effective layer height and traversing speed influence the flow rate. Since the viscosity depends on the production charge, the temperature settings, as well as on the compound itself it varies over a wide range. Additionally, the "die swelling" effect can be observed in the AME process, which can be seen in in literature on other extrusion-based rubber manufacturing technologies [6, 19]. It causes a swelling of the deposited strand after leaving the nozzle.

To correct this, flowrate and ideally strand geometry should be calibrated before and ideally during the printing process.

One possible variant for this is measuring the weight of the extruded material. The volume and the flowrate could be determined from the density. However, this would require a fine resolution of the weight scale, which could be applied under the print bed. Another solution is printing a specified test component and try to evaluate the deviations to target values, like it is often done in FDM printing [20]. For example, a thin wall structure could be printed and the wall thickness could be evaluated. However, this method is not suitable for calibration during the printing process and requires a lot of waste material. A third possibility involves directly measuring the strand geometry and evaluating the cross-section via a laser profile sensor. Both, the strand geometry and the flowrate could be calibrated within this method. The data could aid in adjusting slicer settings or selecting a suitable screw speed factor. Since this mechanism covers both effects, it is recommended for the AME printing process.

2.3 Distortion in the Strand Deposit

Another specific challenge in the AME process are distortion effects of the deposited strands. The consequences can be recognized by the exemplary printed test cuboids shown in Fig. 2. They were printed using a styrene butadiene mixture (SBR), which is commonly used in tire production [21].

Fig. 2 Shape deviations due to force induced deformation of the raw rubber material

The lack of hardening during the process increases the elastic and plastic deformability of the already printed structure. Furthermore, the high viscosity of the filament leads to high transmitted shear forces between the nozzle and the part, resulting in large deformations during printing. In this form, this behavior is unique compared to other 3D printing methods. Due to the deformation, the sides of the cuboid on the left/in the middle warped into a trapezoidal contour by the unidirectional printing direction of the nozzle. Similar effects can be seen on the right. The target dimension of 10 mm edge length of the cube was clearly not reached.

Generally, it is advisable to avoid unidirectional printing strategies, as these enhance the problem. One solution is to reduce the transmitted forces by decreasing the viscosity of the rubber. It can be achieved by an improved temperature control of the nozzle or using shear-thinning effects via induced vibrations. Furthermore, it is possible to reduce the viscosity by adapting the mixture but this changes the properties of the final part.

As a different solution, the travel paths can be adjusted by overshooting or selectively stripping the printed filament strand on the PVA shell. An integrated process control that recognizes the deviations and corrects the path would be necessary. Finally, the vulcanization of the printing material during the printing process could be cited as an additional remedy. The curing reduces the elastic and plastic deformability. However, further research is needed to determine which of these mechanisms works.

2.4 Strand Cut-Off and Extruder Retraction

The next process step is the cut-off of the filament strand. In classic FDM 3D printing it is achieved via an executed "retraction command", in which the filament strand is pulled back into the nozzle by a withdrawal of the extruder. This significantly reduces stinging effects which normally corresponds to fine spider web-like strings between the extrusion stops. However, in rubber 3D printing strand cut off and rubber stringing is particularly challenging, as the strands do not cure. Nozzle movements after the extrusion stops pull entire strands out and leave large residues on the crated layer. These can be identified on the side of the final 3D printing parts, as its visible from the printed rubber front axle bushing shown in Fig. 3. In the current printing technology, this behavior gets intensified since the existing control technology does not support retraction.

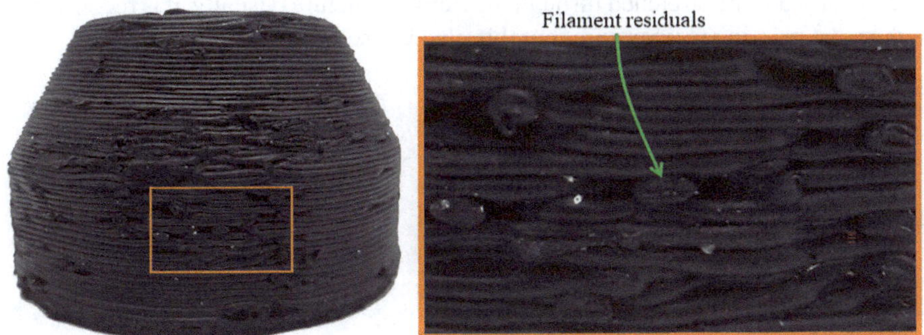

Fig. 3 Filament stringing of the rubber extruder

As a remedy, retraction must be implemented and additional features are required. Especially the nozzle design can affect the separation process. A tapered nozzle bore near the outlet, combined with a nozzle heater, increases the chance of strand separation. Helpful traversing movements in a horizontal or vertical direction can also promote strand breakage. Additionally, a cutter mechanism may tackle the problem. A thin knife-like element could actively cut off the strand. Related systems, that only cover the nozzle but do not cut the strands, are used in rear cases in existing FDM printers [22]. These mechanisms can serve as a basis for adaptation to the AME application.

2.5 Interaction with the Thermoplastic Support Structure

In AME process the surfaces of the two-phase structures interact with each other, as it is highlighted by the tests in Fig. 4. The rubber's surface on the left remains largely intact due to minimal contact with the PVA. In contrast, the right side shows strong interactions, where the negative of the grooved PVA surface is visible on the rubber.

Fig. 4 PVA surface interaction. Intact rubber surface left. Strong interacting surfaces right

The effect is enhanced by overextrusion of PVA or rubber material. Furthermore, an incorrectly set relative position of the extruders causes this behavior due to overlapping path curves. If so, the effect only occurs on one specific component side. As a countermeasure, positions of the extruders should be calibrated in relation to each other. Another remedy is to insert a gap between the component and the support structure.

Figure 5 highlights another PVA interaction. Nozzle dripping from PVA extruder occurs even with functioning retraction on this extruder. Large drops form during rubber printing. The excess material is often whipped by the shell, but sometimes it gets incorporated into new layers, causing gaps in the final part. PVA is also prone to stringing, where fine threads are stretched through the rubber structure. Usually, it is recommended to reduce the printing temperature, but this also worsens the layer adhesion.

Fig. 5 Nozzle dripping and stringing of the PVA extruder

To remedy this, it is important to ensure constant printing conditions for the water-soluble PVA. It should be stored in a filament dryer. Furthermore, a shutter mechanism is helpful, which closes the nozzle when it is not in use. Based on the manual cleaning of the nozzle, a height-adjustable brush could also be placed next to the printing part, stripping of residual material by fine steel wires when the active extruder changes. Attention should also be given to the extruder's travel paths, ensuring minimal interruptions during printing. This could be controlled via the slicer settings. In addition, it is possible to use an ooze shield around the structure. It represents a second thin PVA outer shell, which is manufactured with a gap around the first enclosure. Finally, inline vulcanization would solve these problems, as no PVA enclosure would be needed.

2.6 Part Distortion by Vulcanization Reaction

In the penultimate step of the AME process, components undergo vulcanization, a unique procedure in additive manufacturing. As the exemplary tests with a SBR mixture in Fig. 6 highlight, the chemical reaction during the vulcanization step can lead to an anisotropic shrinkage. The 40 mm × 40 mm × 5 mm plate shown on the left shrinks to approximately 80% of its original length in printing direction, whereas the thickness of 5 mm increases by 20%. Interestingly, there was almost no distortion in the y-direction, which could be described as an orthotropic behavior. Another example is shown on the right. The outer diameter of the concentric printed bushing decreased by 12%. Crossed layer orientations compensate the effect partially by material elasticity and internal stresses. The intensity of this effect clearly depends on polymer morphology and polymer-filler-interaction.

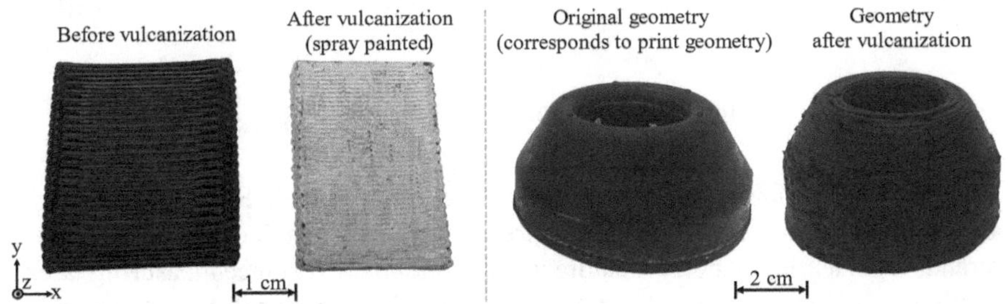

Fig. 6 Shape deviation by vulcanization process

The only viable solution to this problematic is to adapt the raw print geometry, taking the shrinkage effects into account. This can be achieved by iteratively printing and adjusting the model based on geometry deviations. As an alternative, a simulation tool could be developed, which predicts the distortion a priori. Measurements from a test print like the plate in Fig. 6 could be used as reference data to setup the simulation.

3 Conclusion with New 3D-Printer Concept

As a conclusion, the AME process has to face several specific challenges, which can only be overcome with a customized process technology. Based on this analysis, several important technical features can be summarized in one exemplary concept, as shown in Fig. 7. In contrast to the concept in Fig. 1, the movements are not carried out by the extruder. Instead, the print bed is moved in all three dimensions. The static extruder allows to use a more powerful technology, which enables higher throughputs and thus faster printing speeds. Secondly, smaller nozzle diameters could be used which improves part resolution.

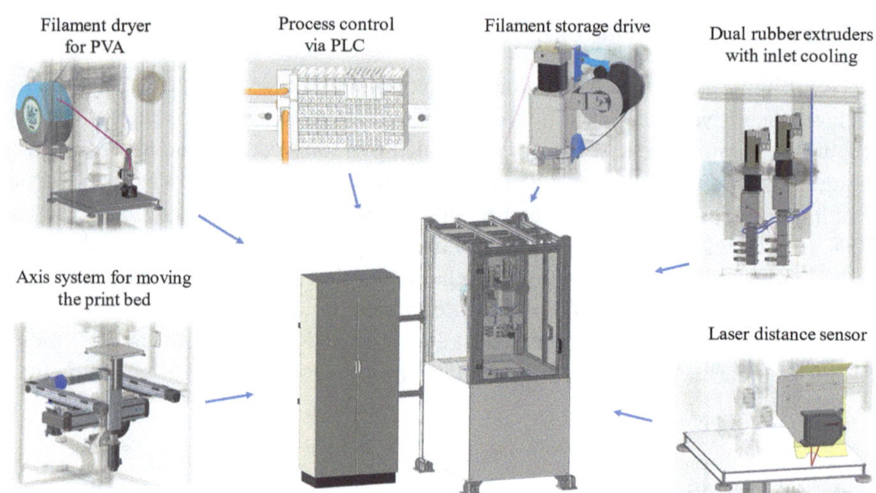

Fig. 7 Adapted rubber 3D printing concept

The printer incorporates a material feed system as outlined in Sect. 2.1, along with a cooling unit to maintain a stable temperature at the extruder's inlet. This should prevent the strands from tearing. Another feature is a side-mounted, distance measuring laser for calibrating the volume flow via the strand geometry. The sensor will be used to calibrate the extruder positions and also serves for levelling the print bed.

The machine is controlled by a programmable logic controller (PLC), which can integrate strand cutters, nozzle shutters, inline process monitoring, heating bed vibrating mechanisms and vulcanization devices. Additionally, it allows to use retraction with the screw extruder.

A dual rubber extruder setup offers enhanced printing speed by using different nozzle diameters. Fine nozzles for perimeters and larger nozzles for filling could be used. It also allows combining various rubber compositions within a single part, which increases the flexibility in rubber 3D printing. While this concept cannot solve all identified problems, it offers an initial foundation for potential extensions and further investigations to improve the part quality in rubber 3D printing.

Acknowledgements. The authors thank the *German Federal Ministry for Economic Affairs and Climate Protection* for their financial support of this research under the project titled: "Additive Manufacturing of Elastomers—AME 2.0" (IGF-Vorhaben Nr.: 01IF23124N / 2).

Competing Interests The author(s) has no competing interests to declare that are relevant to the content of this manuscript.

References

1. Horvath, J.: Mastering 3D Printing; Modeling, printing, and prototyping with reprap-style 3D printers. Apress (2014). https://doi.org/10.1007/978-1-4842-0025-4. ISBN: 978-1-4842-0026-1
2. Savini, A., Savini, G.G.:"A short history of 3D printing; a technological revolution just started" ICOHTEC/IEEE 2015. In: International History of High-Technologies and their Socio-Cultural Contexts Conference, S. 1–8. https://doi.org/10.1109/HISTELCON.2015.7307314. ISBN: 978-1-5090-0065-4
3. Attaran, M.: The rise of 3-D printing: the advantages of additive manufacturing over traditional manufacturing. Bus. Horizons J. **60**, 677–688 (2017). https://doi.org/10.1016/j.bushor.2017.05.011
4. Herzberger, J., Sirrine, J.M., Williams, C.B., Long, T.E.: Polymer design for 3D printing elastomers: recent advances in structure, properties, and printing. In: Progress in Polymer Science, Vol. 97. Elsevier (2019). https://doi.org/10.1016/j.progpolymsci.2019.101144
5. Sundermann, L., Klie, B., Giese, U., Leineweber, S., Overmeyer, L.: Development, construction and testing of a 3D-printing-system for additive manufacturing of carbon black filled rubber compounds. In: *Kautschuk Gummi Kunststoffe Rubberpoint*, Vol. 10, S. 30–35. WIN-Verlag (2020)
6. Drossel, W.-G., Ihlemann, J., Landgraf, R., Oelsch, E., Schmidt, M.: Basic research for additive manufacturing of rubber. In: Polymers, Vol. 12. MDPI (2020). https://doi.org/10.3390/polym12102266
7. Thoguluva, R., Raghavan Vijayaram, T.: A technical review on rubber. In: International Journal on Design and Manufacturing Technologies, Vol. 3, S. 25–37. IAEME Publication (2009). https://doi.org/10.18000/ijodam.70043
8. Röthemeyer, F., Sommer, F.: Kautschuk Technologie; Werkstoffe—Verarbeitung—Produkte. Hanser (2013). ISBN: 978-3-446-43776-0
9. Dasgupta, A., Dutta, P.: Printability of elastomer as a 3D printing material for additive manufacturing. J. Rubber Res. **27**, 137–157. Springer (2024). https://doi.org/10.1007/s42464-024-00241-x
10. Alkadi, F., Lee, J., Yeo, J.-S., Hwang, S.-H., Choi, J.-W.: 3D printing of ground tire rubber composites. J. Precis. Eng. Manuf. Green Technol. **6**, 211–222. Springer (2019). https://doi.org/10.1007/s40684-019-00023-6
11. Quetzeri-Santiago, M.A., Hedegaard, C.L., Castrejón-Pita, J.R.: Additive manufacturing with liquid latex and recycled end-of-life rubber. In: 3D Printing and Additive Manufacturing, S. 149–157. Mary Ann Liebert, Inc. (2019). https://doi.org/10.1089/3dp.2018.0062
12. Toncheva, A., Brison, L., Dubois, P., Laoutid, F.: Recycled tire rubber in additive manufacturing: selective laser sintering for polymer-ground rubber composites. In: Applied Sciences, S. 8778 MDPI (2021). https://doi.org/10.3390/app11188778
13. Costlow, D.B.: Methods and apparatus for additively manufacturing rubber. Patent publication number: US20160185040A1, Bridgestone Americas Tire Operations LLC (2015)
14. Scheungraber, P.: Device and method for the 3D printing of a workpiece made of a rubber like non thermoplastic material. Patent publication number: WO2018206263A1, Rema Tip Top AG (2018)
15. Gerada, I., Hignett, M.: 3d-printer system and 3d-printing method of an elastomerically deformable rubber body, in particular a rubber seal. Patent publication number: WO2020249189A1, Trelleborg Sealing Solutions Germany GmbH (2019)
16. Leineweber, S., Sundermann, L., Bindszus, L., Overmeyer, L., Klie, B., Wittek, H., Giese, U.: Additive manufacturing and vulcanization of carbon black filled natural rubber based components. In: Rubber Chemistry and Technology, Vol. 95, S. 46–57 Lancaster (2021). https://doi.org/10.5254/rct.21.79906

17. Leineweber, S.: Additive Fertigung von Elastomerbauteilen aus rußgefülltem Kautschuk. TEWISS Verlag (2025). ISBN: 978-3-69030-030-8
18. Limper, A.: Verfahrenstechnik der Thermoplastextrusion. Hanser (2012). https://doi.org/10.3139/9783446428669. ISBN: 978-3-446-42866-9
19. Müllner, H.W., Wieczorek, A., Eberhardsteiner, J.: Experimental determination of shear-thinning behaviour during extrusion of rubber blends. In: Strain Volume 47, S. 162–172. Wiley (2011). https://doi.org/10.1111/j.1475-1305.2008.00532.x
20. Prusa Research a.s. "Extrusion multiplier calibration" 2024. Online available at: https://help.prusa3d.com/article/extrusion-multiplier-calibration_2257, Checked on 26.05.2025
21. Abts, G.: Einführung in die Kautschuktechnologie. Hanser (2019). 45461. ISBN: 978-3-446-45855-0
22. Bambu Lab: Replace H2D Quick Change Tool Interface; Step 2: Install the extruder filament guide and the left cutter. n.d. 2025. Online available at: https://wiki.bambulab.com/en/h2/maintenance/replace-quick-change-tool-interface, Checked on 26.05.2025

Production and Optimization of Powder Metallurgical Billets with Lubricant-Infiltrated Cavities for Impact Extrusion

René Laeger(✉), Julius Peddinghaus, Hendrik Wester, Kai Brunotte, and Bernd-Arno Behrens

Institut für Umformtechnik und Umformmaschinen der Leibniz Universität Hannover, Garbsen, Germany
laeger@ifum.uni-hannover.de

Abstract. The lubrication of tools in forming technology primarily serves to reduce wear. Tool lubrication is therefore an essential factor in the optimization of processes with regard to their economic efficiency. There are different approaches that address lubrication on the semi-finished product side. In this contribution, the approach of lubrication through infiltration of porous semi-finished products produced by powder metallurgy is discussed. These semi-finished products should later constantly release lubricant during an extrusion process to enable a constant lubricating film. For this purpose, the porosities serve as reservoirs. The cylinders produced here consist of steel powder mixed with pressing aid to minimize friction during powder pressing. Powder pressing at different pressures was used to produce samples with different densities, which were then sintered. The density was determined by geometric and weight measurement. The semi-finished products were then infiltrated with lubricant. Different lubricants are examined and used to investigate the influence of the viscosity on the infiltration amount. Due to the temperature dependence, the viscosity was determined as a function of temperature. Infiltration was carried out in a lubricant bath with different infiltration times. The amount of the lubricant was measured by weighing to research the correlation between porosity and infiltration quantity. It was found that a lubricant with a high viscosity infiltrates in higher quantity, but retention is lower due to drip-off. In addition, the duration of infiltration at low viscosities has less influence on the lubricant quantity. The results published here can be used to design self-lubricating blanks depending on the desired lubricant quantity for a cold forming process.

Keywords: Powder metallurgy · Forming · Lubrication technology · Wear reduction

1 State of the Art

Tools used in cold forming of metals are exposed to high stresses and abrasive wear due to the low temperatures [1]. Friction reduction is possible on both the tool and workpiece side.

L. Overmeyer and B.-A. Behrens (eds.), *Production at the Leading Edge of Technology*,
Lecture Notes in Production Engineering, https://doi.org/10.1007/978-3-032-19524-1_26

One common tool-side method is forming with coated tools. The coatings are thin layers with application-optimized properties that have a lower friction potential and higher wear resistance than the steel itself, particularly against abrasion [2]. One finding from the investigations in [2] is the strong dependence of roughness on wear. In the case of semi-finished products (SFP) produced by powder metallurgy (PM), increased friction conditions between the SFP and the tool can always be assumed due to open porosities. For this reason, coatings are not effective here.

A common SFP-sided approach is the coating with zinc phosphate coatings [2, 3]. Due to their physical and chemical properties, lubricants can adhere particularly well [4]. A major disadvantage of this coatings is the high ecological footprint and process engineering effort involved. These require several pre-treatment steps, consume large quantities of water and chemicals and generate problematic waste such as sludge containing heavy metals [3]. The aim should be to reduce lubricant consumption, for example through optimized lubrication, in order to improve ecological conditions.

Another approach is primarily aimed at the constant lubricating film. This is maintained on the workpiece side. Porous, impregnated samples are used to create and maintain a lubricating film in situ during forming. The self-lubrication approach originates from sintered plain bearings, which can absorb lubricant due to their porosity and release it in the event of an emergency, for example if there is insufficient lubrication [5]. This provides additional safety for the process. Initial successes in this regard in cold forging were achieved in previous studies [6]. The results showed that with a lower relative density of 0.85, the critical stress in compression tests until material failure can be more than twice as high compared to a relative density of 0.9. This difference in density is caused by a greater absorption of the lubricant. In [6] the infiltration time is always less than 1 h. In addition, only viscosities up to 110 mm^2/s were investigated. The correlations between infiltration duration, viscosity and lubricant amount, which have also been poorly researched in the literature, are taken into account here.

For a potential infiltration of PM SFP made of steel, the porosities must be specifically adjusted. In [7], high porosities of approx. 60 to 70% were achieved when sintering steel. The space holder method is often used for adjusting porosity. This involves introducing a material while powder pressing that is later flushed out, e.g., by rinsing. However, this would be an additional process step compared to other approaches, the expense of which would have to be put into perspective. The porosity can also be adjusted by parameters during powder pressing or sintering. Important parameters are pressures and temperatures, as shown in [8]. The density can be adjusted by varying the pressing parameters, such as the stroke length [6].

While existing research highlights the potential of porous materials and lubricant-impregnated SFPs in cold forming, several key parameters influencing infiltration—such as porosity, infiltration time, and lubricant viscosity—remain insufficiently investigated. Furthermore, many of the methods to adjust porosity involve additional process steps that may not be viable for industrial implementation. This study aims to close this gap by systematically examining how different processing parameters affect lubricant infiltration and evaluating their potential relevance for self-lubricating forming.

2 Goal and Experimental Approach

The desired process chain is shown in Fig. 1. The aim of this investigation is to correlate the density, based on [6], of the porous sintered SFP, the viscosity and the infiltration duration with the infiltration amount. The most suitable process will be used to produce samples, which are intended for later use in an extrusion process to investigate the self-lubricating properties, with the aim of achieving the lowest possible friction.

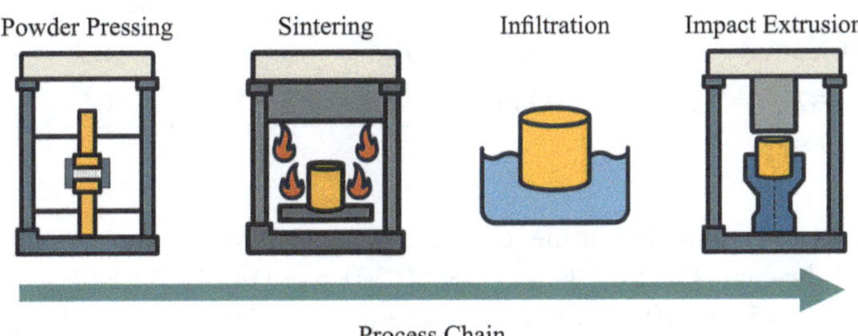

Fig. 1 Process chain of specimen production and application

In contrast to the tests in [6], different viscosities are considered in this work in order to investigate the influence on infiltration. In addition, different infiltration times are applied to clearly determine the influence of duration on the quantity measured by weighing. In studies in [6], due to minor differences in infiltration time (15/30/45 min), only a maximum increase of 31% was observed for longer durations (right side of Fig. 1 in [6]), which is why a larger difference was chosen in this study (45 min to 17 h).

The infiltration of porous structures is governed by complex interactions. A well-established approach to describe the filling of cavities is given by the Washburn Eq. (1) [9]. In the following, this relation will be applied to elucidate the underlying mechanisms and their interdependencies.

$$l = \sqrt{\frac{\gamma * \cos\theta}{\eta * 2} * r * t} \tag{1}$$

with l = depth of penetration, γ = surface tension, θ = contact angle, η = dynamic viscosity, r = capillary radius and t = time.

2.1 Characterization of Materials

Steel Samples. The sinter alloy Atomet 1001 by Rio Tinto was used for the study. The chemical composition is listed in Table 1. Preliminary tests were carried out to produce suitable SFP. These had a diameter of 30 mm and a height of up to 35 mm. A nearly fully compacted SFP with a density of 7.6 g/cm^3 was used to determine relative density. The SFP were geometrically measured and weighed to calculate the density. Two relative densities for each lubricant were chosen for the tests, $\rho_{rel,low}$ = 0.84 with a standard deviation of 0.03 and $\rho_{rel,high}$ = 0.92 with 0.005. These densities are low enough to enable suitable infiltration, but not too low that full compaction is no longer possible in the subsequent extrusion process. The micronized Ethylene-Bis-Stearamide (EBS) wax for sintering processes was added to reduce friction while powder pressing.

Table 1 Chemical composition of Atomet 1001 + copper + graphite [10] + A 2050 M (EBS-wax)

Chemical composition, wt%							
Atomet 1001					Supplements		
C	O	S	Mn	Fe	Cu	C	Wax
0.004	0.087	0.010	0.174	96.230	1.98	0.79	1.00

Lubricants. To investigate the influence of different viscosities on the infiltration of the samples, two lubricants OH 50,199 and OH 50,201 were used. The SFP were infiltrated over durations of 45 and 1020 min to compare the duration of the infiltration with the lubrication amount.

The viscosity is strongly dependent on the temperature. As forming processes do not have a homogeneous constant temperature, the influence of temperature must also be investigated. However, to be able to estimate the viscosities in locally warmer areas during extrusion for later investigations, determined by simulations, the lubricants were examined in an automatic kinematic viscometer (SVM 3001) from Anton Paar GmbH.

2.2 Characterization of the Infiltration

The infiltration tests were carried out at approx. 20 °C. At this temperature, the OH 50,201 has a kinematic viscosity of 151.32 mm^2/s and OH 50,199 of 1804.2 mm^2/s.

The weight of the samples was measured before infiltration, directly after infiltration and after 10 min dripping off using a precision scale. This procedure was used to determine the infiltration quantity and preservation of lubricant in the SFP. Conclusions about the influence of the handling time could be drawn from the analysis of lubricant retention. This knowledge is crucial for the overall process chain design including handling concepts to maintain lubrication in the extrusion process.

3 Results and Discussions

3.1 Viscosity

The kinematic viscosities are shown in the graphs in Fig. 2. Up to a temperature of 80 °C, a significant difference can be observed. The impact extrusion processes described in [11, 12] for example reach temperatures of up to 130 °C in the highest friction areas, while in other areas the temperature is not increased. This would mean approx. the same viscosity for both lubricants in the areas of high friction. However, in areas with low friction and consequently lower temperatures, especially at the beginning of a forming process, the viscosity differs significantly. For this reason, choosing the right lubricant is essential in cold forming.

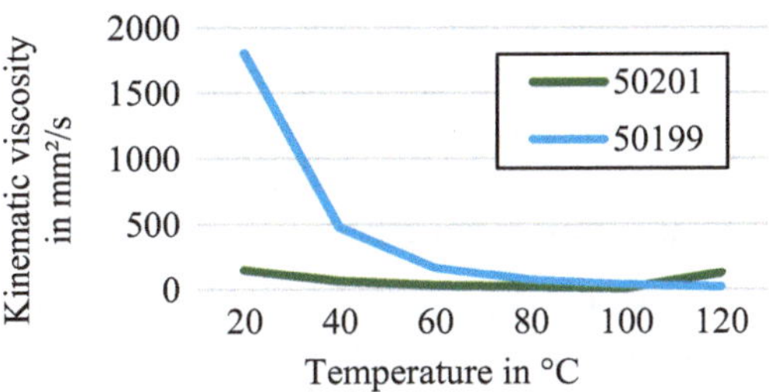

Fig. 2 Comparison of kinematic viscosities of OH 50,199 and OH 50,201 at different temperatures

3.2 Infiltration

Influence of Duration. The amount of lubricant stored is strongly dependent on the time, as evidenced by the increased lubricant amount after 17 h of infiltration compared to 45 min, shown in Fig. 3. After 17 h, the amount is approx. 50% higher than after 45 min. The minor increase in infiltration after such a major increase in infiltration time indicates a nonlinear degressive effect. Smaller pores that form further inside the sample are subject to the Washburn equation [9], which exhibits a degressive progression. This can explain the progression of duration and quantity determined here. With the small differences of 15 min in [6], an increase in lubricant quantity of 31% was already achieved. Assuming similar circumstances and achieving only a 50% increase after a time difference of 975 min, this confirms the degressive behavior and validity of the Washburn equation for these circumstances.

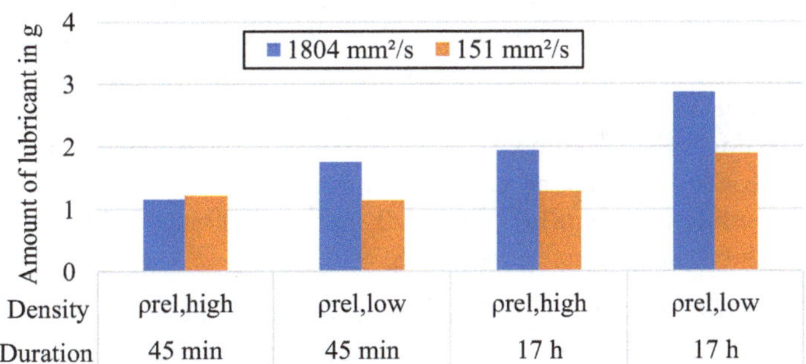

Fig. 3 Lubricant quantity depending on infiltration duration, sample density and viscosity

Influence of Density. A lower density ($\rho_{\text{rel, low}}$) leads to an increased absorption of lubricant. The infiltrated amount is approx. 35% higher compared to the higher density. Only the samples at 45 min with a viscosity of 151 mm^2/s show nearly no difference. Although high viscosity reduces the absolute filling speed, it also increases the sensitivity of the process to structural parameters such as porosity. At low viscosity, the low flow resistance is overcompensated by capillary force, so that differences in porosity are initially negligible. At high viscosity, on the other hand, dissipation has an inhibitory effect from the outset, causing differences in pore structure to lead to clearly distinguishable lubricant quantities even in the early stages of the filling process.

When designing a subsequent cold forming process, it can therefore be assumed that a lower density of SFP allows a greater amount of lubricant to be introduced into the process if a certain infiltration time is exceeded.

Influence of Viscosity. Apart from the specimen with a relative density of $\rho_{\text{rel, high}}$ and an infiltration duration of 45 min, which shows a similar amount, the samples with higher viscosity have an amount of more than 50% of the lubricant, shown in Fig. 4.

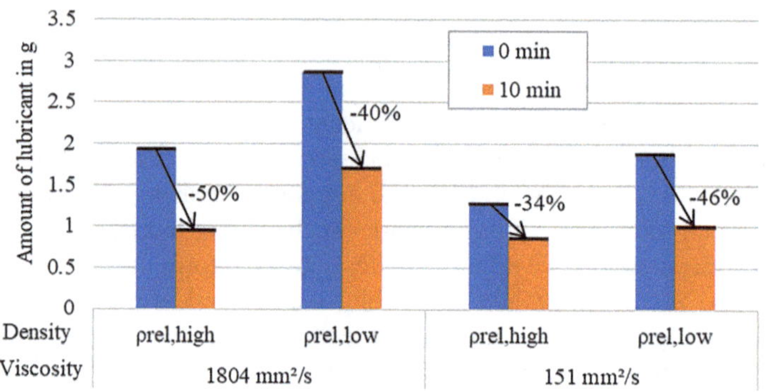

Fig. 4 Lubricant retention over time for different densities and viscosities, 17 h of infiltration.

The hypothesis is that the combination of high viscosity and a complex pore network inhibits the release of the lubricant. In addition, the surface of the sample was not wiped after infiltration. It could be that the increased lubricant volume was strongly influenced by the lubricant adhering to the outside, which remained adhered to the outer pores and roughness due to its viscosity. This is advantageous for reducing friction during subsequent forming, as it allows more lubricant to be introduced into the process. Possible ecological or economic disadvantages must be considered on a case-by-case scenario. Under these conditions, it can be concluded that the lubricant with higher viscosity favors a higher lubricant amount in porous SFP.

Lubricant retention over time. After 10 min, the remaining amount of lubricant in the samples was 50 to 66% of the amount measured immediately after infiltration. The retention in % at a lower viscosity is slightly higher, but the absolute value is still higher for thicker lubricant (Fig. 4). This opposite behavior is probably due to the excess amount of lubricant on the outer surface of the samples. Since more lubricant collects there at high viscosity, the potential for release is also higher.

At higher densities of all samples (45 min and 17 h), slightly more lubricant remains. The larger pore volume in samples with low density also offers more potential for lubricant release. Just one sample is not following the trend. The different trend at high viscosity over 17 h could indicate that porosity plays a different role under these conditions. While at low viscosity, increased porosity primarily leads to faster transmission and thus lower retention, at high viscosity over longer periods of time, the higher number of available pore channels could nevertheless contribute to additional lubricant absorption. This would explain why a higher relative retention rate is observed in more porous samples at high viscosity and 17 h.

4 Conclusion and Outlook

Samples with different densities were infiltrated for two different durations with two different lubricants (with different viscosities). The lubricant quantities were measured by weighing. The most important findings and correlations are listed below:

- Due to the strong dependence of viscosity on temperature, temperature curves should always be determined for lubricants that will be used for infiltration. A thermal analysis of a potentially subsequent cold forming process is also advised.
- An increased lubricant quantity can be achieved through low density of the SFP, high viscosity of the lubricant and long infiltration duration.
- Samples with a high density and lubricants with low viscosity retain the lubricant longer (relative) over time, enabling longer transfer times in a process.

As a result, a long infiltration time, a low density and a high lubricant viscosity are advantageous for the infiltration of as much lubricant as possible if the transfer time in the process chain is short. Low density and high viscosity lead to rapid loss due to dripping.

In further tests, the amount of lubricant required for an extrusion process is identified to determine a suitable combination of parameters for the production of porous SFPs. In addition, the release of the lubricant during forming should be investigated, e.g. in compression tests. In the end, the compaction of the samples should also be determined. This is relevant for the adjustment of the mechanical properties of the product.

Acknowledgements. The presented results are based on the research project "Substitution of conventional mold lubrication by using self-lubricating raw parts in sinter forging II", Project number 264818458. The authors would like to thank the German Research Foundation (DFG) for the financial support.

Competing Interests The author(s) has no competing interests to declare that are relevant to the content of this manuscript.

References

1. Karunathilaka, N., Tada, N., Uemori, T., Hanamitsu, R., Fujii, M., Omiya, Y., Kawano, M.: Effect of lubrication and forging load on surface roughness, residual stress, and deformation of cold forging tools. Metals **9**(7), 783 (2019). https://doi.org/10.3390/met9070783
2. Bobzin, K., Klocke, F., Bergs, T., Brögelmann, T., Feuerhack, A., Kruppe, N.C., Hild, R., Hoffmann, Christopher, D.: Dry forming of low alloy steel materials by full forward impact extrusion with self-lubricating tool coatings and structured workpieces: Universität Bremen
3. Weng, D., Wang, R., Zhang, G.: Environmental impact of zinc phosphating in surface treatment of metals. Met. Finish. **96**(9), 54–57 (1998). https://doi.org/10.1016/S0026-0576(98)81480-3

4. Gariety, M., Ngaile, G., Altan, T.: Evaluation of new cold forging lubricants without zinc phosphate precoat. Int. J. Mach. Tools Manuf **47**(3–4), 673–681 (2007). https://doi.org/10.1016/j.ijmachtools.2006.04.016
5. Beiss, P.: Pulvermetallurgische Fertigungstechnik (2013). https://doi.org/10.1007/978-3-642-32032-3
6. Behrens, B.-A., Brunotte, K., Bohr, D.: Optimised impregnation parameters of self-lubricating powder metallurgical components for cold bulk forming. METAL 2018—27th International Conference on Metallurgy and Materials, Conference Proceedings, pp. 242–248 (2018)
7. Fujii, T., Saito, S., Shimamura, Y.: Fabrication of porous steels via space holder technique and their mechanical properties. Mater. Trans. **64**(7), 1638–1644 (2023). https://doi.org/10.2320/matertrans.MT-Z2023002
8. Mikó, T., Markatos, D., Török, T.I., Szabó, G., Gácsi, Z.: Minimizing Porosity in 17–4 PH stainless steel compacts in a modified powder metallurgical process. J. Compos. Sci. **8**(7), 277 (2024). https://doi.org/10.3390/jcs8070277
9. Washburn, E.W.: The dynamics of capillary flow. Phys. Rev. **17**(3), 273–283 (1921). https://doi.org/10.1103/PhysRev.17.273
10. Rio Tinto Metal Powders. ATOMET 1001 Data Sheet. [online]. n.d. [16.06.2025]. Available from: https://qmp-powders.com/wp-content/uploads/Product%20PDFs/ATOMET_1001.pdf
11. Luca, D.: Finite element simulation and experimental investigation of cold forward extrusion process. MATEC Web Conf. **178**, 2010 (2018). https://doi.org/10.1051/matecconf/201817802010
12. Görtan, M. Okan: Investigation of cold extrusion process using coupled thermo-mechanical FEM analysis and adaptive friction modeling. In: Proceedings of the International Conference of Global Network for Innovative Technology and AWAM International Conference In Civil Engineering (IGNITE-AICCE'17): Sustainable Technology And Practice For Infrastructure and Community Resilience. Penang, pp. 140011 (2017)

A Semi-Analytical Approach for Process Modeling in Cold Ring Rolling

Qinwen Wang[1(✉)], Johannes Seitz[2], Bernd Kuhlenkötter[2], and Alexander Brosius[1]

[1] Institute of Manufacturing Technology, Chair of Forming and Machining Processes, Dresden University of Technology, Dresden, Germany
qinwen.wang@tu-dresden.de

[2] Chair of Production Systems, Ruhr University Bochum, Bochum, Germany

Abstract. Ring rolling is an important incremental forming process for manufacturing seamless rings. However, fast and accurate prediction for process parameters remains challenging, especially in the context of generating large datasets. Therefore, this paper introduces a novel semi-analytical approach, based on the upper bound method, to calculate process parameters incrementally in cold ring rolling. The proposed method is validated against experimental data and finite element (FE) simulations, demonstrating a trade-off between computational speed and accuracy. While slightly less precise than FE simulations, the approach drastically reduces computation time, making it suitable for rapid parameter prediction prior to manufacturing. Incremental calculations further enhance optimization potential, offering a practical solution for improving manufacturing efficiency and sustainability. The development of data-driven approaches, particularly in machine learning (ML), has revealed new possibilities, such as the early identification of form errors. However, training machine learning algorithms requires extensive datasets, which can be obtained in ring rolling from experiments, FE simulations, or (semi-)analytical methods. Therefore, the proposed method enables the generation of large interpolative and extrapolative datasets, which can serve as inputs for ML models.

Keywords: Ring rolling · Incremental forming · Semi-analytical methods · Database generation

1 Introduction and State of the Art

Ring rolling is an incremental forming process widely used in the production of seamless rings for applications such as in the automotive and energy industries. The process can be classified into radial-axial ring rolling (RARR) and tangential ring rolling (TRR), depending on the presence of axial rolls [1]. The former is typically used for manufacturing large rings with diameters up to 16 m at high temperatures above the recrystallization point, while the latter is applied to small rings typically ranging from a few centimeters to a few decimeters at room or semi-hot temperatures starting from the austenitizing temperature [2].

L. Overmeyer and B.-A. Behrens (eds.), *Production at the Leading Edge of Technology*, Lecture Notes in Production Engineering, https://doi.org/10.1007/978-3-032-19524-1_27

To reduce material and energy waste, early prediction of the final ring quality can be beneficial [3]. However, the process involves complex material deformation influenced by coupled effects of geometry, friction and process parameters, which makes accurate quality prediction particularly challenging [4]. With the development of data-driven methods, machine learning (ML) has been increasingly applied to product quality in industrial manufacturing [3]. Seitz et al. explored the potential of transfer learning (TL) in ring rolling. Since data collection in large-scale ring rolling is challenging, TL provides a promising solution by enabling knowledge transfer from small-scale experiments. Moreover, TL can enhance in general machine learning modeling, for cases where no sufficient data is available [5].

Prior research has applied FE simulations and analytical methods to compute process parameters. When properly configured, the FE methodology can provide accurate process predictions. However, it is highly time-consuming, especially for incremental forming processes [4]. (Semi-) analytical modeling can reduce calculational time significantly, though often at some expense of accuracy.

A major class of analytical models in ring rolling focuses on the ring gap. One of the earliest analytical contributions in this area was made by Hawkyard et al., who developed a model to estimate the rolling force and torque based on Hill's slip-line theory for a flat indenter [6]. In this model, the contact length is treated as a straight projection between the tools and the ring, which limits its accuracy depending on the size ratio between the tools and the ring [7]. In 2011 and 2014, Parvizi et al. employed the slab method [8] and later the upper bound method [9] to estimate the radial force. The latter method improved the assumed velocity field based on the work of Yang and Ryoo [10]. However, both methods assume direct control over the ring diameter, limiting their practical applicability. Quagliato et al. introduced the concept of dividing the ring into slices. Based on the deformation of each slice during one rotation of the ring, average strains can be calculated. However, only average strain values for the observed slices are obtained [4]. Notably, no publicly available scripts currently exist for the fast prediction of process parameters in tangential ring rolling.

The aforementioned integration of machine learning modeling through TL, as well as process design, which strikes a suitable balance between accuracy and computational time, has motivated the development of a novel modeling approach. Therefore, this paper introduces a semi-analytical method based on upper bound theory to compute relevant process parameters. The model is capable of predicting the strain, diameter expansion, ring thickness, and forming force. The proposed method provides a quick way to predict process parameters for new configurations.

2 Methodology

In this work, the TRR experiments were conducted at room temperature on a retrofitted UPWS 31,5.2 from Profiroll Technologies GmbH, as presented in Fig. 1a. As the main roll translates towards the mandrel, the ring thickness is reduced, and the ring diameter grows. The process terminates when the main roll and mandrel come into contact, with only the final ring thickness being controlled, as illustrated in Fig. 1b. The machine has been equipped with different sensors. The ring growth is measured using two measuring rolls and a potentiometer. The displacement of the main roll is recorded with a linear displacement transducer. Both rolling force and torque are measured using strain gauges. All measurement signals are acquired via a data acquisition system (HBM QuantumX).

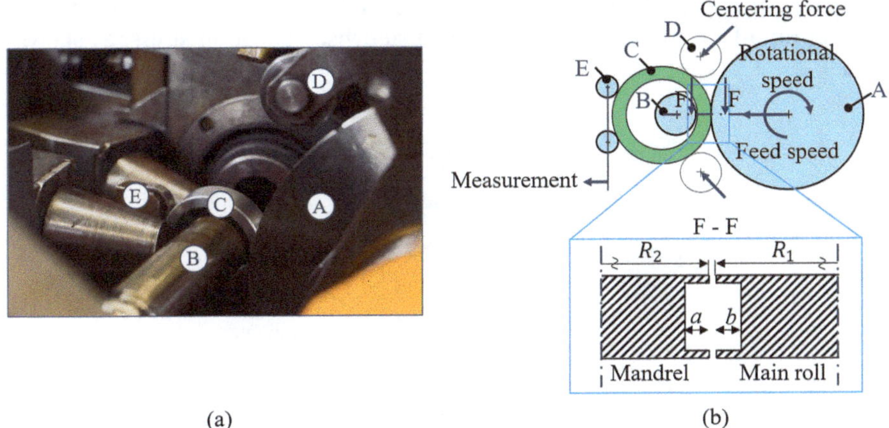

Fig. 1 **a** Photograph of the TPRR machine; **b** Schematic illustration of the process; A—main roll, B—mandrel, C—blank ring, D—guide rolls, E—measuring rolls

The ring material used in this investigation is 100Cr6 (1.3505). Its mechanical property was determined by performing compression tests on cylindrical samples at room temperature. The tests were carried out on a 250 kN universal testing machine from Hegewald and Peschke. Deformation of the compression samples was recorded using an ARAMIS 4M system for digital image correlation. The flow curve, approximated using the Voce extrapolation method with the corresponding coefficients, is presented in Fig. 2. Since the temperature during cold TRR does not exceed 100 °C, thermal effects were neglected.

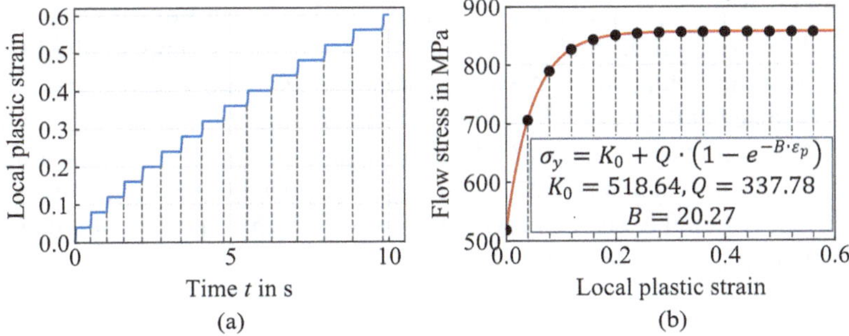

Fig. 2 **a** Local incremental plastic strain; **b** Flow stress curve with corresponding plastic strain points

The following section introduces the method for calculating process parameters, including strain distribution, ring diameter, and rolling force. In the ring's cross-section, strain in the axial direction within the mandrel-main roll gap is significantly smaller than in the other two directions. In the TRR process, the change in the end is within 1%, which justifies the assumption of a plane strain condition. In the proposed model, the ring is divided into n segments, and each segment is further subdivided into m elements, as illustrated in Fig. 3a. Each segment undergoes deformation for a time interval t, as defined in Eq. (4), resulting in a periodic and stepwise change in stress, as shown in Fig. 2a.

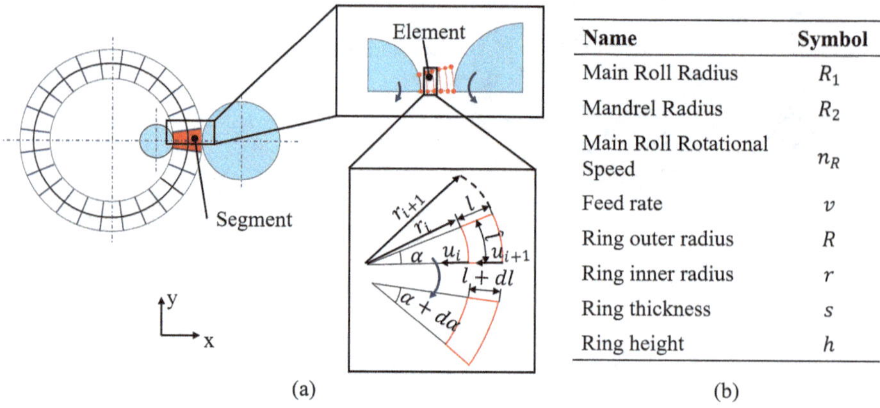

Name	Symbol
Main Roll Radius	R_1
Mandrel Radius	R_2
Main Roll Rotational Speed	n_R
Feed rate	v
Ring outer radius	R
Ring inner radius	r
Ring thickness	s
Ring height	h

Fig. 3 **a** Kinematic model of a ring segment passing through the roll gap; **b** Definition of geometric parameters and symbols

When a segment of the ring passes through the roll gap, each element within it undergoes both radial and circumferential deformation. The subsequent calculations are based on this deforming segment. The model is formulated in a cylindrical coordinate system, where the deformation state of each element is described by two strain components: the radial strain, the circumferential strain. For the i-th element within the segment passing through the roll gap, these strain components are derived from geometric relationships, as illustrated in Fig. 3a, and are mathematically expressed in Eqs. (1–2) as functions of the radial and angular displacement variable u and $d\alpha$. The time required for a ring segment to pass through the roll gap can be calculated using Eq. (4), which accounts for the rotational speed of the main roll n_R, the segment angle α, the current ring radius R and the main roll radius R_1. Correspondingly, the displacement of the main roll during this time interval is formulated in Eq. (5) and is equal to the sum of the radial displacements of all elements within that segment, where v denotes the feed rate of the main roll. All quantities are expressed in SI units.

$$d\varepsilon_r = \frac{\mathrm{d}l}{l} = \frac{u_{i+1} - u_i}{l} \tag{1}$$

$$d\varepsilon_c = \frac{\mathrm{d}\hat{l}}{\hat{l}} = \frac{(r_m + dr_m) \cdot (\alpha + d\alpha) - r_m \cdot \alpha}{r_m \cdot \alpha} \tag{2}$$

$$r_m = \frac{r_{i+1} + r_i}{2} \tag{3}$$

$$t = \frac{\alpha \cdot R}{n_R \cdot R_1} \tag{4}$$

$$u = \sum_{1}^{m} u_i = v \cdot t \tag{5}$$

The equivalent strain can be computed using Eq. (6), from which the equivalent stress is subsequently obtained via the material's constitutive model [11]. In each computational step, following the principles of the upper bound method, the distribution of the displacement u_i for each element is determined such that the total power consumption within the segment is minimized. According to the work of Parvizi [9], this condition is achieved when the internal power of the segment, given in Eq. (7), reaches its minimum. Here, l_i denotes the element length, and $\hat{l}_i$ represents the mean element arc length, as illustrated in Fig. 3a.

$$d\varepsilon_{eq} = \sqrt{\frac{2}{3} \cdot \left(d\varepsilon_r{}^2 + d\varepsilon_c{}^2\right)} \tag{6}$$

$$E_{internal,segment} = \sum_{1}^{m} \varepsilon_{eq,i} \cdot \sigma_{eq,i} \cdot l_i \cdot \hat{l}_i \tag{7}$$

To ensure volume consistency during the forming process, the volume increment described by the strain is incorporated into the internal energy function as a penalty term. To solve the resulting optimization problem, the Broyden–Fletcher–Goldfarb–Shanno (BFGS) algorithm is employed at each incremental step. The complete procedure of the incremental analysis is illustrated in Fig. 4, which outlines the sequence from strain estimation and power minimization to stress evaluation and displacement update.

Using the aforementioned method, the outer and inner radius, R and r, can be predicted throughout the process. These values are also essential for calculating the rolling force via Eq. (8), as proposed by Quagliato et al. [4]. The shear stress k can be obtained from Fig. 2b based on the corresponding equivalent plastic strain. The ring height is denoted by h. The contact length L_C is determined using Eq. (9), proposed by Hawkyard et al. [6], where R_1 and R_2 represent the radii of the tools. s denotes the ring thickness after each rotation, and Δs is the change in thickness, calculated according to Eq. (10). The pressure factor γ, which is required for the force calculation, is derived from the slip-line field solution by Hawkyard et al. [6] and defined in Eq. (11), which was fitted by Quagliato et al. [4].

$$F = 2 \cdot k \cdot \gamma \cdot h \cdot L_C \tag{8}$$

$$L_c = \sqrt{\frac{2 \cdot \Delta s}{\frac{1}{R_1} + \frac{1}{R_2} + \frac{1}{R} - \frac{1}{r}}} \tag{9}$$

$$\Delta s = \frac{2 \cdot \pi \cdot R \cdot v}{n_R \cdot R_1} \tag{10}$$

$$\gamma = -0.0098 \cdot \left(\frac{s}{L_C}\right)^2 + 0.2999 \cdot \left(\frac{s}{L_C}\right) + 0.6909 \tag{11}$$

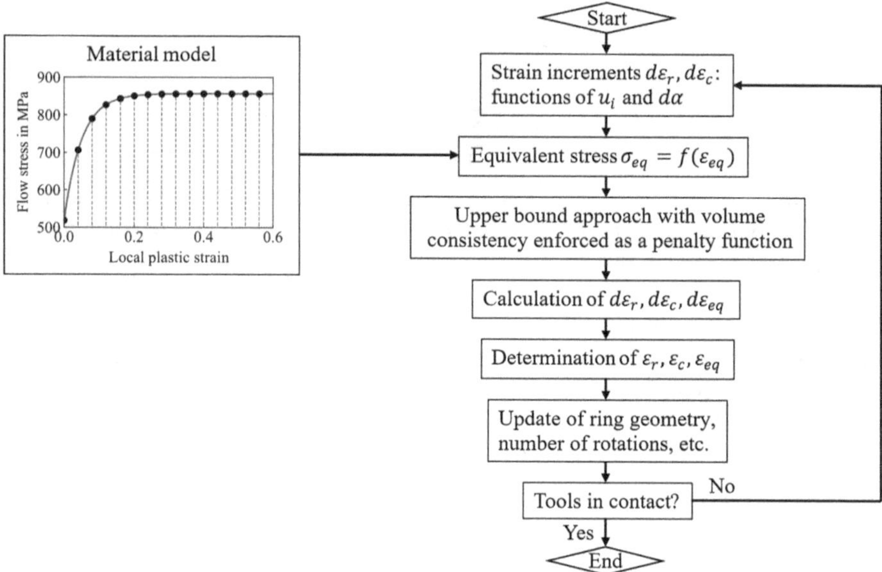

Fig. 4 Flowchart of the incremental ring forming model

3 Model Validation

To obtain comprehensive insights into the process and validate the proposed method, a 3D FE model was developed in LS-DYNA (version 14.0.0) using the standard explicit algorithm, as shown in Fig. 5a. The ring was discretized with 1 mm brick elements with 8 nodes and one integration point (ELFORM 1), while the tools were modeled as rigid parts using shell elements. The material behavior was described by the piecewise linear plasticity law (MAT_24). Contact was modeled with the penalty formulation applying a Coulomb friction coefficient of 0.15, as identified from inverse parameter calibration with experiments. The ring springback was not considered. The initial ring geometry was a square cross-section with an outer diameter of 63 mm, an inner diameter of 50 mm, and a height of 19 mm. The simulation required approximately 330 core-hours, utilizing 32 cores on an Intel® Xeon® Platinum 8470 processor within the high-performance computing system of the Technical University Dresden. Figure 5b, c presents a comparison between experimental and simulation results for one case. The comparison shows a maximum error of approximately 2% for the ring outer diameter and 15% for the rolling force, indicating that the FE simulation results are in good agreement with the experiment and can be considered reliable. As discussed in the previous chapter, the ring expansion in z-direction is within 1%, so that the semi-analytical approach based on a plane-strain assumption is justified and can be validated against the FE results.

Using the upper bound method proposed in the previous chapter, the strain can be calculated incrementally. To illustrate this, an element located in the outermost layer of the ring is selected. The strain obtained using the proposed method is compared with the corresponding result from the FE simulation at the same position, as shown in Fig. 6. The

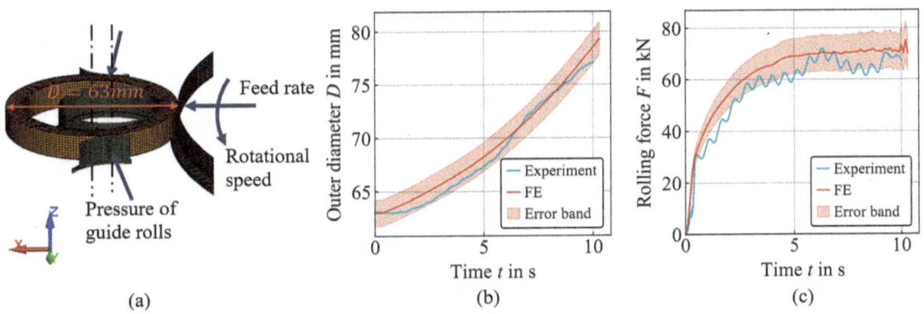

Fig. 5 **a** 3D FE model; **b** Comparison of ring geometry and rolling force between FE and experiment (Feed rate 0.2 mm/s, rotational speed 10.47 rad/s, guide roll force 1500 N)

comparison demonstrates that the local strain of the element can be accurately predicted with a maximum error of 10% to FE results.

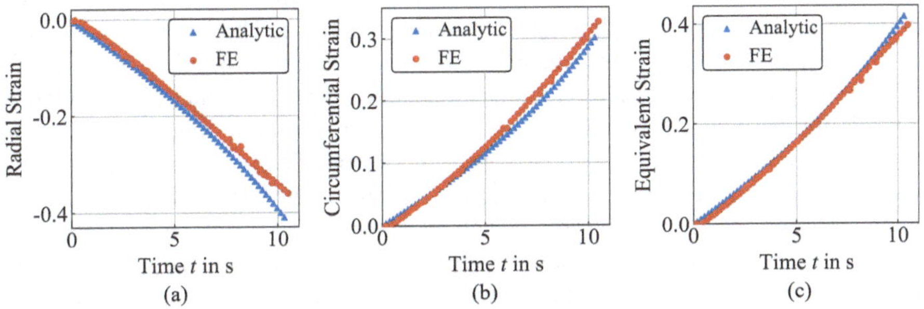

Fig. 6 Strain comparison between semi-analytical method and FE

Both the inner and outer diameters can be predicted, as illustrated in Fig. 7a, b. The inner diameter is compared only to FE results, showing a maximum deviation of 5%, since it was not measured experimentally. Although the maximum deviations of the diameters of the proposed method from experimental results are larger than that of the FE simulation (Fig. 5b), they still remain within a reasonable range. This discrepancy is primarily due to the proposed method neglecting the axial expansion of the ring and the elastic deformation phase. Furthermore, the rolling force is estimated using Eqs. (8–11), with a maximum error of 15% compared to experimental results, as presented in Fig. 7c. Nevertheless, the overall computation time of the proposed method is significantly lower than that of the FE simulation. The entire program, implemented in a Python environment, requires approximately 30 min to run. Future research will include further validation through sensitivity analysis.

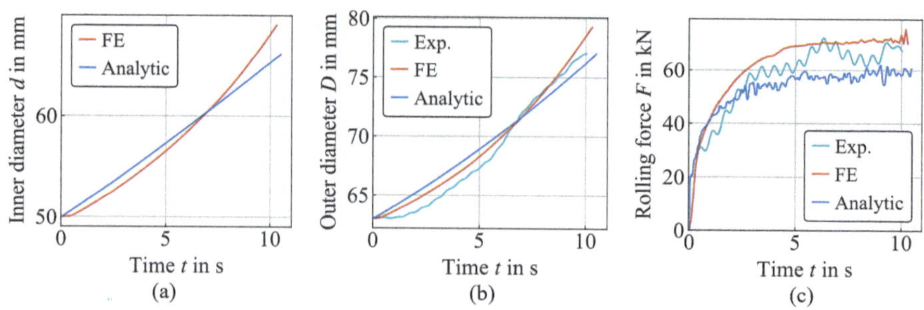

Fig. 7 Validation of the results of semi-analytical method

4 Conclusions and Perspectives

The results of the proposed semi-analytical model, which is based on upper-bound and slip-line techniques, demonstrate that this method is well suited to the preliminary evaluation of machine input configurations. The model incrementally calculates the state of elements to derive local strain, diameter and rolling force. The maximum deviations from the FE results are 10% for strain, 5% for diameter, and 15% for rolling force. It offers significantly reduced computational time compared to conventional FE simulations. While FE models require more than ten hours and reconfiguration for each change in machine input, the semi-analytical approach is more flexible and computationally efficient, completing in 30 min. Future work will focus on further validating the model's accuracy through sensitivity analyses.

The proposed method is particularly suitable for generating large datasets, which can be utilized for machine learning applications. While ML algorithms are effective at interpolating within a predefined Latin hypercube sampling space, the semi-analytical method enables rapid expansion of datasets beyond this space. Moreover, the availability of strain information provides an opportunity to estimate virtual form errors, as it reflects material flow behavior during forming. This information also has the potential to correlate with microstructural evolution.

Acknowledgements. The authors would also like to thank Dr. Luca Quagliato for the discussions during the development of the force model. This work was funded by the Deutsche Forschungsgemeinschaft (DFG, German Research Foundation)—499350001.

Competing Interests The author(s) has no competing interests to declare that are relevant to the content of this manuscript.

References

1. Hua, L., Deng, J., Qian, D.: Precision ring rolling technique and application in high-performance bearing manufacturing. MATEC Web Conf. **21**, 3002 (2015)
2. Wang, Q., Seitz, J., Brosius, A., Kuhlenkötter, B.: Algorithm for a Comparable Calculation of Form Errors based on a Hybrid Database in Cold Ring Rolling. WGP Kongress (2024)
3. Fahle, S., Glaser, T., Kuhlenkötter, B.: Investigation Of Suitable Methods For An Early Classification On Time Series In Radial-Axial Ring Rolling. Hannover: publish-Ing (2021)

4. Quagliato, L., Berti, G.A.: Mathematical definition of the 3D strain field of the ring in the radial-axial ring rolling process. Int. J. Mech. Sci. **115–116**, 746–759 (2016)
5. Seitz, J., Moser, T., Fahle, S., Prinz, C., Kuhlenkötter, B.: Transfer Learning Approaches In The Domain Of Radial-Axial Ring Rolling For Machine Learning Applications (2023)
6. Hawkyard, J.B., Johnson, W., Kirkland, J., Appleton, E.: Analyses for roll force and torque in ring rolling, with some supporting experiments. Int. J. Mech. Sci. **15**(11), 873–893 (1973)
7. Quagliato, L., Berti, G.A., Kim, D., Kim, N.: Slip line model for forces estimation in the radial-axial ring rolling process. Int. J. Mech. Sci. **138–139**, 17–33 (2018)
8. Parvizi, A., Abrinia, K., Salimi, M.: Slab analysis of ring rolling assuming constant shear friction. J. Materi Eng. Perform **20**(9), 1505–1511 (2011)
9. Parvizi, A., Abrinia, K.: A two dimensional upper bound analysis of the ring rolling process with experimental and FEM verifications. Int. J. Mech. Sci. **79**, 176–181 (2014)
10. Yang, D.Y., Ryoo, J.S.: An investigation into the relationship between torque and load in ring rolling. J. Eng. Ind **109**(3), 190–196 (1987)
11. Berg, C.A.: Construction of the equivalent plastic strain increment. Stud. Appl. Math. **51**(3), 311–316 (1972)

Influence and Challenges of Print Infeed on Structural and Optical Properties of Flexo Printed Waveguides

Janka Kirstein(✉), Andreas Evertz, and Ludger Overmeyer

Institute of Transport and Automation Technology, Leibniz University Hannover, Garbsen, Germany
janka.kirstein@ita.uni-hannover.de

Abstract. Flexo printing of optical waveguides enables cost and resource efficient production through the use of polymer-based varnishes and high throughput production. The integration of waveguides into a printed circuit board (PCB) requires temperature resistance of the materials, which necessitates the use of Polyimide (PI) substrates. Therefore, multiple layers of various varnishes need to be printed upon each other for production of cladded waveguides. Each of these aims for highest geometrical quality in order to achieve low loss waveguides. A cladding structure is required for a PI substrate, otherwise optical waveguides cannot be realized due to the material related refractive indices. This work focuses on the relationship between printing parameters and structural and optical properties of rotative printed waveguides. To investigate this, flexo printing on fixed substrates is analyzed according to previously unavailable infeed resolution in single digit micron range. Different infeed settings are considered, as well as deviations in the process that can affect this process parameter. Therefore, the geometrical properties of the printed structures are compared. Further, a correlation is deduced, with the goal of minimizing the total optical attenuation. For fabricated splitters the separation radius is determined from microscopic images. The influence of the waveguide's separation radii, resulting from material accumulation during printing, can be related to optical properties, enabling to connect attenuation with printing parameters for the first time. These results can be used to optimize the printing process in order to use it for the further integration of cladded splitters in electro-optical circuit boards, where they will be utilized to bidirectionally couple into polymer optical fibers (POFs).

Keywords: High-volume additive manufacturing · Flexo-printed optical waveguides · Optical integration and packaging

1 Introduction

The ongoing multimedia trend places enormous demands on the performance of computer networks and electronic devices. However, ever-increasing demands for high-frequency data transmission and high communication bandwidths are causing mounting

L. Overmeyer and B.-A. Behrens (eds.), *Production at the Leading Edge of Technology*,
Lecture Notes in Production Engineering, https://doi.org/10.1007/978-3-032-19524-1_28

problems [1–4]. The development of optical components, particularly optoelectronic printed circuit boards (OPCB), offers a promising solution, since optical connections are insensitive to electromagnetic interference and enable higher data rates. They therefore have potential to significantly improve the electromagnetic compatibility (EMC) of future electronic devices and meet future high-frequency requirements [3–5]. A decisive step is the development of cost-effective OPCBs that can be integrated into the conventional PCB manufacturing process [1–3, 6]. Polymer multimode waveguides promise easy integration into standardized PCBs, offering the alignment tolerances that enable low complexity and cost-effective production. Simple integration can be achieved by an additional optical layer with waveguide structures as a complementary step to conventional PCB production process [5, 6].

Flexo printing is a suitable manufacturing method because it enables high throughput production and low material costs for polymer-based used vanishes [7]. Wolfer has already demonstrated the production of straight and bend optical waveguides using this process by printing layered core structures on a polymethyl methacrylate (PMMA) substrate [8]. Pflieger investigated the use of these uncladded waveguides to create network structures [9] and Pleuß has examined the first printing parameters for use in single-mode range [10]. Evertz has evaluated firstly the integration of optical waveguides into PCBs [11]. Therefore, the structure must be printed on a thermally stable substrate. PI, which is already established as a substrate for flex-PCB, is suitable for this purpose. However, as its refractive index is significantly higher than that of previously used PMMA and respective varnishes used for core printing, a cladding structure becomes necessary [11]. For cladded waveguides, the lower difference in refractive index between core and cladding leads to higher attenuation due to light coupling out. This is due to the critical angle for total internal reflection, which is smaller for smaller refractive index differences [12].

To enable efficient use of waveguides in PCBs, the production and printing parameters have to be optimized to reduce occurring attenuation. For the first time, direct correlations between printing result and printing parameters can be investigated in detail because the used high-precision system enables automated printing delivery with micrometer precision. This study investigates how specifically adjusted printing parameters and non-controllable deviations affect geometric and optical properties. Correlating process parameters with print results enables optimized printing of low-attenuation waveguides and can be used to explain deviations in attenuation results.

2 Method

The goal of this study is to investigate the impact of process parameters and deviations in flexo printing on the geometric and optical properties of waveguides in order to optimize the printing process. Flexo printing is a rotative process in which liquid material is transferred from a printing roller to a flat substrate (roll-to-sheet) via a stamp structure. The printing plate (Kodak Flexcel NX) is applied to a sleeve (ContiTech Axcyl) with adhesive tape (Tesa). The sleeve is located on the outside of the printing roller. Transparent varnishes, which are commercially available printing polymers, are used to print the waveguides. A material combination of Jänecke & Schneemann (JS) 390119 for the cladding (refractive index $n_{clad} = 1.512$) and Weilburger 82F 1530 (formerly 360245) for the core (refractive index $n_{core} = 1.522$) is selected from earlier studies [11]. The stamps are covered with polymer by an anilox roller. The two rollers and the substrate sledge move synchronously and thus enable material transfer at a speed of 30 m/min (see Fig. 1a).

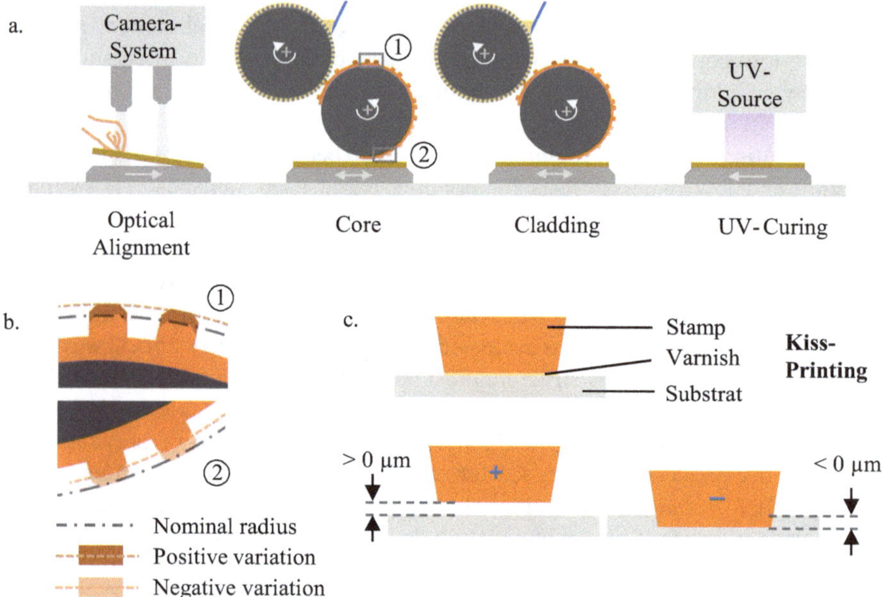

Fig. 1 **a** Schematic view of the flexo printing process in an in-line setup with optical alignment, printing of cladding and core on two separate printing units and curing under a UV-source. **b** Magnified view of the marked areas from a. to illustrate distance deviations between printing form and substrate. **c** Visualization of the stamp position for kiss printing, as well as the signs when changing print infeed

Studies have shown that the distance between the printing roller and the substrate as well as the stamp wide significantly impacts the print result [10]. Both the amount of material transferred and the geometry of the printed structure are affected [13]. Deviations can occur due to concentricity deviations distributed over the print layout (Fig. 1b) or due to inaccurate infeed (Fig. 1c). The modular fixed-bed printing machine Challenger 650 from nsm Norbert Schläfli AG used here offers innovative functions for high-precision production. An integrated camera system enables optical substrate alignment and the printing units can be placed with micrometer precision. Two flexo printing units, available in line with different layouts and varnishes for cladding and core, enable the production of fully cladded waveguides without realigning the substrate or modifying the system. The manufacturing process is carried out layer by layer. Each layer is cured by a broadband mercury vapor UV source.

The waveguide's typical cross-section geometry is as shown in Fig. 2a, whereas the lower cladding turns out flat, the slimmer core builds up height and the top cladding covers both structures. Different process parameters affect the printing result differently. The specific requirements and conditions depend on the application. For example, different line widths require specific settings for kiss printing [10], while integration of waveguides into PCBs requires clad structures to ensure functionality on the temperature-resistant PI substrate with a higher refractive index [10, 13].

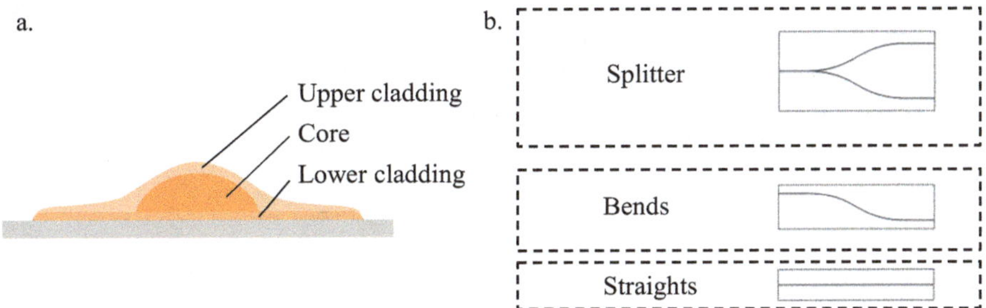

Fig. 2 **a** Cross-section of a printed, fully clad waveguide **b** Print layout structure for cladding and core. The printing form features each structure 15 times, arranged into three rows and five columns

To achieve a core structure with a high aspect ratio, kiss printing is used to print the waveguide core. To determine the kiss-printing area, the print setting must be adjusted iteratively until the printed image is completely visible across the entire substrate without interruptions. When printing the lower cladding, the focus lies on achieving the smoothest, most defect-free surface. This can be achieved by applying pressure (negative sign in Fig. 1c) instead of kiss printing.

As shown in Fig. 2b, the print layout is divided into different areas. The stamps for core printing are 200 µm wide and 2 mm wide structures are provided for cladding. For evaluation purposes a three-layered cladding on PI is analyzed according to line width and defect rate using light microscope measurements. Also, the surface structure is analyzed using a confocal microscope (Objective 20x/0.45). For core layout, deviations in the infeed and resulting changes in waveguide attenuation are analyzed qualitatively using samples with different infeeds on PMMA and using the incoming beam profile on a beam profiler. In addition, the dependence of the separation radius on the print infeed for cladding and core on PI and concentricity deviations of the printing roller with sleeve are analyzed.

3 Results

Three different infeeds are analyzed for printing the lower cladding in order to examine the effects of the infeed on the print result and to determine the best parameters for low attenuation. First, the kiss-printing area is determined and then three layers of the lower cladding are printed on the substrate with infeed values of 0 µm, −100 µm and

−200 μm. The cladding width behaves linearly within the analyzed area. Reducing the infeed by 100 μm widens the printed cladding area by approx. 35 μm. Reducing the infeed significantly decreases the defect rate in the printed cladding area. While varnish transfer is incomplete in kiss-printing area (21% partially or unwetted areas), defects can be significantly reduced by applying pressure between the stamp and substrate. A reduction of 100 μm already leads to a significant improvement with only a few small unwetted pores (0.5%), additional 100 μm leads to a slight further improvement (0.05% unwetted) (see Fig. 3a). Confocal measurements show a height of 10 μm for the lower cladding. Additionally, they also reveal a smoother cladding surface due to reduced infeed, which is consistent with the detection of miswetting at a higher infeed (see Fig. 3b). The cross-sectional profile of the cladding printed at −200 μm infeed already shows crushing edges due to the material suppression on the stamp flanks. Therefore, a compromise must be found between a defect-free, smooth surface and reducing crushing edges.

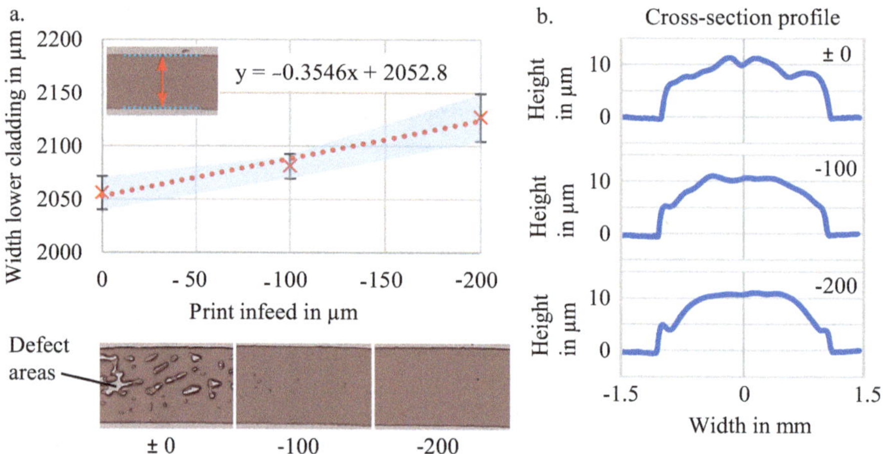

Fig. 3 Results of the investigation of lower clad printing. **a** Dependence of clad width and defects occurring on the print infeed. **b** Cross-section profile for different infeeds

Both print images (core and cladding) show deviations when approaching the kiss-printing position, which can lead to deviations in geometric and optical properties of waveguides if the infeed deviations are too high. However, only the core layout is investigated further, as a significant reduction in the print infeed is necessary for defect-free cladding transfer. This reduction demonstrates the dependency of varnish transfer on stamp width, as shown by Pleuß, as well as further influences and behaviors for such large sizes. Pleuß demonstrates the importance of reduced infeed for stamps smaller than 200 μm [10]. Here, a reduced infeed is also important for wider structures. For core printing, the print infeed is increased stepwise from kiss-printing position to determine where varnish transfer stops and where core printing continues. This procedure reveals that differences in the infeed of at least 40 μm occur during printing over the entire substrate. Figure 4a. illustrates the actual expected print infeeds at various positions of the layout for kiss-printing setting. Areas marked with a reduced print infeed are still fully visible without interruption if the print infeed is increased by the specified value compared to kiss-printing.

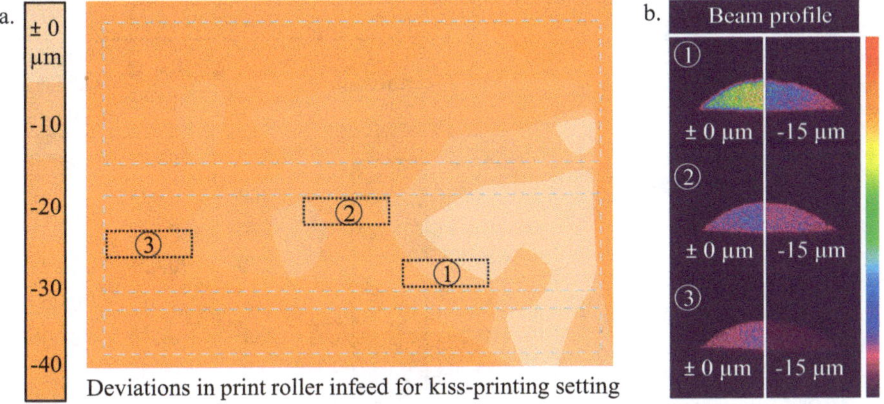

Fig. 4 Influence of print infeed on attenuation **a** Experimentally determined deviations of the infeed for kiss-printing setting distributed across the substrate. Gray dotted lines show areas of the various structures from Fig. 2b. and black dotted lines the positions of the waveguides on the right. **b** Influence of 15 µm reduced infeed on the attenuation of three marked waveguides

A qualitative investigation of waveguide attenuation (core on PMMA) reveals clear differences depending on the position on the substrate and upon reducing the print infeed by 15 µm (see Fig. 4b). The variation across the substrate supports the previous findings of varying infeed in the print layout. The lowest attenuation occurs in waveguides in region of actual kiss-printing area, while the attenuation of other waveguides is already comparatively reduced by the reduced infeed. Comparing positions 2 and 3 reveals that attenuation in position 2 with 15 µm reduced infeed, is similar to the attenuation in position 3 with kiss-printing setting. This also correlates with the deviations determined for the two positions in the print infeed.

Due to qualitative illustration of infeed deviations of at least 40 µm (Fig. 4a) and the influence of infeed on attenuation (Fig. 4b) the deviation is examined quantitatively below. Irregularities in the printing image can be caused by various parameters. These could be clearance in bearings, lack of circularity of printing roller or sleeve, manufacturing tolerances of the printing form or variations in adhesive tape. To determine the tolerance range, distance measurements are taken from the surface of the sleeve next to the printing form and the range of fluctuation is determined. For that measurement a confocal point sensor (confocalDT IFS2403-0.4) from Micro-Epsilon is used. The cylinder is rotated three times around the axis for two measurements to equalize any potential fluctuations or deviations in measured values. The range of fluctuations is calculated from maximum and minimum values for each rotation. The mean value over two sets of three rotations is shown in Fig. 5a for each measurement position. The measurement results show deviations of 37.2 ± 4.1 µm on one side of the printing form and 24.4 ± 1.6 µm on the other side (see Fig. 5b). The measurement data show systematic deviations with a sinusoidal pattern, with one period corresponding to one roller rotation. After the process is set up, the deviation remains consistently across all rotations and prints. Based on the attenuation results for a 15 μm reduction of print infeed, this variance is clearly too high and requires further improvement.

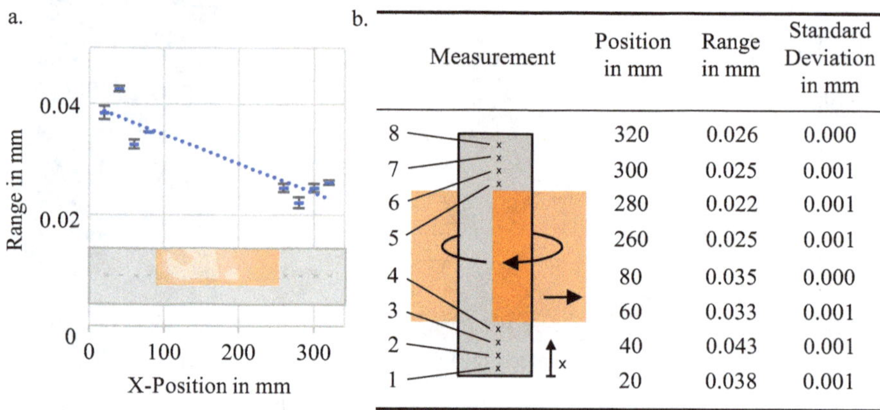

Measurement	Position in mm	Range in mm	Standard Deviation in mm
8	320	0.026	0.000
7	300	0.025	0.001
6	280	0.022	0.001
5	260	0.025	0.001
4	80	0.035	0.000
3	60	0.033	0.001
2	40	0.043	0.001
1	20	0.038	0.001

Fig. 5 Results of the rotation deviation of the printing roller. Mean values of the range from two sets of three rotations, measured on the sleeve next to the printing form. **a** Measurement results in relation to measurement position. **b** Measurement values and visualization of the measurement points on printing cylinder

After demonstrating linear influence of infeed on line width and dependence between infeed and attenuation, the dependence of infeed on the separation radii of splitter structures is also analyzed for core printing. The separation radius is relevant because it depends geometrically on print infeed, but also directly affects splitter attenuation. First, the kiss-printing setting of the core on a raw substrate is determined. For kiss printing on the lower cladding, the print infeed must increases by 10 μm, as the confocal microscope images of the lower cladding show a thickness of 10 μm. To demonstrate a dependency, three layers of core are printed on the lower cladding, ranging from 0 μm (kiss printing on substrate) to +20 μm. Although kiss-printing on the cladding is assumed at +10 μm, these tests can be carried out up to +20 μm because the deviations in the overall layout and the fact that splitters are in an area that is already fully visible with a higher print infeed than the kiss-printing setting.

The results show that the radius decreases with lower pressure between stamp and substrate (see Fig. 6). A variation in the print infeed of 10 μm leads to a radius change of aprox. 7 μm in the examined area. A smaller radius is expected to result in reduced attenuation because there are fewer losses due to decoupling of the light at the separation point.

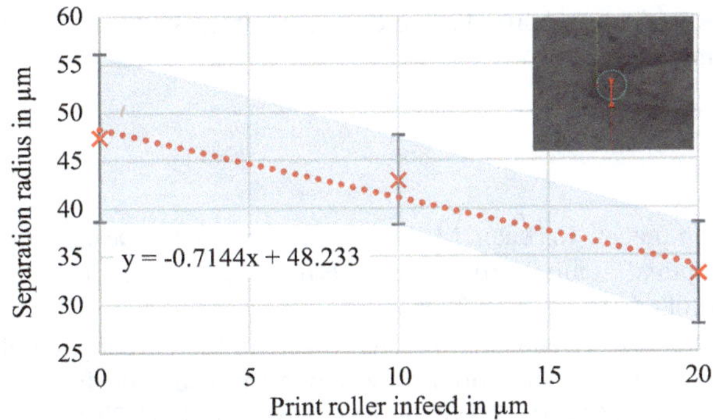

Fig. 6 Results of the investigation of dependency between print infeed and separation radius. Linear relationship between a smaller radius and a larger infeed for three different print infeeds, with 15 determined separation radii each

4 Conclusion

The results show that using reduced infeed offers advantages for a lower cladding. Using a pressure assisted cladding transfer reduces defects and achieves a homogeneous surface. Additionally, a linear correlation between print infeed and line width could be determined in the analyzed area, which can be attributed to a stronger material displacement on the stamp flanks.

Analyses of the core printing revealed a correlation between the infeed process parameter and the resulting separation radius in splitter elements, as well as the general attenuation. Additionally, the limits of the high-precision system were demonstrated. These fluctuations, with a range up to 43 µm, prevent consistent printing over the entire substrate. As changes in the attenuation and in the separation radius are already recognizable for a infeed reduction of 15 µm, deviations below this value should be targeted.

Further analysis of the irregularities in the print image is necessary and requires additional measurements. The circularity of the printing roller, the variance of the printing form and possible deviations of the sledge must also be analyzed using a measurement setup. With smaller print infeed steps and quantitative attenuation measurements, the acceptable tolerance limit should be investigated in more detail to enable better reproducibility across the entire substrate.

In conclusion, the flexo printing system enables high-precision alignment and adjustment of the parameters with limitations due to infeed deviations. In principle, a correlation between the printing parameters and the structural and optical printing results could be demonstrated. Looking ahead, reduced infeed deviations would enable the production of consistent waveguides properties across the entire substrate. The system could then reach its full potential, enabling the production of waveguides with improved, more uniform properties. With consistent properties across the substrate, these waveguides could be integrated into PCBs, all exhibiting the same high quality.

Competing Interests. The author(s) has no competing interests to declare that are relevant to the content of this manuscript.

References

1. Betschon, F., Lamprecht, T., Halter, M., Beyer, S., Peterson, H.: Design principles and realization of electro-optical circuit boards. Proc. SPIE Int. Soc. Opt. Eng./Proc. SPIE **8630**, 86300U (2013). https://doi.org/10.1117/12.2006379
2. Berger, C., Offrein, B.J., Schmatz, M.: Challenges for the introduction of board-level optical interconnect technology into product development roadmaps. Proc. SPIE In.t Soc. Opt. Eng./Proc. SPIE **6124**, 61240J (2006). https://doi.org/10.1117/12.658831
3. Griese, E.: A high-performance hybrid electrical-optical interconnection technology for high-speed electronic systems. IEEE Trans. Adv. Packag. **24**(3), 375–383 (2001). https://doi.org/10.1109/6040.938306
4. Griese, E.: Reducing EMC problems through an electrical/optical interconnection technology. IEEE Trans. Electromagn. Compat. **41**(4), 502–509 (1999). https://doi.org/10.1109/15.809854
5. Shibata, N.T., Takahashi, N.A.: Flexible opto-electronic circuit board for in-device interconnection. 58th Electronic Components and Technology Conference. IEEE, 261–267 (2008). https://doi.org/10.1109/ectc.2008.4549980
6. Bamiedakis, N., Hashim, A., Beals, J., Penty, R.V., White, I.: Low-cost PCB-integrated 10-Gb/s optical transceiver built with a novel integration method. IEEE Trans. Compon. Packaging Manuf. Technol. **3**(4), 592–600 (2013). https://doi.org/10.1109/tcpmt.2013.2242961
7. Assaifan, A.K.: Flexographic printing contributions in transistors fabrication. Adv. Eng. Mater. **23**(5) (2021). https://doi.org/10.1002/adem.202001410
8. Wolfer, T.: Additive Fertigung integrierter multimodaler Polymer-Lichtwellenleiter mittels Flexodruck. TEWISS Verlag (2020)
9. Pflieger, K., Reitz, B., Hoffmann, G., Overmeyer, L.: Layout optimization for flexographically printed optical networks. Appl. Opt. **60**(31), 9828 (2021). https://doi.org/10.1364/ao.420358
10. Pleuß, J., Evertz, A., Overmeyer, L.: Structural limits of flexographic printing towards singlemode dimension waveguides. Proc. SPIE **16**,(2025). https://doi.org/10.1117/12.3041887
11. Evertz, A., Pleuß, J., Reitz, B., Overmeyer, L.: Flexo-printed polymer waveguides for integration in electro-optical circuit boards. Flexible Printed Electron. **9**(3), 035001 (2024). https://doi.org/10.1088/2058-8585/ad5c7b
12. Pflieger, K., Overmeyer, L., Olsen, E.: Flexografically printed optical waveguides for complex low-cost optical networks. Opt. Interconnects XXII, 43 (2022). https://doi.org/10.1117/12.2606364
13. Evertz, A., Fuetterer, L., Overmeyer, L.: PCB-integrated flexo-printed optical networks structures. Proc. SPIE **49**,(2025). https://doi.org/10.1117/12.3041905

Chip Forms in Turning Processes with an Additional Rotational Feed Motion

Berend Denkena, Klaas Maximilian Heide, and Felix Zender(✉)

Institute for Production Engineering and Machine Tools, Hannover, Garbsen, Germany
zender@ifw.uni-hannover.de

Abstract. The performance of turning operations is limited not only by the tool life but also by the form of the resulting chip. Insufficient chip breakage, and thereby long chips, often impedes chip removal. This hampers the automation and often requires time-consuming manual interventions by the operator. Furthermore, these chips can potentially cause damage to the tool, workpiece and machine. Besides the workpiece material and the chip breaker, the resulting chip form is mostly influenced by the form of the undeformed chip, especially the thickness of the undeformed chip. While a greater thickness of the undeformed chip usually results in a more favorable chip breaking behavior, it also leads to a higher load on the tool. A small thickness of the undeformed chip can primarily be achieved by a small tool cutting edge angle and a low feed rate. While the tool cutting edge angle cannot be set arbitrarily during conventional turning processes, novel turning processes allow the angle to be adapted during the process by an additional rotational feed motion. In this paper the effects of the variable cutting edge angle on the form of the resulting chip are investigated. To predict the resulting chip form, various undeformed chip parameters are discussed and it is shown, that these are usable to determine process parameter combinations where favorable chip forms are likely to occur.

Keywords: Turning · Chip · Simultaneous three axis turning

1 Introduction

Insufficient chip breakage and thus large chips are often a major obstacle towards productive and reliable turning processes. First of all, long chips are more difficult to remove from the working zone than short chips. Thus, these chips can cause damage to the cutting tools, workpiece and machine tool and hinder automated machining without human supervision and intervention [10]. To achieve short chips, various influences have to be taken into account. Already the chip formation can lead to favorable chip forms, since chip segmentation often results in a fragmentation during the chip formation or makes the fragmentation easily achievable. This chip formation occurs only in certain materials, e.g., in titanium alloys [9], and during high speed cutting [3]. For continuous chip formation the chip breakage has to be enforced by the secondary chip guidance [6]. In this case, the chip breakage happens when the chip runs upon an obstacle, like

L. Overmeyer and B.-A. Behrens (eds.), *Production at the Leading Edge of Technology*, Lecture Notes in Production Engineering, https://doi.org/10.1007/978-3-032-19524-1_29

the tool or the workpiece, and the strain at the surface of the chip exceeds the fracture strain of the material [12]. Therefore, the form of the resulting chip is especially an issue when machining ductile high-tensile materials, such as the steels used in the aerospace industry. These strains are influenced by the shape of the undeformed chip, particularly the thickness of the undeformed chip, and can be further increased by chip breaker geometries on the rake face of the tool [13]. To be effective, these chip breaker geometries have to be compatible with the engagement conditions during the process, since otherwise either the chip will not flow through the chip breaker geometry, or only insufficient strains will be induced [7].

The shape of the undeformed chip is determined by the feed f, the tool cutting edge angle κ and the geometry of the tool [5]. For conventional turning processes κ is fixed or directly determined by the tool geometry and can thus not be changed at will. This does not apply to novel turning processes with an additional rotational feed motion (simultaneous three axis turning). Here the additional axis enables a flexible adjustment of the tool cutting edge angle according to the geometry of the workpiece and the requirements of the process. First of all, this flexibility allows to machine a greater variety of geometries with the same tool and thus allows a reduction of the necessary tool changes. Additional this can be used to adapt the cutting conditions during the whole process. Degen [2] presents a process design method using the B-Axis of a turn-mill center for both cornered and round cutting inserts (Fig. 1b). In [2] the chip flow, surface and tool wear are taken into account, while the form of the resulting chip is not considered. Whereas this approach uses the B-Axis of a turn-mill center and conventional turning tools, there are also approaches that use the milling spindle of a turn-mill center as an additional rotatory axis (usually referred to as C1-axis) and purpose-built tools with the rake face orientated in the YZ-plane (Fig. 1c). Due to this, the Y-axis of the machine tool is used like the X-axis in conventional turning. A particular advantage compared to the approach of [2] is the lower workspace requirement compared to pivoting the B-axis and the possibility to design every corner of the tool as a different tool.

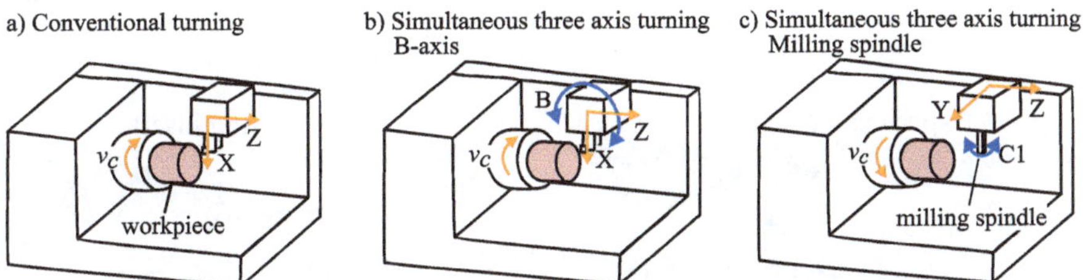

Fig. 1 Conventional turning and simultaneous three axis turning

This additional degree of freedom can be used to optimize the tool path with regard to the shape of the resulting chip. For this, the shape of the resulting chip has to be predicted. For conventional turning processes various models using different approaches have been developed. One group of models is based on finite element simulations. In these models the plasto-mechanic behavior of the workpiece and the thermal effects of the process are estimated based on detailed material models, particularly of fracture behavior [1].

Here, the high computational cost of solving finite element systems has to be considered, especially since during simultaneous three axis turning a multitude of different cutting conditions has to be taken into account. A different approach is to describe the forming of the chip by analytical equations. An early approach for the 2D-case is shown in [11], where the forming chip is described analytically based on the material removal and the resulting strains. In [13] the 3D-case is considered by including the chip flow angle and the chip flow in the chip-breaker. In general, these analytical approaches are limited regarding transferability to the varying cutting conditions occurring during simultaneous three axis turning. Finally there are machine learning approaches, as in [8], where a neuronal network is used to predict the resulting chip form classification for 42CrMo4 during longitudinal turning using tools with different chip breaker geometries. While the feed and depth of cut are varied in the training data, they are not further considered in the model and neither the geometry of the tool or the cutting edge angle are taken into account.

For tool path planning and optimization of simultaneous three axis turning processes rather knowledge about the expected chip form classification is required than about the exact mechanism leading to the resulting chip form. Furthermore, due to the varying engagement conditions during simultaneous three axis turning transferable indicators, like the geometry of the undeformed chip are required. Therefore, this paper discusses various undeformed chip parameters based on turning experiments using varied cutting edge angles.

2 Experimental Setup

The experiments were carried out on a DMG Mori NTX 1000 turn-mill center with the setup shown in Fig. 2a. A tool of the type PSC-50-100-FT15 808,055 with TiCN-Al_2O_3-coated tungsten carbide cutting inserts of the type FT15.808055R08 manufactured by CERATIZIT Austria GmbH was used. The tool used (Fig. 2b) has a corner radius of 0.8 mm and a chip breaker designed as a series of chip breaker mounds along the cutting edge.

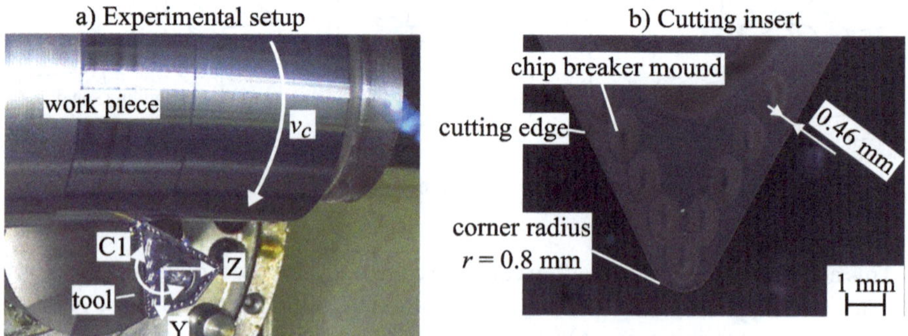

Fig. 2 **a** Experimental setup **b** Cutting insert used

The material machined was 42CrMo4 + QT with a tensile strength of 975 N/mm2. According to the recommendation of the tool manufacturer, a cutting speed v_c of 180 m/min was chosen for dry machining. The chip breaking behavior was investigated full factorial for tool cutting edge angles κ of 110°, 95°, 80° and 60°. The depth of cut a_p was varied between 0.5 mm and 2.5 mm with an increment of 0.5 mm while the feed f was varied from 0.1 mm to 0.4 mm with an increment of 0.1 mm. Whenever a transition in chip-breaking behavior was detected between trial points, additional intermediate parameter steps were examined to capture the change more precisely. The feed travel was 40 mm for each process. To avoid the influence of wear on the chip forming behavior, the cutting inserts were changed before considerable wear occurred. For these process parameters, the workpiece material forms continuous chips.

3 Results and Discussion

3.1 Experimental Results

During the investigations, six distinct chip forms were differentiated. The classification is based on an adaption of the classification proposed in [14] (Fig. 3). Additionally, a distinction was made between long and short chips of the same chip form due to the differences in the chip volume and the risk of entangling with the tool or the workpiece. Here snarled chips and long flat helical chips are classified as unfavorable, short flat helical chips and long cylindrical chips are classified as acceptable, while the chip segments and short cylindrical chips are categorized as favorable forms.

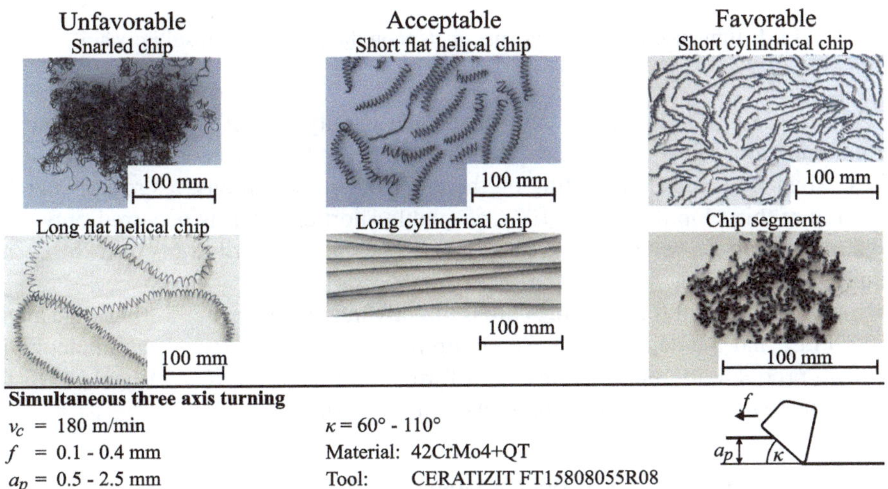

Fig. 3 Chip form classification

In Fig. 4 the resulting chip forms depending on the depth of cut and the feed for the various cutting edge angles are shown. As expected from the literature, more favorable chip forms are found for higher feed rates and higher depths of cut. However, especially for κ of 95° with feeds above 0.3 mm less favorable chip forms occur, while for κ of 60° and 80° favorable chip forms are achieved for these feeds. This might be explained by the distance between the cutting edge and the chip breaker mounds in consideration of the cutting edge angle, since a minimum distance is required to allow a sufficient flow of the chip into the chip breaker geometry of the tool.

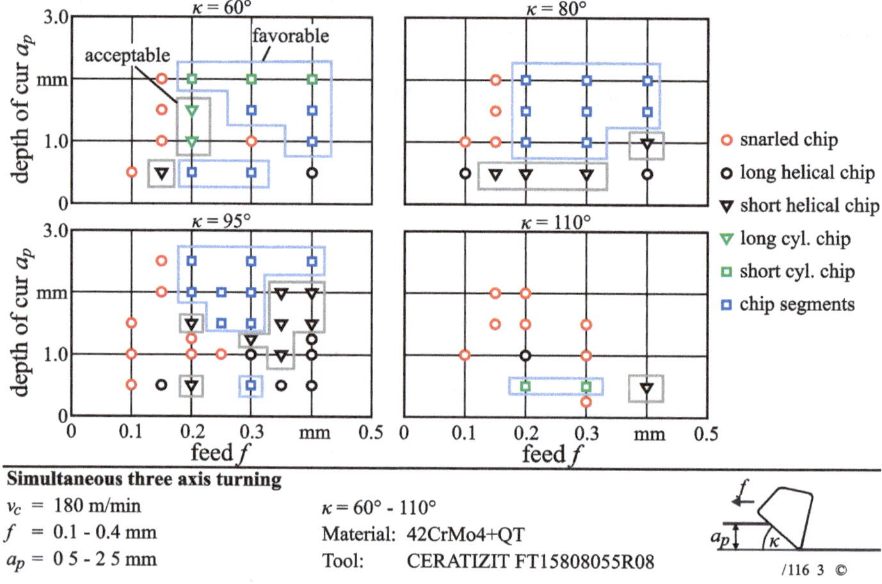

Fig. 4 Influence of the process parameters on the form of the resulting chip

For depths of cut below 1 mm, chip breakage can only be achieved for certain parameter combinations. This might be explained by the chip breaker geometry in the corner radius, which can cause, depending on the engagement conditions, a strong deflection of the rather thin chip. For κ of 110°, acceptable short chip forms occurred only at a_p of 0.5 mm. It should be noted that, in this case, the chip primarily runs over the corner radius of the tool.

In sum, the form of the resulting chip is directly influenced by the cutting edge angle, the feed rate and the depth of cut. So, an optimization of the process parameters with regard to productivity and resulting chip form which take the restrictions due to the unmachined part and the finished part taken into account is desirable.

3.2 Influence of the Shape of the Undeformed Chip

As seen above, an adaption of the cutting edge angle, depending on the intended material removal, is preferable. Since during this adaption a multitude of different cutting edge angles might occur, a description of the engagement conditions based solely on the process parameters is not suitable. An approach describing these engagement conditions relies on the shape of the undeformed chip. For this purpose, various parameters are known. It should be noted, that a wide range of different shapes of the undeformed chip, like shapes with strong local variations of the thickness, for example due to a relatively small depth of cut compared to the corner radius, might occur during the process. For these shapes the conventional undeformed chip thickness is usually no appropriate characterization. Furthermore, the conventional thickness of the uncut chip can be determined analytically, but due to the constant variation of the shape of the undeformed chip, a calculation based on a numerical material removal simulation is desirable. Therefore, the effective undeformed chip thickness h^*, as proposed in [4], is considered as an alternative to the conventional undeformed chip thickness (Fig. 5). h^* is defined as the ratio between the cross-section of the undeformed chip A and the length of the engaged cutting edge b^*.

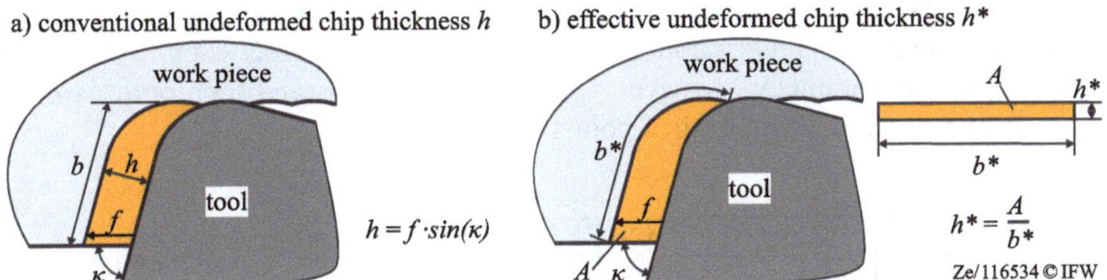

Fig. 5 **a** Conventional undeformed chip thickness, **b** Effective undeformed chip thickness according to [4]

For small depths of cut, the corner radius of the tool has a major influence on the resulting chip form. Furthermore, the effect of chip breaking geometries might change due to its geometry in the corner of the tool or a deviating chip flow angle [7]. Therefore, the length of the straight part of the major cutting edge b_{str} and the relation between the depth of cut and the height of the tool radius $a_p/a_{p,r}$ are taken into account (Fig. 6).

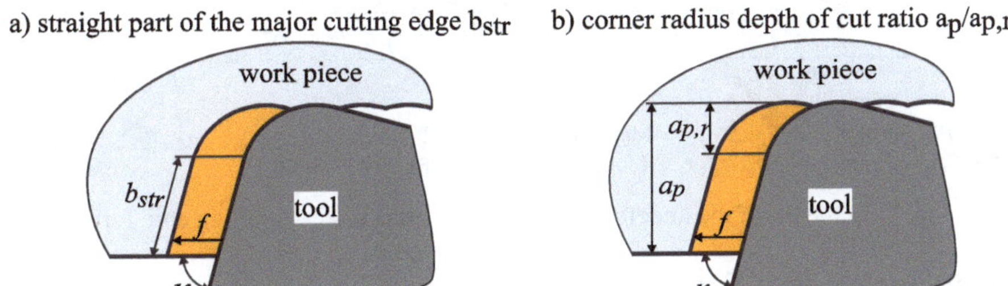

Fig. 6 **a** Straight part of the major cutting edge, **b** Corner radius depth of cut ratio

Figure 7 shows the occurring resulting chip forms depending on these parameters of the undeformed chip. On the one hand, the expected tendency towards more favorable chip forms with a higher thickness of the undeformed chip (in particular for $h > 0.17$ mm or $h^* > 0.14$ mm) and with lower influence of the tool corner radius can be seen. On the other hand, the process parameters with low undeformed chip thickness and large influence of the tool radius, which lead to favorable chip forms, cannot be identified by these undeformed chip parameters. Similarly, the less favorable chip forms occurring at higher chip thicknesses cannot be detected solely based on these indicators. A reason for this might be the geometry of the chip breaker and its influence on the chip flow, which is not represented by these parameters. This might in particular explain the occurrence of the less favorable chip forms at undeformed chip thicknesses close to or greater than the distance between the cutting edge. Thus, these parameters can be used as indicators to determine areas of parameter combinations (e.g., here $h > 0.17$ mm or $h^* > 0.14$ mm combined with a $b_{str} > 0.5$ mm or an $a_p/a_{p,r} > 1.25$), where favorable chip forms are likely to occur. But neither the suitable parameter combinations outside of these areas nor the outliers inside of these areas can be detected. This may be explained by the omission of the chip breaker geometry from the parameters used. Consequently, the prediction of the chip breaking behavior for a tool with a different chip breaker geometry is limited. Still these parameters could be used to optimize the process by first choosing promising process parameters for every point of the process and then optimizing the process further by identifying critical points.

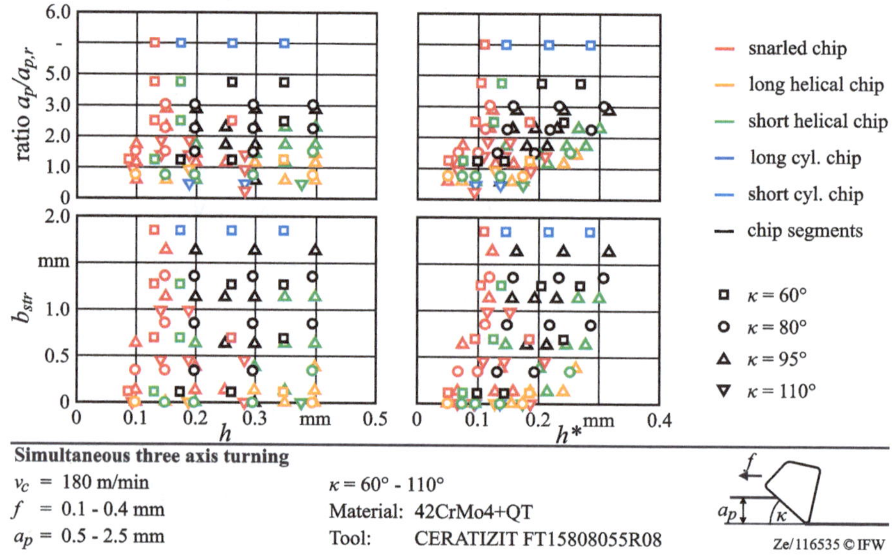

Fig. 7 Influence of the undeformed chip parameters on the resulting chip form

4 Conclusion and Outlook

In this paper the effects of the process parameters and in particular the cutting edge angle on the chip breaking behavior during the turning of 42CrMo4 + QT were investigated. To allow a transfer to varying cutting edge angles, undeformed chip parameters, particularly two variants of the undeformed chip thickness and two parameters describing the influence of the corner radius, were used.

In general, the expected effect of a higher thickness of the undeformed chip and the reduced influence of the corner radius resulting in a more favorable chip form were found, which can be used to define an area where favorable chip forms will likely occur. Thus, sections of the toolpath where probably critical chip forms will occur can be detected. However, since the chip breaker geometry was not included in the parameters used, its specific effects cannot be predicted. This should be addressed in further work by taking the chip flow caused by the chip breaker geometry into account. Further materials and tools with different corner radius and chip breaker geometries should be investigated for this purpose.

Acknowledgements. The authors would like to thank the Federal Ministry for Economic Affairs and Energy for the funding of the project Argonaut (20Y2101E), which enables this investigation. Furthermore, the authors would like to thank CERATIZIT Austria GmbH for providing the cutting tools.

Competing Interests The author(s) has no competing interests to declare that are relevant to the content of this manuscript.

References

1. Buchkremer, S., Klocke, F., Veselovac, D.: 3D FEM simulation of chip breakage in metal cutting. Int. J. Adv. Manuf. Technol. **82**(1–4), 645–661 (2016)
2. Degen, F.: Development of simultaneous three axis turning. Dissertation, Rheinisch-Westfälische Technische Hochschule Aachen (2015)
3. Denkena, B., Ben Amor, R., León-García, L. de, Dege, J.: Material specific definition of the high speed cutting range. Int. J. Mach. Machinability Mater., 176–185 (2007)
4. Denkena, B., Köhler, J.: Consideration of the form of the undeformed section of cut in the calculation of machining forces. Mach. Sci. Technol. **14**(4), 455–470 (2010)
5. Denkena, B., Tönshoff, H.K.: Spanen. Grundlagen. VDI-Buch. Springer, Berlin, Heidelberg, Berlin, Heidelberg (2011)
6. Denkena, B., Bergmann, B., Murrenhoff, M.: Influence of chip breaker geometries on chip breaking behaviour during profile grooving. CIRP J. Manuf. Sci. Technol. **53**, 95–102 (2024)
7. Essig, C.A.: Vorhersage von Spanbruch bei der Zerspanung mit geometrisch bestimmter Schneide mit Hilfe schädigungsmechanischer Ansätze. Ergebnisse aus der Produktionstechnik [2010,23]. Apprimus-Verl., Aachen (2010)
8. Kim, H.-G., Sim, J.-H., Kweon, H.-J.: Performance evaluation of chip breaker utilizing neural network. J. Mater. Process. Technol. **209**(2), 647–656 (2009)
9. Komanduri, R., von Turkovich, B.F.: New observations on the mechanism of chip formation when machining titanium alloys. Wear **69**(2), 179–188 (1981)
10. Liao, Z., Schoop, J.M., Saelzer, J., Bergmann, B., Priarone, P.C., Splettstößer, A., Bedekar, V.M., Zanger, F., Kaynak, Y.: Review of current best-practices in machinability evaluation and understanding for improving machining performance. CIRP J. Manuf. Sci. Technol. **50**, 151–184 (2024)
11. Merchant, M.E.: Mechanics of the metal cutting process. I. Orthogonal cutting and a type 2 chip. J. Appl. Phys. **16**(5), 267–275 (1945)
12. Nakayama, K.: A study on chip-breaker. Bullet. JSME **5**(17), 142–150 (1962)
13. Nedelß, C., Hintze, W., van Luttervelt, C.A.: Characteristic parameters of chip control in turning operations with indexable inserts and three-dimensionally shaped chip formers. CIRP Ann. **38**(1), 75–79 (1989)

14. Verein deutscher Eisenhüttenleute: Spanbeurteilung. Verein deutscher Eisenhüttenleute, Düsseldorf, Stahl-Eisen-Prüfblatt 1178–69 (1969)

High-Pressure Liquid Carbon Dioxide ($HPCO_2$) Jetting as an Alternative Conditioning Method for Diamond Grinding Wheels

Yupeng Li[1(✉)], Arunan Muthulingam[1], and Eckart Uhlmann[1,2]

[1] Institute for Machine Tools and Factory Management (IWF), Technische Universität Berlin, Berlin, Germany
yupeng.li@tu-berlin.de

[2] Fraunhofer Institute for Production Systems and Design Technology IPK, Berlin, Germany

Abstract. Thermal and mechanical loads during the grinding process lead to wear-related changes in the grinding wheel shape and topography which influence the work result of the grinding process. A conditioning process of the grinding wheels is therefore necessary for a reliable and stable grinding process. This study investigates the potential application of high-pressure liquid carbon dioxide ($HPCO_2$) jetting as an alternative method in the truing and dressing of diamond grinding wheels. Experimental results show that the $HPCO_2$ process can be used for profiling, sharpening and cleaning of diamond grinding wheels. Through varying the parameters such as jet pressure p and jet distance s, the adaption of jet characteristics allows $HPCO_2$ jetting to suit different requirements of the jet abrasiveness in different process steps during truing and dressing. These effects on the grinding wheel were evaluated using microscopic and topographical analysis. This study reveals the high potential of $HPCO_2$ conditioning of diamond grinding wheels.

Keywords: Grinding wheel · Dressing · Material removal

1 Introduction

Diamond grinding wheels, characterized by high hardness and efficient cutting performance, are utilized extensively in the grinding process of technical materials with high hardness and brittleness such as cemented carbide and silicon nitride [1]. The intense friction between the grinding wheel and the workpiece during the grinding process results in substantial thermal and mechanical loads on the diamond grinding wheel. This can lead to wear-related changes such as grain breakage, clogging and runout error in the grinding wheel [2]. To restore sharpness and the correct geometric shape of worn grinding wheels, a dressing process is necessary. The severe dressing tool wear and the use of coolant lead to reduced efficiency and increased costs associated with the conventional dressing [3]. Therefore, it is valuable to explore a novel dressing method which operates independently of coolant and tool wear.

L. Overmeyer and B.-A. Behrens (eds.), *Production at the Leading Edge of Technology*, Lecture Notes in Production Engineering, https://doi.org/10.1007/978-3-032-19524-1_30

Based on current investigations regarding the sustainable use of carbon dioxide (CO_2) [4], there has been an increasing focus on research in the field of CO_2 jetting technology. Dunsky and Hashish [5] conducted the first study regarding the feasibility of using high-pressure liquid carbon dioxide ($HPCO_2$) jetting for cutting. As a result, the authors suggested the use of CO_2 as a jetting medium in a residue-free cutting process of aluminium, steel, and brass. Based on this study, Bilz [6] designed a $HPCO_2$ jetting prototype system for cutting under atmospheric conditions. In terms of potential applications, Uhlmann et al. [7] have studied the effect of $HPCO_2$ jetting on cutting AlMg3. Compared with conventional milling tools, the cutting performance of the $HPCO_2$ jet remains constant during the cutting process. Studies mentioned above show that $HPCO_2$ jetting has sufficient material removal performance to process metal materials and the advantages of being coolant-free and independent of tool wear. Therefore, it also has the potential to condition metal-bonded diamond grinding wheels.

This study investigates the potential application of $HPCO_2$ jetting as an alternative method in the truing and dressing of diamond grinding wheels. Jetting experiments were conducted to evaluate the performance of $HPCO_2$ in profiling, sharpening and cleaning under different process parameters. The truing and dressing effects were investigated through microscopic surface analysis of the grinding wheel.

2 Experimental Setup and Method

Figure 1 shows the principal structure of the $HPCO_2$ jetting prototype system developed at the Institute for Machine Tools and Factory Management (IWF), Technische Universität Berlin. To generate a continuous liquid CO_2 flow, gaseous CO_2 is first fed from a gas cylinder into the heat exchanger 1, which is connected to a thermostat PCZEZ 50.06 NEB from National Lab GmbH, Germany, in cooling unit 1. In this step, CO_2 is primarily cooled to a temperature $T = -3$ °C. The CO_2 flow is then fed into a pneumatic high-pressure pump GPD 260-2-NBR from Maximator GmbH, Germany, allowing the CO_2 pressure to be adjusted between $p = 0$ and 4,000 bar. Subsequently, compressed CO_2 is transmitted into heat exchanger 2 and 3, where it is secondarily cooled to a temperature $T = -15$ °C using a circulation chiller Integral T from Lauda Dr. R. Wobser GmbH & Co. KG in cooling unit 2. A low-temperature, high-pressure liquid CO_2 flow is then connected via a high-pressure metal pipe from Allfi AG, Switzerland, to a KUKA KR 10 R900-2 robotic arm. An Allfi type IV 2.0 PWJ 155 jet head is installed on the flange which contains an Allfi type 94 sapphire jetting nozzle with a diameter $d_n = 0.15$ mm. Jetting experiments were conducted on a worktable below the nozzle, with the jet head moving perpendicularly along the surface of fixed grinding tools. The parameters and the vertical serpentine track pattern with a hatch distance of $b = 0.5$ mm are illustrated in Fig. 2. In the truing and dressing process, process forces F required for cleaning, sharpening, and profiling differ. Cleaning requires a low process force F and a large jet influence area A of the $HPCO_2$ jet to improve cleaning efficiency without damage on the grinding tools. This corresponds to a large nozzle distance s. In comparison, sharpening requires a greater process force F to remove the bonds without loosening the diamond grains. Therefore, different jet pressures p are adjusted to determine the appropriate process force F. Since profiling removes both the bonds and diamond grains to precisely process the grinding wheel's geometric profile, it requires the highest process force F and the smallest jet influence area A. To meet the requirements of three sub-processes in the truing and dressing, the experimental parameters in this study were pre-set based on empirical parameters in previous research on $HPCO_2$ jet cutting [6] and is shown in Table 1.

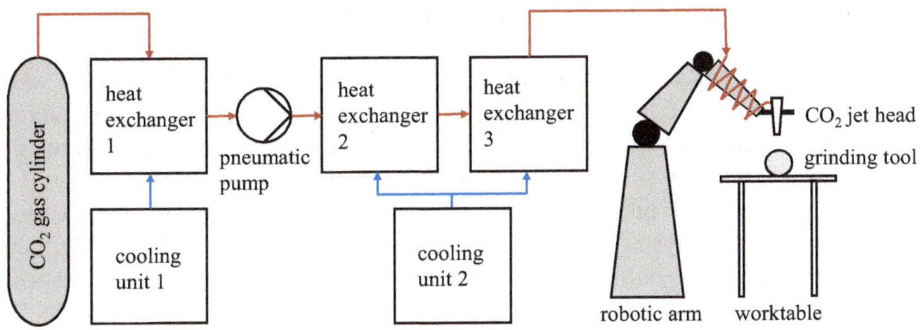

Fig. 1 Schematic of the $HPCO_2$ jetting system

Table 1 Process parameters of cleaning, sharpening and profiling

Subprocess	Nozzle distance s	Feed rate v_f	Jet pressure p	Nozzle diameter b
Unit	mm	mm/min	MPa	mm
Cleaning	56	60	300	0,15
Sharpening	5	150	150, 200, 300	0.15
Profiling	2	150	300	0.15

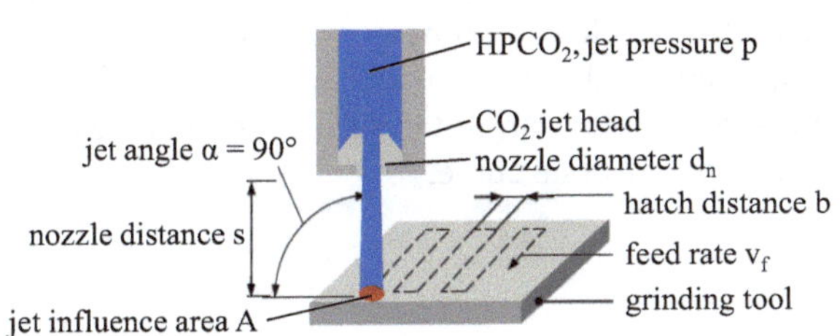

Fig. 2 Parameters of $HPCO_2$ jetting experiments

As an analogue to the investigated diamond grinding wheel, the cleaning experiments were carried out using a metal-bonded diamond grinding belt type FBA-40-2 from Hermes Schleifmittel GmbH, Germany. With a specific island-form macro-structure, changes in the chip space volume of the tool before and after grinding can be observed and calculated directly compared with grinding wheels. Correspondingly, a metal-bonded diamond grinding wheel of type 1A1 D64 C100 from Diamant-Gesellschaft Tesch GmbH, Germany, was used in the sharpening and profiling experiments to evaluate actual dressing effects. To reproduce the wear conditions of grinding tools, pretreatments were carried out before the $HPCO_2$ experiments. Before cleaning experiments, the grinding belt underwent a grinding process of a carbide blank of type EMT100 from Extramet AG, Switzerland, causing the surface to be clogged with chips. Before the profiling tests were conducted, the carbide blank was ground to create a grinding wheel profile error. Only $b_w = 6$ mm of the entire grinding wheel width b_w was engaged until a radius difference of $\Delta r = 0.5$ mm was created between the grinding wheel area that was not engaged, which was subsequently removed by $HPCO_2$ profiling.

After the experiments, an optical surface measurement was carried out with a digital light microscope of the type VHX-5000 from Keyence Deutschland GmbH, Germany, with a magnification ratio of $M = 20$ to 2000 x. Finally, the topography characteristics of processed surface areas on the grinding tools were determined according to DIN EN ISO 25178 [8] using an algorithm tool developed at IWF in MATLAB® from MathWorks, Natick, MA, USA [9].

3 Results and Discussion

3.1 Cleaning

The optical microscope images of the diamond grinding belt are shown in Fig. 3. The grinding belt has an island coating structure (Fig. 3a). After the pretreatment grinding process, the material chips clogged the free chip space between the grains, thereby reducing its free chip space volume (Fig. 3b). Through the cleaning process with $HPCO_2$, the clogged chips were removed and free chip space was exposed. Due to the large distant nozzle distance $s = 56$ mm and the resulting small process force F, neither the substrate of the grinding belt nor the abrasive grains were damaged (Fig. 3c). This indicates that the impact force of the jet exceeded the adhesive force between chips and grinding belt but did not exceed the breaking strength of the metal-bonded diamond coating. The surface texture of the grinding belt was regenerated close to the unused initial state in Fig. 3a, indicating an effective cleaning process with $HPCO_2$. To evaluate the cleaning effects quantitatively, height and volume parameters characterising the grinding belt topography can be derived by measuring the surface morphology of the grinding belt and calculating the Abbott-Firestone curve.

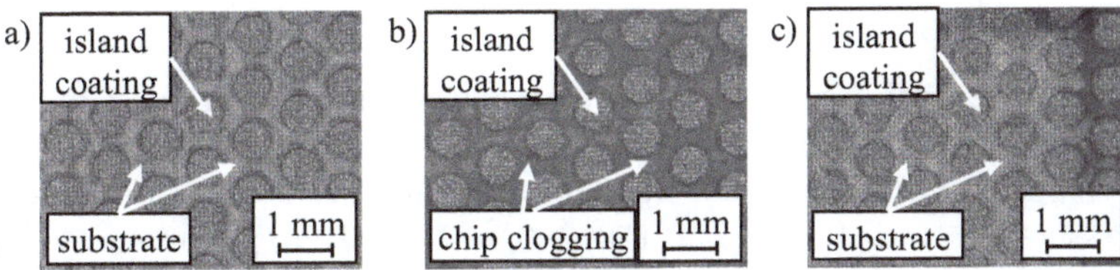

Fig. 3 Surface condition of the diamond grinding belt; **a** unused initial condition; **b** after pretreatment grinding; **c** after $HPCO_2$ cleaning

The specific chip space volume V''_{sp} is defined as the material-free volume normalized by the reference surface area of $A_r = 1\ mm^2$. It serves as a measure of the chip-carrying capacity of the tool. Grain protrusion K describes the average height of the protruding abrasive grains above a defined bonding level. The analysis results are presented in Fig. 4. Compared to the unused initial state, the specific chip space volume V''_{sp} and the grain protrusion K decrease by approximately 27% and 24%, respectively, after grinding. After $HPCO_2$ cleaning, the specific chip space volume V''_{sp} returns to 95.4% of the unused initial state (Fig. 4a), while the average height of the grain protrusion K even increases beyond the initial state to 103% (Fig. 4b). This result emerged as a consequence of the $HPCO_2$ jet's sharpening effect on the grinding belt alongside its intended cleaning effect. In addition to removed chips, the metal bond level of the island-coating was also slightly reduced by the $HPCO_2$ jet, causing more diamond grain protrusion K. Overall, the results show that the investigated parameters V''_{sp} and K differ by less than 5% from the initial state, indicating that the surface topography of the grinding belt can be effectively reset by using $HPCO_2$.

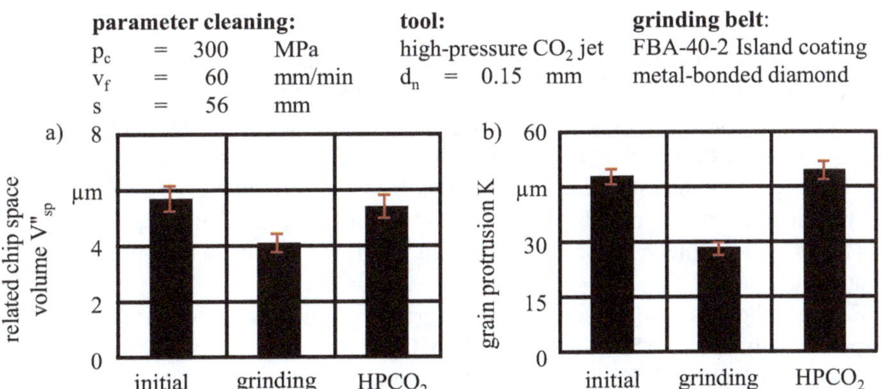

Fig. 4 Analysis of the surface topography of the diamond grinding belt; **a** specific chip space volume V''_{sp}; **b** grain protrusion K

3.2 Sharpening

The purpose of sharpening process is to adjust the surface microstructure without loosening the diamond grains and damaging the macrostructure of the grinding wheel. Accordingly, in sharpening experiments, a smaller nozzle distance s = 5 mm was used, and the jet pressure was varied from p_s = 150, 200, to 300 MPa to preliminarily determine the appropriate sharpening jet pressure p_s regarding the material removal performance of $HPCO_2$ jets. Figure 5 presents the surface topography of the grinding wheel coating with the jet kerfs processed by $HPCO_2$. With a sharpening jet pressure of p_s = 150 MPa, no macroscopic material removal was observed (Fig. 5a), which is appropriate for the sharpening process. At a larger jet pressure of p_s = 200 MPa, the grinding wheel surface exhibits a reset of the bond material and a release of the abrasive grains (Fig. 5b). At p_s = 300 MPa, a large amount of grinding wheel bonding was removed, exposing the metal substrate under the coating (Fig. 5c). Process pressures of p_s = 200 MPa or higher have a profiling effect, changing the macrostructure of the grinding wheel coating and making it unsuitable for sharpening. The following surface analyses were therefore evaluated using a sharpness pressure of p_s = 150 MPa.

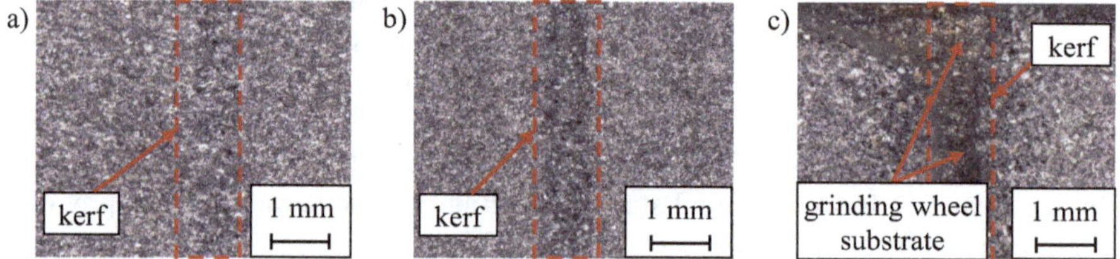

Fig. 5 Influence of jet pressure; **a** p_s = 150 MPa; **b** p_s = 200 MPa; **c** p_s = 300 MPa

According to DIN EN ISO 25178, the topography characteristics of grinding wheels include reduced peak height Spk, reduced valley depth Svk, arithmetical mean height Sa. Reduced peak height Spk is defined as the vertical height of the peak zone in the Abbott-Firestone curve, characterizing the height of abrasive grain tip on the grinding wheel surface. In contrast, reduced valley depth Svk describes the vertical depth of the valley zone and reflects the chip- and coolant-holding capacity of the grinding wheel. Sa is the arithmetic mean of the surface height relative to the mean height and quantifies surface roughness.

Figure 6 shows a comparison of the topography parameters before and after $HPCO_2$ sharpening with p_s = 150 MPa. After $HPCO_2$ sharpening, the reduced peak height Spk increases less than 5% compared to the initial state, indicating that the diamond grains in the surface peak zone were not significantly affected by $HPCO_2$ jet. Process parameters presented in Fig. 6 do not induce fracture, attrition and pull-out of the diamond grains. On the other hand, the reduced valley height Svk is 20% higher than the initial state. This implies a reduced level in bonds in the valley zone, suggesting that the metal bonding is eroded during the dressing process and subsequently removed from the valley zone. The removal of the bond material leads to a rough surface topography which can be reflected by the 16% increase in arithmetic mean height Sa from the initial state. Furthermore, the initial grain protrusion K = 22.5 μm increasing to K = 25.6 μm after $HPCO_2$ sharpening, corresponds to an increase in grain height of ΔK = 3.1 μm by resetting the bond through sharpening. Based on above results, $HPCO_2$ can remove bonds effectively and meanwhile maintain diamond abrasive grains, improving the sharpness of the grinding wheel.

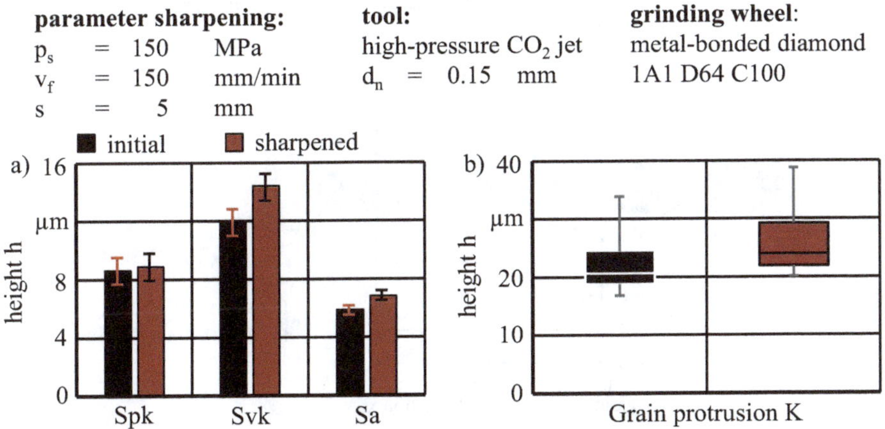

Fig. 6 Analysis of grinding wheel topography in the $HPCO_2$ sharpening; **a** surface characteristics Spk, Svk and Sa; **b** grain protrusion K

3.3 Profiling

In the profiling process, it requires the highest process force F and the smallest jet influence area A in comparison to cleaning and sharpening. Therefore, the smallest nozzle distance s = 2 mm and the maximum jet pressure p_P = 300 MPa was adjusted to ensure a sufficient process force F. In profiling experiments, the radius difference of Δr = 0.5 mm along the grinding wheel width b_w processed in pretreatment was profiled using $HPCO_2$. The macro-scale material removal performance of the $HPCO_2$ jet is investigated. The profiles of the grinding wheel surface are shown in Fig. 7.

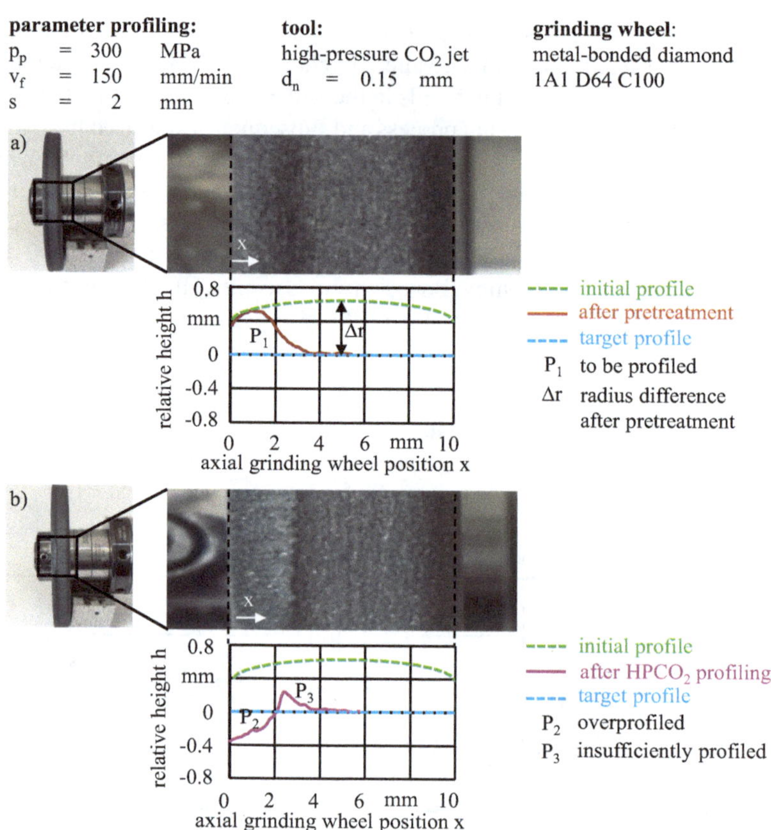

Fig. 7 Profiling with $HPCO_2$; **a** after grinding process; **b** after $HPCO_2$ profiling

After the pretreatment, the grinding wheel radius is reduced by Δr in the range x > 4 mm due to wear during the grinding of a cylindrical carbide blank. However, in the range x < 4 mm, the grinding wheel radius is larger because the grinding wheel was not engaged in this range. In the area for 1.5 mm < x < 4 mm, the carbide block is in smaller depth of engagement with the grinding wheel, resulting in a slope profile and reflecting the circular cross-section profile of the carbide block. To regenerate the grinding wheel into a cylindrical geometry after it lost its original shape during pretreatment, the 0 mm < x < 4 mm area with the protruding profile P_1 needs to be removed to the level of the target profile with the relative height h = 0 mm (Fig. 7a). After the $HPCO_2$ profiling process, the protruding material was partially removed and the area for x < 4 mm was significantly reshaped (Fig. 7b). The process force F of $HPCO_2$ profiling has exceeded the fracture strength of the bond material. The removal of the bond by $HPCO_2$ jet detaches the diamond abrasive particles embedded in the bond matrix, enabling $HPCO_2$ to process the macro-structure of coatings. However, the lowest relative height of the grinding wheel edge decreased to h = −0.36 mm. In this area P_2, more material than necessary was overprofiled. In contrast, insufficiently profiled protruding material remains in the P_3 area. Adjusting the jetting parameters and optimizing the jet shape can lead to an improvement in the $HPCO_2$ profiling process. The result of $HPCO_2$ profiling proves that the $HPCO_2$ jet has sufficient material removal performance to reshape metal-bonded diamond grinding wheels.

4 Summary

This study preliminarily demonstrates the feasibility of using high-pressure liquid carbon dioxide jetting, i. e. $HPCO_2$ jetting, for the truing and dressing of diamond grinding wheels. Experimental results show that by changing process parameters of $HPCO_2$ jetting, such as jet pressure p and nozzle distance s, the material removal performance of the $HPCO_2$ jet can be adjusted to different process requirements of cleaning, sharpening and profiling of diamond grinding wheels. With relatively large nozzle distance s, $HPCO_2$ can effectively remove chips and other contaminants adhering to the tool surface without damaging the grinding tool. As nozzle distance s decreases and jet pressure p increases, the process force F increases and the material removal performance of the $HPCO_2$ jet enhances. Under appropriate parameter combinations, $HPCO_2$ jetting can also remove bond materials without causing diamond grains to detach. By that, a grinding wheel sharpening process is achieved. As the process force F further increases, the $HPCO_2$ jet becomes sufficiently abrasive to remove a significant amount of grinding wheel coating, enabling profiling of the wheel's geometric shape.

In summary, as a flexible, residue-free and dry-processing technology without tool wear, $HPCO_2$ jetting shows potential as a new truing and dressing method in terms of performance and ecological efficiency. However, further research is needed to promote its industrial application. Subsequent research will include investigating the influence of CO_2 phase transition on $HPCO_2$ jet properties and optimizing process parameter combinations to improve truing and dressing results.

Acknowledgements. This publication is based on the results of the research project "Dressing of grinding wheels with hard abrasives by high pressure jets with liquid CO_2" (project number: 510588719), which is funded by Deutsche Forschungsgemeinschaft (DFG, German Research Fundation).

Competing Interests The author(s) has no competing interests to declare that are relevant to the content of this manuscript.

References

1. Biermann, D., Würz, E.: A study of grinding silicon nitride and cemented carbide materials with diamond grinding wheels. Prod. Eng. Res. Devel. **3**, 411–416 (2009)
2. Wegener, K., Hoffmeister, H.W., Karpuschewski, B.: Conditioning and monitoring of grinding wheels. CIRP Ann. **60**(2), 757–777 (2011)
3. Denkena, B., Tönshoff, H.K.: Spanen—Grundlagen. 3rd.edn., Springer, Berlin (2010)
4. Hunt, A., Sin, E., Marriott, R.: Generation, capture, and utilization of industrial carbon dioxide. Chemsuschem **3**(3), 306–322 (2010)
5. Dunsky, C., Hashish, M.: Cutting with high pressure CO_2 jets. Allen, A. G. (Hrsg.). QUEST Integrated, Washington (1994)

6. Bilz, M.: Möglichkeiten und Grenzen des Strahlspanens mittels CO_2-Hochdruckstrahlen. Berichte aus dem Produktionstechnischen Zentrum Berlin. Hrsg.: Uhlmann, E. Dissertation, Technische Universität Berlin. Fraunhofer IRB, Stuttgart (2014)
7. Uhlmann, E., John, P.: Dry cutting with high-pressure liquid CO_2 jets. Adv. Mater. Lett. **10**(1), 2–8 (2019)
8. DIN EN ISO 25178: Geometrische Produktspezifikation, Oberflächenbeschaffenheit: Flächenhaft. Beuth: Berlin, Germany (2012)
9. Uhlmann, E., Muthulingam, A.: Optimizing the sharpening process of hybrid-bonded diamond grinding wheels by means of a process model. Machines **10**(1), 8 (2022)

Autonomously Switching Paraffin-Based Thermal Bridge for Passive Thermal Regulation

Tim Schmitt(✉) and Peter Groche

Institut für Produktionstechnik und Umformmaschinen (PtU), Technische Universität Darmstadt, Darmstadt, Germany
tim.schmitt@ptu.tu-darmstadt.de

Abstract. Thermal management is becoming increasingly important for modern production technology, as climatic conditions cause significant temperature fluctuations that affect production accuracy. Established cooling systems are highly energy-intensive, making them unsustainable. This paper presents the concept of an autonomously switching thermal bridge for passive thermal regulation, based on a paraffin expansion actuator. The system exploits the reversible phase transition to dynamically regulate thermal conductivity without requiring external auxiliary power or active control. Numerical simulations were used to develop the basic principles for utilising the membrane actuators as a thermal bridge. The subsequently realised thermal bridge was tested experimentally in various temperature ranges. It demonstrated good thermal conductivity and autonomous switching capability. The concept is also designed for robustness and production scalability. It therefore holds promise for industrial applications in the field of passive heat management in production environments.

Keywords: Passive thermal regulation · Phase change material · Paraffin expansion actuator · Thermal bridge · Energy efficiency

1 Introduction

In modern production environments, thermal management is becoming increasingly important, as growing climatic fluctuations can impair manufacturing accuracy and process stability [1]. Conventional climate control systems are often energy-intensive and thus limited in sustainability, particularly in large-scale or thermally sensitive industrial applications. Passive thermal regulation systems, which operate without external power or active control, offer the potential to provide a robust, energy-efficient alternative for maintaining thermal stability [2].

In the open state, these systems separate a hot, heat-producing side from a cold, heat-dissipating side, see Fig. 1a. The sink is usually the environment. If

L. Overmeyer and B.-A. Behrens (eds.), *Production at the Leading Edge of Technology*, Lecture Notes in Production Engineering, https://doi.org/10.1007/978-3-032-19524-1_31

the temperature on the hot side exceeds the threshold value, the volume expansion in the phase change of paraffin causes the mechanical displacement of a piston against a bump stop. In this state, heat can flow through the piston into the bump stop. If the temperature of the hot side drops below the threshold value again, the paraffin compresses and the piston is pulled back. In the long term, this results in a passive control behaviour. This is why it is referred to as a thermal bridge instead of a simple heat switch [3]. Due to their independence from auxiliary energy and their robustness, such constructions are so far mainly used in space travel [4]. In these applications the dissipated heat from the power electronics and thrusters must either be dissipated or is used to heat the satellite, depending on the conditions. This requires reliable passive systems [5]. In industrial applications, switchable thermal bridges are not yet common, but they have great potential to improve energy efficiency and reliability.

This work focuses on the use of macroscopic membrane actuators, which were developed at the Institute for Production Engineering and Forming Machines (PtU) and their potential for industrial applications. These actuators, illustrated in Fig. 1b, feature a robust, scalable, and fully metallic design with no moving parts or seals. Advantages include a high actuation force and stiffness, ease of manufacturing and long-term, maintenance-free operation [6]. The actuators are available in various diameters, and the choice of paraffin and the housing properties allows the characteristics of the actuator to be customised. This study examines the potential of these actuators to serve as autonomously switching thermal bridges.

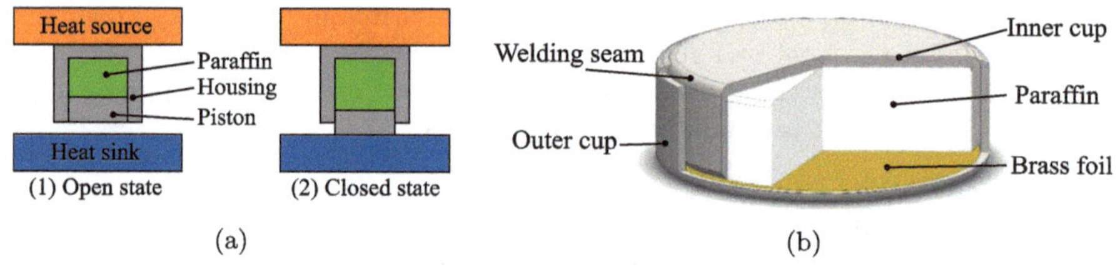

Fig. 1. a Concept and realization of paraffin-based thermal bridges according to [7]. **b** PtU actuator design with welded deep-drawn cups and brass seal according to [8]

2 Approach

High-performance thermal bridges in the aerospace industry are specialised applications designed for specific requirements. In contrast, industrial applications typically face smaller temperature differences, larger available heat-transfer areas, and stringent cost constraints. Under these conditions, moderate switching ratios are already sufficient for reliable passive temperature control. The aim of this work is therefore to develop a thermal bridge concept that meets industrial specifications, particularly in terms of robustness, scalability, and cost-efficiency. Potential fields of application include forming tools, where passive regulation of the operating temperature can improve process stability [9], as well as the temperature management of machine frames, beds, workpieces, or even control units. The actuators developed at the institute with a diameter of 30 mm and a height of 10 mm are used as the central switching element. Their production is process-stable and the characteristics are predictable, see [10], so the focus can only be on the bridge itself.

The switching threshold is determined by selecting the correct type of paraffin. For industrial requirements, the objective is to dissipate heat into the environment at room temperature ($T_m \approx 20\,^{\circ}\mathrm{C}$). Therefore, the paraffin wax needs to be in a liquid state at the upper threshold to fully deploy the actuator, and solid at room temperature in order to insulate from the environment. The paraffin *RT28HC* from *RubiTherm* fulfils the above requirements; its key properties are listed in [11,12]. The melting temperature close to room temperature makes production difficult; the actuators have to be cooled for the welding process, for example.

In order to compare and evaluate different concepts for thermal bridges, quantitative parameters are required. An important parameter for the performance of a thermal bridge is the conductance ratio r [see Eq. (1)], which quantifies the ratio of thermal insulation to conduction. Such state-of-the-art devices for aerospace applications can reach values of $r > 100$ under vacuum environment [5], although performance drops to $r < 50$ in ambient air. Such high values are enabled by sophisticated designs and high-performance materials such as beryllium, copper, and gold [4].

$$r = \frac{\Lambda_{act}}{\Lambda_{dea}} = \frac{\dot{q}_{act} \cdot \Delta T_{dea}}{\dot{q}_{dea} \cdot \Delta T_{act}} \tag{1}$$

The basic concept is to use the membrane actuators as thermal bridges directly, as shown in Fig. 1. The outer cup is connected directly to the hot side. If the temperature exceeds the melting temperature, the actuator expands and connects to the cold side. Apart from the actuator itself, no other components are required, and heat flows directly through it.

2.1 Numerical and Experimental Investigation

The concept is first analysed numerically in order to estimate its performance. The transient simulation is set up in *Abaqus 2024* as a thermo-mechanically coupled and rotationally symmetrical 2D model, see Fig. 2a. The modelling for the expansion actuator is implemented as shown in [10]. The outer surface of the bridge is assumed to be adiabatic, the heat flows on the upper and lower surfaces are specified or analysed as purely conductive boundary conditions.

For an experimental validation approach, the bridge is placed in a test rig to determine heat transfer coefficients, see Fig. 2. The temperatures T_1 and T_4 in the punches can be set independently between 12 and 100°C. The heat flow in

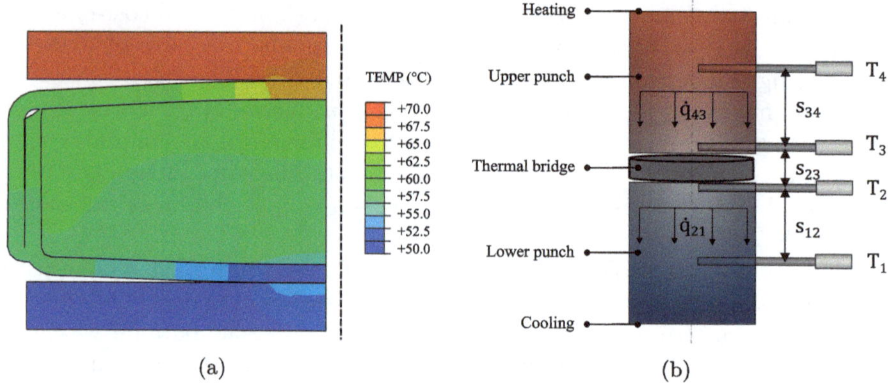

Fig. 2. a Heat distribution in the axisymmetric 2D model, the deformation is shown with exaggerated scaling. In the steady state, only a punctual contact zone is present, limiting the heat flow that can be transferred. **b** Schematic structure of the experimental IHTC test rig according to [13]

the installed bridge results from the temperature difference ΔT and the thermal conductivity Λ, see Eq. (3). Four temperatures per punch are measured using type K thermocouples with 1 Hz and a *QuantumX 840B* measuring amplifier. The detailed setup of the test rig is described in [13].

The heat conduction is assumed to be ideal, static and one-dimensional [14]. Due to the temperature range and the small temperature difference, radiation and convection can be ignored as heat transfer mechanisms [15]. This results in a constant heat flow without losses, see Eq. (2). The heat flow and conductance are calculated according to Eq. (3). To calculate the thermal conductivity of the bridge λ_{23}, the heat flow in the upper punch $\dot{q}_{43}$ and the total height s_{23} of the bridge are used as the reference.

$$\dot{q}_{21} = \dot{q}_{32} = \dot{q}_{43} \tag{2}$$

$$\dot{q} = \frac{s}{\lambda} \Delta T = \Lambda \, \Delta T \tag{3}$$

$$\lambda_{23} = \frac{s_{23}}{\dot{q}_{43}} (T_3 - T_2)$$

3 Results

3.1 Direct Use

The simplest and most obvious approach is to use the membrane actuator itself as a bridge. This concept is proven through simulations. Since no material model is yet available for the paraffin with a switching temperature of 28 °C, the available data for the 58 °C paraffin was used instead [10]. The applied temperature difference is thus comparable to the experimental conditions, see Table 1. The heat on the hot side causes the actuator to expand and fill a 0.5 mm air gap to the housing. This allows the heat flow to pass directly through the actuator to the cold side and dissipative cooling power is transferred. At a temperature difference of 20 K, the actuator expands within 35 s and closes the gap. As soon as there is contact with the cold side, the heat dissipates and heat transfer is favoured by the steel housing. However, the paraffin cools down below its solidification temperature, causing the actuator to collapse. In the steady state only a punctual contact zone is formed, shown in Fig. 2a.

In conclusion, the construction of thermal bridges with this type of actuator requires that the heat flow through the housing exceeds the heat dissipation through the paraffin. Due to the manufacturing route based on sheet metal, it is not possible to increase the wall thickness only in the outer radius. The paraffin with poor thermal conductivity, $0.25\,\mathrm{W\,m^{-1}\,K^{-1}}$ [12], in the paraffin takes up around 78 % of the axial cross-sectional area of the actuator. With an increase in the actuator diameter, this area ratio becomes even worse. Improving the thermal conductivity in the housing by the same thickness, e.g. by using aluminium, has only a minor effect due to the surface ratio between the housing and the paraffin. Consequently, the direct utilisation of such a membrane actuator as a thermal bridge is not effective, see Table 1. Instead, additional heat conducting structures must be used. This means that a heat flux must flow parallel to the actuator, for which the cross-sectional area and the thermal conductivity can be varied.

3.2 Thermal Bridge

If the need for parallel heat conduction is transferred to a technically usable thermal bridge, the concept shown in Fig. 3a results. The actuator is embedded in an aluminium housing on the hot side. When activated, a piston moves and the heat can flow from the housing through radial fits into the piston and from there be transferred to the cold side. A springback is integrated for the piston so that the bridge can also deactivate itself again. Due to the paraffin *RT28* used, the switching threshold is approx. 28 °C. The bridge is validated directly by experiment in the IHTC rig, see Fig. 3b. The run starts from constant temperatures on the hot ($T_3 = 26.3\,°\mathrm{C}$) and cold surfaces ($T_2 = 10.1\,°\mathrm{C}$). The curves are shown in Fig. 4. The heating is switched on at timestamp 1 and the temperature of the hot side rises to 41.8 °C. At timestamp 2, the bridge is activated, a heat flow is transferred and the temperature of the cold side rises to 24.1 °C. After setting a stationary behaviour, the heating is deactivated again at timestamp 3. The heat capacity of the bridge and the experimental setup ensures a gently decreasing heat flow until the bridge is autonomously deactivated at timestamp 4.

3.3 Discussion

As the numerical investigation shows, it is not possible to utilise the membrane actuators directly as a thermal bridge without further ado. In the classic design, heat dissipation from the paraffin is higher than the heat flow through the housing. This leaves only a point-shaped contact surface, severely limiting the heat

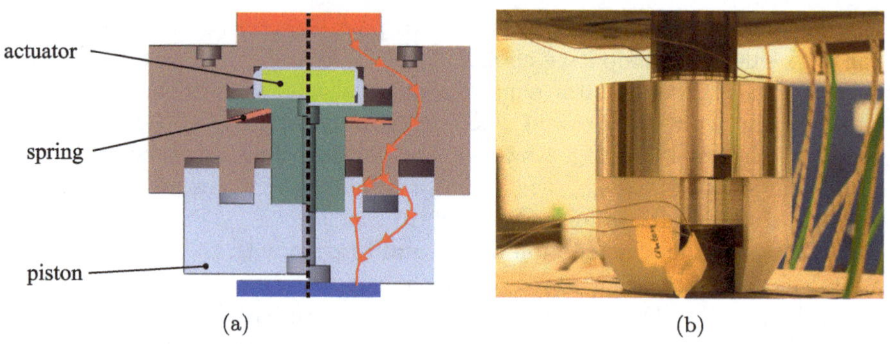

Fig. 3. **a** Sectional view of the thermal bridge in deactivated (left) and activated (right) state. **b** Thermal bridge in the test rig, the cut-out allows visual access

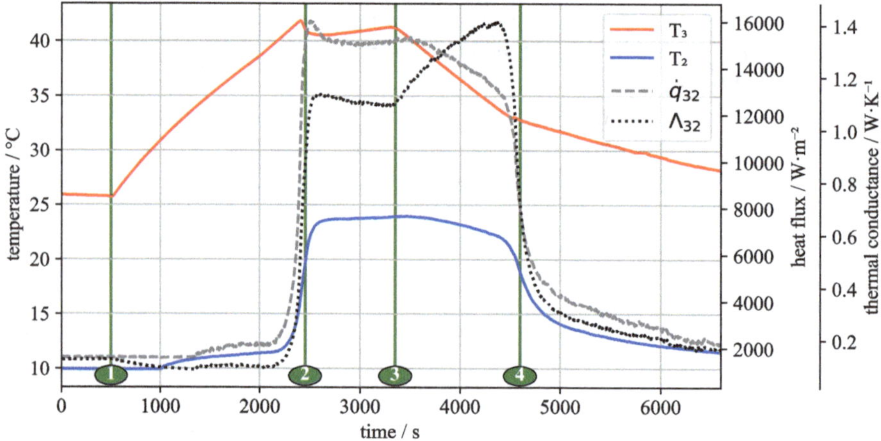

Fig. 4. Experimental investigation of the thermal bridge. At timestamp **1** the heating is enabled, at timestamp **2** the bridge is independently activated, at timestamp **3** the heating is switched off and at timestamp **4** the bridge is autonomously deactivated

dissipation, see Fig. 2a. The transferred thermal fluxes are $\dot{q} = 2316\,\mathrm{W/m^2}$ with a thermal conductivity of $\Lambda_{act} = 0.082\,\mathrm{W\,K^{-1}}$. Due to the simulative simplifications, no heat flow is transferred in the deactivated state; this would correspond to infinite conductivity values. If the thermal conductivity from the experiment is used instead, according to Eq. (1), the conductivity ratio is calculated as $r \approx 1.2$. A bridge with such a low value would not be technically usable. A solution could be enabling heat conduction parallel to the expansion actuator.

Implementing such a concept (see Fig. 3) demonstrates the expected behaviour in the experimental investigation (see Fig. 4). In the sections between timestamp 1 and timestamp 2 and after timestamp 3, there is transient behaviour caused by dynamic effects such as thermal masses. The mean values in the stationary periods are used for the evaluation. This results in a thermal conductance

Table 1. Comparison of the characteristic values of the two thermal bridge concepts

Property	Direct use	Thermal bridge
ΔT	20 °C	17 °C
Λ_{dea}	/	0.105 W K^{-1}
Λ_{act}	0.082 W K^{-1}	1.14 W K^{-1}
r	1.206	13.103

$\Lambda_{32,dea} = 0.105\,\mathrm{W\,K^{-1}}$ in the open state. For the closed state between timestamp 2 and 3, the heat flux is $\Lambda_{32,act} = 1.14\,\mathrm{W\,K^{-1}}$, according to Eq. (1), the conductance ratio for this bridge is $r = 13.1$. While aerospace applications achieve much higher ratios, this comes at the expense of significantly greater effort and cost. For industrial applications, however, such a ratio is already sufficient for reliable passive thermal control, with further optimisation, the conductance ratio can be further increased.

The experiment demonstrated the successful implementation of a thermal bridge using paraffin-based membrane actuators. The ability to switch the bridge independently in both directions enables control behaviour, with a characteristic response time of approximately 1000 s. This provides the basis for temperature control independent of auxiliary energy. Thus, the experiment confirms the fundamental functionality of the concept and offers a first validation of its feasibility for a production environment.

4 Conclusion

This paper demonstrates the concept of a paraffin-based thermal bridge and an approach for its experimental validation, showing the basic suitability of paraffin-based membrane actuators for passive thermal control. Thanks to their low cost, robustness, and reliability, these actuators are promising candidates for use in industrial applications. Their simple design enables their integration into a wide range of systems without the need for control electronics and auxiliary energy. As shown in this paper, the direct use of the actuators as a thermal bridge is not possible in the existing design. To ensure proper function, the heat flow must be transferred mainly in parallel to the actuator. This concept was realised and experimentally tested in the technical bridge. The implemented thermal bridge switches autonomously at a temperature of 28 °C, with a conductivity ratio $r \approx 13$. Future work will focus on transferring this principle into practical implementations and to further optimise the design, for example by improving the switching ratios, reducing the dimensions, and tailoring the switching thresholds through the use of different paraffins. In addition, a more detailed characterisation of the control behaviour is still in progress. Reducing the thermal mass is expected to improve the control dynamics, paving the way for future industrial applications. The presented concept has proven feasible in the experimental validation, and in the future, systematic and statistically validated test series can be conducted to further confirm its reliability.

Acknowledgments. Funded by DFG 1818/65-4 (German Research Foundation).

Competing Interests. The author(s) has no competing interests to declare that are relevant to the content of this manuscript.

References

1. Brecher, C.: Thermoenergetische Gestaltung Von Werkzeugmaschinen: Praxishandbuch. Springer, Wiesbaden (2025). https://doi.org/10.1007/978-3-658-45180-6
2. International Energy Agency: Energy efficiency report 2024 (2024)
3. David G. Gilmore (ed.): Spacecraft Thermal Control Handbook, 2nd edn. Aerospace Press, El Segundo, California (2002)
4. Bevans, J.T., Dan, C.G., Drummond, F.O., et al.: Spacecraft Thermal Control: NASA Space Vehicle Design Criteria Environment, 1st edn. Woodhead Publishing in Mechanical Engineering (1973). https://doi.org/10.1533/9780857096081
5. Shinozaki, K., Nohara, T., Ando, M., Okamoto, A., Maeda, M., Sugita, H., Takada, S.: Research and development of heat switch for future space missions. Trans. Jpn. Soc. Aeronaut. Space Sci. Aerosp. Technol. Jpn. **12**(ISTS 29), 7–11 (2014). https://doi.org/10.2322/tastj.12.Po_4_7
6. Mann, A., Germann, T., Ruiter, M., Groche, P.: The challenge of upscaling paraffin wax actuators. Mater. Des. **190**, 108580 (2020). https://doi.org/10.1016/j.matdes.2020.108580
7. Geismayr, L., Patzwahl, F., Langer, M., Binder, M., Schlick, G.: Thermo-mechanical design and analysis of a multispectral imaging payload using phase change material. In: International Astronautical Congress (IAC), pp. 2–17 (2020)
8. Germann, T., Groche, P.: Dynamic-enhanced macroscopic paraffin wax phase change actuators as a method of process stabilization. In: Behrens, B., et al. (Hg.) 2022—Production at the Leading Edge, pp. 237–258 (2022)
9. Laeger, R., Peddinghaus, J., Rosenbusch, D., Behrens, B.-A.: Reduction of thermally induced wear on a forging tool by heatpipes. In: Bauernhansl, T., Verl, A., Liewald, M., Möhring, H.-C. (eds.) Production at the Leading Edge of Technology. Lecture Notes in Production Engineering, pp. 628–637. Springer, Cham (2024)
10. Mann, A., Bürgel, C., Groche, P.: A modeling strategy for predicting the properties of paraffin wax actuators. Actuators **7**(4), 81 (2018). https://doi.org/10.3390/act7040081
11. Rubitherm, Technologies: Data Sheet RT28HC. https://www.rubitherm.eu/media/products/datasheets/Techdata_-RT28HC_EN_09102020.PDF. Accessed 06 May 2025
12. Ukrainczyk, N., Kurajica, S., Sipusic, J.: Thermophysical comparison of five commercial paraffin waxes as latent heat storage materials. Chem. Biochem. Eng. Quart. **24**, 129–137 (2010). https://doi.org/10.15255/CABEQ.2014.240
13. Schell, L., Sellner, E., Heller, B., Wenzel, T., Groche, P.: Reliable determination of interfacial heat transfer coefficients for hot sheet metal forming. CIRP Ann. **73**(1), 237–240 (2024). https://doi.org/10.1016/j.cirp.2024.03.007

14. Baehr, H.D., Stephan, K.: Wärme- und Stoffübertragung. Springer, Berlin (2016). https://doi.org/10.1007/978-3-662-49677-0
15. Springer-Verlag GmbH: VDI-Wärmeatlas. Springer, Berlin (2013). https://doi.org/10.1007/978-3-642-19981-3

Embossing Optical Structures in Nanometer Range

Dennis Schmiele(✉), Dietmar Friesen, Richard Krimm, and Bernd-Arno Behrens

Leibniz University Hanover, Garbsen, Germany
schmiele@ifum.uni-hannover.de

Abstract. Optical technologies are probably the key technologies of the twenty-first century. Traditional optical systems such as projectors, microscopes or laser optical systems still consist of numerous separate components which generate, manipulate or detect light. By combining various technologies, optical systems can be developed to integrate a large number of functions in a small space at low costs. Optical gratings, for example, are used in waveguides for light coupling or in optical sensors. The structure sizes of such diffractive optical elements are often smaller than one micrometer, which creates major challenges to manufacturing technology. The conflict of objectives between costs, quality, accuracy and functionality can be addressed by embossing in combination with an adaptive process control in real time. In addition to injection molding and laser engraving processes, micro embossing offers a cost-efficient option for the mass production of optical components with structured surfaces. The embossing process also provides a high degree of flexibility to be adapted to new optical materials without the need of extensive adjustments. An embossing device enabling to transfer micro- and nanostructures precisely and reproducibly onto semifinished optical products was developed as part of the cluster of excellence PhoenixD, funded by the German Research Foundation. This paper focuses on demonstrating both the positioning accuracy and the achievable quality of the embossed structures. The innovative combination of embossing processes and real-time process control opens up new chances for the production of high-quality optical components for a wide range of applications.

Keywords: Nanostructures · Embossing · High accuracy

1 Introduction

Until now, the production of highly developed precision optical systems was based on a large number of complex individual components that were assembled in complex, serial processes. This expensive production method still limits their use for daily products, such as in medicine or agriculture. The future vision of the cluster of excellence PhoenixD is the integration of a large number of optical components in a compressed space. One of the main focuses here is on the manufacturability of such novel micro-optical systems and the system technologies required for this. A promising method for achieving cost-effective high-performance optics in mass production is the embossing of micro- and

L. Overmeyer and B.-A. Behrens (eds.), *Production at the Leading Edge of Technology*,
Lecture Notes in Production Engineering, https://doi.org/10.1007/978-3-032-19524-1_32

nanostructures, such as diffractive optical elements (DOEs). DOEs offer an efficient way of manipulating light in a highly specific manner and are therefore ideal for integrated optics.

Research in new manufacturing technologies lead to the academic demonstrator shown in Fig. 1. The intended demonstrator combines functions such as beam guidance via actively controllable mirrors and light coupling via Bragg gratings in a very small space. The opto-electronic platform of the demonstrator is injection molded from polyetheretherketone (PEEK) including the cavities for optical waveguides, filters and photodiodes. After placing the diodes and filters, the fiber optic cavities are filled with polymethylmethacrylate (PMMA) and then a coupling grating is embossed. Even the smallest deviations from the target geometry are critical here, as they can severely affect the function. Embossing such filigree structures requires maximum precision and places high demands on machines and tools. Unavoidable disruptive factors, such as guide tolerances, bearing clearance or thermal expansion, can have a negative impact on precision in the conventional embossing process. For this reason, there is a requirement for an embossing device that compensates for such disturbing factors and thus enables cost-efficient mass production of optics with three-dimensionally structured surfaces.

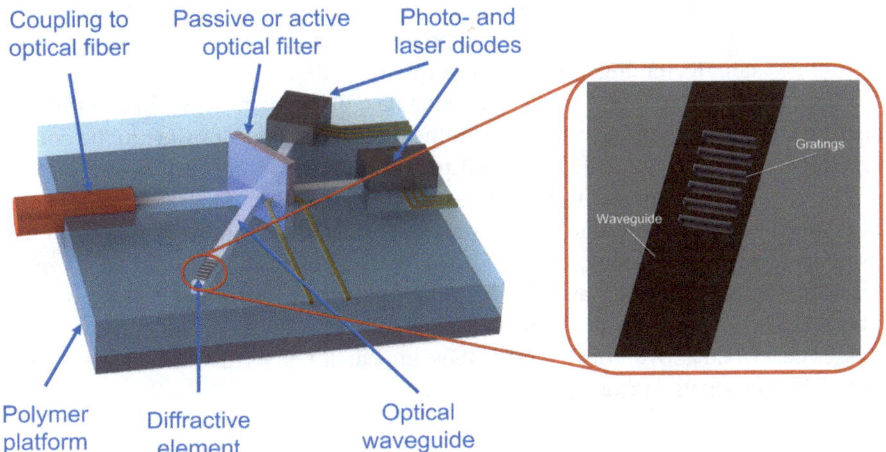

Fig. 1 Demonstrator with several optical functions

2 State of the Art

Currently, there are a variety of processes and system technologies that are used to produce DOEs far beyond the laboratory scale. For example, DOEs of high quality and complexity can be produced using injection molding processes with the appropriate very fine master structures in the tools. In this process, the polymer in question, usually in the form of granulate, is heated in a rotating screw, pressed into the tool under high pressure via a nozzle and formed after the material has solidified (Fig. 2) [1]. This is a primary shaping process, which is not suitable for structuring existing semi-finished products. In addition, it is not possible to adjust the structure (height) without further effort.

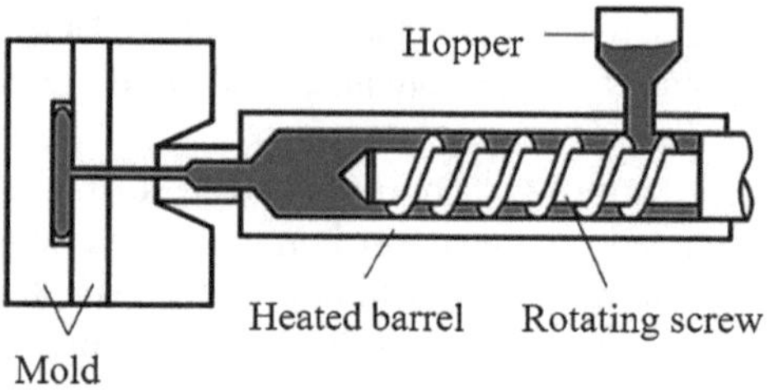

Fig. 2 A sketch of screw injection molding machine [1]

Another variant for the production of micro-/nanostructures is laser direct writing. This is a technology for the direct structuring of materials using a focused laser beam. The laser beam is guided over the material surface to create precise structures or patterns. This process enables the material surface to be changed by vaporization or melting. In contrast to conventional etching, which requires separate masks, laser direct writing offers great flexibility as designs can be quickly adapted digitally. The high precision of the technology enables processing in the micro and nanometer range. Laser direct writing is used in a variety of applications, including semiconductor manufacturing and the production of sensors. It is suitable for various materials such as metals, polymers and glass. A serious disadvantage of this process, in addition to the high acquisition costs of the system technology, is the low output rate [2].

DOEs can also be applied to materials using an embossing process. In contrast to the standard hot embossing process, which is rather expensive and slow due to the complex system technology (vacuum) [3], the roll-to-roll hot embossing process described in Fig. 3 is an efficient manufacturing process. In this process, a decorative or functional layer is transferred to a continuous substrate (e. g. plastic, metal foil or glass [4]) using pressure and heated embossing rollers [5]. This process enables high production speeds, precise transfer and the integration of metallic, optical or functional effects. It is used in the packaging and electronics industries as well as in security printing. Similar to the processes described above, roll-to-roll embossing cannot be used to create structures on semi-finished products in exact positions.

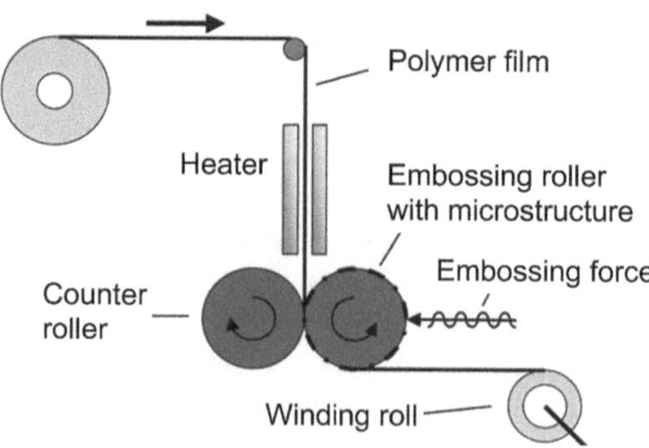

Fig. 3 A sketch of hot roller embossing process [5]

There are already a large number of processes and system technologies, as well as application-specific optimized combinations of these, such as hot embossing and injection moulding [6], for the production of diffractive optical elements. However, none of these manufacturing processes enables DOEs to be produced cost-effectively in large quantities at a precisely defined position on preprocessed components. Therefore, there is a requirement for a new type of system technology that can be used to realize embossing on semi-finished products without damaging the surrounding areas.

3 Achievable Embossing Quality

It has already been shown in [7] that an embossing device consisting of simple and low-cost standard parts can be configured with a system of springs and electromagnets in such a way that vertical positioning accuracy in the sub-micrometer range can be achieved. The precisely adjustable increase of the magnetic force against the spring forces ensures a vertical movement of the embossing stamp. The required positioning accuracy depends on the structure height to be embossed. If a structure height in the micrometer range is required, positioning accuracy in the sub-micrometer range is desirable. This high positioning accuracy is the requirement for the defined compensation of external disturbance variables such as external vibrations, changes in the ambient temperature or dust particles and thus for a precisely controllable embossing process.

For the final assessment of the replica quality, the master structure was first measured optically (Fig. 4). These structures were applied to the embossing die using lithography and subsequent reactive-ion etching. It is a grating structure with a nominal width of 10 μm and a nominal height of 1 μm. To determine the actual dimensions of the master structure, a profile section of the surface is created using a confocal microscope (blue vertical line in Fig. 4, top left) and then the sidewall is approximated linearly to determine the width and the sidewall angle (yellow lines in Fig. 4, bottom). The structure height is determined using horizontal lines (blue dotted lines in Fig. 4, bottom). The averaged real dimensions of the grating are 10.284 μm × 1.117 μm and a sidewall angle of 79.84°. In addition, the process-related phenomenon of microtenching appears [8]. Microcracks appear at the corners of the etched structure, resulting in deeper grooves. For future master structures, the quality and the deviation from the desired target geometry should be minimized by means of process optimization or other methods, e. g. electron beam lithography.

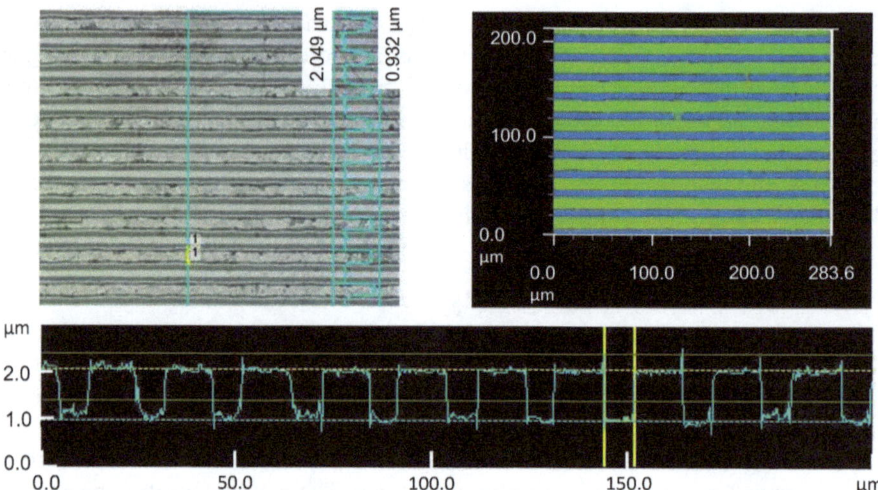

Fig. 4 Analysis of the master structure in the bright field (top, left), as a topography map (top, right) and in the height profile (bottom)

Figure 5 and 6 show the embossing quality that can be achieved with the embossing device described in [7] in foils made of polymethylmethacrylate (PMMA) and OrmoCore®. The two materials investigated are frequently used in optical applications, e. g. waveguides. For the embossing experiments, the embossing die including the grating structure was heated to approx. 120 °C and embossed in 1 mm thick films using the position control of the embossing device. After a holding time of 120 s, cooling of the material was initiated while maintaining the embossing force and the foil was removed from the device after complete cooling. The embossed structures were analyzed by a confocal microscope.

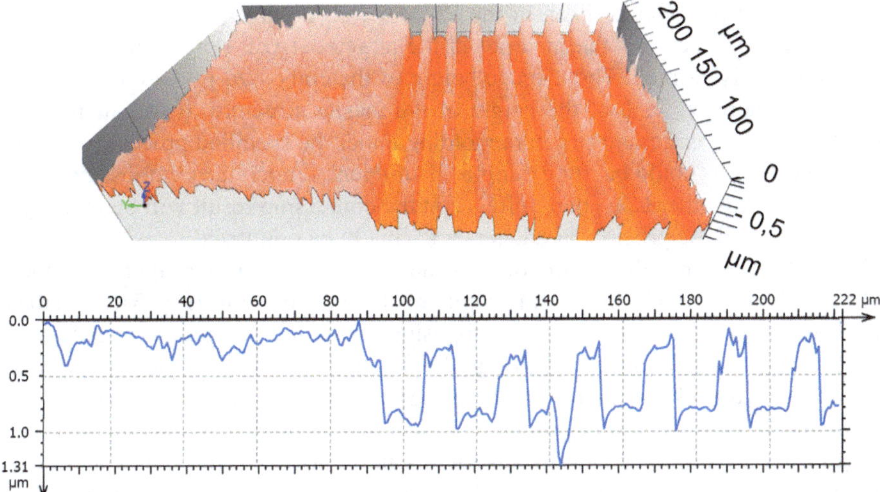

Fig. 5 3D illustration of the topography of the embossed PMMA foil (top) including height profile (bottom)

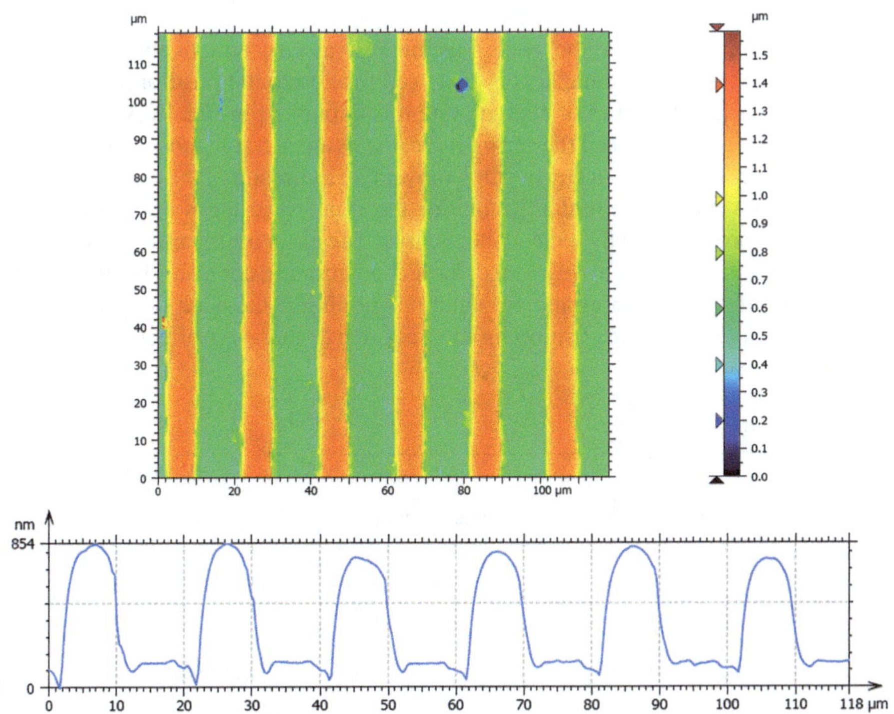

Fig. 6 Topographical illustration of the embossed OrmoCore® foil (top) including height profile (bottom)

When the impression is taken in the PMMA foil, it can be seen (Fig. 5) that the master structure penetrates the material in the targeted 1 μm range, but the structure height is not fully formed. The average height of the embossed grid structure is 0.794 μm and thus approx. 71% of the height of the master structure. It was therefore not possible to achieve complete structure molding under the specified process parameters. The reasons for this could be, for example, insufficient temperature, embossing force or holding time. In further tests, these effects on the embossing result will be investigated in more detail and the process parameters will finally be optimized.

It can also be seen at the bottom of the structure that the surface roughness of the foil can be greatly improved by embossing compared to the initial state (Fig. 5 top, on the left side). The peaks of the structure, on the opposite, still show the original roughness. This could indicate an incomplete transition to the flexible state due to a too low embossing temperature or air inclusions between the master structure and the material and also requires further detailed investigations.

Figure 6 shows that the embossed structure in OrmoCore® has a maximum height of 0.854 μm, which is approximately 76.5% of the height of the master structure. It was also not possible to achieve complete structure molding under the specified process parameters.

4 Horizontal Positioning

In order to be able to provide the system technology as cost-effectively as possible, a system of springs and electromagnets is also used for horizontal positioning, similar to vertical positioning. The setup of such a system is explained in Fig. 7. As soon as the magnetic force begins to exceed the spring force, the guide carriage moves linearly in the direction of the magnet without any twisting movement as a result of the precision guides. By means of its very fine current adjustability, the smallest changes in position can be realized, which are detected by the distance sensor. The basis for the positioning control of the horizontal axis is the already tested vertical positioning control. The difference to vertical, frictionless positioning in horizontal positioning is the need for guides. These can have a negative impact on positioning accuracy due to stick–slip effects, for example.

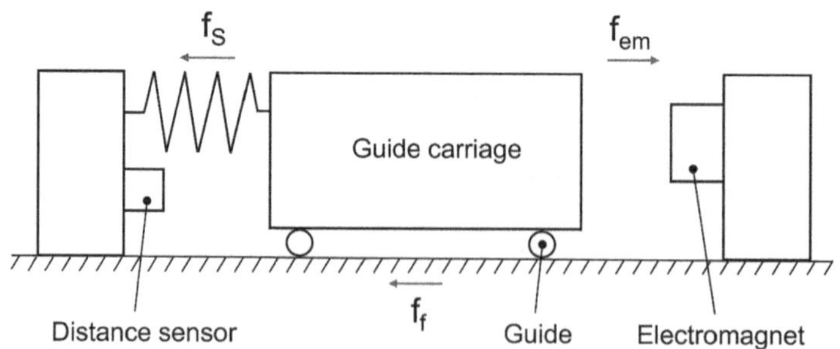

Fig. 7 Schematic structure of a horizontal positioning axis

In initial tests, positioning accuracy of the horizontal axis in the sub-micrometer range was achieved without complex adjustments to the control system. Various positions (red line in Fig. 8, right) in the sub-micrometer range were defined via the GUI and the system behavior was analyzed. The results achieved are shown in Fig. 8. Further optimizations should reduce the initial delay and minimize the overshoot. To maximize the travel range, the system is currently being expanded to include a linear axis for large positioning. This provides the basis for embossing structures on the parts at several points at a large distance from each other (>1 mm). The challenge here is to control this combination of linear axis and spring-magnet system so that positioning accuracy in the sub-micrometer range can be achieved over a positioning distance in the cm range. The linear axis is used to pre-position and hold the target position with an accuracy of 0.01 mm and the fine positioning in the range <0.01 mm is finally achieved by means of the spring-magnet system. The final aim is to combine two such units to form a cross table and thus achieve precise horizontal component positioning.

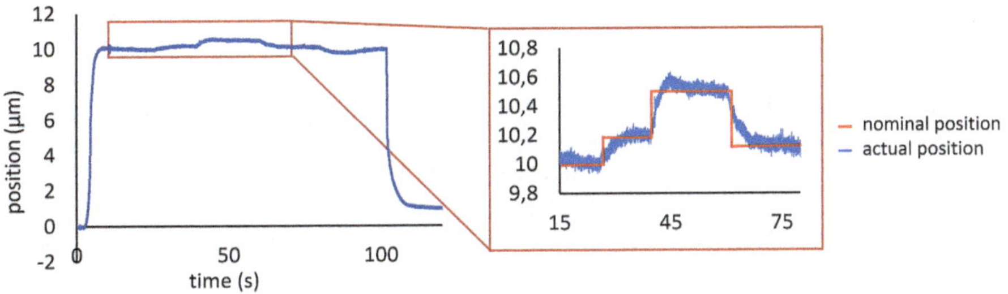

Fig. 8 Position control of a horizontal positioning axis

5 Summary and Future Work

It could be shown that the developed embossing device based on an electromagnetic spring system not only compensates for external disturbances and thus achieves a vertical positioning accuracy in the sub-micrometer range, but is also fundamentally suitable for the embossing process of the very fine grating structures in optical materials. On the basis of this simple drive principle and the control method developed, a concept was then worked out and validated in a test setup, by means of which the horizontal positioning can also be realized with the desired fine accuracy.

The complete and reproducible impression of the structures remains a major challenge. To this end, it is first necessary to identify the exact effects of the individual process parameters on the embossing quality. The expansion of the material spectrum also offers the development of new areas of application and can thus boost cost-efficient production technologies for new types of optical systems.

In addition, a large positioning in horizontal and vertical direction is to be realized in order to maximize the movement range. This step is essential for an automated embossing device for mass production, with which DOEs can be precisely transferred to integrated optical systems.

Acknowledgements. Funded by the Deutsche Forschungsgemeinschaft (DFG, German Research Foundation) under Germany's Excellence Strategy within the Cluster of Excellence PhoenixD (EXC 2122, Project ID 390833453).

Competing Interests The author(s) has no competing interests to declare that are relevant to the content of this manuscript.

References

1. Zheng, R., et al.: Injection Molding. Springer, Heidelberg (2011)
2. Wang, X. et al.: Dielectric geometric phase optical elements fabricated by femtosecond direkt laser writing in photoresists. Appl. Phys. Lett. **110**(18) (2017)
3. Deshmukh, S., Goswami, A.: Hot Embossing of polymers—a review. In: 10th International Conference of Materials Processing and Characterization, pp. 405–414. Taylor & Francis, Mathura, India (2020)
4. Velten, T. et al.: Roll-to-roll hot embossing of microstructures. In: Microsystem Technologies, Vol. 17, pp. 619–627, Springer (2011)
5. Zhu, T., Li, K., Gong, F.: Advances in hot embossing technology for optical glass micor-nanostructures: a review. In: Precision Engineering, Vol. 92, pp. 141–166, Elsevier (2025)
6. Weng, Y.-J.: Development of belt-type microstructure array flexible mold and asymmetric hot roller embossing process technology. In: Coatings, Vol. 9, Issue 4, MDPI (2019)
7. Schmiele, D., Krimm, R., Behrens, B.-A.: Embossing nanostructures. In: Production at the Leading Edge of Technology, WGP 2022, Lecture Notes in Production Engineering, pp. 307–313, Springer (2023)
8. Cui, Z.: Nanofabrication—Principles. Springer, Capabilities and Limits (2008)

Numerical Simulation and Experimental Validation of Quasi-Isothermal Die Forging for Steel-Encased Ti-6Al-4 V Billets

Jytte Möckelmann(✉), Janina Siring, Julian Bosse, Julius Peddinghaus, Hendrik Wester, Kai Brunotte, and Bernd-Arno Behrens

Institute of Forming Technology and Machines, Leibniz Universität Hannover, Garbsen, Germany
moeckelmann@ifum.uni-hannover.de

Abstract. Titanium alloys are frequently manufactured using isothermal forging approach, which is complex in both process and equipment technology. Inert gas is required during forging to prevent oxidation of the titanium. By using a steel casing of AISI 316L around the titanium billet (Ti-6Al-4 V), a quasi-isothermal process is possible without inert gas, whilst sustaining the protection against oxidation and simultaneously reducing cooling. Due to the process-dependant deformation of the steel casing, the simple die design through inversion of the desired forging part geometry no longer applies. To design a die for different casing thicknesses, with a defined titanium forging part, the finite element method was applied. The variation of the initial steel wall thickness and the ram velocities are expected to influence the cooling behaviour during the process and thus formability. This was analysed using the finite-element method. Forging experiments were executed, investigating the aforementioned parameter variations and to evaluate the agreement between the simulation and the experiments. The resulting material distribution of the titanium parts was measured optically. The material distribution of the titanium corresponded well with the simulation results. Independent of the forming speed, thinner wall thicknesses result in cracks in the casing, which can be prevented by using a thicker casing.

Keywords: Titanium · Die forging · Quasi-isotherm

1 Introduction

Titanium alloys are increasingly of interest for lightweight applications in engineering fields due to their strength-to-density ratio and high corrosion resistance [1]. Compared to steel, which has approximately double the density but similar mechanical strengths, the benefits of using titanium alloys become apparent. However, machining titanium is associated with specific limitations, as its low thermal conductivity leads to increased wear of the cutting die [2]. Additionally, the lack of recycling technologies for chips and high process costs poses challenges regarding economic and ecological aspects [3]. Therefore, the focus shifts to forging of titanium alloys. This process enables the production of near-net-shape components through forming, and by using subsequent heat treatment, the mechanical properties can be defined through microstructural changes.

L. Overmeyer and B.-A. Behrens (eds.), *Production at the Leading Edge of Technology*, Lecture Notes in Production Engineering, https://doi.org/10.1007/978-3-032-19524-1_33

Forging is significantly influenced by the forming temperature, as it determines the yield stress and the achievable true plastic strain. Additionally, in cold bulk forming, the presence of the α-phase must be considered, as it offers a more brittle hexagonal close-packed structure. At higher temperatures, a more ductile body-centred cubic β-phase is present. For example, the Ti-6Al-4 V (Ti-64) alloy has a phase transformation temperature of $T_\beta = 995$ °C [2]. Depending on the true plastic strain and the specific forming conditions, different microstructures result from the forging of Ti-64 [4]. Titanium's high reactivity with gases such as oxygen, nitrogen, and hydrogen under conventional forging conditions diminishes the advantages of forming at elevated temperatures. This is because the reaction zone decreases ductility and consequently causes cracks to form in the forged component. The so-called α-case shows crack-prone layers at the surface, as the more brittle α-phase is particularly stabilised by oxygen. Therefore, industrial allowances are factored in, which must be removed through post-processing.

In the conventional forging process, the billet cools down due to contact with the die, which prevents maintaining the necessary temperature window for forging at the surface [5]. Isothermal forging addresses this issue by aligning the die temperature with that of the billet and maintaining it throughout the process. This ensures an even temperature distribution during forging, allowing for more precision, reduced rework, and thus less material loss. This process requires dies to withstand high thermal and mechanical stresses. Consequently, high-temperature-resistant materials such as molybdenum or nickel-based alloys are used, as the dies may be heated to temperatures up to 1,200 °C. Additionally, to minimise sublimation losses and prevent the reaction of titanium or tooling material with oxygen, some procedures use protective gas atmospheres [6]. Overall, the temperature control of the dies also requires additional energy.

Zhang et al. explored how encasing a titanium aluminide billet with different stainless steel foils and glass layers influences its temperature behaviour [7]. They discovered that this technique could halve the cooling rate. The removal of the foils is not discussed in detail. In another approach, Careau et al. examined the forging of steel-encasing Ti-64-powder in a direct powder forging approach [8]. During this process, an intermetallic phase formed between the titanium and the steel capsule made from AISI 316L, which was used to easily remove the encasing due to its brittleness. The discussed studies do not integrate encapsulation into the process design and do not generate complex geometries. The simulation of the deformation of titanium alloys within steel capsules is applied in the processing of powdered titanium alloys in the context of hot isostatic pressing (HIP) [9]. This process primarily considers the densification of the powder particles. Additionally, finite element investigations by Lin et al. have been conducted, in which a TiAl alloy within a steel capsule was intended to improve formability [10]. It was shown that during upsetting, the capsule and the TiAl separate in the middle area.

The encapsulation of Ti-64 billets in an AISI 316L capsule (according to [8]) aims to ensure the necessary temperature window for forming titanium while preventing contact with reactive gases. The encapsulation enables a quasi-isothermal forming process, which eliminates the need for the complex process and equipment technology associated with fully isothermal processes. This approach thus reduces material usage compared to machining and also minimises the necessary machining operations. Typically, there is a predetermined geometry for the finished component that must be adhered to in order to fulfil its functional requirements. In the process of forming steel-encapsulated titanium billets, achieving the die geometry cannot be done through simple inversion of the desired component. In the design of the dies, the steel capsule must be taken into account, as it significantly influences the material flow of the titanium. Therefore, the process is initially designed numerically for the different capsule thicknesses. The iterative adjustment of die dimensions based on the simulated material flow allows for precise process design targeting the desired geometry of the titanium component, despite the encapsulation. Additionally, simulations of the temperature distribution ensure the quasi-isothermal process window of the titanium. Then the process concept was implemented to demonstrate and evaluate the experimental feasibility. Finally, the simulation is validated using the experimental data. Overall, the aim is to develop a method that utilises numerical process design to enable the quasi-isothermal forging of steel-encapsulated titanium billets into desired target geometries.

2 Material and Methods

2.1 Simulation Model

The simulations were carried out using the software Simufact Forming v16.0. The used numerical model is illustrated in Fig. 1d. The dies were modelled as rigid bodies with heat conduction. Material data used in the numerical model such as temperature and strain rate dependent flow curves and CTE for both materials (Ti-64 and AISI 316L) were experimental determined as well as verified in [11]. The simulations were carried out in 2D using rotational symmetry. The steel encased titanium billet was set to a temperature of 1,100 °C. Due to the larger thermal expansion coefficient of Ti-64, there is no interference fit at the forging temperature. Previous studies have demonstrated the presence of an interlayer between titanium and steel after forging when an interference fit was used [12]. This interlayer was predominantly found in the outer edge region of the sleeve, whereas it was less pronounced at the lids. As a result, complex friction and heat transfer coefficients arise, which are neglected within the scope of these investigations. The contact between steel and the dies was modelled through a friction coefficient μ of 0.15 and a friction factor m of 0.3 within a combined friction model of Coulomb and Tresca [13]. Based on the information of Hu et al. [14] the friction was modelled for the contact between titanium and steel with a friction factor m of 0.3. For process design, the heat transfer coefficient was set to a constant value based on the results in [14]. The shell surface (steel wall) was varied between 2 mm (forming path 27 mm) and 4 mm (forming path 28 mm). The press speed was simulated at 20 mm/s and 30 mm/s.

The aim was to produce the same reproducible geometries of the formed titanium product with varying the steel wall thickness, and dies adjusted according to the steel wall thickness used. The aim of the numerical investigations was, on one hand, to achieve a continuous steel encasing to protect the titanium against ambient gases and excessive cooling. During the forming process, the titanium should be kept above the β-transus temperature to prevent phase transformations.

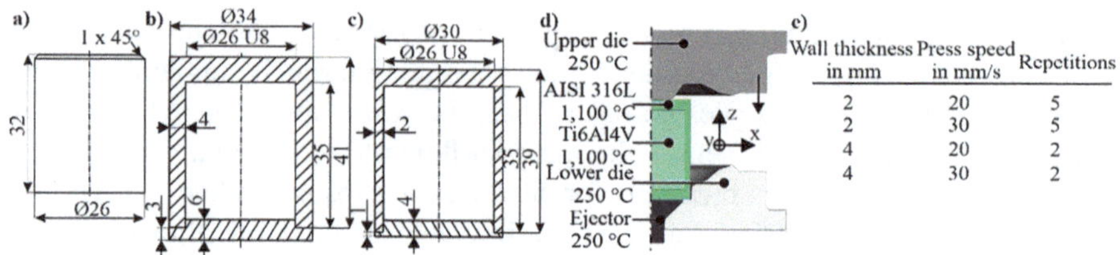

Fig. 1 All dimension in mm. **a** Ti-6Al-4 V billet; Capsule with **b** 4 mm and **c** 2 mm wall thickness; **d** Simulation model; **e** Varied parameters

2.2 Experimental Setup

The wall thickness was varied (2 and 4 mm) to investigate its influence on material flow and die filling while maintaining isothermal process conditions. At the top and bottom area, the thickness was constructed thicker to pretend the titanium against excessive cooling and cracks, which can lead to contact with ambient gases. Dies for different capsule wall thicknesses made from AISI 316L stainless steel are designed numerically to achieve the same component geometry of the Ti-64 core through die forging. The manufacturing of the capsules, cores, and lids followed the specifications in Fig. 1a–c.

Initially the capsules were preheated to 350 °C and afterwards the titanium core was inserted and lid closed by using a hydraulic handpress PJ20H (AC Hydraulic a/S, Viborg, Denmark). The present interference fit between the capsule and the lid ensures an airtight sealing of the titanium core. For the forging experiments, the hydraulic press Schirmer + Plate is used, with the dies made from X38CrMoV5-3 (AISI H11) (Fig. 2a and b) preheated to 250 °C. The forces occurring during the process are recorded using a force transducer, which has been calibrated by the institute between 0 and 500 kN. The sampling rate was set to 500 Hz. Press speed is determined using an external displacement sensor. The encapsulated titanium billets are heated in a convection oven to 1,150 °C for 45 min to ensure a homogeneous forming temperature of 1,100 °C upon transfer to the press. The application of the lubricant CON TRAER G300 by Fuchs to the dies takes place manually. In the experiments, both the wall thickness of the capsule and the forming speed were varied (Fig. 1e). The forged parts are cooled at room temperature. The titanium cores, which were forged within capsules with a wall thickness of 2 mm, are subsequently exemplary removed from the capsules using a conventional lathe. In terms of the experimental results, the capsules with a wall thickness of 2 mm are primarily discussed, as they proved to be the most effective during the experimental procedure. Dimensional measurements of the titanium cores are carried out with a 3D scanning system GOM ATOS Q and compared to the numerical results using Cloud Compare.

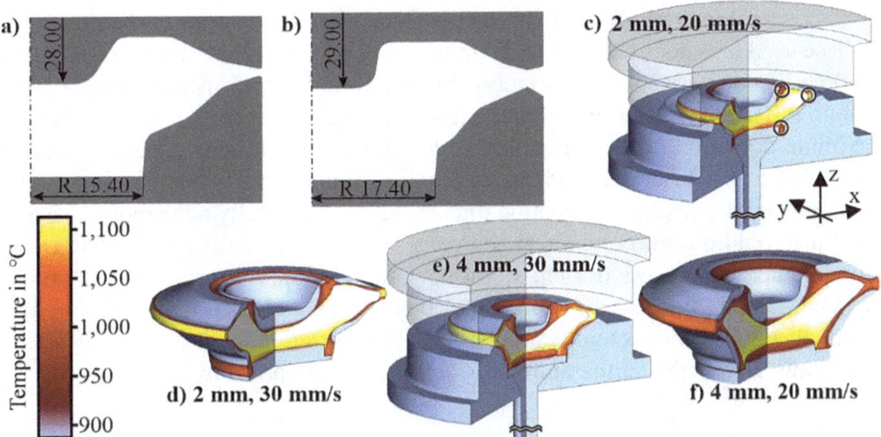

Fig. 2 Dies for a wall thickness of **a** 2 mm and **b** 4 mm; Simulation results of the temperature distribution at an initial temperature of 1,100 °C, with varying steel wall thicknesses (2 mm (**c**, **d**), 4 mm (**e**, **f**)) and press speed (20 mm/s (**c**, **f**), 30 mm/s (**d**, **e**)) at the end of the forging process

3 Results and Discussion

3.1 Numerical Results

The dies for different steel wall thicknesses were iteratively designed through simulations with the aim of achieving the same final titanium geometry. The final dies are shown in Fig. 2a and b. It is noteworthy that the die for the thicker steel wall thickness features an extended mandrel and a generally wider impression in order to produce the same titanium billet. The simulation results of the temperature distribution are presented in Fig. 2c–f. The titanium does not cool below 995 °C in any of the variants and that the steel shows a stronger cooling. Increasing the thickness of the steel affects its temperature and therefore material flow behaviour in the forging process. This was already demonstrated in [11]. For instance, with the slower speed and thinner steel, the temperature cools down to below 900 °C in the steel, with only certain areas along the radius reaching higher temperatures between 900 °C and 950 °C (black circles Fig. 2c). In contrast, with the greater wall thickness, a temperature reduction below 900 °C is calculated only in the upper and lower areas that have prolonged contact time with the dies. The temperature change is influenced by several factors, including the energy input from forming and friction, variations in flow properties, changes in heat transfer to the titanium, and the contact time with the cooler dies. Finally, it is important to emphasize that all variants do not exhibit excessive steel thinning. However, no definitive statement can be made due to the initial omission of a damage model. Overall, the results indicate the effective limitation of titanium cooling, maintaining temperatures above 995 °C during forming.

To evaluate the influence of process parameters such as press speed and steel thickness, the evaluation compared the geometries of numerically simulated formed titanium products based on the simulation results (Fig. 3). In Fig. 3a, the geometric comparison with variations in press speed is shown. Only deviations of less than 1 mm can be observed in the central and flash areas (black circles). In the other areas a deviation of

±0.1 mm is visible. Therefore, it can be concluded that a variation in the press speed with these used dies does not lead to deviations in the formed titanium part. In Fig. 3b), the geometric comparison with variations in steel wall thickness is presented. A deviation of up to 1.1 mm was observed in the mandrel area (black circle). In the other areas a maximum deviation of ±0.5 mm is visible. The numerical results demonstrate that it is possible to produce nearly identical titanium components when varying both the press speed and the steel encapsulation thickness. By specifically adjusting the dies and thus the material flow, the component geometry can be consistently maintained. Despite iterative numerical process design, slight deviations of appr. 1 mm are still expected for varying casing thicknesses. Minor variations in pressing speed have little impact on the formed components, as temperature plays a far more dominant role, partly because speed mainly affects the temperature via contact time, as shown in [11].

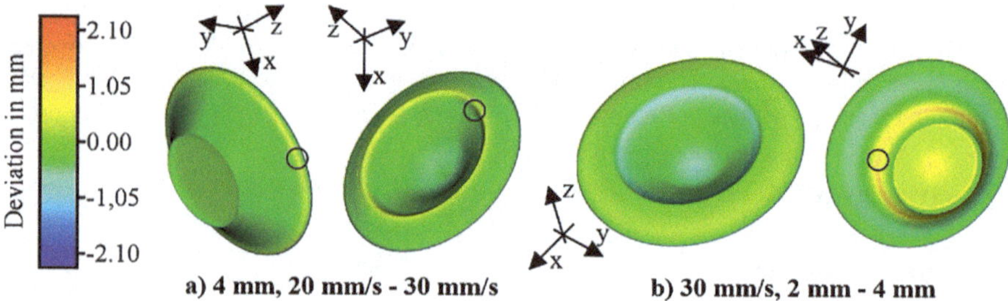

Fig. 3 Exemplary geometry comparison of numerically calculated formed titanium parts (initial temperature 1,100 °C) with **a** a variation of the press speed and **b** a variation of the steel wall thickness

3.2 Experimental Results

First, the encapsulated components are visually inspected (Fig. 4a–d). During the forming of the encapsulated titanium billets with a wall thickness of 2 mm, cracks occur in the steel capsule regardless of the press speed (Fig. 4a and b, red arrows). The cracks mainly appear on the top surface of the capsule with a tendency towards the flash area.

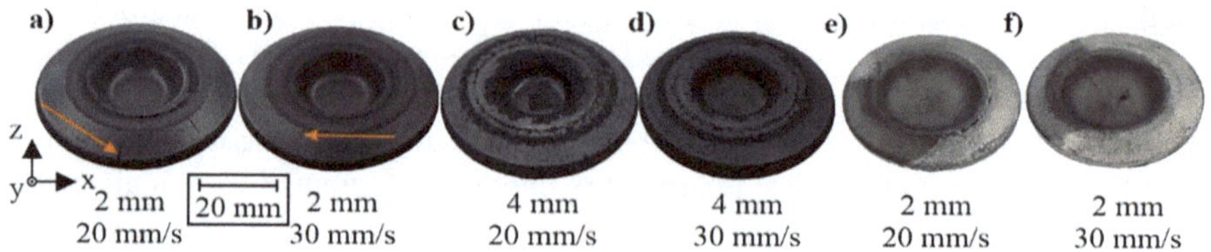

Fig. 4. **a–d** Encapsulated components with varying wall thickness and forming speed **e** top side of the titanium core with a forming speed of 20 mm/s and **f** with a forming speed of 30 mm/s

Reduced temperatures at that location could explain this phenomenon (Fig. 2c and d) and the occurring stresses. In contrast, the capsules with a wall thickness of 4 mm do not show capsule failure. This can be attributed to the higher temperatures of the capsule during forging compared to the thinner walled sample due to slower cooling (Fig. 2c and d). At the same time, increased scale formation is observed on the 4 mm capsule. One possible reason could be humid air in the furnace, this must be investigated in further investigations. Discolorations are visible on the decapsulated titanium cores regardless of the forming speed (Fig. 4e and f). These effects may result from a reaction with oxygen [15], which could have entered the capsule through the lid joint during heating or penetrated through cracks formed during deformation. To verify whether oxygen reaches the titanium through the cracks, the samples could be sectioned at the crack locations and examined for additional α-phase and oxide layers. Therefore, the capsule concept should be further adjusted.

3.3 Validation of the Simulation

To validate the simulation, the forces applied during the forming process are compared with the numerical calculated forces (Fig. 5). At a press speed of 30 mm/s, the comparison between the experimental results and the simulation shows a good agreement (Fig. 5a and b). The experimentally determined maximum force of 373.28 ± 11.59 kN deviates by approximately 6% compared to the simulation (353.74 kN). The standard deviation is determined from the five recorded maximum force values. Small deviations between experiment and simulation can be attributed to the elastic behaviour of the dies, which was not accounted for in the simulation. Due to the partial formation of the interlayer between the titanium and the steel capsule, local variations in friction parameters and heat transfer parameters will result. This was not considered in the simulation and could lead to discrepancies between the simulation and the experimental results. Consequently, there is a good agreement between the simulation and the experimental results. The minor discrepancies can be attributed to variations present in practical conditions.

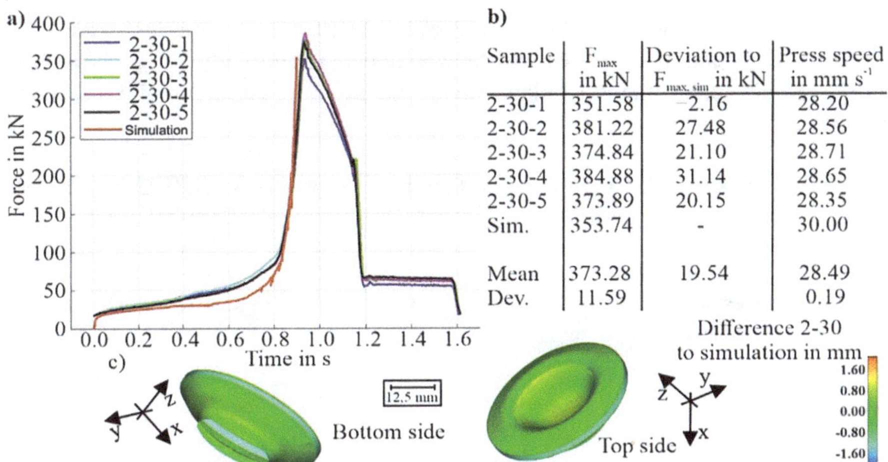

Sample	F_{max} in kN	Deviation to $F_{max,sim}$ in kN	Press speed in mm s^{-1}
2-30-1	351.58	−2.16	28.20
2-30-2	381.22	27.48	28.56
2-30-3	374.84	21.10	28.71
2-30-4	384.88	31.14	28.65
2-30-5	373.89	20.15	28.35
Sim.	353.74	-	30.00
Mean	373.28	19.54	28.49
Dev.	11.59		0.19

Fig. 5. **a** Force–time diagram for a wall thickness of 2 mm at a press speed of 30 mm/s **b** values for 2–30; **c** Comparison of numerically calculated titanium products to the formed titanium core

The comparison of the simulated titanium core with the 3D-scan of encapsulated sample 2–30 reveals deviations ranging from −1.40 mm to 1.20 mm (Fig. 5c). The negative deviations of the decapsulated titanium cores compared to the simulation in the flash area can be attributed to the decapsulation process, during which parts of the titanium were also removed. In further experiments, additional decapsulation strategies and the use of lubricants will be examined in more detail. The positive deviations are primarily due to irregularities in the titanium surface. Overall, there is a clear correlation between the simulated and the forged titanium core. Using simulation-based design of the dies for the process of encapsulated titanium billets successfully produces the components that were designed through simulation.

4 Conclusion and Outlook

In the present study, dies were designed for different steel wall thicknesses that lead to the same titanium geometry by varying the steel wall thickness and pressing speed. The temperature distribution was simulated to ensure the quasi-isothermal condition of the titanium. A numerical comparison of the simulated titanium cores with varying wall thickness and press speed shows overall reproducible titanium cores. Although the experiment and simulation show good agreement, the results for local inhomogeneous temperature distribution and local material flow in particular are based on assumptions and literature data on heat transfer and friction modelling. To further improve the precision of the simulation model regarding temperature distribution experiments are planned to accurately characterise the pressure-dependent heat transfer coefficient as well as friction coefficients. Experimentally, slight cracks are observed in the steel casing with a wall thickness of 2 mm, whereas no cracks occur with a wall thickness of 4 mm. The force–time curves of the experiments and the simulation as well as the comparison of final part geometry of the titanium show a good agreement. In further experiments, the capsule-lid system will be adjusted to prevent the reaction of titanium with oxygen and other decapsulation strategies will be evaluated. Additionally, the process design for steel-encapsulated titanium billets can be extended to more complex component geometries.

Acknowledgements. Funded by the Deutsche Forschungsgemeinschaft (DFG, German Research Foundation)—project number 461918196.

Competing Interests. The author(s) has no competing interests to declare that are relevant to the content of this manuscript.

References

1. Babaremu, K.O., Jen, T.-C., Oladijo, P.O., et al.: Mechanical, corrosion resistance properties and various applications of titanium and its alloys: a review. RCMA **32**, 11–16 (2022)
2. Leyens, C., Peters, M.: Titanium and titanium alloys. Fundamentals and applications. Wiley, Weinheim (2003)

3. Shial, S.R., Masanta, M., Chaira, D.: Recycling of waste Ti machining chips by planetary milling: Generation of Ti powder and development of in situ TiC reinforced Ti-TiC composite powder mixture. Powder Technol. **329**, 232–240 (2018)
4. Bambach, M.D., Seifert, D., Sizova, I.: Intensive forming of grade 5 titanium bars with increased performance for aerospace applications. Procedia Manuf. **47**, 288–294 (2020)
5. Semiatin, S.L., Seetharaman, V., Weiss, I.: The thermomechanical processing of alpha/beta titanium alloys. JOM **49**(6), 33–39 (1997)
6. Klocke, F.: Fertigungsverfahren 4. Springer, Berlin Heidelberg (2017)
7. Zhang, J., Kutzsche, A., Rosenberg, K. et al.: A novel canning technology for forging of gamma-TiAl alloys. MSF, 1421–1426, (2007)
8. Germain Careau, S., Tougas, B., Ulate-Kolitsky, E.: Effect of direct powder forging process on the mechanical properties and microstructural of Ti-6Al-4V ELI. Materials. **14**, (2021)
9. Cui, J., Lv, X., Fu, H.: numerical simulation and hot isostatic pressing technology of powder titanium alloys: A review. Metals, **15**, (2025)
10. Lin, H.Z., Li, S.S., Su, X. et al.: Finite element simulation of canned forging on TiAl alloy containing high Nb. MSF, 813–816 (2005)
11. Siring, J., Brahimi, D., Möckelmann, J., et al.: Numerical process design for hot forging of steel encased titanium workpieces. Material Forming: ESAFORM **54**, 859–867 (2025)
12. Möckelmann, J., Siring, J., Peddinhaus, J. et al.: Isothermes und sauerstofffreies Umformen von Titan. Wt Werkstattstech. Online. **114**, 578–587 (2024)
13. Siring, J., Heine, C., Till, M. et al.: Numerical process design for the production of a hybrid die made of tool steel X38CrMoV5.3 and inconel 718, MRP. **41**, 812–821 (2024)
14. Hu, Z.M., Brooks, J.W., Dean, T.A.: The interfacial heat transfer cofficient in hot die forging of titanium alloy. Proc. Inst. Mech. Eng. C **212**(6), 485–496 (1998)
15. Sun, J., Lu, H., Wang, Z. et al.: High-temperature oxidation behaviour of Ti65 titanium alloy fabricated by laser direct energy deposition. Corr. Sci. **229**, (2024)

Technological Investigations Into Single-Lip Deep Hole Drilling with Large Diameters

Lucas Brause(✉), Sebastian Michel, Mathis Commandeur, and Dirk Biermann

Institute of Machining Technology, TU Dortmund University, Dortmund, Germany
lucas.brause@tu-dortmund.de

Abstract. Single-lip deep hole drilling is a manufacturing process used for the production of deep holes with a length-to-diameter ratio of $l/D \geq 10$ in the diameter range of $D = 0.5 \ldots 80$ mm. It is used, for example, in the production of gear shafts or for cooling bores in the aerospace industry. Single-lip deep hole drilling tools have an asymmetrical tool design. The resulting radial forces at the cutting edges are supported by guide pads on the bore hole surface, which leads to a leveling of feed groves and a high surface quality. In order to optimize the performance and durability of the cutting tool and also ensure the quality of the machined components, it is essential to analyze the interaction between tool, workpiece, and machine. A central aspect of the process analysis is the investigation of the mechanical loads and the dynamic process characteristics caused by the long and slender tool structure and their effects on the workpiece and the cutting tool. In the shown analyses of a single-lip deep hole drill with a diameter of $D = 35$ mm and indexable cutting inserts, the influence of the process parameters on the process forces and the drilling torque in AISI 4140+QT is investigated. In addition, the effect of steady rests as a tool support on the dynamic tool behavior as well as the resulting bore hole quality is analyzed.

Keywords: Single-lip deep hole drilling · Mechanical load · Steady rest · Surface roughness

1 Introduction

In many industrial sectors, components are produced that require the use of deep hole drilling technologies. These technologies usually need special tools and machines and are designed to make high length-to-diameter ratios possible. A representative process is single-lip deep hole drilling. Typical industries in which this manufacturing process is applied are the automotive industry, aerospace industry, and medical technology. Representative components in which these tools are used include crankshafts, camshafts, landing gear components, surgical instruments, and implants [1].

L. Overmeyer and B.-A. Behrens (eds.), *Production at the Leading Edge of Technology*, Lecture Notes in Production Engineering, https://doi.org/10.1007/978-3-032-19524-1_34

Due to the guidance of the tool, single-lip deep hole drilling offers low straightness deviations and high surface qualities. In addition, the deep hole drilling process achieves a high material removal rate. Because of the high surface quality achieved, subsequent machining processes do not have to be carried out, thus reducing time and costs [1–3]. This leads to increased demands on the deep hole drilling process. Therefore, an advanced understanding of the process is necessary for reliable process execution. This is achieved by investigating the process characteristics, such as the mechanical loads and the oscillation behavior of the tool system, which have an influence on the drilling process, the tool, and the machined component [1,3]. In the case of high tool length, support systems are essential elements to reduce bending of the tool and contribute to process stability. Therefore, an understanding of their influence on the deep hole drilling process is also of interest [4].

2 Single-Lip Deep Hole Drilling

According to the VDI guideline 3210 a drilling process is called deep hole drilling if the length-to-diameter ratio exceeds $l/D \geq 10$. Deep hole drilling processes combine several advantages and offer very high cutting performance, ideal cooling and lubrication conditions, and short machining times. It also enables high bore hole qualities in terms of diameter tolerance, surface quality, and geometric shape accuracy, as well as a low straightness deviation [3]. As the asymmetrical cutting edge arrangement leads to radial force components during the drilling process, guide pads are used. These guide pads transfer the process forces to the bore hole wall, which supports and guides the tool in the created bore hole. Another resulting effect is the leveling of the feed grooves, created at the cutting inserts, which leads to a high surface quality [3,5]. Figure 1 schematically shows the mechanical loads and the setup of a single-lip deep hole drilling process.

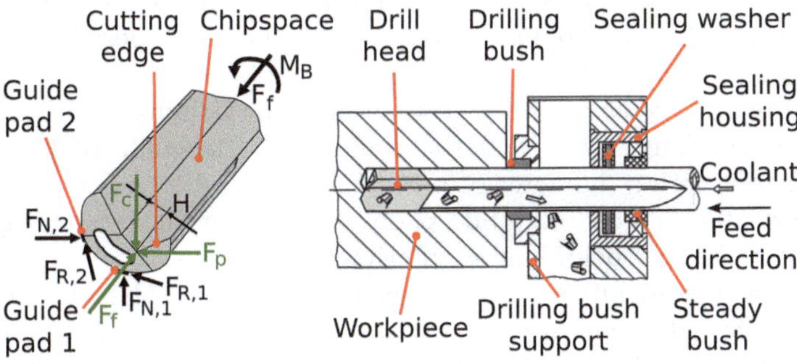

Fig. 1. Mechanical loads and setup for single-lip deep hole drilling according to [3,6]

Mechanical loads occurring during the deep hole drilling process mainly impact the contact areas of the cutting inserts and guide pads of the single-lip drilling tool. The machine spindle generates the torque that is primarily required for the cutting process. The resulting cutting force is spatial and is idealized at a point on the cutting edge that is located outside the rotation axis of the tool. It is composed out of three orthogonally aligned force components: cutting force, feed force, and passive force. Taking a detailed analysis, the cutting force is divided into individual components that impact on the inner and outer main cutting edge as well as the secondary cutting edge. Due to the characteristic asymmetrical tool design, the cutting and passive forces generate corresponding radial forces, which are transferred to the bore hole as normal and frictional forces via the guide pads. The drilling torque corresponds to the sum of the cutting force and frictional torques. In axial direction, the mechanical load on the tool is composed out of the feed force and the cooling lubricant force [1,6,7]. The process parameters, the cooling lubricant concept, the tool design, the cutting edge design, and the workpiece material are considered to be significant influencing variables on the level and distribution of forces and torques that occur during single-lip deep hole drilling [6].

As deep hole drilling tools are generally characterized by a high length-to-diameter ratio, they are particularly susceptible to vibrations. Furthermore, the grooved shank creates a weak point of the tool in terms of tool stiffness due to its torsionally weak shape [3]. Torsional vibrations represent an oscillation of the tool along its own longitudinal axis. Torsional oscillation occurs in the first eigenfrequency or a multiple of the eigenfrequency. They can be heard as chattering during the drilling process. The oscillation also increases tool wear and often affects the quality of the bore hole surface [3,8].

3 Experimental Setup

The tibo KTE40-1000 deep hole drilling machine is used for the experiments. The investigated processes are executed in AISI 4140+QT. Many of the mechanically and dynamically highly stressed, safety-relevant components produced in the aforementioned industries are made from quenched and tempered steels. Figure 2 shows the experimental setup.

To execute the technological investigations, the drilling tool Type 01 with indexable inserts from botek and the cooling oil Blasomill 10 DM are used. The drilling tool with a diameter of $D = 35\,\mathrm{mm}$ has a botek BX coated carbide cutting insert and TiN-coated guide pads. New cutting inserts and guide pads are inserted in the tool before each experiment.

In the deep hole drilling process, steady rests are used to ensure stability and precision. Steady rests are support systems that guide the tool during the drilling process. This is particularly necessary for widely protruding tools, as these tend to deflect due to their asymmetrical structure, which increases the risk of straightness deviation. Support systems also enable higher feeds and extend the service life of the tool [2–4]. Figure 2 shows the two steady rest systems used in the experiments. The steady rest systems are slidable mounted on the guide

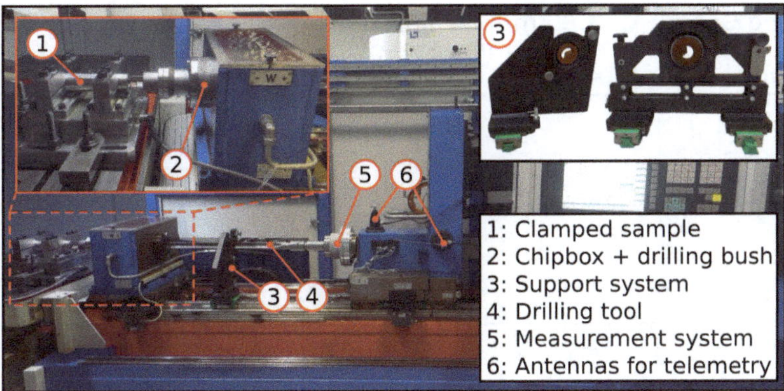

Fig. 2. Experimental setup on the deep hole drilling machine

rails of the machine base. The left of the steady rest systems shown runs one-sided on a guide rail, while the other system is supported by guide rails on both sides.

To record the mechanical process characteristics, a self-designed measuring sleeve is used, which is mounted between the tool and machine spindle; thus it rotates during the process. Strain gauges with a resistance of $\mathrm{R} = 350\ \Omega$ are built into the system with an orientation of $0°$ and $180°$ on the circumference of the sleeve to record axial force and an orientation of $0°$ and $90°$ to record torque due to the deformation of the measuring sleeve. The measured strains are sent to receivers using the internal installed telemetry system Datatel dt12-10DC. The system uses a sampling frequency of $\mathrm{f_s} = 10$ kHz. The measuring range can be adjusted with electrical resistors. The feed force and the bore moment are calculated out of the strains after a calibration of the system.

4 Mechanical Loads, Processdynamics and Roughness

Feed force and bore moment characteristics of a deep hole drilling process are shown in Fig. 3. The experiment presented as an example is executed with the one-sided steady rest, a feed of $\mathrm{f} = 0.15\,\mathrm{mm}$ and a cutting speed of $\mathrm{v_c} = 85\,\mathrm{m/min}$. To obtain comparability, the measured forces are additionally Savitzky-Golay filtered with a third order polynom and a window size of $\mathrm{s_w} = 50$.

The stationary drilling process range $\mathrm{t} \approx 22 \ldots 95\,\mathrm{s}$ is used to compare the feed forces and bore moments occurring under variation of the process parameters. The feed, the cutting speed, and the steady rest are varied. Figure 4 shows the mean of the measured feed forces and bore moments with standard deviation.

In principle, it can be observed that feed forces between $\overline{\mathrm{F_f}} = 4.8 \ldots 6$ kN are measured. The feed force increases as the feed increases. The measured bore moments are in a range between $\overline{\mathrm{M_B}} = 60 \ldots 110$ Nm. An increase in feed also results in an increase in bore moment. Since a higher feed causes greater cutting depth, more shearing work is performed on the cutting edge [5]. Increasing

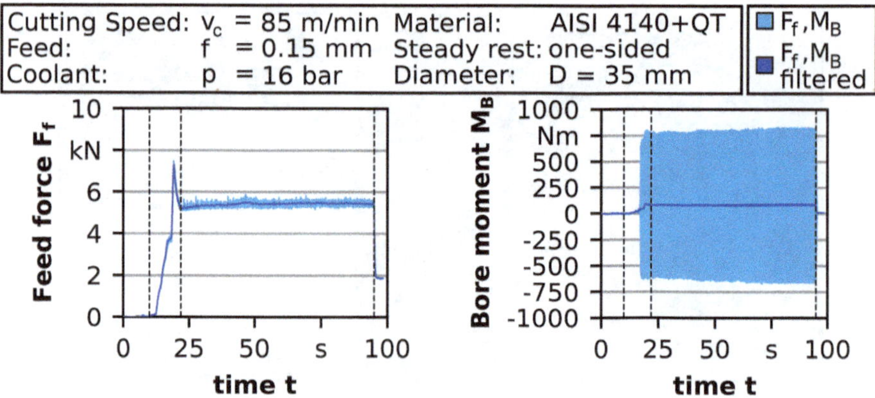

Fig. 3. Typical feed force and bore moment characteristics

the cutting speed has a comparatively small influence on the mechanical process characteristics and leads to a minimal increase of the mechanical load. In addition, the steady rest has no significant influence on the feed force and the bore moment. The standard deviation in the bore moments in Fig. 4 and the bore moment in Fig. 3 provide information on the oscillation behavior of the tool during the deep hole drilling process. Significant torsional oscillations can be recognized during the process. However, no correlation between the oscillation behavior of the tool during the drilling process and the feed, cutting speed, and support can be determined. Nevertheless, it is noticeable that two oscillation levels occur, and some drilling processes oscillate with a higher amplitude. For further analysis, spectrograms of the bore moment of two experiments are presented in Fig. 5 as example. For this purpose, the measurement of the bore moment is transformed into the frequency band.

The left spectrogram shows the torsional oscillation at a cutting speed of $v_c = 75\,\text{m/min}$ and a feed of $f = 0.17\,\text{mm}$. It is evident that the drilling tool oscillates without steady rest at a frequency of $\nu = 494\,\text{Hz}$. The spectrogram on the right shows the torsional oscillation at a cutting speed of $v_c = 85\,\text{m/min}$ and a feed of $f = 0.15\,\text{mm}$. The drilling tool oscillates with the one-sided steady rest at a frequency of $\nu = 659\,\text{Hz}$.

Considering the dominant frequencies as a function of the process parameters and the tool support, it is noticeable that two frequency levels occur. The occurring dominant torsional frequencies are $\nu_1 \approx 500\,\text{Hz}$ and $\nu_2 \approx 660\,\text{Hz}$. According to *Streicher*, the first torsional eigenfrequency of the system is usually energized in single-lip deep hole drilling processes [8]. Comparing the dominant torsional frequencies with the standard deviations of the bore moments indicates correlations. A lower standard deviation in the bore moment is associated with a lower dominant torsional frequency. A higher standard deviation in the bore moment indicates a higher dominant torsional frequency.

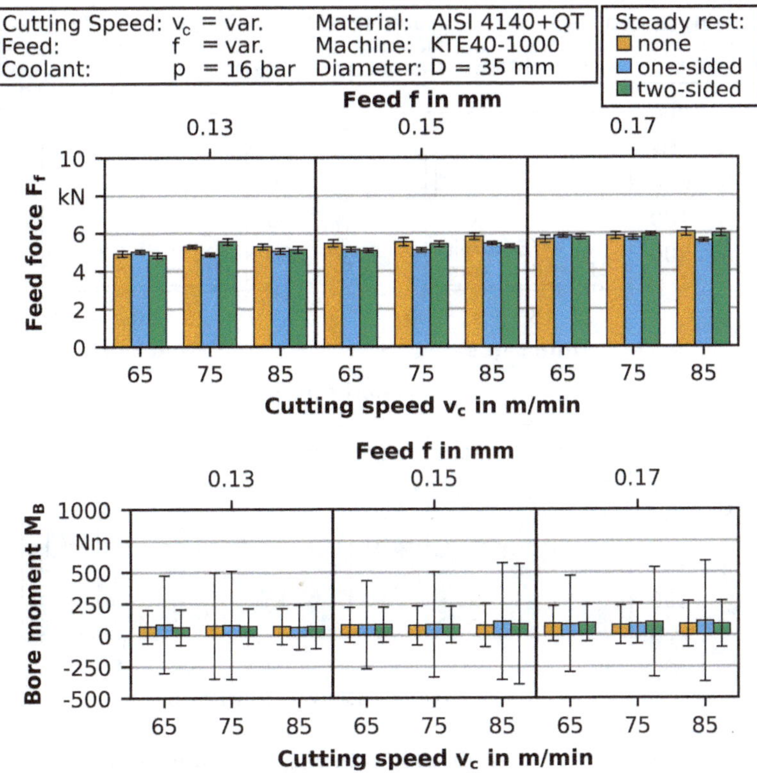

Fig. 4. Feed force and bore moment in the deep hole drilling processes

Also, the achieved surface quality of the bore hole is analyzed under variation of the process parameters and the tool support. Figure 6 shows the roughness characteristics determined in the stationary process range at three measuring points evenly distributed around the bore hole measured with the tactile system MarSurf XR20. For a feed of $f = 0.13\,\text{mm}$, the measurements are executed with measuring sections of $\lambda_c = 0.25\,\text{mm}$ and a measuring length of $l_n = 1.25\,\text{mm}$. To examine the samples drilled with higher feed, $\lambda_c = 0.8\,\text{mm}$ and $l_n = 4\,\text{mm}$ are used.

The steady rest has an influence on the surface quality of the bore hole. In particular, processes without steady rest show higher roughness with increasing cutting speed compared to processes with steady rest. If a steady rest is used, the process parameters have no significant influence on the roughness. In general, the surface quality is very high compared to typical values ($R_z = 5\ldots15\,\mu\text{m}$) for similar single-lip deep hole drilling processes according to VDI 3210 [3].

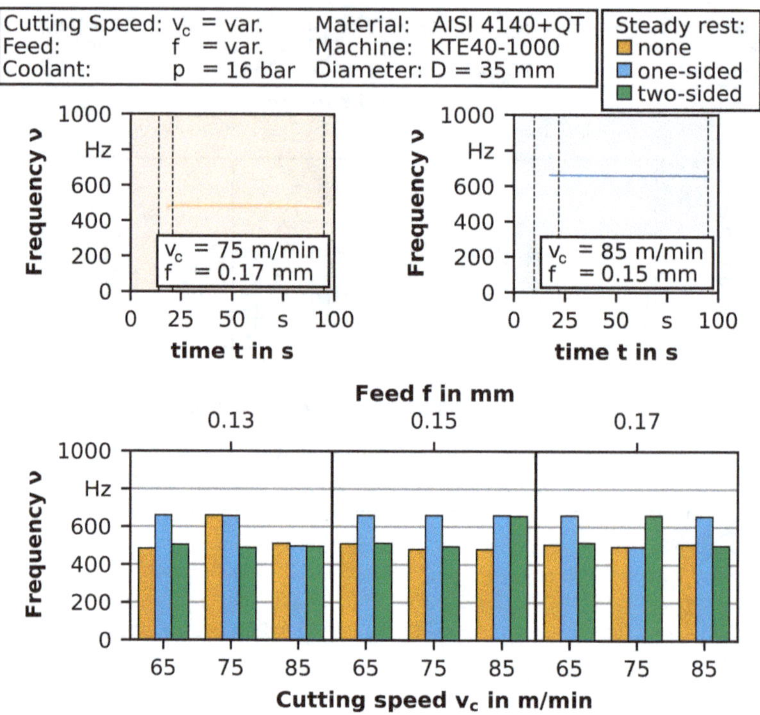

Fig. 5. Oscillation in the single-lip deep hole drilling process

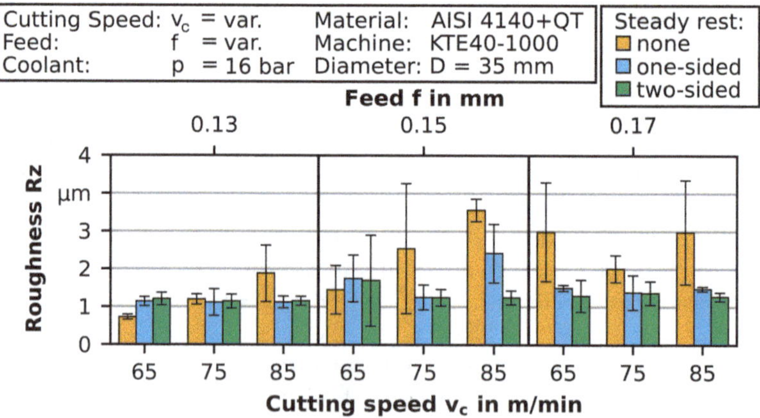

Fig. 6. Roughness of the bore hole surface

5 Conclusion and Outlook

The desire for optimized components and an increase in the performance of the single-lip deep hole drilling process with large diameters needs an enhanced understanding of the process. The investigations presented examine steady rests concerning their influence on the deep hole drilling process. Basic information is also being gathered to develop and design novel sensor systems for the process. The results can be applied in the manufacturing of various components.

During the experiments, variations in process parameters and support systems were made. It can be noted that the examined drilling processes are subjected to oscillations. In analyzing the process oscillations, two dominant frequency levels were identified. It is also noteworthy that the achieved surface roughness with used steady-rests is comparatively low. The influence of tool support on the generated surface quality is particularly evident.

Building on these results, further investigations are of interest. For example, investigations into the relationships between process parameters and dominant torsional frequencies could be made. Additionally, influences from tool bending can be examined more closely. More detailed surface analyses can also be performed, including examinations of roughness profiles. Furthermore, regarding wear studies, examining temperatures generated at the indexable inserts are of interest as they may have an influence on the machined component. The temperatures could be investigated with thin-film sensors, which are integrated into the guide pads. Another aspect to investigate with thin-film sensors are the normal forces on the guide pads. The generated information can provide details about the influence of normal forces and temperatures on the surface integrity of the bore hole [9].

Acknowledgments. The project is funded by the Deutsche Forschungsgemeinschaft (DFG, German Research Foundation) Project number 500498267 and the Fraunhofer Gesellschaft. The authors would also like to thank the project partners Fraunhofer Institute for Surface Engineering and Thin Films (IST), BGTB GmbH, TIBO Tiefbohrtechnik GmbH and botek Präzisionsbohrtechnik GmbH.

References

1. Biermann, D., Bleicher, F., Heisel, U., Klocke, F., Möhring, H.-C., Shih, A.: Deep hole drilling. In: CIRP Annals–Manufacturing Technology, vol. 67, pp. 673-694 (2018)
2. VDI 3208: Deep hole boring with gun drills. Beuth, Berlin (2014)
3. VDI 3210: Deep-hole drilling. Beuth, Berlin (2006)
4. VDI 3211: Deep hole drilling on machining centres. Beuth, Berlin (2015)
5. Klocke, F.: Fertigungsverfahren 1–Zerspanung mit geometrisch bestimmter Schneide. Springer, Berlin (2018)
6. Nickel, J.: Analyse und Modellierung der thermomechanischen Beeinflussung der Randzoneneigenschaften beim Einlippentiefbohren von Bauteilen aus Vergütungsstahl. Dissertation. TU Dortmund University (2023)

7. Griffiths, B.J., Grieve, R.J.: The role of the burnishing pads in the mechanics of the deep drilling process. Int. J. Prod. Res. **23**(4), 647–655 (1985)
8. Streicher, P.: Tiefbohren der Metalle: Verfahrenstechnische und konstruktive Probleme, Dissertation. Universität Stuttgart (1975)
9. Brause, L., Michel, S., Schott, A., Rekowski, M., Herrmann, C., Biermann, D.: Dünnschicht-Sensorsysteme für Führungsleisten. VDI-Z **167**(01-02), 67–69 (2025)

A Conceptual Approach to In-Process Density Monitoring in PBF-LB/M Using High-Speed Thermography and Eddy Current Testing

Lucie Lamprecht(✉), Jork Groenewold, Florian Stamer, and Gisela Lanza

wbk Institute of Production Science, Karlsruhe Institute of Technology, Karlsruhe, Germany
lucie.lamprecht@kit.edu

Abstract. Powder bed fusion by laser beam melting (PBF-LB/M) is widely used in metal additive manufacturing, but maintaining consistent part density remains a challenge due to the lack of in-process sensing technologies that enable feedback control. This study takes the first steps toward an in-process monitoring approach that combines eddy current testing (ECT) and high-speed thermography (HST) to detect and assess density variations in 316L stainless steel samples. A machine learning (ML) model is proposed to predict local density based on the combined ECT and HST data. The data labels can be generated via computed tomography (CT) scans determining reference density distributions. Within this vision, the paper focuses on the analysis of ECT data. Future work involves processing the HST dataset and quantifying porosity from the CT scans to complete the dataset for ML, enabling development and validation of a robust in-process density model.

Keywords: Powder bed fusion by–laser beam melting of metal (PBF-LB/M) · In-process monitoring · High-speed thermography · Eddy current testing · Computed tomography · Machine learning model

1 Introduction and Motivation

Additive manufacturing (AM) processes have made significant advances over the past years evolving from a rapid prototyping tool to a viable manufacturing method in many industries, such as aerospace, healthcare and automotive [1]. Among the various processes, powder bed fusion by laser-beam melting of metals (PBF-LB/M) is the most commonly used AM process for the layer-by-layer production on metals parts, providing complete design freedom [2].

Despite its advantages, PBF-LB/M faces critical challenges in ensuring consistent part quality and reliability. One key factor influencing mechanical performance is the density of the manufactured component, which is significantly reduced by porosity in the form of internal voids or lack of fusion. While relative densities above 99.5% are commonly achieved in industrial practice, even small amounts of residual porosity can compromise the mechanical integrity of components, particularly in safety–critical applications. Optimized process parameters can largely prevent pore formation. However, process parameter optimization is often complex and time-consuming. In addition,

L. Overmeyer and B.-A. Behrens (eds.), *Production at the Leading Edge of Technology*, Lecture Notes in Production Engineering, https://doi.org/10.1007/978-3-032-19524-1_35

inherent process variability, fluctuations in feedstock quality, and machine instabilities still contribute to undesirable porosity levels [3]. This compromises mechanical integrity and limits the broader industrial adoption of PBF-LB/M, especially in safety–critical applications. Due to the high variability in build quality and the strict regulatory requirements in certain high-reliability industries, extensive post-process quality assurance is required. This includes both destructive and non-destructive testing (NDT), contributing significantly to production costs [3, 4]. Computed tomography (CT) remains one of the most commonly used NDT methods for detecting internal porosity, although its applicability is constrained by part size and material density, especially for dense metals [5].

In recent years, there has been increasing interest in in-process monitoring technologies aimed at improving process stability and enabling real-time quality assessment, as shown by Grasso et al. [6]. Eddy current testing (ECT), a widely used non-contact electromagnetic NDT method in post-process inspection, is capable of detecting surface and near-surface defects such as pores and cracks in conductive materials. Integrating ECT sensors into the PBF-LB/M machine by mounting them on the recoater offers the potential for layer-wise, in-process defect detection throughout the build process. Similarly, high-speed thermography (HST) enables in-process monitoring of the melt pool and surrounding thermal behavior, providing indirect insights into energy input and material consolidation [7].

While both techniques offer valuable information, their combination is particularly promising due to their complementary strengths: ECT offers high sensitivity to subsurface defects, while HST provides high spatial resolution of thermal behavior. Conventional detection methods, though technically feasible, are limited in handling the large, time-resolved, and heterogeneous datasets generated in-process. In contrast, machine learning (ML) models are well-suited for such multimodal data, automatically extracting discriminative features and enabling effective fusion of ECT and HST signals, resulting in more accurate and robust defect predictions. Recent studies on sensor data fusion in additive manufacturing further demonstrate that ML-based integration of multiple sensor modalities enhances both the reliability and sensitivity of defect detection [8].

This work introduces a conceptual approach for in-process density monitoring in PBF-LB/M. By correlating layer-wise thermal and electromagnetic signals with post-process X-ray CT data, the feasibility of detecting density deviations during the build is evaluated. The approach is demonstrated using 316L samples manufactured with systematically varying process parameters, resulting in a range of relative densities. The underlying datasets and analysis code will be published following completion of the study.

2 State of the Art

In-process monitoring technologies are increasingly recognized for their potential to enhance process control, improve part quality consistency, and reduce reliance on post-process inspection, as highlighted by Debroy et al. [3].

Among various NDT methods, ECT stands out as a cost-effective option for PBF-LB/M, capable of detecting critical defect types in accordance with ASTM E3166-20 [4].

ECT detects material defects by measuring disturbances in the magnetic field caused by eddy currents deviating around heterogeneities such as cracks, voids, or inclusions. This NDT method is widely used in safety–critical industries, including aerospace, automotive and energy to assess structural integrity and quality [9]. Research on ECT in the context of PBF-LB/M is still limited. Todorov et al. first integrated an ECT sensor array into a lab-scale system, detecting larger surface-visible defects during the build process [10]. More recently, Spurek et al. extended this concept by implementing an ECT system on the recoater of a PBF-LB/M machine to monitor AlSi10Mg parts layer-by-layer. By separating lift-off and conductivity effects, they correlated ECT signals with μCT-derived density, detecting small variations within 99–99.7% relative density [11].

Most other studies focus on off-process setups, where ECT successfully detected near-surface defects and, as shown by Hippert et al., distinguished relative density differences as small as 0.1% through correlations with electrical conductivity [12]. These findings indicate ECT's potential for layer-wise density assessment, offering a promising complement to traditional NDT methods.

Reviewing the literature makes it clear that thermography has also become a prominent NDT technique for in-process monitoring in PBF-LB/M. Dinwiddie et al. [13] used infrared thermography in electron beam-based PBF to monitor beam focus and to differentiate between melted and unmelted regions. Rodriguez et al. [14] investigated temperature uniformity across the powder bed for the implementation of feedback control mechanisms. Similarly, Krauss et al. [15] employed thermographic data to characterize the heat-affected zone and identify internal flaws such as pores. However, only a limited number of studies have employed HST. For instance, Sauer et al. demonstrated that integrating a galvanometer scanner with an HST system enables detailed analysis of subsurface defects [16]. Nevertheless, current research often remains limited to qualitative melt pool observation or single-layer analysis. Consequently, the full potential of HST for quantitative in-process defect detection, particularly in combination with complementary NDT such as ECT, remains insufficiently explored.

3 Materials and Methods

3.1 Experimental Setup and Sensor Integration

The experiments were performed on a SLM280HL PBF-LB/M system. Gas-atomized 316L stainless steel powder (m4p material solutions GmbH, DE), with a particle size distribution of 15–45 μm, was used as feedstock. The powder was applied using a customized recoating system, designed to integrate ECT sensors, as shown in Fig. 1a. The ECT sensors (W1, AMiquam SA, CH) were mounted on the recoater in front of the recoater lip, allowing for layer-wise measurement of the solidified surfaces without interrupting the standard process sequence.

Simultaneously, thermal emission from the process zone was captured using a high-speed infrared ImageIR 9400 camera (InfraTec GmbH, DE). To maintain the closed build environment, the camera was positioned outside the machine and recorded the process through a chamber window and a ceiling-mounted mirror, as illustrated in Fig. 1b. By integrating ECT and HST sensors in a synchronized setup, the methodology allows correlation of electromagnetic and thermal signals for each layer. This multi-modal

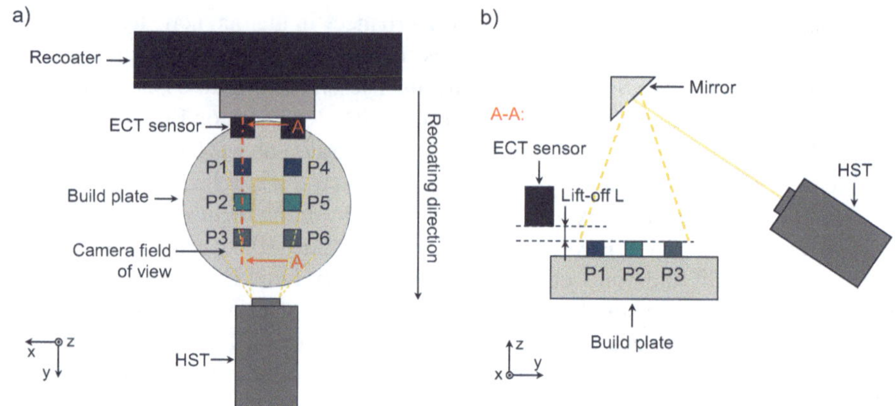

Fig. 1 Schematic overview of the experimental setup and process configuration.**a** Positioning of the samples, ECT sensors and HST camera within the SLM280HL build chamber.**b** Cross-sectional representation showing the relative position of the ECT sensors and the HST camera to the parts. (Adapted from the schematic concept by Spurek et al. [11])

dataset enables the detection of density deviations with higher accuracy and sensitivity than either sensor alone.

A total of six $10 \times 10 \times 9$ mm^3 samples were manufactured directly on a stainless-steel build plate without support structures, using the process parameters summarized in Table 1. The first 30 layers of each sample were monitored for data acquisition to enable a representative initial evaluation. Each pair of adjacent samples was built with identical parameters, allowing for a direct comparison of the ECT signals captured by the two sensors, with each sensor monitoring three samples during the process.

Table 1 PBF-LB/M process parameters

Parameters			P_1 & P_4	P_2 & P_5	P_3 & P_6
Laser power	P_L	[W]	200	225	250
Scanning speed (for 60 layers)	v_s	[mm/s]	700, 800, 900, 1000, 1100		
Hatch spacing	h_s	[mm]	0,12		
Layer thickness	t_l	[µm]	30		
Scan pattern		–	66.7° alternating		
Recoating speed	v_r	[mm/s]	250		
Shield gas		–	Argon		

3.2 Data Acquisition and Processing

ECT data from the monitored layers were recorded as complex signals consisting of real and imaginary components. For targeted analysis, only the signal segments corresponding to the sensor's passage over the samples were extracted from each scan cycle. To facilitate data processing and improve interpretability of signal variations the complex signals were converted into phase angle values $\Phi_{deg}(y)$ using the following equation:

$$\Phi_{deg}(y) = \frac{180^{\circ}}{\pi} \cdot \arctan2(I(y), R(y)) \quad (1)$$

where $I(y)$ and $R(y)$ represent the imaginary and real parts of the signal at position y. To enable layer-wise comparison, the phase angles were mapped to color-coded bars for each layer. By stacking the resulting bars from each layer, a 2D heatmap was created to visually highlight potential density variations or defects across the build height. This visualization, exemplarily shown in Fig. 2a, allows cross-comparison with data obtained from HST and CT.

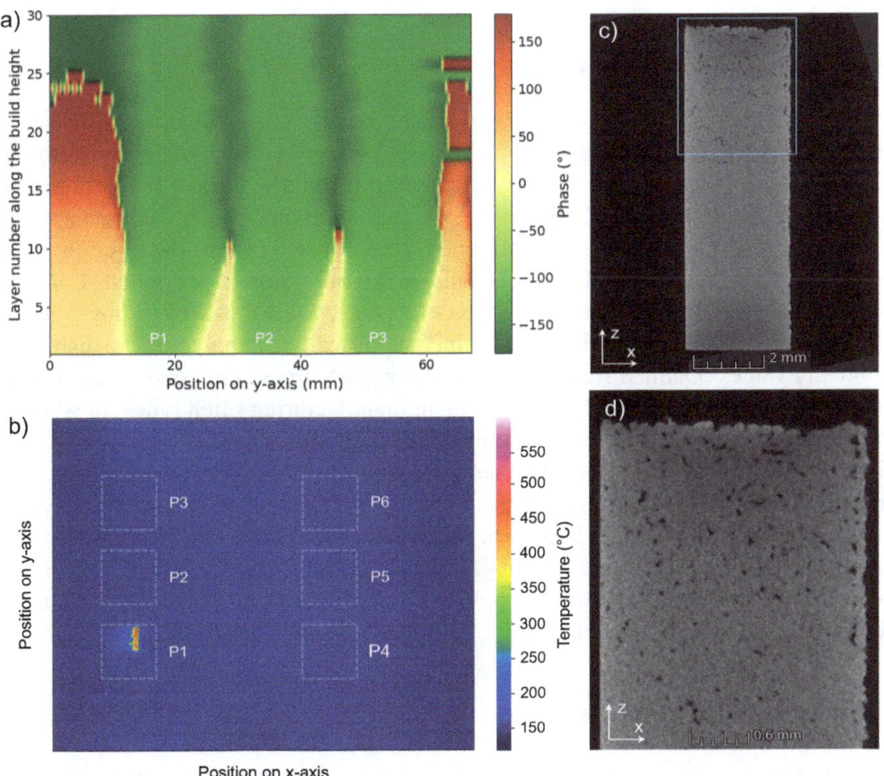

Fig. 2 **a** Phase heatmap of ECT data across 30 layers for three samples. **b** Raw HST thermal image captured for P1, showing all six samples within a single frame. **c** CT scan along the entire build height visualizing porosity distribution for P1. **d** Magnified view of (**c**), highlighting individual pores within the sample

For the HST data, maximum intensity images are generated for each layer, capturing the highest temperature value detected in each frame during laser exposure. This provides a detailed thermal representation of the processed areas. Sample regions are then isolated and converted into temperature vectors along the y-axis, enabling direct comparison with the ECT data. Intensity values are converted to surface temperatures assuming an appropriate emissivity for 316L stainless steel, with camera calibration and application of Planck's law to relate measured radiance to temperature. Although reflections and emissivity variations may slightly affect absolute temperatures, relative differences are sufficient for ML-based analyses. By stacking these vectors across all layers, a layer-wise temperature profile is built, allowing direct comparison between the outputs of the HST and ECT sensors. While the individual temperature images have already been extracted, the construction of the complete layer-wise temperature profile is still ongoing and thus not included in the present evaluation.

To establish reference data for the ML model, $2 \times 2 \times 9\ mm^3$ sections were cut from each sample and high-resolution X-ray CT analysis of these sections using a Zeiss Metrotom 1500 (Carl Zeiss AG, Germany) is currently in preparation.

3.3 ML-Based Evaluation Approach

To enable classification of local porosity based on the sensor data, the CT data are segmented into voxel[1] grids, and for each layer, porosity is calculated by determining the pore volume fraction within the corresponding layer, providing a spatially resolved reference dataset. As an initial approach, a classification model is considered appropriate. In this context, each unit cell is assigned to one of two classes based on its overall porosity. A threshold, defined as a percentage of void volume, is applied. Cells with porosity below the threshold are labeled as defect-free, while those exceeding it are considered defective. The threshold is determined empirically by analyzing the distribution of porosity values obtained from the CT-based reference measurements. This approach is industrially relevant, as quality assurance in manufacturing often relies on whether a critical defect threshold is exceeded, rather than on the precise magnitude of deviation [17]. For data-driven model development, the dataset is split into training, validation, and test subsets. Following standard practice [18], 80% of the data are used for training, while the remaining 20% are divided between validation and testing to ensure reliable model evaluation and prevent overfitting. It should be noted that this section outlines the envisioned methodology, and the machine learning training and evaluation have not yet been carried out at this conceptual stage.

4 First Results and Discussion

The initial results focus primarily on the evaluation of ECT sensor data, as well as preliminary observation of HST data and CT scans. The aim is to illustrate the potential of the proposed approach while highlighting current limitations and challenges. Figure 2a presents a stacked phase heatmap of the first 30 layers, showing three distinct green signal bands corresponding to samples P1 to P3. Yellow regions indicate the initial influence of the metallic build plate, which gradually decreases with each layer. From around layer 12 onward, this results in a color shift to red as the sensor is less affected by the

[1] Voxel = volumetric pixel.

build plate's electromagnetic influence. Minor signal fluctuations between layers are attributed to slight variations in recoater speed leading to non-uniform timing during data acquisition. In the regions between samples, signal disturbances likely result from powder accumulation or mutual electromagnetic interference between adjacent samples, reducing measurement fidelity in these zones. While full processing of the HST dataset is still ongoing, preliminary thermal imaging, as shown in Fig. 2b, displays the melt pool and thermal behavior in sample P1, capturing all six samples in the field of view for spatial orientation and context. These images are intended to form the basis for the temperature profiles described in Sect. 3.2.

Figure 2c shows the porosity distribution across the entire build height as obtained from the CT scan. An initial inspection reveals that the lower regions of the samples, manufactured with a reduced scanning speed of 700 mm/s, exhibit minimal porosity, suggesting sufficient energy input and favorable consolidation. In contrast, porosity increases noticeably in the upper regions, where scanning speeds exceeded 1000 mm/s. These observations are consistent with literature reports linking high scanning speeds to insufficient melt pool stability and lack-of-fusion defects [3, 11]. A magnified CT view in Fig. 2d highlights clustered pores in the upper sample regions, supporting the correlation between increased scan velocity and density degradation.

The preliminary results offer initial insights for further analysis. At this stage, the feasibility of the proposed approach for in-process density monitoring in PBF-LB/M cannot yet be demonstrated conclusively, as direct correlation analyses between the sensor data and reference measurements have not yet been conducted. Nevertheless, the findings provide a conceptual framework and point in a promising direction. Key challenges remain, particularly regarding the spatial alignment of sensor and CT data, as well as the development of a robust feature extraction pipeline for multi-modal fusion. Addressing these challenges is essential for establishing reliable in-process feedback control in future implementations.

5 Summary and Outlook

This study introduced an initial conceptual approach for combining ECT and HST data for in-process density monitoring in the PBF-LB/M process using 316L stainless steel. A tailored experimental setup integrated ECT sensors and a high-speed infrared camera, enabling synchronized acquisition of electromagnetic and thermal signals throughout the build height of the samples. The achieved results at this stage include the preprocessing of ECT signals into spatially resolved phase heatmaps and the demonstration of preliminary HST observations, providing the first evidence of feasibility. In contrast, the vision of this work is to extend these first results by completing HST data processing, constructing full temperature profiles, and correlating the multi-sensor datasets with CT-based porosity measurements and machine learning models for defect recognition and classification.

The next steps involve completing HST data processing and quantifying porosity from the CT scans, as described in Sect. 3.2, to enable training and validation of a ML model for in-process density assessment. While the current results remain preliminary, they establish a proof-of-concept foundation, the methodology demonstrates potential.

Nevertheless, further refinement and systematic validation are required to ensure the approach's robustness, reliability, and practical applicability for in-process quality control in PBF-LB/M.

Acknowledgements. This research was funded by the German Federal Ministry of Research, Technology and Space (BMFTR) through the research project 02J23C100-114 FENI-X.

Competing Interests. The author(s) has no competing interests to declare that are relevant to the content of this manuscript.

References

1. Prashar, G., Vasudev, H., Bhuddhi, D.: Additive manufacturing: expanding 3D printing horizon in industry 4.0. Int. J. Interact. Des. Manuf. **17**, 2221–2235 (2023). https://doi.org/10.1007/s12008-022-00956-4
2. Wohlers, T., Campbell, R.I., Diegel, O., Huff, R., Kowen, J.: Wohlers Report 2020: 3D printing and additive manufacturing state of the industry. Fort Collins, Wohlers Associates (2020)
3. DebRoy, T., Wei, H.L., Zuback, J.S., Mukherjee, T., Elmer, J.W., Milewski, J.O., Beese, A.M., Wilson-Heid, A., De, A., Zhang, W.: Additive manufacturing of metallic components—process, structure and properties. Prog. Mater. Sci. **92**, 112–224 (2018). https://doi.org/10.1016/j.pmatsci.2017.10.001
4. ASTM International: ASTM E3166–20: Standard guide for nondestructive examination of metal additively manufactured aerospace parts after build. ASTM International, West Conshohocken (2020). https://doi.org/10.1520/E3166-20E01
5. Du Plessis, A., Yadroitsev, I., Yadroitsava, I., Le Roux, S.G.: X-Ray micro-computed tomography in additive manufacturing: A review of the current technology and applications. 3D Print. Addit. Manuf. **5**(3), 227–247 (2018). https://doi.org/10.1089/3dp.2018.0060
6. Grasso, M., Remani, A., Dickins, A., Colosimo, B.M., Leach, R.K.: In-situ measurement and monitoring methods for metal powder bed fusion: An updated review. Meas. Sci. Technol. **32**(11), 1–66 (2021). https://doi.org/10.1088/1361-6501/ac0b6b
7. Höfflin, D., Sauer, C., Schiffler, A., Hartmann, J.: Process monitoring using synchronized path infrared thermography in PBF-LB/M. Sensors **22**(16), 5943 (2022). https://doi.org/10.3390/s22165943
8. Shen, T., Li, B., Zhang, J., Xuan, F.: Multi-source information fusion for enhanced in-process quality monitoring of laser powder bed fusion additive manufacturing. Addit. Manuf. **96**, 104575 (2024). https://doi.org/10.1016/j.addma.2024.104575
9. Machado, M.A.: Eddy currents probe design for NDT applications: A review. Sensors **24**(17), 5819 (2024). https://doi.org/10.3390/s24175819
10. Todorov, E.I., Boulware, P., Gaah, K.: Demonstration of array eddy current technology for real-time monitoring of laser powder bed fusion additive manufacturing process. Proc. SPIE **10599**, 1059913 (2018). https://doi.org/10.1117/12.2297511
11. Spurek, M., Spierings, A., Lany, M., Revaz, B., Santi, G., Wicht, J., Wegener, K.: In-situ monitoring of powder bed fusion of metals using eddy current testing. Addit. Manuf. **60**, 103259 (2022). https://doi.org/10.1016/j.addma.2022.103259
12. Hippert, D., Jhabvala, J., Boillat, E., Santi, G., Bleuler, H.: Development of an eddy current testing method as process control for additive manufacturing of metallic components. In: Proceedings of IFToMM world congress 2015, pp. 10–14. IFToMM, Taipei (2015). https://doi.org/10.6567/IFToMM.14TH.WC.OS20.022

13. Dinwiddie, R., Dehoff, R., Lloyd, P., Lowe, L., Ulrich, J.: Thermographic in-situ process monitoring of the electron beam melting technology used in additive manufacturing. In: Proc. SPIE 8705, Thermosense XXXV, 87050K (2013)
14. Rodriguez, E., Medina, F., Espalin, D., Terrazas, C., Muse, D., Henry, C., MacDonald, E., Wicker, R.: Integration of a thermal imaging feedback control system in electron beam melting. In: Proceedings of the solid freeform fabrication symposium, pp. 945–961 (2012)
15. Krauss, H., Eschey, C., Zaeh, M.F.: Thermography for monitoring the selective laser melting process. In: Proceedings of the solid freeform fabrication symposium, pp. 999–1013 (2012)
16. Sauer, C., Höfflin, D., Schiffler, A., Hartmann, J.: Active thermography in PBF-LB/M with the synchronized path infrared thermography. Procedia CIRP **124**, 283–286 (2024). https://doi.org/10.1016/j.procir.2024.08.118
17. Weiser, L.M.: In-process porositätserkennung für den PBF-LB/M-prozess. dissertation, Shaker Verlag, Düren (2023). https://doi.org/10.5445/IR/1000161552
18. Rashidi, H.H., Tran, N.K., Betts, E.V., Howell, L.P., Green, R.: Artificial intelligence and machine learning in pathology: the present landscape of supervised methods. Acad. Pathol. **6**, 1–9 (2019). https://doi.org/10.1177/2374289519873088

Development of a Machine Learning Method for Evaluating Tool Wear in the Gear Skiving Process Based on the Workpiece Topography

Stylianos Tsakiris[1,2](✉), Christopher Janßen[1,2], Mareike Davidovic[1,2], Christian Westphal[1,2], and Thomas Bergs[1,3]

[1] Manufacturing Technology Institute, RWTH Aachen University, Aachen, Germany
s.tsakiris@wzl.rwth-aachen.de

[2] Laboratory for Machine Tools and Production Engineering, RWTH Aachen University, Aachen, Germany

[3] Fraunhofer Institute for Production Technology IPT, Aachen, Germany

Abstract. Manufacturing processes are influenced by external and internal disturbance variables, such as oscillations, clamping errors, and tool wear. At the end of tool life, the maximum wear mark width increases exponentially, leading to tool breakage or higher regrinding costs due to higher material removal. Progressive tool wear also negatively impacts workpiece quality. Gear skiving is a flexible and productive process which enables high quality requirements to be fulfilled, increasing its industrial and scientific relevance. To ensure the quality of manufactured gears, quality control is performed by measuring geometry and roughness of tooth flanks. In this paper a method to evaluate tool wear via gear measurements is presented. In this way, the gear measurement can be used to assess the wear condition of a tool in order to avoid the exponential increase in the maximum wear mark width. For these investigations, the potential of artificial intelligence-assisted analysis, which has not been sufficiently investigated in the field of gear manufacturing, is exploited. To this end a deep neural network is developed which assesses the wear condition based on the gear topography.

Keywords: Gear skiving · Tool wear · Machine learning

1 Motivation and Approach

Manufacturing processes are subject to various disturbance variables that negatively impact the workpiece quality and economic efficiency [1]. One such variable is tool wear, which progresses over the tool's life and leads to an exponential increase of the wear width toward the end of tool life [1]. This can result in tool breakage or significantly higher material removal during regrinding of the tool, reducing the number of possible regrinding cycles and increasing tool costs [2]. Additionally, advanced tool wear impairs the quality of the manufactured workpieces [3]. Monitoring tool wear is therefore crucial but remains challenging in industrial environments [4]. Conventional tool wear measurements are often time-consuming and thus unsuitable for use in series production [5]. In

L. Overmeyer and B.-A. Behrens (eds.), *Production at the Leading Edge of Technology*, Lecture Notes in Production Engineering, https://doi.org/10.1007/978-3-032-19524-1_36

contrast, the geometry and surface quality of the workpiece are regularly measured as part of standard quality control in gear manufacturing [6].

The aim of this paper is to develop a method for evaluating tool wear in gear manufacturing based on the workpiece topography. The manufacturing process considered is gear skiving, which is gaining increasing industrial relevance due to its high productivity and flexibility [7]. For the evaluation of tool wear, a machine learning model (ML) is developed. The data sets used to train and subsequently test and evaluate the model are generated simulatively. The method is then validated using real measurement data. Finally, the influence of the resolution of the generated data on the evaluation quality of the ML model is determined.

2 Development of a Method for Evaluating Tool Wear

In this chapter, the development of a method for enabling the evaluation of tool wear in the gear skiving process based on workpiece topography is described. As the method involves using an ML model, a large amount of data is required for training this model. To this end, the necessary training and testing data is generated via simulation. The input data required for the simulation is also generated automatically. Therefore, the developed method encompasses generating the requisite input data for manufacturing simulations, simulating topographies for training and testing the ML model, and developing the model itself.

The simulations for generating the training and test data sets for the wear evaluation are carried out for a gear with a normal module $m_{n2} = 5$ mm, a number of teeth $z_2 = 21$, an outer diameter $d_{a2} = 116.2$ mm and a normal pressure angle $a_{n2} = 20°$. For the simulative investigation, a right-handed skiving tool with an outer diameter $d_{a0} = 160$ mm and a number of teeth $z_0 = 29$ is used. Similarly to the workpiece, the normal modulus is $m_{n0} = 5$ mm and the normal pressure angle is $a_{n0} = 20°$. Based on the tool and workpiece data, the cross axis angle is set to $\Sigma = -20°$. The axial feed f_a was randomly varied between $f_a = 0{,}1$ mm and $f_a = 0{,}28$ mm.

2.1 Generation of Training and Test Data Sets

A major challenge in the application of machine learning in the field of manufacturing is the acquisition of relevant data [8]. This can also be a limitation of such models, as the availability, quality and composition of the available manufacturing data have an impact on the performance of ML models [8].

For these reasons, the data sets required for training and testing the presented model are generated simulatively. These data sets correspond to simulated tooth flank topographies with 61 profile lines and 45 flank lines, which represent the input variables of the ML model. The SPARTApro manufacturing simulation developed in the Laboratory for Machine Tools and Production Engineering (WZL) at RWTH Aachen University is used for this purpose. SPARTApro is part of the WZL GearToolbox and is based on a plane-based penetration calculation and is used for the simulation of various gear manufacturing processes. Based on the tool, workpiece and process data, the manufacturing process is modeled using the plane-based penetration calculation to generate the tooth gap [9]. Each tooth gap consists of two topographies, one for the left and one for the right tooth flank respectively.

In the manufacturing simulation, the tool is described by an enveloping body, which corresponds to the movement of the tool in the working space in relation to a gear gap. The tool envelope is therefore dependent on both the two-dimensional (2D) tool profile and the kinematics of the simulated process. The tool profile is described by x and y coordinates, which are defined by the reference profile of the tool. The support points describe either the beginning or the end of a rectilinear or circular section. The tool contour is interpolated between these support points [10].

Necessary input for the SPARTApro simulation and thus for the generation of topographies, which are used for training and testing the ML model for wear assessment, are 2D profiles of worn tools. For a realistic representation of the wear on the 2D tool profiles, first, the wear on the rake faces of two tools is measured using a Keyence VHX 6000 digital microscope. Each tool had 45 teeth and had reached the end of their tool life. The worn rake faces were characterized using a Matlab algorithm developed for this method. This algorithm is used to record and save the position, number, shape and size of the existing breakouts. The position of a breakout is determined in relation to the tool tooth root. The distance between the tooth root and tooth tip is standardized for this purpose so that the distribution of the breakouts can also be transferred to other tool profiles. However, the shape of the breakouts is complex and irregular, which is why they cannot be assigned to a simple geometric shape. For this reason, a polynomial function of degree n is defined for each breakout, with which the shape can be approximated. This requires $n + 1$ support points in each case. These are selected equidistantly over the circumference of each breakout. The size of the breakouts is recorded in two directions—along the cutting edge and orthogonal to the cutting edge.

Based on the wear categorization, the characterized breakouts were transferred to the 2D tool profiles required for the simulation. However, a direct transfer of the real breakouts to the 2D tool profiles is not possible. The breakouts of a real worn tool have a three-dimensional (3D) extent and a finite depth perpendicular to the rake face. This does not apply to the 2D profiles that are entered into the simulation. The 2D tool profile is coupled with the gear skiving kinematics to generate the tool enveloping body that penetrates the workpiece planes. Therefore, a wear parameter VK was defined, which determines the depth of each breakout orthogonal to the cutting edge. A comparison between an ideal 2D tool profile and a worn profile is shown schematically in Fig. 1. Furthermore, an expected value μ and a standard deviation σ were specified for each wear class. These are used to define the position and number of breakouts according to a Gaussian distribution along the cutting edge. The wear parameters used to categorize the defined wear classes 'no wear', 'light wear', 'medium wear' and 'heavy wear' are listed in Table 1.

The wear factor VK_{max} combined with the expected value μ and the standard deviation σ enables dimensioning and positioning of the breakouts in such a way that their influence on the quality of a simulated tooth flank is analogous to the quality of a skived gear. These values were defined and adapted iteratively through the following procedure. After recording the wear patterns and defining the wear factors for the first time, worn

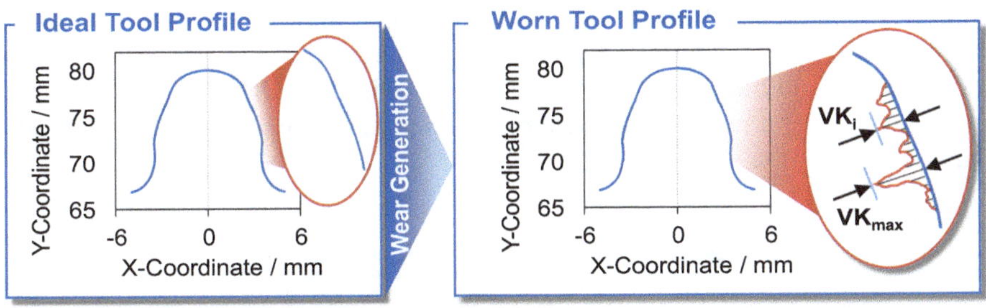

Fig. 1 Comparison of ideal and worn 2D tool profiles

Table 1 Categorization of the wear classes

Wear classes	Wear factor VK_{max}/µm	Expected value µ/–	Standard deviation σ/–
No wear	1	5	0.25
Light wear	5	7	0.30
Medium wear	10	10	0.30
Heavy wear	18	15	0.40

tool profiles are generated for all wear classes using the developed algorithm. Manufacturing simulations are then carried out with the generated profiles. The resulting tooth flanks are compared with an ideal tooth flank using GearAnalyzer, which is also included in the WZL GearToolbox. The flank comparison is used to output both a virtual gear measurement and the topographies with which the ML model is subsequently trained and tested. The gear measurements are used to check whether the wear factors match the associated wear classes. If this is not the case, the factors are adjusted, new tool profiles are generated, and the manufacturing simulation and flank comparison are carried out again. After successfully determining the wear factors for all defined classes, approx. 40,000 simulations and subsequent flank comparisons were carried out using automation realized with Python.

2.2 Development of a Machine Learning Model for Classifying Tool Wear

The ML model, which was developed for the evaluation of tool wear based on the workpiece topography, was generated in Python based on the Keras library. Keras is part of the overall Tensorflow library and is well suited for building deep neural networks (DNN) of high complexity. A deep feed-forward network was set up for the evaluation of tool wear. The evaluation itself was based on a multi-class classification. The DNN is a supervised machine learning approach, as the correct output value is known during the training process [11]. The term feed forward is derived from the direction of the links, which always run in the direction of the output layer. The structure of the developed ML model is shown schematically in Fig. 2. Before the input data is imported into the developed ML model, it is normalized to a value range from zero to one. The aim of normalization is to scale the input data to a uniform order of magnitude. This improves the performance and training stability of a ML model. Two different approaches were investigated for this purpose; min–max normalization and normalization using a standard deviation [11].

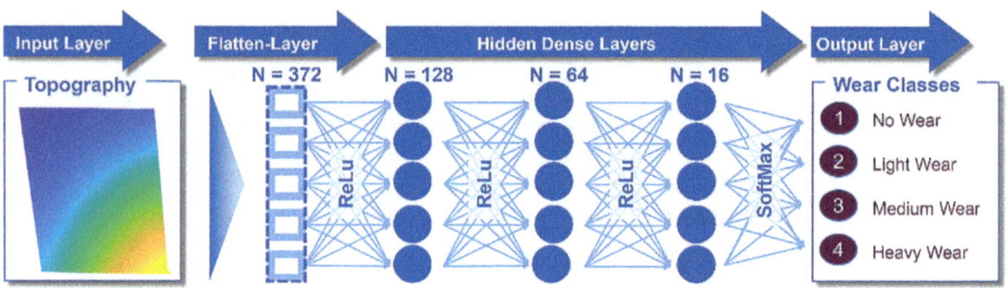

Fig. 2 Structure of the developed ML-model

In the input layer, the simulated topographies of the left and right tooth flank of a gap are imported as a multi-dimensional tensor. It should be noted that the model works with a fixed number of input parameters. The subsequent flatten layer is used to convert a multidimensional tensor into a one-dimensional tensor, which is entered into the next layer. The next layers are called hidden dense layers. Hidden dense layers contain dense connected neurons (N). This means that every single neuron in the layer is connected to every other neuron in the adjacent layers, see Fig. 2. The number of hidden dense layers and the activation functions used were varied during the development of the model. The number of hidden dense layers was set to three and ReLu is used as the activation function in these layers. In the output layer the SoftMax activation function is used, which outputs a percentage probability for each defined class. Four neurons (N = 4) are required in the output layer for four wear classes. To train the model, the batch size was set to the standardized value of 32. The number of epochs, complete runs of all training data through the algorithm, was set to 100. The learning rate, which controls how quickly the DNN converges, was set to 0.0005 using a stochastic optimization algorithm ADAM (Adaptive Moment Estimation).

3 Validation of the Developed Method

80% of the data sets generated were used to train the developed ML model, and the remaining 20% for its testing. The input of a test data set in an ML model serves to assess its performance after completion of the training. The evaluation of the model was then carried out using simulatively generated and real measured topographies.

3.1 Evaluation Based on Simulatively Generated Data

Approximately 30,000 data sets were used to train the model. To compare the influence of normalization, the data sets were normalized using both the min–max and standard deviation methods prior to input. The training was carried out step by step with an increasing amount of training data. After each training process, the model was tested and the classification accuracy achieved was documented. The classification accuracy is defined as the proportion of test data that was assigned to the correct wear class. The percentage accuracy for both normalization approaches is shown in Fig. 3 depending on the number of training data. In all cases 80% of the data was used for training and 20% of the data was used for testing the model. Also, the machine learning parameters used to train this model are shown.

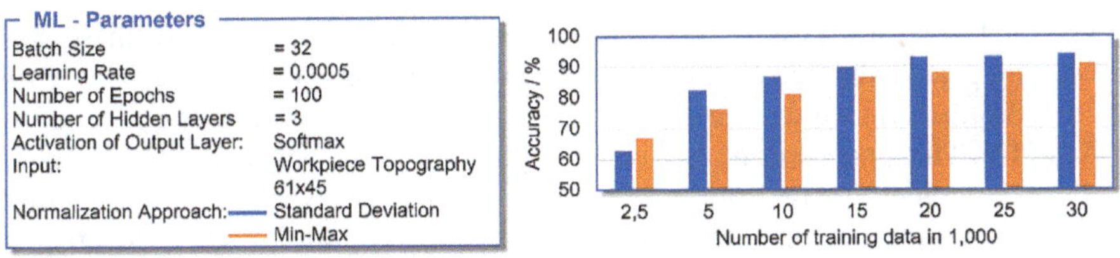

Fig. 3 Accuracy of the classification of tool wear

In Fig. 3 is shown that a higher classification accuracy was achieved by normalizing the input data using the standard deviation when the number of training data was 5,000 or more. Above 15,000 training data sets, an accuracy of 90% was achieved using standard deviation normalization. A comparable accuracy using min–max normalization could only be achieved from using over 27,500 data sets. The maximum accuracy of 94% was achieved with 30,000 normalized training data sets using the standard deviation. For the min–max normalization approach, the classification accuracy after training the same amount of training data is 91.1%. Therefore, it can be concluded that both scaling methods achieved an accuracy of over 90% with enough data. However, the testing results indicated that the normalization of the inputs using standard deviation is more suitable for the classification of tool wear.

3.2 Validation Based on Real Measured Topographies

The validation of the classification of tool wear was carried out using real measured topographies. No real topographies or wear documentation were available for the gear used to train the developed model. For this reason, two different internal gears were measured, which were manufactured for wear tests that had already been carried out. Accordingly, the wear condition of the tool was documented before manufacturing of each of these gears. The gear data is shown in Fig. 4.

A total of three gear gaps of test gear geometry 1 were measured and used as input for the developed model. In each case, the left and right flank of a gear gap were measured. One of the workpieces was manufactured with a new tool, while the other was manufactured with a tool of wear class 3 (medium wear). The final workpiece of test gear geometry 1 was manufactured with a gear skiving tool that had reached the end of its life. All three gears were classified by the model as having been manufactured with a tool of wear class 4 (heavy wear). However, it should be noted that although the tool

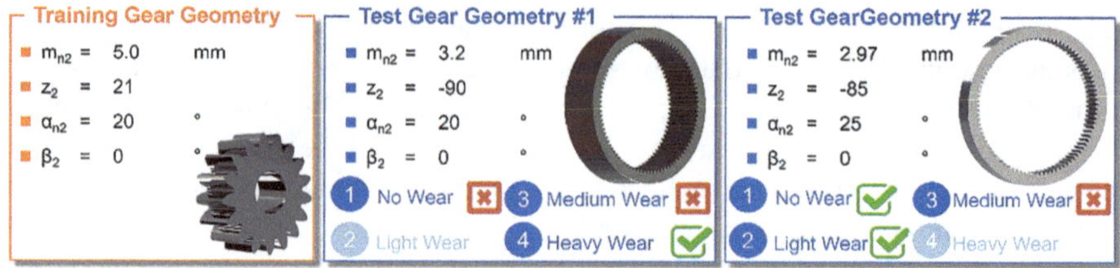

Fig. 4 Data of the training and test gear geometries

was assigned to wear classes 1 and 3 during the manufacturing of two of these gears, the resulting gears exhibited significant geometric deviations. The deviations in profile slope and profile form were more than 50 µm. This corresponds to a gear quality of A12 and A11 [12]. The model was trained considering wear as the only cause for deviations. For test gear geometry 2, two workpieces were manufactured with a new tool and were successfully assigned to wear class 1 (no wear) by the developed model. The other two gears were manufactured with light and medium worn tools respectively. Both of those were assigned to wear class 2—light wear. In total three of the four wear classes were identified correctly. Furthermore, the transferability of the model to internal gears is provided, even though it was trained using topographies of external gears.

Following factors must be taken into account when evaluating the wear classification, the influence of which cannot currently be quantified. The first factor is the pre-processing of the real measured topographies. In a real topography measurement, the coordinates are recorded with a density approximately ten times higher than that of the coordinates output by the GearAnalyzer flank comparison per analyzed tooth flank. Accordingly, the data points of the real topography are initially reduced in order to ensure they match with the dimensions required for input into the model. Another possible factor influencing the evaluation quality of the model is the macrogeometry of the gears. For the investigations carried out, the topographies used to train and validate the model corresponded to gears with different macrogeometries.

4 Influence of Data Resolution on the Evaluation Quality

The influence of data resolution on the evaluation quality is determined by entering a reduced 3×3 grid consisting of 3 profile lines and 3 flank lines into the DNN instead of a complete topography per flank with 61 profile lines and 45 flank lines. While a measurement of a 61×45 topography on a gear measuring machine is time-consuming and is not carried out as standard in series production, a 3×3 grid can already be output of a twist measurement. A twist measurement is primarily carried out on modified gears in addition to the conventional profile and flank line measurement. Accordingly, the application of the developed method could also be integrated into series production to evaluate the tool wear using the profile and flank lines. The 3×3 grid is also output as a virtual gear measurement using the Gearanalyzer flank comparison. The profile and flank lines can be imported both individually and as a grid in order to train and subsequently test the model. Similar to the topographies, the gear measurement data are first normalized and then imported into the DNN. It should be noted that the dimensions of the input layer must be adapted. The remaining structure of the DNN remains unchanged so that the influence of the data resolution can be considered separately. The influence of the data resolution on the quality of the tool wear assessment is depicted in Fig. 5 using the classification accuracy depending on the number of training data.

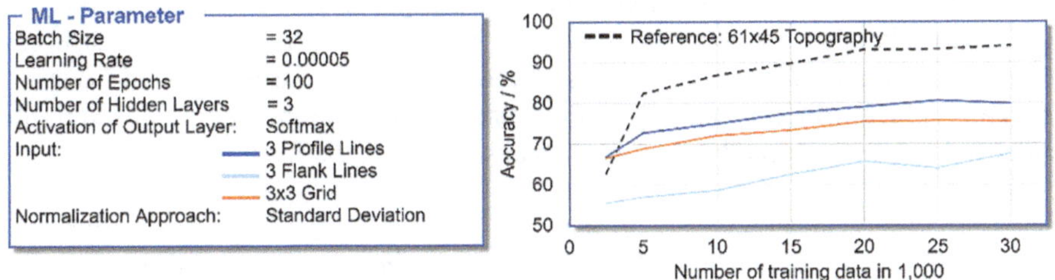

Fig. 5 Influence of data resolution on the quality of wear classification

As a reference, the evaluation accuracy achieved by entering a complete topography is shown with a dashed line. The results shown in the diagram result from data sets normalized by means of standard deviation. For less than 3,500 training data sets, a higher accuracy was achieved with a lower resolution compared to high resolution topographies. This applies to the training of the ML model using three profile lines or a 3 × 3 grid per tooth flank. The maximum evaluation accuracy achieved by the DNN after training using profile lines is 80.8% when using 25,000 training data sets. If the model is trained with the 3 × 3 grid the maximum accuracy is 75.6% when using 30,000 training data sets. If three virtual flank lines are used to train the model, a maximum evaluation accuracy of 67.6% is achieved when using 30,000 training data sets. A lower accuracy for the classification of wear based on flank lines compared to a classification based on profile lines was to be expected, as profile lines are more meaningful for the evaluation of tool wear. However, the accuracy of wear classification based on flank lines decreases when using 25,000 compared to 20,000 training data sets. A possible explanation for this is that the ML model was developed for the input of a 61 × 45 topography. Apart from the learning rate, which was reduced from 0.0005 to 0.00005 based on lower-resolution data using the ADAM stochastic optimization algorithm, no further adjustments were made to the ML parameters. The structure of the model also remained unchanged. The marginal reduction in accuracy can possibly be avoided by adjusting the hyperparameters or even the structure of the model. However, this was not the subject of these investigations.

5 Summary

The aim of this paper was to develop a method for evaluating tool wear in the gear skiving process based on the workpiece topography. This method includes the development of an ML-model, the simulative generation of topographies, which are required for training and testing the model, as well as the generation of the input data necessary for carrying out the simulations. The model was a deep neural network which classifies tool wear in four wear categories. The manufacturing simulations for generating the training and test data were carried out using the manufacturing simulation SPARTApro. Once the model had been trained, it was tested using the simulated data. Two different normalization approaches of the data sets were compared: normalization based on a standard deviation achieved an accuracy of 94%, outperforming Min–Max normalization, which resulted in a maximum accuracy of 91.1%. Furthermore, the developed model was used for the wear classification of real topographies of two gear geometries as input. Three out of the four wear classes were correctly identified. More topographies will be measured in the future to further validate the developed model. Finally, the model will be trained with topographies, which contain deviations caused by different disturbance variables, in order to be able to correctly identify the cause of the deviations.

Competing Interests. The author(s) has no competing interests to declare that are relevant to the content of this manuscript.

References

1. Klocke, F.: Fertigungsverfahren 1. Zerspanung mit geometrisch bestimmter Schneide. Springer, Berlin (2018). https://doi.org/10.1007/978-3-662-54207-1
2. Conradie, P.J.T., Oosthuizen, G.A., Dimitrov, D.: On the effect of regrinding cutting tools for high performance milling of titanium alloys. Int. J. Adv. Manuf. Technol. **90**, (2017). https://doi.org/10.1007/s00170-016-9550-z
3. Hakami, F., Pramanik, A., Basak, A.K.: Tool wear and surface quality of metal matrix composites due to machining: A review. Proc. Inst. Mech. Eng., Part B: J. Eng. Manuf. **231**, (2017). https://doi.org/10.1177/0954405416667402
4. Munaro, R., Attanasio, A., Del Prete, A.: Tool wear monitoring with artificial intelligence methods: A review. J. Manuf. Mater. Process. **7**, (2023). https://doi.org/10.3390/jmmp7040129
5. Jurkovic, J., Korosec, M., Kopac, J.: New approach in tool wear measuring technique using CCD vision system. Int. J. Mach. Tools Manuf. **45**, (2005). https://doi.org/10.1016/j.ijmachtools.2004.11.030
6. Tang, J., Jia, J., Fang, Z., Shi, Z.: Development of a gear measuring device using DFRP method. Precis. Eng. **45**, (2016). https://doi.org/10.1016/j.precisioneng.2016.02.006
7. Guo, E., Hong, R., Huang, X., Fang, C.: Research on the design of skiving tool for machining involute gears. J. Mech. Sci. Technol. **28**, (2014). https://doi.org/10.1007/s12206-014-1133-z
8. Wuest, T., Weimer, D., Irgens, C., Thoben, K.-D.: Machine learning in manufacturing: advantages, challenges, and applications. Prod. & Manuf. Res. **4**, (2016). https://doi.org/10.1080/21693277.2016.1192517
9. Weck, M., Kempa, B., Winter, W.: Analyse des Wälzfräsprozesses mit Hilfe der Durchdringungsrechnung. Arb.Stagung "Zahnrad-Und Getriebeuntersuchungen", (2000)
10. Klocke, F., Brecher, C.: Zahnrad- und Getriebetechnik. Auslegung—Herstellung—Untersuchung—Simulation. Hanser, München (2024). https://doi.org/10.3139/9783446469761
11. Matzka, S.: Künstliche Intelligenz in den Ingenieurwissenschaften. Springer Fachmedien Wiesbaden,Wiesbaden (2021). https://doi.org/10.1007/978-3-658-34641-6
12. DIN ISO 1328–1:2018–03.: Definitionen und zulässige Werte für Abweichungen an Zahnflanken. https://doi.org/10.31030/2773568

Design and Implementation of a Hybrid Position- and Force- Control System for Progressive Dies in a Linear Press

Alexander Müller(✉) and Richard Krimm

Institut Für Umformtechnik Und Umformmaschinen, Leibniz Universität Hannover, Garbsen, Germany
a.mueller@ifum.uni-hannover.de

Abstract. In the production of stamping parts using progressive dies, a decrease in production accuracy and process stability can be observed due to increasing wear of the active elements. The reason for this is that neither the ram trajectory nor the force progression is optimally adapted to the specific requirements of the individual forming processes in progressive dies, which can lead to inefficient material flow and higher stresses in tooling components. This paper explains whether a hybrid force-position control system in a press equipped with linear motors and electromagnets can improve the process stability and quality of sheet metal parts produced in progressive dies compared to conventional presses. This paper describes the implementation of a hybrid drive control system in Simotion Scout for the manufacture of small sheet metal components in a linear press and a progressive die under process-dependent position or force control. The position and speed of the ram are determined using encoders mounted in the guide rails, and the process force is measured with piezoelectric force sensors integrated in the mould. All dimensions are fed back to the control system. Various approaches for force and position controlling were tested. One approach is based on switching between controlling the force and the ram position in dependence of the forming process. In another approach both control loops are superimposed in a cascade control configuration. In addition, controlling the force distribution between the drive components is essential to prevent them from working against each other. A variable control of the ram trajectory and the applied forces enables the production of higher-quality components over a long period of time, even when the die shows signs of wear.

Keywords: Progressive die · Stamping press · Direct drive · Position-force-control

1 Introduction

Flexible and dynamic press drives are becoming more and more important in forming technology, particularly for applications involving a wide range of workpieces, short product cycles, and the need for increasing energy and production efficiency. One particularly challenging application is the operation of progressive dies, wherein several

L. Overmeyer and B.-A. Behrens (eds.), *Production at the Leading Edge of Technology*, Lecture Notes in Production Engineering, https://doi.org/10.1007/978-3-032-19524-1_37

forming processes are executed within a single press stroke, including cutting, bending, and embossing [1]. Such components are produced in large quantities, like in the automotive [2] and electrical industries [3]. In some cases, these processes require divergent ram trajectories or force curves to achieve optimal outcomes, which conventional drive systems (e.g., crank gears, flywheel presses) are often incapable of fully realising. In [4], several simulations based on measurement data have already shown that closed-loop control of the process force on hydraulic press drives can lead to an improvement in component quality.

Our solution to this problem was found in the development of a hybrid direct drive system by IFUM. This system consists of linear motors for precise, dynamic positioning and electromagnetic actuators for the application of additional forces near the bottom dead centre. This combination enables zone-specific control of force and travel, allowing cutting and bending processes to be run with different ram trajectories and embossing processes to be controlled with precise forces. The objective of this work is a hybrid control system for press applications with progressive dies, with a view to evaluating its control and regulation properties.

2 State of the Art

In contemporary manufacturing, eccentric presses, knuckle-joint presses and servo presses are utilised for the production of small sheet metal parts with progressive dies [5]. Mechanical eccentric presses are the most frequently employed, but despite their capacity for high stroke rates, their adaptability is considerably constrained. Furthermore, the bottom and top dead centre may undergo slight alterations following the initial setup. These variations are a consequence of operating time and temperature. This can result in inconsistent embossing depths during the embossing process. Servo presses with direct servo motor drives enable flexible motion profiles and precise process control. In [6], simulations showed that fuzzy intelligent control can control the press stroke and process force precisely and flexibly. Direct-drive presses with linear motors do not require mechanical transmission elements and are presented in [7] using the "Stanzrapid" press as an example. In [8], a press with a linear motor is analysed based on its dynamic properties and forces. Unlike established presses, linear motor presses have a relatively low forming force. Due to its high efficiency and dynamic capabilities, the demonstrator was constructed with a linear press that incorporates an electromagnet, which applies additional process force [9].

3 Components and Tools of the Linear Press

A press demonstrator with two different electric drive types was developed at IFUM (see Fig. 1). This press is equipped with four linear motors and two electromagnets. The linear motors reach 20 kN of force at maximum along the entire stroke path. The arrangement of the electromagnets was constructed in a way that their effective range begins 1.5 mm in advance of the bottom dead centre of the ram. With an air gap of 0.5 mm, the electromagnets reach a maximum force of 100 kN. Both position and force control can be realised with the linear motors, whereas the electromagnets can only be used in the force control mode. In addition, two measuring systems are installed. Process forces can be calculated partially via the linear motor current.

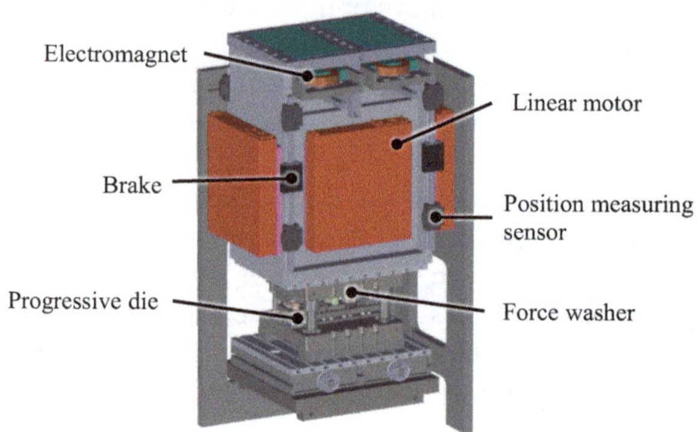

Fig. 1 Hybrid press demonstrator with progressive die

3.1 Sensor, Control and Power Electronics

The actuators are controlled via power electronics from Siemens®. The topology of the control system has already been presented in [10]. The system consists of the motion control system (Simotion D445-2), which performs the higher-level control with position control and actual value processing. Here also precise motion sequences and trajectories can be programmed. The drive-orientated control systems are connected to the motion control system via the ProfiBus and ProfiNet communication interfaces. A Sinamics S120 with 4 frequency inverters is used to operate the linear motors. The electromagnets are each operated by means of a Sinamics DCM current controller. The measured actual position of the ram is transferred from the encoder to the controller via the Sinamics sensor module. The process force measured by the force washer is first preconditioned in charge amplifiers and then forwarded to the control system's analogue inputs via a voltage signal proportional to the force.

3.2 Progressive Die

A progressive die was designed and manufactured for the hybrid press demonstrator, which is specially designed for the demonstrator's drive zones (see Fig. 2). The progressive die is comprised of five stages. One bending stage, three cutting stages and one embossing stage. The engagement heights of the cutting and bending stages are within the pure effective range of the linear motors, and these forming processes should never exceed their maximum force. The embossing punch is positioned at the greatest distance from the embossing die. In the process of establishing the press, it is imperative that the bottom dead centre is meticulously aligned with the engagement height of the embossing stage. This ensures that the electromagnets are situated within their effective range during the embossing phase.

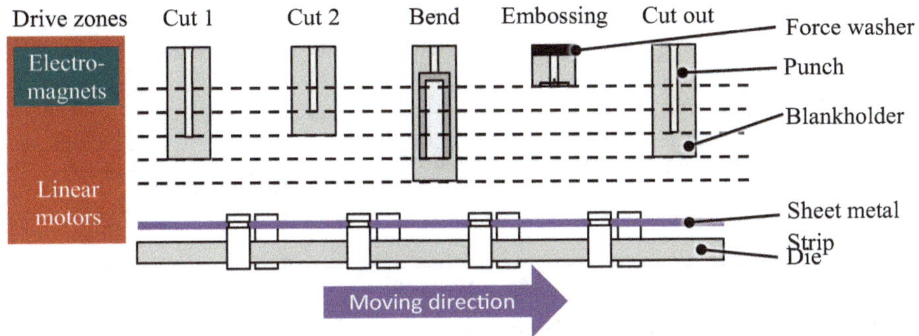

Fig. 2 Engagement heights of the forming processes in the progressive die

4 Control Loops

The objective of the control system is to facilitate the position- and force-controlled movement of the press ram with control parameters based on previously defined set values. The control system is confronted with three superior issues. Firstly, there are two manipulated variables but only one mass that is to be moved within one degree of freedom. Secondly, the movement of a mass with multiple coupled actuators that have to work together. Last but not least, the movement of a hanging load that is solely driven by electromagnetic fields requires protection in case of a power loss.

Figure 3 illustrates the sequence of drive control over the entire stroke in relation to the forming processes. During the cutting and bending phase, the ram is moved in position-control mode with the linear motors. The electromagnets are activated immediately following the final cutting process, shortly in advance of the bottom dead centre. The force-controlled stamping process is initiated. Finishing this, the linear motors start the return stroke with the electromagnets deactivated.

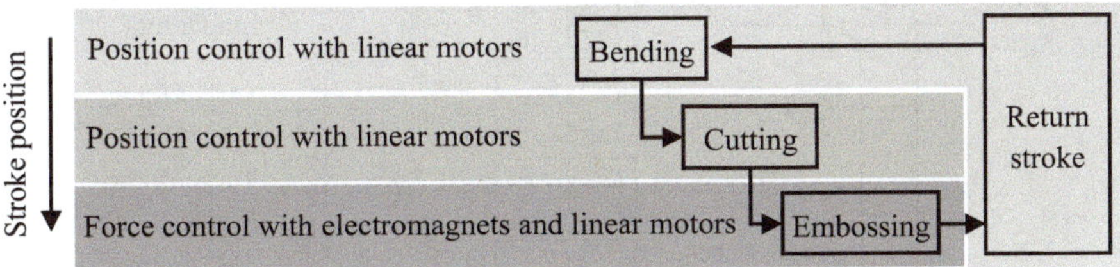

Fig. 3 Drive control sequence over the stroke in relation to the forming processes

4.1 Linear Motors

If the linear motors are activated, they are each controlled by a cascade controller consisting of position, speed and current control (see Fig. 4). The target position is specified by the higher-level Simotion controller and processed directly by a position controller.

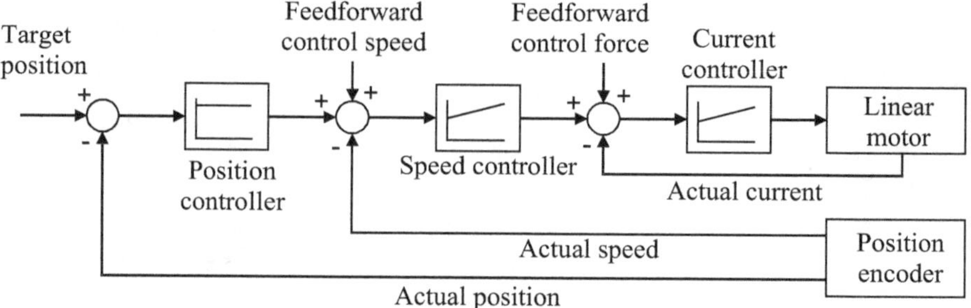

Fig. 4 Structure of a cascade controller with current-, speed- and position-controller

A desired stroke trajectory can be stored in a table of setpoints. In this case the target speed is forwarded from the position controller output to the drive control of the linear motors via the integrated ProfiBus interface. The drive controller uses the speed information from the specified motion profile to feedforward control the speed controller. The controller calculates the linear motor force from the actual motor current. This force feedforward control takes place directly before the linear motors are activated and the brakes are deactivated. By this, the moment of inertia of the press ram is also taken into account as an additive force [11]. This allows the linear motors to apply the required holding force triggered by the ram weight directly after the axes are released.

Between the drives occur interactions because all four linear motors are mechanically coupled to each other via the ram. Without load distribution between the motors, it can happen that one motor alone provides the force for ram positioning or, in the worst case, the motors work against each other. In this case, the linear motors fail to reach their maximum performance. To address this challenge, a control system for force sharing, as outlined in [12], was implemented between the linear motors (see Fig. 5).

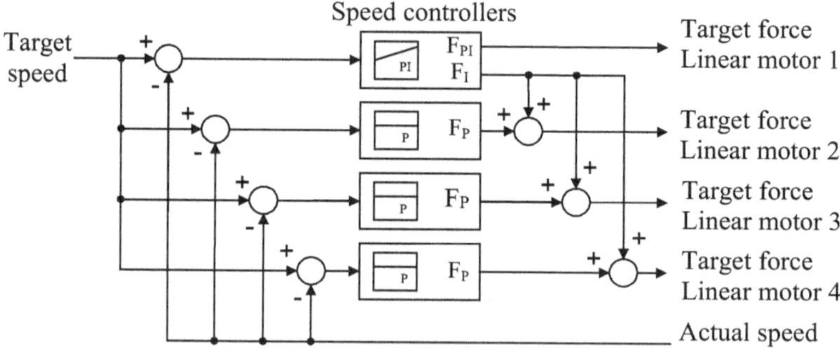

Fig. 5 Force sharing control loop between the linear motors

For this purpose, one linear motor was initially configured as the leading positioning axis and the other three motors as synchronised axes. All synchronised axes are connected as following axes. To ensure that the slave axes do not work dynamically against the master axis, the speed controller of the master axis was designed as a PI-controller and the controllers of the slave axes as P controllers.

4.2 Electromagnets

As soon as the effective range of the electromagnets is reached, control is initiated via the electromagnets. Depending on the control deviation, the DC controllers can adjust the magnetic current in the armature circuit in order to achieve the desired coining force.

In contrast to the linear motor control, the force coupling is replaced via a parallel switching interface, whereby the control is implemented as in Fig. 6 on one of the two current converters, while the second simply takes over the armature control set of the master, and thus both electromagnets receive the same current. In the Sinamics DCM controller, a speed controller is reused as a force controller, and the analogue force signal from the charge amplifier is connected to the actual speed value. In the case of force coupling between electromagnets and linear motors, the force exerted by the linear motors is supplied to the electromagnet's force controller as a feedforward loop.

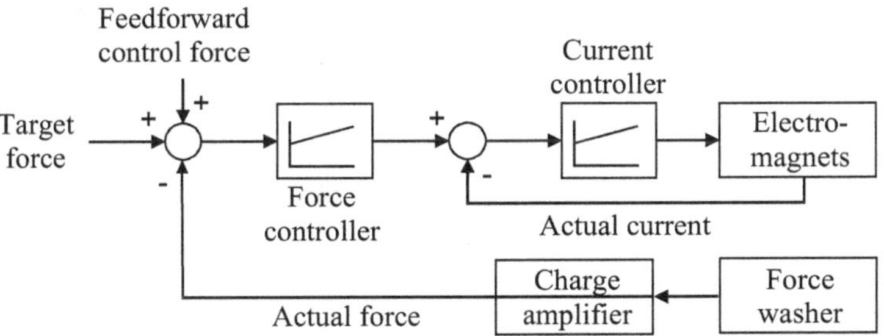

Fig. 6 Structure of the control loop for the electromagnets

5 Discussion

5.1 Experimental Results

The hybrid drive control of the linear press with progressive die was experimentally tested with a cosine-shaped ram trajectory, a stroke height of 12 mm and a stroke rate of 15 strokes per second, as well as an embossing force of 20 kN and a break of 1.4 s at the bottom dead centre for the emboss force control. The current stroke position, linear motor forces and the stamping force were measured in dependence of time with and without electromagnetic force control (see Fig. 7). The initial positive motor force demonstrates the initial holding of the ram in position by the linear motors against its weight. As the ram moves downward, the blankholder forces set in, causing the linear motor force to change sign. Slight deviations of the ram's movement from the targeted trajectory can be identified at Fig. 7, Pos. 1 and 2. These deviations are attributable to the interaction between the cutting punch and the sheet metal. The rupture of the sheet metal results in a cutting impact. At this point, the control system of the linear motors cannot work fast enough against the sudden change in force, so the stamping die hits the sheet unintentionally (see Fig. 7, Pos. 3). Subsequently, the machine switches to force-controlled mode, with the electromagnets remaining in this state for a duration of 1.4 s (see Fig. 7, Pos. 4). After the embossing process the electromagnets are deactivated, and the linear motors start again with the return stroke. As the cutting punches rise from the sheet, the linear motors experience a brief positive increase in force at Fig. 7, Pos. 5 due to the sheet stripping.

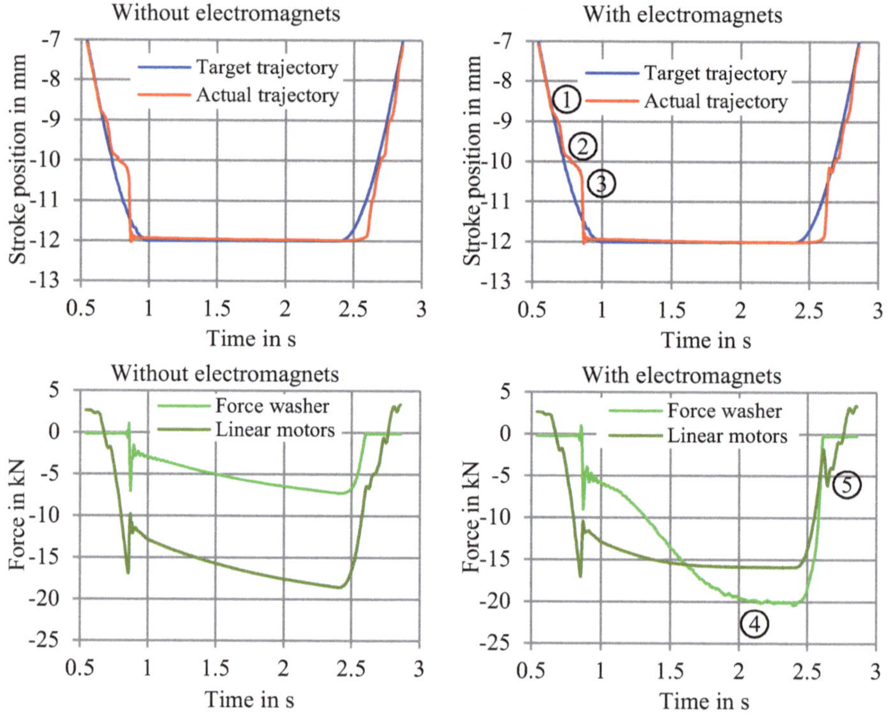

Fig. 7 Stroke position and force curves with and without electromagnets

The force curves show that the linear motors are very close to their performance limits without the electromagnets and that the necessary embossing force is not achieved due to the hold-down forces. This also indicates that the requisite embossing depth has not been attained. The electromagnets allow the embossing force to be precisely controlled, ensuring consistent embossing depth.

5.2 Outlook

With the implemented control system, small sheet metal components can be produced in position- and force-controlled forming processes. However, the control system can still be expanded and improved. On the one hand, the transition between position and force control should be further improved. Another aim is to increase the stroke rate. To achieve this objective, it is necessary to further optimise the control parameters. In addition, the position and force target trajectory must be tuned in order to achieve a better match with the progressive die and its engagement height.

In conclusion, plans are in place to analyse the produced components using optical surface measurements and compare the embossing quality between components produced with the hybrid press and a conventional eccentric press.

Acknowledgements. Funded by the Deutsche Forschungsgemeinschaft (DFG, German Research Foundation)—Project ID: 239414277

Competing Interests. The author(s) has no competing interests to declare that are relevant to the content of this manuscript.

References

1. Birkert, A., Haage, S., Straub, M.: Umformtechnische Herstellung komplexer Karosserieteile. Springer, Berlin, Heidelberg (2013)
2. Moghaddam, M.J., Soleymani, M.R., Farsi, M.A.: Sequence planning for stamping operations in progressive dies. J. Intell. Manuf. **26**(2), 347–357 (2015)
3. Xu, Z., Li, Z., Zhang, R., et al.: Fabrication of micro channels for titanium PEMFC bipolar plates by multistage forming process. Int. J. Hydrogen Energy **46**(19), 11092–11103 (2021)
4. Havinga, J., van den Boogaard, T., Dallinger, F., et al.: Feedforward control of sheet bending based on force measurements. J. Manuf. Process. **31**, 260–272 (2018)
5. Doege, E., Behrens, B.-A.: Handbuch Umformtechnik. Springer, Berlin, Heidelberg (2016)
6. He, Y., Luo, X., Wang, X.: Research and simulation analysis of fuzzy intelligent control system algorithm for a servo precision press. Appl. Sci. **14**(15), 6592 (2024)
7. Schepp, F.: Linearmotorgetriebene pressen für die stanztechnik. Shaker, Aachen (2002)
8. Gao, J., Zhao, S., Li, J., et al.: Study on dynamic characteristic and forging capacity of the direct drive press with symmetrical toggle booster mechanism driven by TPMLM. Proc. Inst. Mech. Eng., Part B: J. Eng. Manuf. **235**(11), 1800–1809 (2021)
9. Krimm, R., Behrens, B.-A., Reich, D.: Hybridaktorischer Pressenantrieb*/Hybrid Actoric Press Drive - Press drive made of linear motors and electromagnets for the manufacturing of complex components. Wt Werkstattstech. Online. **105**(10), 747–752 (2015)
10. Krimm, R., Behrens, B.-A.. Reich, D.: Concept of a linear hybrid press drive. Konf. Zum Symp. Autom. Syst. Technol., 67–71 (2016)
11. Siemens: SIMATIC S7-1500: Hanging loads with SIMATIC axis technology object and SINAMICSS. https://support.industry.siemens.com/cs/ae/en/view/109807260. Last accessed 23 May 2025
12. Siemens: SINAMICS S: DCC load sharing. https://support.industry.siemens.com/cs/ae/en/view/38470057. Last accessed 21 May 2025

Experimental Investigations on the Influence of Geometric Features and Process Parameters on the PBF-LB/M Manufacturability of Strut-Based Lattice Structures

Lukas Melzig(✉), Dominik Rauner, and Michael F. Zaeh

Technical University of Munich, TUM School of Engineering and Design, Institute for Machine Tools and Industrial Management (iwb), Boltzmannstraße 15, 85748 Garching, Germany
lukas.melzig@iwb.tum.de

Abstract. The efficient combustion of hydrogen in gas turbines is considered a key aspect of a sustainable power generation. However, the challenging combustion behavior of hydrogen requires adjustments to the burner design. The integration of lattice structures, which can be manufactured by powder bed fusion of metals using a laser beam (PBF-LB/M), has been demonstrated to stabilize the flame and mitigate emissions. However, the use of lattice structures manufactured by PBF-LB/M is restricted due to the constrained part quality. In this study, the influence of geometric features and process parameters on the PBF-LB/M manufacturability of strut-based lattice structures was investigated. In accordance with a design of experiments, lattice specimens were manufactured from the nickel-based superalloy Inconel 718. The dimensional accuracy and the powder removal were evaluated non-destructively by applying micro-computed tomography, while digital microscopy was employed to analyze the surface roughness. The results indicated that the geometric characteristics of the lattice structures and the process parameters have a significant influence on the manufacturability. It was shown that decreasing strut diameters and increasing volume fractions of the lattice structures resulted in a reduced dimensional accuracy, an increased surface roughness, and, thus, in an insufficient powder removal.

Keywords: Metal additive manufacturing · Inconel 718 · Combustion of hydrogen · Micro-computed tomography · Tailored process adaptations

1 Introduction

Lattice structures can be defined as the periodic arrangement of unit cells, forming a porous structure of interconnected faces or struts. They exhibit a broad spectrum of applications and are progressively being used in the energy sector [1]. In this case, the integration of lattice structures within the combustion chamber of gas turbines can enable the secure and efficient combustion of hydrogen, thereby reducing greenhouse gas emissions within the energy sector [1, 2].

L. Overmeyer and B.-A. Behrens (eds.), *Production at the Leading Edge of Technology*, Lecture Notes in Production Engineering, https://doi.org/10.1007/978-3-032-19524-1_38

Powder bed fusion of metals using a laser beam (PBF-LB/M) allows for the fabrication of intricate lattice structures tailored for their applications, due to its considerable design freedom [3]. However, process-induced limitations, specifically the diminished dimensional accuracy, the elevated surface roughness, and the impaired powder removal, prevent the full-scale industrial adoption of lattice structures manufactured by PBF-LB/M [4, 5]. The lack of support structures and the usually small volume of solid material within the lattice structures lead to an insufficient dissipation of the heat induced by the laser beam. This causes an increase in the size of the melt pool, resulting in a thickening of the lattice struts. In addition, the down-skin of the inclined struts, which typically have overhang angles of less than 45°, in particular is susceptible to overheating, whereby particles from the surrounding powder bed are sintered to the downside of the lattice struts [5–7]. Consequently, efforts are being made in industry and academia to enhance the PBF-LB/M manufacturability of lattice structures through modified build strategies.

In their study, Ulff et al. [8] demonstrated that the dimensional accuracy of body-centered cubic (BCC) lattice structures can be enhanced through local adjustments to the laser power. The authors also found that different strut diameters resulted in various and differently pronounced geometric deviations, which require lattice structure-specific adjustments to the process parameters. However, a targeted adjustment of the PBF-LB/M process parameters necessitates a comprehensive investigation into the respective influence of these parameters on the manufacturability of diverse lattice types.

Di Prima et al. [9] examined the impact of the laser power and the scan speed on the thickness of the struts and the static mechanical properties of regular truncated dodecahedron lattice structures. They showed that the thickness of the lattice struts increased with an elevated laser power and a decreased scan speed, thereby identifying a clear influence of these parameters on the dimensional accuracy. The authors were also able to correlate differences in the mechanical properties due to different thicknesses of the lattice struts with the applied process parameters.

Großmann et al. [10] exhibited that, in addition to the process parameters, the exposure strategy significantly influences the manufacturability of lattice structures. The hatch exposure was found to be less suitable for the manufacturing of filigree lattice structures compared to the contour exposure due to necking effects at the nodal points.

Moj et al. [11] utilized micro-computed tomography (μCT) measurements to examine the influence of geometric features of different lattice structures on their respective PBF-LB/M manufacturability. For this purpose, six distinct lattice types, each comprising two different unit cell sizes and volume fractions, were examined. The actual volume fraction (VF) and dimensional deviations were determined. The VF is defined as the proportion of solid material in the volume of a unit cell. For periodic lattice structures, the VF of a unit cell can be transferred to the entire structure. The authors identified variations in the measured VF (MVF) compared to the designed VF (DVF) for all examined lattice types, with the MVF predominantly exceeding the DVF. These deviations were more evident in smaller unit cells than in larger ones.

It can be concluded that both the process parameters and the geometric features of the respective lattice structure have a significant influence on the manufacturability. However, recent studies have only investigated either the influence of the process parameters on a specific lattice structure or the influence of different lattice types for given process parameters. To leverage the full potential of lattice structure-specific adjustments to the build strategy, it is imperative to consider the influence of the process parameters and the lattice structures in conjunction. In this study, therefore, the influence of geometric features and process parameters on the manufacturability of different strut-based lattice structures (SBLS) manufactured from the nickel-based superalloy Inconel 718 via the PBF-LB/M process was investigated.

2 Materials and Methods

2.1 Experimental Setup

In preliminary experiments, a parametrized specimen design (see Fig. 1) was derived, which ensures a good suitability for the non-destructive testing using µCT and for the examination using a digital microscope. The geometry consisted of two main components, a base, constructed with an isosceles triangular ground plane, which was utilized to ensure the precise positioning of the specimens during the µCT scan, and the respective SBLS placed on top of it. The different SBLSs were specified by their respective lattice type, the size of the unit cell s_c, and the diameter d_s of the lattice struts. To determine the influence of geometric features and the process parameters on the manufacturability of different SBLSs, a two-part design of experiments (DoE) was employed. Both partial DoEs were full-factorial.

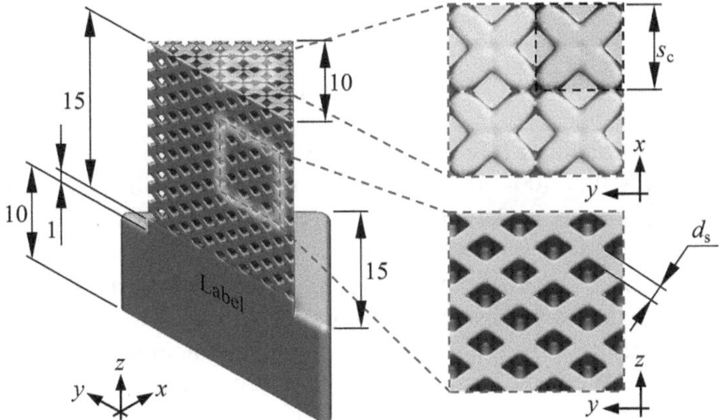

Fig. 1 Schematic of the parametrized specimen with the strut-based lattice structure specified by the lattice type, the unit cell size s_c, and the strut diameter d_s; dimensions in mm; z: build direction

The first part of the DoE examined the impact of varying strut diameters and VFs on the part quality, specifically considering the dimensional accuracy and the surface roughness, for BCC, Diamond, and Fluorite SBLSs. Strut diameters d_s of 0.3 mm, 0.6 mm, and 0.9 mm, along with VFs of 30%, 50%, and 70%, were examined. To obtain the desired VF for a given strut diameter, the size of the respective unit cell s_c was selected accordingly. Uniform process parameters were employed for the manufacturing of the different SBLSs with the PBF-LB/M process, whereby a laser power P of 285 W and a scan speed v of 960 mm/s was set. These values are referred to as standard parameters P_0 and v_0 in the further course of this study.

In the second part of the DoE, BCC SBLSs with $d_s = 0.6$ mm and a VF of 50% were manufactured with varying process parameters to determine the influence of these parameters on the dimensional accuracy and the surface roughness. The base of the specimens shown in Fig. 1 was consistently manufactured with the standard parameters. The laser power was reduced in two stages, commencing from P_0, while the scan speed was increased in two stages, starting from v_0. The values of the laser power and the scan speed (see Table 1) were selected to ensure that the combination of the lowest laser power and the highest scan speed resulted in a 50% reduction in the volumetric energy density E_v compared to the standard parameters (i.e., $E_v = 67.5$ J/mm^3). Given the laser power P, the scan speed v, the hatch distance h, and the layer thickness d, E_v can be calculated using the following equation [12]:

Table 1 Process parameters adjusted according to the design of experiments

Parameter	Levels		
Laser power P in W	285 (P_0)	235 (P_1)	185 (P_2)
Scan speed v in mm/s	960 (v_0)	1110 (v_1)	1260 (v_2)

$$E_v = P/(vhd) \tag{1}$$

The two-part DoE comprised a total of 35 specimens, which were manufactured in a single pass on an EOS M400-1 PBF-LB/M machine (EOS GmbH, Germany) from the nickel-based superalloy Inconel 718C (Oerlikon Metco Europe GmbH, Germany). The machine was equipped with a 1 kW continuous-wave Yb fiber laser with a wavelength of 1064 nm and a laser spot size of 90 μm. The specimens were randomly distributed on the build platform, which was kept at a constant temperature of 353 K throughout the manufacturing process. Argon 5.0 was used to maintain an inert atmosphere inside the process chamber. While the laser power and the scan speed were modified, the hatch distance and the layer thickness remained at $h = 110$ μm and $d = 40$ μm, respectively. Given that the influence of the scan pattern was not part of this study, a non-patterned line exposure with a rotation of 67° between two subsequent layers was selected. The powder recoating was executed using a rubber lip, ensuring a consistent powder application despite minor thermally induced distortions of the SBLSs. Subsequent to the PBF-LB/M and the powder removal, the specimens were cut from the build platform using an ABS 460 L band saw (KNUTH Werkzeugmaschinen GmbH, Germany). Prior to the various examinations, all SBLSs were thoroughly cleaned in a SONOREX SUPER RK 106 ultrasonic bath (BANDELIN electronic GmbH & Co. KG, Germany).

2.2 Evaluation

The non-destructive testing of the SBLSs was carried out using X-ray μCT. The μCT scans were generated with a YXLON Precision CT system (Comet Yxlon GmbH, Germany). Due to the size of the specimens and the targeted positioning of the scanner unit (60 mm) and the detector (1000 mm) relative to the position of the lattice structures, a voxel size of 12 μm was achieved. The μCT scans were performed uniformly with a voltage of 170 kV, a current of 120 μA, and a copper filter with a thickness of 1 mm.

The analysis of the μCT scans was conducted using the software Dragonfly (Object Research Systems Inc., Canada). The initial phase of the μCT evaluation required the three-dimensional (3D) reconstruction from two-dimensional (2D) gray value image stacks. The design and the dimensioning of the specimens, as well as the choice of the scan parameters, should serve to circumvent the occurrence of beam hardening, a phenomenon that is typically observed in X-ray μCT. The resolution of the image data necessitated only minor post-processing. Contrast-limited adaptive histogram equalization was applied to enhance the contrast of the images. To minimize image artefacts and for improved contrast, the specimens were tilted by 10° around the *x*- and *y*-axis.

The dimensional accuracy was evaluated by comparing the computer-aided design (CAD) geometry of the respective SBLS with the 3D reconstruction of the corresponding μCT scan. This required the conversion of the 3D reconstruction into a contour mesh. The necessary image segmentation, which was executed based on the Otsu method [13], was carried out automatically. Following a manual preliminary alignment of the contour mesh to the CAD geometry, an automated fine alignment based on an internal fitting algorithm was employed. The calculation of the volume enclosed by the contour mesh allowed for the comparison of the designed and the actual VF of the SBLSs. By comparing the contour mesh with the nominal CAD geometry, areas of geometric deviation were identified.

To validate and supplement the evaluation of the μCT scans, a qualitative analysis of the surface roughness of the down-skin and up-skin areas of the lattice struts was conducted using the digital microscope VHX-7000 (KEYENCE Deutschland GmbH, Germany). For this purpose, the front side (denoted by the label in Fig. 1) of the specimens, which were embedded in epoxy resin molds, was subjected to a sequential grinding process using SiC paper from grit sizes 320 to 2500. This was followed by a polishing with a diamond paste of 3 μm and 1 μm granularity. The magnification levels were selected individually for the different lattice structures to ensure a good resolution.

3 Results and Discussion

3.1 Dimensional Accuracy

Corresponding to the DoE, the MVF with the DVF for the different SBLSs and process parameters were compared in the scatter plots shown in Fig. 2. As illustrated in Fig. 2a to Fig. 2c, the MVF for all lattice types is notably higher than the DVF when a small strut diameter of 0.3 mm is selected. The highest value for the MVF at a DVF of 30% and $d_s = 0.3$ mm was observed for the BCC lattice structure at 39.9%. This was followed by the Fluorite structure at 64.6% at a DVF of 50%, and for the Diamond structure at

79.8% at a DVF of 70%. The latter revealed the highest absolute deviation across all strut diameters and VFs. As the strut diameter increased, a significant decrease in the MVF for all unit cells was observed, until the MVFs for some of the SBLSs fell below the DVF. The BCC and the Fluorite specimens with $d_s = 0.9$ mm fell the furthest below the DVF of 70%, with 68.5% and 68.7%, respectively. It can be seen that the relative change in the MVF with a growing strut diameter was reduced as the DVF increased. However, the influence of the VF was found to be significantly less pronounced than that of the strut diameter. It is assumed that, as the strut diameter decreased, the heat induced by the laser beam could no longer be dissipated sufficiently, as the cross-section available for the heat dissipation became smaller. This led to overheating, particularly in the down-skin areas, which caused areas of the surrounding powder to be melted, resulting in a pronounced dross formation and sintering of surrounding powder particles. The deviation of the MVF from the DVF is particularly attributed to the strong dross formation in the down-skin areas. As the strut diameter increased, the process heat was dissipated more effectively, leading to reduced overheating, less dross formation, and, consequently, a smaller deviation of the MVF from the DVF. However, a further investigation is necessary to determine mechanisms resulting in the undercutting of the DVF of some lattice structures.

As the VF increases at a constant strut diameter, the size of the unit cell s_c has to become smaller. This leads to an augmentation of the number of unit cells within the lattice structure, consequently leading to an increased area exposed to the laser beam. This results in an increased heat input per layer, leading to a considerable dross formation, even with larger strut diameters. The effect of the strut diameter and of the DVF on the MVF was observed to a similar extent for all unit cells. However, no clear trend was identified, confirming that adjustments to the process parameters to enhance the manufacturability should be tailored to the specific lattice structure.

Figure 2d shows the influence of the laser power and the scan speed on the MVF of a BCC lattice structure with a strut diameter of 0.6 mm and a DVF of 50%. It can be seen that the deviation of the MVF from the DVF increased with an elevation in the laser power and a decreasing scan speed. With an MVF of 51.9%, the highest discrepancy was determined for the standard process parameters. The lowest laser power of 185 W and the highest scan speed of 1260 mm/s resulted in the lowest deviation of the DVF with an MVF of 50.1%. It is therefore evident that the laser power and the scan speed have a quantifiable impact on the actual VF, consequently affecting the dimensional accuracy of the lattice structures. As the laser power is reduced and the scan speed is increased, E_V decreases, leading to a reduced heat input, which in turn causes less pronounced overheating in the down-skin areas. As a result, there is less dross formation and there are fewer sintered powder particles. A more detailed analysis of the influence of the respective process parameters will be part of further investigations.

The comparison of the 3D reconstruction and the CAD geometry revealed that dimensional deviations primarily occurred in the down-skin areas and are predominantly attributable to dross formation (see Fig. 2e). An increase in the dross formation in the build-up direction could not be determined. Under standard process parameters, the BCC lattice structure with $d_s = 0.3$ mm and a DVF of 30% exhibited the highest dimensional deviation of 0.03 ± 0.05 mm on average. The Diamond lattice structure,

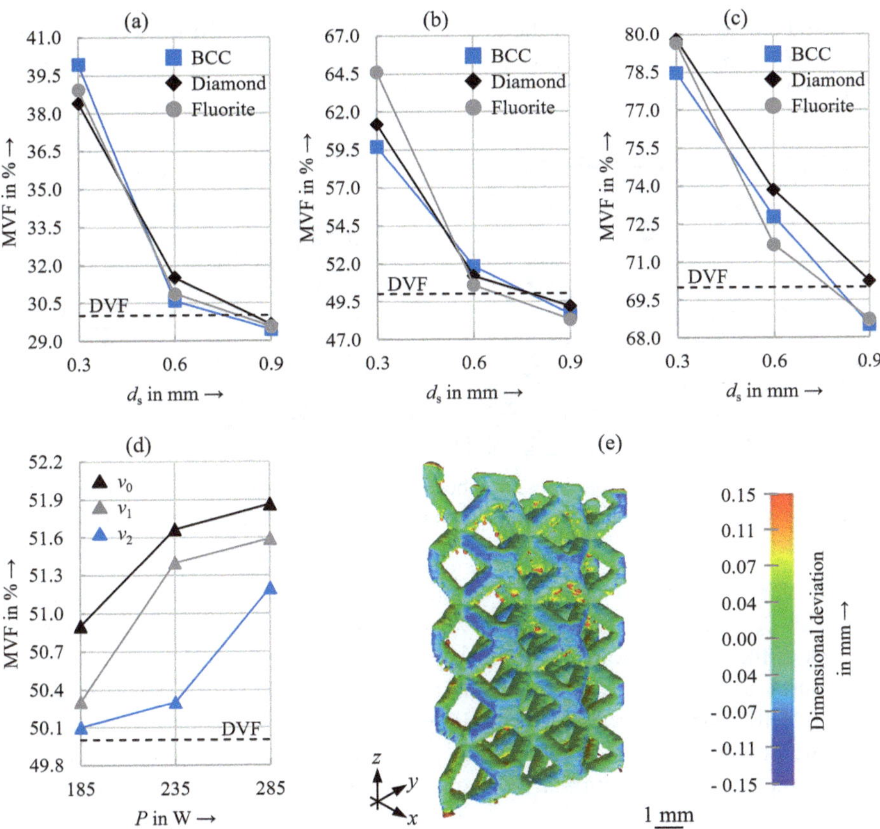

Fig. 2 Influence of the strut diameter d_S, the designed volume fraction (DVF), and the lattice types body-centered cubic (BCC), Diamond, and Fluorite on the measured volume fraction (MVF): **a** DVF of 30%, **b** DVF of 50%, and **c** DVF of 70%; **d** influence of the laser power P and the scan speed v ($v_0 = 960$ mm/s, $v_1 = 1110$ mm/s, and $v_2 = 1260$ mm/s) on the MVF for a BCC lattice structure with $d_S = 0.6$ mm and a DVF of 50%; **e** visualization of dimensional deviations of a BCC lattice structure with $d_S = 0.9$ mm and a DVF of 30%; z: build direction

with $d_S = 0.9$ mm and a DVF of 30%, deviated the least from the CAD geometry, with an average of 0.01 ± 0.06 mm.

3.2 Surface Roughness and Powder Removability

As shown in Fig. 3, the pronounced dross formation and sintered powder particles in the down-skin areas of the depicted BCC lattice structure with $d_S = 0.9$ mm and a VF of 30% resulted in an elevated surface roughness compared to the up-skin areas. The qualitative comparison of the microscopy images revealed that with a decreasing strut diameter and an increasing VF, the surface roughness increased in both the down- and the up-skin areas. This was observed for all examined unit cells. Lattice structures manufactured with a reduced E_v compared to the standard parameters (i.e., $E_v = 67.5$ J/mm^3) exhibited a decreased surface roughness in the down- and up-skin areas. To fully determine the influence of the process parameters, a quantitative analysis of the surface roughness is required, which is the subject of further investigations.

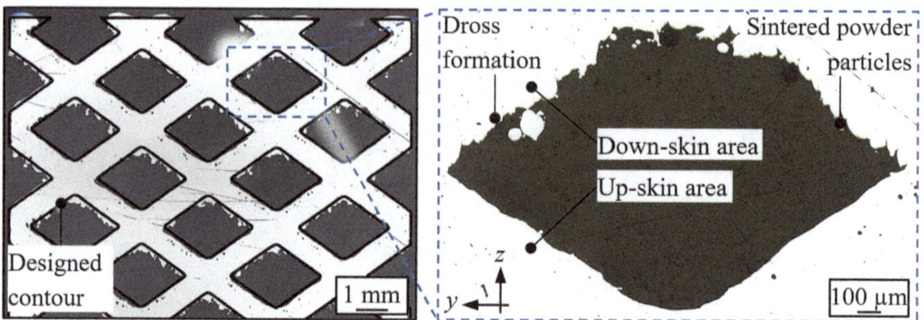

Fig. 3 Analysis of the surface roughness of a body-centered cubic (BCC) lattice structure with a strut diameter d_S of 0.9 mm and a designed volume fraction (DVF) of 30%; z: build direction

The evaluation of the μCT scans revealed the presence of powder residues in all SBLSs with $d_S = 0.3$ mm and VFs of 50% and 70%. For these specimens, the highest deviations of the VF and from the nominal geometry, as well as the qualitatively most elevated surface roughness, were observed. It is assumed that in these cases, the actually open-pored lattice structure was partially clogged due to the pronounced dross formation, making the powder removal infeasible. In addition to closed pores, the high surface roughness increased the frictional resistance, which further complicated the powder removal.

4 Conclusions and Outlook

In this study, the influences of geometric features and process parameters on the PBF-LB/M manufacturability of SBLSs were investigated. It was shown that the lattice type, the strut diameter, and the VF, as well as the laser power and the scan speed influence the dimensional accuracy, the surface roughness, and subsequently the powder removability. Decreasing strut diameters and increasing VFs caused overheating in the down-skin areas of the lattice struts for all lattice types. This resulted in a pronounced dross formation and, thus, led to an elevated surface roughness and a diminished dimensional accuracy. Reducing the E_V through a decrease in the laser power and an increased scan speed resulted in less overheating, less pronounced dross formation, and, thus, in a reduced surface roughness and an improved dimensional accuracy. It was shown that modifications to the process parameters result in an enhanced manufacturability. However, such adaptations need to be tailored to align with the distinct characteristics of the respective SBLS.

To exploit the full potential of lattice structure-specific adjustments to the build strategy, the influence of the scan pattern on the dimensional accuracy and the surface roughness of SBLSs will be determined in future studies. A quantitative analysis of the surface roughness will be conducted to support this objective.

Acknowledgements. The results of this experimental study were obtained as part of the project ENERGIZE (523881008), funded by the German Research Foundation (DFG). The authors would like to express their sincere gratitude towards the DFG for the trustful cooperation.

Competing Interests. The author(s) has no competing interests to declare that are relevant to the content of this manuscript.

References

1. Boretti, A., Huang, A.: AI-driven DfAM of aeronautical hydrogen gas turbine combustors. Int. J. Hydrogen Energy **77**, 851–862 (2024)
2. Ahmad, S., Ullah, A., Samreen, A., Qasim, M., Nawaz, K., Ahmad, W., Alnaser, A., Kannan, A.M., et al.: Hydrogen production, storage, transportation and utilization for energy sector: A current status review. J. Energy Storage **101**, 113733 (2024)
3. Blakey-Milner, B., Gradl, P., Snedden, G., Brooks, M., Pitot, J., Lopez, E., Leary, M., Berto, F., et al.: Metal additive manufacturing in aerospace: A review. Mater. Des. **209**, 110008 (2021)
4. Sun, C., Wang, Y., McMurtrey, M.D., Jerred, N.D., Liou, F., Li, J.: Additive manufacturing for energy: A review. Appl. Energy **282**, 116041 (2021)
5. Horn, M., Koch, L., Schafnitzel, M., Schmitt, M., Schlick, G., Schilp, J., Reinhart, G.: Parametric compensation scheme for increasing the geometrical accuracy of lattice structures in medical implants produced by powder bed fusion. Procedia CIRP **104**, 839–844 (2021)
6. Vrána, R., Koutný, D., Paloušek, D., Pantělejev, L., Jaroš, J., Zikmund, T., Kaiser, J.: Selective laser melting strategy for fabrication of thin struts usable in lattice structures. Materials **11**, 1763 (2018)
7. Bagheri, Z.S., Melancon, D., Liu, L., Johnston, R.B., Pasini, D.: Compensation strategy to reduce geometry and mechanics mismatches in porous biomaterials built with selective laser melting. J. Mech. Behav. Biomed. Mater. **70**, 17–27 (2017)
8. Ulff, N., Leingang, E., Schubert, J., Zanger, F.: Improvement of manufacturing accuracy of graded Ti-6Al-4V BCC lattice structures by local laser power adaption. Procedia CIRP **124**, 808–813 (2024)
9. Di Prima, M., van Belleghem, S., Badhe, Y., Snodderly, K., Porter, D., Burchi, A., Gilmour, L.: Build parameter influence on strut thickness and mechanical performance in additively manufactured titanium lattice structures. J. Mech. Behav. Biomed. Mater. **151**, 106369 (2024)
10. Großmann, A., Gosmann, J., Mittelstedt, C.: Lightweight lattice structures in selective laser melting: Design, fabrication and mechanical properties. Mater. Sci. Eng., A **766**, 138356 (2019)
11. Moj, K., Owsiński, R., Robak, G., Gupta, M.K., Scholz, S., Mehta, H.: Measurement of precision and quality characteristics of lattice structures in metal-based additive manufacturing using computer tomography analysis. Measurement **231**, 114582 (2024)
12. Meiners, W.: Direktes Selektives laser Sintern einkomponentiger metallischer Werkstoffe. Shaker, Aachen (1999)
13. Otsu, N.: A threshold selection method from gray-level histograms. Automatica **11**, 62–66 (1975)

Aritificial Intelligence in Production

Comparison of Unsupervised Learning Methods for Anomaly Detection of Green Compacts in Powder Pressing

Nils Niedernostheide(✉), Olcay Özgün, Christopher Prinz, and Bernd Kuhlenkötter

Chair of Production Systems, Ruhr-University Bochum, Bochum, Germany
niedernostheide@lps.ruhr-uni-bochum.de

Abstract. Powder metallurgy is a basic technology capable of producing complicated parts close to their final contours. As a result, in many cases, complex post processing is not required, making the process a cost-effective production process, especially in mass production. The process step of powder pressing, in which the green compacts are shaped for subsequent sintering, has proven to be the main factor influencing quality deviations in the final component. This study analyzes a pressing process of an industrial partner that produces green compacts simultaneously in several cavities. Defective green compacts are to be identified and sorted out to enhance transparency and resource efficiency after pressing. For this purpose, deviating process sequences are to be recognized using anomaly detection. Acoustic data and process data, such as force curves, are analyzed to detect anomalies. Due to the spatial conditions, process parameters are recorded collectively across all cavities on the upper and lower slides. Unsupervised learning methods such as envelope curve, isolation forest, dynamic time warping, and autoencoder are used for anomaly detection. These approaches are compared with each other and evaluated in terms of their suitability for recognizing incorrectly pressed green compacts. The aim is to test which approach is best suited to recognizing anomalies and preventing defective green compacts from passing through the subsequent sintering process.

Keywords: Anomaly detection · Unsupervised learning · Powder pressing

1 Introduction

Powder metallurgy (PM) is a processing technique for the synthesis of alloys and composite materials [1] and is being investigated in science primarily for its versatility in the production of materials in the powder state that cannot be synthesized in the liquid state [2]. In industry, PM is used in the mass production of uniform products with high demands on component precision [3]. Depending on the manufacturing and reprocessing process, the metal powder must be pressed into a solid mass. In sintering processes, as considered in this paper, such pre-pressing of the powder is necessary [4]. During powder pressing, the powder is compacted into a green compact and thus influences the size, shape, and properties of the end product [1]. This makes powder pressing a critical step in the sintering process with regard to component quality [5].

L. Overmeyer and B.-A. Behrens (eds.), *Production at the Leading Edge of Technology*,
Lecture Notes in Production Engineering, https://doi.org/10.1007/978-3-032-19524-1_39

This paper examines a powder press of an industrial partner in which green compacts are pressed simultaneously in up to four cavities and components for the automotive sector are produced, which are subsequently sintered. The aim is to detect anomalies in the pressing process to identify and sort out defective green compacts after the pressing process and thus prevent them from passing through the further sintering process. Acoustic data, range of movement, and force curves of the upper and lower press punches are available for this purpose. The quality of the green compacts was not recorded, so no labels are available, and the quality must be determined as 'good' or 'bad' based on the process data using an unsupervised anomaly detection.

Anomaly detection can be used to identify abnormal operational and system faults [6]. Anomalies are data patterns that show deviating characteristics from normal instances [7], so that they were presumably generated by a different mechanism [8]. Accordingly, the task of anomaly detection (also known as outlier detection) is to identify the deviating data patterns [9]. This makes it possible to recognize anomalies in time series data at an early stage and thus detect errors [6]. A variety of methods are available for the detection of anomalies [7], whereby unsupervised methods are most frequently used where no labels are available.

Various applications can be found in the literature that detect anomalies based on vibration data. For example, envelopes are a widely used tool for detecting anomalies or faults in rolling bearings [10]. Weiss et al. use a dynamic time warping (DTW) and non-linear regression approach to monitor the cracking of green compacts in hydraulic metal powder presses [11]. There are also several approaches in which autoencoders (AE) are applied to acoustic data for anomaly detection [12, 13]. Oliveria et al. compare different unsupervised anomaly detection methods, such as AE and Isolation Forest (IF), for the identification of faults in heavy-duty railway operations [14]. Brito et al. show that IF can be used for anomaly detection of rotating machines, such as bearings and gearboxes [9].

2 Dataset and Methods

This chapter presents the data set and the pre-processing steps. The methods for anomaly detection and their implementation are also described.

2.1 Dataset

The data is recorded on a hydraulic press from komage (SE120-DL), which records the movement data and the force curves using its built-in movement sensors from MTS-Temposonics (RHM0250MD701S1G8100 /RHM0150MD701S1G8100) and force sensors from Wika (S-20 (14106998)). The Optimizer4D system from QASS GmbH is used to record the acoustic data. The data set contains data from 900 press strokes and contains only numerical time series. Each press stroke is approximately three seconds long and the data is recorded every millisecond, resulting in an approximate number of 3000 time steps. The acoustic data is compressed using a Fast Fourier Transform and consists of 512 frequency bands, with each frequency band corresponding to a time series. The 512th frequency band corresponds to a frequency of approximately 655 kHz and all other bands below are equidistant to 0 Hz. A time series is produced for the range of movement and the force curve. This means that for each press stroke, 514 time series over 3000 time steps are recorded for the upper and lower punches. Prior to training, the time series are adjusted to a uniform length of 3000 time steps so that all methods can process the data and the data are normalized to eliminate the influence of different value ranges (see Fig. 1), e.g. the range of movement from –73 to 19 mm and the force from 0 to 745 kN. In addition, the 900 press strokes are divided into 700 training data and 200 test data. Furthermore, the entire data set was manually checked for anomalies to ensure that the training data set consisted exclusively of normal data and, on the other hand, to be able to check in the test data set whether the ten anomalies present were correctly detected by the various methods after training. The fact that the training data set only consists of normal data gives the model a priori knowledge that needs to be considered. An anomaly in the present application is characterized by an additional deflection in the time series, which indicates, for example, an additional pressing (see Fig. 2), and can lead to deviations in the size, shape, and properties of the end product.

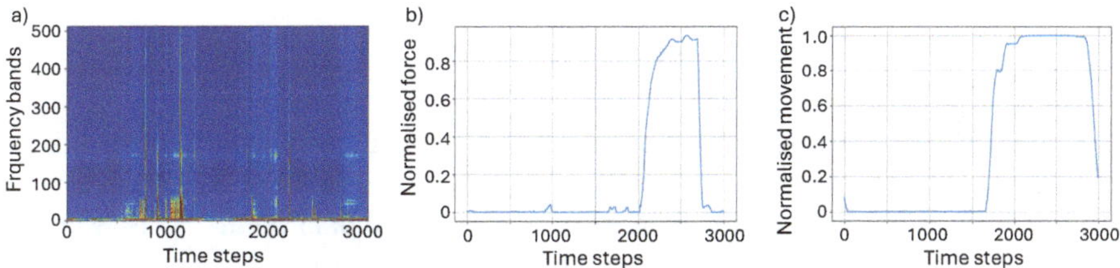

Fig. 1 **a** Example of acoustic data with a color scale for better visualization, **b** example of a force curve, and **c** an example of the range of movement

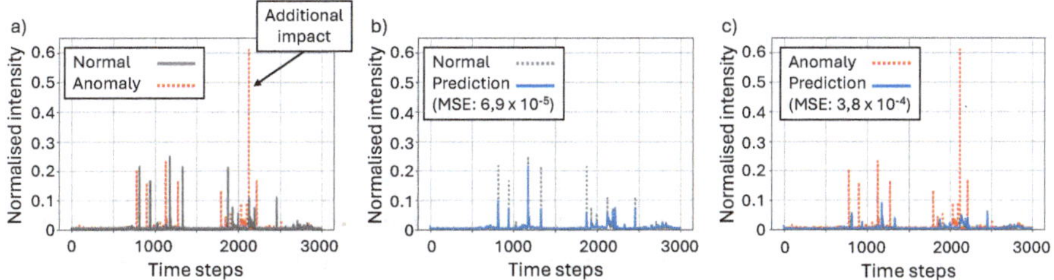

Fig. 2 **a** Comparison of a normal and abnormal course of the 400th frequency of the acoustic data, **b** comparison of the normal course and its prediction by the autoencoder and **c** comparison of the abnormal course and its prediction. The low MSE values result from normalization and must be considered in relation to each other

2.2 Methods

Envelope. For anomaly detection using envelope curves, two envelope curves are derived from the training data set, which specify the upper limit UE_t and the lower limit LE_t, where the individual data points of the time series may lie without being detected as an anomaly. These result from the mean value μ_t of the training data, the standard deviation

σ_t and selectable factor k according to formula 1:

$$UE_t = \mu_t + k \cdot \sigma_t, \ LE_t = \mu_t - k \cdot \sigma_t \tag{1}$$

For anomaly detection, the number of data points that lie outside the range of the two envelopes is then determined. A threshold value is set to determine the number above which an anomaly is present.

Dynamic Time Warping. DTW is used to establish a similarity relationship between two time series and can be used to measure the similarity between two sequences. For the similarity measurement, two input sequences are used and a matrix is generated that calculates the distances d_{x_i,y_j} of the value x_i of the first sequences for each time step to every other value y_j of the second sequences. The individual distances in the matrix from the first individual distance $d_{1,1}$ to the last individual distance $d_{n,m}$ are then used to form the path p for which the sum of the distances D of the path p is minimal (formula 2) [11]. For anomaly detection, a threshold value must be defined from which distance D an anomaly is present.

$$D = \sum d_{x_i,y_j} \ \forall \ d_{x_i,y_j} \ \in \ p = min[d_{1,1}, \ \ldots \ , d_{n,m}] \tag{2}$$

Isolation Forest. The IF approach offers an alternative to distance-based anomaly detection. With IF, no distances are calculated, but the lengths of random tree structures are determined. Each tree represents an attempt to isolate the data points. This is done by randomly dividing the data set, with each division leading to an enlargement of the tree structure. If anomalies have greater deviations from the normal data points, this means that isolating the anomalous data points requires fewer splits than for normal data points and the tree path for anomalous data points is therefore shorter. An anomaly score s is required for anomaly detection. For an instance x, the anomaly score results from formula 3 as follows [15].

$$s(x, n) = 2^{-\frac{E(h(x))}{c(n)}} \tag{3}$$

Here, n is the number of instances in the data set, $h(x)$ is the path length of an isolation tree, $E(h(x))$ is the average path length over several trees, and $c(n)$ is the average path length. The anomaly score is therefore in a value range of $0 < s \leq 1$. If this is approximately 1, it is an anomaly and if the value is less than 0.5, it is a normal instance. The advantage of this algorithm is that no complex and computationally intensive distances need to be calculated, resulting in a short calculation time [15].

Autoencoder. Another approach to detect anomalies that neither calculates the distance nor splits the data set is the use of an autoencoder. An autoencoder is a neural network that consists of two sub-models, an encoder and a decoder. The encoder compresses the input data into a latent representation with the aim of extracting the most important features. The latent space is characterized by a smaller dimension of the input data, which contains the most relevant information. The decoder aims to reconstruct the compressed data with minimal difference from the original input data [16]. If an autoencoder is trained on normal data and assuming that anomalies deviate from the normal data, the reconstruction error on anomalous data is larger than on normal data. Therefore, the reconstruction error can be used for anomaly detection [17]. The mean squared error (MSE) can be used as a metric for the reconstruction error (formula 4) [18].

$$MSE = \sum_{i=1}^{M} \sum_{j=1}^{N} (x_{ij}^{o} - x_{ij}^{re})^2 \quad (4)$$

Here, i describes the current time series, j the current time step, M the maximum value for the number of time series considered, N the maximum value for the time steps, x^O the original value, and x^{re} the reconstructed value. Figure 2 shows an example of anomaly detection based on acoustic data using an autoencoder.

2.3 Implementation

For the use of the envelope approach, three is selected for the selectable factor k. For DTW, fastdtw [19] is used and average time series are formed from the training data, which serve as the first input sequence. The respective time series of the test data serves as the second input sequence. The Isolation Forest Model from scikit-learn [20] with its default settings is used for the IF approach. A separate IF model is trained for each time series. The autoencoder used consists of two Conv2D layers, a BatchNormalisation layer, a Dropout layer, two LSTM layers, and two Conv2D Transpose layers from TensorFlow [21]. The hyperparameters used are 20 epochs, a learning rate of 0.0001, a dropout rate of 0.3, and a batch size of four.

3 Results and Discussion

In this chapter, the results of the anomaly detections are compared with each other. The accuracy, precision, recall, and F1-score are used as the basis for evaluation [22]. Due to the unbalanced test dataset (ten anomalies in 200 press strokes), the model achieves an accuracy of 95% if no anomalies are detected. The area under the curve (AUC) of the receiver operating characteristic curve (ROC) is also used as an evaluation metric and can vary between 0 and 1 [22], with random guessing leading to an AUC of 0.5 [23]. In this application, all anomalies must be detected to identify unusual pressing strokes and to exclude potentially faulty green compacts from further sintering process. The recall is therefore set to 1 so that all anomalies are recognized, even if normal time series are also identified as anomalies. This reduces the precision, which indicates how many detected anomalies are actually anomalies, and the F1-score, which is the harmonic mean of precision and recall. The values for the individual evaluation metrics are listed in Table 1.

Table 1 Evaluation of the anomaly detection of the acoustic, movement, and force data

	Recall = 1	Upper punch				Lower punch			
		Envelope	IF	AE	DTW	Envelope	IF	AE	DTW
Acoustic	AUC	**0.99**	**0.99**	0.91	0.61	**0.99**	**0.99**	0.94	0.82
	Accuracy	**0.97**	**0.97**	0.80	0.32	**0.98**	0.97	0.80	0.37
	Precision	**0.67**	**0.67**	0.02	0.05	**0.71**	0.67	0.02	0.07
	F1	**0.80**	**0.80**	0.33	0.10	**0.83**	0.80	0.33	0.14
Movement	AUC	**0.99**	0.19	**0.99**	0.46	**0.99**	0.50	**0.99**	0.52
	Accuracy	**0.99**	0.05	0.97	0.13	**0.98**	0.05	0.97	0.08
	Precision	**0.91**	0.05	0.67	0.05	**0.71**	0.05	0.67	0.05
	F1	**0.95**	0.10	0.80	0.10	**0.83**	0.10	0.80	0.10
Force	AUC	**0.99**	0.25	**0.99**	0.34	**0.99**	0.21	**0.99**	**0.99**
	Accuracy	**0.97**	0.05	**0.97**	0.05	**0.97**	0.05	**0.97**	**0.97**
	Precision	**0.67**	0.05	**0.67**	0.05	**0.67**	0.05	**0.67**	**0.67**
	F1	**0.80**	0.10	**0.80**	0.10	**0.80**	0.10	**0.80**	**0.80**

The anomaly detection using envelopes achieves an AUC of 0.99, an accuracy of at least 0.98, and an F1-score of at least 0.80 for all data. The IF approach achieves an AUC of 0.99 and an F1-score of 0.80 for the acoustic data and thus achieves the same results as the envelope approach for the upper punch. In the case of anomaly detection on the range of movement and the force curves, the IF achieves worse results than with the randomized frame, with an AUC of 0.50 and lower. Anomaly detection using AE achieves an AUC of over 0.90 on all data and achieves the same values for accuracy (0.97), precision (0.67), and F1-score (0.80) as the envelope curve approach for the force curves, but with a precision of 0.02 achieves poorer results on the acoustic data. DTW achieves the same values as AE and as the envelope curve approach on the force curve data of the lower punch. Otherwise, the anomaly detection using DTW achieves a precision and F1-score of less than 0.2 for the other data.

The differences in the performance of the IF on the acoustic data compared to the force curve and range of movement data could be due to the fact that 512 time series are available for the acoustic data and an incorrectly classified time series in the acoustic data does not directly lead to an anomaly, as the total number of frequencies that are classified as an anomaly are considered. This is not possible for the force curve and the range of movement, as only one time series is considered in each case. This could lead to greater robustness in the acoustic data and could be decisive, as anomaly detection using IF is based on the random division of the data set. If this is the case, increasing the number of isolation trees can be used in the future to improve the performance of force displacements and movements. One possible explanation for the poor results achieved by DTW is that DTW shifts the anomalies in time so that the total distance is lower and thus the similarity is increased. This can lead to local anomalies not being recognized, as only the calculated total distance is considered for anomaly detection. As a result, individual or a small number of large individual distances cannot be recognized and therefore the local anomalies cannot be detected either. The poorer results of the AE on the acoustic data compared to the force curves and the range of movement data are because the difference in the reconstruction error between normal and abnormal data is small for the acoustic data, and therefore more normal data are falsely detected as anomalies.

In the future, it should be checked whether an adjustment of IF and AE leads to better results than the envelope curve approach. It should also be checked whether the same results can be achieved with other powder press data and other pressing processes to validate the transferability of the results. Furthermore, it should be noted that the training data set was manually analyzed for anomalies in advance. The performance of the various methods can therefore decrease if anomalies are present in the training data set. For this reason, the influence of anomalies in the training data set on the performance of the methods must be analyzed.

4 Conclusion

This paper presents a comparison between envelope curve, dynamic time warping, isolation forest, and autoencoder for anomaly detection during powder pressing of green compacts. For this purpose, a data set of 900 press strokes is available that contains the acoustic data, range of movement and force curve of the upper and lower punches of the powder press. The ROC-AUC, accuracy, precision, recall, and F1-score are used as evaluation metrics, whereby the recall corresponds to the value 1, as it is crucial for the present application that all anomalies are recognized to sort out defective green compacts after the pressing. As a result of the comparison, it is shown that the envelope curve approach achieves the best values of the evaluation metrics on the available data compared to Dynamic Time Warping, Isolation Forest, and Autoencoder, whereby the Isolation Forest approach on the acoustic data of the upper punch and the Autoencoder on the force curves show comparable results to the envelope curve approach. The results obtained should be validated in future work and the influence of anomalies in the training data set on the performance of the methods should be investigated.

Competing Interests. The author(s) has no competing interests to declare that are relevant to the content of this manuscript.

References

1. Herrera Ramirez, J.M., Perez Bustamante, R., Isaza Merino, C.A., Arizmendi Morquecho, A.M.: Unconventional techniques for the production of light alloys and composites. Springer International Publishing, Cham (2020)
2. Ghods, S., Schultz, E., Wisdom, C., Schur, R., et al.: Electron beam additive manufacturing of Ti6Al4V: Evolution of powder morphology and part microstructure with powder reuse. Materialia **9**, 100631 (2020)
3. Akhtar, S., Saad, M., Misbah, M., Sati, M.: Recent Advancements in powder metallurgy: A review. Materials Today: Proceedings **5**, 18649–18655 (2018)

4. Mendoza-Duarte, J., Sagarnaga-Fernandez, M., Moreno-Resendiz, E., Medrano-Prieto, H., et al.: Aluminum-lithium alloy prepared by a solid-state route applying an alternative fast sintering route based on induction heating. Mater. Lett. **263**, 127178 (2020)
5. Krüger, M., Vogel-Heuser, B., Weiß, I., Trunzer, E.: Data-driven product quality monitoring in quality-critical forming processes. IFAC-PapersOnLine **54**, 220–225 (2021)
6. Du, S., Ma, X., Li, X., Wu, M. et al.: Anomaly detection based on principle of justifiable granularity and probability density estimation. In: 2023 China Automation Congress (CAC), pp. 1364–1367. IEEE, (2023)
7. Ding, Z., Fei, M.: An anomaly detection approach based on isolation forest algorithm for streaming data using sliding window. IFAC Proceedings Volumes **46**, 12–17 (2013)
8. Hawkins, D.M.: Identification of outliers. Springer, Netherlands, Dordrecht (1980)
9. Brito, L., Susto, G., Brito, J., Duarte, M.: An explainable artificial intelligence approach for unsupervised fault detection and diagnosis in rotating machinery. Mech. Syst. Signal Process. **163**, 108105 (2022)
10. Lee, D.-H., Hong, C., Jeong, W.-B., Ahn, S.: Time-frequency envelope analysis for fault detection of rotating machinery signals with impulsive noise. Appl. Sci. **11**, 5373 (2021)
11. Weiss, I., Vogel-Heuser, B., Trunzer, E., Kruppa, S.: Product quality monitoring in hydraulic presses using a minimal sample of sensor and actuator data. ACM Trans. Internet Technol. **21**, 1–23 (2021)
12. Chinnasamy, M., Sumbwanyambe, M., Hlalele, T.: Acoustic anomaly detection of machinery using autoencoder based deep learning. In: 2024 32nd Southern African Universities Power Engineering Conference (SAUPEC), pp. 1–6. IEEE, (2024)
13. Coelho, G., Matos, L., Pereira, P., Ferreira, A., et al.: Deep autoencoders for acoustic anomaly detection: experiments with working machine and in-vehicle audio. Neural Comput. & Applic. **34**, 19485–19499 (2022)
14. Oliveira, D.F.N., Vismari, L.F., de Almeida, J.R., Cugnasca, P.S., et al.: Evaluating unsupervised anomaly detection models to detect faults in heavy haul railway operations. In: 2019 18th IEEE international conference on machine learning and applications (ICMLA), pp. 1016–1022. IEEE, (2019)
15. Liu, F., Ting, K., Zhou, Z.-H.: Isolation forest. In: 2008 eighth IEEE international conference on data mining, pp. 413–422. IEEE, (2008)
16. Zhang, J., Creighton, D., Lim, C., Rolfe, B. et al.: Anomaly detection in the automotive stamping process: An unsupervised machine learning approach. IOP Conf. Ser.: Mater. Sci. Eng. **1307**, 12035 (2024)
17. Yang, S., Liu, S., Shang, P., Wang, H.: An overview of methods of industrial anomaly detection. In: 2024 7th international conference on Robotics, Control and Automation Engineering (RCAE), pp. 603–607, IEEE, (2024)
18. Engida, Z., Neto, H., Slonimer, A., Bedard, J. et al.: Anomaly detection in complex data: a practical application when outliers are few. In: OCEANS 2022, Hampton Roads, pp. 1–7. IEEE (2022)
19. PyPI: fastdtw. https://pypi.org/project/fastdtw/. Last accessed 24 April 2025
20. Scikit-learn developers: Isolation Forest. https://scikit-learn.org/stable/modules/generated/sklearn.ensemble.IsolationForest.html. Last accessed 24 April 2025
21. TensorFlow: tf.keras.Layer. https://www.tensorflow.org/api_docs/python/tf/keras/Layer
22. Yong, L., Nugroho, H.: Acoustic anomaly detection of mechanical failure: Time-distributed CNN-RNN deep learning models. In: Wahab, N.A., Mohamed, Z. (eds.) Control, instrumentation and mechatronics: Theory and practice, pp. 662–672, Springer Nature Singapore, Singapore (2022)
23. Ruff, L., Kauffmann, J., Vandermeulen, R., Montavon, G., et al.: A unifying review of deep and shallow anomaly detection. Proc. IEEE **109**, 756–795 (2021)

Towards Dynamic Resource Adaptation for Real-Time Containers

Rebekka Neumann(✉), Armin Lechler, and Alexander Verl

Institute for Control Engineering of Machine Tools and Manufacturing Units, University of Stuttgart, Stuttgart, Germany
rebekka.neumann@isw.uni-stuttgart.de

Abstract. The growing demand for flexible use of computing resources in automation has led to increased adoption of container-based virtualization. To achieve real-time behavior, system models and analysis are applied before the containers are deployed. However, not all system properties can be determined and modeled in detail at design time due to changing workloads, dynamic task placement, and task interference. To compensate for these uncertainties, resources for real-time containers are statically assigned and over-provisioned. This paper proposes an orchestration extension for adapting the computing resources of real-time containers at runtime to use system resources more efficiently. The architecture is designed conceptually and validated through analytical evaluation, indicating increased CPU availability for non-realtime containers under dynamic quota adaption.

Keywords: Container-based virtualization · Real-time containers · Resource allocation

1 Introduction

Control systems are evolving from rigidly connected, predefined shop floor devices to complex, responsive, interconnected systems [4]. Thus, flexible approaches, such as service-oriented architecture (SOA) are necessary to implement this dynamic environment. To enable hardware-independent software deployment, container-based virtualization is increasingly being incorporated into automation technology to provide these software services [8]. Technically, containers provide operating system (OS)-level virtualization for Linux processes, providing isolation from the host and other containers [3]. This is achieved using kernel features such as `namespaces` for process isolation and `cgroups` for resource management.

Despite the use of container-based virtualization, the potential for flexible resource provisioning has not yet been exploited. Although control applications can be deployed hardware-independently, the allocated resource budgets are predefined and pessimistically chosen, since these container placements are calculated at design time. To compensate for uncertainties in system load and shared

L. Overmeyer and B.-A. Behrens (eds.), *Production at the Leading Edge of Technology*, Lecture Notes in Production Engineering, https://doi.org/10.1007/978-3-032-19524-1_40

access to resources, such as network interfaces and caches, resources of control applications are over-provisioned. For instance, a CPU core, or even an entire compute node may be reserved for an application that does not require all its computing power. Tasks may be given more quota than necessary, which leads to inefficient resource utilization [5].

To counteract static resource reservation, available container resources must adapt to demand at runtime. This allows for unused resource capacities to be used by other applications. This work presents an extension to an orchestration framework that enables dynamic resource adaptation for RT control containers. he architecture is designed conceptually and evaluated analytically, demonstrating potential improvements in CPU availability for non-real-time workloads compared to conventional static allocation.

The remainder of this paper is structured as follows. In Sect. 2, related work regarding the orchestration, the process of resource allocation and deployment, of real-time containers is presented. In Sect. 3, the concept of the dynamic adaptation orchestration extension is outlined. In Sect. 4, the results of this paper are summarized, and an outlook for future work is given.

2 Related Work

In the field of RT container orchestration, the assignment of RT containers to nodes based on specific criteria is a widely discussed topic. In Wang et al. [13] and Dai et al. [2], deployment processes for RT containers are presented. An orchestration framework for control containers is presented in Walker et al. [12]. The open-source container orchestrator Kubernetes[1] is commonly used to manage containers in distributed systems. However, since Kubernetes was designed for web applications, it must be adapted to support the deployment of RT containers Sollfrank et al. [7]. Both Fiori et al. [3] and Struhár et al. [9] present Kubernetes extensions for deploying RT containers with a RT interface to scheduling using the Hierarchical Constant Bandwidth Server (HCBS) scheduling extension, presented in Abeni et al. [1]. In Walker et al. [11] a Kubernetes-based orchestration tool is presented for deploying RT containers with RT communication interfaces.

The aforementioned publications focus on container placement, considering different aspects, such as the latency of services, load balancing, and minimum number of nodes assigned. The performance of the tasks is monitored to ensure that the previously defined RT requirements are met; however, neither minimizing nor flexibly allocating resources at runtime is considered.

3 Conceptual Orchestration Extension for Resource Adaption

The aim of the presented orchestration extension is to increase the resource utilization flexibility of compute nodes while preventing performance degradation

[1] https://kubernetes.io.

for RT tasks. The reduction of CPU bandwidth is based on response time analysis. The response time is defined as the time between when a task becomes ready and when it is completed. In contrast to the execution time, the response time takes scheduling delays into account.

The framework adapts the CPU quota for RT containers and allocates the freed CPU time to compute-intensive, non-critical tasks. As shown in [10], the response time of a RT container is influenced by other co-located applications, due to context switches and cache interference. A distributed edge environment consisting of several compute nodes connected via RT network is assumed. Each compute node, referred to as *Node*, is real-time capable as defined in Neumann et al. [6]. The high-level architecture of the orchestrator is shown in Fig. 1 and the functionality of the core components are described below.

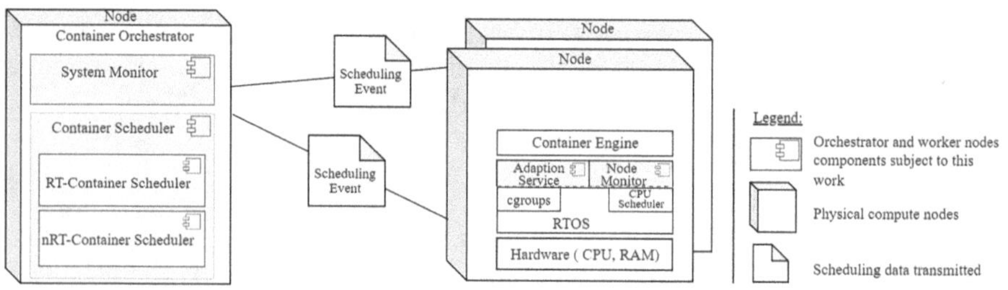

Fig. 1. Overview of the orchestration extension architecture

The *Container Orchestrator* includes services to deploy containers to nodes and monitor nodes. The *RT-Container Scheduler* selects the nodes to which the RT containers will be deployed. RT containers are assigned an initial CPU quota based on a specifically calculated worst case execution time (WCET), which will be assumed to be the maximum execution time a container will need even with influencing co-located containers. The WCET which will be referred to as the "initial quota" in the following. The initial quota depends on the resource access patterns and with this the interference of the co-located applications. In this work the initial quota is assumed to be pessimistic, as resources are usually in automation domain.

Optimizing the node selection process for RT containers is beyond the scope of this work, as it has been widely discussed in the literature. Non-RT (nRT) containers are scheduled by the *nRT-Container Scheduler* component. RT containers are given higher priority than nRT containers. Therefore, the nRT containers are scheduled after all RT containers have been placed. Both *Container Schedulers* generate scheduling events that trigger the deployment process. During operation, the *Node Monitor* component of the respective node continuously monitors node- and container-level metrics, such as resource consumption, deadline misses, and response time. Note that deadline misses are not tolerable and must be monitored to ensure they do not occur. The *Node Monitor* comprises the actual CPU time utilized by RT containers for every CPU core and period which are monitored by the *CPU Core Monitor* of each core. Based on response time analysis and actual resource utilization, the *Adaption Service* reduces the allocated CPU time. The *Adaption Service* is responsible for increasing and decreasing CPU quotas, as well as migrating services between cores. When the CPU quota is adapted, the*Adaption Service* directly changes the resource limits in the `cgroup` the corresponding container is assigned to, which then influences the scheduled limits. It must be confirmed that the applications do not exceed the requirements for other resources, such as RAM and I/O. The *System Monitor* of the *Container Orchestrator* keeps track of the actual utilization of every node (and every core) as well as the sum of the initial quotas of all RT tasks running on the nodes. The orchestration decisions of the *RT-Container Scheduler* are based on the sum of the initial quotas of all deployed RT containers, since these are assumed to be safe for predictable execution of RT containers. A new RT container can only be scheduled if enough quota is left to satisfy the worst-case demands of all RT containers including the potential new one on the node. Meanwhile, the *nRT-Container Scheduler's* orchestration decisions are based on actual CPU utilization because non-critical applications can use freed quota and can be interrupted any time.

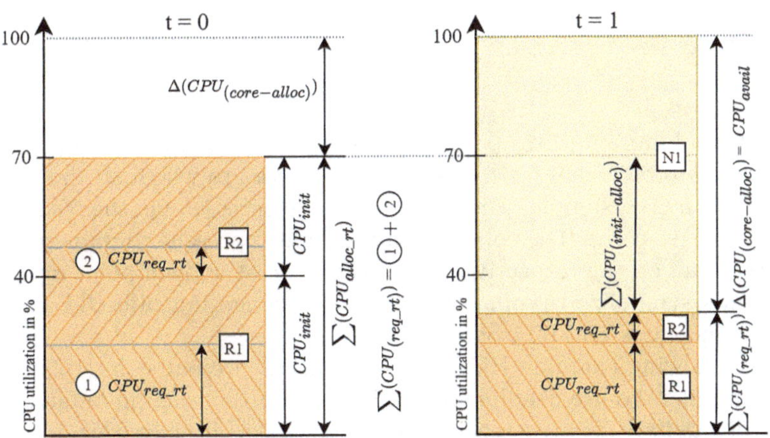

Fig. 2. Visualization of the *Node Monitor* properties of core 2

In automation, even a simple control application typically runs on an isolated CPU core or an entire node to ensure timing predictability and avoid runtime interference between applications that share resources.

The following example, divided into different subsequent scenarios, shows the potential for downsizing resources at runtime under strict monitoring and the benefits in terms of CPU utilization that can be used for other applications. The *CPU Core Monitor* attributes, aggregated and made available by the *Node*

Monitor, are visualized in Fig. 2. The values of the *Node Monitor* for the below described example are summarized in Table 1.

Table 1. Example scenarios for dynamic CPU quota adaption with multiple RT containers and one nRT container

NodeMonitor: CPU core 2		RT Container 1 (R1) $CPU_{init} = 40\ \%$	RT Container 2 (R2) $CPU_{init} = 30\ \%$	RT Container 3 (R3) $CPU_{init} = 20\ \%$	nRT Container 1 (N1)
t = 0	$\Sigma(CPU_{alloc\_rt}) = 70\ \%$ $\Sigma(CPU_{init\_rt}) = 70\ \%$ $\Sigma(CPU_{req\_rt}) = 31\ \%$ Quota adaption ↓	$CPU_{req\_rt} = 22\ \%$	$CPU_{req\_rt} = 9\ \%$	$CPU_{req\_rt} =$	$CPU_{avail} =$
t = 1	$\Sigma(CPU_{alloc\_rt}) = 31\ \%$ $\Delta(CPU_{(init-alloc)}) = 39\ \%$ $\Delta(CPU_{(core-alloc)}) = 69\ \%$	$CPU_{req\_rt} = 22\ \%$	$CPU_{req\_rt} = 9\ \%$	$CPU_{req\_rt} =$	$CPU_{avail} = 69\ \%$
t = 2	$\Sigma(CPU_{alloc\_rt}) = 90\ \%$ $\Sigma(CPU_{init\_rt}) = 90\%$ $\Sigma(CPU_{req\_rt}) = 63\ \%$ Quota adaption ↓	$CPU_{req\_rt} = 40\ \%$	$CPU_{req\_rt} = 30\ \%$	$CPU_{req\_rt} = 20\ \%$	$CPU_{avail} =$
t = 3	$\Sigma(CPU_{alloc\_rt}) = 63\ \%$ $\Delta(CPU_{(init-alloc)}) = 27\ \%$ $\Delta(CPU_{(core-alloc)}) = 37\ \%$	$CPU_{req\_rt} = 30\ \%$	$CPU_{req\_rt} = 21\ \%$	$CPU_{req\_rt} = 12\ \%$	$CPU_{avail} = 37\ \%$
t = 4	$\Sigma(CPU_{alloc\_rt}) = 50\ \%$ $\Sigma(CPU_{init\_RT}) = 50\ \%$ $\Sigma(CPU_{req\_rt}) = 11\ \%$ Quota adaption ↓	$CPU_{req\_rt} =$	$CPU_{req\_rt} = 8\ \%$	$CPU_{req\_rt} = 3\ \%$	$CPU_{avail} = 50\ \%$
t = 5	$\Sigma(CPU_{alloc\_rt}) = 11\ \%$ $\Delta(CPU_{(init-alloc)}) = 39\ \%$ $\Delta(CPU_{(core-alloc)}) = 89\ \%$	$CPU_{req\_rt} =$	$CPU_{req\_rt} = 8\ \%$	$CPU_{req\_rt} = 3\ \%$	$CPU_{avail} = 89\ \%$

The first column indicates points in time, t. The second column provides the *CPU Core Monitor* values for the isolated CPU core to which all the containers are allocated in the example. These values can be accessed via the *Node Monitor* for the respective cores. The columns on the right represent containerized applications that run on the respective core of the node at a given point in time. The colored fields symbolize running applications. Orange indicates RT containers, and yellow indicates non-RT containers.

In the following example HCBS scheduling, as presented in [1], is used for the RT containers to use deadline scheduling with fixed maximum quota at container level and fixed-priority scheduling for the tasks of the containers, to realize setting CPU bandwidths for containers.

Initial deployment of RT containers: Two RT containers, R1 and R2, are assigned to the (isolated) second CPU core of node one with the following scheduling metrics for initial quota: Period $P_1 = 1\ ms$, CPU quota $Q_1 = 0.4\ ms$ and $P_2 = 1\ ms$, $Q_2 = 0.3\ ms$. Implicit deadlines are assumed. During operation, the execution metrics are tracked for every running service and CPU core, then aggregated in the *Node Monitor*. At $t = 0$, 70 % of the quota of the isolated CPU core is allocated by RT processes, as represented by the metric $\sum(CPU_{alloc\_rt})$. The sum of the initial CPU quotas of all RT containers assigned to the respective core $\sum(CPU_{init\_rt})$ is at 70 % at this time. Based on the response time monitoring, the actual required bandwidth CPU_{req_rt} of R1 and R2 is determined. The sum of the actual required CPU time, denoted by $\sum(CPU_{req\_rt})$, is equal to 31 %, resulting in 39 % less CPU bandwidth necessary than initially allocated. The values for $\sum(CPU_{req\_rt})$ are given with safety factors to ensure predictable execution of a given set of periodic applications with defined input parameters, if the system configuration remains unchanged. To reduce the difference between the allocated and the necessary quotas, the quota adaptation process is triggered. The *Adaption Service* reduces the CPU limits for R1 and R2 accordingly. After the adaptation process, the applications run with the reduced quota and are continuously monitored. Completion of the adaptation process is reported to the *System Monitoring* component of the *Container Orchestrator*. At $t = 1$, the $\sum(CPU_{alloc\_rt})$ equals the $\sum(CPU_{req\_rt})$, leaving a difference of 39 % CPU time per period. This is expressed by the value $\Delta(CPU_{(init-alloc)})$. The free quota of the core is described with $\Delta(CPU_{(core-alloc)})$. It can be used by non-critical, non-real time applications.

Adding another RT container: At $t = 2$ a third RT container, R3, is be allocated on the same core as R1 and R2. Immediately after the scheduling event is received by the node, N1 is stopped and the quota of R1 and R2 are re-increased to their initial values. R1 and R2 will most likely need more CPU bandwidth than before with the co-located RT container R3. This is due to different effects, such as context switches and interference between co-located containers that share resources, such as the cache. The re-adaptation process must be at least as fast as the deployment process, including pulling the container image and starting the container, if necessary. Future work must detail this process to enable the deterministic re-adaptation of the quota to the initial values and to determine a viable solution for creating the new container with only the available resources due to other RT processes executing.

In this example, the initial quota Q_3 is $0,2\ ms$, therefore initially reserving 20 % of the CPU time. The sum of the three RT containers is $\sum(CPU_{alloc\_rt}) = 90$ %. As before, the actual required quota is lower than the initial quota. Note that the required CPU times, R1 and R2, are higher than in the previous scenario. As before, the CPU time is adjusted accordingly, and N1 is restarted.

Removing a RT container: If a RT application, for example, a CNC application used to manufacture a product, is no longer needed, the respective container is terminated and removed. In this example R1 is stopped and removed. As before, the CPU time for R2 and R3 can be further reduced, due to less scheduling-related overhead and interference between the applications. The *Adaption Service* adapts the quota, and $CPU_{avail} = 89$ %, which now can be used for N1.

4 Conclusion

This work presents a conceptual framework for dynamically adjusting CPU quotas at the node level. This framework increases the flexibility and resource utilization of automation while avoiding the overprovisioning of computing resources. Shortcomings in existing solutions were identified, and a concept to address these gaps was developed. The framework's components were introduced, and a typical use case was presented to demonstrate the benefits of the solution.

However, some details, such as the modeling of the monitoring process indetail, remain subject to future work. First, the factors influencing the necessary monitoring time to achieve reliable values must be evaluated. Then, the lower limits of the reduced quota necessary for predictable execution must be identified.

Acknowledgments. This work was supported by the Deutsche Forschungsgemeinschaft (DFG, German Research Foundation) - Project-ID 544536759.

References

1. Abeni, L., Balsini, A., Cucinotta, T.: Container-based real-time scheduling in the linux kernel. SIGBED Rev. **16**(3) (2019)
2. Dai, W., Zhang, Y., Kong, L., Christensen, J.H., Huang, D.: Design of industrial edge applications based on iec 61499 microservices and containers. IEEE Trans. Ind. Inform. **19**(7) (2023)
3. Fiori, S., Abeni, L., Cucinotta, T.: Rt-kubernetes. In: Hong, J., Bures, M., Park, J.W., Cerny, T., (eds.) Proceedings of the 37th ACM/SIGAPP Symposium on Applied Computing. New York, NY, USA, ACM (2022)
4. Grosch, B., Fuhrländer-Völker, D., Stock, J., Weigold, M.: Cyber-physical production system for energy-flexible control of production machines. Procedia CIRP **107**, 221–226 (2022). Leading manufacturing systems transformation–Proceedings of the 55th CIRP Conference on Manufacturing Systems 2022.
5. Lelli, J., Faggioli, D., Cucinotta, T., Lipari, G.: An experimental comparison of different real-time schedulers on multicore systems. Journal of Systems and Software **85**(10), 2405–2416 (2012)
6. Neumann, R., von Arnim, C., Neubauer, M., Lechler, A., Verl, A.: Requirements and challenges in the configuration of a real-time node for opc ua publish-subscribe communication. In: 2023 29th International Conference on Mechatronics and Machine Vision in Practice (M2VIP) (2023)
7. Sollfrank, M., Loch, F., Denteneer, S., Vogel-Heuser, B.: Evaluating docker for lightweight virtualization of distributed and time-sensitive applications in industrial automation. IEEE Transactions on Industrial Informatics **17**(5), 3566–3576 (2021)
8. Struhár, V., Behnam, M., Ashjaei, M., Papadopoulos, A.V.: Real-time containers: a survey. Schloss dagstuhl–leibniz-zentrum für informatik. OASIcs, Fog-IoT 2020. vol. 80 (2020)
9. Struhar, V., Craciunas, S.S., Ashjaei, M., Behnam, M., Papadopoulos, A.V.: React: Enabling real-time container orchestration. In: 2021 26th IEEE International Conference on Emerging Technologies and Factory Automation (ETFA). IEEE (2021)

10. Struhár, V., Craciunas, S.S., Ashjaei, M., Behnam, M., Papadopoulos, A.V.: Hierarchical resource orchestration framework for real-time containers. ACM Trans. Embed. Comput. Syst. **23**(1) (2024)
11. Walker, M., Imhoff, N., Neumann, R., Neubauer, M., Lechler, A., Verl, A.: Methodology for the combined orchestration of real-time containers and time-sensitive networks. In: 18th CIRP Conference on Intelligent Computation in Manufacturing Engineering, CIRP (2024)
12. Walker, M., Rupietta, L., Lechler, A., Neubauer, M., VerI, A.: Orchestration of heterogeneous, distributed, real-time control containers. In: IEEE 22nd International Conference on Industrial Informatics (INDIN), IEEE (2024)
13. Wang, Z., Goudarzi, M., Aryal, J., Buyya, R.: Container orchestration in edge and fog computing environments for real-time iot applications. In: Buyya, R., Hernandez, S.M., Kovvur, R.M.R., Sarma, T.H., (eds.) Computational Intelligence and Data Analytics. Lecture Notes on Data Engineering and Communications Technologies. vol. 142, Springer Nature Singapore, Singapore (2023)

Human-Centered Evaluation Framework for AI-Powered User Interfaces for CNC Machine Tools in Production

Gilbert Ely Engert(✉), Ann-Kathrin Bischoff, and Matthias Weigold

Institute for Production Management, Technology and Machine Tools (PTW), Technical University Darmstadt, Darmstadt, Germany
g.engert@ptw.tu-darmstadt.de

Abstract. Future user interfaces (UIs) for CNC machine tools must evolve to meet increasing industry demands and address the growing shortage of skilled workers. Enhancing usability, efficiency, and accessibility is key to making machine operation more attractive and reducing cognitive workload. Therefore, there is a need to redesign the user interface for operating machine tools to be more user-friendly in the future. However, a suitable evaluation framework is essential to ensure that new user interfaces are truly user-centered and intuitive for production workers, ultimately improving working conditions in modern manufacturing. This paper presents a human-centered evaluation framework designed to assess CNC machine tool interfaces, focusing on the potential of AI-powered UIs compared to existing conventional UIs. The evaluation framework is based on a comparative user study, where participants perform machining tasks using a conventional UI and a newly developed AI-powered UI. Data is collected using a questionnaire during the evaluation and includes task performance, usability, user experience, and cognitive workload to enable a user-focused assessment. Task performance is defined by completion time and success rate, usability is measured using the System Usability Scale (SUS) and interaction principles from DIN EN ISO 9241–110:2020, while cognitive workload is assessed using the NASA Task Load Index (NASA-TLX). Additionally, open feedback and the Thinking Aloud method provide qualitative insights for further user experience improvements. Participants' background information, expectations, and post-task reflections are also collected in the questionnaire. The study results reveal usability gaps, expectation mismatches, and cognitive workload differences, emphasizing the potential benefits of AI-powered user interfaces. As part of this research, the AI-powered UI was developed, which integrates intelligent assistance functions based on Large Language Models (LLMs) to simplify machine operation for production workers. Utilizing LLMs enables natural language interactions, assisting users in generating G-Code, troubleshooting errors, and optimizing workflows.

Keywords: Operator assistance systems · Large language models · Usability

L. Overmeyer and B.-A. Behrens (eds.), *Production at the Leading Edge of Technology*, Lecture Notes in Production Engineering, https://doi.org/10.1007/978-3-032-19524-1_41

1 Introduction

Machine tools are a key driver of economic growth [1]. In a survey conducted in 2023, 45% of companies reported being unable to fully utilize their production capacity due to a shortage of skilled personnel. One promising strategy to address this issue is the creation of more attractive workplaces [2]. At the same time, increasing task complexity leads to a higher workload for machine operators. As Sonderegger and Sauer emphasize, it is no longer sufficient to enhance only the speed or precision of machine tools; operator comfort and interaction quality needs to be improved through well-designed user interfaces (UIs) [3]. Existing machine tool UIs are frequently criticized for their complexity. According to Brecher et al., the number of physical control elements has increased by 400% since the 1960s, significantly complicating operation [4]. In a study by Brandl et al., 72.7% of users reported that error messages were unclear, and 68.2% criticized the menu structure as confusing [5]. In response, modern UIs increasingly incorporate ergonomic and usability principles [6]. However, to assess their actual impact on usability, user experience, and user acceptance, a systematic and human-centered evaluation framework is required.

The **research objective** of this paper is to develop and apply a suitable framework for the human-centered evaluation of UIs for machine tools, including a comparative analysis of conventional UIs—referring to long-established interfaces used in industrial practice—and AI-powered UIs. The **research question** is how can user interfaces for machine tools be evaluated in a human-centered manner, and how can AI-powered UIs be systematically compared to conventional UIs?

2 Evaluation of User Interfaces

This chapter provides an overview of the fundamental concepts and established approaches for the evaluation of UIs, forming the basis for the human-centered evaluation framework developed in this study.

To conduct a human-centered evaluation, DIN EN ISO 9241–210 and Butz et al. propose a set of suitable methods, including user-based evaluation, inspection-based evaluation, questionnaires, observation, interviews, focus groups, and long-term observation [7, 8]. In addition, ISO/TR 16,982 outlines further methods and provides guidance on how to select and adapt them, based on the specific evaluation context [9].

Usability and User Experience (UX) are considered key performance indicators (KPIs) for the evaluation of UIs. The definition of usability has evolved over time. According to ISO 9241–11:2010, usability is defined as "*the extent to which a system can be used by specified users to achieve specified goals with effectiveness, efficiency, and satisfaction in a specified context of use.*" [7] Usability is commonly operationalized through aspects such as simplicity, consistency, controllability, and user confidence, offering a practical and reliable measure of perceived usability across different systems and contexts. The System Usability Scale (SUS) provides a standardized measurement ranging from 0 to 100 and can be interpreted across five usability levels: Excellent usability (85–100), good usability (70–84), ok/acceptable usability (50–69), poor usability (25–49) and awful usability. (0–24) [10]. User Experience (UX) encompasses the overall emotional, aesthetic, and psychological experience of the user [11]. It includes both positive and negative aspects, such as satisfaction, enjoyment, motivation, aesthetic appeal, creativity and emotional engagement. In contrast to usability, which is well-defined and measurable, UX aspects are typically more subjective and less clearly structured [12].

In addition to usability and UX, interaction principles represent another key dimension in interface evaluation. These principles define fundamental design criteria that support effective and intuitive user interaction. According to DIN EN ISO 9241–110:2020, they include: suitability for the user's task, self-descriptiveness, conformity with user expectations, learnability, controllability, user error robustness, and user engagement. [7]. Another evaluation dimension is workload, which can be measured using the NASA Task Load Index (NASA-TLX) developed by Hart and Staveland in 1988. The NASA-TLX captures perceived workload through six dimensions rated on a scale from 0 (low workload) to 100 (high workload): mental demand, physical demand, temporal demand, perceived performance, effort, and frustration [13].

Although a variety of evaluation methods and metrics are established, their effective combination and context-specific application remain challenging—particularly in industrial environments. The complexity of production systems and the diversity of users demand a structured and human-centered evaluation approach. Therefore, a suitable evaluation framework is required, to systematically integrate the dimensions of usability, user experience, workload, and interaction principles into a coherent assessment of UIs for machine tools.

3 Evaluation Framework

The objective of the developed framework is to enable a structured evaluation of UIs for machine tools, including a comparative analysis between established conventional UIs and AI-powered UIs. The framework focuses on usability and user experience, especially for individuals without specialized expertise in machine tool operation. Within the framework, participants sequentially interact with both a conventional and a new developed AI-powered UI for CNC machine tools in a real production environment. The specific characteristics of both interfaces are described in Sect. 4. To minimize order effects, the sequence in which participants interacted with the two UIs was systematically counterbalanced using a Latin square design. This means that participants alternated in which UI they started with, ensuring that both UIs appeared equally often in the first position. Each assessment session follows a standardized procedure and integrates quantitative and qualitative methods across three core phases, as illustrated in Fig. 1. A diverse group of participants is selected to ensure variation in age, gender, experience level, and professional background, capturing a broad spectrum of user perspectives. Each participant completes a one-hour evaluation session, which includes pre-task questionnaires, task-based interaction with both UIs, and post-task assessments. The evaluation is conducted by a two-person experimental team with clearly defined roles. One team member acts as the facilitator, introducing the study, guiding participants through the procedure, answering any questions or resolving uncertainties. The second team member serves as a silent observer, documenting the evaluation process and systematically recording verbalized thoughts expressed by the participants.

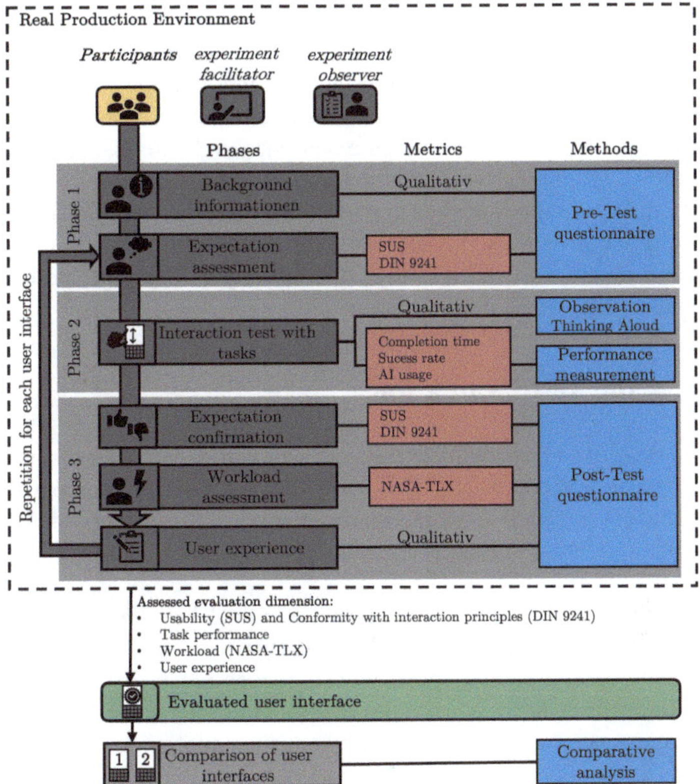

Fig. 1 Framework for evaluating user interfaces for CNC machine tools in production

Phase 1—Pre-Evaluation

At the beginning of the experiment, sociodemographic data is collected to provide contextual background, along with self-reported prior experience in machine tool operation and the frequency of using Large Language Models (LLMs), such as ChatGPT. Participants are also asked to evaluate their expectations regarding usability and interaction quality of the respective UI. For this purpose, the SUS and the interaction principles are used. A five-point Likert scale, ranging from *"strongly agree"* to *"strongly disagree"*, is applied in the pre-test questionnaire.

Phase 2—Task Based Interaction

Following the pre-evaluation participants complete five representative tasks involving typical CNC machine tool operations. The tasks include: *(1) initializing the machine, (2) moving the axis to the workpiece, (3) working with existing G-codes, (4) creating new G-code for a specific operation,* and *(5) resolving a simulated machine error.* During task execution, participants are observed using the Thinking Aloud method, while the task completion time, success rate, and AI usage are recorded.

Phase 3—Post-Evaluation

Following the interaction test with tasks, participants re-evaluated the respective UI to assess expectation fulfillment with a post assessment questionnaire regarding actual usability, conformity with the interaction principles and perceived workload, measured using the NASA-TLX. Participants are also invited to provide open-ended feedback, offering qualitative insights into their user experience and potential areas for improvement. After the three phases were completed by a participant, the entire process is repeated for both UIs. In this study, each participant therefore completed the procedure twice: once for the conventional UI and once for the AI-powered UI. The Framework can be used to evaluate a UI individually or to directly compare the performance and usability outcomes between UIs.

4 Results of the Test Evaluation

To validate the developed framework and ensure a human-centered assessment, a preliminary test evaluation was conducted. This study enabled an initial comparison between conventional and AI-powered UIs highlighting the potential benefits of AI integration in CNC machine operation and offering valuable insights for further refinement of the framework.

The evaluation is conducted using two nearly identical 3-axis milling machines manufactured by Gebr. Heller Maschinenfabrik, positioned side by side in a real production environment. One machine is equipped with a conventional UI (Siemens Sinumerik 840D), while the other features a novel developed AI-powered interface. The AI-based UI integrates assistance functions powered by LLMs, enabling natural language interaction for tasks such as G-code generation, machine specific explanations and trouble shooting. Users interact with the AI via a text-based interface—similiar to a chatbot—enabling intuitive and natural language communication. The system responds accordingly, making complex technical knowledge more accessible and supporting the user in production-related tasks. It is specifically designed to simplify machine operation and reduce cognitive workload for users. This experimental setup enables a direct side-by-side comparison between the two interfaces under realistic operating conditions, thereby avoiding artificial or laboratory-based test environments. The setup is illustrated in Fig. 2.

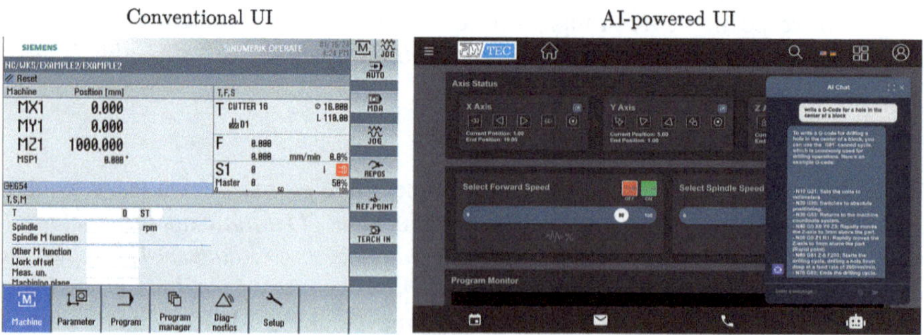

Fig. 2 Comparison of the conventional UI with Siemens Sinumerik 840d (left) and new developed AI-powered UI (right)

The test evaluation involved nine participants (six male, three female), aged 23–78 years (mean = 38), with diverse professional backgrounds. While most participants reported limited theoretical and practical experience with machine tools, they indicated a high frequency of using LLMs in their daily routines. The distribution is shown in Table 1.

Table 1 Participant distribution based on the collected background informations

Background	Participant distribution
Professional background	Mechanical Engineering (3), Design (2), Human Ressources (1), IT (1), Chemistry (1), Economic (1)
Theoretical experience	No to low level (5), intermediate (3), high level (1)
Practical experience	No to little (8), skilled (1), very experienced (0)
LLM usage	Never (0), once a month (1), once a week (4), daily (4)

Usability and Conformity with Interaction Principles

While participants initially had lower expectations regarding the usability of the conventional UI compared to the AI-powered UI, these expectations were also significantly disappointed during actual use. In contrast, the AI-powered UI not only exceeded prior expectations already in the pre-task assessments but also surpassed them during hands-on interaction. The conventional UI fell short across all interaction principles, particularly in conformity with user expectations, task suitability, and user engagement. Conversely, the AI-powered UI showed a strong match between expectation and experience in five out of seven categories. Notably, controllability even exceeded expectations. The overall usability is rated with the SUS and the overall conformity with the interaction principles is rated with the 5-point Likert scale ranging from strongly agree (5) to strongly disagree (1) as shown in Table 2.

Table 2 Usability rated with the SUS and conformity rated with 5-point Likert scale. Higher is better

Metric	Conventional UI		AI-powered UI	
	Expected	Actual	Expected	Actual
Total SUS	55.28 (Ok)	12.22 (awful)	74.17 (good)	83.33 (good)
Standard deviation	29	10	13	16
Total conformity	2.97	1.68	4.00	4.00
Standard deviation	0.74	0.35	0.32	0.87

Workload Based on NASA-TLX

Regarding perceived workload, the AI-powered UI again yielded substantially more favorable results. Participants reported more than a 50% reduction in mental and temporal workload, alongside decreased physical effort and frustration. Subjective performance perception improved significantly. The aggregated NASA-TLX score confirmed the overall lower cognitive load experienced with the AI-powered interface. The dimensions of the workload are shown in Table 3.

Table 3 Workload rated with the NASA-TLX. Lower is better (except of performance)

Metric	Conventional UI	AI-powered UI
Mental demand	85.00	33.33
Physical demand	16.67	6.11
Temporal demand	60.00	21.11
Performance	20.00	75.00
Effort	68.33	34.44
Frustration level	88.89	20.56
Total TLX	56.48	31.76
Standard deviation	6.52	9.09

Task Performance

Analysis of task performance revealed in significant differences in efficiency and success rate between the two interfaces. With the AI-powered UI, nearly all tasks were completed successfully, especially more complex ones, where all participants actively utilized the AI system. In contrast, more than half of the participants were unable to complete four tasks when using the conventional UI. Additionally, the AI-based UI demonstrated remarkably higher efficiency, resulting in average time savings between 73 and 197 s per task compared to the conventional interface. The task performance is shown in Table 4.

User Experience

These quantitative findings were further supported by qualitative feedback. The AI-powered UI was consistently described as more intuitive, structured, and user-friendly. In particular, the integrated LLM-based AI system was highlighted as a key strength, offering targeted support during task execution. In contrast, the conventional interface was criticized for being non-intuitive, cluttered, and difficult to understand.

Due to the small sample size, non-parametric significance test and corresponding effect size analyses were conducted. The significance test determine whether the observed differences are likely due to chance or represent systematic effects [14, 15]. Therefore the Wilcoxon signed-rank test assessed paired differences for SUS, Conformity, NASA-TLX, and task times, while an Exact Sign Test was applied for task success. In addition, effect sizes (Cohen's *dz*, Cliff's delta) were calculated to evaluate the practical relevance of these differences[16, 17]. Results show statistically significant

Table 4 Task performance based on success rate, completion time and AI usage

Tasks	Conventional UI		AI-powered UI		
	Mean success rate (%)	Mean completion time (s)	Mean success rate (%)	Mean completion time (s)	Mean AI usage rate (%)
Task 1	22.22	265	100	125	0
Task 2	22.22	286	88.89	142	0
Task 3	22.22	286	100	90	0
Task 4	11.11	298	77.78	225	100
Task 5	33.33	270	100	89	100

differences with large effect sizes, highlighting clear benefits of the AI-powered UI as shown in Table 5.

Table 5 Statistical significance tests and effect sizes for evaluation metrics

	Significance test	Effect size	Interpretation
Total SUS	$p < 0.01$	Cohen's dz = 1,12	Large effect
Total conformity	$p < 0.05$	Cliff's delta = 0,82	Large effect
Total TLX	$p < 0.05$	Cohen's dz = 0,95	Large effect
Task success rate	$p < 0.05$	Cliff's delta = 0,75	Large effect
Task completion time	$p < 0.01$	Cohen's dz = 1,15	Large effect

5 Conclusion and Outlook

The developed framework proves to be effective for conducting human-centered evaluations of machine tool operation UIs. By integrating quantitative performance metrics with qualitative methods—such as open feedback, think-aloud protocols, and observations—it enables a detailed assessment of usability and user experience, identifying both strengths and weaknesses of conventional and AI-powered UIs. However, the application is resource-intensive, requiring approximately one hour per participant and two trained facilitators. Recruiting participants with industrial experience also proved challenging, limiting the generalizability of some findings. Despite these constraints, the framework demonstrates high methodological quality and supports comprehensive evaluations under real production conditions. The test evaluation further shows that AI-based UIs can be effectively used by less experienced users, offering practical benefits such as reduced training requirements, increased process reliability, and efficient use by non-specialists.

Future work will focus on expanding and refining the evaluation framework to increase its generalizability and practical relevance. A key objective is to involve a larger number of participants, with an emphasis on including more experienced users from the industrial domain, to better reflect real-world operator behavior. In addition, the user tasks will be further adapted to represent more realistic and comprehensive production processes. To enhance user engagement and motivation during the study, improvements in the overall user experience of the evaluation process itself are also planned. Furthermore, future evaluations should include machine tools with modern UIs, allowing a more detailed benchmarking. Finally, the flexibility of the developed framework makes it a promising tool, not only for CNC UIs, but also for the evaluation of other machine systems and AI solutions in industrial environments.

Acknowledgements. This research was part of the project "KompAKI" (02L19C151) which is funded by the German Federal Ministry of Eduaction and Research and Projektträger Karlsruhe PTKA. The authors are responsible for the content of this publication. The authors would like to thank the "KompAKI" project partners for their content support.

Competing Interests. The author(s) has no competing interests to declare that are relevant to the content of this manuscript.

References

1. Brecher, C., Weck, M.: Werkzeugmaschinen Fertigungssysteme 1: Maschinenarten und Anwendungsbereiche. Springer, Berlin Heidelberg (2019)
2. Verein Deutscher Werkzeugmaschinenfabrik: Die deutsche Werkzeugmaschinenindustrie und ihre Stellung im Weltmarkt. Verein Deutscher Werkzeugmaschinenfabriken, Frankfurt (2024)
3. Sonderegger, A., Sauer, J.: The influence of design aesthetics in usability testing: Effects on user performance and perceived usability. Appl. Ergon. **41**, 403–410 (2010)
4. Brecher, C., Kolster, D., Herfs, W.: Innovative Benutzerschnittstellen für die Bedienpanels von Werkzeugmaschinen. Z. Für Wirtsch. Fabr. **106**, 553–556 (2011). https://doi.org/10.3139/104.110607
5. Brandl, C., Brecher, C., Czerniak, J., Hellig, T., Mertens, A., Schlick, C., Sittig, S.: Multimodale, aufgabenorientierte Bediensysteme zur flexiblen und nutzerzentrierten Mensch-Maschine-Interaktion an Produktionsmaschinen. Rheinisch-Westfälische Technische Hochschule Aachen Fakultät 4—Maschinenwesen Werkzeugmaschinenlabor WZL
6. Sgarbossa, F., Grosse, E.H., Neumann, W.P., Battini, D., Glock, C.H.: Human factors in production and logistics systems of the future. Annu. Rev. Control. **49**, 295–305 (2020)
7. DIN EN ISO 9241–210: DIN EN ISO 9241–210. Ergonomie der mensch-system-interaktion, (2020)
8. Butz, A., Krüger, A., Völkel, S.T.: Mensch-maschine-interaktion. De Gruyter Studium, Oldenburg (2022)
9. ISO/TR 16982: ISO/TR 16982: Ergonomics of human-system interaction—usability methods supporting human-centred design, (2024)
10. Bangor, A.: Determining what individual SUS scores mean: Adding an adjective rating scale. **4**, (2009)
11. Bill, A., Tullis, T.: Measuring the user experience: Collecting, analyzing, and presenting UX metrics (Interactive Technologies). Elsevier (2022)
12. Rogers, Y., Sharp, H., Preece, J.: Interaction design beyond human-computer interaction. John Wiley & Sons Inc., New Jersey (2007)

13. Phil, S., Gore, B.: TLX @ NASA Ames—home, https://humansystems.arc.nasa.gov/groups/TLX/. Last accessed 02 June 2025
14. Field, A.: Discovering statistics using IBM SPSS statistics. SAGE, Los Angeles London New Delhi Singapore Washington DC Melbourne (2018)
15. Wilcoxon, F.: Individual comparisons by ranking methods. Biom. Bull. **1**, 80–83 (1945)
16. Cohen, J.: Statistical power analysis for the behavioral sciences. L. Erlbaum Associates, Hillsdale, N.J (1988).
17. Cliff, N.: Dominance statistics: Ordinal analyses to answer ordinal questions. Psychol. Bull. **114**, 494–509 (1993). https://doi.org/10.1037/0033-2909.114.3.494

A Simulation-Based Deep Learning Approach for Predicting Part Quality in Thermoplastic Injection Moulding

André Hürkamp(✉), Tom Hartmann, Manuel Megnet, and Klaus Dröder

Technische Universität Braunschweig, Institute of Machine Tools and Production Technology, Braunschweig, Germany
a.huerkamp@tu-braunschweig.de

Abstract. In this work a deep learning approach for real-time warpage prediction in injection moulding is presented. A simplified geometry derived from an industrial HUD top cover is investigated as an example geometry. Simulation data from Autodesk Moldflow serve as the basis for training surrogate models. Two model variants are compared: a direct, fully data-driven prediction of the full displacement field by neural networks (NN) and a reduced-order approach combining Proper Orthogonal Decomposition with NN (POD+NN). Both approaches reach a $R^2 \geq 0.94$ and have prediction times below 50 ms. The results show that the POD+NN model achieves high accuracy with reduced complexity, supporting efficient integration into digitalised manufacturing workflows. A prototype dashboard enables interactive analysis.

Keywords: Injection moulding · Simulation · Machine learning

1 Introduction

Injection moulding is a widely used manufacturing process for plastic parts, offering high design flexibility and material versatility as well as short cycle times. Identifying optimal process settings typically relies on expert knowledge and extensive trial-and-error procedures [1,2]. Hence, ensuring part quality at minimal cost requires a deep understanding of material behaviour, process physics, and their impact on final part performance. To achieve this, mathematical models are developed and used to optimise mould design and performance indicators with controllable process and design parameters [3].

In the processing of thermoplastic polymers, especially the thermal behaviour can cause significant shrinkage and warpage, which poses a challenge in achieving dimensional accuracy [4]. Despite the initial conditions being controlled, the melt state within the mould cavity varies due to environmental influences, leading to dimensional deviations in the final part. Therefore, online quality monitoring is required [5]. Sophisticated numerical methods implemented in commercial software such as Autodesk Moldflow or Moldex3D are well suited for analysing

L. Overmeyer and B.-A. Behrens (eds.), *Production at the Leading Edge of Technology*, Lecture Notes in Production Engineering, https://doi.org/10.1007/978-3-032-19524-1_42

and optimising process and part. However, these simulations are often computationally expensive and time-consuming. In order to use it more efficiently for optimisation and monitoring, it is necessary to reduce calculation time by implementing surrogate models that run significantly faster, yet still provide sufficient accuracy. In this regard, surrogate-based optimization offers a computationally efficient alternative to costly physics-based simulations [6]. In the field of sheet metal forming, Link et al. [7] used convolutional neural networks with a RES-SE-U-Net architecture to determine the quality of a deep drawn part based on the local lubrication. The model was trained based on finite element simulations using a 2D projection of the blank. In the context of injection moulding, Uglov et al. [8] developed a framework for warpage prediction, also projecting the simulation results onto a two-dimensional space using a U-Net like convolutional encoder. In [9], a principal component analysis on sheet metal parts was applied to perform a variance-based sensitivity analysis in low-dimensional space.

In this contribution, a simulation-based deep learning approach for prediction of 3D-warpage is presented. As an example part, a simplified geometry based on an industrially available head-up display (HUD) top cover is investigated. The HUD top cover retains key geometric features, while reducing complexity for efficient simulation and analysis. The top cover of the HUD assembly serves both as a protective and optical interface. Manufacturing induced distortion can change the optical path, leading to image misalignment, unwanted reflections, and reduced visibility. Therefore, warpage is a key quality criterion and potential safety risk for the driver.

2 Modelling of Polymer Flow and Warpage

The flow of the polymer melt during injection moulding is modelled as a transient, incompressible, non-isothermal, and highly viscous free-surface flow. For a very low Reynolds number (Re $\ll 1$) as for the polymer flow, the governing equations reduce to the Stokes momentum balance together with the continuity constraint

$$-\nabla p + \nabla \cdot \boldsymbol{\tau} + \rho \mathbf{g} = \mathbf{0}, \quad \nabla \cdot \mathbf{v} = 0 \tag{1}$$

with pressure p and deviatoric stress $\boldsymbol{\tau}$. Assuming a generalized Newtonian fluid, the constitutive relation for the deviatoric stress

$$\boldsymbol{\tau} = \eta(\dot{\gamma}, T)\left(\nabla \mathbf{v} + \nabla \mathbf{v}^{\top}\right) \text{ with } \dot{\gamma} = \frac{1}{2}\left\|\nabla \mathbf{v} + \nabla \mathbf{v}^{\top}\right\|. \tag{2}$$

is given by the shear- and temperature-dependent viscosity $\eta(\dot{\gamma}, T)$.

Due to non-uniform cooling and pressure gradients, residual stresses accumulate during filling. For unfilled, amorphous polymers, these are modelled as elastic reactions to incompatible, stress-free strains constrained by the surrounding material. The total free strain is given by

$$\boldsymbol{\varepsilon}^{\text{free}} = \underbrace{\boldsymbol{\alpha}_T : \Delta T}_{\text{thermal}} + \underbrace{\boldsymbol{\beta}\,\Delta p}_{\text{pressure-induced}}, \tag{3}$$

where α_T denotes the thermal expansion, ΔT the temperature drop from the stress-free reference, and $\boldsymbol{\beta}$ the pressure shrinkage coefficient. The residual stress is computed via the linear elastic constitutive relation

$$\boldsymbol{\sigma}^{\text{res}} = \mathbb{C} : \left(\boldsymbol{\varepsilon} - \boldsymbol{\varepsilon}^{\text{free}}\right), \tag{4}$$

with $\mathbb{C}$ being the fourth-order elasticity tensor and $\boldsymbol{\varepsilon}$ the total strain. After cooling, $\boldsymbol{\sigma}^{\text{res}}$ acts as an internal load to compute the final part deformation $\mathbf{u}$ in a structural finite element analysis.

3 Deep Learning Workflow for Warpage Prediction

The objective of this work is to develop a deep learning architecture based on finite element simulations to predict the deformations due to warpage. Two model variants are investigated: (i) a direct neural network (NN) that maps process parameters to the full displacement field, and (ii) a POD+NN model, in which a neural network predicts POD coefficients that are linearly decoded to reconstruct the full displacement field $\mathbf{u}$.

3.1 Direct NeuralNetwork Formulation

The objective of the machine learning task is to learn a surrogate that maps process parameters $\mathbf{p} \in \mathbb{R}^d$, where d denotes the number of input parameters of the injection-moulding cycle to the resulting displacement field $\mathbf{u}$ (i.e. warpage). A feed-forward network

$$\hat{\mathbf{u}} = \mathcal{N}_\theta^{\mathbf{u}}(\mathbf{p}) : \mathbb{R}^d \to \mathbb{R}^{3n}, \tag{5}$$

parametrised by weights θ, acts as a universal approximator for the displacement components at the n nodes.

Given a data set $\mathcal{D} = \{\mathbf{u}_i\}_{i=1}^N$, where $\mathbf{u}_i$ is the reference displacement obtained from the high-fidelity simulations, the network parameters are obtained by minimising the mean-absolute-error (MAE)

$$\mathcal{L}_{\text{MAE}}(\theta) = \frac{1}{N} \sum_{i=1}^{N} \left\| \mathbf{u}_i - \mathcal{N}_\theta^{\mathbf{u}}(\mathbf{x}_i) \right\|_1. \tag{6}$$

3.2 POD-Based Surrogate Modelling

To enable fast and accurate prediction of high-dimensional displacement fields in injection-moulding simulations, we adopt a combined model-reduction and machine-learning strategy [10–12]. For model-reduction, Proper Orthogonal Decomposition (POD) is applied to extract the principal components of the solutions, enabling a compact representation and linear reconstruction of the displacement field. From the data set $\mathcal{D} = \{\mathbf{u}_i\}_{i=1}^N$, the snapshot matrix

$$\mathbf{U} = \left[\mathbf{u}^{(1)} \cdots \mathbf{u}^{(N)}\right] = \mathbf{W}\boldsymbol{\Sigma}\mathbf{V}^\top, \tag{7}$$

is assembled and decomposed via singular value decomposition (SVD) $\mathbf{W}\boldsymbol{\Sigma}\mathbf{V}^{\top}$. By performing the SVD of $\mathbf{U}$, the POD modes are directly obtained as the columns $\boldsymbol{\varphi}_i = \mathbf{W}_{\mathrm{i}}$ of $\mathbf{W} = \boldsymbol{\Phi}$. The corresponding POD coefficients are given by $\mathbf{A} = \mathbf{W}^{\top}\mathbf{U}$ with columns $\mathbf{a}^{(j)} = \boldsymbol{W}^{\top}\mathbf{u}_j$. A truncated basis $\boldsymbol{\Phi}_m$ retaining the first m modes defines a low-dimensional latent space. Finally. the displacement field

$$\hat{\mathbf{u}} = \sum_{i=1}^{m} a_i(\mathbf{p})\,\boldsymbol{\varphi}_i(\mathbf{x}) = \boldsymbol{\Phi}\mathbf{A} \tag{8}$$

can then be approximated. Because the coefficients $a_i(\mathbf{p})$ depend only on the process parameters, we can train a feed-forward neural network

$$\hat{\mathbf{a}} = \mathcal{N}_{\theta}^{\mathbf{a}}(\mathbf{p}) : \mathbb{R}^{p} \to \mathbb{R}^{m} \tag{9}$$

by minimizing the empirical loss $\mathcal{L}(\theta) = \frac{1}{N}\sum_{j=1}^{N} \left\|\mathbf{a}^{(j)} - \mathcal{N}_{\theta}^{\mathbf{a}}(\boldsymbol{\mu}^{(j)})\right\|_2^2$. Substituting $\hat{\mathbf{a}}$ into (8) reconstructs the full displacement field via the POD modes acting as a linear decoder.

4 Results

For both model architectures (NN and POD+NN), a simulation database obtained from the Moldflow model shown in Fig. 1 was used for the training. The model consists of $n = 147{,}191$ nodes and $n_{\mathrm{el}} = 796{,}413$ tetrahedral elements.

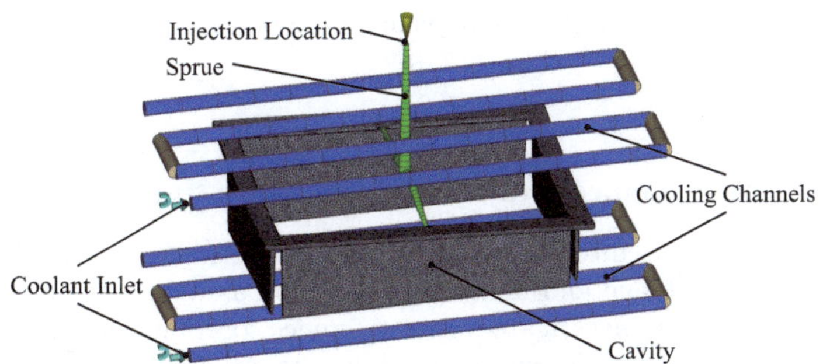

Fig. 1. Illustration of the simulation model in Autodesk Moldflow including mesh, runner system, and cooling channels (top); sampled process parameter space for surrogate model training (bottom)

As material, a polycarbonate (Makrolon® 2405 from Covestro [13]) was utilized using a Cross-WLF model for the viscosity η. For each training sample i the $d = 5$ scalar process parameters $\mathbf{p}^i = \left(T_c^i,\ T_m^i,\ t_c^i,\ \dot{q}^i,\ p_p^i\right)^{\top} \in \mathbb{R}^5$ are collected within the given parameter range of Fig. 1. Hereby, T_c is the coolant temperature, T_m the melt temperature, t_c the cycle time, $\dot{q}$ the volumetric injection rate, and p_{pack} the packing pressure as function of the injection pressure p_{in}.

Latin hypercube sampling was used to define 332 parameter combinations to perform individual injection moulding simulations. The simulations were carried out using Autodesk Moldflow Insight 2023 (workstation with two AMD EPYC 7643 48-Core processors with 2.30 GHz, 256 GB RAM). Each simulation required approximately 14 min of wall-clock time, resulting in a total elapsed time of 76 h for all simulations. Due to parallelisation across multiple cores, this corresponds to a cumulative CPU time of approximately 14 h per simulation. The maximum warpage values of all simulations are between 0.64 mm and 1.95 mm (see Fig. 2). For both architectures, the dataset was split into 80% training and 20% test data.

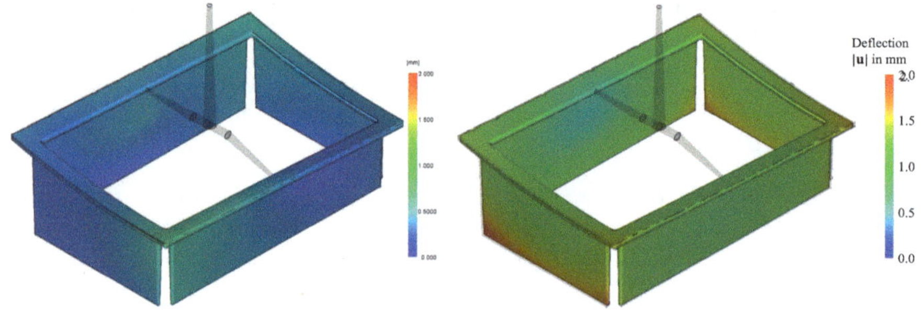

Fig. 2. Minimal warpage (left) and maximum warpage (right) of all samples

4.1 Model Performance Evaluation

For evaluating both architectures, performance metrics such as coefficient of determination R^2, average and maximum deviation as well as training and prediction times have been investigated. As a reference solution, the direct approach (NN) was used. The model learns the nonlinear mapping $\mathcal{N}_\theta^{\mathbf{u}} : \mathbb{R}^5 \rightarrow \mathbb{R}^{3\times 147191}$. Hence, in total 441,573 displacement components will be computed. The network architecture consists of ten hidden layers, each comprising 20 neurons. A constant learning rate of 10^{-3} was used, with the mean absolute error (MAE) (6) as the loss function and the Adam optimizer for training. To ensure robustness and exclude statistical outliers, the model was trained three times independently.

For the POD-NN approach, a low-dimensional latent space with $m = 140$ POD-modes is retained, capturing 99.9 % of the total snapshot energy. A neural network is then trained to learn the nonlinear mapping $\mathcal{N}_\theta^{\mathbf{a}} : \mathbb{R}^5 \rightarrow \mathbb{R}^{140}$ for the $m = 140$ POD coefficients. The full displacement field is reconstructed via a linear decoder using the POD basis $\mathbf{\Phi} \in \mathbb{R}^{441573\times 140}$.

Table 1. Characteristic results and performance metrics of the surrogate model

Characteristic value	Direct NN	POD+NN
Average deviation [mm]	0.0166	0.0178
Coefficient of determination R^2	0.954	0.94
Maximum deviation in mm	0.131	0.101
Relative error at max. deviation in %	13.2	13.0
Average maximum deviation in mm	0.049	0.043
Approx. training time in s	12,000	80
Approx. prediction time in ms	3	40

metrics of both approaches are shown in Table 1. For the direct NN approach, the average deviation across all test data is 0.0166 mm. The coefficient of determination averages R^2=0.954 on the test data. The absolute maximum error on all nodes of the test samples is 0.131 mm which corresponds to a relative error of 13 %. However, the average maximum error per sample is much lower with 0.049 mm.

The POD+NN model achieves a prediction accuracy comparable to the direct NN approach. While the average errors remain comparable between both approaches, the POD+NN model yields a slightly lower maximum error. A relevant advantage of the POD+NN approach is the significantly faster model training. The training time with POD+NN averaged 78 s, while without POD it averaged around 12,000 s (3.33 h). In contrast, the prediction speed for direct NN was significantly shorter at approx. 3 ms than for POD+NN, where the prediction time was approx. 40 ms when using 140 modes. By reducing the number of modes for the reconstruction, the prediction time can be further reduced (approx. 8 ms using only 20 modes).

4.2 Influence on Training Data Size

To evaluate the influence of training data quantity on model performance, the amount of training data was gradually increased. For each training data size, three surrogate models were trained. The maximum deviation and coefficient of determination in Fig. 3 show that the model accuracy increases with an increasing amount of training data. Although model accuracy continues to improve with increasing training data, a good approximation is already achieved with as few as 120 samples. Hence, by using efficient sampling methods, the effort for data generation can be reduced. However, the decreasing standard deviations indicate greater robustness and stability of the training process with larger datasets.

4.3 Graphical User Interface for Model Deployment

To support the integration of the surrogate model into an industrial workflow, a prototype dashboard was developed for interactive visualisation and evalu-

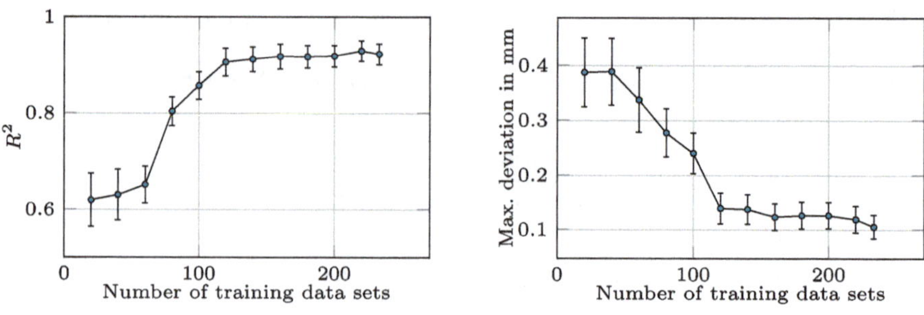

Fig. 3. Approximation accuracy of POD+NN in dependence of training data sets: R^2-score (left); maximum deviation (right)

ation of simulation and prediction results. The current implementation shown in Fig. 4 is based on a Python environment and integrates PyTorch for deploying the trained neural network models, PyQt for the design of a responsive graphical user interface (GUI), and Vedo for 3D visualisation. It enables real-time exploration of different process settings and their predicted effects on part quality. Deploying the surrogate and visualising the results takes in this example approx. 60 ms. Thus, it provides immediate visual feedback by dynamically updating the displayed part geometry in response to user input via interactive sliders facilitating intuitive analysis and rapid decision-making. Currently, the surrogate model is fed by user inputs in the dashboard. In future applications, the input may come directly from an manufacturing execution system (MES) as depicted in Fig. 4 to enable quality monitoring.

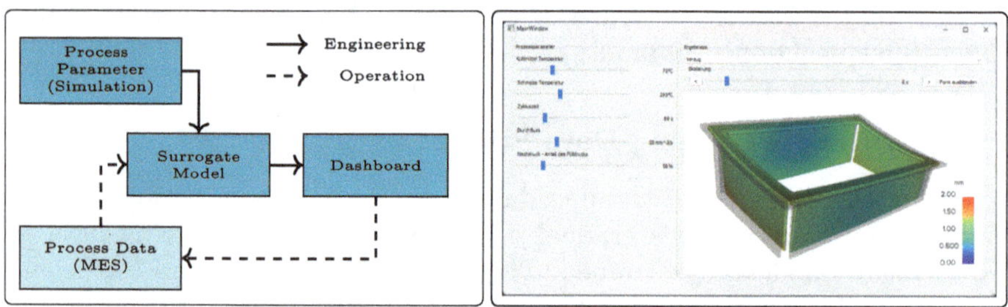

Fig. 4. Integration of surrogate model into prototype of dashboard application

5 Conclusion

This study presents a simulation-driven deep learning workflow for predicting warpage in injection-moulded plastic components, demonstrated on a simplified geometry of a HUD top cover. Two surrogate modelling strategies were compared: a direct, fully data-driven neural network approach, and a reduced-order model using POD-based dimensionality reduction with a linear decoder. Both approaches yield accurate warpage predictions in comparison to high-fidelity simulations, with the POD+NN model achieving comparable results while significantly reducing output dimensionality. This reduction facilitates faster training while preserving physical interpretability. The results demonstrate that deep learning models trained on simulation data can predict deformation fields in milliseconds, offering a promising tool for process optimization and for real-time quality control in injection moulding without a large number of experimental data. The fast prediction time below 50 ms also qualifies the approach for integration into digital twin systems and CPPS environments. Although training the model for a specific part design currently requires considerable computational effort (approx. 76 h in this study), future research should aim at improving generalisation across different product geometries. This would reduce the need for extensive simulations for each new product and enable broader applicability in industrial settings. Future work will therefore also focus on linking the surrogate with inputs from an existing MES and validate the quality monitoring directly on the shopfloor.

Acknowledgments. This work is part of the research project "DIAZI Digitalisierung des Industrialisierungsprozesses in der Automobil- und Zulieferindustrie" (Grant No. 13IK008D). The authors gratefully acknowledge the funding provided by the Federal Ministry for Economic Affairs and Energy (BMWE).

References

1. Shen, C., Wang, L., Li, Q.: Optimization of injection molding process parameters using combination of artificial neural network and genetic algorithm method. J. Mater. Process. Technol. **183**(2–3), 412–418 (2007). https://doi.org/10.1016/j.jmatprotec.2006.10.036
2. Bibow, P., Dalibor, M., Hopmann, C., Mainz, B., Rumpe, B., Schmalzing, D., Schmitz, M., Wortmann, A.: Model-driven development of a digital twin for injection molding. Lect. Notes Comput. Sci., 12127 LNCS, 85–100 (2020). https://doi.org/10.1007/978-3-030-49435-3_6
3. Fernandes, C., Pontes, A.J., Viana, J.C., Gaspar-Cunha, A.: Modeling and optimization of the injection-molding process: a review. Adv. Polym. Technol. **37**(2), 429–449 (2018)https://doi.org/10.1002/adv.21683
4. Hopmann, C., Xiao, C., Kahve, C.E., Fellerhoff, J.: Prediction and validation of the specific volume for inline warpage control in injection molding. Polym. Test. **104**, 107393 (2021).https://doi.org/10.1016/j.polymertesting.2021.107393
5. Gao, R.X., Tang, X., Gordon, G., Kazmer, D.O.: Online product quality monitoring through in-process measurement. CIRP Ann. **63**(1), 493–496 (2014). https://doi.org/10.1016/j.cirp.2014.03.041
6. Pfrommer, J., Zimmerling, C., Liu, J., Kärger, L., Henning, F., Beyerer, J.: optimisation of manufacturing process parameters using deep neural networks as surrogate models. Procedia CIRP **72**, 426–431 (2018). https://doi.org/10.1016/j.procir.2018.03.046

7. Link, P., Penter, L., Rückert, U., Klingel, L., Verl, A., Ihlenfeldt, S.: Real-time quality prediction and local adjustment of friction with digital twin in sheet metal forming. Robot. Comput. Integr. Manuf. **91**, 102848 (2025). https://doi.org/10.1016/j.rcim.2024.102848
8. Uglov, A., Nikolaev, S., Belov, S., Padalitsa, D., Greenkina, T., Biagio, M.S., Cacciatori, F.M.: Surrogate modeling for injection molding processes using deep learning. Struct. Multidisc. Optim. **65**(11), 1–13 (2022). https://doi.org/10.1007/s00158-022-03380-0
9. Schwarz, C., Ackert, P., Mauermann, R.: Principal component analysis and singular value decomposition used for a numerical sensitivity analysis of a complex drawn part. Int. J. Adv. Manuf. Technol. **94**(5–8), 2255–2265 (2018). https://doi.org/10.1007/s00170-017-0980-z
10. Swischuk, R., Mainini, L., Peherstorfer, B., Willcox, K.: Projection-based model reduction: Formulations for physics-based machine learning. Comput. Fluids **179**, 704–717 (2019). https://doi.org/10.1016/j.compfluid.2018.07.021
11. Hürkamp, A., Lorenz, R., Ossowski, T., Behrens, B.A., Dröder, K.: Simulation-based digital twin for the manufacturing of thermoplastic composites. Procedia CIRP **100**, 1–6 (2021). https://doi.org/10.1016/j.procir.2021.05.001
12. Fresca, S., Manzoni, A.: POD-DL-ROM: Enhancing deep learning-based reduced order models for nonlinear parametrized PDEs by proper orthogonal decomposition. Comput. Methods Appl. Mech. Eng. **388**, 114181 (2022). https://doi.org/10.1016/j.cma.2021.114181
13. Covestro Solution Center. https://solutions.covestro.com/en/products/makrolon/makrolon-2405_000000000000945088?SelectedCountry=US

Development of a Decision-Making Method for Using Large Language Models in Production-Related Processes

Ann-Christine Gerchel[1(✉)] and Thomas Gartzen[2]

[1] TH Augsburg, Fakultät Für Elektrotechnik, Augsburg, Deutschland
ann-christine.gerchel@tha.de
[2] TH Köln, Fakultät für Anlagen, Energie- und Maschinensysteme, Köln, Deutschland
thomas.gartzen@th-koeln.de

Abstract. Large language models (LLMs) have demonstrated significant potential across various industries. Despite their success in areas such as customer service, sales and marketing, their application in manufacturing and production environments often lacks clear, practical guidance for identifying suitable use cases and making effective implementation decisions. This study addresses this gap by developing a structured method to identify production-related processes suitable for LLMs and to guide implementation decisions. The research begins by outlining the fundamentals of LLMs. Practical examples from literature and industry are presented to illustrate the current state of LLM use in production-related contexts. Based on these insights, a two-part decision-making method is introduced. The first part focuses on identifying processes that could benefit from LLM integration. Criteria are developed through qualitative content analysis of the presented examples, and an evaluation process inspired by utility value analysis is applied. The second part employs a structured assessment framework to guide implementation decisions, assessing factors such as the appropriate deployment type (e.g., on-premises or cloud-based), the optimal size of the LLM, and knowledge transfer requirements. The questions are based on the previously outlined fundamentals of LLMs. The practical application of the method is demonstrated through a detailed case study, showing its effectiveness in identifying suitable processes and supporting practical decision-making as well as the need for further refinement.

Keywords: Large language models · Production processes · Knowledge management · Generative artificial intelligence · Decision-making method · Natural language processing

1 Introduction

The ongoing digitalization of manufacturing increasingly incorporates artificial intelligence (AI) to optimize operations, improve quality, and reduce costs [1]. While applications such as machine vision [2] and predictive maintenance [3] are well established within production environments, the systematic use of language-based AI technologies remains comparatively less mature.

L. Overmeyer and B.-A. Behrens (eds.), *Production at the Leading Edge of Technology*,
Lecture Notes in Production Engineering, https://doi.org/10.1007/978-3-032-19524-1_43

Large Language Models (LLMs) are AI systems designed to understand and generate human language and are used across business domains such as customer service, marketing, and software development due to their capabilities in natural language processing and automation [4–6]. So far, there's no structured approach to help select an LLM to integrate it into production and production-related processes. This research addresses the gap by developing a structured decision method to support the integration of LLMs in production environments and is guided by the following research questions:

1. How can a production-related process suitable for LLM optimization be identified?
2. How can an LLM be effectively deployed in a production business?

The goal is to provide a proposal for a practical framework to support decision-making for the use of LLMs in production settings. It focuses on identifying relevant processes and evaluating key implementation aspects, such as integration, data availability, and knowledge transfer. Therefore, this paper covers existing use cases of LLMs, followed by the proposed methodology, a case study, and concluding insights.

2 Background and Applications of Large Language Models in Production Contexts

LLMs are mainly built on transformer architectures introduced by Vaswani et al. [7] and use self-attention mechanisms to process input sequences in parallel. Recent advances in computational power and training methods have enabled the development of large-scale models such as GPT, LLaMA, and Gemini. Performance is driven by the volume and diversity of training data, model architecture, parameter count, and training configurations [8]. Despite their potential, LLMs pose notable risks in production environments. Sensitive company data may be exposed if not handled securely [9]. Furthermore, LLMs can generate plausible yet incorrect outputs ("hallucinations"), which could lead to operational errors in manufacturing [10, 11]. Their opaque reasoning processes (black-box behavior) limit transparency and accountability [12], and training data biases may result in skewed or discriminatory outputs [13].

Nevertheless, LLMs have been tested and applied across various production contexts, including production planning, programming, quality, and human–machine interaction. In academic research, Gkournelos et al. [14] and Fan et al. [15] used LLMs for assembly planning and toolpath generation. While models supported planning tasks well, challenges arose in handling structured outputs and domain-specific tasks. Xia et al. [16] integrated an LLM into a digital twin system to generate process plans, while Zhao et al. [17] demonstrated LLM-based job scheduling that outperformed traditional heuristics in some cases. Fakih et al. [18] introduced an iterative feedback loop for LLM-supported PLC programming. Xia et al. [19] fine-tuned transformer models on manufacturing documents for more accurate code generation. In knowledge management, LLMs were evaluated using real documentation and failure logs [11, 20], supporting employees with operational knowledge. In additive manufacturing, LLMs assisted with G-code generation and troubleshooting printing issues [21]. In manual assembly, digital assistants based on LLMs reduced cognitive load for workers [22].

Industrial deployment is also increasing. Automotive OEMs such as Mercedes-Benz, Toyota, and Stellantis use LLMs for voice assistance, customer interaction, and documentation analysis [23–25]. In manufacturing, Bosch applies generative AI for defect image generation [26], while Siemens' Industrial Copilot supports engineers with automation tasks and maintenance operations [27]. ABB and Microsoft integrate LLMs into IoT platforms to enhance decision-making and knowledge conservation [28].

3 Decision Method for LLM Integration in Production

This research follows a Design Science Research (DSR) methodology to address the lack of a structured method for integrating LLMs into production. Following the framework by March and Smith, this paper's core artifact is a decision-making method for LLM integration. The research process involved building this method by synthesizing existing knowledge and evaluating it through a case study to demonstrate its utility and provide a foundation for future refinement [29].

3.1 Part 1: Identifying LLM-Suitable Processes

The first part of the decision method develops criteria for identifying processes suitable for LLM integration. This approach is based on the iterative process of qualitative content analysis inspired by Kuckartz [30]. In this study, the chapters of experimental design, execution, and results from literature examples and all the industry examples were coded using an inductive coding system developed during text analysis. The categories emerging from this coding process were subsequently adopted as evaluation criteria without further subcategorization. The texts underwent multiple rounds of analysis. The criteria for process evaluation resulting from the qualitative content analysis include:

1. Process Complexity (PC): Degree of ambiguity, variability, and number of non-standard decision points that lack a clear, predefined set of rules or instructions.
2. Data Availability (DA): Accessibility and readiness of structured and unstructured operational data (e.g., sensor readings, production logs, system data via APIs) necessary for the LLM to understand context or perform calculations.
3. Automation Potential (AP): Extent to which the LLM can automate process steps.
4. Employee Acceptance (EA): Willingness of personnel to adopt LLM support.
5. Documentation (D): Quality and accessibility of existing explicit knowledge, such as manuals, SOPs, and historical records, suitable for LLM consumption.
6. Frequency of Support (FoS): How often the LLM can be applied or the process occurs.
7. Fault Tolerance (FT): Robustness against errors and impact of incorrect LLM responses.
8. Materially Critical Process (MCP): Risk of significant material damage upon process failure.
9. Time-Critical Process (TCP): Urgency of the process and the impact of delays on downstream operations, considering the potential for timely LLM intervention when human support is constrained.

In total 9 scientific and 5 practical Use Cases were analyzed, and the absolute occurrence of the individual criteria were summed up. A weighting was then derived from the absolute number of occurrence, normalized to a total weight of 100%. The resulting weights can be found in the weight column in Fig. 1.

Criteria	Weight		5	4	3	2	1		Value
Process Complexity	18,18	very high	☑	☐	☐	☐	☐	very low	90,91
Data Availability	12,73	much data	☑	☐	☐	☐	☐	no data	63,64
Automation Potential	7,27	high potential	☑	☐	☐	☐	☐	no potential	36,36
Employee Acceptance	12,73	very high	☑	☐	☐	☐	☐	none	63,64
Documentation	16,36	detailed	☑	☐	☐	☐	☐	none	81,82
Frequency of Support	3,64	frequent	☑	☐	☐	☐	☐	very low	18,18
Fault Tolerance	9,09	very high	☑	☐	☐	☐	☐	none	45,45
Materially Critical Process	9,09	uncritical	☑	☐	☐	☐	☐	very critical	45,45
Time Critical Process	10,91	uncritical	☑	☐	☐	☐	☐	very critical	54,55
	100							Final Score	500

Fig. 1 Decision-method part 1 with best possible final score

A Utility Value Analysis (UVA) is employed to score each process on a 5-point Likert scale per criterion. The UVA is employed for its suitability in multi-criteria decision-making, allowing for the systematic aggregation of diverse factors into a single score. This provides a transparent and comparable basis for prioritizing processes for LLM integration. The score C_i is multiplied by its weight W_i, and summed, resulting in a maximum of 500 points. The formula to calculate the final score is expressed in (1):

$$Final\ Score = \sum(C_i * W_i) \tag{1}$$

The final score aids in prioritizing processes for LLM integration and is presented as a normalized percentage related to the minimum possible score of 100 points. Against this background, a score of 300 points, for example, translates to 50%. This limit can also be used as a threshold of the suitability of a process. The higher the final score, the more suitable a process for LLM integration is. Figure 1 illustrates the decision model for part one. Evaluation also includes process details, affected business areas, and necessary data access. Metrics to measure actual LLM benefit should be established upfront.

Part 2: Determining the Type of LLM Deployment

The second phase introduces a structured assessment framework inspired by the Analytic Hierarchy Process (AHP). While the framework is inspired by the AHP for structuring decision criteria, it does not strictly adhere to all its foundational assumptions, notably the use of pairwise comparisons. This pragmatic approach was chosen to ensure usability and practicality in the rapidly evolving LLM landscape, avoiding the complexity of full AHP for numerous potential LLM options. Acknowledging the inherent complexity and difficulty of identifying and quantifying the interdependencies among all criteria in a practical manner, this framework pragmatically treats each criterion as an independent factor in the scoring process. This approach ensures the method remains both usable and accessible for practitioners. The design of the questionnaire, where not every question influences every criterion of the framework, further helps to mitigate potential overlaps and maintain this conceptual independence.

This framework is not intended to be a final, prescriptive solution but rather a preliminary proposal designed to activate discussion among stakeholders. The criteria are not meant to be mutually exclusive but instead serve as a structured starting point to highlight key considerations for LLM deployment. Final decisions require expert judgment to navigate the specific complexities of the business environment.

The table in Fig. 2 presents the corresponding recommendations based on the evaluation scores. A high score indicates, for example, that an on-premise deployment strategy and fine-tuning for knowledge transfer may be more appropriate. Conversely, a low score suggests that a cloud-based solution and prompt engineering are more suitable. It is important to note that not every question influences every aspect of the framework. For instance, employee acceptance is independent of the available budget.

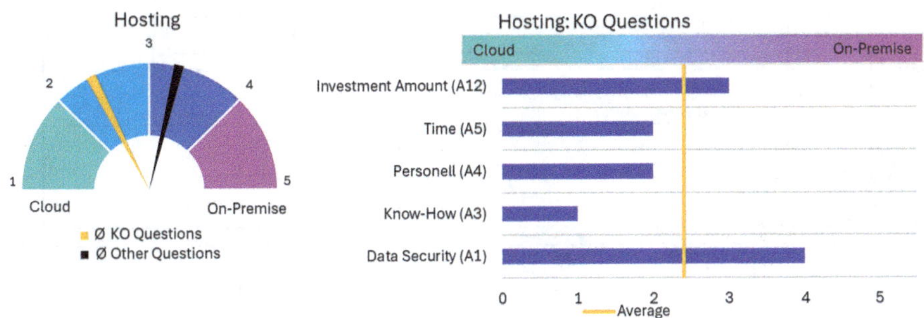

Fig. 2 Example of a gauge and bar chart

The 'KO' questions, highlighted in red in Fig. 2, are defined as critical, non-negotiable criteria. These criteria emerged from the research as key factors derived from an analysis of industrial challenges, data security, and resource availability. The significance of these questions lies in their impact on the feasibility or acceptability of various deployment strategies. By prioritizing these foundational limitations, the framework ensures that viable options are identified from the outset. For instance, a lack of internal expertise in LLM maintenance (Question A3) directs the recommendation toward vendor-provided cloud solutions and Prompt Engineering.

The results are presented using gauge charts that display the mean values for both KO and non-KO questions, with scores calculated only from the relevant questions for each category. For instance, only questions like A1, A3, and A4 contribute to the hosting strategy recommendation. The example in Fig. 3 illustrates this concept. A separate bar chart for KO questions provides more detailed insight. Due to their overriding nature, the KO questions push the recommendation towards Cloud Hosting. However, the bar chart shows that with increased support time, personnel capacity, and know-how, an on-premise "build-while-buy" strategy can be considered long-term. Therefore, the framework's results are not exclusive but rather exploratory proposals for starting an LLM implementation. For organizations new to AI, Part 2 should only be applied after identifying viable processes in Part 1. Defining baseline KPIs and including all interest groups is crucial for addressing concerns and establishing feedback mechanisms.

OP: On-Premise
RAG: Retrieval Augmented Generation
PE: Prompt Engineering
FT: Fine-Tuning

Category	Subcategory	Code	Question/Statement	Rating (1 to 5)	Hosting: highest rating	Hosting: lowest rating	Knowledge Transfer: highest rating	Knowledge Transfer: lowest rating	Model Size: higher rating = bigger model (hr = bm)	Integration Option	Staff Acceptance & Training Needs: "employee readiness"	Data Basis: "data readiness"
Concerns	Data privacy and legal considerations	A1	To what extent is the data required for the process proprietary, particularly sensitive data?		OP	Cloud	RAG	PE				
		A2	How willing is the company to take responsibility for the results of the LLM?								Acceptance	
	Support	A3	How extensive is the internal know-how for the support and maintenance of an LLM?		OP	Cloud	FT	PE				
		A4	What are the personnel capacities that can be used for the support of the LLM?		OP	Cloud	FT	PE				
		A5	How much time can be invested in the support of the LLM?		OP	Cloud	FT	PE				
	Acceptance	A6	What is the overall AI maturity level of the company?		OP	Cloud	FT	PE			Acceptance/Training	
		A7	Obstacles such as demographic limitations will not be a problem.								Acceptance	
		A8	How high is the employees' trust in an LLM?								Acceptance	
		A9	How high is the willingness for training?								Acceptance/Training	
		A10	How high is the need for training?								Training	
	Costs	A11	How high is the willingness to invest?		OP	Cloud	FT	PE			Acceptance	
		A12	What is the available investment sum?		OP	Cloud	FT	PE	hr = bm			
Companys technical ressources	IT architecture	B1	How are existing systems already connected?							adapt to system		
		B2	The connection of an LLM to the existing systems is permitted and easy to implement.							Implementation effort		
	Hardware	B3	Large capacities of processor power are available for the on-premise application.		OP	Cloud			hr = bm			
		B4	Large, free storage capacities are available for the model and further data.		OP	Cloud			hr = bm			
		B5	The internal network infrastructure is extensive and resilient.		OP	Cloud			hr = bm			
Data availability and basis		C1	Aggregating the necessary data is easy to implement. (e.g. do data from multiple sources have to be processed?)									good
		C2	Access to internal company knowledge is important.				RAG	PE				good
		C3	A structured knowledge database exists to which the LLM can access.				RAG	PE				good
		C4	The quality of the existing data is high.									good
		C5	The cleaning and preparation of the data that the LLM is to use is simple.									good
		C6	The data that the LLM is to process is complete.									good
Requirements for the LLM	Performance	D1	The LLM will be used by many employees.						hr = bm			
		D2	Only a few requests per day and employee are made.						hr = bm			
		D3	A lot of data is processed per request (e.g. documents).						hr = bm			
		D4	Conversations with the LLM will be longer (context window).						hr = bm			
		D5	The LLM does not need a high output speed (tokens per second).						hr = bm			
		D6	The latency (seconds to the first token) can be longer.				FT	PE	hr = bm			
		D7	Hallucinations/misinformation can have very critical consequences.				RAG	PE	hr = bm			
		D8	A failure of the LLM would hardly affect the processes.						hr = bm			
	Flexibility and adoptability	D9	Industry/domain-specific language is important for the application of the LLM.				FT	PE	hr = bm			
		D10	The location independence of the LLM is unimportant (e.g. for home office).		OP	Cloud						
Future prospects		E1	An expansion, e.g. to other locations, is not to be expected.		OP	Cloud						
		E2	Through the use of the LLM, the concentration can be focused on the core business in the long term.								Acceptance	

Fig. 3 Recommendations for the second part of the decision model with KO-questions

4 Case Study

This chapter applies the developed decision-making methodology to an anonymized case study at PrintCNC Dynamics, a metal parts manufacturer specializing in metal 3D printing and CNC machining. The company operates multiple 3D printers and CNC machines and serves a varied customer base from prototyping to serial production. With about 30 employees working in three shifts, PrintCNC Dynamics is exploring LLM integration partly because of anticipated workforce reductions. The IT landscape includes cloud-based MES and ERP systems, integrated CRM and QMS via APIs, and an on-premise server hosting CAD software and a growing local knowledge database.

4.1 Part 1: Identifying Use-Cases

The first part of the decision method was applied to two processes, described further in the following paragraphs, at PrintCNC GmbH to identify and evaluate potential LLM use cases. The scores for these two distinct processes are documented in Fig. 4.

	Criteria									Final Score
	Process Complexity	Data Availability	Automation Potential	Employee Acceptance	Documen-tation	Frequency of Support	Fault Tolerance	Materially Critical Process	Time Critical Process	
Repair work	4	5	1	5	4	1	2	1	2	55% (329 P)
Maintenance	2	3	2	2	5	3	3	2	2	39% (262 P)

Fig. 4 Ratings of decision method part 1 for PrintCNC dynamics

Repair of 5-axis CNC machine: A critical machine failure during the night shift demands immediate repair, but operators lack experience and the technician is unavailable. Evaluation scored 55% (Fig. 4), indicating meaningful LLM support potential. The LLM could provide step-by-step repair guidance using documentation, prior maintenance records (via RAG) and data from a digital twin.

Routine maintenance: Regular maintenance follows a well-established checklist, taking 1–2 h for experienced staff. Evaluation scored only 39% (Fig. 4), suggesting low benefit from LLM assistance unless less experienced staff take over these tasks and the LLM could serve as training tool. The LLM could serve as an interactive digital knowledge assistant. It could provide instant clarifications on steps, tools, or safety from documentation.

These two use cases show exemplary the use of the decision method. While the LLM shows clear value in supporting complex, unplanned repairs, its impact on routine procedures appears limited under current conditions. This highlights the importance of also evaluating irregular or ad hoc tasks, where LLMs may provide significant support despite low frequency.

4.2 Part 2: Deployment Strategy Evaluation

For PrintCNC Dynamics, the second part of the framework (Fig. 5) suggests a cloud-based LLM deployment due to limited in-house expertise. A phased strategy is recommended, starting with cloud services and potentially moving on-premise as internal capabilities grow. However, data security concerns call for careful vendor selection. For knowledge transfer, RAG and prompt engineering are preferred over fine-tuning to ensure data security and manage costs. A medium-sized model is deemed best for balancing performance and reliability. While the company's IT infrastructure supports API-based data retrieval, a vector database is needed for knowledge documents. Employee acceptance is high, and integration with existing ERP and MES systems is feasible.

Category	Concerns											
Subcategory	Data privacy and legal considerations		Support			Acceptance					Costs	
Code	A1	A2	A3	A4	A5	A6	A7	A8	A9	A10	A11	A12
Rating (1 to 5)	4	5	1	2	2	2	4	3	4	5	5	3

Category	Companys technical ressources				
Subcategory	IT architecture		Hardware		
Code	B1	B2	B3	B4	B5
Rating (1 to 5)	MW & API	3	1	2	3

Category	Data availability and basis					
Subcategory						
Code	C1	C2	C3	C4	C5	C6
Rating (1 to 5)	5	4	3	4	4	4

Category	Requirements for the LLM									
Subcategory	Performance								Flexibility and adoptability	
Code	D1	D2	D3	D4	D5	D6	D7	D8	D9	D10
Rating (1 to 5)	2	4	3	3	2	3	2	4	3	5

Category	Future prospects	
Subcategory		
Code	E1	E2
Rating (1 to 5)	5	4

Fig. 5 Ratings of decision method part 2 for PrintCNC dynamics

5 Reflection and Future Directions

This paper presents an exploratory proposal for a decision-making framework, not a fully validated one. Its current design includes speculative assumptions and has not yet been externally validated, except for PrintCNC Dynamics. This lack of empirical feedback from practitioners is a significant limitation. Furthermore, simplifications exist in the assessment questions, which do not cover all criteria comprehensively. For instance, the deployment strategy doesn't fully consider factors like model availability (open vs. closed-source) or specific corporate strategic motivations.

The first part of the method effectively balances comprehensiveness with practicality by using a weighted scoring system to prioritize potential LLM applications. While this approach is easy to apply, its reliance on manufacturing-specific examples limits its generalizability. Additionally, the inductive coding process, performed by a single researcher, could introduce bias, and the weighting scheme would be more robust with expert validation and validation through more real-world applications.

The second part uses a questionnaire-based approach to guide implementation. While it provides valuable, quantifiable insights, its recommendations can be rigid. The direct mapping of scores to single strategies may overlook important contextual nuances and the potential for hybrid solutions (e.g., combining RAG with prompt engineering). The method is intended as a structured guideline for discussion, not a final, prescriptive solution. Its usability is also limited using subjective Likert scales, which can lead to conflicting results. The reliance on KO questions is a temporary solution that needs further development.

Future work must prioritize rigorous empirical testing and validation through more researchers in diverse industrial settings to transform this proposal into a proven framework. Obtaining direct practitioner feedback and benchmarking the method against actual deployment outcomes will be essential for validating its utility. For Part 1, future research should focus on expanding and validating criteria using expert interviews and pilot studies. More sophisticated weighting methods like the Delphi Method or machine learning approaches could improve accuracy and universality. For Part 2, the framework needs to expand its scope to include more complex factors like data sensitivity and regulatory requirements. The binary nature of KO questions could be replaced with alternatives like Fuzzy Logic for more nuanced suitability scores or Constraint Programming to model feasible options. Developing dedicated software, perhaps with an LLM-powered bot, is also suggested to enable more complex analysis and ensure the method remains relevant and accessible.

References

1. Gao, R.X., Krüger, J., Merklein, M., et al.: Artificial Intelligence in manufacturing: State of the art, perspectives, and future directions. CIRP Ann. **73**, 723–749 (2024). https://doi.org/10.1016/j.cirp.2024.04.101
2. Haffner, O., Kučera, E., Rosinová, D.: Applications of machine learning and computer vision in industry 4.0. Applied Sci. **14**, 2431 (2024). https://doi.org/10.3390/app14062431
3. Ucar, A., Karakose, M., Kırımça, N.: Artificial intelligence for predictive maintenance applications: key components, trustworthiness, and future trends. Appl. Sci. **14**, 898 (2024). https://doi.org/10.3390/app14020898
4. Kurian T. https://nilleow9.medium.com/customers-are-putting-gemini-to-work-e5ab7a9f9e4f, last accessed 2025/06/15
5. OpenAI.: https://openai.com/de-DE/stories/. Last accessed 15 June 2025
6. Anthropic.: https://www.anthropic.com/customers. Last accessed 15 June 2025
7. Vaswani, A., Shazeer, N., Parmar, N. et al.: Attention is all you need, (2017)
8. Engelke, U., Engelke, B.: ChatGPT—Mit KI in ein neues Zeitalter. 1st edn. mitp Verlags GmbH & Co.KG, Frechen (2023)
9. Li, H., Chen, Y., Luo, J. et al.: Privacy in large language models: Attacks, defenses and future directions, (2023)
10. Huang, L., Yu, W., Ma, W. et al.: A survey on hallucination in large language models: Principles, taxonomy, challenges, and open questions, (2023)
11. Kernan Freire, S., Foosherian, M., Wang, C. et al.: Harnessing large language models for cognitive assistants in factories. In: Lee, M., Munteanu, C., Porcheron, M. et al. (eds.), Proceedings of the 5th international conference on conversational user interfaces, pp. 1–6. ACM, New York, USA (2023)
12. Mockenhaupt, A.: Digitalisierung und künstliche intelligenz in der produktion. Springer Vieweg, Wiesbaden, Heidelberg (2021)
13. Touvron, H., Martin, L., Stone, K. et al.: Llama 2: Open foundation and fine-tuned chat models, (2023)
14. Gkournelos, C., Konstantinou, C., Makris, S.: An LLM-based approach for enabling seamless Human-Robot collaboration in assembly. CIRP Ann. **73**, 9–12 (2024). https://doi.org/10.1016/j.cirp.2024.04.002
15. Fan, H., Liu, X., Fuh, J.Y.H., et al.: Embodied intelligence in manufacturing: leveraging large language models for autonomous industrial robotics. J. Intell. Manuf. **36**, 1141–1157 (2025). https://doi.org/10.1007/s10845-023-02294-y
16. Xia, Y., Zhang, J., Jazdi, N. et al.: Incorporating large language models into production systems for enhanced task automation and flexibility. VDI-Berichte., **2437**, 375–390 (2024). https://doi.org/10.51202/9783181024379
17. Zhao, Z., Tang, D., Zhu, H. et al.: A large language model-based multi-agent manufacturing system for intelligent shopfloor, (2024)
18. Fakih, M., Dharmaji, R., Moghaddas, Y. et al.: LLM4PLC: Harnessing large language models for verifiable programming of PLCs in industrial control systems, (2024)
19. Xia, L., Li, C., Zhang, C., et al.: Leveraging error-assisted fine-tuning large language models for manufacturing excellence. Robot. Comput.-Integr. Manuf. **88**, 102728 (2024). https://doi.org/10.1016/j.rcim.2024.102728
20. Kernan Freire, S., Wang, C., Foosherian, M., et al.: Knowledge sharing in manufacturing using LLM-powered tools: user study and model benchmarking. Front Artif Intell **7**, 1293084 (2024). https://doi.org/10.3389/frai.2024.1293084
21. Badini, S., Regondi, S., Frontoni, E., et al.: Assessing the capabilities of ChatGPT to improve additive manufacturing troubleshooting. Adv. Ind. Eng. Polym. Res. **6**, 278–287 (2023). https://doi.org/10.1016/j.aiepr.2023.03.003

22. Colabianchi, S., Costantino, F., Sabetta, N.: Assessment of a large language model based digital intelligent assistant in assembly manufacturing. Comput. Ind. **162**, 104129 (2024). https://doi.org/10.1016/j.compind.2024.104129
23. Ott, L.: ToyoGPT: Das sichere ChatGPT für Toyota Deutschland. objective partner AG. https://objective-partner.de/toyogpt-ugpt/. Last accessed 15 June 2025
24. Mercedes-Benz Group.: https://group.mercedes-benz.com/innovation/digitalisation/industry-4-0/chatgpt-in-vehicle-production.html. Last accessed 15 June 2025
25. Hoffmann, D., Schmiedchen, R., Tiedemann, Y.: Wie Generative AI die Automobilindustrie verändert. https://www.automotiveit.eu/technology/kuenstliche-intelligenz/wie-genai-die-zukunft-der-automobilindustrie-vorantreibt-667.html. Last accessed 15 June 2025
26. Bosch Global.: https://www.bosch.com/de/stories/ki-bilderkennung-fertigung/. Last accessed 15 June 2025
27. Siemens.: https://www.siemens.com/de/de/produkte/automatisierung/themenfelder/ki-industrie/industrial-copilot.html. Last accessed 15 June 2025
28. ABB.: https://new.abb.com/news/detail/104829/abb-and-microsoft-collaborate-to-bring-generative-ai-to-industrial-applications. Last accessed 15 June 2025
29. March, S.T., Smith, G.F.: Design and natural science research on information technology. Decis. Support Syst. **15**, 251–266 (1995). https://doi.org/10.1016/0167-9236(94)00041-2
30. Kuckartz, U.: Qualitative inhaltsanalyse. Methoden, Praxis, Computerunterstützung. 4th edn. Beltz, Weinheim (2018)

Causal Inference for Quality Enhancement in Injection Molding

Daeyeop Na(✉), Stefan Schulte, David Kiviriga, and Joachim Metternich

Institute of Production Management, Technology and Machine Tools, Technical University of Darmstadt, Darmstadt, Germany
D.Na@ptw.tu-darmstadt.de

Abstract. Many production processes exhibit complex and partially unknown cause-and-effect relationships, posing challenges to effective process control aimed at enhancing product quality. Some processes involve complex interactions, making precise quality optimization challenging. In injection molding, product quality prediction algorithms have traditionally relied on machine learning models utilizing feature selection or deep learning techniques. However, these methods have limitations in identifying the true causal relationships between process parameters and quality outcomes. As a result, it is challenging to accurately determine the key parameters that have the most significant impact on quality. To address this issue, this study uses causal inference, which encompasses causal discovery techniques, to establish clear causal links between process parameters and product quality by integrating domain knowledge in injection molding. By applying causal inference techniques to injection molding, this approach enables the precise identification of process parameters that have a significant impact on product quality. This method enables effective process control, ultimately contributing to quality improvement in injection molding products.

Keywords: Causal inference · Injection molding · Quality improvement

1 Introduction

In the era of Industry 4.0, technologies such as the Internet of Things (IoT), cloud computing, and big data have enabled production environments that generate large volumes of process-related data.

This data availability paves the way for advanced artificial intelligence (AI)-based quality control systems [1]. Early detection of deviations during production can help reduce the occurrence of defective products and improve overall manufacturing efficiency [2]. Injection molding is a widely used process for mass-producing high-precision plastic components, where product quality is critical. However, machine parameters may not accurately reflect the actual molding behavior, and quality can still vary due to other factors even when settings remain unchanged, making it difficult for conventional models to identify their causal impact [3]. To overcome these limitations, recent research has been focusing on developing machine learning and deep learning-based algorithms

L. Overmeyer and B.-A. Behrens (eds.), *Production at the Leading Edge of Technology*,
Lecture Notes in Production Engineering, https://doi.org/10.1007/978-3-032-19524-1_44

that can predict product quality directly from process data in injection molding. These predictive approaches aim to enable accurate and efficient quality assessment for each molded part without relying on exhaustive physical inspections. While conventional deep learning and machine learning models can accurately predict product quality by learning correlations between input variables and outcomes, they struggle to explain the causal impact of individual process parameters [4]. Therefore, Identifying the root causes of quality degradation remains a challenge. To address this issue, this study applies causal machine learning techniques to sensor data and machine setting/parameter data collected from the injection molding process.

The following sections first review prior research on quality prediction in injection molding and the use of causal machine learning, and then present the proposed causal learning framework, including its methodological details and application results for identifying key parameters that directly influence product quality.

2 Literature Review

2.1 Predictive Quality in Injection Molding and Remaining Research Gap

In recent years, various techniques have been studied to predict the quality of injection-molded products, with a particular focus on machine parameters collected from injection molding machines. Previous studies have applied machine learning techniques by using various process parameters as input data. For example, features extracted from mold pressure curves such as peak pressure, pressure gradient, viscosity index, and energy index have been utilized for quality prediction [5]. Similarly, key time points in the injection molding process such as filling start, switchover point, maximum pressure, end of packing, and end of cooling have been used as input variables for predictive models [6]. In addition, deep learning models such as backpropagation neural networks have been employed to model and optimize product quality based on key process variables, thereby improving quality through parameter optimization [7].

While these studies have contributed to improving the accuracy of quality prediction in injection molding processes, fundamental limitations remain. Specifically, the causal relationships between process parameters and product quality have not yet been fully clarified. Most conventional machine learning models rely primarily on correlations, making it difficult to distinguish between cause and effect [8]. This reveals a clear research gap: understanding which parameters truly influence product quality, and to what extent, is essential for accurate prediction and effective process optimization. To overcome this limitation, causal machine learning is employed to uncover the underlying causal relationships between process parameters and product quality.

2.2 Causal Machine Learning

Causal inference goes beyond simple correlation between variables and aims to determine whether one variable actually influences another, thus demonstrating a causal relationship. It plays a central role in data-driven decision-making, policy design, and scientific analysis across various fields [9]. This aligns with the well-known phrase "correlation does not imply causation" as a statistical association between two variables may arise from various underlying factors and does not necessarily indicate a direct causal link [10].

To better understand and model such causal relationships, graphical model–based approaches are commonly applied. In particular, Directed Acyclic Graphs (DAGs) are widely used in causal inference, where nodes represent variables and edges (arrows) indicate directed causal relationships between them.

These causal graphs not only provide a visual representation of the causal structure among variables but also serve as a theoretical and mathematical foundation for causal inference. In practice, the underlying causal structure is often not known in advance, and one must infer it directly from observational data, a process referred to as causal discovery.

In recent years, a variety of causal discovery algorithms have been developed to learn the structure of causal graphs directly from observational data. Traditional constraint-based methods such as the Peter-Clark (PC) algorithm and the Fast Causal Inference (FCI) algorithm assume full observability and are primarily designed for independent and identically distributed (i.i.d.) data. While FCI accounts for latent variables, it cannot adequately model temporal dependencies, which limits its use in time series applications [11–13]. Linear non-Gaussian methods such as DirectLiNGAM and VARLiNGAM offer explicit causal directionality but rely on assumptions like linearity and non-Gaussian noise, which may not hold in complex production data [14].

To overcome these limitations, time-series-specific methods such as PCMCI+ were introduced. PCMCI+ handles strong autocorrelation, non-linear dependencies, and partial observability by combining conditional independence testing with flexible causal graph construction. In particular, it provides a statistically robust framework for modeling causal relationships in time series while maintaining low false positive rates [15].

Once a causal graph is constructed, causal inference typically proceeds in two stages [16]. The first stage is causal estimation, where the goal is to quantitatively estimate the magnitude and direction of the causal effect of a specific treatment variable (X) on an outcome variable (Y). In this context, a confounder may introduce a backdoor path between X and Y, creating statistical dependence that does not reflect a true causal effect. Such spurious associations can be blocked by conditioning on observed variables that satisfy the backdoor criterion. In certain cases, frontdoor adjustment may also be applicable, provided specific conditions are met. This method leverages a mediator variable that lies between X and Y and enables causal identification when it satisfies the necessary properties. Once the causal effect is identified through either the backdoor or frontdoor criterion, the magnitude of the effect is typically estimated using regression-based methods appropriate to the identification strategy and the characteristics of the data [17].

The second stage is causal refutation, which evaluates the credibility and robustness of the estimated causal effect. Commonly used refutation techniques include:

- **Placebo Refutation**: This method replaces the treatment variable with a randomly generated placebo variable to assess whether the originally estimated effect could have arisen by chance. The p-value ("probability value") is the probability of obtaining the observed result, or a more extreme one, if the null hypothesis were true. If the refutation yields a low effect size and a high p-value, it suggests that the original causal effect is unlikely to be spurious and is therefore considered statistically reliable.

- **Random Refutation**: In this approach, the analysis is repeated using randomly selected control variables to test the sensitivity of the causal estimate to the choice of covariates. If the causal effect remains consistent in both magnitude and direction, and the p-values remain high, this indicates that the estimation is robust to variations in covariate selection and is not likely the result of overfitting or selection bias.
- **Subset Refutation**: This technique involves re-estimating the causal effect on randomly chosen subsets of the data to examine whether the effect is dependent on specific portions of the dataset. If the estimated causal effect remains stable across different subsets and continues to exhibit statistical significance, it provides evidence that the result is generalizable and not sensitive to the specific sample used.

These refutation methods play a critical role in ensuring that the model captures genuine causal relationships rather than reflecting random statistical associations, data artifacts, or sampling noise.

However, industrial systems such as those found in manufacturing are often large and complex, making causal structure learning a challenging task. Even domain experts may struggle to fully define all causal relationships, and relying solely on manual construction of causal graphs may result in incomplete or biased representations [18]. To address these limitations, recent research has explored hybrid approaches that combine domain expertise with automated causal discovery methods.

In this study, such a hybrid approach is proposed to enable more effective causal identification of key process parameters influencing product quality in complex injection molding processes.

3 Methods: Causal Machine Learning Framework

In this study, a causal machine learning framework is applied to identify the causal factors influencing the quality label in injection molding processes. The analysis is based on approximately 290 sensor and control variables collected from injection molding machines at a German automotive components manufacturer. The data is gathered from a pilot production line that has been specifically established for this project. The overall structure of the applied framework is illustrated in Fig. 1.

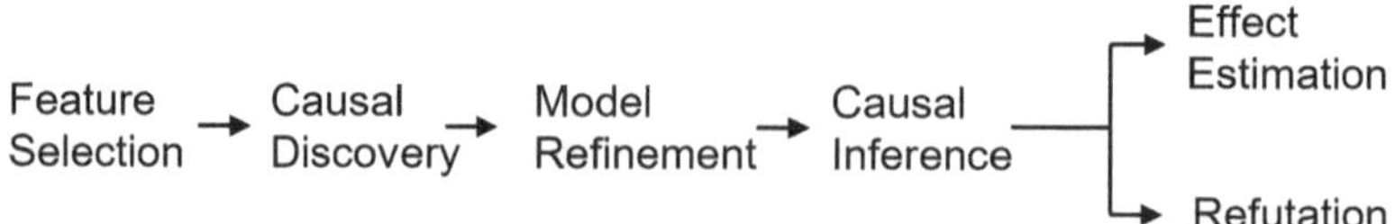

Fig. 1 Overview of the applied causal machine learning framework in this study

To reduce the computational complexity involved in analyzing the complete set of variables, a machine learning-based feature importance analysis is conducted first. In this step, the relative importance of the features is assessed using the SHAP-Algorithm

(SHapley Additive exPlanations) [19]. The variables with high importance scores are then subjected to causal discovery techniques to infer the underlying causal structure, which is represented as a Directed Acyclic Graph (DAG).

Given that the multivariate sensor data from injection molding are time-resolved and characterized by complex, partly nonlinear dependencies and potential unobserved confounders, PCMCI + is chosen as the causal discovery method due to its suitability for such time series data. The initial DAG is subsequently refined by removing edges that contradict domain knowledge or violate established physical constraints (Fig. 2).

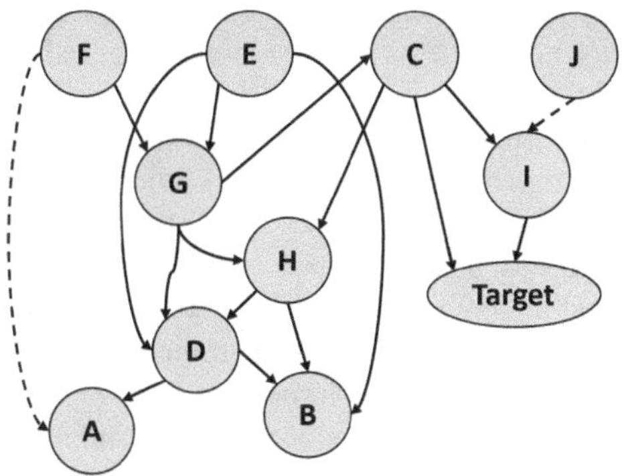

Fig. 2 The DAG constructed from causal discovery and domain knowledge with focus on the quality label. Dashed edges indicate those removed following causal discovery and solid black lines represent edges that remained unchanged after the discovery step

Finally, causal inference procedures are carried out based on the refined graph shown in Fig. 2. This process begins with the formal identification of causal effects, such as estimating the causal impact of specific treatment variables (e.g., temperature) on the product quality label. Standard criteria such as the backdoor or frontdoor criterion are used to check whether the causal effect can be identified, and the appropriate adjustment set is determined to block potential confounding paths.

Once identification is achieved (e.g., via the backdoor or frontdoor criterion), the strength and direction of the causal effects are estimated using regression-based methods such as linear regression that are standard tools for causal effect estimation. These methods are adapted to the identification strategy and the characteristics of the data, allowing for the quantification of how changes in treatment variables affect the product quality label. To evaluate the reliability and robustness of the estimated causal effects, the refutation methods introduced in Sect. 2.2 are applied, including Placebo, Random and Subset Refutation.

4 Results and Discussion

As shown in Fig. 2, based on the results of causal discovery and domain knowledge, hydraulic pressure at switch-over, dosing volume, peak injection pressure, temperature, and injection volume were identified as the key variables that may exert a causal influence on the quality label. The quality label represents the measured distance from the center to the edge of the molded product, serving as a critical quality indicator. Temperature refers to the surface temperature of the product immediately after injection. Hydraulic pressure at switch-over is the pressure at the point where the process transitions from injection to holding pressure to compensate for shrinkage. Dosing volume is the amount of molten plastic prepared before injection, which can influence the switch-over pressure depending on its size. Peak injection pressure is the highest pressure reached during cavity filling, influenced by injection speed, melt properties, and mold design.

Table 1 presents the results of causal effect estimation and refutation analysis, both conducted as part of the causal inference stage, for these five variables. The estimated effect values, computed using the DoWhy Python library for causal inference, indicate the magnitude and direction of change in the quality label associated with changes in each variable [20]. As shown in Table 1, the estimated causal effects are as follows: hydraulic pressure at switch-over: −0.011, temperature: −0.026, peak injection pressure: −0.005, injection volume: 0.195, and dosing volume: −0.177. As observed in the causal estimation results, the preceding SHAP-based feature importance analysis identified temperature as one of the most influential factors for the quality label. However, its direct causal effect was relatively small compared to the other four parameters. In addition, the causal inference analysis indicated that injection volume and dosing volume exhibited relatively higher causal effects. This suggests that the importance rankings in SHAP primarily reflect mediated correlations through other process variables, whereas causal inference excludes indirect paths and estimates only the pure direct effects. Based on the criteria described earlier, the analysis shows that the refutation tests demonstrated consistent stability across all variables. In the Placebo Refutation, all effects disappeared completely (0.0, p-value = 1.0), confirming that the original estimates are not due to random chance. In the Subset Refutation, effect sizes remained nearly unchanged and p-values ranged from 0.9 to 1.0, indicating strong robustness against sample variation. In the Random Refutation, the effect sizes were also consistent; however, the p-value for injection volume was relatively lower at 0.68, suggesting slightly reduced robustness compared to the other variables with p-values above 0.84. Overall, these results indicate that the causal estimates for these variables are statistically reliable, with injection volume potentially showing slight variation in its estimated effect depending on which other variables are controlled for in the analysis.

Table 1 Causal effect estimation and refutation results

Edge	Estimation	Placebo refutation		Random refutation		Subset refutation	
		New effect	p-value	New effect	p-value	New effect	p-value
C → Target	−0.011	0.0	1.0	−0.011	0.88	−0.011	0.98
I → Target	−0.026	0.0	1.0	−0.026	0.96	−0.026	1
G → Target	−0.005	0.0	1.0	−0.005	0.98	−0.005	0.96
E → Target	0.195	0.0	1.0	0.195	0.68	0.194	0.9
F → Target	−0.177	0.0	1.0	−0.177	0.98	−0.178	1

5 Conclusion

This study applied causal machine learning to identify and analyze the causal relationships between process parameters and product quality in complex injection molding processes, without relying on conventional machine learning or deep learning approaches. Through this method, key parameters affecting quality were identified, and a comparison with conventional machine learning showed meaningful results.

The analysis was conducted for a single quality metric only, and the scope of causal discovery was limited as a wide range of algorithms was not explored. Future work will include analyzing causal relationships involving multiple quality metrics and a broader set of process parameters, as well as developing an automated hybrid approach that integrates causal discovery algorithms with domain knowledge for real-time quality monitoring and process optimization. This will enable rapid identification and mitigation of quality degradation causes in injection molding operations, and early detection of changes in key variables with causal impacts on quality, allowing for automatic process parameter adjustments to reduce defect rates and improve productivity.

Acknowledgements. The research leading to these results has received funding from the German Federal Ministry for Economic Affairs and Energy (BMWE) through the research project Endi-QM - Energieeffizienz durch intelligentes in-Prozess Quality Monitoring (FKZ. 03EN4032D).

Competing Interests. The author(s) has no competing interests to declare that are relevant to the content of this manuscript.

References

1. Hiou, M.Y., Akroum, H.: Leveraging AIoT for advanced quality control in production lines. In: 2023 IEEE International Conference on Artificial Intelligence & Green Energy (ICAIGE), p. 1. IEEE (2023)
2. Singh, S., Batra, R., Rai, K., Sujai, S.: Proactive quality evaluation: a novel strategy-assisted early detection in manufacturing. Proc. Eng. Sci. **5**, 343 (2024)
3. Ke, K.-C., Huang, M.-S.: Quality classification of injection-molded components by using quality indices, grading, and machine learning. Polymers (Basel) **13** (2021)
4. Vowels, M.J.: Trying to outrun causality with machine learning: limitations of model explainability techniques for identifying predictive variables (2022). arXiv: arXiv:2202.09875
5. Chen, J.-Y., Yang, K.-J., Huang, M.-S.: Online quality monitoring of molten resin in injection molding. Int. J. Heat Mass Transf. **122**, 681 (2018)
6. Gim, J., Rhee, B.: Novel analysis methodology of cavity pressure profiles in injection-molding processes using interpretation of machine learning model. Polymers (Basel) **13** (2021)
7. Yin, F., Mao, H., Hua, L., Guo, W., et al.: Back Propagation neural network modeling for warpage prediction and optimization of plastic products during injection molding. Mater. Des. **32**, 1844 (2011)
8. Altman, N., Krzywinski, M., 2015. Association, correlation and causation. Nat Methods *12*, p. 899.
9. Sudeepa Roy, B.S.: Causal inference in data analysis with applications to fairness and causal inference in data analysis with applications to fairness and explanations. In: 18th International Summer School 2022, Berlin, Germany, September 27–30, 2022, Tutorial Lectures, in Reasoning Web. Causality, Explanations and Declarative Knowledge, p. 114 (2022)

10. Yao, L., Chu, Z., Li, S., Li, Y., et al.: A survey on causal inference. ACM Trans. Knowl. Disc. Data **15**, 1 (2021)
11. Entner, D., Hoyer, P.O.: On causal discovery from time series data using FCI, p. 121 (2010)
12. Shimizu, S., Hoyer, P., Hyvärinen, A., Kerminen, A.: A linear non-Gaussian acyclic model for causal discovery. J. Mach. Learn. Res. **7**, 2003 (2006)
13. Spirtes, P., Glymour, C.: An algorithm for fast recovery of sparse causal graphs, **9**, 62 (1991)
14. Shimizu, S., Inazumi, T., Sogawa, Y., Hyvarinen, A., Kawahara, Y., Washio, T., Hoyer, P.O., Bollen, K.: DirectLiNGAM: a direct method for learning a linear non-Gaussian structural equation model (2011)
15. Runge, J.: Discovering contemporaneous and lagged causal relations in autocorrelated nonlinear time series datasets (2020)
16. Sharma, A., Kiciman, E.: DoWhy: an end-to-end library for causal inference (2020)
17. Pearl, J.: Causal inference in statistics: an overview. Stat. Surv. **3** (2009)
18. Marazopoulou, K., Ghosh, R., Lade, P., Jensen, D.: Causal discovery for manufacturing domains (2016). arXiv: arXiv:1605.04056
19. Lundberg, S., Lee, S.-I.: A unified approach to interpreting model predictions (2017)
20. Sharma, A., Kiciman, E.: DoWhy: an end-to-end library for causal inference (2020). arXiv arXiv:2011.04216

Moving Towards Continuous Virtual Commissioning Using Open Interfaces for Industrial Robot Integration and Control

Simon Nowinski(✉), Lars Klingel, and Alexander Verl

Institute for Control Engineering of Machine Tools and Manufacturing Units (ISW), University of Stuttgart, Stuttgart, Germany
simon.nowinski@isw.uni-stuttgart.de

Abstract. The increasing demand for flexible control and simulation concepts in the industry is driven by competitive pressures and shorter product life cycles. Virtual Commissioning (VC) offers the advantage of identifying design and planning errors before the system is commissioned in a real environment. In VC, the test configurations are referred to as Model-in-the-Loop (MiL), Software-in-the-Loop (SiL), and Hardware-in-the-Loop (HiL). The current approach of the different test configurations requires in addition to the organizational preparation and integration, technical efforts such as manual switching and setup. Therefore, this study applies a switching, conversion and transfer mechanism with open interfaces, which enable Continuous Virtual Commissioning (CVC). These mechanisms are tested on an industrial robot cell. This lays the technological foundation for flexible switching between the configurations of VC. Furthermore, robotic joint and cartesian position deviations are observed across the configurations of VC, indicating configuration-specific alterations.

Keywords: Continuous virtual commissioning · Open interfaces · Robotics

1 Introduction

Due to the demand of flexible production systems, the number of robot manufacturers and available robot models is increasing. As a result, users have to work with various systems more frequently. For these reasons, there is a need to simplify the programming and integration of systems, considering progressively changing tasks and environments [1]. Besides conventional Robotic Control (RC), robots can be operated via Computer Numerical Control (CNC) with an CNC-Kernel integration [2,3]. Nevertheless CNC systems are designed specifically for machining and material removal processes, whereas RC systems are primarily used for handling and assembly tasks. CNC systems use standardized

L. Overmeyer and B.-A. Behrens (eds.), *Production at the Leading Edge of Technology*, Lecture Notes in Production Engineering, https://doi.org/10.1007/978-3-032-19524-1_45

programming languages, such as G-Code according to ISO 6983 and often rely on Computer Aided Manufacturing (CAM) tools, which automatically generate complex paths based on CAD data. In contrast, robots are typically based on vendor-specific languages, supported by common robot simulation environments and programmed manually on-site using teach pendants, editors [2] or other teaching methods like *Programming by Demonstration* (PbD) [1,4]. CNC-based robot teaching is researched with the Direct Robot Control (DRC) mechanism [5], or with the use of extended dynamic behavior to improve the trajectory planning [6]. Furthermore it is still necessary to bridge the gap between CNC and RC already in early development stages. Virtual Commissioning (VC) as method is suitable for virtual teaching without the need of the real machine or robot. In combination, an open platform architecture is needed for independent proprietary programming languages and controls (see Fig. 1). This paper shows possibilities for Continuous Virtual Commissioning (CVC) with Model-in-the-Loop (MiL)-teaching as promising starting point, streamlining the development of robotic system applications.

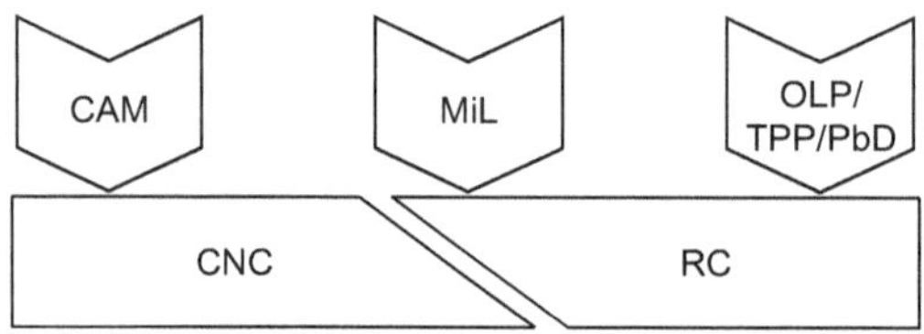

Fig. 1. MiL-Teaching as early development tool to build the bridge between CNC and RC

The paper is structured as follows. After the introduction, the state of art with related work is presented (Sect. 2). The following section outlines the concept of CVC (Sect. 3). Subsequently the implementation and realization are shown (Sect. 4). Section 5 presents the results of the evaluation. Finally, an conclusion and outlook are given (Sect. 6).

2 State of Art and Related Work

VDI/VDE 3693 [7] defines the XiL-configurations as follows. In the MiL simulation, prototypically and simple implemented control algorithms are tested. Actors and sensors of the machine describe the machine model. An abstract control model language is used as common simulation model networked on one computer. The following Software-in-the-Loop (SiL) simulation runs on a simulation computer with a machine model via virtual communication with an emulated (virtualized) control connected. The virtualized control runs production code. The Hardware-in-the-Loop (HiL) simulation is an advanced stage, for integration and overall testing the finished control code. This stage uses a simulation computer with machine model via real-time communication with a physical control connected. Manufacturer-specific control and simulation solutions without open interfaces, hinder the integration of complex, multi-variant, and highly linked systems and components. For end-to-end engineering, open control and simulation platforms are necessary. A case study in [8] shows the requirements on VC with continuous approach. It formed a foundation for further concept outlines. Thus, intuitive component-based modeling and supported teaching of movements in the 3D model is possible. From such control models, configurations and programs should automatically derive for motion control and the programmable logic controller (PLC). Furthermore, virtual methods should be used to optimize cycle times, robot positioning and configurations. For the latter, its necessary to map the dynamics of the robot system in the simulation, which will later also be used in motion control. Z. Liu [9] presents a method for the interoperability of component models between different simulation tools. An extended International Electrotechnical Commission (IEC) 61131–3 library tested by the author is integrated into the modeling process. The developed component model is exported in a PLCopen XML format (XML - eXtensible Markup Language) and imported into the respective VC simulation tool. At this stage, different component behavior models are created to form a system model. Plant models available in a proprietary format can be translated into the neutral exchange format by the simulation model developer. This enables the integration of simulation components into the desired simulation environment using an import function. This method allows the use of neutral exchange formats, however only elementary control components are validated in this case. Nibert et al. use a modular approach to system design to maintain the different configurations MiL, SiL and HiL. This approach decouples functional components from interface components. Thereby a single functional model can be used in different development stages. Only the interface components are changed as required [10]. This approach allows for a integrated transition between configurations, however the approach is only tested on a vehicle design and controller level, not specific on industrial production systems. Additionally neither the Input/Output (I/O)-management or classification is presented. Another publication by Tinsel et al. [11] shows an XML-based transformation mechanism to perform a continuous transition between X-in-the-Loop (XiL)-configurations during VC with industrial controls. An extension in the submodels adapts their structure to configuration changes. A converter then applies the structural changes defined in the extension. In addition, automatic code generation accelerates the XiL cycle by modifying the model structure. The resulting configuration rule is stored in the model extension for reuse in future configurations. However, this method is only applied to XML-file transformation. Studies show potential, restrictions, limitations and challenges for VC integration at the machine, cell, and plant levels either with the use of manual control, basic integrated scripts without a dynamic plant model, or with SiL as the earliest test stage considered [3] while directly using PLC I/O-links [12,13]. Likewise a vehicle-based test approach with ROS

(Robot Operation System) but by using external ROS-nodes which correspond to SiL specifications [14].

Previous work outlines the challenges with XiL-configurations of continuous coupling and adaption in the engineering process. From that point, derived mechanisms and experimental investigations on XiL-configuration management for CVC are shown in [8] and [15]. An further elaborated and extended concept single-tool-solution is shown in the next chapter.

3 Concept Overview

Following the examination of initial work on practical CVC and challenges [15], this article proposes an approach and breakdown on CVC mechanisms. To be able to initialize and change simulation and control states between the different XiL-configurations in the specific environments, a switch-, convert- and transfer-mechanism are defined [15]. Figure 2 shows the architecture of the mechanisms with numbers as reference.

(1) Switch between XiL-configurations with necessary setup, supported by Single Source of Truth (SSOT) modular approach [10]. The change and utilization of the Real-Time Operation System (RTOS) besides the Non-Real-Time Operation System (NRTOS) and parallelization of model calculations [16] are considered in the mechanism. The test requirements, input and output interfaces, and functional components in the simulation environment's block diagram can be adapted during switching.
(2) Convert and reuse control programs from the MiL-configuration for every following XiL-configuration. Due to differences in syntax, the parameters may need to be adapted in the conversion process.
(3) Transfer control programs between simulation and control using the server-client principle with websockets as communication protocol and a buffer to compensate for connection problems or time-delayed accesses by the client.

4 Mechanism Implementation and Realization

Compatible software tools are used to ensure the successfully implementation of various test configurations with real-time capabilities. The simulation tool used is ISG-virtuos [17]. It contains a motion and teach tool for MiL control. Beckhoff TwinCAT 3 is used for the control environment. ISG-virtuos enables the solver configuration to be switched between Windows- and TwinCAT-Solver. This allows the interfaces to be addressed between ISG-virtuos and TwinCAT 3. A Graphical User Interface (GUI) allows the user to switch between the different XiL test configurations (MiL, SiL, and HiL). This GUI communicates via the ISG-virtuos "Remote Interface" in order to execute changes within the simulation environment and allow switching. In addition, control with the GUI is accessible via the "TwinCAT 3 Automation Interface" to initialize the TwinCAT

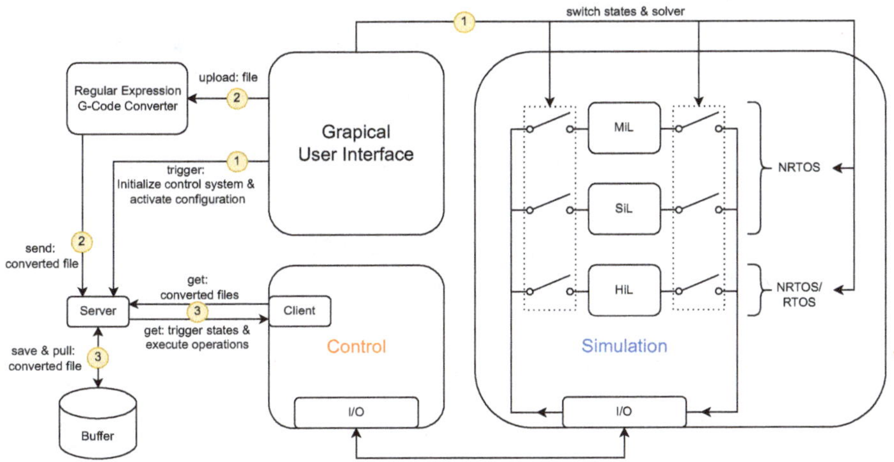

Fig. 2. XiL-Mechanisms: (1) switch-, (2) convert-, and (3) transfer-mechanism (based on [15])

control project, among further functionalities. ISG-virtuos enable real-time simulation and deterministic communication via the TwinCAT solver. In this case, an I/O-EtherCAT simulation is created in order to establish communication via EtherCAT between the simulation and control PC.

5 Mechanisms Evaluation

Figure 3 shows the iterative structure of the test configurations defined in Sect. 3 and 4. In the virtual and real test cell a "Stäubli TX2-40" robot is used. Starting with the MiL configuration, in which only the simulation with internal motion and teach tool is used. Following the SiL configuration with emulated interfaces, in this case a TCP/IP (Transmission Control Protocol/Internet Protocol) connection between ISG-virtuos simulation and Beckhoff TwinCAT 3 control. The TwinCAT 3 control environment can run in this configuration either on the same PC as the simulation, or remotely on a different PC. The ending HiL configuration uses the real-time TwinCAT-solver and EtherCAT interfaces between the simulation and control environment to ensure real-time communication. The integrated ISG CNC-Kernel in ISG-virtuos and TwinCAT is used for the calculation of forward kinematics and inverse kinematics. The MiL-Controller is used for teaching the forward kinematics of the robot with joint angles (marked in red) and inverse kinematics in cartesian coordinates (marked in green).

Initially only the flange-tool-center-point is calculated, therefore kinematic offset of the tool-CS (Coordinate System) from the flange-CS needs to be considered in simulation and real control. The illustrated pose is teached invers with the MiL-Controller. The cartesian pose is then converted and transferred to the control for the SiL and HiL configuration. The cartesian calculations of the MiL-simulation and SiL- and HiL-control are mostly identical (measured a cartesian

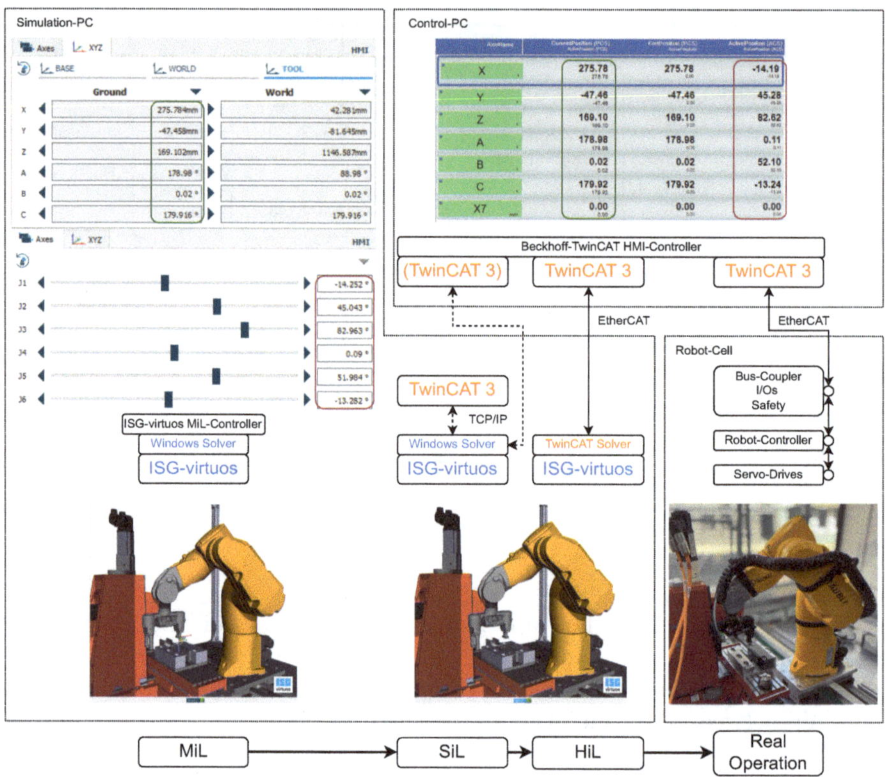

Fig. 3. Hardware and software level assignment for the iterative XiL test configurations unto real operation with "Stáubli TX2-40" robot in ISG-virtuos and Beckhoff TwinCAT 3

deviation of $-4\mu m \leq \Delta \leq 4\mu m$). The system relies solely on the manufacturers calibration and no additional external sensors are employed for error compensation. The overall system uncertainty includes potential deviations caused from different kernel versions between simulation and control PC as well as platform dependent floating-point resolutions. The joint angle solutions in inverse kinematics indicate a measured deviation of $-0.35° < \Delta < 0.25°$ between the XiL-stages. These effects can occur due to multiple possible solution branches of the robot arm [18]. Specifically in this case, through the axis positioning of the chosen pick-and-place operation of the robot. This study is limited by the gap between simulated joint behavior and real-world robot execution, particularly due to configuration-specific uncertainties in control signal interpretation, sensor accuracy, and physical interaction dynamics. These discrepancies can affect the fidelity of joint deviation analysis. High-precision measurement systems, such as laser trackers, to validate joint positions in real environments are necessary.

6 Conclusion and Outlook

In this study CVC was introduced and multiple CVC mechanisms were applied. The mechanisms are derived based on agile principles with the effort reduction through automation of processes to decrease production downtimes in the MiL-stage and subsequent SiL- and HiL-stage. The mechanisms are designed through the concept of open interfaces and allow CVC of industrial robots. The mechanisms contribute to the increase of commissioning flexibility, with high accuracy between different XiL-stages, as the results show. This study contributes to a deeper insight of XiL-specific joint deviations in robotic systems by comparing simulation with real measurements. The identification of deviation intervals across XiL-stages and emphasizing the influence of uncertainties, provides a foundation for improving the consistency between simulated and actual robot behavior. However, limitations such as the absence of external measurements, real-time correction mechanisms, and reliance on one specific robot platform and controller, constrain the observability of the results. Future research will focus on integrating high-precision measurement systems, compensation approaches, and developing feedback loops from diverse physical robots and controllers back into the virtual control system and simulation. These steps will enhance model accuracy and control fidelity, ultimately supporting more robust deployment of robotic systems.

Acknowledgments. The authors thank the German Federation of Industrial Research Associations (AiF) for funding the project: "SIRob" (KK5311205RL4).

References

1. Lehmann, C. et al.: Anwendungsbeispiele zur integration hetero-gener Steuerungssysteme bei robotergest utzten Industrieanlagen. In: Vogel-Heuser, B., Michael ten Hompel, Bauernhansl, T. (eds.) Hand- buch Industrie 4.0. Berlin, Heidelberg, Springer Berlin Heidelberg, pp. 29–42 (2024) ISBN: 978-3-662-58527-6. https://doi.org/10.1007/978-3-662-58528-349
2. Hagele, M. et al.: Industrial Robotics. In: Siciliano, B., Khatib, O. (eds) Springer handbook of robotics 2nd edn. Springer Hand-books, Berlin and Heidelberg Springer, pp. 1385–1422 (2016). ISBN 978-3- 319-32550-7. https://doi.org/10.1007/978-3-319-32552-154
3. Manescu, G.I. et al.: Behavioral algorithms for mechatronic sys-tems: a digital twin perspective. In: 2025 26th International Carpathian Control Conference (ICCC). (Stary Smokovec, High Tatras, Slovakia). IEEE, pp. 1–5 (2025) ISBN 979-8-3315-0127-3. https://doi.org/10.1109/ICCC65605.2025.11022913.
4. Billard A.G., Calinon, S., R udiger Dillmann. Learning from Humans . In: Siciliano, B., Khatib, O. Springer handbook of robotics 2nd edn. Springer Handbooks. Berlin and Heidelberg: Springer, 2016 pp. 1995 (2014). ISBN 978-3-319-32550-7. https://doi.org/10.1007/978-3-319-32552-174
5. Wojtulewicz, A., Chaber, P.: Industrial robot control sys- tem with a predictive maintenance module using IIoT technology. eng. In: Sensors (Basel, Switzerland) 25.4 (2025). Journal article the authors declare no conflict of interest. https://doi.org/10.3390/s25041154 eprint: 40006383.

6. Pfeifer, D., Scheifele, C., org Fehr, J.: Engineering platform for the efficient design and operation of cnc-controlled robots reusing digital twins from virtual commissioning. In: Procedia CIRP 134 (2025). PII: S2212827125004524, pp. 19–24. ISSN 22128271. https://doi.org/10.1016/j.procir.2025.03.028
7. VDI, ed. Virtuelle Inbetriebnahme—Modellarten und Begriffe. VDI/VDE 3693 Blatt 1. Version DE88985735. DIN Media GmbH.
8. Lars Klingel et al. Durchg angige Steuerung und Simulation/Continuous control and simulation Open platforms for automated production sys-tems with industrial robots. In: wt Werkstattstechnik online 114.05 (2024), pp. 183 188. ISSN 1436-4980. https://doi.org/10.37544/1436-4980-2024-05-19
9. Zheng Liu. Methode zur Entwicklung und für den Test der Komponenten-modelle für die virtuelle Inbetriebnahme. ger. (2022). https://doi.org/10.25673/92084
10. Nibert, J., Herniter, M., Chambers, Z.: Model-based system design for MIL, SIL, and HIL . In: world electric vehicle journal 5.4 (2012). PII: wevj5041121, pp. 1121–1130. https://doi.org/10.3390/wevj5041121
11. Tinsel, E.F., Dr.-Ing Alexander Verl.: A simulation model exten- sion to enable continuous control tests during the Virtual Commissioning. In: Procedia CIRP 112 (2022). PII: S2212827122012008, pp. 97–102 ISSN 22128271. https://doi.org/10.1016/j.procir.2022.09.046
12. Chiara Nezzi et al.: Virtual commissioning of a mechatronic plant for insu-lating material processing: a company application towards Digital Twin. In: Procedia Computer Science 253 (2025). PII: S1877050925001085, pp. 384–392. ISSN 18770509. https://doi.org/10.1016/j.procs.2025.01.100
13. Raza, M., Bilberg, A., de Oliveira Hansen, J.P.: Multiple-level virtual commissioning in manufacturing systems: possibil-ities and challenges for integration. In: Procedia CIRP 134 (2025). PII: S2212827125004950, pp. 396–400. ISSN 22128271. https://doi.org/10.1016/j.procir.2025.02.136
14. Roger Zeits et al., eds. SAE Technical Paper Series. WCX SAE World Congress Experience (Detroit, Michigan, United States). SAE Technical Paper Series. SAE International400 Commonwealth Drive, Warrendale, PA, United States, (2024)
15. Simon Nowinski et al.: Durchg angige virtuelle Inbetriebnahme/Continuous virtual commissioning Challenges and solution approaches for the contin-uous engineering of production systems . In: wt Werkstattstechnik online 115.06 (2025), pp. 507–516. ISSN 1436-4980. https://doi.org/10.37544/1436-4980-2025-06-141
16. Scheifele, C.: Plattform zur Echtzeit-Co-Simulation für die virtuelle Inbetriebnahme. de. (2019). https://doi.org/10.18419/OPUS-10742
17. ISG Industrielle Steuerungstechnik GmbH. ISG-kernel / ISG-virtuos. Homepage. ISG Industrielle Steuerungstechnik GmbH. 2025. url: https://www.isg-stuttgart.de/ (visited on 06/04/2025)
18. Kevin M. Lynch and Frank C. Park. Modern Robotics. Cambridge Univer- sity Press, (2024) ISBN 9781316661239. https://doi.org/10.1017/9781316661239

Surrogate Models for TCP Displacement in Milling

Matthäus Loba(✉), Felix Roenneke, Marcel Fey, and Christian Brecher

Laboratory for Machine Tools and Production Engineering (WZL), RWTH Aachen University, Aachen, Germany
m.loba@wzl.rwth-aachen.de

Abstract. During milling operations, process forces cause displacements at the Tool Center Point (TCP). Depending on the selected tools and process parameters, different engagement conditions arise, resulting in a high number of load cases. To determine the resulting displacements, analytical beam models or finite element (FE) models can be used. While analytical models offer high computational efficiency, FE models provide higher accuracy but require significantly longer computation times depending on the number of elements. This makes real-time calculation using FE in process-monitoring simulations infeasible. Surrogate models based on neural networks offer a way to overcome this limitation. In this paper, an approach is presented that uses ANNs to predict the tool deflection during milling. Based on FEM simulations, multiple ANNs were trained. The best ANN approximates the FE simulation with a median absolute percentage error (*MdAPE*) of less than 1% on the training data. On the test data, a different ANN reaches an *MdAPE* of 16.54%. In addition, the computation time remains below 0.2 ms.

Keywords: Surrogate models · Milling · Stiffness modelling

1 Introduction

Modern machine tools offer the possibility to trace position and current of each axis. Based on the encoder difference [1] or motor current [2], process forces can be estimated. This enables the prediction of workpiece quality in a process-parallel manner [3]. To estimate the deviations caused by process forces, the resulting deflection in the cutting zone of both the tool and the workpiece must be known. Therefore, different models can be used, which must fulfill two requirements—accuracy and computational efficiency—to enable valuable real time process monitoring. While excessive simplifications lead to inaccuracies, detailed models such as finite element models (FEM) can result in high computational times [4]. Surrogate Models based on neural networks (NN) have the potential to fulfill both requirements [5, 6].

This paper investigates the extent to which these models can be used to predict deflection due to process forces during milling. First, a brief state-of-the-art overview of force and stiffness modeling in milling, as well as surrogate modeling, is provided to

L. Overmeyer and B.-A. Behrens (eds.), *Production at the Leading Edge of Technology*, Lecture Notes in Production Engineering, https://doi.org/10.1007/978-3-032-19524-1_46

identify gaps in current research. Next, feature selection is performed to determine the relevant influences on the tool deflection and to avoid overfitting. Based on this analysis and the tool geometry, a method is presented to reduce the number of FEM simulations required for NN training. Subsequently, multiple NNs are trained and compared in terms of efficiency and accuracy. Finally, conclusions are drawn.

2 State of the Art

The resulting deflection of the milling tool during machining depends on the actual engagement situation (force distribution) as well as the spindle speed (excitation frequency). Approaches to calculate the force amplitude are based on the undeformed chip thickness h and width b. The most commonly used model in literature for milling was developed by Armarego [7, 8] and later generalized by Altintas and Lee for end mills [9]. According to this model, three force components occur on each infinitesimal element: tangential dF_t, radial dF_r and axial dF_a. For an end mill, the elemental force in direction is given by:

$$dF_q(\theta, z) = \left[K_{qe} + K_{qc}dh\right]db;\ q \in t, r, a \tag{1}$$

For equal-spaced mills h can be approximated along the tool axis:

$$dh(z) = \left|\mathrm{f}_z \cdot \sin\left(\varphi_{z=0} + \frac{z \cdot \tan(\gamma)}{r}\right)\right| \forall \varphi_a \leq \varphi \leq \varphi_e \ \text{ else } \ h = 0\text{ mm} \tag{2}$$

Krüger identifies three approaches to modelling tool deflection in milling [4]:

1. Linear, analytical models (e.g. equivalent spring stiffness)
2. Discrete analytical beam models (e.g. Bernoulli Beam or Finite Differences)
3. Discrete numerical models (e.g. FEM or boundary element method (BEM)).

The model types differ essentially in their complexity and accuracy (cf. Fig. 1). In end milling, the tool-side stiffness results from a serial combination of the stiffnesses of the machine tool, the tool holder, and the end mill. Substructuring techniques enable the prediction of the resulting stiffness by coupling each component [10–12]. The machine tool stiffness here is modeled as an equivalent spring damper element. It can be identified using quasi-static methods like load–deflection measurements or via dynamic testing methods [13], for measuring frequency response functions by using impact hammer and accelerometers. For measurements dummy tools are used and afterwards decoupled. Subsequently, the tool including its holder is coupled to the machine tool, modeled as beam-model [10–12] or FEM [14]. If beam models are used, several simplifications are applied, e.g. tapered elements are divided into multiple cylindrical elements or using an equivalent diameter for fluted segment [10, 15]. Nevertheless, these assumptions often lead to inaccuracies. To predict surface quality the material removal simulation must take into account both, the tool deflection and the dynamically shifting point between tool and workpiece caused by spindle rotation and the helix angle γ [16]. The high variety of engagement situations and number of load cases for one engagement situation leads that FE models have not yet been used in real-time simulations to map tool deformation.

Surrogate models aim to reduce the complexity of a model to achieve lower computation time while maintaining sufficient accuracy [5, 6]. In terms of modeling deflections caused by forces, various work exists (Fig. 2).

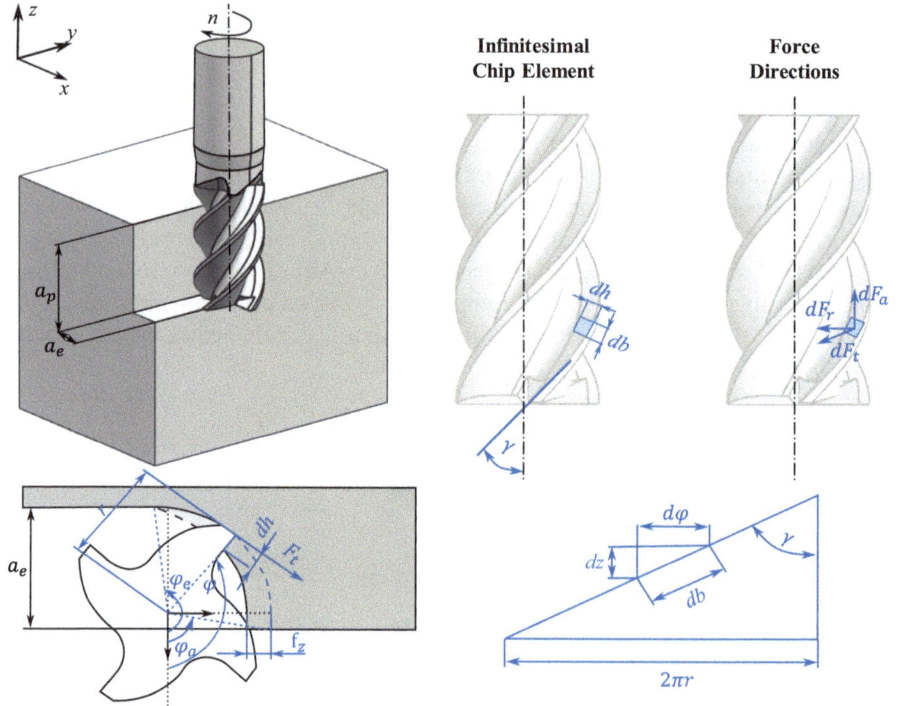

γ	Helix Angle [deg]	r	Radius [mm]		
a_p	Cutting Depth [mm]	n	Spindle Speed [rpm]	b	Uncut Chip Width [mm]
a_e	Cutting Width [mm]	f_z	Feed per Tooth [mm]	h	Uncut Chip Thickness [mm]
F_a	Axial Force [N]	F_r	Radial Force [N]	F_t	Tangential Force [N]
φ_e	Material Entry Angle [deg]	φ	Actual Tooth Angle [deg]	φ_a	Material Exit Angle [deg]

Fig. 1 Geometric and mechanical variables in milling acc. [7–9]

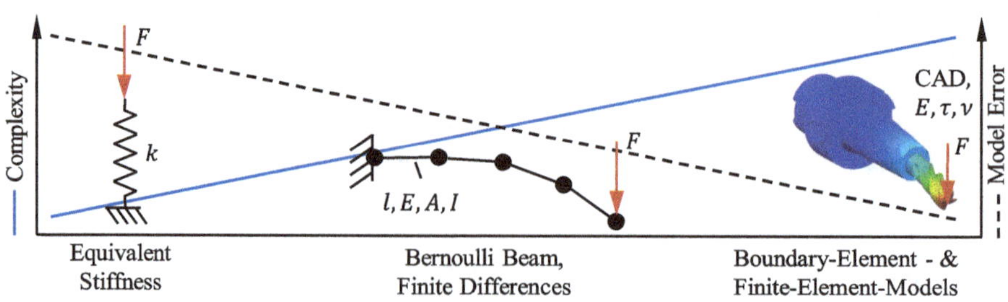

Fig. 2 Overview stiffness modeling acc. [4]

Papadopoulos et al. show in [17] that ANNs are suitable for approximating non-linear beam elements. Based on a carbon nanotube, they validate their mode and decrease the computation time by more than 98%. Suto et al. use a radial basis function (RBF) and a multilayer perceptron (MLP) to model different 2D beam structures [18]. They also could decrease the computation time. Here the RBF shows the best result according to performance as well as accuracy. Greve and van de Weg propose in [19] recurrent NNs for surrogate modeling of parametric FE simulations. They use long-short term memory models (LSTM) to predict time-dependent structural responses like displacement and equivalent plastic strains. Both modeling strategies can reach an obtained coefficient of determination, R2, of 0.9993. In summary, it can be said that machine learning algorithms can help to perform model order reduction. A transfer to the modeling of TCP displacements due to cutting forces therefore seems possible.

NNs in milling are mostly used to predict surface roughness (SR) [20–22], cutting force [22–25] or tool wear [26] with cutting speed, feed rate and cutting depth as input [27, 28]. Training algorithms such as Levenberg–Marquardt and gradient descent are used, with mean square error (MSE), root mean square error (RMSE) and mean absolute percentage error (MAPE) as statistical performance metrics [27]. Denkena et al. apply LSTM networks to predict cutting forces from machine internal signals [23, 24], using the predicted forces to compensate the tool deflection [29]. Brosius et al. predict cutting force coefficients with NN to parametrize a (Kienzle-)force model [25]. Gouarir et al. employ convolutional NN to predict tool wear and reach a accuracy of 90% [26]. Lin et al. predict SR using ANN with cutting parameters and triaxial acceleration sensor data as inputs [21]. Compared with regression, the ANN achieves higher prediction accuracy. These works show that it is possible to predict the cutting force and its influencing factors, as well as the resulting SR, but not the intermediate tool deflection. Predicting this deflection is therefore necessary, for example, to estimate the finishing allowance in real-time simulations and enable a more accurate root cause analysis in machining.

3 Feature Selection

The accuracy of surrogate models depends on their input, output and training data as well as the model structure. In a first step, therefore the relevant values must be identified. As output data, the tool center point (TCP) deflection can be chosen. The resulting deflection depends on the material properties (e.g., Young's modulus E), the geometry (e.g., moment of inertia I) of the tool holder system, as well as the load case, all of which are modeled in the FE simulation. In terms of this work is assumed, that the properties of the tool holder system are not changing e.g. due to different clamping forces [12]. By implying forces on each node depending on the actual infinitesimal chip element based on the actual engagement situation (cf. Figure 1 and Eqs. 1 and 2) the load case is defined. Therefore, as an input for the NN, variables that influence the load must be chosen. In milling the resulting load case depends on the engagement situation. For an equally spaced mill, five variables influence the resulting force amplitude: the cutting depth ap, entry φ_e and exit φ_a angle, the angular position of the cutting edge φ_i, the spindle position φ_{sp} and the chosen feed per tooth fz. Here, fz is the only variable that scales the forces but does not affect load case (cf. Eq. 2). By normalizing (linear elasticity) fz it can be neglected as an input. Furthermore φ_i has a constant shift to φ_{sp}, which changes along the height due to γ but is constant due the defined geometry in the FE model. Moreover, in this work, effects caused by ploughing or chip formation are initially neglected.

In the next step, an NN structure must be defined. In the literature, milling tools are often modeled using linear beam elements [10, 11]. To enable interpolation between different load cases and to compute the deflection at a specific time in real-time, a feedforward network is chosen. The final concept of the surrogate model structure is shown in Fig. 3.

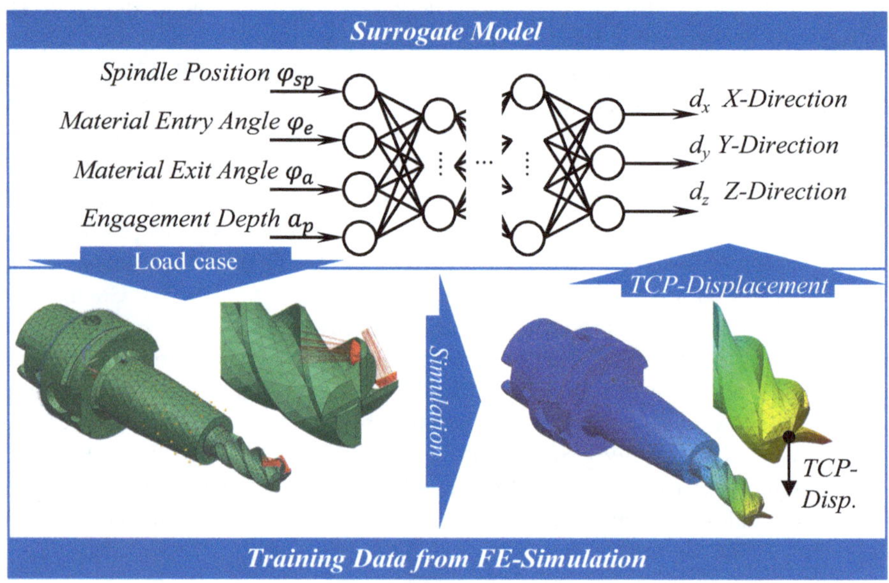

Fig. 3 Concept of the surrogate model

4 Methodology for Generating Training Data

As mentioned in the last chapter, the load case depends on the tool geometry as well as the engagement situation (a_p, φ_e, φ_a, φ_{sp}). The number of cutting edges engaged along the tool axis ($n_{t,a}$) is influenced by γ, the total number of cutting edges n_t and the axial depth of cut a_p:

$$n_{t,a} = \left\lfloor \frac{a_p \cdot n_t \cdot \tan(\gamma)}{2\pi r} \right\rfloor \tag{3}$$

The maximum number of cutting edges engaged in the cutting plane ($n_{t,c}$) occurs at $\varphi_e = \pi$ and $\varphi_a = 0$ ($n_{t,c} = n_t/2$) and can be zero if $|\varphi_e - \varphi_a| < \frac{2\pi}{n_t}$. The number of angular positions of φ_i can be reduced by using the spindle position φ_{sp} as a variable and just taking the positions from zero to the spacing angle ($\varphi_{sp} \in \left[0, \frac{2\pi}{n_t}\right]$). Nevertheless, the evaluation points of φ_i, φ_e, and φ_a must be reduced. Optimal values are here a divisor and multiples of the tool spacing angle. In milling usually is categorized into down ($0 \leq \varphi_i \leq \frac{\pi}{2}$) and up ($\frac{\pi}{2} < \varphi_i \leq \pi$) milling. Therefore, two strategies to create the training data are chosen:

1. Up milling: $\varphi_e = \pi \qquad \varphi_a = \pi - \varphi_{grid}$
2 Down milling: $\varphi_e = \varphi_{grid} \qquad \varphi_a = 0$
   with

$$\varphi_{grid} = i \cdot \varphi_{step} \;\; with \;\; i \in \mathbb{N}, \;\; 1 \leq i \leq \left\lfloor \frac{\pi}{\varphi_{step}} \right\rfloor \tag{4}$$

   φ_{step} specifies the step size for the discretization in terms of material beginning and ending. Furthermore, a third strategy can be added to increase variety:
3. Center milling: $\varphi_e = (\pi + \varphi_{grid})/2\varphi_a = (\pi - \varphi_{grid})/2$
   Depending on the number of chosen strategies $\mathrm{n_{strat}}$, $\varphi_{step,m}$ for the material and $\varphi_{step,sp}$ for the angular spindle position discretization, this leads to a minimum number of training data $\mathrm{n_{train}}$ of:

$$\mathrm{n_{train}} = \mathrm{n_{t,a}} \cdot \mathrm{n_{strat}} \cdot \frac{\pi}{\varphi_{\mathrm{step,m}}} \cdot \frac{2\pi}{\mathrm{n_t} \cdot \varphi_{\mathrm{step,sp}}} \tag{5}$$

   Figure 4 summarizes the influence of the individual variables on the support points for generating the training data.

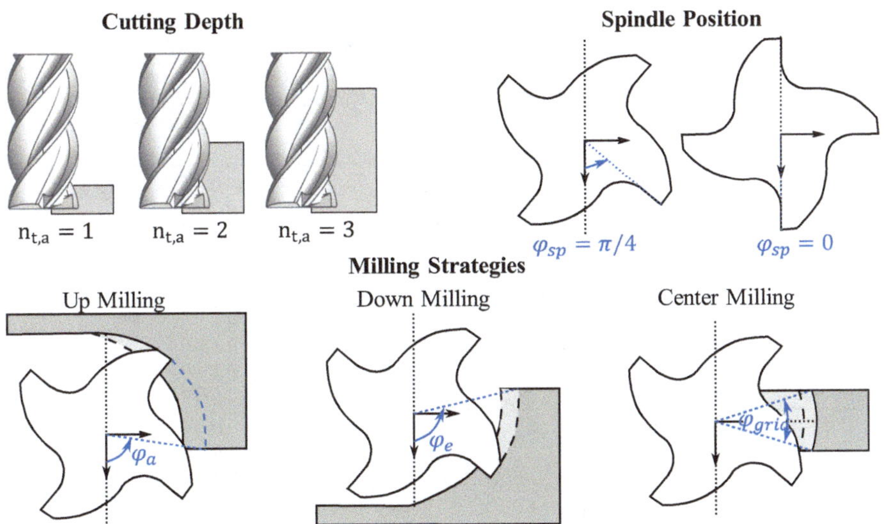

Fig. 4 Generation of evaluation points for the training data

5 Surrogate Model Results

The aim of the surrogate model is to reduce computation time while achieving sufficient accuracy. Modern machine tools allow to trace internal signals at the frequency of the position control loop (500–1000 Hz) [30]. Consequently, the computation time must remain below 1–2 ms. In terms of accuracy, a relative error deviation of up to 10% is considered acceptable. The model architecture from Fig. 3 was trained with different algorithms, training functions and number of hidden layers and neurons. To assess whether NNs are suitable for surrogate modeling in milling, only down milling is considered. Future work should investigate the extent to which a single NN can be extended to multiple milling strategies.

A Ø16 mm tool with four cutting edges, $\gamma = 45°$. To get robust results, first over 20,000 different net parametrizations are trained. The training algorithms, activation functions, and the number of hidden layers and neurons are varied. Afterwards the best 100 NNs get trained five times and the best result gets stored. Here the maximum time (15–20 s) and number of epochs (1000–10,000) is increased and the minimum performance decreased (10e-8 to 10e-10) to improve performance. To ensure reproducibility and comparison, the random seed is initialized at the beginning of each run, so all networks use the same seed in each run [31]. The five best results are shown in Table 1 in terms of the *RMSE*. The computation time is reduced significantly. An average FEM run needs up to 10 s per load case; by contrast, the surrogates need less than 1 ms, making them real-time capable. Nevertheless, the accuracy of the prediction on the test data is not sufficient (rel. error >10%). With an increased number of training data this error could be reduced. In Fig. 5 this gets visible by showing the prediction in feed normal direction. The target values are visualized by 'o'-marker and the prediction by 'x'-marker. The different colors symbolize the different nets.

Table 1 Training results

Net	Training algorithm	Activation function	Hidden layers	Neurons per Layer	*RMSE* (μm²)	*MdAPE* (Train, Val., Test data)	Computation time (ms)
1	Bayesian regularization	Elliot Symmetric Sigmoid	6	7	0.1241	0.0032 0.1188 0.5677	0.1586
2	Bayesian regularization	Elliot Symmetric Sigmoid	4	6	2.8774	0.004 0.0654 0.2277	0.1571
3	Levenberg-Marquart	Log-Sigmoid	5	8	8.3933	0.1316 0.1396 2.6269	0.1501
4	Bayesian regularization	Elliot symmetric Sigmoid	3	6	8.5716	0.0017 0.0306 0.4458	0.1194
5	Bayesian regularization	Log-Sigmoid	3	5	8.8800	0.0206 0.0269 0.1653	0.1123

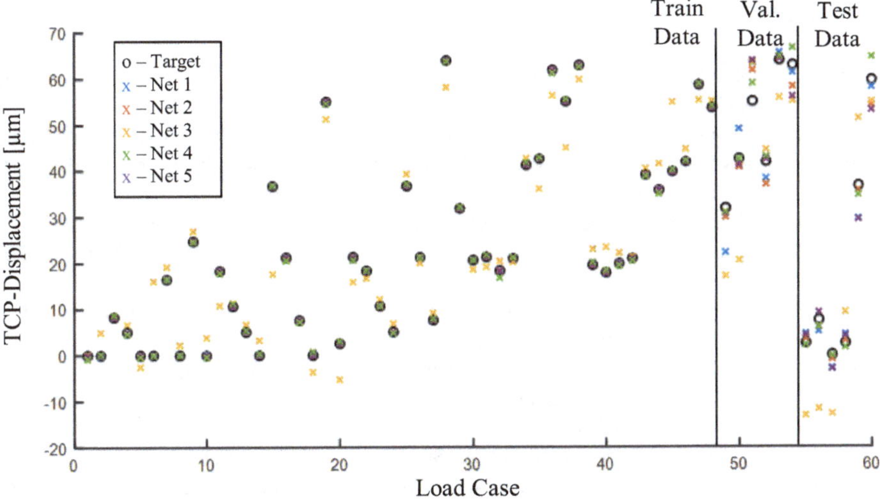

Fig. 5 Surrogate prediction

6 Conclusion

This paper presents a methodology for designing surrogate models for milling tools. A strategy for systematically generating training data is developed, enabling a reduction in the number of required FE simulations. The resulting surrogate model can reconstruct the load case and calculate the resulting TCP displacement with a relative error of 0.3% for training data. The trained neural networks can perform the calculation in less than 0.2 ms. This demonstrates that model order reduction using a feedforward network is feasible, for the training data. For the test data, the best ANN achieves a relative error of 16.5%. Further investigations are necessary to determine how far this error can be reduced, for example by changing the initial seed [31]. Moreover, it should be investigated to what extent stiffness losses, e.g. a reduction in clamping force of the holder [12], can be taken into account. In future work, neural networks are intended to reduce uncertainties in quality prediction during the process-parallel material removal simulation [3]. As an input the engagement situation and the relative spindle will be used as input, to predict the tool deflection during milling.

Acknowledgements. Funded by the Deutsche Forschungsgemeinschaft (DFG, German Research Foundation)—Projektnummer 524834456.

Competing Interests. The author(s) has no competing interests to declare that are relevant to the content of this manuscript.

References

1. Fey, M., Epple, A., Kehne, S., et al.: Verfahren zur Bestimmung der Achslast auf Linear- und Rundachsen. G01L 1/04 (2016)
2. Yamada, Y., Kakinuma, Y.: Sensorless cutting-force estimation for full-closed controlled ball-screw-driven stage. Int. J. Adv. Manuf. Technol. **87**, 3337–3348 (2016). https://doi.org/10.1007/s00170-016-8710-5
3. Königs, M., Brecher, C.: Process-parallel virtual quality evaluation for metal cutting in series production. Procedia Manuf. **26**, 1087–1093 (2018). https://doi.org/10.1016/j.promfg.2018.07.145
4. Krüger, M.: Modellbasierte Online-Bewertung von Fräsprozessen. Dissertation. Berichte aus dem IFW, vol. 7. IFW, Hannover (2014)
5. Kudela, J., Matousek, R.: Recent advances and applications of surrogate models for finite-element-method computations: a review. Soft. Comput. **26**, 13709–13733 (2022). https://doi.org/10.1007/s00500-022-07362-8
6. Tahkola, M., Keranen, J., Sedov, D., et al.: Surrogate modeling of electrical-machine torque using artificial neural networks. IEEE Access **8**, 220027–220045 (2020). https://doi.org/10.1109/ACCESS.2020.3042834
7. Armarego, E., Whitfield, R.C.: Computer-based modelling of popular machining operations for force and power prediction. CIRP Ann. **34**, 65–69 (1985). https://doi.org/10.1016/S0007-8506(07)61725-9
8. Armarego, E.: UNESCO/CIRP seminar on manufacturing technology. In: Proceedings of the UNESCO/CIRP Seminar on Manufacturing Technology, pp. 167–173. CIRP, Paris (1982)
9. Altintaş, Y., Lee, P.: A general mechanics and dynamics model for helical end mills. CIRP Ann. **45**, 59–64 (1996). https://doi.org/10.1016/S0007-8506(07)63017-0
10. Brecher, C., Chavan, P., Fey, M.: Efficient joint identification and fluted-segment modelling of shrink-fit tool assemblies by updating extended tool models. Prod. Eng. Res. Devel. **15**, 21–33 (2021). https://doi.org/10.1007/s11740-020-00999-0
11. Schmitz, T.L., Duncan, G.S.: Three-component receptance-coupling substructure analysis for tool-point dynamics prediction. J. Manuf. Sci. Eng. **127**, 781–790 (2005). https://doi.org/10.1115/1.2039102
12. Denkena, B., Wienrich, S., Krüger, M.: Receptance-coupling substructure analysis for modelling the dynamic process behaviour in milling processes with hydraulic chucks. Forsch. Ingenieurwes. **89**, 1–14 (2025). https://doi.org/10.1007/s10010-025-00838-7
13. Brecher, C., Weck, M.: Thermo-elastisches Verhalten von Werkzeugmaschinen. In: Brecher, C., Weck, M. (eds.) Werkzeugmaschinen Fertigungssysteme – Konstruktion, Berechnung und messtechnische Beurteilung, 9th edn., pp. 577–594. Springer, Berlin (2017)
14. Albertelli, P., Goletti, M., Monno, M.: An improved receptance-coupling substructure analysis to predict chatter-free high-speed cutting conditions. Procedia CIRP **12**, 19–24 (2013). https://doi.org/10.1016/j.procir.2013.09.005
15. Kops, L., Vo, D.T.: Determination of the equivalent diameter of an end mill based on its compliance. CIRP Ann. **39**, 93–96 (1990). https://doi.org/10.1016/S0007-8506(07)61010-5
16. Dépincé, P., Hascoët, J-Y.: Active integration of tool-deflection effects in end milling. Part 1. Prediction of milled surfaces. Int. J. Mach. Tools Manuf. **46**, 937–944 (2006). https://doi.org/10.1016/j.ijmachtools.2005.08.005
17. Papadopoulos, V., Soimiris, G., Giovanis, D.G., et al.: A neural-network-based surrogate model for carbon nanotubes with geometric nonlinearities. Comput. Methods Appl. Mech. Eng. **328**, 411–430 (2018). https://doi.org/10.1016/j.cma.2017.09.010

18. Suto, K., Sakai, Y., Tanimichi, K., et al.: Surrogate model of elastic large-deformation behaviours of compliant mechanisms using co-rotational beam elements. In: 15th World Congress on Computational Mechanics (WCCM-XV) & 8th Asian Pacific Congress on Computational Mechanics (APCOM-VIII), pp. 1–10. CIMNE, Porto (2022)
19. Greve, L., van de Weg, B.P.: Surrogate modeling of parametrized finite-element simulations with varying mesh topology using recurrent neural networks. Array **14**, 100137 (2022). https://doi.org/10.1016/j.array.2022.100137
20. Pontes, F.J., Ferreira, J.R., Silva, M.B., et al.: Artificial neural networks for machining-process surface-roughness modeling. Int. J. Adv. Manuf. Technol. **49**, 879–902 (2010). https://doi.org/10.1007/s00170-009-2456-2
21. Lin, Y.-C., Wu, K.-D., Shih, W.-C., et al.: Prediction of surface roughness based on cutting parameters and machining vibration in end milling using regression method and artificial neural network. Appl. Sci. **10**, 3941 (2020). https://doi.org/10.3390/app10113941
22. Yeganefar, A., Niknam, S.A., Asadi, R.: The use of support-vector machine, neural network and regression analysis to predict and optimise surface roughness and cutting forces in milling. Int. J. Adv. Manuf. Technol. **105**, 951–965 (2019). https://doi.org/10.1007/s00170-019-04227-7
23. Denkena, B., Bergmann, B., Stoppel, D.: Reconstruction of process forces in a five-axis milling centre with an LSTM neural network in comparison to a model-based approach. JMMP **4**, 62 (2020). https://doi.org/10.3390/jmmp4030062
24. Denkena, B., Klemme, H., Stoppel, D.: Machine learning-based reconstruction of process forces. In: Valle, M. (ed.) Advances in System-Integrated Intelligence: Proceedings of the 6th International Conference on System-Integrated Intelligence (SysInt 2022), LNNS, vol. 546, pp. 23–32. Springer, Cham (2023)
25. Arnold, F., Hänel, A., Nestler, A., et al.: New approaches for the determination of specific values for process models in machining using artificial neural networks. Procedia Manuf. **11**, 1463–1470 (2017). https://doi.org/10.1016/j.promfg.2017.07.277
26. Gouarir, A., Martínez-Arellano, G., Terrazas, G., et al.: In-process tool-wear prediction system based on machine-learning techniques and force analysis. Procedia CIRP **77**, 501–504 (2018). https://doi.org/10.1016/j.procir.2018.08.253
27. Chakraborty, S., Chakraborty, S.: Applications of artificial neural networks in machining processes: a comprehensive review. Int. J. Interact. Des. Manuf. **18**, 1917–1948 (2024). https://doi.org/10.1007/s12008-024-01751-z
28. Nasir, V., Sassani, F.: A review on deep learning in machining and tool monitoring: methods, opportunities, and challenges. Int. J. Adv. Manuf. Technol. **115**, 2683–2709 (2021). https://doi.org/10.1007/s00170-021-07325-7
29. Denkena, B., Bergmann, B., Stoppel, D.: Tool-deflection compensation by drive-signal-based force reconstruction and process control. Procedia CIRP **104**, 571–575 (2021). https://doi.org/10.1016/j.procir.2021.11.096
30. Wellmann, F.: Datengetriebene, kontextadaptive Produktivitätssteigerung von NC-Zerspanprozessen, 1st edn. Apprimus Wissenschaftsverlag, Aachen (2019)
31. Lucic, M., Kurach, K., Michalski, M., et al.: Are GANs created equal? A large-scale study. https://doi.org/10.48550/arXiv.1711.10337

Enabling Data-Driven Process Monitoring for Small Batch Sizes Through Cross-Component Learning

Berend Denkena, Jonas Becker, and Maximilian Krüger(✉)

Institute of Production Engineering and Machine Tools (IFW), Leibniz University Hannover, Garbsen, Germany
krueger@ifw.uni-hannover.de

Abstract. Process errors, such as tool breakage, can occur during manufacturing with machine tools. These errors lead to rejects, machine damage, and downtime. Process monitoring identifies abnormal behavior and enables timely intervention, thereby ensuring a machine tool's safety and preventing rejects. Data-driven monitoring methods achieve the highest detection rates while triggering only a small number of false alarms. However, these methods are currently deployed only in series production, as a high number of recurring processes is required for the training phase. Therefore, in small batch size manufacturing, the potential of data-driven methods cannot be realized, and process reliability cannot be guaranteed. In this paper, a novel data processing method is proposed to overcome this hurdle and enable data-driven process monitoring even for small batch sizes. With the introduced approach, process signals recorded during the machining of two components are first segmented into shorter subsequences. Subsequently, similarities between these subsequences are identified by k-means time series clustering. During this process, different segment lengths are compensated for through a combination of Dynamic Time Warping (DTW) and DTW Barycenter Averaging (DBA). Finally, monitoring thresholds specifically tailored for one of the components are generated based on the data from the other. Thus, cross-component learning is achieved. This work, therefore, constitutes an important step towards safer and more efficient machining of small batch sizes.

Keywords: Process monitoring · Dynamic time warping · Machine learning

1 Motivation

In response to increasing competition and the pursuit of a sustainable society, companies must continually enhance productivity. This requires that manufacturing becomes more efficient [1]. Consequently, process monitoring systems (PMS) that reliably detect process errors are needed. PMS enable timely interventions that prevent rejects, costly machine damage, and downtime, thereby ensuring process reliability and increasing efficiency and productivity [2]. Furthermore, because they ensure machine safety, PMS

L. Overmeyer and B.-A. Behrens (eds.), *Production at the Leading Edge of Technology*,
Lecture Notes in Production Engineering, https://doi.org/10.1007/978-3-032-19524-1_47

are a prerequisite for achieving higher levels of automation. PMS typically rely on data-driven monitoring methods, as they achieve comparatively high detection rates while minimizing false alarms [3]. However, deploying these data-driven methods requires a significant amount of process data, which limits their applicability to large batch sizes. Therefore, for smaller batch sizes, research has explored simulation-based monitoring [4]. However, these methods present several drawbacks, including the requirement for considerable computational resources and the development of intricate models [5, 6]. Furthermore, they necessitate machine-specific parameterization to accurately reflect real-world conditions [7]. Despite extensive efforts, simulation-based methods fall short of replicating the same level of accuracy as data-driven approaches in series production [8]. Consequently, recent research aims to alleviate the data requirement for the successful deployment of data-driven approaches [9, 10].

To this end, this paper introduces a novel data processing method that enables effective data-driven process monitoring for small batch sizes. The method relies on a similarity-based cross-component learning approach and is demonstrated using milling of different components as an example. Similar to the methods described in [10, 11], process signals are first segmented into recurring subsequences. Subsequently, similar subsequences from other processes are identified and used as references. To maximize the quantity of referenceable data, non-linear time variations are compensated for in a similar manner, as proposed in [12]. Finally, to highlight the advantages of the proposed method, monitoring limits are calculated based exclusively on the data of one component and applied to successfully monitor a second component. This allows, for example, the real-time monitoring of pockets during milling, regardless of when they occur in a machining process or what aspect ratio they have.

2 Experimental Setup

For this work, experiments were conducted on a DMG Mori HSC 30 linear milling center equipped with a Siemens Sinumerik 840D solution line control system. The tools used were solid carbide end milling cutters, specifically the FORTIS Alligator Steel Type 1,954,510. The material machined was 42CrMo4 steel, supplied as block-shaped blanks measuring 170 mm × 100 mm × 100 mm. For data collection, a Beckhoff Industrial PC (IPC) and TwinCAT 3 software were employed to acquire machine internal and external signals. The machine's internal signals, such as axis position, speed, acceleration, and current, along with the spindle's speed and current, were recorded at 100 Hz using the Axis Data Stream (ADAS) software option provided by Siemens, while the external data from a triaxial acceleration sensor (KS943B100 by MMF) was recorded at 20 kHz.

For data acquisition, ten workpieces of a test component and three workpieces of a training component were machined. The test component comprised two edge trimming operations along two L-shaped contours, as illustrated by the tool path in Fig. 1. For this, the blank workpiece was prepared by milling a groove with two 90° angles. Following this preparation, the recorded milling operation was executed. The first L-shaped contour (tool path and plot section a) is error-free, whereas the second (tool path and plot section b) includes a hole that partially overlaps the machined area, thus simulating a process error ε. The exact position of the hole was varied during data acquisition to introduce different degrees of process error. This way, ten error-free L-shaped trimming processes, as well as five with a smaller process error and five with a larger one, were recorded. After each measurement, the block was face-milled with a second tool to minimize wear and the process was repeated afterwards with a Z-axis offset.

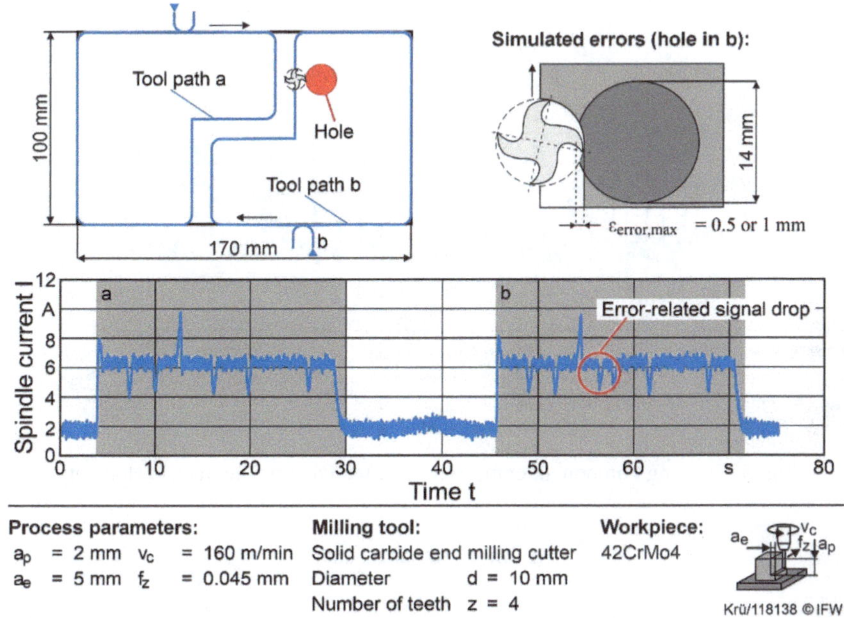

Fig. 1 Test component comprised of L-shaped geometries with a simulated error

The training component, on the other hand includes several distinct geometric elements, as illustrated in Fig. 2, by the tool path and observed spindle current. The machining process consists of two edge trimming operations, two fully open grooves, six grooves open on one side, two closed grooves milled from the top with a zigzag entry, a circular pocket with a helix entry from the top, and two rectangular pockets with rectangular entries from the top. After each run, the block was face-milled as described before.

An overview of the collected data is provided in Table 1.

Throughout all milling experiments the condition of the tool was monitored by using a Keyence VHX-970F microscope to visually inspect the cutting edges. It was observed that during the recording of the used data the average flank wear of the tool increased by approximately 2 µm. It was therefore concluded that no significant wear occurred.

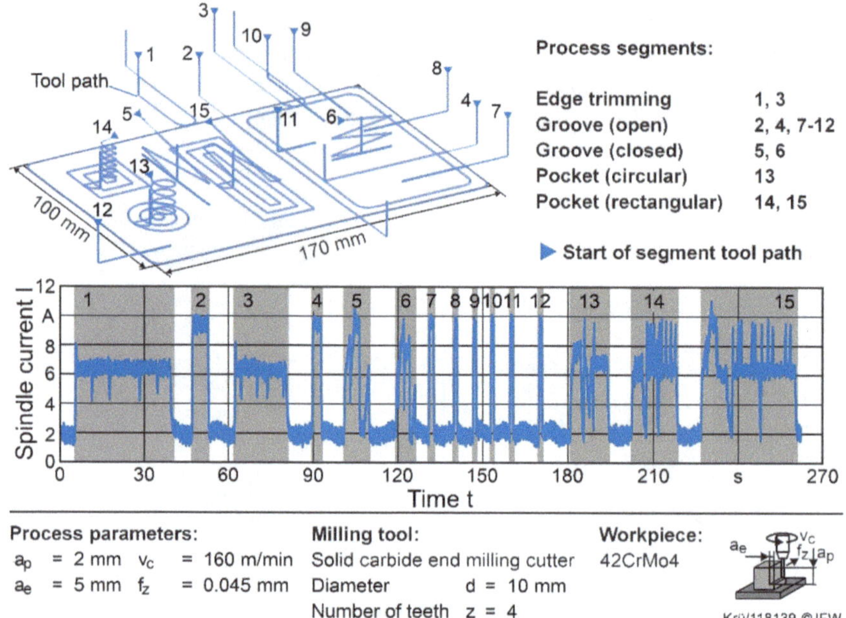

Fig. 2 Training component comprised of fifteen distinct geometric elements

Table 1 Acquired data

	Purpose	Quantity (pcs)
Training component	Providing similar reference data	3
Test component	Monitoring target and data for a comparison with statistical approaches	10
Error cases	Testing with $\varepsilon_{error,max} = 1.0$ mm	5
	Testing with $\varepsilon_{error,max} = 0.5$ mm	5

3 Monitoring Approach

The initial step in the proposed cross-component learning method involves segmenting a process's signals into repeating subsequences. This is accomplished by segmenting processes based on the active G-code line. To trigger segmentation, the active line number, a parameter available in the machine control, is used. This leads to a finer segmentation than presented in [12] and allows for cross-component learning even for new geometries like the here-presented L-shape, which is not present in the training data.

Additionally, to ascertain whether a segment actually involves milling, a simple tool engagement detection was implemented. For this purpose, the signal from the acceleration sensor is filtered using a bandpass filter configured to ±1 Hz of the expected frequency at which the teeth engage with the workpiece. Significant increases in the energy of the filtered signal are then used as indicators of tool engagement. Segments without any tool engagement are discarded, resulting in a total of fifteen segments for the signals of one L-shaped trimming. Notably, the first and last segments primarily involve movement in the air without tool engagement; however, brief tool engagement was detected at their ends or starts. This may be due to the blank's allowance, as the block exceeded the nominal size by up to a several hundred micrometers. Subsequently, each segment's spindle current is downsampled to 100 data points and smoothed to reduce noise and extract the segment's characteristic motif. The resulting spindle current is shown in Fig. 3, where all segments are stretched to the same length, regardless of their actual duration.

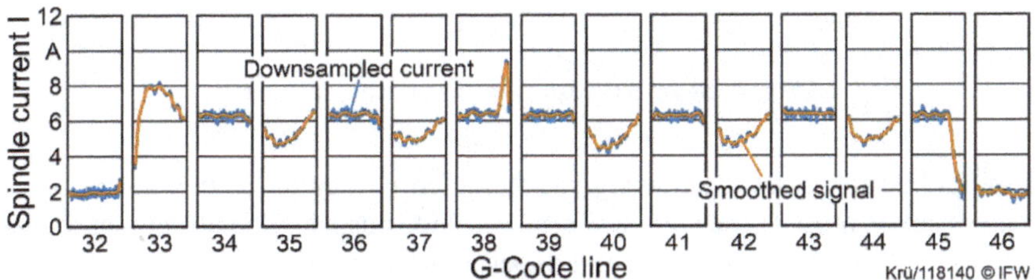

Fig. 3 Segments of the current signal of the first L-shaped trimming

For the sake of comprehensibility, the subsequent steps are explained using the 150 segments from the first ten L-shaped trimmings as an example. The smoothed segments are initially clustered using the k-means clustering algorithm in conjunction with DTW. DTW is well-regarded for its capability to measure similarities between temporal sequences that may vary in speed, thus accommodating accelerations and decelerations. It is widely employed in automatic speech recognition due to its effectiveness in handling variations in speaking speeds [13, 14]. For implementation, the tslearn machine learning toolkit for time series analysis in Python [15] is utilized. To determine the optimal number of clusters, the heuristic Elbow method is employed. For this, the total distance between all aligned time series across all clusters, known as "inertia", is calculated. The cutoff point is selected as the last increase in the number of clusters that results in a decrease in inertia of more than 20% compared to the previous increment.

In the given example, the clustering results in six distinct clusters, corresponding to air movement, material entry, straight path sections, outer corners, inner corner, and material exit, as illustrated in Fig. 4 (top left to bottom right). In addition to the raw current signals, the barycenters of the clusters are also displayed. They represent the centroids and, consequently, the characteristic motifs of the clusters. This is particularly evident in the third cluster, where the drop-off observed at the end of each time series results in a single drop-off at the end of the barycenter.

Through the described clustering process, a database of reference clusters can be created from different components. In this work, this database is constructed using data from the training component, which consists of 110 milling segments.

For this purpose, the warping paths of the smoothed signals within a cluster are calculated to align them with the barycenter of the cluster using DBA. These warping paths

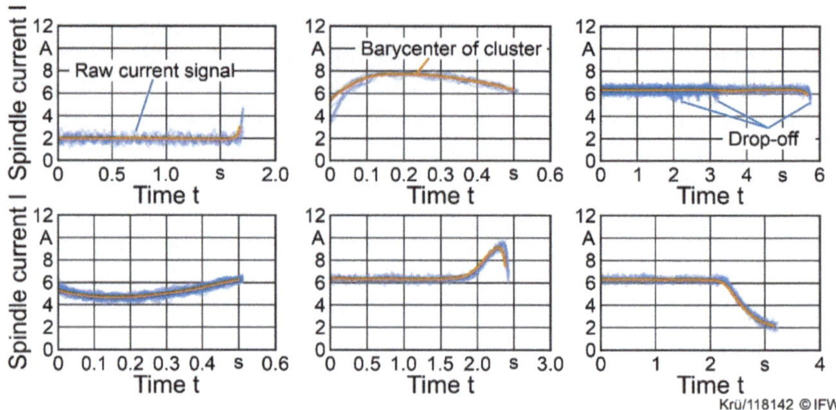

Fig. 4 Segment clusters of the new component

are then applied to adjust the raw signals to match the length of the barycenter. This method preserves most of the signals' variance while ensuring the alignment of characteristic motifs. Subsequently, lower and upper monitoring thresholds are calculated for each cluster. The entire process is illustrated in Fig. 5 for the segments corresponding to the rectangular tool paths of the two pockets in the training component (elements 14 and 15 in Fig. 1). In this case, the benefits of the approach are particularly evident; the signals not only vary in length, but the characteristic peak at the end also represents a different proportion of each segment's total duration. This varying proportion necessitates non-linear warping to maximize data efficiency and provide monitoring thresholds that are statistically reliable.

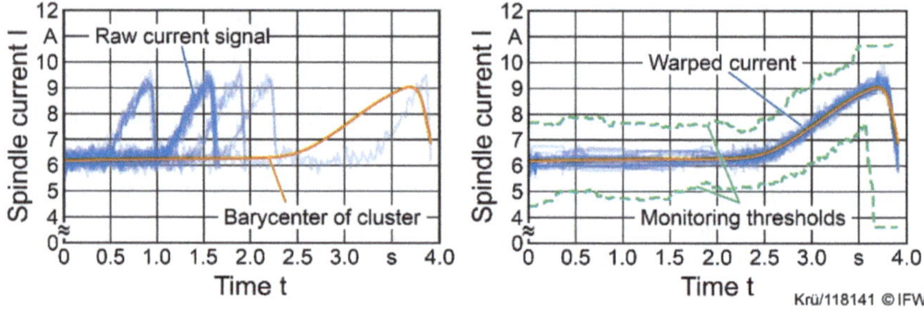

Fig. 5 Creation of a reference cluster out of raw signals of varying length

The monitoring thresholds, as illustrated, are calculated based on the mean and standard deviation of the maximum or minimum within a sliding window of 25 ms, as described in Eqs. 1 and 2. The constants used are derived from the central 99.9% range of a normal distribution, along with a safety factor of two.

$$\text{upper threshold} = \text{mean}(\text{max}_{\text{window}}) + \text{std}(\text{max}_{\text{window}}) * 3.29 * 2 \quad (1)$$

$$\text{lower threshold} = \text{mean}(\text{min}_{\text{window}}) - \text{std}(\text{min}_{\text{window}}) * 3.29 * 2 \quad (2)$$

To apply the proposed monitoring approach to a new batch, the segmentation process described earlier is applied to the current signal recorded during the machining of the first workpiece of a new component. Subsequently, similar segments are identified in the database by calculating the DTW-distance between the characteristic motifs of the segments and the barycenters of the clusters. The thresholds from the reference clusters are then warped to fit the motifs of the segments from the new component. For this, the warping paths calculated during the initial identification of the suitable reference clusters are inverted. Finally, by combining all limits, a single signal envelope tailored specifically for the new component is formed. This allows for process monitoring using data-driven insights derived from a previous component, thereby achieving cross-component learning.

4 Results

To validate the proposed method, the signal envelopes generated for the test component were evaluated with the b section of the test component by shifting the signal in time. In the test, the monitoring system was able to successfully detect all simulated errors without causing any false alarms. An example of this is shown in Fig. 6, in which the process error is detected due to the signal falling below the lower threshold.

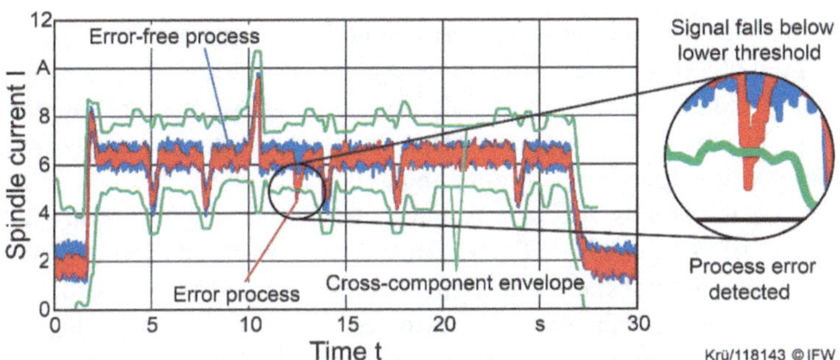

Fig. 6 Evaluation of the proposed method

To further evaluate the proposed cross-learning approach, a comparison is drawn with the confidence limit estimator presented in [16]. This conventional statistical approach utilizes data from machining multiple workpieces of an identical component. As a continuously self-learning method, its performance is limited by the amount of available data, which is why it is well-suited to quantifying the advantage of the proposed cross-learning approach. The learning rate of the statistical approach is set to 0.4, as recommended for the first ten workpieces of a new component. The other hyperparameters, such as the size of the sliding window, the parameter derived from the central 99.9% range of a normal distribution, and the safety factor of two, are kept identical to ensure consistent boundary conditions. For the comparison, the statistical approach is trained with an increasing number of signals from the error-free section of the test component.

Since the performance depends on the order in which signals are provided, the evaluation is repeated several times with a randomized order.

The test revealed that the conventional approach is, on average, able to detect the larger error from the seventh workpiece onwards and the smaller error from the ninth workpiece onwards, while causing one or two false alarms during the monitoring of the ten workpieces, depending on the order in which the signals are provided. An overview of the observed performances is provided in Table 2.

Table 2 Comparison of the performance

Approach	Earliest detection of $\varepsilon_{error,max} = 1.0$ mm	Earliest detection of $\varepsilon_{error,max} = 0.5$ mm	False alarms observed
Cross-component learning	2nd workpiece	2nd workpiece	0
Conventional statistical	7th workpiece	9th workpiece	1–2

5 Conclusion and Outlook

In this work, a novel cross-component learning method for process monitoring has been presented. The method involves segmenting signals of milling processes based on the active G-code line and employing k-means clustering, DTW, and DBA to identify similarities among recurring subsequences. This approach creates a database of reference clusters from previously machined components. The database is then utilized to generate monitoring thresholds specifically tailored for a new component, based on the characteristic motif extracted from the first machined workpiece of a new batch. A comparison with a conventional statistical approach demonstrated how the proposed method addresses the data deficit inherent in small batch size production, paving the way for enhanced process monitoring. Therefore, this work represents an important step towards safer and more efficient manufacturing of small batch sizes.

In further research, the influence of process parameter variations on the method will be evaluated. To enhance effectiveness, a new approach is needed to estimate signal characteristics for new process parameters based on the most similar reference clusters. Additionally, since the current method can only provide monitoring thresholds from the second workpiece onwards, a method to characterize new components based on the available G-code is necessary to extend the applicability to single-item production.

Acknowledgements. The research presented was conducted as part of the project RoPro. This Project is supported by the Federal Ministry for Economic Affairs and Energy (BMWE) on the basis of a decision by the German Bundestag.

Competing Interests. The author(s) has no competing interests to declare that are relevant to the content of this manuscript.

References

1. Blaeser-Benfer, A., Schröter, W., Vollborth, T., Rießelmann, J.: Produktivität für kleine und mittelständische Unternehmen – Teil I: Handlungsleitfaden für den industriellen Mittelstand. RKW Rationalisierungs- und Innovationszentrum der Deutschen Wirtschaft e. V., Eschborn (2012)
2. Klocke, F., Pritschow, G.: Autonome Produktion. Springer, Heidelberg (2004). https://doi.org/10.1007/978-3-642-18523-6
3. Kim, D.-H., et al.: Smart machining process using machine learning: a review and perspective on machining industry. Int. J. Precis. Eng. Manuf.-Green Technol. **5**(4), 555–568 (2018). https://doi.org/10.1007/s40684-018-0057-y
4. Tidriri, K., Chatti, N., Verron, S., Tiplica, T.: Bridging data-driven and model-based approaches for process fault diagnosis and health monitoring: a review of researches and future challenges. Ann. Rev. Control. **42**, 63–81 (2016). https://doi.org/10.1016/j.arcontrol.2016.09.008
5. Denkena, B., Koeller, M.: Simulation based parametrization for process monitoring of machining operations. Procedia CIRP **12**(1013), 79–84 (2013). https://doi.org/10.1016/j.procir.2013.09.015
6. Altintas, Y., Aslan, D.: Integration of virtual and on-line machining process control and monitoring. CIRP Ann. **66**(1), 349–352 (2017). https://doi.org/10.1016/j.cirp.2017.04.047
7. Radu, P., Schnakovszky, C.: A review of proposed models for cutting force prediction in milling parts with low rigidity. Machines **12**(2), 140 (2024). https://doi.org/10.3390/machines12020140
8. Denkena, B.;,Bergmann, B., Neff, T., Witt, M.: Einzelteilüberwachung in Beschleunigungsphasen. wt Werkstattstechnik online **109**(5), 347–351 (2019). https://doi.org/10.37544/1436-4980-2019-05
9. Ströbel R., et al.: Intelligente Prozessüberwachung für die flexible Produktion – Integration von Physics-Informed Machine Learning und Active Learning. ZWF Zeitschrift für wirtschaftlichen Fabrikbetrieb **120**(2025), 224–231. De Gruyter (2025). https://doi.org/10.1515/zwf-2024-0154
10. Netzer, M.: Intelligente Anomalieerkennung für hochflexible Produktionsmaschinen – Prozessüberwachung in der Brownfield Produktion. Shaker Verlag GmbH, Düren (2023)
11. Denkena, B., et al.: Time Series Search and Similarity Identification for Single Item Monitoring. Production at the Leading Edge of Technology. WGP 2021. Lecture Notes in Production Engineering. Springer, Cham (2022). https://doi.org/10.1007/978-3-030-78424-9_53
12. Denkena, B., Klemme, H., Krüger, M.: Sicher in kleiner Serie fertigen. WB Werkstatt + Betrieb Das Portal für spanende Fertigung. Carl Hanser Verlag, Münschen (2024). https://www.werkstatt-betrieb.de/a/fachartikel/sicher-in-kleiner-serie-fertigen-5462413
13. Sakoe, H., Chiba, S.: Dynamic programming algorithm optimization for spoken word recognition. IEEE Trans. Acoust., Speech, Signal Process. **26**(1), 43–49 (1978). https://doi.org/10.1109/TASSP.1978.1163055
14. Jiang, S., Chen, Z.: Application of dynamic time warping optimization algorithm in speech recognition of machine translation. Heliyon **9**(11) (2023). https://doi.org/10.1016/j.heliyon.2023.e21625
15. Tavenard, R., et al.: Tslearn, A machine learning toolkit for time series data. J. Mach. Learn. Res. **21**(118), 1–6 (2020). http://jmlr.org/papers/v21/20-091.html
16. Brinkhaus, J.-W.: Statistische Verfahren zur selbstlernenden Überwachung spanender Bearbeitungen in Werkzeugmaschinen. PZH Produktionstechnisches Zentrum GmbH, Garbsen (2009)

Machine Learning for Sales Forecasting in Industrial Environments: A Systematic Literature Review of Input Factors and Applied Methods

Ferenc Wolter(✉), Joschua Prieß, Valentin Rösler, Alexander Rokoss, and Florian Stamer

Institute for Production Technology and Systems (IPTS), Leuphana University Lueneburg, Lüneburg, Germany
ferenc.wolter@leuphana.de

Abstract. Accurate sales forecasting is crucial for manufacturing companies, as it significantly impacts inventory, capacity utilization, lead times, and delivery performance. However, these forecasts are subject to considerable uncertainty driven by economic, political, and environmental dynamics. Under such conditions, the reliability of established forecasting approaches is reduced. Concurrently, digitalized supply chains yield large data sets that can enhance the prediction quality of sales forecasts. Machine learning (ML) facilitates the extraction of underlying patterns from these diverse data sets, enabling data-driven and more accurate predictions. The existing scientific literature introduces numerous ML-based approaches for sales forecasting. This study presents the findings of a systematic literature review on the characteristics of ML-based sales forecasts, utilized input factors, applied methods for both feature selection and regression analysis, as well as interpretability approaches from the field of explainable artificial intelligence. Thus, this study highlights existing research gaps, particularly concerning the selection and quantification of relevant input factors for ML-based sales forecasting in industrial environments.

Keywords: Sales and demand forecasting · Machine learning · Industrial environments

1 Introduction

Accurate sales forecasting is crucial for manufacturing companies, as it significantly impacts numerous logistical objectives. Overestimating sales volumes can result in low capacity utilization or avoidable inventories, while underestimation can lead to longer order lead times and insufficient delivery performance [1]. However, sales forecasting depends on a variety of different influencing factors and faces significant uncertainties driven by economic, political and environmental dynamics [2]. Under such conditions, heuristic and conventional time series approaches have not been able to generate a

L. Overmeyer and B.-A. Behrens (eds.), *Production at the Leading Edge of Technology*,
Lecture Notes in Production Engineering, https://doi.org/10.1007/978-3-032-19524-1_48

satisfactory prediction quality, since they are limited in handling high-dimensional or non-linear input structures [3]. Concurrently, the increasing digitalization of supply chains results in the generation of large data sets that can enhance the prediction quality of sales forecasts. Machine learning (ML) represents a promising approach for sales forecasting as the corresponding models can incorporate numerous and heterogenous input factors in the decision-making process. In recent years, various ML approaches for sales forecasting have been presented in the scientific literature [4, 5]. Since the input factors utilized, the ML regression models applied, and the approaches for identifying the most relevant input factors for the forecast differ greatly, the comparability of the existing studies is affected. Thus, the aim of this study is to provide a comprehensive overview of ML-based sales forecasting and guide further research as well as transfer into industrial practice by conducting a systematic literature review. The analysis focuses on the following three research questions:

- RQ1: Which input factors are utilized for ML-based sales forecasting?
- RQ2: Which ML regression models are applied for sales forecasting?
- RQ3: Which approaches are used to quantify the influence of individual input factors on the prediction quality of the ML regression model?

This paper is structured as follows. Section 2 covers the theoretical background on sales forecasting as well as ML methods and outlines related work. Section 3 describes the methodology, which is based on the systematic literature review by Tranfield et al. [6]. Section 4 presents the results, covering a descriptive analysis of the examined studies, the characteristics of sales forecasts, utilized input factors, applied ML regression models and approaches on identifying the most important input factors. The paper concludes with a summary and an outlook.

2 Theoretical Background

2.1 Machine Learning in Sales Forecasting

Sales forecasting refers to the process of estimating future sales volumes based on historical data and explanatory variables to support financial, capacity, and inventory planning [1, 7]. Closely related to sales forecasting is the concept of demand forecasting, which typically adopts a more market-oriented or customer-centric perspective [8]. Since both aim to predict future product outflows and rely on similar forecasting objectives, data, and methodological approaches, this study analyzes them jointly [8, 9]. Those sales forecasts vary by horizon (ranging from short-term operational to long-term strategic) and in their aggregation level (e.g., product, customer, or regional) [10]. Conventional approaches are typically based on time series models such as exponential smoothing and autoregressive-moving-average, as well as heuristics and judgmental methods applied by domain experts [10]. Although effective in stable contexts, these methods often degrade under volatile, nonlinear, or multi-causal demand patterns [3].

In recent years, the increasing availability of granular and high-dimensional data has led to growing interest in ML-based approaches for enhancing the prediction quality of sales forecasts [3, 8]. Supervised ML regression models leverage labeled data to learn dependencies between input and output variables and generate predictions for new data instances [11]. Previous studies show that ML regression models are particularly applicable for sales forecasting [4, 5]. These studies highlight multiple linear regression (MLR), support vector regression (SVR), decision trees (DT), random forests (RF), and artificial neural networks (ANN) as suitable regression models. As the performance of these ML regression models depends on the utilized input factors, selecting the most informative ones can improve prediction quality and reduce computing time. Feature selection addresses this by removing irrelevant or redundant input factors and retaining only those that are most relevant for modeling [12]. The feature selection methods segment into three main categories: filter, wrapper and embedded methods. Explainable artificial intelligence aims to enhance the transparency of ML regression models by providing insights into the prediction process. Common approaches include feature importance analysis, model-agnostic methods such as Shapley additive explanations (SHAP) and inherently interpretable models [13]. These methods allow for determining the relevance of individual features and their impact on model predictions.

2.2 Related Work

Several systematic literature reviews examine the application of ML methods in sales forecasting. Ahaggach et al. conduct a broad mapping study, identify research trends, and classify forecasting methods at an aggregate level [4]. In addition, Saha et al. provide a cross-industry overview of sales forecasting approaches from the field of artificial intelligence and discuss corresponding business challenges and the lack of interpretability of complex models [14]. Bender et al. analyze methodological developments in the industrial environment and classify both data characteristics and regression models [5]. Swaminathan and Venkitasubramony examine sales forecasts in the fashion industry and emphasize the relevance of context-related input factors, without systematically categorizing them [15]. Finally, Eglite and Birzniece focus on deep learning regression models in retail and highlight their performance based on selected studies [16]. However, these systematic literature reviews are limited by their industry-specific focus, a lack of systematic analysis of input factors, a narrow coverage of regression models, and the omission of methods for assessing feature relevance or interpreting model decisions, respectively. This study aims to close this gap by systematically categorizing the input factors utilized, covering the whole range of ML-based regression models, and including feature selection methods and explainable artificial intelligence approaches.

3 Methodology

For the identification of utilized input factors and applied ML methods for sales forecasting, the method of systematic literature review according to Tranfield et al. was selected [6]. This method enables a holistic overview of relevant research and complies with the exploratory research objectives. For the literature review, the two established databases Scopus and Web of Science were searched for suitable publications [17]. The search was limited to journal articles and conference papers in English from the research fields of computer science, engineering as well as business, management and accounting. The search string was defined based on similar literature reviews that had already been conducted and was further optimized in a previous scoping study [4, 5, 15]. The research objective resulted in four main parts that formed the search string. The first part combined "sales" and "demand" with terms related to the prediction character using a proximity operator without expanded range to make the search for these terms independent of their order. The second part comprised terms that represent the concept and models of ML. The third part consisted of terms that refer to industrial environments. The fourth part of the search string was designed to ensure that the publications address the actual execution of forecasts rather than conceptual aspects. The following search string was applied to the title, abstract and keywords for the Scopus literature database and its notation was adapted accordingly for Web of Science.

(TITLE-ABS-KEY (("sales" OR "demand") W/0 ("forecast*" OR "predict*" OR "estimat*")) AND TITLE-ABS-KEY ("machine learn*" OR "supervised learn*" OR "regression" OR "support vector" OR "decision tree" OR "random forest" OR "neural network") AND TITLE-ABS-KEY ("manufactur*" OR "production" OR "industr*") AND TITLE-ABS-KEY ("case stud*" OR "empiric* stud*" OR "use case*" OR "application")) AND (LIMIT-TO (DOCTYPE,"ar") OR LIMIT-TO (DOCTYPE,"cp")) AND (LIMIT-TO (SUBJAREA,"COMP") OR LIMIT-TO (SUBJAREA,"ENGI") OR LIMIT-TO (SUBJAREA,"BUSI")) AND (LIMIT-TO(LANGUAGE,"English")).

The corresponding search on 28 May 2025 found 456 publications in Scopus and 125 publications in Web of Science. After removing 99 duplicates, 482 publications remained to be analyzed. For the content analysis of these publications, the following three inclusion criteria were derived from the research objective:

1. Does the publication present a case study on sales forecasting?
2. Is at least one ML regression model applied for the forecasting task?
3. Is the forecast object an industrially manufactured product?

These criteria were used to determine the relevance of the publications by excluding publications that relate, for example, to general overviews, models that cannot process input factors or solely monetary forecasts. For the remaining 482 publications, a title and abstract analysis was conducted considering the three inclusion criteria. This reduced the number of potentially relevant publications to 166. These publications were examined more closely in a full-text analysis, which ultimately resulted in 116 relevant publications meeting the inclusion criteria [18]. To classify the case studies on ML-based sales forecasting, information on the industry, market segment, aggregation level of the forecast object, granularity as well as time horizon of the forecast was extracted. Regarding the research questions, the data origin, input factors, ML-based regression models, and methods for quantifying the influence of input factors on the forecast (evaluation metrics, feature selection, explainable artificial intelligence) were also examined.

4 Results

4.1 Descriptive Analysis of Case Studies and Sales Forecasts Characteristics

The number of case studies published annually indicates that the research field of ML-based sales forecasting has gained significant relevance in recent years. Although Fig. 1 shows no distinct trend, the number of publications considerably increased between

2018 and 2019. While a maximum of five publications per year were found for the period between 2008 and 2018, each subsequent year recorded 9–23 publications, from nearly twice to more than four times the earlier annual maximum.

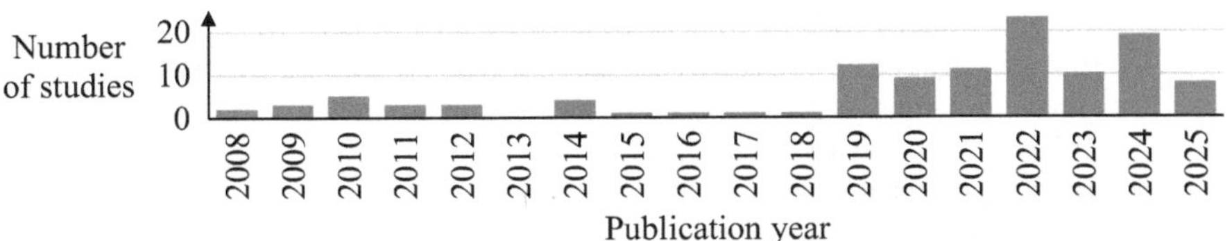

Fig. 1 Annual publications related to ML-based sales forecasting

The subsequent sections evaluate the content of the case studies. It should be noted that the relevant information was not consistently available for all 116 publications, and individual case studies may exhibit multiple characteristics. The case studies presented originated from various industries. The automotive industry (29 publications) alongside retail and consumer goods (23) were the most frequently examined. These were followed by electronics (13), food and beverages (10), mechanical engineering (9), health and pharmaceuticals (8) as well as clothing (6). In addition, information on market segmentation was available for 52 case studies. Accordingly, most case studies focused on a national (32 publications) or international (13) sales market. In contrast, regional (5) or local (2) sales markets were less frequently addressed. Depending on the circumstances of the company and sales market, different characteristics were chosen for sales forecasts. For example, sales forecasts for the majority of case studies were conducted at the product level (71 publications). In other case studies, the forecast object was aggregated at the level of a product group (24) or the entire company (7). Differences in forecast granularity also appeared in the case studies. In this regard, monthly sales (64 publications) were forecasted more frequently than weekly (23) or daily (12) sales. Coarser degrees of granularity, such as quarterly (5) or annual (5) sales forecasts, were far less common. Apart from the case studies that forecasted sales for the next day (4 publications), forecast horizons were more evenly distributed. Sales were forecasted for more than one day and up to one month (30), more than one month and up to three months (13), more than three months and up to one year (27) or beyond one year (6).

4.2 Data Origin and Input Factors

Of the total 116 case studies, 114 utilized real data exclusively and two used a combination of both real and simulated data for ML-based sales forecasting. To answer RQ1, each case study was examined to determine whether it included at least one input factor from a certain category. A wide variety of input factors were utilized in the scientific literature (see Fig. 2). Historical sales data was by far the most frequently considered input factor with a total of 82 case studies. This indicates that the time series principle also applies to ML regression models for sales forecasting. Moreover, external input factors in the form of calendar data (e.g., season, month, or holidays) and economic indicators (e.g., gross domestic product, inflation, or currency exchange rates) were used in 49 and

36 case studies, respectively. Internal input factors such as product information (e.g., product type, lifespan, or color) as well as prices and discounts (sales price, manufacturing costs, or price of complementary goods) were utilized almost as frequently with 30 case studies each.

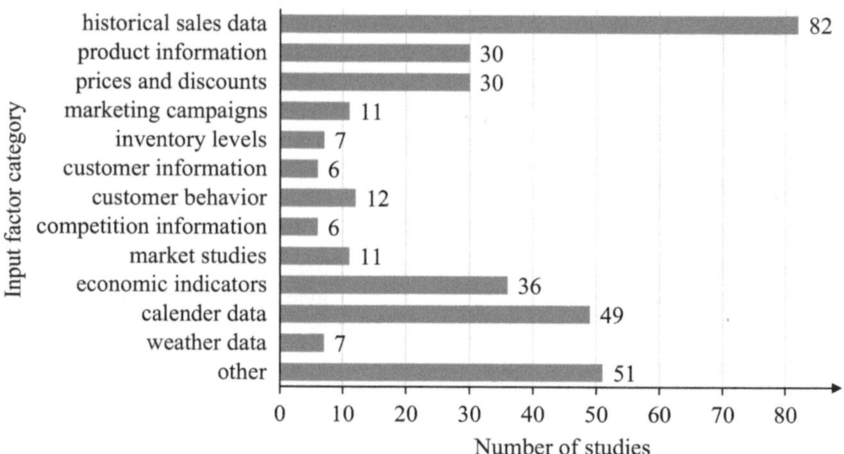

Fig. 2 Utilized input factor categories of the examined case studies

Two examples of case studies that incorporate a wide range of input factors for ML-based sales forecasting were presented by Gao et al. and Sharma et al., covering nine and eight different input data categories, respectively [19, 20].

4.3 Machine Learning Based Regression Models and Evaluation Metrics

To address RQ2, ML regression models were evaluated both by their number of applications and by how often they were reported as the most accurate model in comparative case studies, in which at least two different ML regression models were applied. Multilayer perceptron, a variant of artificial neural networks, was the most prominent ML regression model for sales forecasting, since it was applied in 62 case studies (see Fig. 3). This was followed by multiple linear regression (MLR, 32), support vector regression (SVR, 27), long short-term memory (LSTM, 24), and random forest (RF, 21). However, the most applied ML regression models only slightly correlated with the ones that most frequently showed the highest prediction quality in comparative case studies.

In this regard, the reviewed case studies indicated a dominance of hybrid models, in which at least two (ML) regression models were integrated, as they performed best in 11 out of 14 comparisons (79%) with other ML regression models. Blended models showed similar superiority, demonstrating the highest prediction quality in 9 out of 13 comparative case studies (69%) by combining multiple forecast results of different (ML) regression models. A possible reason for this is that hybrid and blended models were often only compared with the respective individual models as benchmarks. Although it has not been applied extensively to date, extreme gradient boosting (XGB) appears

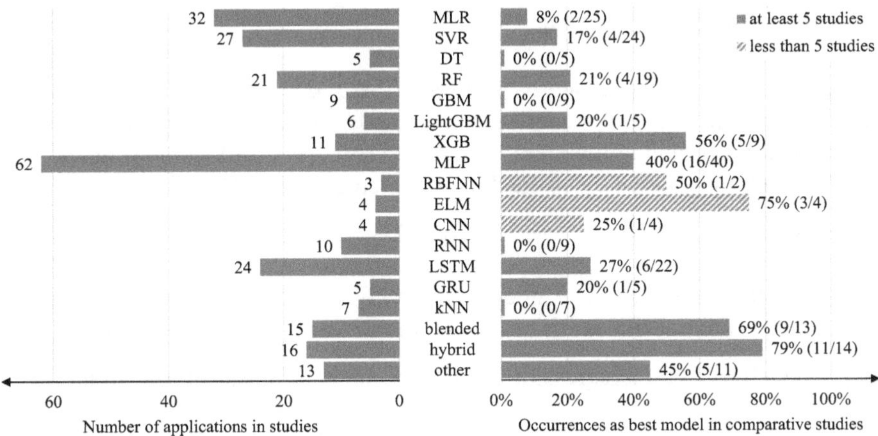

Fig. 3 Number of applications and occurrences as best model for ML regression models

to be a promising approach, demonstrating the highest prediction quality in 5 out of 9 comparisons (56%) with other ML regression models. Two variants of artificial neural networks, multilayer perceptron (MLP) and long short-term memory (LSTM), prevailed in numerous comparative case studies, with 40 and 27% respectively. The evaluation metrics root mean square error (RMSE, 58 publications), mean absolute percentage error (MAPE, 55), mean absolute error (MAE, 41), mean squared error (MSE, 30) and coefficient of determination (R^2, 22) were most frequently used to quantify the prediction quality of ML regression models. Mean absolute deviation (MAD, 8) and the correlation coefficient (R, 6) were evaluated less often.

4.4 Feature Selection and Explainable Artificial Intelligence

Regarding feature selection, 31 of 116 reviewed case studies applied methods to reduce the number of input factors for the final model to the most influential. The filter method was used in 18 case studies and the wrapper method in ten case studies, typically based on correlation coefficients or stepwise regression, respectively. Both approaches were employed, for example, by Hasheminejad et al. [21]. Embedded feature selection was applied in five case studies, while other case studies occasionally preselected input factors based on domain knowledge. Explainable artificial intelligence was rarely applied to quantify the importance of individual input factors for the ML regression model. In 19 corresponding case studies, this was primarily achieved either by selecting inherently interpretable ML regression models, such as multiple linear regression, or by applying feature importance approaches. In four recent case studies since 2023, including Upadhyay et al., Shapley additive explanations were applied to reverse-engineer the most influential input factors for the ML regression model [22].

5 Conclusion and Outlook

This study examines existing approaches for ML-based sales forecasting with a focus on input factors, ML regression models and methods for identifying the most important input factors. A systematic literature review of 116 papers, following Tranfield et al., yielded the subsequent results for the three research questions [6]:

- RQ1: Which input factors are utilized for ML-based sales forecasting?

The case studies utilize diverse input factors, with most using historical sales data, calendar data, economic indicators, product information as well as prices and discounts.

- RQ2: Which ML-based regression models are applied for sales forecasting?

ML-based sales forecasting is dominated by hybrid and blended ML regression models as well as artificial neural networks, such as multilayer perceptron and long short-term memory. In addition, extreme gradient boosting demonstrates high prediction quality compared to other ML regression models.

- RQ3: Which approaches are used to quantify the influence of individual input factors on the prediction quality of the ML-based regression model?

About one-fourth of case studies apply feature selection methods to identify the most relevant input factors for sales forecasting with the majority relying on filter methods. Alongside inherently interpretable ML regression models, Shapley additive explanations are used in some recent case studies to explain sales forecasts.

Overall, this study shows that ML has been applied in several variations for sales forecasting. However, the existing case studies do not yet allow a reliable conclusion regarding which specific input factors are most relevant for sales forecasting. Consequently, future research should systematically identify the most important input factors across industries, both to reveal company- or industry-specific drivers and to establish broader generalizations about the relevance of input factors. Closing this research gap will enable companies to develop domain-specific ML-based models for sales forecasting in a targeted and resource-efficient manner.

Acknowledgements. The project "ML-based sales forecasting using internal and external context data" leading to this paper is funded by the European Union and the State of Lower Saxony within the European Regional Development Fund program (87006251).

Competing Interests. The author(s) has no competing interests to declare that are relevant to the content of this manuscript.

References

1. Schmidt, M., Nyhuis, P.: Produktionsplanung und -steuerung im Hannoveraner Lieferkettenmodell. Innerbetrieblicher Abgleich logistischer Zielgrößen. Springer, Berlin, Heidelberg (2021)
2. Sa-ngasoongsong, A., Bukkapatnam, S.T., Kim, J., Iyer, P.S., Suresh, R.P.: Multi-step sales forecasting in automotive industry based on structural relationship identification. Int. J. Prod. Econ. **140**, 875–887 (2012). https://doi.org/10.1016/j.ijpe.2012.07.009
3. Tripathi, S., Yadav, S., Kumar, H.: A comparative analysis of traditional time series models and deep learning approaches in sales prediction: strengths and limitations. Pharma Innov. **8**, 20–24 (2019). https://doi.org/10.22271/tpi.2019.v8.i2Sa.25244
4. Ahaggach, H., Abrouk, L., Lebon, E.: Systematic mapping study of sales forecasting: methods, trends, and future directions. Forecasting **6**, 502–532 (2024). https://doi.org/10.3390/forecast6030028
5. Bender, B., Bretschneider, S., Fattah-Weil, J.: Advances in demand forecasting: a systematic review of methods, the role of AI, and data strategies in manufacturing. In: Proceedings of the 30th Americas Conference on Information Systems, Paper 7 (2024). https://aisel.aisnet.org/amcis2024/stratcompis/stratcompis/7/. Accessed 15 Sep. 2025
6. Tranfield, D., Denyer, D., Smart, P.: Towards a methodology for developing evidence-informed management knowledge by means of systematic review. Br. J. Manag. **14**, 207–222 (2003). https://doi.org/10.1111/1467-8551.00375
7. Schönsleben, P.: Handbook integral logistics management. Springer, Berlin, Heidelberg (2023)
8. Usuga Cadavid, J.P., Lamouri, S., Grabot, B.: Trends in machine learning applied to demand and sales forecasting: a review. ILS Proc. (2018). https://hal.science/hal-01881362v1. Accessed 15 Sep. 2025
9. Syntetos, A.A., Babai, Z., Boylan, J.E., Kolassa, S., Nikolopoulos, K.: Supply chain forecasting: theory, practice, their gap and the future. Eur. J. Oper. Res. **252**, 1–26 (2016). https://doi.org/10.1016/j.ejor.2015.11.010
10. Chopra, S., Meindl, P.: Supply chain management. Strategy, planning, and operation. Pearson, Boston (2016)
11. Suthaharan, S.: Machine learning models and algorithms for big data classification. Thinking with examples for effective learning. Springer, New York (2016)
12. Guyon, I., Elisseeff, A.: An introduction to variable and feature selection. J. Mach. Learn. Res. **3**, 1157–1182 (2003)
13. Lundberg, S., Lee, S.-I.: A unified approach to interpreting model predictions. In: 31st Conference on Neural Information Processing Systems, pp. 4768–4777. Curran Associates, Red Hook, New York (2017). https://doi.org/10.48550/arXiv.1705.07874
14. Saha, R., Shofiullah, S., Faysal, S.A., Happy, A.T.: Systematic literature review on artificial intelligence applications in supply chain demand forecasting. Acad. J. Bus. Admin. Innov. Sustain. **4**, 109–127 (2024). https://papers.ssrn.com/sol3/papers.cfm?abstract_id=5062817. Accessed 15 Sep. 2025
15. Swaminathan, K., Venkitasubramony, R.: Demand forecasting for fashion products: a systematic review. Int. J. Forecast. **40**, 247–267 (2024). https://doi.org/10.1016/j.ijforecast.2023.02.005
16. Eglite, L., Birzniece, I.: Retail sales forecasting using deep learning: systematic literature review. Complex Syst. Inform. Model. Quart. **30**, 53–62 (2022). https://doi.org/10.7250/csimq.2022-30.03
17. Martín-Martín, A., Orduna-Malea, E., Thelwall, M., López-Cózar, E.D.: Google Scholar, Web of Science, and Scopus: a systematic comparison of citations in 252 subject categories. J. Inform. **12**, 1160–1177 (2018). https://doi.org/10.1016/j.joi.2018.09.002

18. Wolter, F.: Literature repository of studies on machine learning based sales forecasting. CSV dataset. Zenodo (2025). https://doi.org/10.5281/zenodo.17122547
19. Gao, T., Ji, Z., Niu, D., Xi, Y.: Research on product sales forecasting based on multi-value chain collaborative data management system in manufacturing industry. In: 2022 IEEE International Conference on e-Business Engineering (ICEBE), pp. 86–93. IEEE, Piscataway, New Jersey (2022). https://doi.org/10.1109/ICEBE55470.2022.00024
20. Sharma, H., Sahu, N., Singhal, R.K., Garg, A., Singhal, R., Tripathi, S.: Data-driven forecasting and inventory optimization using machine learning models and methods. In: 2024 1st International Conference on Advanced Computing and Emerging Technologies (ACET), pp. 1–6. IEEE, Piscataway, New Jersey (2024). https://doi.org/10.1109/ACET61898.2024.10730235
21. Hasheminejad, S.A., Shabaab, M., Javadinarab, N.: Developing cluster-based adaptive network fuzzy inference system tuned by particle swarm optimization to forecast annual automotive sales: a case study in Iran market. Int. J. Fuzzy Syst. **24**, 2719–2728 (2022). https://doi.org/10.1007/s40815-022-01263-6
22. Upadhyay, H., Shekhar, S., Vidyarthi, A., Prakash, R., Gowri, R.: Sales prediction in the retail industry using machine learning: a case study of BigMart. In: 2023 International Conference on Electrical, Electronics, Communication and Computers (ELEXCOM), pp. 1–6. IEEE, Piscataway, New Jersey (2023). https://doi.org/10.1109/ELEXCOM58812.2023.10370313

Hybrid Safety Concept for Collaborative Robots in Thermal and Mechanical Hazard Environments: Case Study in Automotive Adhesive Applications

Bsher Karbouj(✉) and Jörg Krüger

Department of Industrial Automation Technology, Technical University of Berlin, Berlin, Germany
karbouj@tu-berlin.de

Abstract. The integration of collaborative robots (cobots) in industrial settings requires safety concepts that address multiple hazards. However, a key gap persists in aligning biomechanical safety standards with task-specific risks such as thermal exposure and mechanical collisions. This paper addresses that gap through a case study in automotive adhesive applications, where cobots and operators share workspaces involving high-temperature tools and dynamic tasks. A hybrid safety concept was developed, combining active measures (inclination sensors, virtual safety zones) with passive protections (heat shields, rounded tool edges). Biomechanical thresholds per ISO/TS 15066 were validated using CoboSafe sensors, while thermal risks were assessed through controlled testing. A mixed-methods approach evaluated speed effects, contact forces, and incident reduction across 90 operational cycles. At 0.35 m/s, collision forces remained within ISO limits ($\leq$65 N for facial regions), and no incidents were observed with the hybrid safety system. In contrast, 8 safety-relevant events occurred in a baseline setup without protective measures. While thermal exposure reduction was not directly quantified in percentage terms, qualitative observations confirmed lower surface temperatures and minimized contact risk through shielding and controlled motion. The proposed concept demonstrates a scalable and practical approach for enhancing safety in high-risk human–robot collaboration where mechanical and thermal hazards coexist.

Keywords: Human–Robot collaboration · Hybrid safety systems · Adhesive manufacturing · Biomechanical compliance · Thermal risk mitigation

1 Introduction

Collaborative robots (cobots) are designed to operate safely alongside human workers, often in shared workspaces, without the need for traditional safety barriers [1]. This paradigm shift necessitates robust safety concepts to manage the complex interactions between humans and robots [2]. To address safety concerns in human–robot collaboration (HRC), the International Organization for Standardization introduced ISO/TS 15066,

L. Overmeyer and B.-A. Behrens (eds.), *Production at the Leading Edge of Technology*, Lecture Notes in Production Engineering, https://doi.org/10.1007/978-3-032-19524-1_49

which provides guide-lines for assessing and mitigating risks associated with physical interactions between humans and cobots [3]. This standard emphasizes biomechanical limits, such as permissible force and pressure thresholds, to prevent injuries during unintended contacts. However, ISO/TS 15066 primarily focuses on mechanical hazards and does not comprehensively address other potential risks, such as thermal hazards, that may arise in specific industrial applications [4]. In certain manufacturing processes, such as adhesive application in the automotive industry, cobots are required to handle high-temperature tools, introducing thermal risks alongside mechanical ones. This gap underscores the need for an integrated safety approach that considers the multifaceted nature of risks in collaborative work environments.

This paper aims uniquely to develop and validate a hybrid safety concept tailored for HRC scenarios involving both mechanical and thermal hazards. By conducting a case study on adhesive application processes in automotive manufacturing, we systematically analyze biomechanical forces, thermal exposures, and the effectiveness of implemented safety measures. The proposed concept integrates active monitoring systems, such as real-time inclination sensors and virtual safety boundaries, with passive safeguards, including heat-resistant shields and ergonomic tool designs. Through empirical validation using CoboSafe measurement devices and controlled experiments, we assess the concept's ability to maintain safety standards without compromising productivity. The findings of this research contribute to the advancement of safety protocols in industrial applications where cobots and human workers interact in environments characterized by complex, multi-risk scenarios.

2 Related Work

Human–Robot Collaboration safety strategies are typically classified as either active or passive, depending on their behavior relative to the hazard event. Active measures operate in real time to detect and mitigate risks—e.g., sensors, emergency stops, or speed-limiting controls. In contrast, passive measures provide built-in protection by design, such as physical guards or heat shields, without relying on system feedback. This definition aligns with safety engineering conventions, where anticipatory systems like seatbelt pretensioners are considered active despite being mechanical. Key active strategies include Speed and Separation Monitoring (SSM) [5], Safety-Rated Monitored Stop (SRMS) [6], and Power and Force Limiting (PFL) [7]. These techniques, defined in standards like ISO10218, primarily address mechanical safety [8]. ISO/TS 15066 further refines these guidelines by specifying biomechanical limits for contact forces across 29 body regions, derived from empirical pain studies [9]. However, this standard focuses on mechanical hazards and does not cover thermal or electrical risks common in tasks like welding or adhesive bonding. Additionally, its reliance on idealized human body models may not reflect real-world variability. Thermal hazards in HRC remain underexplored. Surveys indicate that only 19% of implementations include thermal safeguards, while 25% neglect safety measures altogether [10]. Most existing systems rely on mono-risk strategies centered on SSM and SRMS (32 out of 46 reviewed studies) [8], leaving critical protection gaps in high-temperature processes. Recent work in welding automation highlights the need for hybrid safety concepts. Kovács and Tick [11] propose a risk-based safeguarding model for robotic gas metal arc welding (GMAW), integrating physical protections with cybersecurity measures to address both thermal and operational risks. Their approach reinforces the value of layered safety strategies in hazardous HRC environments. Given the growing complexity of collaborative tasks, the literature underscores the need for integrated safety concepts that combine biomechanical compliance with application-specific mitigation strategies.

3 Hybrid Safety Concept

This section presents the design and implementation of a hybrid safety concept for collaborative robots operating in environments with both mechanical and thermal hazards. The concept was developed based on a real-world glue application scenario involving high-temperature tools and human–robot collaboration. The concept is composed of two main layers: (1) active safety measures including real-time sensor-based monitoring, virtual safety zones, and emergency stop mechanisms, and (2) passive physical protections such as heat shielding, compliant tool design, and mechanical guards. This generalizable approach aims to ensure compliance with biomechanical standards (e.g., ISO/TS 15066) while accommodating task-specific hazards like thermal exposure. The remainder of this paper demonstrates the concept's implementation and validation in a real-world automotive adhesive application.

3.1 Use Case Scenario

The use case involves a semi-automated adhesive process in the automotive sector, where foam parts are bonded to structural panels using hot glue. Previously executed manually, the task posed risks including burns, glue overflow, and ergonomic strain due to tool temperatures of up to 190 °C. To address these hazards, a UR10e collaborative robot was introduced for glue dispensing, while operators now focus on loading, positioning, and inspection. This created a shared workspace with new safety demands related to dynamic motion and thermally active tooling. Figure 1 shows the layout of the collaborative cell, including the robot, glue gun with inclination sensor, and spatial segmentation of the work area. Operator-accessible (green) and robot-exclusive (orange) zones are defined virtually, enabling safe collaboration without physical barriers.

Risk assessment identified two primary hazard domains:

- Mechanical collision during transitions or unintended movement.
- Thermal injury from contact with the glue gun or glue spray.

These findings motivated a hybrid safety approach combining real-time monitoring and physical protection tailored to task-specific risks.

3.2 Active and Passive Safety Measures

The concept integrates both active (sensor/digital) and passive (mechanical/physical) safety layers. The key active component is a dual-axis inclination sensor mounted on the glue gun, monitoring tool orientation. Exceeding a $\pm 45^\circ$ tilt triggers a PLC interrupt for

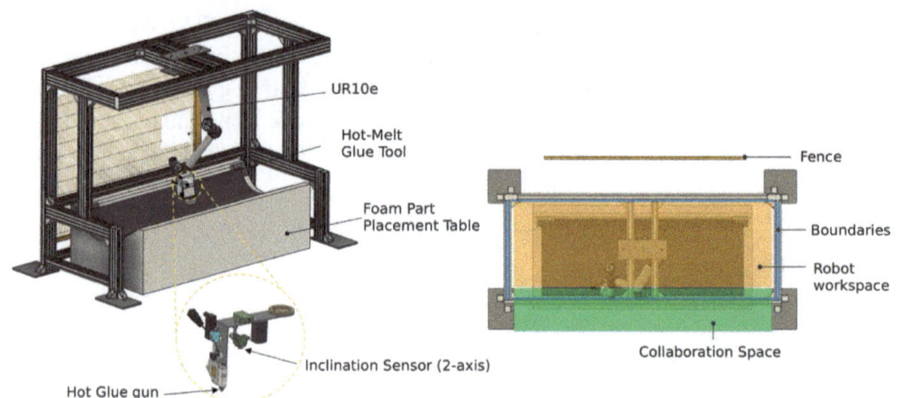

Fig. 1 Collaborative robotic cell for adhesive application

emergency stop, preventing off-target glue spray. Although the robot's pose is available through its internal kinematics, it may not accurately represent the real-time orientation of the glue gun due to potential mechanical play, flexible mounting, or dynamic deviations. The inclination sensor provides direct, tool-mounted feedback to ensure accurate monitoring of tilt angles and enables reliable emergency stop triggering when thresholds are exceeded.

A second active safeguard is the definition of virtual boundaries via the robot teach pendant, limiting motion to a calibrated envelope. These were defined by jogging the end-effector to safe extremes and storing them in the controller.

Passive protections include:

- A heat shield made of low-conductivity material to ensure touch-safe surface temperatures.
- An anti-drip guard restricting glue ejection to a safe zone, blocking hand intrusion.
- Rounded tool geometries and soft materials to reduce collision impact.

Together, these measures provide redundant coverage for biomechanical and thermal risks. Table 1 summarizes the components and their safety roles.

Table 1 Summary of hybrid safety components and their target risk domains

Safety layer	Type	Component	Target risk
Passive	Physical	Heat shield	Thermal contact
Passive	Mechanical	Rounded tool geometry	Collision
Active	Sensor	Inclination sensor (2-axis)	Glue deviation
Active	Digital	Virtual safety boundaries	Workspace overlap

3.3 End-Effector Safety Design

While Sect. 3 outlined the key passive components within the hybrid safety architecture, this section details their mechanical implementation, focusing on the end-effector as the primary human–robot contact point.

The end-effector (see Fig. 3) represents the closest point of contact between robot and operator, requiring enhanced safety design due to hot-melt adhesive use. A commercial glue gun was mounted on a modular aluminum plate, and custom components were added:

- Heat Shield: Multi-layer enclosure with polymer and foil, reducing surface temperature.
- Anti-Drip Guard: Polymer housing that confines glue ejection to a narrow path.
- Rounded Geometry: Filleted edges reduce pressure points, supporting ISO/TS 15066 compliance.
- Cable Routing: Internal wiring and tubing prevent entanglement or disconnection.
- Mounting Interface: Custom adapter for secure robot-tool connection.

To ensure systematic identification and mitigation of thermal hazards, the thermal risk assessment was conducted in accordance with the general risk assessment procedure defined in ISO 12100 [12]. Furthermore, contact temperature thresholds were evaluated based on ISO 13732-1 [13], which provides criteria for assessing human skin burn risks due to hot surfaces. These standards guided the selection of materials and the design of passive protective elements such as the heat shield.

4 Experimental Evaluation and Results

This section presents the experimental validation of the hybrid safety concept under realistic conditions. A robotic cell simulating an automotive adhesive task was used to assess biomechanical compliance, thermal risk mitigation, and system robustness via observations, ISO-based force testing, and thermal measurements.

4.1 Experimental Setup and Testing Procedure

To evaluate the effectiveness of the hybrid safety concept, a series of experimental tests were conducted in a pilot robotic cell replicating an automotive adhesive bonding scenario (see Fig. 2). The setup and measurement protocol were designed to assess both biomechanical compliance and thermal risk mitigation under realistic operating conditions. Robot Platform: The system was built around a Universal Robots UR10e collaborative arm with a 10 kg payload and 1300 mm reach. The robot was mounted overhead on an aluminum frame to replicate spatial constraints common in industrial adhesive workstations. End-Effector: The customized glue gun assembly described in Sect. 3.3 was mounted on the robot flange and connected to a temperature-controlled hot-melt adhesive supply, operating at up to 190 °C. The end-effector integrated the dual-axis inclination sensor, heat shield, and anti-drip guard. Measurement Devices: Force and pressure measurements were conducted using CoboSafe CBSF devices from

GTE Industrieelektronik GmbH. These sensors simulate different human body regions and conform to the ISO/TS 15066 guidelines. The following sensor pads were used:

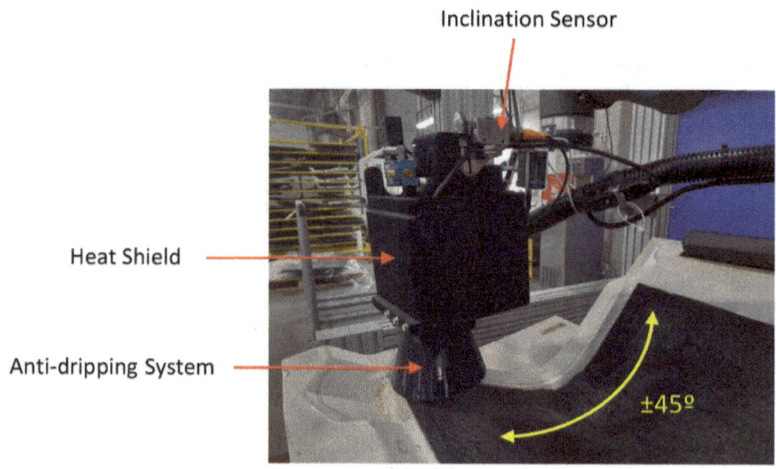

Fig. 2 End effector with integrated safety components

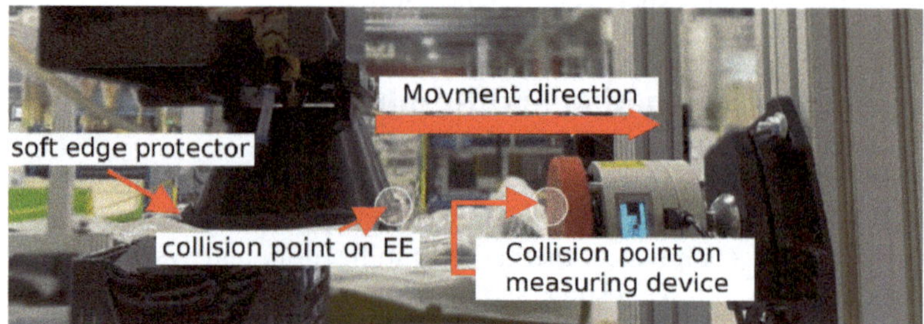

Fig. 3 Test setup showing the measuring device, movement direction and collision points

- CBSF 35 (K2 = 35 N/mm) for sensitive areas such as the face.
- CBSF 75 (K2 = 75 N/mm) for torso and upper arms.
- CBSF 150 (K2 = 150 N/mm) for less sensitive regions like the back.

Test Scenarios: A total of 90 cycles were performed under consistent environmental conditions (22 °C room temperature, 190 °C glue gun, 0.35 m/s robot speed), simulating a realistic production workflow., divided equally across three human proxy configurations. Each operational cycle consisted of the robot executing the full adhesive application task, including tool activation, glue dispensing along a predefined path, and return to the home position. Safety mechanisms (e.g., inclination interrupts, safety zone limits) were triggered under controlled conditions to assess responsiveness and coverage.

Thermal Testing: Temperature measurements were recorded using a non-contact IR sensor to evaluate the effectiveness of the heat shield. Measurements were taken before and after shield integration at multiple locations on the end-effector surface.

Data Logging and Analysis: All force, pressure, and temperature data were logged and analyzed to verify compliance with ISO limits. Additionally, qualitative observations were made during each cycle to track near-miss events, unintended contacts, or glue-related hazards.

4.2 Force and Pressure Validation

To evaluate compliance with biomechanical safety standards, a series of force and pressure measurements were conducted using certified CoboSafe CBSF sensors (see Sect. 4.1). The robot operated at 0.35 m/s over 90 cycles across three contact setups; both transient and quasi-static contacts were tested. Table 2 shows the ISO/TS 15066 thresholds alongside our measured forces/pressures by body region. The most critical transient event with CBSF-35 (face) yielded a peak force of 63 N, below the ISO/TS 15066 force limit of 65 N. For non-facial regions (CBSF-75/150), transient peak forces ranged from 93.5 to 131 N, all below the respective regional limits. Pressures were derived from the measured forces using each pad's calibrated effective contact area; in the hybrid configuration no ISO/TS 15066 pressure-limit exceedances were observed, whereas baseline trials (safeguards off) showed 15–20% force exceedances and occasional pressure exceedances. Quasi-static contacts produced lower peaks; for example, a back/shoulder test with CBSF-150 measured 58 N (static), within the applicable static limit. The compliant, rounded end-effector geometry and elastic interfaces helped reduce tool-surface pressure.

Table 2 ISO/TS 15066 force and pressure thresholds and measured values for the evaluated body regions under transient and quasi-static contacts

Body region	Thresholds				Measured results			
	Transient		Static		Transient		Static	
	Force (N)	Pressure (N/cm^2)	Force (N)	Pressure (N/cm^2)	Force (N)	Pressure (N/cm^2)	Force (N)	Pressure (N/cm^2)
Skull and forehead	130	110	0	0	117	234	116	233.5
Face	65	110	0	0	93.5	182	93.5	182
Nape	300	280	150	140	93.5	182	93.5	182
Chest	280	340	140	170	93.5	182	93.5	182
Underarm and wrist	320	360	160	180	93.5	182	93.5	182
Hand and finger	280	380	140	190	93.5	182	93.5	182

(continued)

Table 2 (*continued*)

Body region	Thresholds				Measured results			
	Transient		Static		Transient		Static	
	Force (N)	Pressure (N/cm^2)	Force (N)	Pressure (N/cm^2)	Force (N)	Pressure (N/cm^2)	Force (N)	Pressure (N/cm^2)
Back and shoulders	420	320	210	160	131	273	58	271
Upper arm and elbow	300	380	150	190	131	273	58	271

4.3 Thermal Risk Assessment

In addition to biomechanical safety, thermal hazards were analyzed due to the presence of high-temperature components on the glue dispensing end-effector. The glue gun operates at an internal temperature of up to 190 °C, which poses a significant risk of burns if exposed surfaces are left unprotected. To mitigate this, a multi-layer heat shield was integrated into the tool design, as shown in Fig. 3. The shield was constructed from a combination of thermally insulating polymer and reflective aluminum foil to minimize conductive and radiative heat transfer.

Surface temperature measurements were conducted using a non-contact infrared thermometer at critical points around the end-effector housing. Without shielding, the outer nozzle surface reached temperatures between 110–125 °C, depending on the duration of operation. After installing the heat shield, the accessible surface temperatures were reduced to below 50 °C, consistent with touch-temperature criteria in ISO 13732-1 [13] for short-duration contact (touch-safe). In addition to thermal insulation, an anti-drip guard was added to the nozzle tip to prevent glue from being dispensed outside the intended zone. This passive safety element further minimized the risk of hot glue contact with the operator's hands during manual handling tasks. The combined effect of the heat shield and anti-drip guard effectively eliminated all observed thermal safety events during the 90 test cycles. No operator-related incidents or contact with hazardous hot surfaces were reported throughout the testing phase. In contrast, during baseline tests without thermal shielding, glue overspray occurred in 8 out of 30 cycles. This comparison supports the thermal effectiveness of the passive protection system under real operating conditions.

4.4 System Responsiveness and Safety Effectiveness

To evaluate the system's real-time responsiveness, the inclination sensor was tested under repeated 45° threshold violations. It consistently detected excessive tilt and triggered emergency stops via the PLC. The average response time was 210 ms—sufficient to prevent glue ejection beyond the work zone. A qualitative study across 90 robot cycles confirmed the system's effectiveness: no injuries, glue contact, or hazardous robot–operator interactions occurred. Virtual safety zones limited motion, and the anti-drip guard effectively contained the adhesive.

To assess impact, a baseline test (30 cycles, identical toolpath and parameters) was conducted without safety measures. Overspray occurred in 8 cycles (26.7%), often during upward motions exceeding 45° before interruption could occur—highlighting the need for real-time monitoring. The hybrid system, by contrast, recorded zero incidents.

5 Discussion and Conclusion

The results validate the proposed hybrid safety concept as an effective approach for mitigating both mechanical and thermal risks in collaborative robotics. Across 90 test cycles, the integration of inclination sensing, virtual safety zones, and passive protections (e.g., heat shields, anti-drip guard) ensured compliance with ISO/TS 15066 and prevented all safety-relevant incidents. Compared to the baseline (8 incidents in 30 cycles), the hybrid system improved safety without reducing productivity. Notably, ISO-compliant force limits alone were insufficient for thermal hazards, highlighting the value of task-specific measures. Some limitations remain, including occasional pressure exceedances in baseline tests and static zone definitions. Future improvements may involve adaptive sensing, vision-based detection, and learning-based risk models to enhance responsiveness in dynamic environments.

Competing Interests. The author(s) has no competing interests to declare that are relevant to the content of this manuscript.

References

1. Karbouj, B., Rashwany, K.A., Alshamaa, O., Krüger, J.: Adaptive behavior of collaborative robots: review and investigation of human predictive ability. Procedia CIRP **130**, 952–958 (2024)
2. Keshvarparast, A., Battini, D., Battaia, O., Pirayesh, A.: Collaborative robots in manufacturing and assembly systems: literature review and future research agenda. J. Intell. Manuf. **35**, 2065–2118 (2023)
3. Villani, V., Pini, F., Leali, F., Secchi, C.: Survey on human–robot collaboration in industrial settings: safety, intuitive interfaces and applications. Mechatronics **55**, 248–266 (2018)
4. Dhanda, M., Rogers, B.A., Hall, S., et al.: Reviewing human-robot collaboration in manufacturing: opportunities and challenges in the context of industry 5.0. Robot Comput.-Integr. Manuf. **93**, 102937 (2025)
5. Karagiannis, P., Kousi, N., Michalos, G., et al.: Adaptive speed and separation monitoring based on switching of safety zones for effective human robot collaboration. Robot Comput.-Integr. Manuf. **77**, 102361 (2022)
6. Rubagotti, M., Tusseyeva, I., Baltabayeva, S., et al.: Perceived safety in physical human–robot interaction—a survey. Robot Auton. Syst. **151**, 104047 (2022)
7. Svarny, P., Rozlivek, J., Rustler, L., Hoffmann, M.: 3D collision-force-map for safe human-robot collaboration (2021)

8. Lucci, N., Lacevic, B., Zanchettin, A.M., Rocco, P.: Combining speed and separation monitoring with power and force limiting for safe collaborative robotics applications. IEEE Robot. Autom. Lett. **5**, 6121–6128 (2020)
9. Behrens, R., Pliske, G., Umbreit, M., et al.: A statistical model to determine biomechanical limits for physically safe interactions with collaborative robots. Front. Robot. AI **8** (2022)
10. Arents, J., Abolins, V., Judvaitis, J., et al.: Human–robot collaboration trends and safety aspects: a systematic review. J. Sens. Actuat. Netw. **10** (2021)
11. Kovács, T.A., Tick, A.: Safeguarding human-robot collaboration in gas metal arc welding: a risk assessment approach for welding automation. In: 2024 IEEE 22nd Jubilee International Symposium on Intelligent Systems and Informatics (SISY), pp. 000541–000546 (2024)
12. International Organization for Standardization: ISO 12100:2010—Safety of machinery—General principles for design—Risk assessment and risk reduction (2010)
13. International Organization for Standardization: ISO 13732-1:2006—Ergonomics of the thermal environment—Methods for the assessment of human responses to contact with hot surfaces—Part 1: Hot surfaces (2006)

Setup Assistance for Transfer Presses Based on AI

Niyazi Ayaz(✉), Dennis Schmiele, Malte Nagel, Lennart Hinz, and Richard Krimm

Leibniz University Hanover, Garbsen, Germany
ayaz@ifum.uni-hannover.de

Abstract. Due to complex interactions with not completely identified interdependencies, the creation and maintenance of stable process conditions during production on multi-stage presses can currently only be realized with a great amount of set-up effort and by using implicit knowledge. Changes of the process parameters in one stage affect the process flow in other stages. So the restoration of good part production is a lengthy process, depending on the complexity of the tool set. As part of the German Research Foundation's Priority Program 2422 "Data-driven process modelling in forming technology", AI-based methods are used to achieve a deeper understanding of the relationships between process variables and their impact on workpieces quality. To this end, a representative demonstrator part is analysed in simulations with regard to manufacturability and geometric quality. Characteristics are defined and a multi-stage forming process is elaborated. The quality of the process is quantified by comparing measured and nominal quality parameters. An approach consisting of two linked AI models to predict the system interrelationships is in development. First, system-inherent interactions are identified, then geometric quality characteristics are predicted, depending on the system configuration. Currently training data are being generated. Objective is to generate recommendations for machine operator actions for a convenient setup to restore good part production. Such recommendations lead to a considerable reduction of time in changeover processes on multi-stage presses and a corresponding increase in productivity.

Keywords: Multi-stage presses · Setup assistance · Quality assurance

1 Introduction

Sheet metal forming is one of the most widely used manufacturing processes across various industrial sectors. In addition to meet high economic and technical demands, manufacturers face growing challenges due to globalization, increasing competition, shorter product life cycles and pressure to reduce costs. A key strategy to enhance efficiency lies in the use of multi-stage presses. These systems allow complex forming operations to be carried out within a significantly smaller workspace, enabling more compact production layouts. At the same time, they reduce process time by combining multiple forming steps into a single, streamlined sequence. By integrating space-efficient multi-stage presses, companies can produce components more economically without

L. Overmeyer and B.-A. Behrens (eds.), *Production at the Leading Edge of Technology*,
Lecture Notes in Production Engineering, https://doi.org/10.1007/978-3-032-19524-1_50

compromising quality. Additional efficiency gains can be achieved by means of process optimization, further supporting competitive and cost-effective manufacturing. However, the setup of such multi-stage processes is very complex and time consuming. Comparing the desired geometry with the real geometry shows that newly installed or readjusted toolsets can lead to discrepancies. Having multiple individual stages with individual force-path progressions inside the workspace can result in an eccentric load distribution [1]. Those effects can change the way of closing of the machine resulting in bad shape and geometric characteristics. A mean to affect the forces between the ram and the bolster plate is to change the installation height of each stage inside the press [1].

The installation process can take multiple days for multi stage presses with up to 20 stages. The reason therefore is that the adjustment is carried out on implicit knowledge of the operator. The operator incrementally changes the installation heights of the individual stages based on his own experience. There is no domain specific knowledge which could help the operator in making his decisions [1].

The aim of this project is to create an AI-based setup assistant which will be able to reduce the amount of time needed to install or readjust the tools by directly telling the operator the required heights. Therefore, first of all a lot of data is required which will be used to train and validate the AI. To create this amount of data a forming toolset with 7 stages has been designed. Each stages height can be adjusted individually, by changing the amount of underlying foils, therefore manipulating the part quality. The meanwhile manufactured toolset is equipped with extensive sensory to acquire relevant system parameters. At the end of the forming process the final part will be geometrically captured to determine if the shape and geometry values lie between the defined tolerances. If they don't the required changes in each height will be suggested by the setup assistant.

2 State of the Art of Setup Assistants in Production

As early as in the 1990s Neugebauer and Thamm described the reinstallation procedure of a tool as inherently noncausal. Even with precise tool presetting and carefully configured machine parameters, a consistently stable process could not be ensured. This result suggests that the underlying physical interdependencies weren't sufficiently understood. The knowledge, experience and skills of the responsible operators remained a decisive factor in achieving satisfactory results. With the increasing improvement of sensors and computing power with topics like machine learning this may change in the future [2].

The following presents some relevant examples of assistance systems in science and industry related to setup of machines with the aim of process optimization resulting in setup time reduction, among other things.

Barrenetxea et al. present a model based assistant tool called "SUA" for the efficient setup and optimization of centerless grinding processes. Setup times are reduced by substituting manual trial and error with guided, simulation driven configuration. Operators are guided toward parameter configurations that maximize stability and surface integrity [3].

Torres-Treviño et al. present a hybrid expert system designed to automate the selection of machining parameters for turning and milling operations. The system integrates symbolic regression—using alpha–beta operators—with an evolutionary algorithm called "Evonorm". This regression technique produces interpretable mathematical models derived from experimental or historical process data. The models are constructed using a set of predefined mathematical functions, enabling compact and analytically tractable representations. Model validation is carried out through residual analysis to ensure high predictive reliability. Comparative evaluations reveal that symbolic regression achieves superior performance over both linear regression and genetic programming. The expert system focuses on minimizing surface roughness but is adaptable to other criteria. It requires only basic data inputs and user-defined process goals, making it applicable across a range of machining contexts [4].

Gräler et al. present a progressive development of an intelligent setup assistant designed for forming processes involving progressive tools. In their 2018 work, the authors address the high variability in initial setup conditions due to stochastic disturbances such as material inconsistencies or a variation of lubrication. To reduce the reliance on manual, experience-driven setup procedures, they introduce a two-stage compensation strategy: first, a model-based calculation of optimal adjustment screw positions using second-order regression models derived from Design of Experiments. Second, an iterative correction based on deviations between actual and target part geometries [5]. They propose an extended system architecture that explicitly integrates non-adjustable but measurable disturbances, such as tool wear, material thickness variation, temperature shifts, into the setup assistant. This is achieved through a hybrid modeling approach: while DoE remains the basis for modeling tool behavior under controlled conditions, machine learning algorithms—such as regression trees or neural networks—are used to model the effects of disturbances during real production. Simulation studies demonstrate that this extended approach significantly improves setup quality and reduces iteration steps, especially under variable production conditions [6].

3 Demonstrator Workpiece and Tool Design

A demonstrator workpiece will be used to create the training and validation datasets for the AI, by intended geometry changes during the manufacturing process. During the design of the demonstrator workpiece the relevance to real industrially produced parts played a big role. Manufacturing the chosen workpiece should require multiple forming steps and be relevant for the future. Therefore a design similar to an assembly plate has been decided on. In advance of production of the designed workpiece, each individual production process was validated by means of simulations. Numerous finite elements simulations with varying geometrical parameters were carried out, analyzing the strain, the displacement and the material flow [7]. The designed tool set is shown in Fig. 2. As described earlier it consists of 7 modular stages, each ones height being able to be adjusted individually by the usage of underlying foils. A description of each process is displayed in the following Table 1.

Since the sequence of the forming processes had not been predefined externally, simulations were carried out in which the order of the forming steps was varied. The results demonstrate that the sequence does in fact influence the component properties. When deep drawing was performed in advance of stretch forming (Fig. 1a), waviness occurred in the flange area. Conversely, when the order was reversed, bulging appeared in the base area (Fig. 1b).

Table 1 Description of the tool stages

Stage no.	Description
1	Pre cut of a square blank from the coil with 200 mm × 200 mm dimensions
2	Round blank Cut of the square blank with additional gripping points
3	Deep drawing with a depth of 15 mm (varying 10–20 mm)
4	Form stretching four areas with a depth of 3 mm (varying 2–5 mm)
5	Embossing of two dies with a depth of 0.2 mm (varying 0–0.2 mm)
6	Piercing of four holes, three of them with a diam. of 5 mm the other with 20 mm
7	Round cut of the final part diameter of 136 mm discarding the gripping points

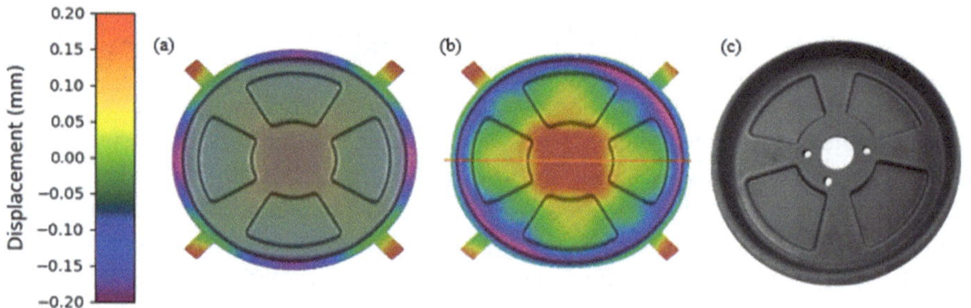

Fig. 1 Demonstrator part **c** and exemplary finite element simulation results **a** and **b** for different operations at stage 2 and 3

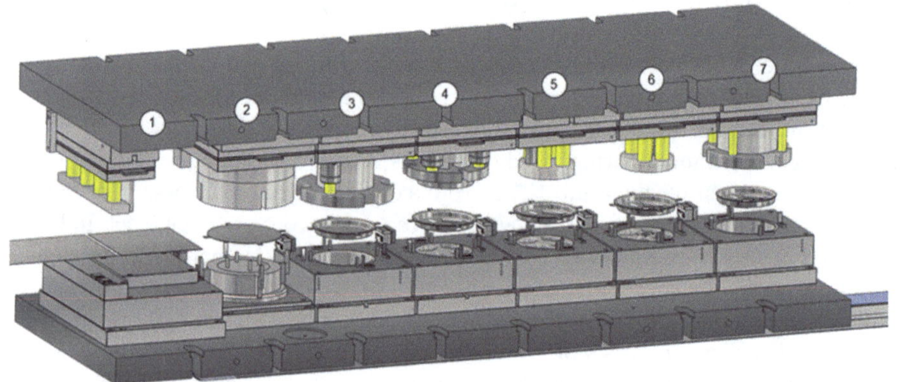

Fig. 2 Modular multi-stage toolset

4 Proof of Concept and First Results

The study will be conducted on a servo-electrically directly driven toggle joint press with a nominal press force of 2000 kN. DC06 will be used as material for the workpiece, a conventional deep drawing steel with a tensile strength of approx. 270–330 N/mm^2 and a thickness of 0.75 mm. Currently the force–time-curves and trajectories are measured in each stage of the tool set. Additionally laserscanners are used to create clouds of points of the parts. Those clouds of points can be used to derive characteristics to evaluate the part quality.

The concept of measurement has been validated during forming first demonstrator workpieces. Figure 3 shows the measured path-time-curves (trajectories) of each stage. The distances in height between the stages 1 and 2 and the other stages result from constructive restrictions in the shear cutting stages. To reduce the amount of wear the cutting processes start directly before the bottom dead center. It can be seen (Fig. 3, right) that the measuring system is able to detect small vibrations proving it is able to quantify dynamic system behavior.

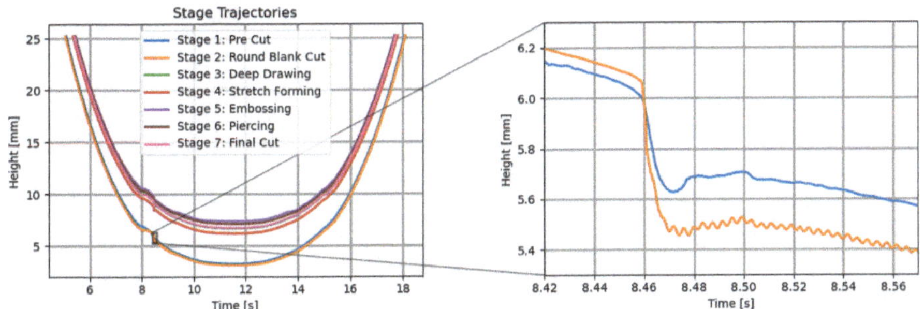

Fig. 3 Travel path in dependence of time for each stage and detected vibration in stage 2

Figure 4 shows the deflections of the stages under load due to the forming processes. It is determined by calculating the difference between a reference trajectory, measured without any sheet metal in the press (without any process forces) on the one hand and an operational stroke on the other hand. These calculations will play a big role in determining the effects of the underlying foil and how it will be able to compensate undesired changes in the geometry of the produced part, by the AI.

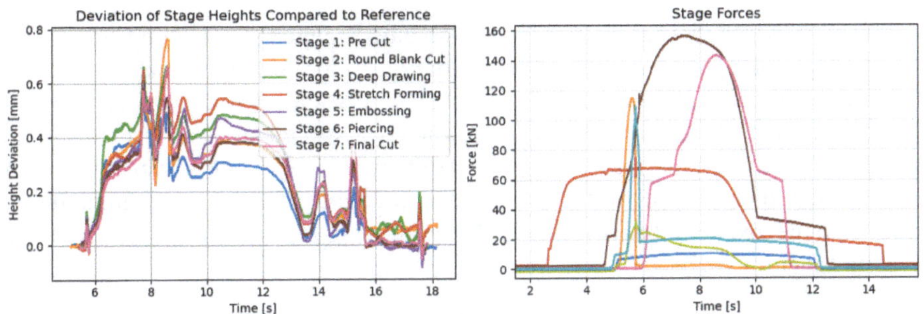

Fig. 4 Measured deflection (left) and forces (right) in dependancy of time for each stage

Figure 5a shows the calculated differences of the geometry in regards to height difference h_2 between the two layers shown in Fig. 5b for a produced demonstrator workpiece. The height differences can be measured in micrometric range. The result illustrates a slight tilt in the xy-plane, which results from a tilt in the tool set or the machine. Additionally, some systematic observations can be made, such as the bulging in the area of stretch forming (black circles), as expected from the simulation results. However, any attempt at interpretation should be postponed until a larger dataset is available.

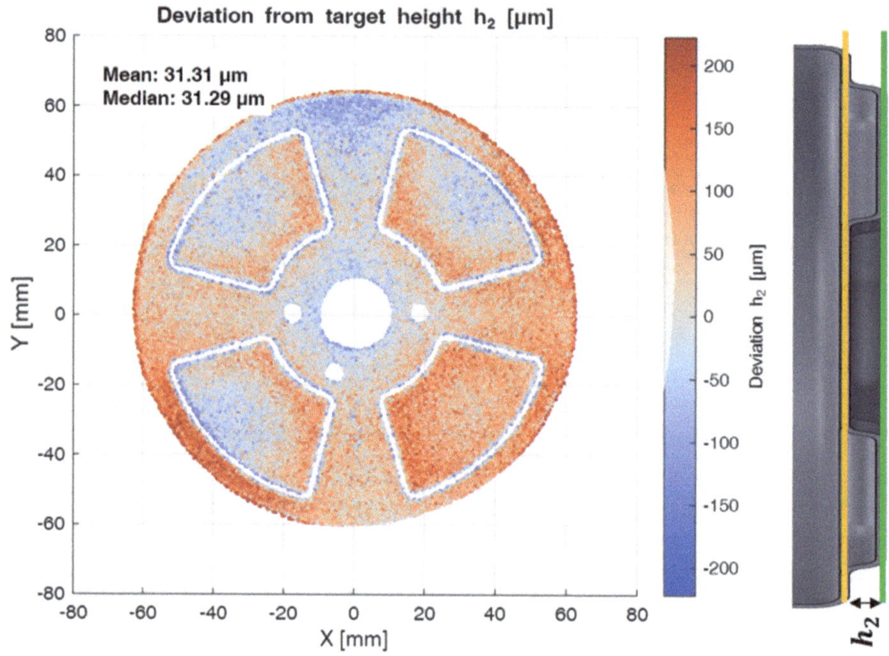

Fig. 5 Height deviations derived from the cloud point

Figure 6 shows the first application of the AI-model, aimed at predicting height h_2 for a predefined set of input parameters as shown in Fig. 7. The current evaluation was based on a limited dataset of 20 measurement runs (16 parts each), with controlled variations restricted to stages 3–5. While the dataset is not yet sufficient for statistically robust conclusions, it enabled an initial validation of the model's predictive capability with a MSE of 0.022 and a RMSE of 0.148 mm. The predictive framework is implemented as a fully connected neural network (FCNN) employing a SiLU (Sigmoid Linear Unit) activation function.

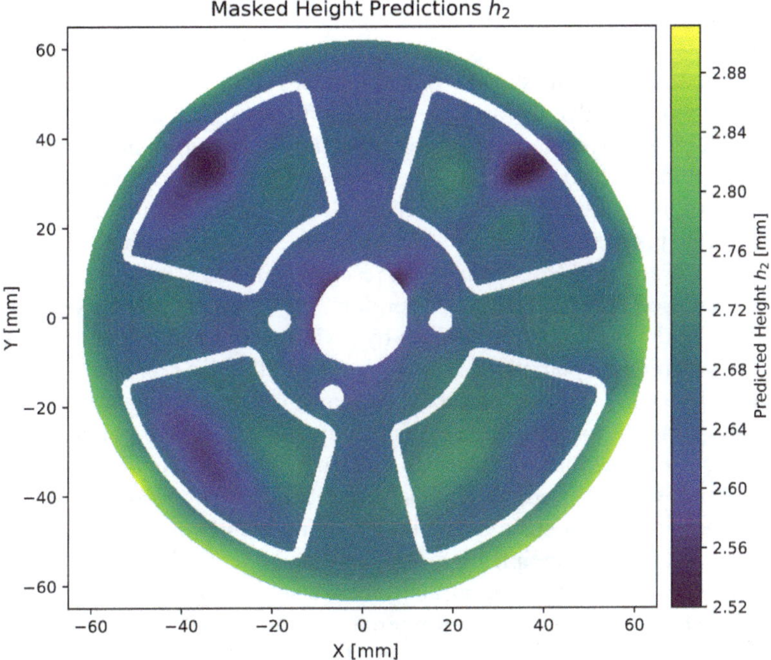

Fig. 6 Exemplary height prediction by the AI-model for a predefined set of input parameters

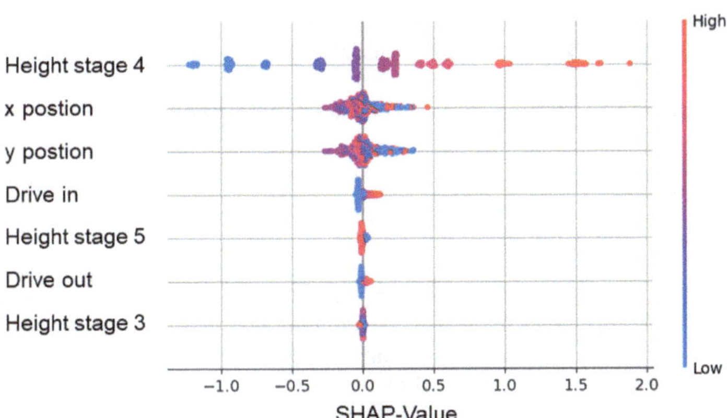

Fig. 7 SHAP Analysis for the input values of the neural network

To interpret the model's predictions, SHAP (Shapley Additive Explanations) values were calculated; these quantify the contribution of each input feature to a specific prediction, enabling a transparent assessment of parameter influence [8]. SHAP analysis indicated stage 4 as the most influential factor (see Fig. 7).

5 Summary

Workpieces produced in multi-stage forming processes are susceptible to deviations from their intended geometry due to various external influences such as machine deflection, temperature fluctuations, material thickness variations, or changes in lubrication conditions. These influences lead to discrepancies between the actual geometry and the target specifications of the workpiece. To avoid these deviations and obtain the desired geometry, manual readjustments of the forming machine are often required. However, this adjustment process is time-consuming and resource-intensive, resulting in production delays and increased operational costs. Therefore, the development of a system capable of compensating the mentioned influences by giving recommendations for installation heights, holds significant potential for operators of multi-stage presses.

Looking ahead, the focus is on the development of an intelligent system that can proactively support the forming process. Currently, training data are being generated to enable the application of machine learning concepts in multi-stage forming operations. The goal is to develop a predictive model capable of recommending optimal height settings for each individual stage of the tool set. Importantly, the model will be based on explainable AI (XAI) principles. This means that the reasoning behind the system's recommendations will be transparent and interpretable, allowing operators and engineers to understand the logic behind the proposed adjustments. For future test series, additional sensors for measuring lubrication, sheet thickness, and temperature will be implemented in order to enable a more detailed analysis of the interaction mechanisms between external process influences and the achievable workpiece quality.

Acknowledgements. The research project with the title 'AI based setup assistance system for multi-stage presses' was promoted under KR 3718/11-1 by the German Research Foundation (DFG). The authors would like to thank the German Research Foundation (DFG) for their support.

Competing Interests. The author(s) has no competing interests to declare that are relevant to the content of this manuscript.

References

1. Krimm, R.: Berechnung der lastabhängigen Maschinenauffederung zur Verkürzung der Anlaufzeit neuer Transferwerkzeugsätze, Dissertation, Universität Hannover (2006)
2. Neugebauer, R., Thamm U.: Maßnahmenkatalog für den Wiederanlauf von Pressen, Blech Rohre und Profile 43 (1996) 4, pp. 178-183
3. Barrenetxea, D., et al.: Model-based assistant tool for the setting-up and optimization of centerless grinding process. Mach. Sci. Technol. **16**(4), 501–523. Taylor & Francis (2012)
4. Torres-Treviño, L.M., et al.: An expert system for setting parameters in machining processes. Expert Syst. Appl. **40**(17), 6877–6884. Elsevier (2013)
5. Gräler, M., et al.: Assisted setup of forming processes: architecture for the integration of non-adjustable disturbances. Procedia CIRP, **81**, 1348–1353. Elsevier (2019)

6. Gräler, M., et al.: Assisted setup of forming processes: compensation of initial stochastic disturbances. Procedia Manuf. **25**, 358–364. Elsevier (2018)
7. Krimm, R., et al.: Method planning for deep drawing in the production of workpieces with variable shapes. In: MATEC Web Conference, vol. 408. MATEC Web of Conferences (2025)
8. Lundberg, S. M., Su-In L.: A unified approach to interpreting model predictions. Adv. Neural Inform. Process. Syst. **30** (2017)

Bridging Autonomization and Human–Machine Interaction: An AI-Based Framework for Smart Manufacturing

Ann-Kathrin Bischoff(✉), Gilbert Ely Engert, and Matthias Weigold

Department of Mechanical Engineering, Institute for Production Management, Technology and Machine Tools (PTW), Technical University of Darmstadt, Darmstadt, Germany
a.bischoff@ptw.tu-darmstadt.de

Abstract. In this paper, we propose a novel framework that integrates Human–Machine Interaction into autonomous production environments by leveraging Artificial Intelligence (AI), particularly Large Language Models (LLMs). The increasing shortage of skilled workers in the manufacturing sector and the ongoing autonomization of production, demand more intuitive and adaptive workflows for machine operators. As production moves towards a higher level of autonomy, operators will no longer manage only individual machines but rather interact with a superordinate system. Within this proposed framework the superordinate system is an AI-driven system which enables the interaction with the production process. The AI-driven system extracts knowledge from data provided by existing digitalization tools, e.g. error reporting or tooling equipment documentation. By implementing LLMs, production workers are enabled to interact with the extracted knowledge via text or speech to retrieve knowledge and receive decision-making support through reasoning capabilities. The AI system will further gather specific knowledge from the machine operators, which will be looped back into the AI system to continuously learn from user inputs. The LLM integration allows for a more flexible, efficient, and human-centric production environment as proposed within Industry 5.0. While primarily conceptual, our work includes initial insights that demonstrate the feasibility and potential benefits of integrating LLMs into the manufacturing environment. We outline the key components of the framework, discuss its technical feasibility, and evaluate its impact through simplified use cases. Our findings highlight the potential of LLMs to transform Human–Machine Interaction in manufacturing settings, bridging the gap between autonomization and human expertise.

Keywords: Autonomous manufacturing · Agentic AI · Large language models

1 Introduction

While agentic AI, as an autonomous AI system, is already impacting industries like the energy, transportation, healthcare or finance sectors, its application in the manufacturing sector is still at an early stage [1]. Despite the advancements introduced by

L. Overmeyer and B.-A. Behrens (eds.), *Production at the Leading Edge of Technology*,
Lecture Notes in Production Engineering, https://doi.org/10.1007/978-3-032-19524-1_51

the Industry 4.0 – such as digitalization, Internet of Things and data mining – progress towards fully autonomous systems remains low [2]. As manufacturing systems evolve towards autonomous systems, the role of the human operator shifts from directly operating individual machines to interacting with complex, higher-level systems, such as the production itself [3]. Further, with the skilled labor shortage facing the industry [4], intelligent systems offer a solution to document and transfer knowledge. This evolution highlights the need for an intuitive interaction that adapts to the workers and allows humans oversight while leveraging the full potential of intelligent systems.

The proposed framework of this paper outlines a way to incorporate humans in autonomous production systems as an orchestrator, rather than eliminating humans completely, making high-labor cost countries more attractive for production [4]. The following sections outline the theoretical foundations, the architecture of the AI-driven system and insights on its feasibility and potential advantages.

2 Background

The proposed framework of this paper builds upon three interconnected fields of research. First, the concept of autonomy is defined in the context of manufacturing. Second, within the field of AI, Large Language Models (LLMs), Agentic AI and Multi-Agent-Systems (MAS) are examined as key technologies driving new forms of autonomy. Third, to derive the role of humans within autonomous production, insights from the field of human–machine interaction are considered.

2.1 Autonomy in Manufacturing

Autonomous systems are characterized by their ability to reach predefined goals independently without the need of human intervention or detailed programming, to adapt to humans during human–machine interaction and to obtain decisions [2, 4].

Bauer et al. are proposing a five-stage model to describe and characterize autonomous productions. The stages are defined by different maturity levels of twelve production features, such as maintenance or worker integration. For example, the maturity levels reach from “machine operation and control by worker” to “autonomous machines and self-organizing system” for the feature worker integration. The resulting five stages of autonomous production are *Stage 0 – Analog Production* (no automation or digitalization, worker does manual work), *Stage 1 – Transparent Production* (connected entities provide data, worker knows statuses), *Stage 2 – Flexible Production* (partial automation, modular set-up and data-driven insights), *Stage 3 – Semi-Autonomous Production* (systems act independently, worker monitors and optimizes processes) and *Stage 4 – Autonomous Production* (fully autonomous and adaptive, worker becomes orchestrator) [3]. A comparison between earlier and recent data reveals a noticeable development in the adoption of autonomous technologies in manufacturing. In 2019, 25% of experts rated their current state as stage 0 and 67% as Stage 1 [3]. According to a 2024 industry report by the National Association of Manufacturers (NAM) 12.5% indicate their operations as fully integrated and automated, 36.3% as partially autonomous [5]. This indicates a clear shift from early-stage implementation toward more advanced and operational forms of autonomy. Given the current industry progress, implementing the proposed framework of this paper stages 2 and 3 come into focus. However, advancing towards stage 4 poses additional challenges, such as more adaptive data and information handling [3]. Further, human–machine interaction becomes less intensive and the variability of lot sizes increases [3].

2.2 Large Language Models, Agentic AI and Multi-agent-Systems

Recent advances of LLMs towards Agentic AI and MAS have opened new opportunities to reach higher degrees of autonomy [6]. To outline their respective contributions for autonomous systems, they will be described in the following. LLMs build the technological basis for Agentic AI, which can autonomously execute tasks, whereas MAS are a collaboration of autonomous agents with the goal to solve complex tasks [7].

Large Language Models (LLMs)

LLMs are advanced neural networks trained on vast amounts of data to understand human-like language [8]. They are capable of performing a wide range of tasks such as question answering, summarization, information extraction, and reasoning [9]. Different optimization methods are Prompt Engineering, Retrieval Augmented Generation (RAG) or Fine-Tuning [10]. Distilled LLMs are compressed and accelerated LLMs with specific knowledge [11].

Agentic AI

The key aspects of Agentic AI are its autonomous and goal-oriented behavior, which allows interaction with the environment, the capability of learning and decision-making [12] and the optimization of workflows [1]. Compared to traditional agents, Agentic AI rapidly adapts in uncontrolled environments, becoming more resilient [13]. Furthermore, Agentic AI has the capability of memorizing provided information and utilizing it through self-supervised learning [13]. Different Agentic-AI modules are *specialized agents*, *advanced reasoning and planning*, *persistent memory* and *orchestration* [6]. Architectural approaches for Agentic AI are MAS, Hierarchical Reinforcement Learning and Goal-oriented Modular Architectures [13].

Multi-agent Systems (MAS)

As individual entities within MAS, LLM agents are defined as "autonomous, intelligent systems powered by […] LLMs that integrate modular components – reasoning, memory, cognitive skills, and tools – to solve complex tasks in dynamic and evolving environments" [7]. Multiple Agents operating as a group form the MAS. The MAS architectures can be network-based, centralized, hierarchical or distributed [12, 14].

2.3 Human–Machine Interaction

The goal of human–machine interaction is to design efficient, user-friendly systems by connecting the user, its tasks and tools with the environment [15]. Within manufacturing, LLMs enable more efficient workflows, precision and innovation [16]. The human-centered design process for interaction systems DIN EN ISO 9241–210 ensures human needs are met [17] with increasing degrees of autonomy. With a higher degree of autonomy the role of the workers evolves from manual machine operator to orchestrator [3]. Due to its ability to support workers, AI is classified as a human-centric technology [18], making it suitable for integration in human–machine interaction.

3 Conceptual Framework

The following conceptual framework provides the architecture of the AI-driven system. The key components, the functionality and the technical feasibility are explained below.

3.1 Framework Development and Key Components

The framework was developed using a deductive methodology, based on a systematic literature analysis of (1) existing Agentic AI architectures and paradigms and (2) specific requirements for utilizing LLMs in industrial productions for human–machine interaction. Based on the findings the custom framework for agentic AI for industrial productions was derived and contextualized.

The key Agentic AI modules *specialized agents*, *persistent memory* and *orchestration* were selected, as they enable the execution of complex tasks, context-aware interaction based on memory and the coordination of the task execution (derived from (1)) [6]. The overall architecture follows the MAS approach due to its suitability for dynamic environments, such as industrial productions (derived from (1)) [13]. The proposed system architecture consists of two main components: the User Interface (UI) and the AI-driven system. The UI allows users to interact with the system – input a prompt and receive output. Four sub-components constitute the AI-driven system:

1. Control Agent as *orchestration*: Responsible for classifying the user prompt
2. Enablers as *specialized agents*: Different tools for optimizing the LLM
3. Memory Database as *persistent memory*: Access to multiple databases, such as the knowledge used in the RAG tool or sensor data from machines
4. Switch Agents as *specialized agents*: Different agents containing distilled LLMs.

The Enablers are chosen as they represent common optimization methods for LLMs: Prompt Engineering, RAG Retrieval and Reasoning-Capabilities (derived from (1)) [10]. To specialize the AI-driven system for industrial production the components MES and ERP and SCADA, SPS and sensors of the Memory Database are chosen in accordance to the automation pyramid as defined by the ISA-95 standard (derived from (2)) [19]. The integration of the components User Feedback and RAG Knowledge further enhance the AI-driven system by incorporating additional knowledge of the machine operator and documents such as machine manuals (derived from (1) and (2)). The selection of the Switch Agents is based on general LLM applications, such as the Q&A Agent and Decision-Support Agent which can provide different decision actions (derived from (1)), and on specified LLM applications in the industrial sector such as the Energy Efficiency Agent (derived from (2)). Further Switch Agents can be implemented depending on specific requirements.

3.2 Framework of the AI-Driven System

Figure 1 illustrates the proposed framework using a Unified Modeling Language (UML) component diagram. The UML "lollipops" signify a provided interface by a component [20]. The corresponding required interfaces show a components need for an interface [21]. The AI-driven system is not specified for one specific role of an employee. For example, users can be production workers or production managers. Users can also work on multiple machines throughout the production, as the AI-driven system is not specified to a singular machine, but rather for the production system as a whole.

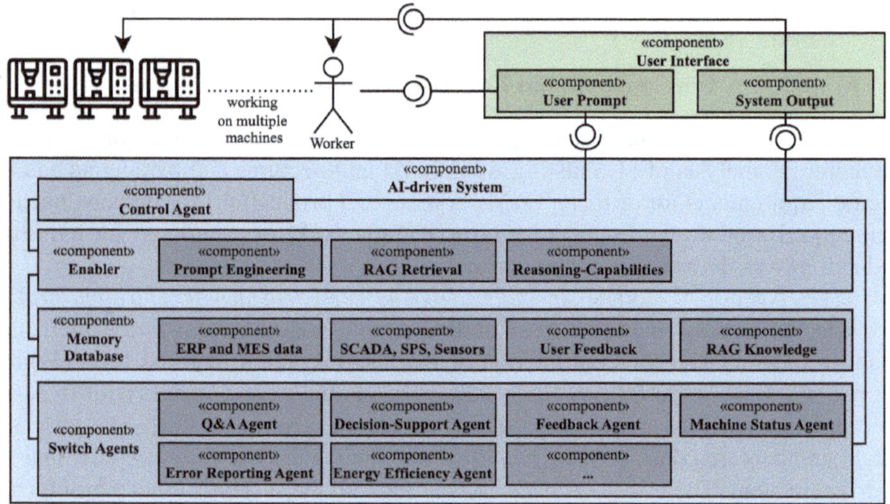

Fig. 1 Framework of the AI-driven system

Once the user provides an interface, which is the user prompt in this framework, to the UI «component», it is forwarded to the AI-driven system «component». The Control-Agent then classifies the prompt [14], chooses an Enabler and the needed Memory Databases and finally orchestrates it to a specified Switch Agent. The agents are able to communicate with each other through the Control Agent [14]. The Switch Agents generate the system output and forward it to the corresponding component. Finally, the user receives the system output via the UI and is provided to the machines if, for example, the decision to adjust machine parameters is made.

With the proposed structure of the framework, it is ensured that firstly answers are generated efficiently with the best optimization method (Enabler), secondly that the answer is enriched with relevant data (Memory Database) and lastly with the context of the data the right specializes agents can generate an accurate and precise response (Switch Agents). The sequence is essential because each step provides the necessary foundation for the next, ensuring efficiency, factual relevance and domain-specific and contextual accuracy.

All the Switch Agents are distilled LLM Agents that are optimized for their respective task. By using distilled LLM agents, the energy consumption can be reduced and efficiency enhanced [11]. As distilled LLMs require less computing power, they present a viable solution for manufacturing companies with limited computing power and an opportunity to run the system locally, ensuring data security [22]. The Enablers allow the AI-driven system to adjust the Switch Agents to the user prompt. For example, a simple user input can be answered with the Q&A Agent with an optimized system prompt by the enabler Prompt Engineering, while a more complex user input requires reasoning-capabilities. Each Switch Agent can be optimized with each Enabler. The Memory Database can provide the Switch Agents with further information for generating a reliable system output. For example, a user prompt asking about the status of the production process of a specific order needs access to data about the order from the ERP-database. Inquiring the system with a reoccurring error of a specific machine, the User Feedback data might be needed. Each Switch Agent can be fed with data from multiple Memory Databases. The annotation of metadata in the Memory Database allows for more transparency in the generated answer, since every generated response provides the corresponding metadata. Malfunctions of the LLMs, such as hallucinations, can be reduced by applying tailored optimization methods through the Enabler Agents.

Unlike general-purpose LLMs like ChatGPT, the Agentic AI framework exceeds simple prompt answering – it tracks user goals, integrates domain-specific knowledge, and executes multi-step tasks autonomously.

3.3 System Architecture and Technical Feasibility

The system architecture of the proposed framework is based on a hierarchical MAS structure [14] as the Control Agent orchestrates the Switch Agents. The modules of the Agentic-AI system are *specialized agents* («component» Switch Agents), *persistent memory* («component» memory database) and *advanced reasoning* («component» reasoning-capabilities). This allows the system to autonomously retrieve, process, and adapt knowledge based on the user input and the gathered data, enabling dynamic and context aware decision support.

The AI-driven system is implemented in Python with pertinent libraries. At the core of the AI-driven system, the LangChain framework from LangGraph is used for building the MAS workflows and integrating the tools and memory database [23]. Semantic search and long-term memory is realized using vector databases such as FAISS or chroma [24]. Communication between the interfaces of the components are realized through REST APIs using HTTP for the bidirectional Client–Server-Communication [25]. For agent orchestration, the LangChain Supervisor Agent is implemented [26]. Data for the memory database is retrieved through standardized protocols such as Open Platform Communications Unified Architecture (OPC UA) and Asset Administration Shell (AAS) [27].

4 Initial Insights and Evaluation

The insights and evaluation of the AI-driven system are based on the conceptual framework, respective to preliminary stage of the system realization. An empirical evaluation through practical application of the AI-driven system is planned, as the framework has to date been evaluated solely through analytical methods due to its conceptual nature.

4.1 Evaluation of Impact Through Simplified Use Cases

To evaluate the impact of the AI-driven system in the production environment, four simplified use-cases are assessed. Table 1 shows the evaluation overview of the system.

Table 1 Evaluation of AI-driven system through simplified use-cases

Use-case		Without AI-driven system	With AI-driven system	Utilized sub-components
1. Knowledge management and documentation	WP	Manual search through handwritten or typed documentation, fragmented data sources	Natural language prompts to retrieve machine- or task-specific information	*Enabler*: RAG retrieval *Memory Database*: User Feedback and RAG knowledge *Switch Agent*: Q&A
	PE	Medium cognitive load, time-intensive, error-prone	Low cognitive load, timely, error-tolerant	Agent and feedback agent
2. Energy efficiency optimization	WP	Data analysis and optimization depend on periodic expert reviews, no real-time feedback	Continuous monitoring and analysis of data, real-time suggestions	*Enabler*: Reasoning-capabilities *Memory Database*: SCADA, SPS, sensors *Switch Agent*: Energy
	PE	High cognitive load, time-intensive, reliable	Low cognitive load, timely, error-tolerant	Efficiency agent
3. Error reporting automation	WP	Manual error logging, inconsistent terminology, troubleshooting through trial and error	Error classification, suggestions based on historical data, guided troubleshooting	*Enabler*: RAG retrieval *Memory Database*: User feedback and RAG knowledge *Switch Agent*: Error re
	PE	High cognitive load, time-intensive, error-prone	Low cognitive load, time-consuming, error-tolerant	porting agent

Since the proposed framework is primarily conceptual, the evaluation is based on the qualitatively modeled work processes (WP) and the projected effort (PE) for the worker. The PE is based on the cognitive load (high, medium, low), amount of time (time-intensive, time-consuming, timely) and error-proneness (error-prone, reliable, error-tolerant) as these are the critical aspects for workers in the production environment. Throughout all use cases there is potential for the cognitive load, needed time and error-proneness to be reduced, leading to a more resilient and efficient production [13].

4.2 Insights on the Feasibility and Benefits

The integration of LLMs in production demonstrates promising feasibility, especially in enhancing human–machine interaction and decision support. Our initial implementation of LLMs indicates they can effectively assist in specific knowledge transfer, question answering and assisting the machine worker with general tasks, such as generating G-Code. The potential benefits include increased flexibility, reduced operator workload, and improved process efficiency. However, challenges such as system reliability, real-time performance, and domain-specific adaption need addressing to fully realize the advantages.

5 Conclusion and Outlook

While primarily conceptual, this work provides a technical framework for the incorporation of agentic AI for a more autonomous production. By leveraging LLMs, specialized agents and memory databases, the proposed system facilitates knowledge transfer, decision support and options and context-aware interaction through natural language. The framework contributes to the progression along the autonomy stages in production as it implements autonomous agents, allowing the human to orchestrate the shopfloor [3]. The key contribution of the Agentic AI framework lies in its modular, agent-based structure tailored to production environments combining adaptive optimization, context-aware enrichment and domain-specific response generation across the production system as a whole. By implementing dynamic agents that can coordinate tasks and pass on information, rather than a rigid system, the AI-driven system allows for a higher stage of autonomy in the production, depending on the depth and extent to which machines and systems are integrated.

In future work, the framework will be implemented, validated and iteratively refined through empirical user studies, involving different roles of production employees. The evaluation will focus on usability, system acceptance, and cognitive load, as well as technical aspects such as robustness and error-resistance of the AI-driven system. The approach holds potential for a human-centric, flexible and more efficient manufacturing landscape with higher autonomy.

Acknowledgements. This research was part of the project "etaGPT" which is funded by the German Federal Ministry of Economic Affairs and Energy and Projektträger Jülich and managed by etalytics GmbH. The authors are responsible for the content of this publication. The authors would like to thank the "etaGPT" project partners for their content support.

Competing Interests. The author(s) has no competing interests to declare that are relevant to the content of this manuscript.

References

1. Hosseini, S., Seilani, H.: The role of agentic AI in shaping a smart future: a systematic review. Array **26**, 100399 (2025). https://doi.org/10.1016/j.array.2025.100399
2. Mo, F., Monetti, F.M., Torayev, A.: A maturity model for the autonomy of manufacturing systems. Int. J. Adv. Manuf. Technol. **126**, 405–428 (2023). https://doi.org/10.1007/s00170-023-10910-7
3. Bauer, D., Schumacher, S., Gust, A., Seidelmann, J., Bauernhansl, T.: Characterization of Autonomous Production by a Stage Model. Procedia CIRP. **81**, 192–197 (2019). https://doi.org/10.1016/j.procir.2019.03.034
4. Behrens, B.-A., Groche, P., Krüger, J., Wulfsberg, J.P.: WGP-STANDPUNKT INDUSTRIEARBEITSPLATZ 2025, **30** (2018)
5. Brousell, D.R.: Survey: smart factories are still a work in progress. https://manufacturingleadershipcouncil.com/survey-smart-factories-are-still-a-work-in-progress-2-35893/. Accessed 12 Aug. 2025
6. Sapkota, R., Roumeliotis, K.I., Karkee, M.: AI Agents vs. Agentic AI: a conceptual taxonomy, applications and challenges (2025). https://doi.org/10.48550/arXiv.2505.10468
7. Bousetouane, F.: Agentic systems: a guide to transforming industries with vertical AI agents (2025). https://doi.org/10.48550/arXiv.2501.00881
8. Nazir, A., Wang, Z.: A comprehensive survey of ChatGPT: advancements, applications, prospects, and challenges. Meta-Radiol. **1** (2023). https://doi.org/10.1016/j.metrad.2023.100022
9. Kalyan, K.S.: A survey of GPT-3 family large language models including ChatGPT and GPT-4. Nat. Lang. Process. J. **6**, 100048 (2024). https://doi.org/10.1016/j.nlp.2023.100048
10. Gao, Y., Xiong, Y., Gao, X.: Retrieval-augmented generation for large language models: a survey (2024). https://doi.org/10.48550/arXiv.2312.10997
11. Liu, J., Zhang, C., Guo, J.: DDK: Distilling domain knowledge for efficient large language models (2024). https://doi.org/10.48550/arXiv.2407.16154
12. Horne, D.: The agentic AI mindset—a practitioner's guide to architectures, patterns, and future directions for autonomy and automation. (2025)
13. Acharya, D.B., Kuppan, K., Divya, B.: Agentic AI: autonomous intelligence for complex goals—a comprehensive survey. IEEE Access **13**, 18912–18936 (2025). https://doi.org/10.1109/ACCESS.2025.3532853
14. Ghorbani, J., Fallah, Y.P., Choudhry, M.A.: Investigation of communication media requirements for self healing power distribution systems. In: 2013 IEEE Energytech. pp. 1–7. IEEE, Cleveland, OH, USA (2013). https://doi.org/10.1109/EnergyTech.2013.6645325
15. Butz, A., Krüger, A., Völkel, S.T.: Mensch-Maschine-Interaktion. De Gruyter Oldenbourg, Berlin, Boston (2022)
16. Li, Y., Zhao, H., Jiang, H.: Large language models for manufacturing (2024). https://doi.org/10.48550/arXiv.2410.21418
17. DIN EN ISO 9241-210. Ergonomie der Mensch-System-Interaktion (2020)
18. European Commission: ERA industrial technologies roadmap on human-centric research and innovation for the manufacturing sector. Publications Office, LU (2024)
19. Martinez, E.M., Ponce, P., Macias, I., Molina, A.: Automation pyramid as constructor for a complete digital twin, case study: a didactic manufacturing system. Sensors **21**, 4656 (2021). https://doi.org/10.3390/s21144656

20. Donohoe, P. (ed.): Software Architecture. Springer US, Boston, MA (1999). https://doi.org/10.1007/978-0-387-35563-4
21. Bjerkander, M., Kobryn, C.: Architecting systems with UML 2.0. IEEE Softw. **20**, 57–61 (2003). https://doi.org/10.1109/MS.2003.1207456
22. Cantini, R., Orsino, A., Talia, D.: Xai-driven knowledge distillation of large language models for efficient deployment on low-resource devices. J. Big Data **11**, 63 (2024). https://doi.org/10.1186/s40537-024-00928-3
23. LangChain: Overview. https://langchain-ai.github.io/langgraph/concepts/multi_agent/. Accessed 3 June 2025
24. LangChain: Vector stores | LangChain, https://python.langchain.com/docs/integrations/vectorstores/. Accessed 3 June 2025
25. Kumar, S.: A Review On Client-Server Based Applications and Research Opportunity. Int. J. Sci. Res. 10, 33857–33862 (2019). https://doi.org/10.24327/ijrsr.2019.1007.3768.
26. LangChain: Multi-agent | LangChain. https://langchain-ai.github.io/langgraph/agents/multi-agent/. Accessed 3 June 2025
27. Busboom, A.: Automated generation of OPC UA information models—a review and outlook. J. Ind. Inf. Integr. **39**, 100602 (2024). https://doi.org/10.1016/j.jii.2024.100602

Quality Assurance for the Serial Production of Metallic Bipolar Plates Using Artificial Neural Networks

H. Krauss(✉), K.R. Riedmüller, and M. Liewald

Institute for Metal Forming Technology,, University of Stuttgart, Stuttgart, Germany
heiko.krauss@ifu.uni-stuttgart.de

Abstract. The availability of powerful and clean energy sources for the electrification of passenger and commercial vehicle transportation represents a crucial requirement for achieving current climate targets. Fuel cell systems are one example of such energy sources for modern drive systems, which is particularly important in the field of heavy-duty applications. In fuel cell systems, bipolar plates (BPPs) take on a crucial function by supplying reaction gases and discharging the produced water. However, even slightest fluctuations of process parameters can lead to typical forming defects during the production of the BPPs such as tearing and springback. Consequently, there is an ongoing need to develop methods for detecting and preventing these defects during the production process. Against this background, this paper presents a system for quality assurance and process control for the serial production of metallic BPPs. This involves a digital twin of the forming process using an artificial neural network (ANN) to control the forming process by providing specific recommendations for process parameters that will lead to a higher dimensional accuracy of the formed BPPs. Multiple simulations of the embossing process of BPPs with varying process parameters such as the friction coefficient and blank holder force were performed to first create a process database. This database was subsequently used to train the ANN of the digital twin. Finally, it was shown that the ANN can predict process parameters and is therefore suitable to be used by a control algorithm to recommend process parameter adjustments.

Keywords: Artificial neural networks · Foil forming · Metallic bipolar plates · Single-stage embossing

1 Introduction

Fuel cells convert hydrogen and oxygen into electrical energy and heat, producing water as a byproduct. A conventional proton-exchange membrane fuel cell (PEMFC) consists of two electrodes separated by an electrolyte, which is usually a membrane, and a BPP enclosing the membrane electrode assembly (MEA). Finally, several of such PEMFCs are combined to fabricate PEMFC stacks. A

L. Overmeyer and B.-A. Behrens (eds.), *Production at the Leading Edge of Technology*, Lecture Notes in Production Engineering, https://doi.org/10.1007/978-3-032-19524-1_52

BPP has several functions, including the electrical contacting of the individual MEAs, the supply of the reaction gases and the discharge of the resulting water. Because of these functions, BPPs have to meet high requirements in terms of gas impermeability, electrical conductivity as well as corrosion resistance [1]. In order to meet these requirements, BPPs must offer special properties, particularly in regard to their susceptibility to corrosion resistance, so that graphite, carbon composites, stainless steel, titanium or aluminum alloys, for example, are currently considered as BPP materials. On the other hand, a relatively high dimensional accuracy is required, since the BPPs significantly influence the dimensions and costs of the final PEMFC stack [2].

Today, BPPs from stainless steel are predominantly used in PEMFCs, which are manufactured by combining two monopolar plates (MPPs). The MPPs are first produced separately by hollow embossing, hydroforming, hollow embossing rolling, electromagnetic forming, incremental micro forming or milling and then joined by welding [1]. Here, hollow embossing is particularly economical for high-volume production due to its short cycle times. However, producing the high-precision MPPs from metallic foils in large quantities poses serious challenges. The embossing process reacts sensitively to small fluctuations in material or process parameters owing to the small thickness of the metallic foil and the complex geometry of the MPP. If not controlled, these fluctuations can lead to cracks or wrinkles in the MPPs or cause poor dimensional accuracy [1,3]. Therefore, extensive quality assurance is required for the serial production of metallic BPPs, which leads to longer production times and higher costs if done manually.

Against this background, the aim of the present work was to introduce a system for quality assurance and process control for the serial production of metallic BPPs. However, process parameters that cannot be measured directly, such as the coefficient of friction, are particularly difficult to monitor and therefore to control. In addition to contact forces, temperature, actual contact surface and relative speed, friction is primarily influenced by the amount of lubrication. Therefore, an indirect determination of the coefficient of friction using these measurable variables and its subsequent adaptation by adjusting the amount of lubrication is generally possible. For this purpose, this paper proposes an ANN that predicts actual friction coefficients based on the full-surface measured geometry of formed MPPs. Following this prediction, ideal friction conditions for the process can be set by adjusting the amount of lubricant using a control algorithm.

2 Related Work

The use of ANNs instead of finite element (FE) simulations for numerical modelling of metal forming processes is a current area of research. ANNs are of interest in this context as they provide faster predictions than FE simulations. Faster predictions in turn enable shorter design iterations, inline control and therefore a better optimization of the metal forming process.

Previous research work using ANNs for metal forming process displacement field and material thinning field predictions relied on network architectures incorporating convolutional neural networks (CNNs). In [4], a CNN architecture was used to predict images of material thinning fields and displacement fields for a simple cup geometry. The die, blank and blank holder geometries were projected onto a plane and encoded as an image. Process parameters like temperature and punch velocity were also encoded in images and used with forming tool images as inputs of the CNN. The training dataset was generated from FE simulations.

ANNs can also be used to optimize the process design or process parameters. In [5], for example, the geometric parameters of a textile draping forming process were optimized by using two CNNs sequentially. For this purpose, the surfaces of the forming tool were encoded as images, from which the first CNN predicted strain field images. Subsequently, a second CNN used these generated strain field images to predict optimal rubber pad positions. Furthermore, [6] introduced a digital twin of a deep drawing process that incorporated a CNN and an optimization algorithm to improve the part quality by adjusting local friction inline. The CNN predicts, similar as in [4], images of displacement and thinning fields. It was trained by FE simulations that were validated in real experiments. This CNN was then used as a surrogate model of the metal forming process in an optimization algorithm that applied a loss function to penalize thinning, additional lubrication and changes in blank holder force. However, since changes of the blank holder force were disfavored here, only the local lubrication was optimized. Another limitation was that the digital twin used thinning as part quality metric. But the thinning was only measured manually with a thickness gauge to validate the FE simulation, which is not feasible for complex geometries of BPPs.

In summary, previous research has employed ANNs as a replacement for FE simulations to enable faster predictions. These predictions were then used with an optimization algorithm to determine optimal process parameters offline. To the best of the authors knowledge, there is currently no work that predicts unmeasurable process parameters such as the friction coefficient inversely from formed geometries to control parameter fluctuations inline. Since previous research has shown, that CNNs are particularly suited for predictions with surfaces, the ANN employed in this work utilizes a network architecture based on CNNs.

3 Method

The quality assurance and control system proposed in this work is based on [7] and shown in Fig. 1. A holographic sensor array measures the geometry of a produced MPP within $100ms$ and an accuracy of $1\mu m$ in height [8]. The measured geometry is used to calculate the dimensional deviation to the reference geometry for quality assurance. The dimensional deviation is then used by an ANN to predict actual process parameters. This enables the developed system to adjust the lubrication with a conventional control algorithm to possible deviations in friction conditions that may arise, for example, due to temperature changes or minor fluctuations in material properties during the forming process. The ANN is based on data from a digital twin of the MPP manufacturing process, which is generated from FE simulations with varying friction coefficients and blank holder forces. This contribution only focuses on these simulation data in order to demonstrate the general effectiveness of the quality assurance and control system.

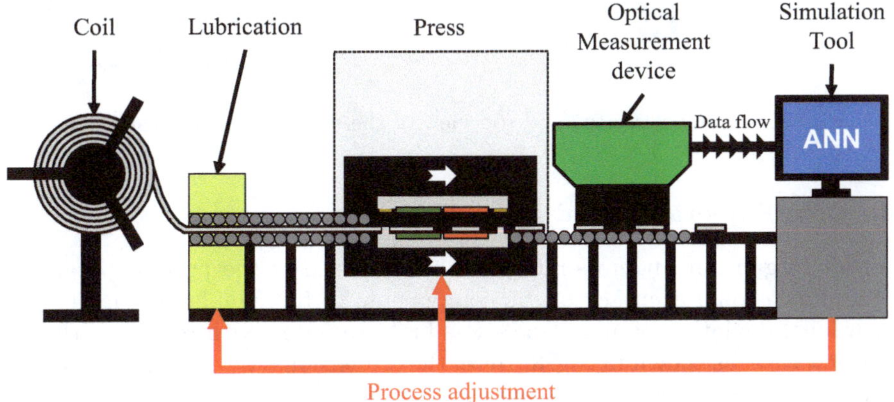

Fig. 1. Schematic representation of the quality assurance and control system [7]

3.1 FE Simulation

To create a dataset for the ANN training, FE simulations of the hollow embossing forming process were carried out with varying parameters. The simulation model used was based on a model developed in previous work of the co-authors [7,9] and already validated with experimental tests. The 3-Parameter Barlat material model from [9] was chosen to model the stainless steel 1.4404 with a thickness of $100\mu m$ as the MPP material. Tool surfaces and the academic plate design of the MPP are shown in Fig. 2. The academic plate design incorporates key features of bipolar plates, but does not possess any specific functional characteristics and serves primarily for validation and testing purposes [9]. As a supplement to [9], a blank holder was added to gain a further process parameter with the blank holder force, which can also be adjusted to improve the dimensional accuracy of the formed MPP. FE simulations of MPPs generally are computationally intensive due to the fine meshing required to model the complex surface geometry with small radii of $0.1mm$. Therefore, only the two process parameters friction coefficient and blank holder force were varied in this study in the ranges of $0.0 - 0.3$ and $0kN - 100kN$. In total, 130 simulations were carried out using the LS-OPT Software and space filling sampling.

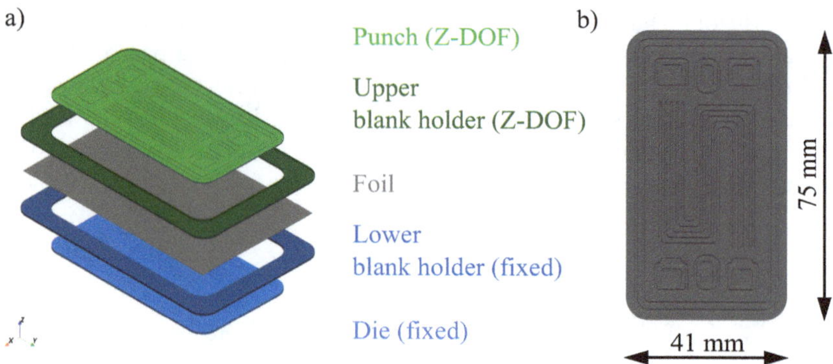

Fig. 2. Tooling setup (**a**) and top view of the academic plate design (**b**)

3.2 Data Exploration and Preparation

In order to gain an understanding on the influence of the process parameters on the dimensional accuracy of the components to be manufactured, the mean dimensional deviation of all samples was first evaluated with regard to the best and worst case, as shown in Fig. 3. Here, the dimensional deviation is the nearest distance between nodes of the simulated and the specified reference surface. Since the hollow embossing process was simulated with shells, the true surface had to be calculated from the shell thicknesses and the nodal coordinates of the mesh. The nearest distances were then determined by finding the nearest reference nodes with a k-d tree and calculating the Euclidean distances to the respective simulation nodes. All samples are triangulated so that dimensional deviations can be interpolated linearly, providing a clearer visualization of the response surface. Figure 3 shows that the dimensional deviation rises with very low friction and blank holder force. A possible explanation for this is that a low coefficient of friction and blank holder force result in a low retention force, leading to a higher draw-in of the metal foil. With a higher draw-in in turn, internal stresses arise due to tangential stresses, which are higher near the edges of the blank. These stresses lead to springback, as can be seen in the worst case of Fig. 3. Furthermore, very high retention forces lead to higher thinning at sharp edges of the MPP and thus also to a higher dimensional deviation.

3.3 Modeling

In the present work, a network architecture was used that employs CNNs to predict process parameters based on images of dimensional deviations. The ResNet18 [10] applied for this purpose is a commonly used network architecture well suited for image classification tasks. To make the network suitable for regression, the softmax activation function was replaced by a sigmoid activation function. The sigmoid activation function was chosen to constrain all outputs between zero and one, thus preventing the network from predictions outside the

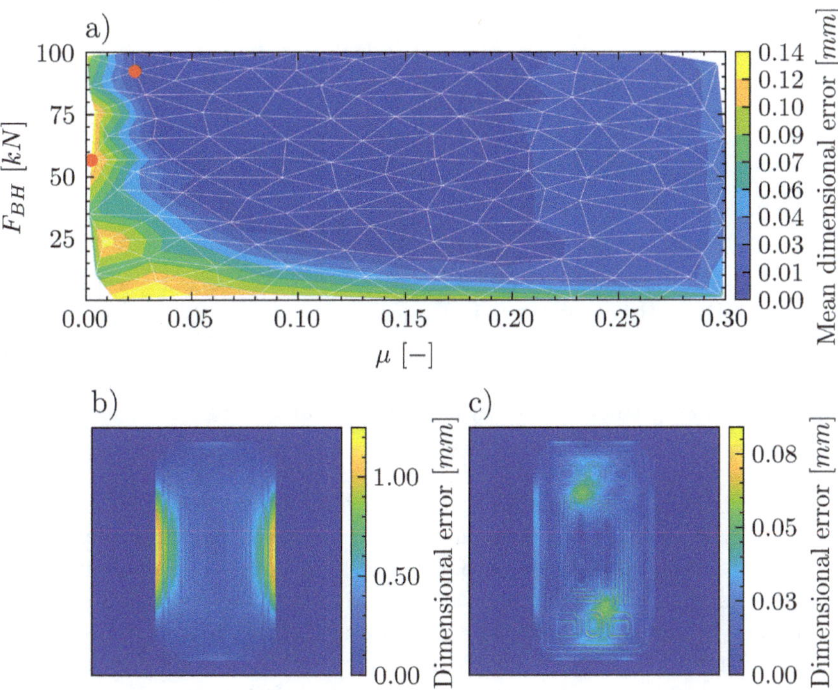

Fig. 3. Mean dimensional deviations of the simulated academic plate design (**a**) and the respective worst case (**b**) and best case (**c**)

specified process parameter ranges. Consequently, the process parameters were scaled with a MinMaxScaler to the sigmoid output range. The Adam optimizer was used with a learning rate of 1e-5 and mean squared error (MSE) as the loss function instead of categorical cross-entropy. The model was trained with an 80:20 validation split and a batch size of 16 for 500 epochs.

4 Results

Fig. 4 shows the predictive performance of the derived ResNet18-Regressor model. The mean absolute error (MAE) of the predicted static friction coefficient is 0.051 (1.7%) for the training dataset and 0.029 (9.5%) for the test dataset. The MAE of the predicted blank holder force is 3.1kN (3.1%) for the training dataset and 12kN (12%) for the test dataset.

5 Discussion

As can be seen in Fig. 4, the proposed model is able to predict process parameters based on detected dimensional deviations. However, due to the small size of the training dataset, the model tends to overfit. In the current state, the

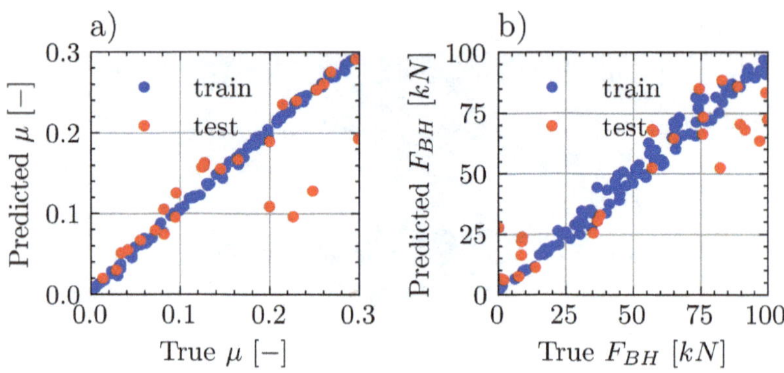

Fig. 4. Model predictions of static friction (**a**) and blank holder force (**b**)

proposed ANN is only applicable to the plate design used for model training. This means that separate models have to be trained for other plate designs. To incorporate characteristics of other plate designs, the model input could be extended by images of the reference plate design or forming tool surfaces. This would further increase the demand for training data, which is bottlenecked by long simulation times of MPPs. With the current state of MPP FE simulation it is therefore advisable to train models for specific MPP designs instead of training a generalizable model. But in principle, the proposed method or model could be applied to other forming processes, like deep drawing, where the dimensional deviation can be encoded in images. The decisive factor is that the resulting part of the forming process has a unique dimensional deviation that is influenced by friction conditions. This is necessary since inverse predictions are not possible if the same dimensional deviation relates to different process parameters.

6 Conclusion

This work introduces a quality assurance and control system that uses an ANN to predict actual process parameters from dimensional deviations. These parameters can then be used by a control algorithm to stabilize the hollow embossing process and ensure optimal part quality. A dataset consisting of formed MPP surfaces as well as their corresponding static friction coefficients and blank holder forces was obtained through 130 FE simulations and space-filling sampling. The ANN, which is used to predict process parameters, is based on a modified ResNet18 architecture for regression. The ANN can predict process parameters from dimensional deviations but tends to overfit due to the small training dataset. Future work will involve experiments using the quality assurance and control system to investigate whether adjusting lubrication based on predicted friction coefficients can stabilize the hollow embossing process. Data generation strategies, such as using partial MPP surfaces, sampling based on dimensional deviations, or replacing FE simulations by an ANN, could be explored to generate more data samples and reduce overfitting. Additionally, the dataset could be extended to include material parameter variations if experiments show that these fluctuations affect the dimensional accuracy of the formed MPP surface significantly.

Acknowledgments. This work was supported within the "DFG-Fraunhofer Transfer Programme", established by the German Research Foundation (DFG-Project 460294948). Authors feel grateful for the financial support.

References

1. Porstmann, S., Wannemacher, T., Drossel, W.G.: A comprehensive comparison of state-of-the-art manufacturing methods for fuel cell bipolar plates including anticipated future industry trends. J. Manuf. Processes **60**, 366–383 (2020). https://doi.org/10.1016/j.jmapro.2020.10.041
2. James, B., Huya-Kouadio, J., Houchins, C., Gomez, M.: Mass Production Cost Estimation of Direct H_2 PEM Fuel Cell Systems for Transportation Applications (2017 - 2021) (Final Report) (2021). https://doi.org/10.2172/2309724
3. Alo, O.A., Otunniyi, I.O., Pienaar, H.: Manufacturing methods for metallic bipolar plates for polymer electrolyte membrane fuel cell. Mater. Manuf. Process. **34**(8), 927–955 (2019). https://doi.org/10.1080/10426914.2019.1605170
4. Attar, H.R., Zhou, H., Foster, A., Li, N.: Rapid feasibility assessment of components to be formed through hot stamping: A deep learning approach. J. Manuf. Processes **68**, 1650–1671 (2021). https://doi.org/10.1016/j.jmapro.2021.06.011
5. Zimmerling, C., Poppe, C., Kärger, L.: Estimating optimum process parameters in textile draping of variable part geometries - a reinforcement learning approach. Procedia Manufacturing **47**, 847–854 (2020). https://doi.org/10.1016/j.promfg.2020.04.263
6. Link, P., Penter, L., Rückert, U., Klingel, L., Verl, A., Ihlenfeldt, S.: Real-time quality prediction and local adjustment of friction with digital twin in sheet metal forming. Rob. Comput. Integr. Manuf. **91**, 102,848 (2025). https://doi.org/10.1016/j.rcim.2024.102848
7. Beck, M., Riedmüller, K.R., Liewald, M., Bertz, A., Aslan, M.J., Carl, D.: Investigation on the Influence of Geometric Parameters on the Dimensional Accuracy of High-Precision Embossed Metallic Bipolar Plates, pp. 427–438. Springer International Publishing (2023). https://doi.org/10.1007/978-3-031-18318-8_44
8. Fratz, M., Beckmann, T., Anders, J., Bertz, A., Bayer, M., Gießler, T., Nemeth, C., Carl, D.: Inline application of digital holography [invited]. Appl. Optics **58**(34), G120 (2019). https://doi.org/10.1364/ao.58.00g120
9. Beck, M., Karadogan, C., Cyron, P., Riedmüller, K.R., Liewald, M.: Study on the numerical prediction quality of material models regarding springback of hollow embossed metallic bipolar half-plates. In: Material Forming: ESAFORM 2024, vol. 41, pp. 1472–1481. Materials Research Forum LLC (2024). https://doi.org/10.21741/9781644903131-163
10. He, K., Zhang, X., Ren, S., Sun, J.: Deep residual learning for image recognition. In: 2016 IEEE Conference on Computer Vision and Pattern Recognition (CVPR), pp. 770–778. IEEE (2016). https://doi.org/10.1109/cvpr.2016.90

Concept of a Process Model for Data-Driven Root Cause Analysis in Manufacturing

Dominik Pietsch[1,2(✉)], Uwe Wieland[1,3], and Steffen Ihlenfeldt[2]

[1] Volkswagen AG, Wolfsburg, Germany
dominik.pietsch@volkswagen.de
[2] TUD Dresden University of Technology, Dresden, Germany
[3] HTWD-University of Applied Sciences, Dresden, Germany

Abstract. In manufacturing data-driven root cause analysis (RCA) is a complex process due to the spatial and temporal separation of causes and effects and the difficulty in identifying corresponding data patterns. In addition, dynamic changes in manufacturing processes constantly alter cause-and-effect relationships (CERs), further complicating analysis. To support the implementation and development of data-driven RCA projects in manufacturing, this paper develops a dedicated process model (PM). This includes creating a role, artifact, activity, and phase model, and integrating these submodels into a single, coherent PM. The PM streamlines the implementation of data-driven RCA projects in manufacturing through clear responsibilities, artifacts, and activities. The PM is validated through an automotive final assembly RCA project.

Keywords: Process model · Root cause analysis · Data-driven · Manufacturing

1 Introduction

The increasing digitalization of manufacturing, particularly since the emergence of Industry 4.0, has made it possible to optimize costs, time, and product quality by using intelligent data analysis methods [2]. Despite the significant potential of using machine learning (ML) for intelligent data analysis in manufacturing, the success rate of data science projects remains low [10]. This also affects the execution of projects in the area of data-driven RCA. RCA, traditionally a knowledge-driven process, is undergoing a digital transformation towards data-driven approaches in recent years [11]. In this context, data-driven RCA aims to identify the root causes of manufacturing errors using data. The combination of low success rates in data-driven projects on the one hand, and the increasing shift towards data-driven RCA in manufacturing on the other, complicates the realization of projects in this domain. Therefore, this work aims to develop a dedicated PM for data-driven RCA projects in manufacturing. This PM clarifies responsibilities, objectives, and methods to facilitate project execution.

L. Overmeyer and B.-A. Behrens (eds.), *Production at the Leading Edge of Technology*, Lecture Notes in Production Engineering, https://doi.org/10.1007/978-3-032-19524-1_53

The rest of the paper is structured as follows. Section 2 describes the components of PMs and discusses related work. Section 3 describes the concept of a PM for data-driven RCA in manufacturing. Section 4 presents a validation of the developed PM through a use case in the automotive industry. Finally, Sect. 5 discusses the key findings and areas for further research.

2 Process Models for Data-Driven Projects

PMs enable the repeatable and quality-assured execution of projects [4], aiming to reduce costs, ensure high-quality outcomes, and enhance communication with stakeholders [3]. They provide a structured framework that defines project progression and includes the necessary artifacts, activities, and roles [3]. From this, five sub-models can be derived: artifact model, activity model, phase model, role model, and the model for methods, tools, and standards [7].

An artifact is a primary or secondary work result that is created, modified, or used through activities, with specific roles responsible for them [1]. In PMs, roles represent skill profiles rather than individuals. This allows one person to assume multiple roles, and vice versa [1]. Artifacts are processed using methods. A method is a set of activities that support project execution within the PM. Phases represent the temporal grouping and arrangement of activities and artifacts [7].

In industry and science, there exists a plethora of PMs for executing data-driven projects. The cross-industry standard process for data mining (CRISP-DM) [16] is the de facto standard in this regard [14]. Several extensions have been developed to adapt CRISP-DM for manufacturing and engineering. For instance, Huber et al. [6] introduced the data mining methodology for engineering applications (DMME), adding phases such as technical understanding, realization, and implementation to better address engineering contexts. Building on DMME, the data analytics production line optimization model (DAPLOM) [5] restructures the DMME phases and then links them to specific methods for production line optimization. Schoch [13] proposed a PM that focuses on predictive quality management in the automotive sector and therefore adds a data development phase to CRISP-DM. Lundén et al. [9] enhanced CRISP-DM by integrating domain expert (DE) validation during data preparation and data acquisition in manufacturing. Kornas [8] developed a quality management system for battery cell production that combines expert knowledge with process data in order to reduce scrap. The presented data analytics self-service tool for process experts is based on CRISP-DM. Furthermore, CRISP-DM has been integrated into Six Sigma's DMAIC cycle by Wieland et al. [15].

Although there are numerous PMs for data science projects in manufacturing, there is no dedicated, process-independent PM for data-driven RCA in manufacturing. Existing PMs fail to fully define roles, artifacts, activities, or phases and offer limited methodological support for RCA. They also inadequately address dynamic manufacturing environments with evolving CERs, which result in model degradation over time. Furthermore, current PMs lack guidance on integrating explainable ML methods and only partially integrate domain knowledge. Both are essential for a robust and trusted ML model for data-driven RCA. To fill this gap, this paper proposes a PM dedicated to data-driven RCA in manufacturing that emphasizes explainability, domain knowledge integration, and adaptability to changing manufacturing conditions.

3 Process Model Development

Before developing the PM for data-driven RCA in manufacturing, it is essential to outline the requirements and constraints. Based on the related work discussed in Sect. 2, three key requirements emerge: the integration of domain knowledge (R_1), the incorporation of explainable ML methods (R_2), and the representation of the entire ML model lifecycle, from conception and development to deployment, monitoring and maintenance to ensure adaptability in dynamic manufacturing environments (R_3).

In addition to these three requirements, the study introduces four constraints that narrow the scope. First, the PM only describes ML-relevant artifacts, activities, and phases (C_1). Consequently, project management activities are not considered. Second, the focus is limited to three core roles (C_2). Roles such as project managers or software developers are excluded. Third, the current version of the PM concept only partially describes individual phases, activities, and artifacts, and it is formulated in a method-neutral way (C_3). This limitation will be addressed in future work. Lastly, the PM is specifically tailored to continuous RCA problems in manufacturing (C_4). Unlike one-time RCA, the focus is on continuous manufacturing processes that repeatedly produce errors. The goal is to use ML models to identify the root causes of these errors and employ them as a decision support system for process experts.

The following sections introduce the concept of a PM for data-driven RCA in manufacturing. Section 3.1 presents the role model. Sections 3.2, 3.3 and 3.4 describe the three main parts of the PM, their activities, and their artifacts.

3.1 Role Model

Three core roles are defined for the execution of data-driven RCA projects in manufacturing. The first role is the data scientist (DS), whose primary objective is to develop the data-driven root cause analysis model (RCAM). This role requires expertise in statistics, ML, model selection, training, and optimization, as well as programming skills (e.g., Python), data visualization and interpretation, and strong analytical thinking capabilities. The second role is the ML operations engineer (MLE), who is responsible for provisioning, integrating, and preparing relevant manufacturing data, as well as deploying ML models into runtime environments. The MLE role encompasses competencies in extract, transform, load (ETL) processes, ML operations (MLOps), as well as continuous integration and continuous delivery (CI/CD). The final role is that of the DE, who contributes process and manufacturing knowledge and can validate analytical results. This role requires an in-depth understanding of manufacturing processes and relevant CERs for RCA, as well as a problem-solving mindset.

3.2 Preliminary Clarification

As illustrated in Fig. 1, the preliminary clarification phase encompasses the definition of the problem for a specific RCA use case in manufacturing, the execution of a proof of concept (PoC), and the scoping of the project. Initially, a problem analysis is conducted, from which the use case profile ($A_{1.1}$) is derived using a machine learning canvas. The DE is responsible for creating the use case profile, which outlines the problem statement. This phase is designed to ensure that the DS and MLE gain a thorough understanding of the manufacturing use case by leveraging domain knowledge (R_1). It also serves to consolidate all relevant information, along with an initial target vision, into a comprehensive use case profile.

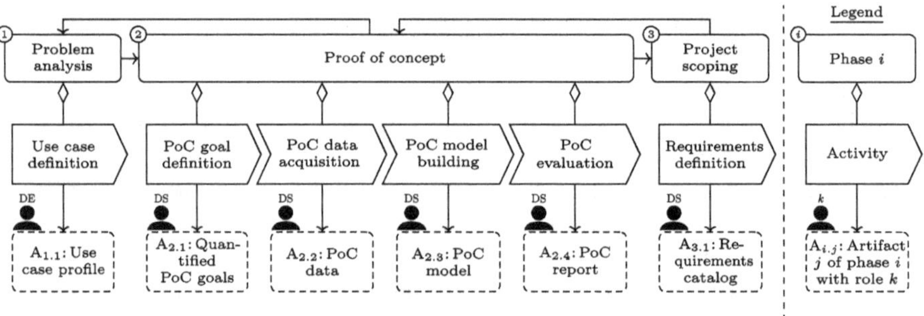

Fig. 1. Process model part 1: preliminary clarification

Following the problem analysis phase, the PoC phase is initiated. The PoC verifies whether CERs that are relevant to the specified RCA use case can be represented by existing manufacturing data and ML models. Within the scope of the PoC, four key artifacts are developed: Quantified PoC goals ($A_{2.1}$), PoC data ($A_{2.2}$), PoC model ($A_{2.3}$), and PoC report ($A_{2.4}$). The DS is responsible for these artifacts. At the conclusion of the PoC, an evaluation is conducted to assess the suitability of ML methods for addressing the RCA problem in the manufacturing use case. If any missing information from the problem analysis phase is identified during PoC execution, the process loops back to the problem analysis phase. This feedback mechanism allows for the extension of the use case profile ($A_{1.1}$) and ensures the provision of the necessary input for successful PoC completion.

Once the PoC has been successfully completed and evaluated, the requirements definition phase begins under the leadership of the DS. This phase aligns the initial problem description, documented in the use case profile ($A_{1.1}$), with the insights obtained from the PoC report ($A_{2.4}$). The outcome is the development of artifact $A_{3.1}$, the requirements catalog, which defines quantifiable objectives for the project execution. These include model architectures, interpretability levels, minimum performance thresholds, and both functional and non-functional requirements.

If the defined objectives and requirements deviate from the initial problem analysis or the findings of the PoC, the process may return to earlier phases for revision. This ensures that the resulting requirements accurately reflect the domain-specific context of the RCA use case. In cases where misalignment occurs, re-execution of the respective phases is necessary. Such iterative adjustments are essential to maintain coherence between the defined requirements and the real-world RCA scenario.

3.3 Development

During the development, as shown in Fig. 2, the three phases of data acquisition, RCAM training, and RCAM evaluation are carried out.

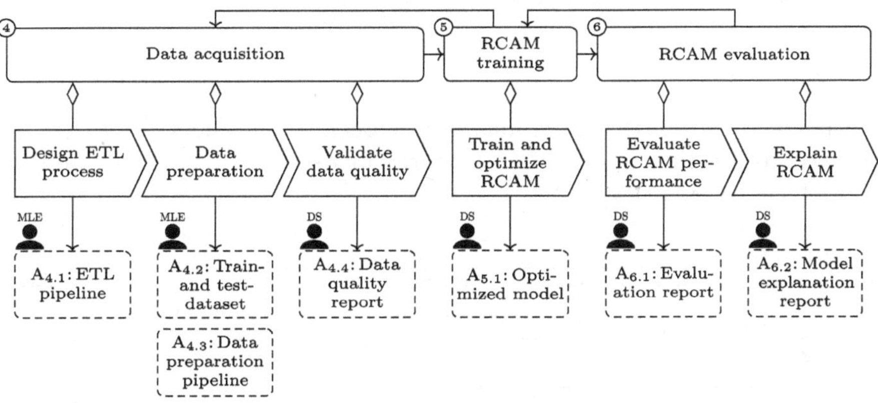

Fig. 2. Process model part 2: development

The data acquisition phase begins with the creation of an ETL pipeline ($A_{4.1}$), which is developed by the MLE and builds upon insights gained from the PoC data ($A_{2.2}$) acquisition. This pipeline ensures access to all relevant data required for RCA. Subsequently, the activity of data preparation leads to the generation of training and test datasets ($A_{4.2}$) and the development of a data preparation pipeline ($A_{4.3}$). Following this, the data quality is evaluated and summarized in a data quality report ($A_{4.4}$). If the data quality is found to be insufficient, adjustments to previously created artifacts, such as the ETL pipeline ($A_{4.1}$), may be necessary.

The RCAM training phase is characterized by a highly iterative process, led by the DS, and focuses on selecting and developing an ML model optimized based on the training data. To incorporate domain knowledge, human-in-the-loop ML approaches, such as active learning, are employed during this phase (R_1). Throughout training, challenges may arise, such as insufficient quantity or quality of the training and test data ($A_{4.2}$). In such cases, adjustments to previously developed artifacts, such as the ETL pipeline ($A_{4.1}$), may be required. This can result in a partial or complete repetition of the data acquisition phase.

In the sixth phase, the DS evaluates the optimized RCAM using test data, followed by a comprehensive explanation. The RCAM explanation is a central component of this PM, aiming to fulfill R_2. At this activity, local and global interpretation methods from explainable ML are applied to interpret the RCAM. These methods are used by the DS to collaboratively evaluate the identified patterns with the DE, for example, through feature importances or partial dependence plots. This process enhances trust in the optimized RCAM while simultaneously assessing the CERs learned by the RCAM. If deficiencies are identified during evaluation or interpretation, previous phases may be revisited and adjusted accordingly. For instance, a different model architecture may be selected during the RCAM training phase, or modifications to data preparation may be required in the data acquisition phase. Such iterative refinements are essential to ensure alignment between the RCAM and the real-world RCA scenario.

3.4 Operation

Following deployment, the operationalization begins, as illustrated in Fig. 3. This process comprises three phases that primarily focus on the deployment and monitoring of the RCAM (R_3).

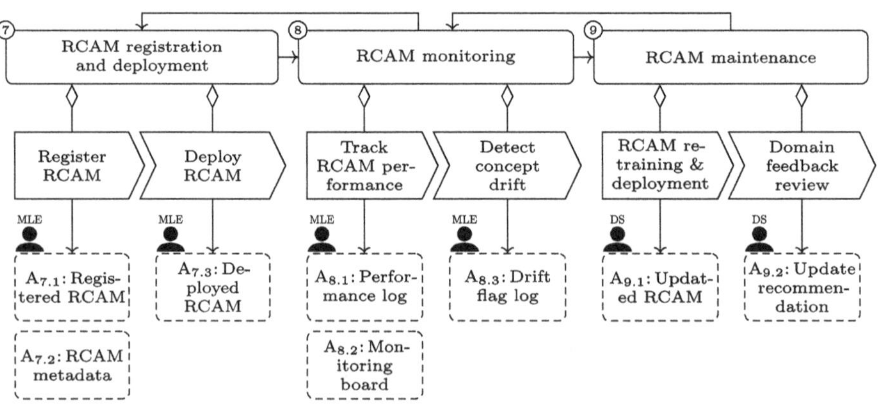

Fig. 3. Process model part 3: operation

First, in the RCAM registration and deployment phase, the MLE registers the optimized RCAM from phase five in a model registry ($A_{7.1}$) and documents all relevant model information, including training and test performance, in an

RCAM metadata file ($A_{7.2}$). Next, the MLE deploys the model into a runtime environment ($A_{7.3}$).

In the eighth phase, RCAM monitoring is implemented during operational deployment. With support from the MLE, the model's performance is continuously tracked and recorded in a performance log ($A_{8.1}$), and visualized through a monitoring dashboard ($A_{8.2}$). As part of this process, the MLE establishes a concept drift detection mechanism, resulting in the creation of a drift flag log ($A_{8.3}$).

If a degradation in model performance is observed due to concept drift or other factors, the RCAM maintenance phase is initiated. This phase involves retraining and redeploying the RCAM to operationalize an updated version ($A_{9.1}$). Additionally, the DS collaborates with the DE to collect and review real-world domain feedback (R_1) and derive further RCAM update recommendations ($A_{9.2}$). At this stage, returning to earlier phases may be necessary, for example, if domain feedback points to required adjustments in the monitoring dashboard or drift detection methods.

4 Case Study in Automotive Manufacturing

A real-world use case from an automotive final assembly process was used to validate the PM introduced in Fig. 3. The objective was to conduct RCA of errors that arise during electronic commissioning of vehicles. These errors are often caused by incorrect software or hardware versions, or improper installations.

The described PM was fully executed within the scope of the project, with close collaboration maintained between the DS, MLE, and DE roles throughout. During the problem analysis and PoC phases, the feasibility of data-driven RCA for electronic commissioning was demonstrated using manually exported data from various source systems. This data was subsequently prepared, and used for an initial ML model training. A PoC dataset containing 12, 444 examples with 2164 unique errors, and 142 unique causes was collected. ML models such as random forest, k-nearest neighbors, decision tree, and an association rule mining classifier were trained on this dataset. The random forest model achieved the highest classification F_1-score of 88.6% on test data. Notably, the highly interpretable rule mining classifier achieved comparable performance, with only 3.9% lower F_1-score, using just 348 extracted associations or CERs used as classification rules (for more details, see Pietsch et al. [12]).

The requirements derived from the PoC served as guiding constraints for the development phase. Based on PoC insights, ETL and data preparation pipelines were implemented. Data quality was validated, and corrective measures were taken to adjust the ETL and data preparation pipelines accordingly. In alignment with the PoC results, the model selection favored tree-based methods, such as random forest, to classify error data into root causes. To incorporate expert knowledge, active learning was tested and successfully reduced labeling efforts by up to 85%, and actively involving the DE in the RCAM training process. The classificiation results were further validated and interpreted using explainable

ML techniques, such as Shapley values, which enriched the understanding of the RCAM and the learned CERs.

This was followed by the operational phase, where the model was deployed into a runtime environment and became operational in manufacturing using services from cloud-based hyperscalers. In parallel with development, concept drift detection methods based on statistical process control were validated for future model monitoring. Drift detection approaches were tested using supervised classification metrics (precision, recall, F_1-score) and unsupervised soft-vote entropy. Entropy-based drift detection achieved a true positive detection rate of 83.5%, making it a promising unsupervised method for the RCA use case in the future. The retraining process was automated via a pipeline, and domain feedback is systematically collected and evaluated by the DS to identify potential improvements to the RCAM.

5 Conclusion and Future Directions

This work presents a PM for data-driven RCA in manufacturing. The PM is structured around roles, artifacts, and activities related to specific phases and consists of three parts: preliminary clarification, development, and operation. The PM integrates domain knowledge into ML model development and includes an activity focused on model explainability. This activity builds trust with DEs and allows the use of existing CERs to validate the RCAM's learned relationships. The PM covers the entire ML model lifecycle, including monitoring and maintenance. While not explicitly addressed here in detail, the model supports high agility and multiple feedback loops.

In future work, the constraints described in Fig. 3 will be addressed. The partial description used in this study and the method-neutral formulation (C_3) provide opportunities to further develop the PM. This includes fully specifying all activities, artifacts, and phases and enriching them with appropriate methods, tools, and standards. Additionally, the PM will be validated in further manufacturing use cases, and the insights gained will be incorporated into the refinement of the presented PM.

Acknowledgments. The results, opinions, and conclusions expressed in this publication are not necessarily those of Volkswagen Aktiengesellschaft. The authors declare that they have no known competing financial interests. This research was funded by the Federal Ministry for Economic Affairs and Climate Protection on the basis of a resolution of the Deutscher Bundestag with funding code 13IK020I. Funded by the European Union. During the preparation of this work the authors used DeepL Write and GPT-4 based chatbots in order to rephrase and translate. After using these tools, the authors reviewed and edited the content as needed and take full responsibility for the content of the publication.

Competing Interests. The author(s) has no competing interests to declare that are relevant to the content of this manuscript.

References

1. Broy, M.: Projektorganisation und Management im Software Engineering, 1st edn. Xpert. press ser. Springer, Berlin (2013)
2. Clancy, R., Bruton, K., O'Sullivan, D.T., Cloonan, A.J.: The HyDAPI framework: A versatile tool integrating Lean Six Sigma and digitalisation for improved quality management in Industry 4.0. IJLSS **15**(5), 1127–1154 (2024). https://doi.org/10.1108/IJLSS-12-2021-0214
3. Gnatz, M.A.J.: Vom Vorgehensmodell zum Projektplan. Ph.D. thesis. Technical University of Munich (2005)
4. Hammerschall, U.: Flexible Methodenintegration in anpassbare Vorgehensmodelle. Ph.D. thesis. Technical University of Munich (2008)
5. Harman, D., Buschmann, D., Scheer, R., Hellwig, M., Knapp, M., Schmitt, R.-H., Eigenbrod, H.: Data analytics production line optimization model (DAPLOM)—a systematic framework for process optimizations. In: Behrens, B.-A., Brosius, A., Drossel, W.-G., Hintze, W., Ihlenfeldt, S., Nyhuis, P. (eds.) Production at the Leading Edge of Technology. Lecture Notes in Production Engineering, pp. 412–420. Springer International Publishing, Cham (2022). ISBN 978-3-030-78423-2. https://doi.org/10.1007/978-3-030-78424-9_46
6. Huber, S., Wiemer, H., Schneider, D., Ihlenfeldt, S.: DMME: data mining methodology for engineering applications–a holistic extension to the CRISP-DM model. Proc. CIRP **79**, 403–408 (2019). https://doi.org/10.1016/j.procir.2019.02.106
7. Kalus, G.: Projektspezifische Anpassung von Vorgehensmodellen: Feature-basiertes Tailoring. Ph.D. thesis. Technical University of Munich (2013)
8. Kornas, T.: Qualitätssicherungssystem für die Anlaufphase und den Serienbetrieb einer Batteriezellenproduktion. Ph.D. thesis. Technical University of Braunschweig (2021)
9. Lundén, N., Bekar, E.T., Skoogh, A., Bokrantz, J.: Domain knowledge in CRISP-DM: An application case in manufacturing. IFAC-PapersOnLine **56**(2), 7603–7608 (2023). https://doi.org/10.1016/j.ifacol.2023.10.1156
10. Martinez, I., Viles, E., Olaizola, I.G.: A survey study of success factors in data science projects. In: 2021 IEEE International Conference on Big Data (Big Data), pp. 2313–2318. IEEE Computer Society, Los Alamitos, CA, USA (2021). https://doi.org/10.1109/BigData52589.2021.9671588
11. Pietsch, D., Matthes, M., Wieland, U., Ihlenfeldt, S., Munkelt, T.: Root cause analysis in industrial manufacturing: a scoping review of current research, challenges and the promises of AI-driven approaches. Journal of Manufacturing and Materials Processing **8**(6), 277/1–19 (2024). https://doi.org/10.3390/jmmp8060277
12. Pietsch, D., Wieland, U., Stoll, M., Ihlenfeldt, S.: Interpretable and Adaptive Data-Driven Root Cause Analysis in Manufacturing through Association Rule Mining. Procedia CIRP **134**, 31–36 (2025). https://doi.org/10.1016/j.procir.2025.03.006
13. Schoch, A.: Datenzentrierte Künstliche Intelligenz für ein prädiktives Qualitätsmanagement in der Automobilindustrie (DZKI-PQ). PhD thesis. University of Kassel (2023)
14. Schröer, C., Kruse, F., Gómez, J.M.: A systematic literature review on applying CRISP-DM process model. Proc. Comput. Sci. 181:526–534 (2021). ISSN 18770509. https://doi.org/10.1016/j.procs.2021.01.199
15. Wieland, U., Fischer, M., Hilbert, A.: Menschliches Expertenwissen und Prozessdaten im Verbund: six sigma und data mining im Kontext von Industrie 4.0. HMD Praxis der Wirtschaftsinformatik **50**(6), 66–75 (2013). https://doi.org/10.1007/BF03342070

16. Wirth, R., Hipp, J.: CRISP-DM: towards a standard process model for data mining. In: Proceedings of the Fourth International Conference on the Practical Application of Knowledge Discovery and Data Mining, vol. 1, pp. 29–40 (2000)

Towards Safe Neural Networks for Autonomous Mobile Systems: Literature Review and Research Agenda

Patrick Ziegler(✉), Jonathan Hilbinger, Jörg Franke, and Sebastian Reitelshöfer

Institute for Factory Automation and Production Systems (FAPS), Erlangen, Germany
patrick.ziegler@faps.fau.de

Abstract. Autonomous driving has achieved notable success in structured environments like highways and urban roads, where deterministic control systems and well-defined safety protocols ensure reliable operation. In controlled settings such as warehouses, fully automated vehicles rely on static sensor fusion techniques to navigate in predictable conditions. However, off-highway domains, such as agriculture and construction, present substantial challenges due to variable terrain, unpredictable obstacles, and environmental factors including dust, rain, and vegetation. These complexities necessitate advanced perception systems that utilize sophisticated data-processing pipelines, incorporating neural networks and multimodal sensors. Despite their functional potential, neural network architectures raise concerns in safety–critical applications, necessitating innovative solutions to ensure robust operation. In this paper, we address these challenges with a systematic literature review that explores the evolving landscape of safety–critical AI applications, focusing on promising approaches to enhance AI-based perception for off-highway autonomous mobile systems (AMS). Our findings highlight the role of mid-level sensor fusion, Out-of-Distribution (OoD)-detection, and explainable AI (XAI) techniques in improving reliability and functional safety. Drawing on these insights, we propose a research agenda for developing a reliable perception system tailored to off-highway applications. This agenda emphasizes the reliance on multiple research streams for handling diverse environmental conditions and enabling real-time integration of different validation architectures.

Keywords: Artificial Intelligence · Neural networks · Functional safety

1 Introduction

The rapid advancement of AMS has transformed various industries in recent years. Ranging from transportation to logistics, autonomous driving has achieved significant milestones in structured environments such as highways and factories. In these structured settings normative safety protocols using deterministic safety systems and static sensor technologies suffice to enable reliable and interpretable environment perception [1]. This is demonstrated, for instance, by various manufacturers commercially deploying mobile robots for intralogistics material transport [2]. In the automotive domain, manufacturers

L. Overmeyer and B.-A. Behrens (eds.), *Production at the Leading Edge of Technology*,
Lecture Notes in Production Engineering, https://doi.org/10.1007/978-3-032-19524-1_54

like Mercedes-Benz have achieved SAE Level 3, which is currently feasible only in controlled conditions with deterministic algorithms and lidar-based mechanisms or through extensive testing and failure analysis not feasible for smaller batch size applications [3, 4].

Extending autonomous capabilities to increasingly unstructured and complex environments presents challenges, as these situations are characterized by variable terrain, unpredictable obstacles, and dynamic environmental factors, including dust, rain, and vegetation. These demand more diverse and flexible perception systems capable of integrating different physical measurement principles and process significant amounts of sensor data efficiently. Additionally, interpreting relations between different sensor modalities requires understanding of complex relations and sensor specific characteristics across different information spaces [5]. Currently the realization of these challenging tasks is increasingly conducted by artificial neural networks that are already used for similar complex data processing tasks in less safety–critical environments [6]. However, the deployment of neural networks in safety–critical applications raises significant reliability concerns, due to their general black-box character and tendency to display overconfident predictions and hallucinations in unknown scenarios. These challenges necessitate innovative new approaches to ensure robust operation and functional safety, while not limiting the potential of the neural network itself. To address these challenges this paper conducts a systematic literature review that explores the current research landscape of artificial intelligence in safety–critical applications, especially focusing on the development of AMS in unstructured environments. Subsequently the key contributions of this paper are:

- A literature review on artificial intelligence in safety-critical applications
- A taxonomy for characterization of existing technologies and research directions
- A proposed research agenda outlining a robust perception system for AMS.

This paper is organized as follows: Initially, chapter "Solution Space Management for the Development of Ecologically Sustainable Products in the Early Innovation Phase" outlines the methodology used to conduct this literature review, detailing the initial conceptualization, development of search strings, as well as the subsequent systematic review process. Next, chapter "Initiating the Corporate Sustainability Transformation: Internal Drivers for Emission Reduction" presents the findings, highlighting the field's recent significance and organizing the identified literature into a taxonomy that categorizes relevant research streams. Chapter "Consolidation Studies on Solid-State-Recycling by Forming Aluminum Chips" synthesizes these findings to propose a research agenda, based on our taxonomy that guides future research. Finally, chapter "Evaluation of Measurement Techniques for Assessing Corroded Bolted Joints in Remanufacturing: A Reference Measurement for a Data-Driven Approach" concludes this study and provides an outlook on future developments in this field.

2 Materials and Methods

To ensure a systematic literature review, we designed a structured review process based on the works of Vom Brocke et al. [7] and Webster et al. [8]. Fundamentally, this review process is divided into a two-stage process. Firstly, an initial search-string development phase explores the topic and identifies a comprehensive collection of relevant keywords. In the second phase, a search string-based literature review process is conducted, and the resulting literature is distilled into a taxonomy. Figure 1 illustrates the key steps of our systematic literature review process.

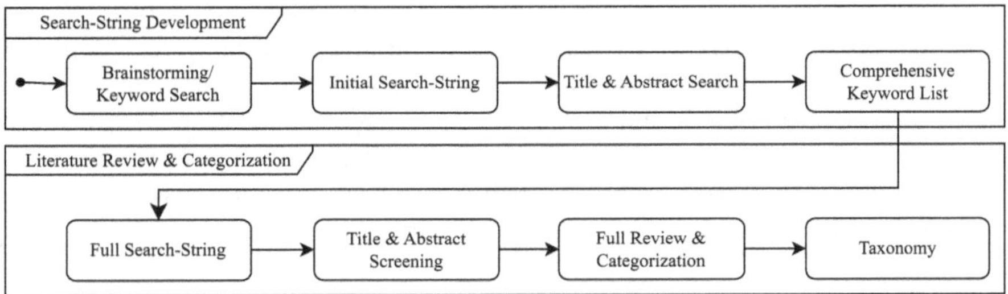

Fig. 1 Structured two-stage literature review process, beginning with initial keyword formulation and concluding with a taxonomy categorization of relevant literature

In the search string development phase, we begin with a brainstorming to identify relevant keywords related to the topic. These keywords are used to conduct an initial search in the Scopus database using a broadly defined search string.

TITLE-ABS-KEY ((“neural network” OR “artificial intelligence”) AND (“functional safety”)). This search yields 220 publications, which we reduce to 56 through title screening. A detailed abstract review of the identified literature provides a comprehensive overview of the topic, as well as refined keyword list for detailed search string definition. Based on this overview, we additionally formulate research questions and establish inclusion and exclusion criteria to guide the literature selection process. The research questions employed in this literature review are as follows:

1. What methods can mitigate functional safety risks in neural networks?
2. How can safety mitigation methods ensure functional safe robotic applications?

In addition to the relevance defined by the research questions, we establish inclusion and exclusion criteria to ensure methodological rigor and address formal requirements, as detailed in Table 1.

Table 1 Inclusion and exclusion criteria applied to filter scientific relevance and applicability of the reviewed literature

Inclusion criteria	Exclusion criteria
Methodological depth	Lack of relevance to the research topic
Accessibility via research channels	Non-scientific sources
Comparability and transferability	Preliminary or non-validated research

The inclusion criteria ensure only literature with sufficient methodological depth that goes beyond position papers is included in this review. Additionally, literature must be accessible through typical research channels and in or transferable to our research domain. The exclusion criteria eliminate studies lacking topical relevance, preliminary or non-validated research, and non-scientific sources. Using these foundational considerations, we develop a comprehensive search string based on the identified keyword list. As outlined in Table 2, the defined search string consists of five core components. Firstly, we focus the search on relevant properties that we want the identified approach to cover. Second, we narrow the context of the search to actual functional safety, excluding general model improvement with the keywords shown in line 2. In sections three and four we limit the search to the relevant technology and area of application. Finally, during search string development, we identified an important distinction between hardware and software faults (e.g. bit flips in accelerator hardware). To account for this distinction and focus on software, the fifth component excludes hardware-related publications.

Table 2 Final search string used in the systematic literature review. Based on a comprehensive keyword list and composed of five logical components to identify relevant literature

Component	Search string	Operator
Approach	(trustworthy* OR redundan* OR transparen* OR interpret* OR robust* OR explain*)	AND
Context	("Safety–Critical System*" OR "Functional Safety" OR "Safe Operation" OR "Safety Verification")	AND
Technology	("Neural Network" OR "Deep Learning" OR "Machine Learning" OR "Artificial Intelligence")	AND
Application	("driving" OR "robots" OR "autonomous")	AND
Focus	("hardware")	AND NOT

This search string results in 366 initial findings in the Scopus database. To reduce this set, we screen publication-titles against the defined inclusion and exclusion criteria, eliminating out-of-scope contributions and reducing the number of relevant publications to 118. A subsequent in-depth abstract examination further applies these criteria, narrowing the collection to 81 publications. Based on this literature collection, the next step involves a comprehensive review of the publications, identifying key concepts and categorizing them based on their individual focus areas.

3 Findings and Taxonomy

As outlined in chapter "Initiating the Corporate Sustainability Transformation: Internal Drivers for Emission Reduction", the total findings of our literature collection encompass 81 scientific publications selected for detailed examination. As shown in Fig. 2, the majority of these publications were published within the past five years, highlighting the present relevance of the research topic.

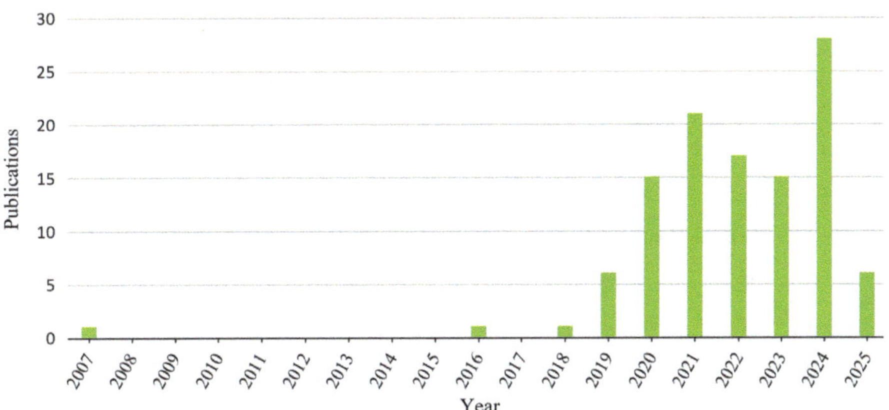

Fig. 2 Publication year distribution of the identified literature (n = 81) from 2007 to 2025, showing significant recent research interest between 2020 and 2025

3.1 Categorization and Taxonomy

To structure the identified body of literature, we propose a taxonomy that categorizes relevant publications into fundamental technological concepts within the research domain. The purpose of this taxonomy is to identify the current landscape of research directions and thereby identify potential entry points for further research. As illustrated in Fig. 3, the taxonomy initially divides the literature into two main categories: **Theoretical Frameworks** and **System Architectures**.

Within the **Theoretical Frameworks,** we group supporting literature that does not directly introduce technical implementations but instead contributes conceptual foundations. This includes **Development Strategies**, which outline processes and guidelines aimed at ensuring functional safety via rigorous development standards. Publications within the **Certification Frameworks** subcategory encompass strategies proposing adaptions of existing normative and regulatory frameworks to facilitate the integration of machine learning components. This research is crucial for enabling future easier certification of learning enabled components. Finally the **Theoretical Frameworks** category includes **Surveys** that cover the current state of the art in adjacent domains and application scenarios.

The **System Architecture** category encompasses research focused on enhancing functional safety through innovative technical approaches. Firstly, a large number of research emphasizes the role of **Redundant Design** for reliable coverage of diverse failure scenarios and environmental conditions. Secondly, many publications emphasize the integration of **Deterministic Concepts** to guarantee predictable system behavior. Although these concepts facilitate straightforward certification and validation, they inherently limit the flexibility of neural networks. As manually designing flexible deterministic implementations of complex behaviors is a significant challenge, recent research has focused on automating the abstraction process, by extracting deterministic structures from "black-box" functionality.

The **Neural Network** subcategory summarizes research on training, analysis, monitoring or design of the fundamental neural network components. Literature identified

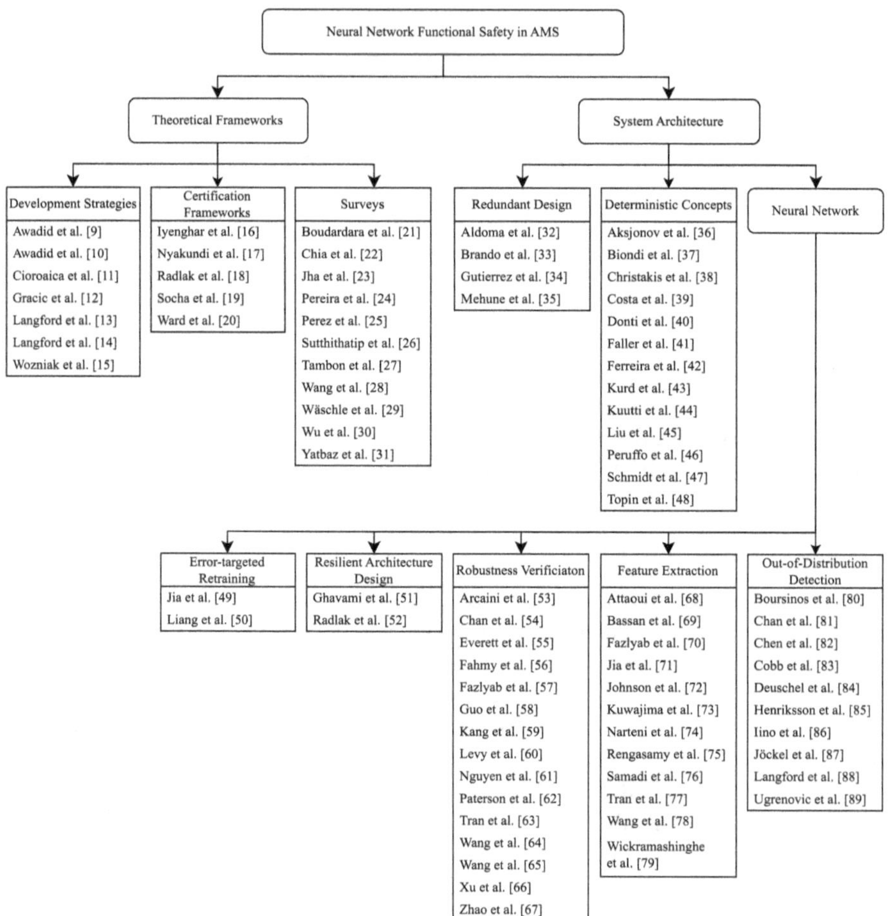

Fig. 3 Taxonomy of the identified literature, organized into two major research domains: theoretical frameworks and system architectures

within the **Error-targeted Retraining** branch focuses on data selection strategies for error-specific retraining towards known fault scenarios. This includes approaches for selectively retraining specific neural network segments to address specific fault scenarios, while preserving verified functionality. **Resilient Architecture Design** explores methods to improve neural network architectures themselves with techniques, like in-network noise reduction filters or specific activation functions to increase neural network resilience. Research within **Robustness Verification** generates different testing scenarios that prove a neural networks robustness. This includes small input perturbations, as well as novel approaches including generative AI or vector space approaches.

The **Feature Extraction** category summarizes approaches that enhance the transparency of neural network decision-making. This includes different approaches to evaluate intermediate features of an architecture to ensure and monitor expected behavior. Lastly, neural networks exhibit a tendency to display overconfident predictions in the face of uncertainty. Hence, for neural networks to operate safely, algorithms need to identify scenarios that are out of their designed operational domain. This **Out-of-Distribution** research uses additional neural network structures, such as autoencoders, latent vector space representations or entropy maximizing retraining to identify neural network anomalies and uncertainties.

3.2 Evaluation of Technological Concepts

The proposed taxonomy classifies research into Theoretical Frameworks and System Architectures. While Development Frameworks and Certification Strategies within the Theoretical Frameworks offer formal verification, they do not enhance neural network robustness. Consequently, to identify a technological solution, we focus on discussing the potential of System Architecture concepts.

In order to continually prepare for disturbances in any individual system component, the core of any safety–critical system is based on **Redundant Design**. This is addressed by the authors in this category, specifically highlighting the importance of covering the entire data processing pipeline from physical measurement principles to result propagation. Furthermore, this must be achieved with sufficient architectural diversity to ensure that specific hazards do not cause simultaneous failures in both system instances. To improve the explainability of neural networks, the **Deterministic Concepts** category provides a reliable solution. However, the functionality of the concepts targets simpler applications and does not allow expansion to complex environment perception tasks. **Error-targeted Retraining** only considers error-handling after errors occur. While this can greatly benefit the development process, it does not improve the inherent safety of the networks. Similarly, **Resilient Architecture Design** contributes to neural network performance but does not provide any verifiable safety assurance. In contrast, the research directions **Robustness Verification** and **Feature Extraction,** originating from the XAI domain, can be used to monitor a network's reasoning process and ensure that small input perturbations do not cause unexpected changes in the output. This supports both robustness and transparency, contributing to the safety assurance of neural network architectures. Finally, **Out-of-Distribution** detection concepts enable the introduction of a two-state system, where in one state reliable neural network predictions are expected, while in the other a safe state is defined. This significantly increases flexibility in functionality development, as reliable functionality is only required in a limited target domain.

4 Research Agenda: Reliable Neural Network Operation in Safety–Critical AMS

This review has identified several research directions that contribute to advancing the functional safety of learning-enabled components. The diversity of these approaches underscores that no single method is sufficient to guarantee the reliable operation of neural networks in safety–critical systems. Instead, enhancing the reliability and transparency of such systems requires a multi-layered strategy that integrates advancements from different research domains. To accelerate the safe adoption and certification of neural networks in these environments, existing methods must be intelligently combined and further advanced within their respective domains. Moreover, this strategy must be tailored to the specific characteristics of the target application, leveraging task-specific properties for reliability and safety assurance. Based on the systematic literature review and the evaluation presented in chapter "Initiating the Corporate Sustainability Transformation: Internal Drivers for Emission Reduction", we propose a research agenda, as shown in Fig. 4, that synthesizes the most promising concepts into a comprehensive safety architecture tailored for mobile robot perception in challenging environments.

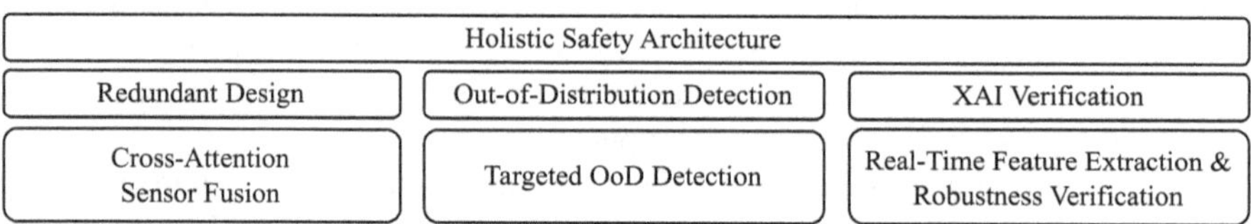

Fig. 4 Three-pillar research agenda proposing aresearch focus on intelligent sensor fusion, targeted OoD detection and real-time XAI verification

A key element of this architecture is **Redundant Design**. This is first realized through network redundancy by parallelizing the data flow across duplicated neural networks with slightly heterogeneous architectures. To extend this redundancy across different physical measurement principles, we propose the integration of cross-attention mechanisms into the network architecture to enable learnable sensor data fusion. These mechanisms dynamically leverage individual sensor modalities by learning sensor-specific relationships. Another crucial component for ensuring trust in the system's decisions is **Out-of-Distribution** detection. These mechanisms identify inputs that fall outside the training distribution, effectively mitigating the risk of overconfident predictions in unfamiliar scenarios. For robust OoD detection in multimodal settings, we will propose a method that leverages the existing multimodal feature space generated by the primary detection pipeline. To further enhance system reliability, we plan to incorporate XAI methods for **Feature Extraction** and **Robustness Verification**. These methods allow for systematic analysis of intermediate features and verification of robustness against input perturbations during inference. However, a major challenge lies in the formalization of a method that incorporates such concepts while maintaining real-time performance. To address this, we will propose a method for selectively validating relevant parts within the inference process with appropriate XAI methods. Finally, we combine these insights on network robustness, feature activations, and environment familiarity to derive a metric that robustly monitors the network's operational health. Combined with redundant design across the entire processing pipeline, this approach targets key failure modes and supports the development of a more robust and trustworthy perception system for off-highway applications.

5 Conclusion

This survey provides a comprehensive overview of current literature on enhancing the reliability of neural network components. It establishes a foundation for systematically improving functional safety in neural network-enabled applications and highlights the growing relevance and research activity in this field. This reflects the widespread adoption of neural networks across technological domains and emphasizes their increasing importance in safety–critical systems. To conduct our survey, we first defined two research questions to guide the review. We then introduced a taxonomy that categorizes the identified publications into distinct technological methods, thereby addressing the first research question and providing a systematic framework to guide future research. Based on these insights, we also propose a research agenda that outlines key research concepts for developing robust, learning-enabled perception systems and, in answering our second research question, demonstrates how the identified methods can support safe robotic applications.

Competing Interests. The author(s) has no competing interests to declare that are relevant to the content of this manuscript.

The references section includes in text references. For a full list of the identified body of literature see: https://github.com/paziegler/TowardsSafeAMS

References

1. ISO 3691-4—Industrial trucks—Safety requirements and verification—Part 4: Driverless industrial trucks and their systems, 53.060(3691-4)
2. Fragapane, G., de Koster, R., Sgarbossa, F.: Autonomous Mobile Robots for Material Handling in Intralogistics, pp. 251–274. Springer International, Cham (2024)
3. Lieret, M., Fertsch, J., Franke, J.: Fault detection for autonomous multirotors using a redundant flight control architecture. In: 2020 IEEE 16th International Conference on Automation Science and Engineering (CASE), pp. 29–34. IEEE (2020)
4. Schwindt-Drews, S., Storms, K., Peters, S., et al.: Acceptance and trust: drivers' first contact with released automated vehicles in naturalistic traffic. Universitäts- und Landesbibliothek, Darmstadt (2025)
5. Huang, K., Shi, B., Li, X., et al.: Multi-modal sensor fusion for auto driving perception: a survey (2022)
6. LeCun, Y., Bengio, Y., Hinton, G.: Deep learning. Nature **521**, 436–444 (2015). https://doi.org/10.1038/nature14539
7. vom Brocke, J., Simons, A., Niehaves, B., et al:. Reconstructing the giant: on the importance of riguor in documenting in the literature search process (2009)
8. Webster, J., Watson, R.T.: Analyzing the past to prepare the future: writing a literature review. MIS Quart. **26** (2002). https://doi.org/10.2307/4132319

Predictive Quality Control of Rubber Injection Molding

Alexander H. Olbrich[1(✉)], Ludger Overmeyer[2], and Ulrich Giese[1]

[1] Deutsches Institut für Kautschuktechnologie e. V, Hannover, Germany
alexander.olbrich@freudenberg.com

[2] Institute of Transport and Automation Technology (ITA), Leibniz University Hannover, Garbsen, Germany

Abstract. This study presents a predictive quality control (PQC) approach for rubber injection molding (IM), aiming to reduce molding related defects of rubber products via facilitating machine learning (ML) based quality modelling. Predictive quality of cyclical processes like injection molding is a fast-emerging field of research with a predominant focus on thermoplastic. Integrating predictive quality models into control strategies offers potential savings by reducing waste. However, the complex interactions between measured process variables and setting parameters place additional demands on the controllability of these models. While previous research (Olbrich et al. in 15th Fall Rubber Colloquium. KHK, Hannover. 2024) was focused on modelling molding related quality properties with process data, the aim of this work was to integrate ML models into closed-loop control approaches to control process deviations. The integrated model serves as a quality observer, which substitutes quality measurements. If the predicted quality deviates from the target value, the controller adjusts cyclically the setpoints of the IM process to countermeasure deviations. As part of this research a demonstrator was developed and tested to predict the weight of an elastomer test plate as quality property. Due to the long cycle times caused by vulcanization, the opportunities for experimental trials are considerably restricted. For this, the controller was designed and tuned virtually, facilitating the predictive quality models. The proposed predictive quality approach was afterwards successful validated on differed defined disturbance scenarios, like material change overs between rubber batches of different viscosities. The proposed approach therefore demonstrates significant potential for reducing waste while enhancing quality assurance, in rubber IM.

Keywords: Machine learning · Control of quality properties · Rubber injection molding

1 Introduction

This work focuses on controlling the product properties of injection molded elastomer components based on machine learning (ML). Previous research was focused on predictive quality for injection molded rubber products [1]. The injection molding (IM) process passes through several nonlinear phases within a production cycle, making

L. Overmeyer and B.-A. Behrens (eds.), *Production at the Leading Edge of Technology*, Lecture Notes in Production Engineering, https://doi.org/10.1007/978-3-032-19524-1_55

direct physical modeling of product properties complex or impossible. However, due to the cyclic nature of the process, it is possible to model the process behavior with ML using representative data [2]. Therefore, the use of ML for predictive quality and control in thermoplastic IM has been a focus of research [3]. In contrast, rubber IM remains unexplored due to the more complex material behavior, influenced by factors such as reaction kinetics affecting flow behavior or phenomena like wall slip [4]. In the area of quality attributes, a distinction is mainly made between the prediction of dimensional product properties using regression, and the classification of defects [5, 6]. Due to its ease of measurement and qualitative significance, part weight is often used as the target variable for prediction [6]. In this context, artificial neural networks (ANN) are a suitable model class for modeling the nonlinear process behavior [7]. Besides ANNs, other ML methods such as Random Forests (RF), Support Vector Machines, and Gaussian Processes Regressors (GPR) are also applied in research. However, comparing these models based on metrics like accuracy or training data requirements is difficult due to varying experimental setups, including tools, quality characteristics, and sensor configurations [5]. Nevertheless, compared to ANNs, RFs offer interpretability via feature importance, while GPRs enable local uncertainty estimation [8, 9]. A predictive quality model can be integrated into a control loop as observer for calculating product properties [10]. Alternative approaches in IM aim to achieve quality through statistical process control when product properties are not available [4]. These process variables, such as maximum injection pressures, can be predicted by ML models and subsequently used as inputs for predictive quality models [11]. Another focus of research is the use of ML as a controller for the IM process, to stabilize and optimize the process by adjusting setpoints. Different control approaches include the integration of ML methods into control concepts such as model predictive controllers or fuzzy controllers, as well as their use as inverse process models to predict new set points to countermeasure deviations [10, 12]. The mathematical inversion of neural networks is complex due to the activation functions, making direct integration not feasible [11]. This problem can be avoided by directly training an inverse model, in which the setpoints can be predicted directly out of quality deviations [13, 14]. The dynamic effects of the IM process can be represented in the modeling for controller design by internalizing them through computationally intensive recurrent ANNs [15]. Global optimization methods minimize the error between ANN predictions and targets to find an optimal operating point without model inversion [16]. Liang et al. combine the approach of a quality observer and an inverse ANN as a control unit, using two separate models to regulate the predicted quality, showing a predictive quality control (PQC) approach [13]. Statistical experimental designs are commonly used for the experimental generation of training data [6]. Simulations in the case of thermoplastics can be used to generate training data for ML models or to design controllers, thereby reducing experimental effort [13, 17]. This study addresses a PQC approach for the prediction and control of product properties of injection-molded elastomer components. The presented approach includes three of the mentioned ML model categories, product quality, inverse model, and process model. The quality model serves as an observer, and the inverse model is combined with a PID controller. The controller is tuned through simulation using the process model, with the observer integrated in a combined surrogate model of the IM process and the developed control loop. Finally, the PQC approach is experimentally validated using practical disturbance scenarios.

2 Approach and Experimental Setup

This chapter presents the approach, and the selected application example used in this work. Figure 1 illustrates the schematic overview of the exemplary application used in this study. The IM machine is connected to an edge device via an OPC UA client-server setup. The individual modules are implemented in Python on an industrial computer. The retrieved process data can be categorized into setpoints, time series like pressure profiles, and cyclically aggregated measurements of process parameters derived from the time series (e.g., average injection pressure).

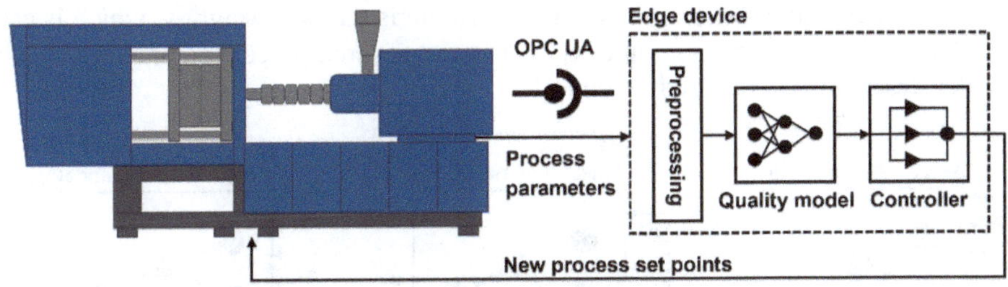

Fig. 1 High level overview of system architecture

Process data is preprocessed, stored in a database, and transferred to a machine learning model. A predictive quality model calculates the relevant product properties based on the setpoint parameters and aggregated process values. The ML models used in this work are ANNs with a feedforward architecture, implemented in Keras [18].

This model class was chosen due to the exceptionally high accuracies achieved by ANNs, which are essential given the generally tight tolerance bands required in industrial applications. Based on the quality prediction from the previous cycle, the PQC adjusts the setpoints to achieve a target quality value. The controller and the model operate discretely in the cycle of the IM process. The controller combines a classical PID approach with a downstream ANN and is described in detail in Chap. 4. The adjusted setpoints are written back to the machine for the next cycle. Figure 2 shows the used exemplary product: An elastomer test plate and the filling related defects. The part weight serves as a representative metric to describe defects such as overfilling (flash) and underfilling. A design of experiments with varying process and material settings was conducted to generate representative data for model training.

In a second experimental phase, defined disturbance scenarios were executed to enforce typical production-related viscosity deviations, such as those caused by batch changes, and subsequently compensated for them. This is further discussed in Chap. 5.

Fig. 2 Used product geometry. Elastomer testing plates made from carbon black filled EPDM with a hardness of 70 Shore. The used tool contains a single cavity with a cold runner system

3 Modelling of Process and Quality Based on ML

Chapter 3 covers the ML models that were used and integrated into the PQC approach. The models used are schematically illustrated in Fig. 3. They are categorized into models (ANN_A) for predicting product quality, inverse models (ANN_B) for predicting the required setpoints, and process models (ANN_C) for predicting critical process variable. In the chosen example, the predicted target variable is the part weight y, which is estimated based on the process setpoints $x_{1...6}$, u and measured process parameters $x_{7.}$ The variables descriptions are listed in Table 1.

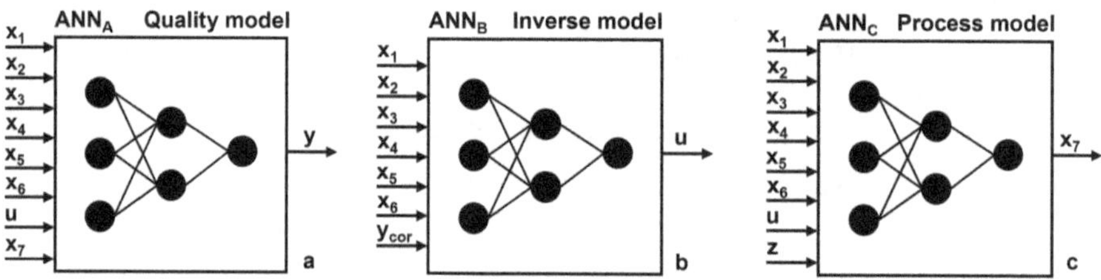

Fig. 3 Schematic overview of the models used, including input and output variables

The separation of the models is due to two key aspects. ANN_A can predict product quality from the process conditions with high accuracy. To determine the necessary correction in the event of a disturbance, it is possible to use optimization algorithms to minimize the error between the predicted and target weight by computing an optimal set of setpoint parameters. The challenging invertibility of an ANN requires the use of computationally intensive optimizers. Due to the time constraints imposed by short cycle times (< 60 s), it is therefore more practical to facilitate a second inverse ANN_B to reach response times within a cycle. ANN_B receives an additional input variable: a corrected weight previously calculated by a controller to achieve the desired target weight (see Chap. 4). It then predicts a setpoint which is applied to the IM machine in the next cycle. In this example, the predicted control variable u is the switch over volume between injection and holding phase, which is influencing the amount of injected material. Another limitation of ANN_A is the dependency of process variable x_7 on setpoints. To achieve decoupling, a third model ANN_C is used to predict the measured process variables in advance. ANN_C calculates the shot volume, which is determined by subtracting the measured material cushion volume from the dosing volume. The cushion volume is the measured volume that remains in the cylinder after the holding pressure

Table 1 Overview of the conducted Design of Experiments as the basis for model training

ID	Parameter	Type	Unit	Min.	Max.	R(y)
x_1	Dosage volume	Set value	cm^3	11	14	−0.06
x_2	Injection speed	Set value	cm^3/s	0.4	0.8	−0.03
x_3	Cold runner temperature	Set value	°C	95	110	0.46
x_4	Holding pressure	Set value	Bar	1950	2150	0.22
x_5	Holding time	Set value	s	2.7	4	0.47
x_6	Back pressure	Set value	Bar	70	110	0.24
u	Switch over volume	Set value	cm^3	0.5	4.2	−0.19
x_7	Shot volume	Process value	cm^3	8.59	9.98	0.95
x_8	Residual cushion volume	Process value	cm^3	0	3.84	−0.32
y	Product weight	Quality property	g	7.82	8.83	1
y_{cor}	Corrected weight	State variable	g	–	–	–
z	Normalized batch viscosity	Disturbance	–	−1	1	−0.06

phase, allowing direct monitoring of process fluctuations. Since this variable is measured directly on the machine, integrating ANN_C into the controller is not required.

However, ANN_C can be used for virtual controller design to reduce the need for physical trials (see Chap. 4). Table 1 provides an overview of the training data used. In the experimental design, setpoints and batches with varying viscosities were considered. The set values x_1 to x_6 were varied by 5 equidistant steps, while u as preferred control variable was varied by 13 steps to achieve a higher resolution. The material deviations were considered by 3 batches with different viscosities. The dataset comprises 649 production cycles with 118 parameter combinations. The data was split into training, test, and validation sets in a 70/15/15 ratio, ensuring strict separation of unique setpoint combinations. Table 1 shows the Pearson correlation coefficients R(y) between the parameters and the weight, represented in the data set. The cushion exhibits only a weak correlation with part weight due to nonlinear interactions with the set parameters. The physically motivated conversion of the cushion into shot volume, shows a correlation of 0.95 to the part weight, which is a good prerequisite for a monitoring variable. However, with an RMSE of 65 mg and R^2 of 0.9, the resulting linear model would remain too inaccurate for a PQC. Due to the complex interactions within the considered process window, the setting parameters show no correlation with part weight, although local linearities around specific operating points are conceivable. The modeling results are presented in Table 2, along with achieved performance metrics on the test data.

ANN_A achieves a high accuracy (R^2 = 0.992) and thus will be integrated as a quality observer in the control loop. With an accuracy of $R^2 = 0.980$, model ANN_B can capture the complex interactions between setting parameters and is therefore suitable for integration as an actuator in the control loop, which is described in the next section.

Table 2 Training results including architecture (number of neurons per layer) as well as performance metrics based on the test data

ID	Description	Input	Out	Architecture	RMSE	R^2
ANN_A	Quality model	$x_1, \ldots, x_7, u$	y	8–(7–2–2-1)-1	27 mg	0.992
ANN_B	Inverse model	$x_1, \ldots, x_6, y_{cor}$	u	7–(3–2–2-1)-1	0.057 cm^3	0.980
ANN_C	Process model	$x_1, \ldots, x_6$, u, z	x_7	8–(20–20-1)-1	0.160 cm^3	0.981

4 Controller Architecture and Tuning on Surrogate Model

This chapter describes how the ANNs are integrated into the PQC system and tuned through simulation. Figure 4 shows the block diagram of the control loop. Based on the process parameters and setpoints of the current cycle, the ANN_A predicts the part weight y (n) for the discrete cycle. From the calculated deviation e (n) from the target value w(n), a corrected weight y_{cor} is determined using a PID controller. The inverse model ANN_B then calculates a new switchover volume based on this correction and the current process setpoints, which is transmitted to the machine as the new control variable u (n + 1) for the next cycle. The neural network enables consideration of the nonlinear transfer behavior, eliminating the need to linearize the operating point.

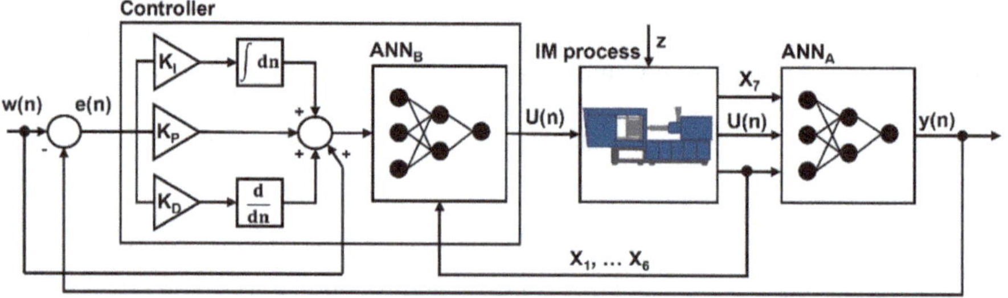

Fig. 4 Block diagram of the control loop with integrated ANNs

The additional integral component of the controller allows for the elimination of steady-state errors over the course of multiple cycles. For virtual tuning of the PID parameters, the surrogate model shown in Fig. 5 is used, in which the real IM process is replaced by ANN_C. For this purpose, multiple IM cycles are simulated, and viscosity jumps are introduced. In the presented example, 1.000 disturbance scenarios with randomly selected setpoints were simulated, and the PID variables were adjusted to prevent overshooting and ensure disturbance compensation.

Figure 6 shows a virtual disturbance scenario in which two viscosity disturbances, introduced as step functions, are compensated. The controller compensates deviations of 100–150 mg and reaches in both cases a stable state after 10 cycles.

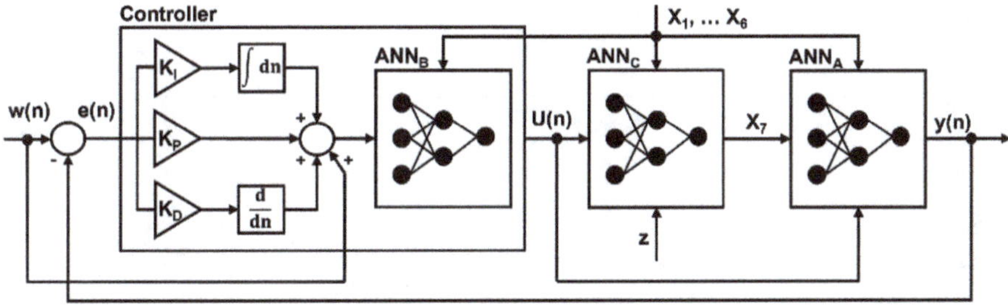

Fig. 5 Block diagram of the surrogate model with integrated process model in the control loop

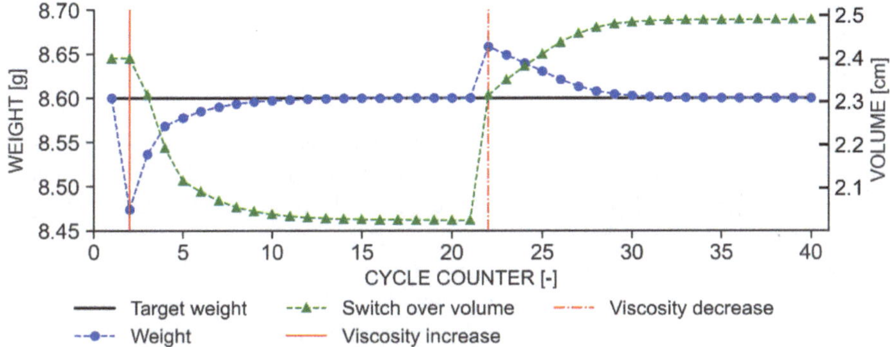

Fig. 6 Simulated step responses of viscosity changes on the surrogate model

5 Experimental Results and Discussion

This chapter presents the results of the experimental validation of the designed controller. Figure 7 shows a batch change to a material with higher viscosity. Due to the volume of the cold runner, the increase in viscosity occurs gradually over 20 cycles rather than as a sudden step change. The PQC continuously compensates for the increase in viscosity and corrects unintended disturbances such as a material feed interruption, via the adjustment of the switch over volume. The weight deviation is reduced from 142 mg to 5 mg compared to reference parts with the original settings.

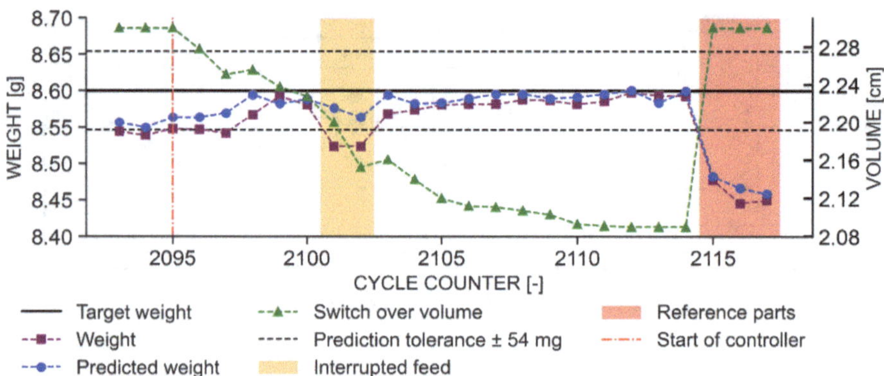

Fig. 7 Results of compensating the change to a higher-viscosity batch, including predicted and measured weight and control variable adjustments of switch over volume on the IM machine

Table 3 provides an overview of the experimental test runs, including the absolute correction of the switchover point in the steady-state condition. In all cases, fluctuations were quickly compensated, and the weight deviations from the target value were significantly reduced from approximately 150 mg to 5–25 mg. The total set point adjustments of the switch over volume Δu reaching form—0.48 cm^3–0.15 cm^3. Additionally, it was possible in test no. 6 to operate the system under a much lower cold runner temperature of −15 °C—well outside the temperature range represented in the training data—and still regulate it successfully, demonstrating the extrapolation capability and robustness of the PQC.

Table 3 Results of validation trials for multiple validation scenarios

ID	Description	Δu (cm$^{3)}$	Error ref. (mg)	Error (mg)
1	Material change from low to high viscosity	−0.21	−142	−5
2	Material change from low to high viscosity	−0.27	−213	−20
3	Material change from low to high viscosity	−0.21	−133	4
4	Material change from low to high viscosity	−0.38	−139	5
5	Material change from high to low viscosity	0.15	71	24
6	Set runner temperature from 100 to 80 °C	−0.48	−412	−6

The PQC significantly reduces material waste during material changes. Moreover, viscosity-related process fluctuations can be automatically compensated during ongoing production without the need for operator intervention. One key advantage of PQC is the ability to predict part quality directly during the process, eliminating the delay between production and quality inspection. In the context of rubber IM, PQC applications targeting mechanical properties would be particularly valuable, as parts often require several hours of resting time before measurements can be taken, by which point corrective intervention is no longer possible. It should be noted that the models are tailored to the specific reference part and must be adapted for other products. Furthermore, due to production-related drifts such as tool wear, model performance must be continuously monitored and retraining performed as needed [19].

6 Conclusion

The developed PQC approach enables the control of metric product properties in cyclic processes using machine learning. Initial limitations of modeling with neural networks were overcome by introducing various sub-models, each serving a strictly defined purpose. The model used for predicting product properties enables quality feedback without the need for additional inline measurement equipment. PQC combines an inverse ANN, which maps the nonlinear transfer behavior to calculate the control variable, with a classical PID approach that uses the integral component to compensate for steady-state errors. The approach is complemented by an ANN for predicting quality attributes, enabling the virtual tuning of the controller using a surrogate model. The experimental validation of this approach was carried out using the challenging example of rubber injection molding. The deployed ANNs achieve high prediction accuracies, thereby enabling the compensation of viscosity disturbances. In the experimental validation, PQC enables nearly scrap-free material changes and reduces the tolerance deviations of the part weight from approximately 150 mg to 5–25 mg. The compensation of operating points in the extrapolation range of the models also demonstrates the robustness of the developed approach.

Acknowledgements. The authors would like to thank the company Freudenberg Technology Innovation SE & Co. KG for funding this research and for providing the equipment and machines used for this work.

Competing Interests The author(s) has no competing interests to declare that are relevant to the content of this manuscript.

References

1. Olbrich, A., Overmeyer, L., Giese, U.: Predictive quality for injection molded elastomer products. In:15th Fall Rubber Colloquium. KHK, Hannover. (2024)
2. Kenig, S., et al.: Control of properties in injection molding by neural networks. Eng. Appl. Artif. Intell. **14**, 819–823 (2001)
3. Selvaraj, S., et al.: A review on machine learning models in injection molding machines. Adv. Mater. Sci. Engineering. Wiley. (2022)
4. Röthemeyer, F., Sommer, F.: Kautschuktechnologie, 3rd edn. Hanser, Munich (2013)
5. Polenta, A.: A comparison of machine learning techniques for the quality classification of molded products. Information. **13**(6), 272 (2022)
6. Krauß, J., Borchardt, I.: Prediction and control of injection molded part weight using machine learning–a literature review. Procedia CIRP. **118**, 867–872 (2023)
7. Sun, X., et al.: Prediction of quality index of injection-molded parts by using artificial neural networks. Adv. Mater. Res. **291–294**, 432–439 (2011)
8. Tayalati, F., Azmani, A., Azmani, M.: Application of supervised machine learning methods in injection molding process for initial parameters setting: prediction of the cooling time parameter. Prog. Artif. Intell. (2024)

9. Ren, Z., et al.: A learning-based model predictive control scheme for injection speed tracking in injection molding process. Complex Intell. Syst. **10**, 7845–7861 (2024)
10. Chen, Z., Turng, L.: A review of current developments in process and quality control for injection molding. Adv. Polym. Technol. **24**(3), 165–182 (2005)
11. Mustafa, M.: Modellbasierte Ansätze zur Qualitätsregelung beim Kunststoffspritzgießen. Shaker Verlag, Aachen (2000)
12. Zhao, P., et al.: Intelligent injection molding on sensing optimization, and control. In: Advances in Polymer Technology. Wiley (2020)
13. Liang, J., Wang, P.: Self-learning control for injection molding based on neuronal networks optimization. J. Inject. Molding Technol. **6**(1), 58–71 (2002)
14. Lukas, M., Overmeyer, L., et al.: Real-time temperature control in rubber extrusion lines: a neural network approach. J. Adv. Manuf. Technol. **133**, 5233–5241 (2024)
15. Rehmer, A., et al.: An internal dynamics approach to predicting batch-end product quality in plastic injection molding using recurrent neural networks. IEEE Conf. Control. Technol. Appl. CCTA, 1427–1432 (2022)
16. Meiabadi, M., et al.: Optimization of plastic injection molding process by combination of artificial neural network and genetic algorithm: journal of optimization in industrial. Engineering. **13**, 49–54 (2013)
17. Lockner, Y., Hopmann, C.: Induced network-based transfer learning in injection molding for process modelling and optimization with artificial neural networks. J. Adv. Manuf. Technol. **112**, 3501–3513 (2021)
18. Collet F., et al: Keras. (2015). https://keras.io. Last accessed 20 July 2025
19. Schulze-Struchtrup, A.: Ganzheitliche Formteil-Qualitätsprognose für das Spritzgießen thermoplastischer Kunststoffe auf der Basis maschineller Lernverfahren. Universität Duisburg-Essen (2021)

Enhancing Key Performance Indicator Definition Using Large Language Models

Hans Joachim Groß(✉), Muriel Schnelle, Jonas Barth, and Joachim Metternich

Technical University of Darmstadt, Institute of Production Management, Technology and Machine Tools (PTW), Otto-Berndt-Str. 2, 64287 Darmstadt, Germany
J.Gross@ptw.tu-darmstadt.de

Abstract. Modern production systems face growing complexity due to globalization, customer-specific demands, and sustainability requirements. Existing Performance Management Systems (PMS), including methods like the Balanced Scorecard (BSC) or Hoshin Kanri, often lack a systematic approach to deriving strategy-aligned Key Performance Indicators (KPIs). This results in inconsistencies, redundancies, and unclear indicator relationships that hinder operational transparency. This paper presents a structured framework supported by Large Language Models (LLMs) for strategy-driven KPI definition in Shop Floor Management (SFM). By combining driver tree logic with the interpretive power of LLMs, KPIs are derived from corporate strategy. The approach enables transparent, scalable, and sustainable KPI systems while ensuring alignment between local performance measures and corporate objectives.

Keywords: Performance management · Key performance indicator · Large language model · Framework

1 Introduction

Production systems are undergoing rapid transformation driven by globalization, economic growth, and technological innovation [1]. Manufacturers face increasing challenges due to rising demand for customized products, shorter product life cycles, and growing ecological and social requirements [2]. A lack of transparency and alignment between local performance targets and corporate strategy can cause inefficiencies [3]. This growing complexity necessitates strategic performance management and consistent KPI structures across organizational levels. Especially when KPIs are inconsistently defined or not systemically connected, the increasing volume and heterogeneity of measured production data hinder timely and targeted decision-making [4]. Recent developments in LLMs enable improvements in digital SFM by processing natural language and deriving insights from unstructured data [5]. Additionally, it has been demonstrated that Retrieval Augmented Generation (RAG) approaches can effectively enrich LLMs with contextual data, improving response quality without the need to modify model parameters [6]. This work combines the advantages of LLMs to

L. Overmeyer and B.-A. Behrens (eds.), *Production at the Leading Edge of Technology*, Lecture Notes in Production Engineering, https://doi.org/10.1007/978-3-032-19524-1_56

define consistent KPIs for tackling the growing complexity in modern production systems. Furthermore, a Framework is developed to guide the user through the KPI-definition-process under consecration of established approaches like BSC.

The article is structured as follows: Sect. 2 outlines challenges in transforming production systems and introduces performance management as a measurement tool. Section 3 derives requirements for a KPI definition framework and presents its conceptual development. A proof of concept illustrates the frameworks' functionality in Sect. 4. Section 5 summarizes the key findings and suggests future research directions and development.

2 Current State of Research

2.1 Performance Management Systems

SFM is a key element of lean production that promotes continuous improvement through transparency, short-cycle communication, and employee involvement [7]. The Darmstadt SFM model defines five core elements: performance management, problem-solving, lean leadership, visual management, and competency development [8]. PMS use structured KPI frameworks to align operational activities with strategic goals [9]. Through the top-down cascade of objectives and bottom-up feasibility assessments by employees, individual efforts are linked to corporate strategy [7,10]. In production, PMS enable transparency, early bottleneck detection, and targeted resource control [7]. Effective KPI selection requires a systematic approach and linkage to target values. The SMART (specific, measurable, achievable, relevant, time-bound) principle is a common guideline for formulating effective KPI-goals [11]. A consistent KPI structure across departments is essential for transparency and comparability [12]. Prominent PMS approaches include the BSC, which translates strategy into four perspectives: financial, customer, internal processes, and learning and growth [13]. It is further supported by four management processes: translating the vision, communicating and linking, business planning, and feedback and learning [14]. Hoshin Kanri combines top-down goal setting with iterative bottom-up alignment using the catchball method [15]. The performance pyramid connects strategic and operational levels through top-down strategy deployment and bottom-up KPI aggregation [16]. KPIs should be derived from strategic goals through clearly defined success factors and visualized using driver trees to ensure continuity from corporate vision to operational metrics [17]. Finally, KPIs and targets should drive the operational process improvement rather than isolated departmental optimization [7].

2.2 Weaknesses in Existing PMS

Despite the widespread use, studies reveal weaknesses in existing PMS (table 1), particularly the lack of a consistent process for KPI identification [11,18]. These shortcomings are exacerbated by the increasing complexity in production systems [18]. Recent research has increasingly focused on data-driven decision-making in SFM, particularly in KPI analysis, to enhance transparency and support operational decisions [19]. Advanced approaches apply intelligent methods such as anomaly detection and correlation analysis to identify cross-process KPI deviations [18], or leverage machine learning for performance evaluation in complex small-batch production [20]. Despite these advances, consistent and process-oriented KPI systems remain underdeveloped. Current research emphasizes analysis of existing indicators but lacks structured approaches for end-to-end KPI definition [19]. Multidimensional KPI networks covering time, quality, efficiency, and flexibility can enhance alignment across global production networks [3], yet guidance for defining such systems is scarce. In the context of Circular Economy (CE), for example, the absence of harmonized KPIs due to inconsistent metrics and definitions hampers the integration of lean and CE goals [21]. A holistic KPI system based on the triple bottom line (economic, environmental, social) is essential for balancing efficiency with sustainability [21].

Table 1. Deficiencies of PMS according to [11,18]

ID	Deficit
D1	Large number of recorded KPIs that obscure focus on the essentials
D2	Overemphasis on financial KPIs
D3	KPIs, often arbitrarily chosen; no clear link to corporate strategy
D4	Unclear, missing, or insufficiently communicated KPI definitions
D5	Relationships between individual indicators are often unknown
D6	Lack of future orientation of KPIs
D7	Missing connection of the PMS to the actual causes of problems
D8	Employees lack knowledge to interpret KPIs
D9	Missing integration with existing management and manufacturing systems
D10	Lack of real-time performance evaluation
D11	High effort required for data collection
D12	KPIs not broken down to individual employee level
D13	Misaligned KPIs across and within departments
D14	KPIs provide distorted view due to outdated or selective data

These challenges are showing, there is no existing approach for a systematic and consistent approach to deriving KPIs from corporate strategy, with a clear focus on process view and process improvement. Additionally, they intensify existing challenges for performance measurement, as shown in Tabel 1. Against this backdrop, it is necessary to develop a framework that supports the structured, strategy-driven, cross-functional definition of KPIs, providing a basis for transparency, goal alignment, and holistic performance management.

2.3 Capabilities of Large Language Models

Recent developments in the field of LLMs offer promising opportunities for enhancing digital SFM [5]. Their core capabilities are memorization, reasoning, generalization, and diversification, enabling the interpretation of natural language and structured extraction of information from unstructured sources such as failure logs, manuals, or historical data [22]. Memorization helps overcome challenges like missing expert knowledge or time constraints by recalling domain-specific facts and procedures, while reasoning allows LLMs to perform step-by-step analyses and support operational decisions logically [22]. However, limitations persist: LLMs trained predominantly on public data struggle in domains requiring access to private or context-specific knowledge [6]. Additionally, probably incorrect generated outputs (hallucinations) can impair reliability [23]. To mitigate these issues, prompt engineering and retrieval-augmented generation (RAG) approaches have gained traction [23]. These methods extend the model's context through task-specific prompts or embedded document chunks, enabling more relevant responses even to short, vague user inputs [6]. In complex production environments, especially under time pressure and incomplete user queries, LLMs must still navigate domain-specific semantics and extract actionable insights. Despite these challenges, their ability to generalize from examples and adapt to prompt refinement makes them valuable tools for detecting deviations, identifying causes, and proposing corrective actions across the shop floor [6].

3 Development of KPI Definition Framework

To address the challenges, presented in Sect. 2.2, PMS must support strategy-oriented and multidimensional KPI definition. Therefore, PMS must evolve toward supporting structured and strategy-oriented KPI definition. Enhanced transparency across hierarchical levels enables a consistent pursuit of corporate goals in day-to-day operations and fosters a shared understanding of value creation across the organization. To this end, a KPI definition framework has been developed. First, the requirements are formulated, followed by a description of the framework and a demonstration of its benefits using a fictitious use case. The deficits D1-D5, D7-D9, and D11-D14 of existing PMS (Table 1) can be avoided through a consistent consideration in the framework. To address them in a targeted manner, the requirements shown in Table 2 are placed on the framework. Deficits D6 and D10, related to forecasting and real-time evaluation, are outside the scope, as the framework focuses on KPI selection and system assessment.

The goal of the framework (Table 1) is to support Operational Excellence employees (users) in defining suitable and harmonized KPIs, thereby ensuring transparency in production processes despite their complexity and interdependencies. To this end, the concept for the framework is based on the advantages of an LLM to support the user in interpreting the company strategy and guiding through the KPI-definition process, taking into account the requirements in Table 2. A three-stage process is chosen to enable an initial contextualization followed by a structured derivation of KPIs (Stage 1), thereby ensuring their strategic alignment. Based on the identified KPIs, an iterative coordination with relevant stakeholders takes place in Stage 2. This early involvement

Table 2. Strategic, content-related, and methodological requirements for framework development and addressed deficits from Table 1

ID	Requirement	Addressed
R1	Clear linking of corporate strategy and operational KPIs	D3, D12, D13
R2	Balanced consideration of economic, ecological, and social and circular objectives	D2
R3	Adaptability of the definition and selection of KPIs to the specific corporate division under consideration	D4, D7, D13
R4	The number of KPIs used should be limited through strategically justified selection processes	D1, D3
R5	Consideration of established approaches for KPI definition	D5
R6	Clear definitions and communication structures for KPIs	D4, D8
R7	Representation of systemic interrelationships between KPIs	D5
R8	Avoidance of redundant and contradictory KPIs	D1, D5
R9	Active involvement of employees in the KPI development process	D4, D8
R10	Clear identification of the data required for KPI formation	D9, D11, D15
R11	Enabling focus on the most important improvements	D7, D12

aims to foster transparency during the KPI definition phase and increase acceptance among stakeholders. Finally, Stage 3 focuses on the technical validation and feasibility of the KPIs. This ensures that the defined KPIs are not only meaningful but also technically measurable and implementable.

After entering the company strategy, organisation chart, and business area considered, an LLM extracts keywords and generates a driver tree in Stage 1. For consideration of potential cause-effect relationships between corporate levels and departments with regard to KPI definition, the LLM analyses the organizational chart first (addressing R1, R3). These relationships form the starting point for building the KPI cascade and aligning objectives across hierarchical levels (R1, R7). The LLM interprets unstructured strategy texts (e.g., vision, goals, existing KPIs, regulatory requirements) and transforms them into structured KPI suggestions by interpreting natural language and extracting success factors from the company strategy out of a text field (R1). The step-by-step approach based on driver tree logic is intended to encourage the strategy-compliant, targeted and consistent definition of KPIs with a clear focus on process view and process improvement (R1, R6, R8, R11). Target prompts such as *What success factors influence goal X?* or *What sub-goals lie between goal Y and the measurable result?* guide the derivation of sub-goals and success factors, supported by a curated KPI database (R2, R4, R8, R9). To this end, the KPI database contains information such as the definition, description, and calculation of KPIs and will be connected via RAG, which means that training the model is not necessary. The output of *Stage 1*, is a by the LLM generated cascaded KPI structure with contextual information for implementation (R6, R7). In *Stage 2*, the KPIs are refined in coordination with stakeholders and top management, new insights feed into a feedback loop (R9, R11). In *Stage 3*, the LLM assesses the availability of the required data for the KPI calculation defined in *Stage 2* by interfacing with

MES, ERP, and other systems (R10). If data exists, the LLM uses available tools to assist KPI calculation by proposing SQL queries or providing visualisation suggestions. If not, it suggests appropriate data collection measures. The final output is a KPI sheet containing source data, responsibilities, cascade position, and scope. The structured approach ensures alignment with strategic goals, reduces redundancy, and clarifies interdependencies, avoiding user overload.

4 Demonstration of the Proposed Framework

To demonstrate the functionality of the developed KPI definition framework, a prototype was implemented and validated using a fictitious use case. The foundation is a curated KPI database based on SQLite, comprising 125 indicators across twelve categories (e.g., quality, productivity, sustainability). The KPI database was created from a literature review and lists textually described KPIs, along with their explanations and calculation formulas. Each KPI was assigned to one of three organizational levels (production, management, top management). The database is integrated into the system using a RAG approach. The KPI descriptions were embedded using the text-embedding-model Sentence Transformers and stored in a vector database using FAISS index. In order to input the corporate strategy into the LLM, a user interface based on the Streamlit Python package was developed. After submission and initiation of the analysis, the strategy is passed as a user prompt together with a system prompt to a LLM. The LLM used was Llama 3.3:70b, deployed through an Ollama instance. The system prompt specifies the role of the LLM as a lean expert and additionally provides an introduction to the driver tree logic according to [17]. For the functional test, three hierarchical levels (production, management, top management) were assumed, with the requirement that KPIs can be aggregated across these levels. Furthermore, it was defined that the LLM should primarily retrieve KPIs from the connected database and explicitly mark any additional, model-generated KPIs. For evaluation purposes, the framework was tested using the strategy as the user prompt input: *"Our machines are designed to set the standard for the semiconductor industry through shorter innovation cycles and maximum precision."* First, the LLM derived the driver tree logic and subsequently generated sub-goals, success factors, and their descriptions. In addition, relevant KPIs from the database as well as further KPIs proposed by the LLM were provided. Table 3 summarises the LLM's output in tabular form under consideration of the reasoning structure of the LLM. Appropriate KPIs are derived from the database based on success factors by converting the corporate strategy into a vector.

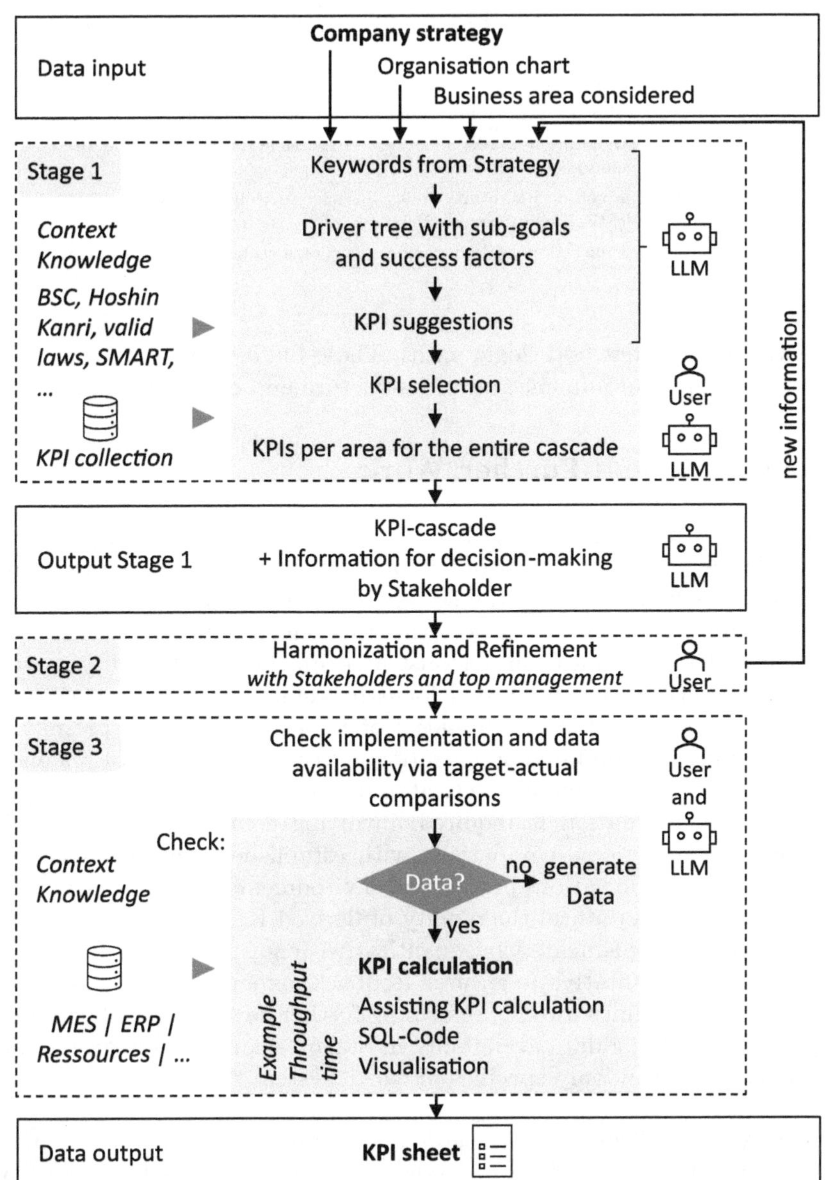

Fig. 1. KPI definition Framework

As shown in Table 3, strategy-aligned sub-goals were identified. The success factors from driver tree 1 for reducing innovation cycles and their KPIs require critical evaluation regarding their relevance. In contrast, driver tree 2 shows a clearer structure and yields more interpretable results. The three additional KPIs

Table 3. Summarized output of the KPI definition Framework

Provided driver trees	Driver tree 1	Driver tree 2
Sub-goals	Shorten innovation cycles	Increase precision
Success factors	Machine flexibility, changeover efficiency	Process stability, quality control
KPIs from database	Machine Flexibility Index (MFI), changeover efficiency	reproducibility, defect density
Additional KPIs	Innovation cycle duration	process stability index, quality rate

consistently complement both logic paths. These findings indicate that varying the system prompt can influence output structure and database integration.

5 Conclusion and Further Work

This paper presents a framework that uses an LLM to address key deficits of current PMS, particularly unclear indicator definitions and weak strategic alignment. The framework supports the selection and definition of KPIs to enable more transparent, strategy-driven performance management in complex production systems. A first application demonstrated its core functionality as well as current limitations. Future work will focus on refining the framework components, assessing the automation potential of LLMs (e.g., trough prompt engineering), defining interpretable organisational structures for the LLMs and validating the framework through practical application. To this end, a long-term, practice-oriented evaluation is required, involving continuous use in strategy reviews and KPI revisions, comparison with established methods such as the BSC, and application in various production environments. The evaluation should also consider user acceptance, the quality of derived KPIs, and their impact on decision-making, complemented by quantitative (e.g., precision, coverage, efficiency gains) and qualitative (e.g., user feedback, expert reviews, case studies) assessments, with findings directly informing further development. Furthermore, the module for verifying data availability in Stage 3 should be integrated, as the KPI database currently only specifies in which system the data may reside.

Acknowledgments. This work was carried out within the research project CLiCE-DiPP (funding code: 01MN23023F). Special thanks go to the Federal Ministry for Economic Affairs and Energy (BMWE) for funding the project and to the project management agency DLR (DLR Project Management Agency) for their support.

References

1. Vidal, G.H., Coronado-Hernáandez, J.R., Minnaard, C.: Measuring man- ufacturing system complexity: a literature review. J. Intell. Manuf. **34**, 2865–2888 (2023)

2. Romero, D. et al.: In: Goedicke, M., Neuhold, E., Rannenberg, K. (eds.) Advancing Research in Information and Communication Technology, pp. 194–221. Springer International Publishing, Cham (2021)
3. Verhaelen, B., Mayer, F., Peukert, S., Lanza, G.: A comprehensive KPI network for the performance measurement and management in global production networks. Prod. Eng. Res. Devel. **15**, 635–650 (2021)
4. Raoufi, K. et al.: Current state and emerging trends in advanced manufacturing: smart systems. Int. J. Adv. Manufact. Technol. (2024)
5. Müller, M., Alexandi, E., Metternich, J.: Digital shop floor management enhanced by natural language processing. Procedia CIRP **96**, 21–26 (2021)
6. Cicek, E., Wang, Y., Metternich, J. Enhancing Deviation Management Using Knowledge Graph based Retrieval-Augmented Generation System: accepted for publication. Procedia CIRP (2025)
7. Suzaki, K.: The New Shop Floor Management: Empowering People for Continuous Improvement. The Free Press, New York (1993)
8. Hertle, C., Tisch, M., Metternich, J., Abele, E.: Das Darmstäadter Shopfloor Management-Modell. Zeitschrift füur wirtschaftlichen Fabrikbetrieb **112**, 118–121 (2017)
9. Neely, A., Gregory, M., Platts, K.: Performance measurement system design. Int. J. Oper. Prod. Manag. **15**, 80–116 (1995)
10. Ohlig, J., et al.: Performance management on the shop floor - an investigation of KPI perception among managers and employees. Int. J. Qual. Serv. Sci. **12**, 461–473 (2020)
11. Brenner, J.: Shopfloor Management und seine digitale Transformation: Die besten Werkzeuge in 45 Beispielen, Hanser, München (2019)
12. Meißner, A., Grunert, F., Metternich, J.: Digital shop floor management: A target state. Procedia CIRP **93**, 311–315 (2020)
13. Kaplan, R.S., Norton, D.P.: The balanced scorecard: measures that drive performance. In: Harvard Business Review, pp. 71–79 (1992)
14. Kaplan, R.S., Norton, D.P.: Strategic learning & the balanced scorecard. Strategy & Leadership **24**, 18–24 (1996)
15. Serdar Asan, Ş, TanyaŞ, M.: Integrating Hoshin Kanri and the balanced scorecard for strategic management: the case of higher education. Total Quality Manag. Bus. Excellence **18**, 999–1014 (2007)
16. Ravelomanantsoa, M.S., Ducq, Y., Vallespir, B.: A state of the art and comparison of approaches for performance measurement systems definition and design. Int. J. Prod. Res. **57**, 5026–5046 (2019)
17. Gottmann, J.: Produktionscontrolling: Wertströome und Kosten optimieren. Springer Fachmedien, Wiesbaden (2016)
18. Longard, L.P.: Datenbasierte Entscheidungsunterstüutzung füur das wertstromübergreifende Performance Management Ph.D. thesis, TU Darmstadt (2024)
19. Eichenseer, P., Winkler, H.: A data-oriented shopfloor management in the production context: a systematic literature review. Int. J. Adv. Manufact. Technol. **134**, 4071–4097 (2024)
20. Cupek, R., Ziebinski, A., Drewniak, M.: Fojcik, M. In: Nguyen, N.T., Hoang, D.H., Hong, T.-P., Pham, H., Trawiánski, B. (eds.) Intelligent Information and Database Systems, pp. 661–673. Springer International Publishing, Cham (2018)
21. Wittine, N., Kruthaup, B., Stolipin, J., Wenzel, S.: Combining Lean Management And Circular Economy: A Literature Review. Hannover (2024)

22. Li, J. et al.: Fundamental Capabilities of Large Language Models and their Applications in Domain Scenarios: A Survey. In: Ku, L.-W., Martins, A., Srikumar, V (eds.) Proceedings of the 62nd Annual Meeting of the Association for Computational Linguistics (Volume 1: Long Papers), pp. 11116–11141. Association for Computational Linguistics, Stroudsburg, PA, USA (2024)
23. Gao, Y. et al.: Retrieval-augmented generation for large language models: A survey (2023). arXiv:2312.10997 2

Towards Data-Driven Closed-Loop Control of Product Properties in Progessive Die Forming

Ciarán-Victor Veitenheimer(✉), Robin Krämer, Felix Georgi, and Peter Groche

Institute for Production Engineering and Forming Machines, Technical University of Darmstadt, Darmstadt, Germany
ciaran.veitenheimer@ptu.tu-darmstadt.de

Abstract. Progressive die forming is a highly productive method in sheet metal forming, enabling the manufacturing of complex components through multiple sequential forming stages. However, forming processes are typically influenced by a multitude of sources of uncertainty, including variations in the properties of semi-finished products and tool wear. This often results in fluctuating product properties and unstable forming processes. Currently, compensating for these variations requires extensive knowledge from experienced operators. Automated adaptation to varying process conditions presents multiple challenges: In the process of selecting and integrating sensors, it is essential to ensure that relevant process variables are being measured, which is typically associated with proximity to the forming zone. Conversely, the process should be unaffected by sensor integration. Additionally, in progressive dies, signals from multiple stages can interfere with each other. Therefore, the objective of this work is to investigate which sensor technology, in combination with data-driven models, is best suited for closed-loop control of forming processes in progressive dies. The present study proposes a methodology for the control of part angles in an adaptive bending stage. The actuator for the adaptive bending stage is a combination of motor and wedge gear, allowing for continuous adjustment of the bending angle. To model the influence of uncertainty in metal forming, the process is deliberately disturbed by varying the properties of the semi-finished product. Therefore, different types of sensors are integrated into the tool system and used in combination with machine learning models to control the part properties, with the aim of reducing the dependence on manual expertise.

Keywords: Smart manufacturing · Closed-loop control · Progressive die

1 Introduction and State of the Art

Progressive dies represent a highly efficient class of forming tools, that has been widely adopted in industrial practice thanks to their ability to perform multiple

L. Overmeyer and B.-A. Behrens (eds.), *Production at the Leading Edge of Technology*, Lecture Notes in Production Engineering, https://doi.org/10.1007/978-3-032-19524-1_57

operations in a single press stroke. While their productivity is undisputed, their operational rigidity and lack of feedback mechanisms render them vulnerable to process disturbances. Deviations in material properties, tool wear, or temperature gradients cannot be compensated in real time, which often results in quality deviations or increased scrap, especially in sensitive forming stages [4]. The concept of integrating closed-loop control into forming processes has gained traction in recent years, with the aim of transitioning from observation to closed-loop control. Allwood et al. [2] outlined the theoretical and practical requirements for such systems, emphasizing observability, controllability, and model-based decision-making. Their framework has since been used to guide the design of sensor-integrated, feedback-capable forming setups. To render forming systems observable, various sensor types have been explored. Force and strain sensors yield direct measurements of contact conditions and tool loads, while acceleration sensors provide dynamic signatures suitable for detecting vibration-related anomalies [14]. While the integration of force sensors often necessitates extensive adaptation work on the tool, the attachment of acceleration sensors to dies is typically a relatively straightforward process. However, all data-driven approaches require modeling to map sensor data to actionable variables [13]. In the literature, modeling typically focuses on material characterization and wear monitoring ([6,11]). Nevertheless, approaches related to process control can also be found and will be outlined in the following. Data-driven models have proven effective in this domain, enabling predictive estimation of bending angles or sheet deformation ([5,12]). Unterberg et al.[10] used deep convolutional neural networks trained on force signals from blanking operations to classify material grades in situ. Molitor et al.[7] implemented a model for online springback compensation by combining camera-based angle detection with a neural network trained on force signals. Similarly, Sharad and Nandedkar [9] used a neural network to predict springback in U-bending operations. In high-speed forming, Schenek et al. [8] demonstrated that material properties could be extracted from punching force profiles using shallow neural networks. This offers an in-line alternative to tensile testing. These examples illustrate the increasing industrial relevance of data-driven modeling in achieving robust process understanding and control. Yet, the complexity of progressive dies–with their high actuation frequency, spatial constraints, and superposing tool interactions–makes closed-loop control particularly challenging.

2 Methodology and Experimental Setup

As described in Chap. 1, progressive dies are of high industrial relevance due to their high productivity. Conversely, if the process deviates from its stable range due to uncertainty, there is a significant risk of high scrap rates. The objective of this study is threefold: to develop a flexible design for the bending stage of a progressive die; to integrate heterogeneous sensor technology at various points in the tool; and to control the bending angle of a tab based on this integration. The effectiveness of the process is to be evaluated by processing different materials, sheet thicknesses and target angles. The progressive die used in this study is depicted schematically in Fig. 1. As illustrated in Fig. 1a, the general structure is as follows: The sheet metal is inserted into the left side of the tool, and two pilot holes with a diameter of $d = 6$ mm are punched into the strip edges in the first stage. In order to characterize the semi-finished product, a force measuring ring of type 9051C from Kistler (illustrated in red in Fig. 1; F_{p}) is installed directly in the force flow of the punch on the operator side. Moreover, two acceleration sensors of the 352C03 type from PCB Piezotronics have been incorporated into this area (see blue section in Fig. 1). The first sensor is attached to the upper ($a_{\mathrm{p,t}}$), the second is attached to the lower plate ($a_{\mathrm{p,b}}$).

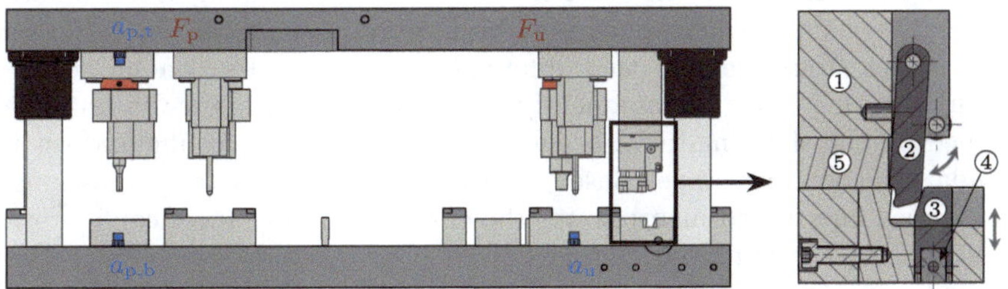

(a) Progressive die with two force (red) and three acceleration (blue) sensors

(b) Detailed view of adaptive bending stage

Fig. 1. Schematic representation of the progressive die with integrated sensors and adaptive bending stage

The secondary stage of the tool is exclusively designated for guiding the coil, with no additional forming operations being conducted at this stage. In the subsequent stage, a U-shaped hole is punched out. This hole is ultimately the prerequisite for the tab that is finally bent out. The punching force is measured using the same type of force sensor as in the first stage (F_{u}). An additional acceleration sensor is also situated on the lower tool plate for the purpose of monitoring vibrations (a_{u}). The final stage of the process is the adaptive bending stage, which involves the variable adjustment of the tab angle. Figure 3 illustrates the bending angle by way of example. The operating principle of the adjustable bending angle is schematically illustrated in Fig. 1b: The bending punch (2) is connected to the upper tool (1) with a rotational degree of freedom. The blank holder (5) is spring-mounted and also attached to (1). As the upper tool is lowered by the press ram during operation, the blank holder clamps the sheet strip at the bending edge. Subsequently, the bending punch contacts the pre-cut tab and bends it downward. A wedge (3) tilts the bending punch further in the bending direction during operation, forming the component to the specified angle. Variability in the bending angle is achieved by adjusting the height of wedge (3). A servomotor (Bosch MS2N03-B0BYN-CMSG1) and spindle move a wedge (4) in and out of the drawing plane. Motor control and sensor signal acquisition are handled by a measurement computer running LabVIEW, with a National Instruments cRIO 9047 synchronizing sensor data via hardware trigger at 25 kHz. The recorded time series, which may contain information about the sheet material, form the basis for data-driven bending angle control. To compensate springback, the final part angle must be estimated within a single stroke from the sensor signals. The use of 1D convolutional neural networks (1D-CNNs) appears to be a promising approach for this purpose, as their parameter-shared convolutional kernels enable efficient extraction of time invariant feature patterns, thereby enhancing generalization with limited data and improving robustness to temporal shifts [3]. In contrast, reinforcement learning, despite its inherent suitability for control-oriented tasks, exhibits significant data inefficiency in real-world settings. Moreover, in the absence of extensive pretraining in a simulation environment, its application appears to be largely impractical. An additional challenge arises from the fact that different bending angles can be achieved within the bending stage. Consequently, the final part geometry is not solely determined by the sheet material properties but also by the configured setting of the adaptive bending stage. Based on this, a 1D-CNN that processes various time series inputs and incorporates the current bending stage configuration via motor position is employed. Sensor data is preprocessed by removing offsets and drifts, normalizing, and interpolating to a uniform length of $l_{\text{sig}} = 14700$ points. Acceleration signals can additionally be filtered and transformed into the frequency domain. Since the signals are significantly longer than the scalar information provided by the motor position, this positional data is fed into the model after feature extraction through the convolutional layers (concept describe in [1]). Figure 2 shows the structural layout of the model used.

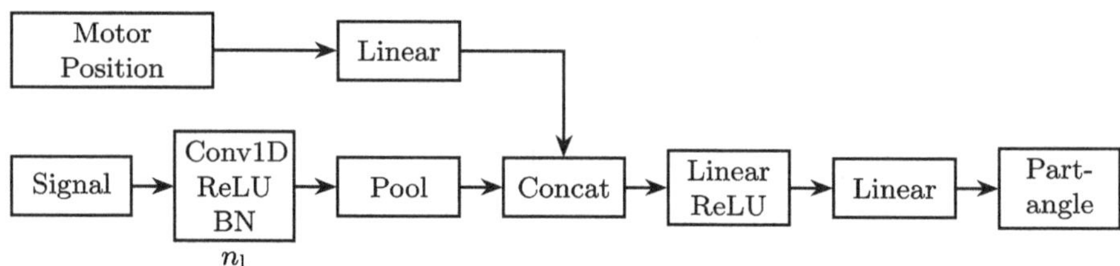

Fig. 2. Schematic structure of the CNN used

For each signal, Bayesian hyperparameter optimization is performed using the Python library Optuna. The varied hyperparameters and the corresponding search space are listed in Table 1. The data is consistently split into training, validation, and test sets with a ratio of 70:15:15.

The experiments are conducted on a Bruderer (BSTA 810) at a stroke rate of 80 strokes per minute. The initial experiments are carried out using DC04 sheet metal with a thickness of 0.95 mm, supplied from coil. To simulate the influence of material fluctuations and to evaluate the prediction performance

Table 1. Hyperparameters varied during optimization and their respective ranges

Description	Hyperparameter	Search space
Number of layers	n_l	[1, 5]
Learning rate	lr	$[1 \times 10^{-5}, 1 \times 10^{-2}]$
Batch size	bs	[8, 128]
Kernel size	$s_k, k \in [1, n_l]$	[8, 64]
Filter size	$s_f, f \in [1, f_l]$	[8, 64]
Filter type	-	{High-pass, Low-pass}
Cutoff frequency	f_{cutoff}	{500, 1000, 5000, 10000}

of the model, additional tests are performed using DC01 with sheet thicknesses of 0.8 and 1.0 mm. The resulting datasets are labeled accordingly as $\mathbf{m}_i$ with $i \in$ {DC04, DC01,0.8, DC01,1}. For training and validation purposes, the final unloaded bending angle of all tabs must be measured. Preliminary experiments revealed that angle detection using a camera system is insufficiently accurate in this context. This inaccuracy is primarily attributed to factors such as limited accessibility, lighting conditions, and burr formation inherent to the process. Therefore, all components were measured using a 3D laser scanner (Hexagon Absolute Arm AS-1). Figure 3 shows a scanned sheet metal section, clearly illustrating the burr formation around the U-shaped punching. To assess measurement accuracy, three of the depicted tabs were scanned and evaluated five times using the system. The results of these measurements are presented in Table 2 and are deemed sufficiently accurate for the purposes of this study.

Fig. 3. Exemplary 3D scan of final part geometry

Table 2. Corresponding measurement statistics for three tabs, each measured five times

Part	Mean [°]	σ [°]
1	104.376	0.071
2	104.708	0.053
3	104.732	0.059

3 Results

Before initiating model training, the influence of the adjustable bending stage is investigated. To this end, 30 experiments were conducted with DC04 at each of seven different motor positions. The results were evaluated as described above and are shown in Fig. 4a. It is clearly visible that the motor position has a distinct and approximately linear effect on the unloaded part angle, demonstrating the potential for targeted adjustment. For DC04, the resulting angles range on average from 100.81° to 109.03°.

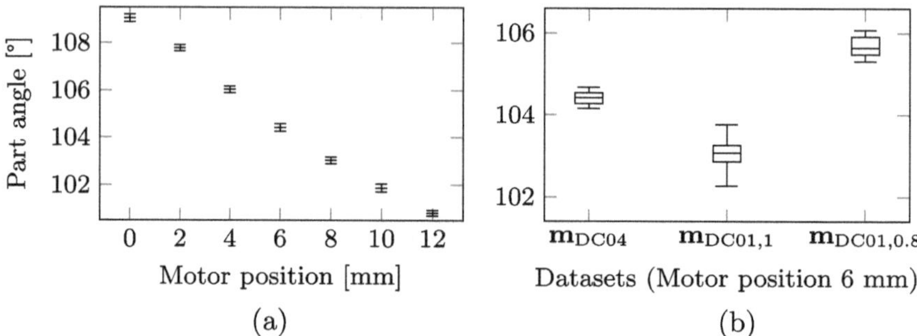

Fig. 4. Boxplots of uncontrolled part angles for various motor positions and DC04 (**a**) and single motor position (6 mm) with different materials (**b**)

To investigate the influence of material fluctuations, additional experiments were conducted using different materials and sheet thicknesses, as described in Chap. 2. The corresponding results for a fixed motor position (6 mm) are shown in Fig. 4b. The differences in springback behavior are clearly visible: $\mathbf{m}_{\mathrm{DC01,1}}$ reaches a median angle of 103.07°, exhibiting 1.35° less springback compared to components in $\mathbf{m}_{\mathrm{DC04}}$. Notably, a higher variation in the resulting part angles was observed. One potential explanation for this phenomenon is the wider permissible strength tolerance observed in DC01 in comparison to DC04. Components from the dataset $\mathbf{m}_{\mathrm{DC01,0.8}}$ reach a median angle of 105.48°, which corresponds to 1.05° more springback than the reference dataset. This appears plausible due to the reduced sheet thickness and the associated increase in elastic bending contribution. Figure 5 presents the test loss of the final trained and evaluated models as box plots. The loss measures the deviation between the predicted and the actual part angle, directly reflecting prediction accuracy. Each plot is based on 30 models trained and evaluated using the methodology described in Chap. 2. The results for sensor a_{u} are not shown in Fig. 5, as the variation is significantly larger compared to the other models. All model results are summarized in tabular form (see Table 3). Based on the hyperparameter optimization, all acceleration signals are low-pass filtered with a cutoff frequency of $f_{\mathrm{cutoff}} = 500$ Hz and transformed into the frequency domain. Figure 5 clearly shows that models using $a_{\mathrm{p,t}}$ as input achieve the highest prediction accuracy, with a maximum deviation of 0.228°. This demonstrates that the signal allows for precise characterization of the sheet material properties related to springback. The median difference between the predicted and actual bending angle based on $a_{\mathrm{p,t}}$ is only 0.1°. In contrast, models trained using $a_{\mathrm{p,b}}$ show considerably greater variance.

Models that process the a_u signal demonstrate a significant degree of variability and are deemed to be unsuitable, with deviations of up to 14.3°. One potential explanation for this phenomenon is the close proximity of the sensor to the adaptive bending stage. At this point, vibrations caused by the spring-loaded blank holder, the bending punch and the adjustment mechanism may overlay the signal, thereby disrupting the extraction of meaningful features. Given that sensor $a_{p,b}$ is mounted on the same tool plate, this may also provide a potential explanation for the higher variance observed in models based on this signal compared to those using $a_{p,t}$. The two models that utilise force data as an input demonstrate greater consistency in their performance when compared to those that are based on acceleration data. The median deviations of F_u and F_p are 0.24 ° and 0.17 °, respectively, with maximum deviations of 0.82 ° and 0.79 °. Whilst the prediction accuracy is lower in comparison to models that utilize $a_{p,b}$ as an input, the preprocessing is considerably more efficient, as no filtering or transformation operations are necessary. In addition, these models demonstrate reduced sensitivity to alterations in hyperparameters and positioning.

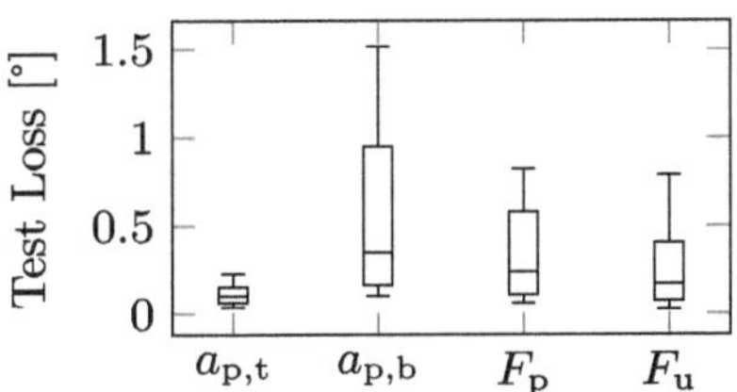

Fig. 5. Boxplots of test loss for models trained with different sensor signals

Table 3. Test loss statistics for models trained with different sensor signals

Sensor	Median [°]	Min [°]	Max [°]
$a_{p,t}$	0.101	0.037	0.228
$a_{p,b}$	0.349	0.105	1.515
a_u	0.877	0.085	14.30
F_p	0.239	0.061	0.821
F_u	0.172	0.029	0.789

4 Conclusion and Outlook

A progressive die with an adaptive bending stage was successfully commissioned in this work, and the adjustability of the bending angle was reliably demonstrated. In addition, force and acceleration sensors were integrated at various positions within the tool, serving as the foundation for property control. Based on the integrated sensor signals and the target angle, an accurate prediction of the bending angle was achieved using a 1D-CNN. The highest prediction accuracy was obtained using an acceleration sensor mounted on the upper tool in close proximity to the pilot punch ($a_{p,t}$). Models using acceleration signals as input show high sensitivity to the sensor placement within the progressive die. Furthermore, extensive preprocessing in the form of filtering and transformation operations is essential, along with comprehensive hyperparameter optimization. In contrast, models trained with force signals demonstrate significantly higher robustness with respect to sensor placement and hyperparameter settings but do not reach the same level of prediction accuracy.

Future work should investigate the extent to which sensor fusion can improve both the robustness of the sensing system with respect to integration location and prediction accuracy. Additionally, the proposed approach must be validated in real time application across a broader range of materials and sheet thicknesses.

References

1. Akilan, T., Wu, Q.M.J., Safaei, A., Jiang, W.: A late fusion approach for harnessing multi-cnn model high-level features. In: 2017 IEEE International Conference on Systems, Man, and Cybernetics (SMC). pp. 566–571 (2017). https://doi.org/10.1109/SMC.2017.8122666
2. Allwood, J.M., Duncan, S.R., Cao, J., Groche, P., Hirt, G., Kinsey, B., Kuboki, T., Liewald, M., Sterzing, A., Tekkaya, A.E.: Closed-loop control of product properties in metal forming. CIRP Ann. **65**(2), 573–596 (2016). https://doi.org/10.1016/j.cirp.2016.06.002
3. Cao, J., Bambach, M., Merklein, M., Mozaffar, M., Xue, T.: Artificial intelligence in metal forming. CIRP Ann. **73**(2), 561–587 (2024) https://doi.org/10.1016/j.cirp.2024.04.102.=
4. Farioli, D., Kaya, E., Fumagalli, A., Cattaneo, P., Strano, M.: A data-based tool failure prevention approach in progressive die stamping. J. Manuf. Mater. Process. **7**(3). article 92 (2023). https://doi.org/10.3390/jmmp7030092
5. Havinga, J., van den Boogaard, T., Dallinger, F., Hora, P.: Feedforward control of sheet bending based on force measurements. J. Manuf. Process. **31**, 260–272 (2018). https://doi.org/10.1016/j.jmapro.2017.10.011
6. Huang, C.-Y., Dzulfikri, Z.: Stamping monitoring by using an adaptive 1D convolutional neural network. Sensors **21**(1), 262 (2021). https://doi.org/10.3390/s21010262
7. Molitor, D.A., Arne, V., Kubik, C., et al.: Inline closed-loop control of bending angles with machine learning supported springback compensation. Int. J. Mater. Form. **17**. article 8 (2024). https://doi.org/10.1007/s12289-023-01802-y
8. Schenek, A., Görz, M., Liewald, M., Riedmüller, K.R.: Data-driven derivation of sheet metal properties gained from punching forces using an artificial neural network. Key Eng. Mater. **926**, 2174–2182 (2022). https://doi.org/10.4028/p-41602a
9. Sharad, G., Nandedkar, V., Springback in sheet metal U bending–FEA and neural network approach. Procedia Mater. Sci. **6**, 835–839 (2014). https://doi.org/10.1016/j.mspro.2014.07.100
10. Unterberg, M., Niemietz, P., Trauth, D., Wehrle, K., Bergs, T.: In-situ material classification in sheet-metal blanking using deep convolutional neural networks,. Prod. Eng. Res. Dev. **13**, 743–749 (2019). https://doi.org/10.1007/s11740-019-00928-w
11. Unterberg, M., Becker, M., Niemietz, P., Bergs, T.: Data-driven indirect punch wear monitoring in sheet-metal stamping processes. J. Intell. Manuf. **35**(4), 1721–1735 (2024). https://doi.org/10.1007/s10845-023-02129-w
12. Viswanathan, V., Kinsey, B., Cao, J.: Experimental implementation of neural network springback control for sheet metal forming. J. Eng. Mater. Technol. **125**(2), 141–147 (2003). https://doi.org/10.1115/1.1555652

13. Volk, W., Groche, P., Brosius, A., Ghiotti, A., Kinsey, B.L., Liewald, M., Madej, L., Min, J., Yanagimoto, J.: Models and modelling for process limits in metal forming. CIRP Ann. **68**(2), 775–798 (2019). https://doi.org/10.1016/j.cirp.2019.05.007
14. Yang, M.: Sensing technologies for metal forming. Sens. Mater. **31**(10), 3121 (2019). https://doi.org/10.18494/SAM.2019.2399

Improved Machine Learning Based Flank Wear Identification by Incorporating 3D-Height Maps

Sebastian Schibsdat(✉) and Jan H. Dege

Institute of Production Management and Technology, Hamburg University of Technology, Hamburg, Germany
sebastian.schibsdat@tuhh.de

Abstract. Coated cutting tools are very important in the field of machining, but predicting the temporal evolution of tool wear is still a major challenge due to the high complexity of the wear process. Although machine learning (ML) shows promise in this respect, traditional supervised learning models rely on labour-intensive and error-prone manual measurement of tool wear to generate training data, which is a process that is time-consuming and inefficient. This paper uses a machine vision approach that automates the identification of flank wear on coated indexable inserts for turning applications and extends these techniques by integrating 3D tool surface data via height maps into the input. For this purpose, images of indexable inserts are taken during external longitudinal turning of C45 + N and X5CrNi18–10 using a focus variation microscope. This enables a height map of the wear area to be generated. Image annotation is performed on the 2D images to maintain comparability with traditional methods. Furthermore, the images and height maps are automatically cropped to focus on the flank wear region. This reduces the class imbalance and allows the model to maintain higher image resolution, despite the limited input. Three different U-Net models are trained on this dataset: two with height maps and one without. A comparison of the three models focusing on prediction accuracy and learning efficiency shows that including height maps improves prediction performance and reduces the required dataset size for comparable accuracy.

Keywords: Machine vision · Convolutional neural networks · Tool Wear detection · Turning

1 Introduction

Machining is a key manufacturing process. It enables the production of complex parts with great precision. Despite major advances in technology, tool wear continues to be a major problem. Predicting the wear of coated tools is challenging due to the multi-layered, process-dependent effects during cutting. These are difficult to model analytically. A new approach to these problems is the use of machine learning (ML) techniques, which shows promising results [1]. Many approaches used in the literature are based on supervised ML models that use regression or classification, for which labelled data is required [1]. Wear prediction is based on the tool's measured wear, which is usually

L. Overmeyer and B.-A. Behrens (eds.), *Production at the Leading Edge of Technology*, Lecture Notes in Production Engineering, https://doi.org/10.1007/978-3-032-19524-1_58

measured manually by experts and can be influenced by human error. To reduce the human factor in this laborious process, machine vision models have been suggested. Miao et al. and Holst et al. use a U-Net to create a segmentation of flank wear on a turning tool and a ball end mill [2, 3]. Bergs et al. expands this by incorporating tool type identification and comparing general trained U-Net segmentations with specialised versions [4]. Yoo et al. use a similar approach, but with different CNN models [5]. Chen et al. address the problems of imbalanced datasets and blurry boundaries for wear areas with an augmentation process involving diffusion and a two-step segmentation process using a fuzzy clustering technique [6]. In Zhu et al. the U-Net tool segmentation is used to predict surface roughness [7]. This study also compares the performance of different U-Net types. All of the aforementioned approaches use a 2D image for prediction and modelling. Adhesions of the workpiece material on the tool surfaces can make identification with just colour information difficult. Integrating additional information, such as surface geometry, could improve the model's accuracy. Additionally, increasing the input dimension by one more channel representing the topology increases the available training data per instance and could reduce the amount of training data needed for the same level of prediction accuracy.

This paper presents a machine vision model that has been trained using images of coated indexable inserts for turning applications. The model takes height maps of the tools as an additional input alongside the image. A separate model that uses only the image as input is compared against it. In order to ascertain the impact of incorporating height maps on the size of the training data, new models will be trained and evaluated using subsets of the original data. To focus on improvements through height maps, the classical U-Net [8] was selected as an ML model, since it is a proven CNN network used in various segmentation tasks, characterised by efficient and stable learning. This is a similar approach to that taken by Miao et al. and Holst et al., as their research has already shown that the base U-Net is capable of achieving an acceptable level of prediction accuracy [2, 3].

2 Materials and Methods

2.1 Experimental Setup

Images of the various stages of tool wear were gathered during external longitudinal turning experiments on a CNC lathe Max Mueller Gildemeister MD5S using custom-coated, indexable inserts. The setup had an orthogonal clearance of $\alpha_0 = 16°$, a tool orthogonal rake angle of $\gamma_0 = 6°$, and a cutting edge inclination of $\lambda_s = -6°$. Indexable inserts of the CNMG160608-MM5 type, manufactured by Walter AG, were used. They were made of fine-grained tungsten carbide (WC grain sizes ≤ 1 µm) and contained a binder of 10 wt% cobalt. A custom TiAlN coating was applied. This coating had a thickness of $t_{Coating} = 3.5 \pm 0.3$ µm. Some tools also had an additional decor layer containing TiN with a thickness of around $t_{Decor} = 50$ nm. The workpiece materials used were C45 + N (1.0503) and X5CrNi18–10 (1.4301). Each workpiece was cylindrical, with a usable length of $l_W = 500$ mm and a diameter ranging from $d_{Wmax} = 150$ mm to $d_{Wmin} = 100$ mm. Additionally, one side was flattened to enable temperature measurements of the rake face for use in other research. The process parameters used are listed in

Table 1. These process parameters are taken from a larger, fractional factorial design of experiments.

Table 1 Cutting process parameter used

Workpiece material	Cutting depth a_p in mm	Cutting speed v_c in m/min	Cutting feed f in mm	Cutting edges used	Wear measurements
X5CrNi18–10	1.1–1.6	170–290	0.15–0.35	11	52
C45 + N	1.2; 1.6	130; 180	0.15; 0.25	7	134

The cutting process was interrupted at intervals of $t_c = 30$ s in order to take a microscope image of the cutting tool. To facilitate this, the microscope was mounted on a gantry above the lathe, allowing the camera to be lowered into the machine chamber. With this setup, the insert could be measured in situ. Furthermore, the images obtained were highly homogeneous due to the precisely reproducible position of the camera, which simplified the preparation of the dataset. Each cutting edge was used until it reached a maximum flank wear width of $VB_{max} = 300$ μm, tool breakage, or a significant change in chip formation. For the dataset used in this paper, 18 cutting edges were used. This resulted in n = 186 (C45: 134, X5CrNi18–10: 52) measurements of tool wear.

Images and height maps were captured using a microscope VHX-7020, Keyence Corporation, including an adjustable magnification lens. Used in conjunction with the microscope controller VHX-X1F, Keyence Corporation, this allows the usage of focus variation for greater depth of field and 3D surface mapping. For these images, the microscope was oriented orthogonally to the flank face at an angle of $\psi = 35°$ to the primary cutting edge, Fig. 1. This angle was selected to also capture the secondary cutting edge. Magnification of 150x was selected, which leads to a field of view of approximately 2.34 × 1.75 mm. If the tool wear was too large to be captured completely at this magnification, an additional image was taken at 100x magnification (3.54 × 2.65 mm).

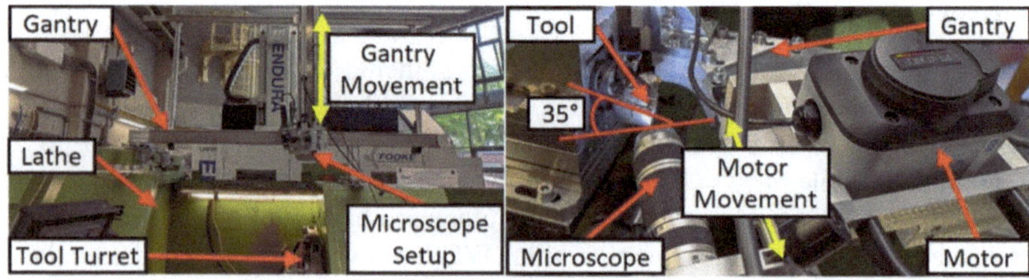

Fig. 1 Microscope setup

2.2 Hardware and Software

Image annotation was carried out manually using the MATLAB R2024b, The MathWorks Inc., 'Image Labeller' tool. Dataset preparation, hyperparameter study and final model training were conducted in Python. The main libraries used are listed in Table 2. Dataset creation, augmentation and model training were carried out on a computer with two Nvidia RTX 3090 GPUs, 128 GB of 2133 MHz RAM, an AMD Ryzen Threadripper 3970X processor and an SSD with 2 TB of storage.

Table 2 Main python libraries used for the study

Basic	Augmentation	ML training and model
Python 3.12.10 Pandas 2.2.3 Pillow 11.2.1 Numpy 2.2.5	Albumentations 2.0.7 Opencv-python 4.11.0.86	Torch 2.7.0 + cu128 Torchvision 0.22.0 + cu128 Scikit-learn 1.6.1 Optuna 4.3.0

3 ML Methodology

3.1 Database

As a result of the experiments, 404 images (.jpg format) with a resolution of 2048x1536 pixels were obtained. These included various stages of tool wear, ranging from new state to tool breakage, Fig. 2a, b. Accompanying height maps were extracted and saved as a .csv file for all images. The images were sorted into two categories according to the material used. To make creating a consistent mask easier, the tool images were first sorted by time used, and then annotated. First, the entire visible tool was segmented into the 'Tool' category, including the worn areas. Next, all types of flank wear and adhesion on the cutting edge were segmented as a single 'Wear' class, which overrode the previous 'Tool' declaration for these pixels. The mask was saved as a .png file.

Fig. 2 Example of flank wear and image pre-processing

The extracted height maps revealed an artificial surface in areas where parts of the tool were not visible. This appears to be an artefact resulting from processing through the microscope controller. To correct this, the background area of the image was identified and the corresponding areas in the height maps were set to zero.

To increase the resolution of the restricted input in the ML model, the images should be cropped to the region of interest. The first step is to rotate the images, accompanying height maps and masks so that the cutting edge aligns with the horizontal edge of the image. Additionally, the areas above and below the cutting edge are cropped. This reduces the image size to 1898×936 pixels, a total reduction of 43.5%, without any loss of information about wear. Figure 2c, d shows this process applied to one example, with the same changes applied to the accompanying mask created through the annotation process.

To allow for an accurate comparison of the different models, data augmentation was performed offline. First, the dataset was split into training and test data in a 70/30 ratio. The test data was saved directly without any additional modification and includes 121 images. Unless stated otherwise, all performance evaluations in this paper are based on these 121 images. To increase the available training data, multiple augmented versions of a single training image were created using the Albumentations Python library. The following augmentations were applied: Rotate, Horizontal Flip, Vertical Flip, Grid Distortion, Elastic Transform and Colour Jitter (only on image). Geometric and pixel-level augmentation were performed identically on the accompanying height map and mask. Colour space augmentation, however, could only be performed on the image.

Each augmentation had a 50% probability of applying a randomised value within a predetermined range. Multiple augmentations could be applied simultaneously. This increased the size of the training set of 282 images tenfold to a total of 2820 images.

3.2 Training Setup

The ML training was implemented with the PyTorch library. To incorporate both available graphics cards, the DataParallel wrapper of PyTorch was used. The basic U-Net [8] was selected as the model. As in the original proposed U-Net, two convolution layers with a 3×3 kernel and stride 1, with a rectified linear unit (ReLU) activation function, are implemented at encoder and decoder levels, with feature sizes of 64, 128, 256, 512 and 1024. Unlike the original proposed U-Net, a batch normalisation layer is implemented behind every convolution layer, as well as padding. Batch normalization was introduced to improve training speed and stability [9]. The padding ensures that the image size remains unchanged at each level. This means that the image size in the decoder is the same as in the encoder. Not only does this improve the output resolution, it also makes it easier to implement the forwarding of the encoder stage to the decoder stage and the comparison of the ground truth with the model's output.

The height map is implemented as an additional input in two ways: as an extra channel to the input image (RGB-H) and as another input (RGB + H). Including the additional input in the U-Net requires a change to the first encoder stage. Here, both inputs go through a convolution layer, a batch normalisation layer, and a ReLU layer. The two feature maps of depth 64 are concatenated to create a single feature map with depth of 128, which is then reduced to 64 using a convolution layer. A batch normalisation and ReLu layer follow. The rest of the model remains unchanged. An overview of the model structure is given in Fig. 3.

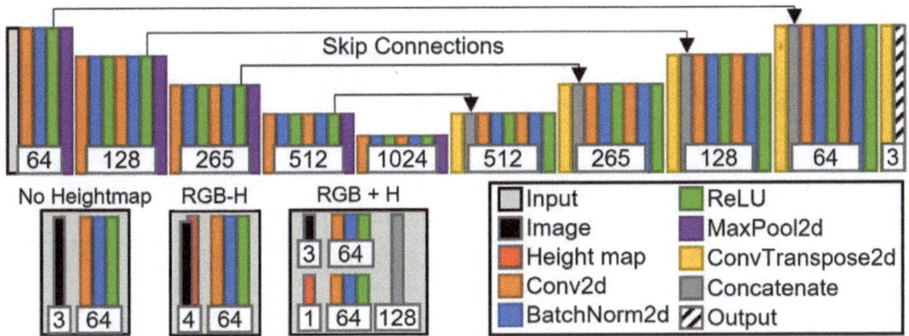

Fig. 3 Overview of the modified U-Net and input scenarios

For the model, the input size was chosen as 50% of the original image's resolution, which was 949 × 468 pixels. Normalisation of images and height maps was between 0 and 1. Cross Entropy Loss was used as the loss function and Adam as the optimiser. The model's predictions were evaluated using the Intersection over Union (IoU) score, which was calculated as the mean on the models predictions of the test set.

An initial hyperparameter study with the Optuna library was performed to determine the learning rate and batch size. The input scenario with no height map was trained 50 times with different learning rates in the range [1 × 10–5, 1 × 10–2] and batch sizes in the range [4, 24]. Validation was performed using the IoU score on a hold-out set with a 70/30 split from the training data. A hold-out sample was randomly selected and used consistently. The Tree-structured Parzen Estimator (TPE) was used as a sampler, and the Hyperbandpruner for pruning underperforming hyperparameters. Models were trained for in range [2, 10]. As the top ten models used a learning rate between 2.3×10^{-5} and 9.3×10^{-5}, so 5×10^{-5} was selected for use in this study. The hyperparameter study showed a small improvement with smaller batch sizes. However, the impact of a larger batch size was deemed acceptable in order to reduce training time. The batch size was set to 24, as this was the maximum using the available hardware.

Each of the three input scenarios (no height map, RGB-H and RGB + H) was trained for 30 epochs using the entire training dataset. A validation on the test set was performed for each epoch. This process was repeated using 80%, 60%, 40% and 20% of the training set. The reductions were achieved by excluding two to eight augmentations from all images, respectively. As there are no non-augmented images in the training set, the random distribution of test set images remains unchanged in these cases.

4 Results

For the three different models, the top plot of Fig. 4 shows the mean and standard deviation of the mean IoU score of the prediction on the test set over 30 epochs, based on five training runs. Over all training runs the model with no height map achieves its best prediction at epoch 30, with an IoU score of 0.705; the RGB-H model achieves its best prediction at epoch 30, with an IoU score of 0.7501; and the RGB + H model achieves its best prediction at epoch 28, with an IoU score of 0.7504.

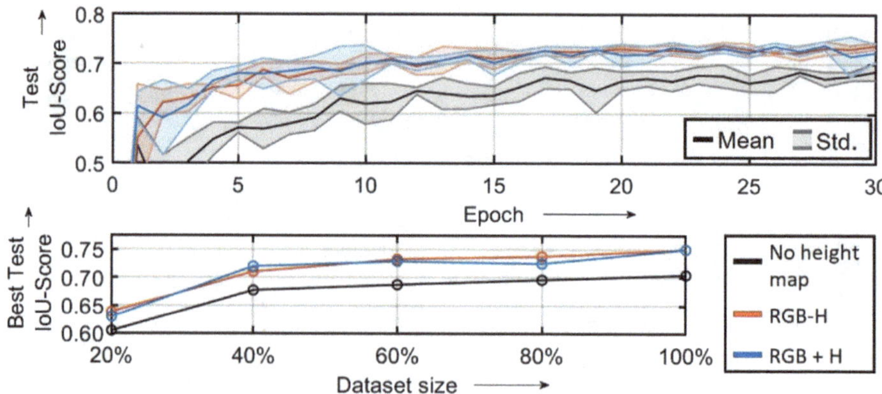

Fig. 4 Mean IoU score of the test set prediction of the three different models over the course of training on the complete dataset and smaller subsets

As the validation loss increased or stagnated from around epoch 20 in all three models, further improvement with a higher epoch count is unlikely. As the figure shows, at all epochs, the models with height maps outperformed the model that used only the image as input. Furthermore, at epochs 6 (RGB-H) and 5 (RGB + H), the models with height maps achieved a mean IoU scores within 0.05 of the final mean value, whereas the model without a height map did so at epoch 12. This demonstrates that the height map improves both the quality and speed of the model's convergence to the optimum.

At the bottom of Fig. 4, the plot shows the mean IoU score for all three models across all epochs in subsets of different sizes of the dataset. The dataset corresponding to 100% refers to the best runs of the models shown at top of Fig. 4. For datasets smaller when 100% only one training run was performed. Across different dataset sizes, the models with hight map as input both performed similar good, while the model with no height map performed worst overall. Decreasing the dataset size decreases the IoU score of the model with no height map and the model with combined RGB-H input equally. The model with split RGB + H input shows stable IoU scores between 40% and 60%. All models demonstrate a significant drop in performance when using only 20% of the dataset. As the IoU score does not plateau at or before 100% of the dataset, it can be concluded that the dataset is still too small to reach the theoretical maximum of model performance, as more training data would increase the performance further.

Fig. 5 shows an example prediction output for all models at epoch 30 for a single test image. In the model without the height map, some areas of discolouration caused by the adhesion of workpiece material were incorrectly identified as wear areas. The model with the combined RGB-H input also misidentified some discolouration, but to a lesser extent. Finally, the RGB + H split input model had an even smaller problem with discolouration, but ignored the adhesion of workpiece material to the cutting edge. As can be seen, all models produce predictions that resemble the manually created mask, albeit with an overestimation that is reduced by adding height maps. Predictions based on additional height map inputs are very similar for both models. Both models can accurately estimate wear progression along the cutting edge.

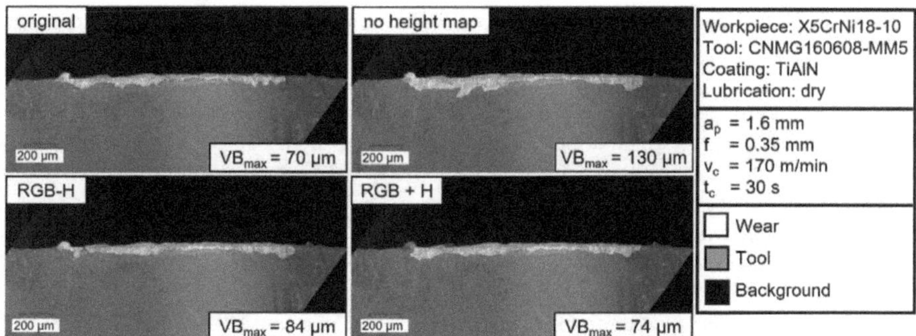

Fig. 5 Example prediction of a test image shown as overlay over the input image

5 Conclusion

This study presents a method for automatically evaluating flank wear width on indexable inserts. A dataset of 404 microscope images with height maps was created, showing varying levels of tool wear, with manual annotations for segmentation. A modified U-Net was trained to predict flank wear using three input methods: images only; images with height maps as a fourth dimension (RGB-H); and images with height maps as parallel inputs (RGB + H). Results showed incorporating height maps improved the IoU score by up to 0.045 (6.4%) and required up to 7 (58%) epochs less to achieve an IoU score close to the final value. This improvement was retained in smaller datasets.

As next steps, the implementation of wear evaluation after ISO 3685 is required so that the predicted mask can be transferred to comparable characteristic wear values. To further improve the model, different loss algorithms and models, such as the Segment Anything Model (SAM) [10] or newer variations of the U-Net [11], will be tested.

Acknowledgements. This work was funded by the Deutsche Forschungsgemeinschaft (DFG, German Research Foundation)—521385417.

Competing Interests The author(s) has no competing interests to declare that are relevant to the content of this manuscript.

References

1. Colantonio, L., Equeter, L., Dehombreux, P., Ducobu, F.: A systematic literature review of cutting tool wear monitoring in turning by using artificial intelligence techniques. Machines. **9**, 351 (2021). https://doi.org/10.3390/machines9120351
2. Miao, H., Zhao, Z., Sun, C., Li, B., Yan, R.: A U-Net-based approach for tool wear area detection and identification. IEEE Trans. Instrum. Meas. **70**, 1–10 (2021). https://doi.org/10.1109/TIM.2020.3033457
3. Holst, C., Yavuz, T.B., Gupta, P., Ganser, P., Bergs, T.: Deep learning and rule-based image processing pipeline for automated metal cutting tool wear detection and measurement. IFAC PapersOnLine. **55**, 534–539 (2022). https://doi.org/10.1016/j.ifacol.2022.04.249

4. Bergs, T., Holst, C., Gupta, P., Augspurger, T.: Digital image processing with deep learning for automated cutting tool wear detection. Procedia Manuf. **48**, 947–958 (2020). https://doi.org/10.1016/j.promfg.2020.05.134
5. Yoo, Y., Yang, G., Park, K., Hyun, Y., Jeong, S.: Extendable machine tool wear monitoring process using image segmentation based deep learning model and automatic detection of depth of cut line. Eng. Appl. Artif. Intell. **135**, 108570 (2024). https://doi.org/10.1016/j.engappai.2024.108570
6. Chen, H., Cheng, C., Hong, J., Huang, M., Kong, Y., Zheng, X.: An on-machine tool wear area identification method based on image augmentation and advanced segmentation. J. Manuf. Process. **132**, 558–569 (2024). https://doi.org/10.1016/j.jmapro.2024.10.085
7. Zhu, X., Chen, G., Ni, C., Lu, X., Guo, J.: Hybrid CNN-LSTM model driven image segmentation and roughness prediction for tool condition assessment with heterogeneous data. Robot. Comput. Integr. Manuf. **90**, 102796 (2024). https://doi.org/10.1016/j.rcim.2024.102796
8. Ronneberger, O., Fischer, P., Brox, T.: U-Net: convolutional networks for biomedical image segmentation. (2015).
9. Ioffe, S., Szegedy, C.: Batch normalization: accelerating deep network training by reducing internal covariate shift. (2015).
10. Kirillov, A., Mintun, E., Ravi, N., Mao, H., Rolland, C., Gustafson, L., Xiao, T., Whitehead, S., Berg, A.C., Lo, W.-Y., Dollár, P., Girshick, R.: Segm. Anything. (2023). https://doi.org/10.48550/arXiv.2304.02643
11. AL Qurri, A., Almekkawy, M.: Improved UNet with Attention for Medical Image Segmentation. Sensors. **23**, 8589 (2023). https://doi.org/10.3390/s23208589

Digital Transformation in Manufacturing

Evaluating Deep Learning Approaches for Industrial Process Monitoring

Paulina Merkel[1(✉)], Mareile Kriwall[1], Bernd-Arno Behrens[2,1], and Malte Stonis[1]

[1] IPH—Institut für Integrierte Produktion Hannover gGmbH, Hannover, Germany
merkel@iph-hannover.de

[2] Institute of Forming Technology and Machines (IFUM), Leibniz University Hannover, Garbsen, Germany

Abstract. Industrial process monitoring has a significant impact on ensuring quality and efficiency, particularly in complex manufacturing processes such as cross-wedge rolling. This study evaluates the performance of four deep learning approaches—Artificial Neural Networks (ANN), Convolutional Neural Networks (1D-CNN), Autoencoders, and Recurrent Neural Networks (RNN) using Long Short-Term Memory (LSTM)—for detecting deviations during the cross-wedge rolling of hybrid workpieces. Given the challenge of a small dataset, data augmentation techniques are applied to enhance model training. The investigation aims to determine whether deep learning can effectively classify process deviations and which architecture performs best. The results indicate that 1D-CNN models achieve the highest classification accuracy within the test group. Additionally, various hyperparameters are analyzed to identify those with the most significant impact on model performance. The findings contribute to the optimization of deep learning-based process monitoring, highlighting the potential of 1D-CNN and Autoencoder models for industrial applications.

Keywords: Deep learning · Process monitoring · Cross-wedge rolling

1 Introduction

As industrial processes grow more complex, traditional monitoring methods often fall short in detecting subtle deviations that may compromise product integrity [1]. One intricate process is cross-wedge rolling—a technique used to preform workpieces with high precision [2]. Recent advances in Artificial Intelligence (AI), particularly deep learning, offer promising solutions for real-time process monitoring [3]. Deep learning models can uncover complex patterns in process data, enabling the early detection of anomalies that might otherwise go unnoticed [3]. However, their successful application in industrial settings is often challenged by limited datasets [4]. This study explores the use of four prominent deep learning architectures—Artificial Neural Networks (ANN), Convolutional Neural Networks (CNN), Long Short-Term Memory (LSTM), and Autoencoders for detecting and classifying deviations during the cross-wedge rolling of hybrid workpieces. Four requirements for the research work were defined: the use of a small data

L. Overmeyer and B.-A. Behrens (eds.), *Production at the Leading Edge of Technology*,
Lecture Notes in Production Engineering, https://doi.org/10.1007/978-3-032-19524-1_59

set, classification of process deviations, application to the cross-wedge rolling process, and real-time capability. In light of the small dataset, data augmentation techniques are employed to improve model robustness. Through systematic evaluation and hyper-parameter optimization, the study aims to identify the most effective architecture for reliable process monitoring of cross-wedge rolling processes. The forming of hybrid workpieces takes approximately nine seconds. To ensure real-time capability, only the first two seconds of process data are monitored, allowing classification results to be used for in-process control. Additionally, algorithm inference times must remain below the threshold of 0.5 seconds.

2 State of the Art and Current Research

This chapter contains relevant knowledge about cross-wedge rolling, AI, and current work in the field of process monitoring.

Cross-wedge rolling is a bulk forming process used primarily for the near-net-shape production of axially symmetric components [5]. In cross-wedge rolling, a billet is plastically deformed between two contoured tools, which progressively reduce the diameter and increase the length of the workpiece [5]. The method enables a high material utilization by reducing machining allowance, and improved mechanical properties due to favorable fiber flow and grain refinement [6, 7]. Within the framework of the Collaborative Research Center 1153 "Tailored Forming" cross-wedge rolling is used to form hybrid workpieces composed of two or more joined materials with differing mechanical or thermal properties [8]. The objective is to develop and investigate novel process chains that enable the production of functionally graded, hybrid components by first joining dissimilar materials and subsequently forming them. In this context, cross-wedge rolling is employed to plastically deform previously joined workpieces into elongated, near-net-shape geometries. The process is therefore designed to account for the differing flow behaviors, forming limits, and thermal responses of the materials involved.

AI refers to computational methods that enable machines to perform tasks that typically require human intelligence, such as pattern recognition, decision-making, and learning from data [9]. Within AI, machine learning allows systems to learn patterns from historical data and make predictions or classifications without being explicitly programmed [9]. Deep learning is based on neural networks with multiple layers forming a deep architecture that can model complex, non-linear relationships [9]. Models consist of layers of interconnected neurons, which perform weighted computations and pass the results through non-linear activation functions [9].

Process monitoring has a significant impact on ensuring the quality and reliability of hybrid components [1]. Due to inherent process variations during the initial joining steps, batches of nominally identical hybrid billets can exhibit slight differences in geometry, material properties, or joint quality. While these variances may be within acceptable tolerances for the joining stage, they can have significant consequences during subsequent forming processes such as cross-wedge rolling. In particular, small deviations can lead to material flow instabilities, joint failures, or surface defects during rolling, resulting in reduced process robustness and lower yield. To address this challenge and improve the overall process reliability, the Collaborative Research Center 1153 is developing real-time process monitoring and inline control systems.

Process monitoring in metal forming typically involves the acquisition and analysis of sensor data such as force, torque, temperature, displacement and acoustic emissions [10–12]. These signals are monitored in real time and threshold-based evaluations, statistical process control, and signal trend analysis are commonly used to detect deviations, tool wear, or emerging defects [13]. AI systems can learn to recognize subtle patterns and correlations that precede process deviations or product defects [3]. Detecting anomalies in datasets is an important task to identify unusual patterns or outliers that deviate from normal data [14]. In contrast to the detection of anomalies, the classification of process deviations in this use case refers to the additional derivation of the underlying error on the basis of the available data.

CAO ET AL. identified four key areas for AI applications in metal forming: process simulation, process design, process control, and quality control [15]. TRETYAKOV ET AL. and PETRIK ET AL. applied AI methods to predict and model microstructural parameters, while BRANDAO AND SHIGAKI demonstrated advancements in digital twin technologies for metal forming [16–18]. FAHLE developed a machine learning model to enhance quality in radial-axial ring rolling [19], and DOEDE ET AL. implemented a process monitoring system for screw presses based on the CRISP-DM framework [20].

3 Experimental Procedure

A schematic of the cross-wedge rolling process and resulting data is shown in Fig. 1. Only two seconds of data are used for the development of AI models. The relevant section is marked in the Figure. The forming process begins with induction heating of the workpieces. They are then manually transferred to the cross-wedge rolling machine, where the upper tool is lowered onto the workpiece, causing slight compression. Finally, the upper and lower tools move transversely toward each other. The wedges of the tool shape the workpiece. For the training of the AI, a data set was created consisting of a reference and six deviations from the reference. The experimental setup was thoroughly described by DENKENA ET AL. [21]. The deviations included errors in the rolling process (rolling temperature lowered by 100 °C, distance between tools lowered 5 mm) and errors from the joining process (layer deviations in height, position, width, and a material deviation). Each time series consists of a measurement of the force for the horizontal movement of the upper tool and temperature measurements on the workpieces. The force sensor of the upper tool is integrated into the hydraulic cylinder used for the movement of the tool. The pyrometer used for temperature measurement is placed at the end of the tool directed at the bearing seat of the workpiece. Each deviation was repeated so that six time series were recorded. The tests with reference parameters were repeated twelve times.

Features were extracted from the time series data. The features were selected through data visualization and the identification of relevant points in the time series. Eight of the features captured characteristic points, such as: average temperatures between 0–0.25 s and 0.75–1.25 s; local force maximum between 0.25–0.5 s; force values at 0.75 s, 1 s, and 2 s; the span of force values between 0.5–1.5 s; and a local force minimum between 0.38–0.6 s. Additionally, six statistical features were extracted, including the average, median, and maximum of the force, interquartile range, and the number of maxima and minima.

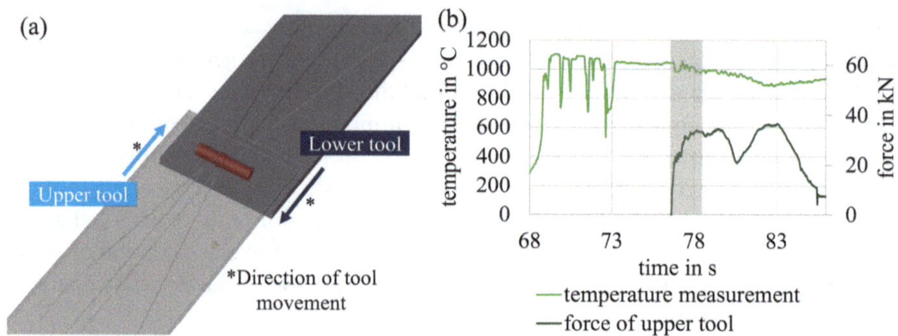

Fig. 1 **a** Schematic representation of a cross-wedge rolling process, **b** resulting data

A data augmentation was performed to enhance the number of training data. To create augmented data the averages of two datasets were calculated and used as artificial data for training. Testing was only done with not-augmented data. In order to minimize the computational effort, a fixed training and test data set was defined, which was used for the training and evaluation of the algorithms.

To investigate deep learning approaches for this use case, different variations of neural networks are investigated: ANN, 1D-CNN, LSTM, and Autoencoder. Each model type is evaluated for its ability to detect anomalies and classify process outcomes. These algorithms are selected due to their complementary strengths. ANN serve as a baseline for comparison and are suitable for general pattern recognition. 1D-CNNs are effective in extracting local features from sequential or structured one-dimensional data, making them well-suited for analyzing time series or process signals. LSTM are designed to process sequential data and capture temporal dependencies, which is beneficial for time-dependent process signals. Autoencoders are typically used for anomaly detection by learning compact representations of normal data and identifying deviations.

Hyperparameter were tuned using grid search. Grid search is a systematic method for optimizing the hyperparameters of a model. It works by defining a set of possible values for each hyperparameter and then evaluating the models performance for all possible combinations of these values [22]. For each deep learning method a different subsample of hyperparameters was chosen (see Table 1). A varying number of hyperparameters was adjusted depending on the architecture of each deep learning model. For example, training and evaluating an ANN requires configuring multiple architectural parameters, such as the number of neurons and hidden layers. In contrast, LSTM networks were used with predefined configurations (e.g., "nano", "small"), which bundle several architectural choices into preset options.

To evaluate the performance of AI algorithms validation metrics are essential. Accuracy is defined as the proportion of correctly classified instances (both true positives and true negatives) over the total number of predictions. It offers a general overview of the algorithm's performance and is particularly useful for assessing the overall correctness of a model across all classes. Accuracy alone can be misleading, since a model could achieve high accuracy simply by predicting the majority class, while failing to detect faults. To address this limitation, the False Positive Rate is also considered. False-positive rate is

Table 1 Hyperparameter for training and evaluation

		Level 1	Level 2	Level 3	Level 4
ANN	Activation function in output & hidden layer	ReLU	Tanh	Sigmoid	Softmax
	Number of Neurons in first & second layer	32	64	128	-
	Number of layers	2	3	-	-
	Learning rate	0.0001	0.001	0.01	-
1D-CNN	Filters	32	64	-	-
	Fully connected units	64	128	-	-
	Kernel size	3	5	7	-
	Dropout	0.3	0.5	-	-
	Learning rate	0.0005	0.001	-	-
	Epochs	50	100	-	-
LSTM	Activation functions	ReLU	Tanh	Sigmoid	Linear
	Model size	Nano	Small	Large	XL
	Dropout	0.05	0.1	0.2	0.5
	Learning rate	0.001	0.0001	0.00001	-
Autoencoder	Activation functions	ReLU	Tanh	Sigmoid	-
	Number of neurons in hidden layer	32	64	128	-
	Layer depth	2	3	-	-
	Dropout	0	0.2	-	-
	Learning rate	0.01	0.001	0.0001	-
	Epochs	100	200	-	-

the proportion of normal (non-defective) cases that are incorrectly classified as faults. In industrial applications, a high false-positive rate leads to unnecessary interventions, increased downtime, and reduced trust in the monitoring system.

4 Results and Discussion

The results of training and evaluation are presented in Table 2. The table presents the results of anomaly detection and classification using the described deep learning methods. For each model, the average accuracy across different hyperparameter settings is shown, along with the highest achieved accuracy and the corresponding false positive rate. Additionally, the frequency with which the highest accuracy was achieved across multiple hyperparameter configurations was analyzed to ensure that strong results are not due to chance, but rather reflect the algorithm's effective learning capability. The results show that the use of ANN leads to relatively low average accuracy. An analysis of the corresponding hyperparameters reveals that particularly low accuracy is observed when the activation functions Linear or Tanh (short for: hyperbolic tangent) are used. Among all tested parameters, the choice of activation function had the greatest impact on model performance. This was especially evident in the output layer, where the use of Sigmoid or Softmax significantly improved the results.

Table 2 Results of Deep Learning Algorithms

		ANN	1D-CNN	LSTM	Autoencoder
Anomaly detection	Average accuracy	59 %	94 %	75 %	83 %
	Highest accuracy	95 %	100 %	100 %	95 %
	False positive rate	0 %	0 %	0 %	0 %
	Number of best models	67 out of 864	4 out of 96	1 out of 192	18 out of 216
Classification	Average accuracy	35 %	81 %	59 %	52 %
	Highest accuracy	76 %	86 %	81 %	81 %
	False positive rate	18 %	0 %	0 %	0 %
	Number of best models	10 out of 864	31 out of 96	1 out of 192	4 out of 216

1D-CNN showed classification accuracies of up to 86%. Thereby, CNNs achieved the highest result in this study. Importantly, this performance was consistently observed across multiple hyperparameter configurations, indicating that the results are robust and not due to random chance. Autoencoders also performed comparatively well, achieving an anomaly detection accuracy of 95% and a classification accuracy of 81%. Notably, 18 out of 216 trained models reached an anomaly detection accuracy of 95%, further supporting their reliability. LSTM networks achieved a maximum accuracy of 100% in anomaly detection; however, this result was obtained by only one out of 192 models and is therefore likely due to random variation. A similar trend was observed in classification, where the best-performing LSTM model reached 81% accuracy. For both autoencoders and LSTM networks, the learning rate emerged as the most influential hyperparameter. Across all algorithms, the architecture of the layers—specifically the number of layers and neurons—had relatively little effect on performance. The false positive rate for classification is 0% for CNN, LSTM, and Autoencoder. Only the usage of ANN leads to a high false-positive-rate of 18%.

1D-CNNs worked best for this use case and effectively captured local patterns in structured time series data. Autoencoders worked well for anomaly detection due to their ability to learn compact representations of normal behavior. LSTMs were sensitive to hyperparameters and less robust, likely because they require more data to model temporal dependencies effectively. ANNs lacked structural biases like locality or sequence modeling, making them less capable of capturing relevant patterns in the data.

To evaluate the impact of data augmentation, the training process was repeated using the original, non-augmented dataset. The results show that data augmentation led to an increase in average accuracy across all hyperparameter configurations—by approximately 1% for LSTM and 2% for the other algorithms. For 1D-CNN, Autoencoder, and LSTM, the highest accuracy was equal to or higher when using augmented data. In the case of ANN, the highest accuracy was achieved using non-augmented data.

5 Conclusion

The evaluation of different deep learning algorithms for anomaly detection and classification reveals clear differences in performance and reliability. ANN showed limited effectiveness, particularly due to a high false positive rate. 1D-CNN consistently achieved strong and reliable results, with the highest classification accuracy (86%) and a false positive rate of 0%, making them the most robust model in this study. Autoencoders demonstrated an anomaly detection with up to 95% accuracy. LSTM networks achieved the highest individual accuracy (100%), but this result was not reproducible across different hyperparameter settings, suggesting limited robustness. Overall, CNN and Autoencoders emerged as the most promising approaches for this use case. From a practical perspective, the results indicate that the investigated deep learning algorithms are well-suited for anomaly detection in cross-wedge rolling processes and can serve as a solid foundation for the development of a real-time process monitoring system. Further research will focus on adapting the results to other datasets, investigating reproducibility through additional experiments, implementing the monitoring system in a real-world process, and examining the generalizability of the findings to other cross-wedge rolling applications.

Acknowledgement. This research was funded by the Deutsche Forschungsgemeinschaft (DFG, German Research Foundation)—CRC 1153, subproject B01—252662854.

Competing Interests The author(s) has no competing interests to declare that are relevant to the content of this manuscript.

Data Availability. Datasets related to this article can be found at https://doi.org/10.25835/z0zrg4km, hosted at Institutional Repository of Leibniz Universität Hannover.

References

1. Chukwunweike, J.N., Anang, A.N., Adeniran, A.A., et al.: Enhancing manufacturing efficiency and quality through automation and deep learning: addressing redundancy, defects, vibration analysis, and material strength optimization. World J. Adv. Res. Rev. **23**(03), 1272–1295 (2024). https://doi.org/10.30574/wjarr.2024.23.3.2800

2. Li, J., Chu, W., Feng, P., et al.: Numerical and experimental research on shafts with elliptical sections formed by cross wedge rolling. Int. J. Adv. Manuf. Technol. **127**, 2969–2978 (2023)
3. Zhao, C.: Perspectives on nonstationary process monitoring in the era of industrial artificial intelligence. J. Process Control. **116**, 255–272 (2022)
4. Brigato, L., Iocchi, L.: A close look at deep learning with small data. In: 25th International Conference on Pattern Recognition (ICPR), pp. 2490–2497, Milan, Italy (2021). https://doi.org/10.1109/ICPR48806.2021.9412492
5. Pater, Z.: Development of cross-wedge rolling theory and technology. Steel Res. Int. **81**(9), 25–32 (2010)
6. Li, Q., Lovell, M.: Cross wedge rolling failure mechanisms and industrial application. Int. J. Adv. Manuf. Technol. **37**, 265–278 (2008). https://doi.org/10.1007/s00170-007-0979-y
7. Wei, Y., Shu, X., Han, S., et al.: Analysis of microstructure evolution during different stages of closed-open cross wedge rolling. Int. J. Adv. Manuf. Technol. **95**, 1975–1988 (2018). https://doi.org/10.1007/s00170-017-1359-x
8. Merkel, P., Budde, L., Grajczak, J., et al.: Feasibility study for the manufacturing of hybrid pinion shafts with the cross-wedge rolling process. Int. J. Mater. Form. **16**, 45 (2023). https://doi.org/10.1007/s12289-023-01761-4
9. Goodfellow, I., Bengio, Y., Courville, A.: Deep Learning. MIT Press, Cambridge (2016)
10. Heß, B., Groche, P.: Untersuchungen zum oszillierenden Verzahnungsdrücken, pp. 38–39. Schmiede-Journal, März (2014)
11. Usamentiaga, R., Venegas, P., Guerediaga, J., et al.: Infrared thermography for temperature measurement and non-destructive testing. Sensors. **14**(7), 12305–12348 (2014). https://doi.org/10.3390/s140712305
12. Behrens, B.-A., Bouguecha, A., Buse, C., et al.: Potentials of in situ monitoring of aluminum alloy forging by acoustic emission. Arch. Civ. Mech. Eng. **16**, 724–733 (2016). https://doi.org/10.1016/j.acme.2016.04.012
13. Kriwall, M., Schellenberg, D., Niemann, C., et al.: Optical measurements and force measurements as a basis for predicting the tool life of forging dies. Prod. Engineering. (2024). https://doi.org/10.1007/s11740-024-01282-2
14. Schmidl, S., Wenig, P., Papenbrock, T.: Anomaly detection in time series: A comprehensive evaluation. Proc. VLDB Endow. **15**, 1779–1797 (2022). https://doi.org/10.14778/3538598.3538602
15. Cao, J., Bambach, M., Merklein, M., et al.: Artificial intelligence in metal forming. In: CIRP Annals, vol. 73, (2024). https://doi.org/10.1016/j.cirp.2024.04.102
16. Tretyakov, D., Bylya, O., Shitikov, A., Gartvig, A., Stebunov, S., Biba, N.: The prospects of implementation of artificial intelligence for modelling of microstructural parameters in metal forming processes. Material forming Esaform. **2024** (2024). https://doi.org/10.21741/9781644903131-238
17. Petrik, J., Bambach, M.: DeepForge: leveraging AI for microstructural control in metal forming via model predictive control. J. Manuf. Process. **121** (2024). https://doi.org/10.1016/j.jmapro.2024.05.023
18. Brandao Júnior, P.S., Shigaki, Y.: Advancements in digital twin application in the metal-forming industry: state of the art and challenges. In: Proceedings of the 15th International Multi-Conference on Complexity, Informatics and Cybernetics (IMCIC 2024) (2024). https://doi.org/10.54808/IMCIC2024.01.46
19. Fahle, S.: Entwicklung eines maschinellen Lernansatzes zur Qualitätsverbesserung im Radial-Axial Ringwalzen durch Zeitreihenklassifikation (2022). https://doi.org/10.13154/294-9409
20. Doede, N., Merkel, P., Kriwall, M., et al.: Implementation of an intelligent process monitoring system for screw presses using the CRISP-DM standard. Prod. Eng. **19** (2024). https://doi.org/10.1007/s11740-024-01298-8

21. Denkena, B., Behrens, B.-A., Overmeyer, L., et al.: Sensitivity of process signals to deviations in material distribution and material properties of hybrid workpieces. Int. J. Adv. Manuf. Technol. **130** (2023). https://doi.org/10.1007/s00170-023-12807-x
22. Passos, D., Mishra, P.: A tutorial on automatic hyperparameter tuning of deep spectral modelling for regression and classification tasks. Chemom. Intell. Lab. Systems. (2022). https://doi.org/10.1016/j.chemolab.2022.104520

Defreezing Horizons for Resilient Production Network Configuration Planning

Frederik Rincke(✉), Moritz Hörger, Martin Benfer, and Gisela Lanza

wbk Institute of Production Science, Karlsruhe Institute of Technology (KIT), Karlsruhe, Germany
frederik.rincke@kit.edu

Abstract. Today's volatile and dynamic world pushes historically-grown production networks to their limits. Market pressure requires production network planning to design highly efficient network configurations. Short-term shocks cause frozen decisions to be outdated. Consequently, there is a necessity for re-evaluating decisions within frozen horizons to address short-term shocks in efficient network configurations. This paper introduces dynamic rolling-horizon network configuration planning, which is less vulnerable to shocks. The approach introduces the possibility to reevaluate decisions in a frozen horizon, and "defreeze" those frozen decisions for a penalty cost, if substantial adjustments become necessary. Therefore, the concept utilizes updated forecast data in shock scenarios in a network configuration planning by adjusting network decision considering the recent system state and relevant disruption scenarios. Our conceptual approach represents an important extension of production network configuration planning, which supports practitioners with increased flexibility to manage today's volatile world. Due to its generic properties, the defreezing mechanism can further be implemented in other rolling horizon planning processes.

Keywords: Production network · Network configuration · Resilience · Rolling horizon planning

1 Introduction

In recent years, product life cycles in the automotive industry have become increasingly short [1]. The resulting increasing number of ramp-ups, in conjunction with new competition from emerging markets, necessitates companies to reduce manufacturing costs [2] by distributing production activity to global production networks (GPN) [3]. For cost-efficient GPN, a lean network configuration with low redundancies must be in place [4], highlighting the conflict between resilience and cost efficiency in network configuration [5]. As the adjustment of production networks requires considerable time [6], decision problems for network configuration utilize rolling horizon planning with frozen horizons to stabilize the planning for the upcoming periods [2]. However, today's volatile and uncertain world requires a faster pace of network configuration planning and the possibility to adapt to disruptions and changed circumstances within the frozen

L. Overmeyer and B.-A. Behrens (eds.), *Production at the Leading Edge of Technology*, Lecture Notes in Production Engineering, https://doi.org/10.1007/978-3-032-19524-1_60

horizons [7]. Consequently, practitioners must possess the capacity to assess prior decisions in the context of new circumstances, with a view to modifying unfavorable-turned decisions at the earliest opportunity. A particular example to this can be seen in automotive OEM and supplier production networks. As these are often globally distributed, the VUCA world (volatile, uncertain, complex, ambiguous) has enormous impact, since external and internal shocks can cause the short-term adjustment of production plans [8]. One recent example for necessary adjustments of plans is the US-EU tariffs, which stir up trade relations between these countries and further advance decoupling strategies [9]. Assuming these shifts are long-lasting, they would cause production lines and factories to shift out of the EU towards the US [10]. Therefore, production planning decisions within a frozen horizon need to be rethought. However, adjustments of taken decisions come at a price, if actions occurred to implement it. The goal of this paper is to develop a mechanism which enables the reversal of taken decision at an early stage in the frozen horizon by evaluating sunk planning cost. The remainder of the paper is structured as follows: Sect. 2 provides an overview over related work, including production planning processes in global production networks (GPN) and rolling horizon planning approaches. Section 3 introduces the conceptual defreezing mechanism framework, which is adaptable for all optimization problems which include a frozen horizon. Lastly, Sect. 4 discusses and contextualizes the defreezing approach, and provides an outlook for further research.

2 Related Work

The following section introduces related work for production planning in GPN and rolling horizon planning approaches. Production planning in GPN assigns activities in industrial value creation processes to ensure that resources within the GPN are utilized efficiently [11]. Rolling horizon planning involves making short-term decisions on an ongoing basis based on a forecast horizon that is periodically shifted forward and updated as necessary in order to plan optimally in dynamic, uncertain environments, to flexibly respond to changes in the forecast [12].

2.1 Production Planning in GPN

Ahn et al. [13] develop a generative probabilistic planning tool for complex and uncertain global production networks. Offline-Deep-Reinforcement Learning, graph neural networks and policy simulations simulate probabilistic demand, lead time and production conditions to calculate a robust optimal solution. Bruetzel et al. [2] use Monte Carlo simulation for scenario generation and stochastic modelling to increase the robustness of GPN planning. Bhutta et al. [14] propose an integrated production-distribution model for GPN, optimizing site selection and strategic capacity considerations, as well as distribution strategies, accounting for freight cost, exchange rates and tariffs. Azaron et al. [15] design a multi-objective, two-stage stochastic programming model, including optimizing warehouse and retailer site selection, production and inventory levels as well as shipping quantities within the GPN. While these approaches yield optimal solutions in the context of GPN planning, the optimization models are intended for a single utilization and do not permit the update of data at a later point in time.

2.2 Rolling Horizon Planning Approaches

Sethi and Sorger [12] develop a framework for rolling horizon decision making, in which uncertainty in the future can be reduced through costly forecasts. As an early adopter of the rolling horizon planning, Sridharan and Udayabhanu [16] investigate the influence of the characteristics of frozen horizons on the stability of master production schedules in material requirements planning systems. They find that freezing orders instead of full periods has the most impact on the stability of the MPS. Glomb et al. [17] develop a generic rolling horizon approach for a multi-period optimization model. They segment large planning problems into overlapping intervals, enhancing scalability and solution quality for large optimizations. For further information, Chand et al. [18] provide an overview over rolling horizon approaches and their applications. While these approaches provide stability in planning processes, they cannot handle short-term disruptions and requirement shifts occurring within the frozen horizon, since decisions are frozen and irrevocable. Therefore, a method to assess the adjustment of decisions within the frozen horizon must be developed.

3 Defreezing a Frozen Horizon

The following section describes the defreezing mechanism motivated in Sect. 1. First, in Sect. 3.1, sunk cost for planning decisions are addressed, which serve as input for the defreezing framework described in Sect. 3.2.

3.1 Sunk Cost for Planning Decisions

The cost occurring to change frozen horizons are always specific to a planning problem. There are certain types of cost which typically occur between the start of planning for a set action (the "start" of the frozen horizon) and the finalization of the stated action. For example, in a network configuration decision, after a production line is decided upon, costs for planning the production line and its setup occur in the periods between the decision and the set-up/decommission of the production line. These may include investment costs, personnel costs, costs for service providers, and more [19]. If the dismantling of a plant or machine is planned, these costs also include decommissioning costs [19]. As these costs occur to implement an initial decision, the realized costs can be considered as sunk costs if the decision gets overthrown within the frozen horizon. Therefore, these costs can serve as penalty costs when re-evaluating decisions in the frozen horizon ("defreezing"). An overview for costs occurring in the set-up or decommissioning of a new production line are described in Table 1.

Since the costs describe the actions taken after an initial decision, they typically increase over time; however, they can do this steadily or rise sharply. Therefore, the costs increase with late plan changes after the initial decision, and thus early adaptions in the frozen horizon cause lower costs than late adoptions. We further assume that the costs can be measured (e.g., direct allocation of FTE to planning tasks), and that they include external factors (e.g., interest rates/labor cost increases for later periods). Exemplary set-up and decommissioning costs over time can be found in Fig. 1.

Table 1 Exemplary set-up (S) and decommissioning (D) costs for production lines

Cost type	Example	Type
Investment costs	Purchasing of machines and (IT-)infrastructure elements	S
	Purchasing of software tools	S
Personnel costs	Salary for production planners	S/D
	Coordination costs for project management	S/D
	Costs for personnel expansion / reduction	S/D
	Training costs for employees	S
Service provider costs	Costs for external service providers	S/D
	Costs for external audits/certifications	S
Contractual costs	Down payments for machines or services	S/D
Decommissioning costs	Costs for line dismantling	D
	Costs for disposal of line	D

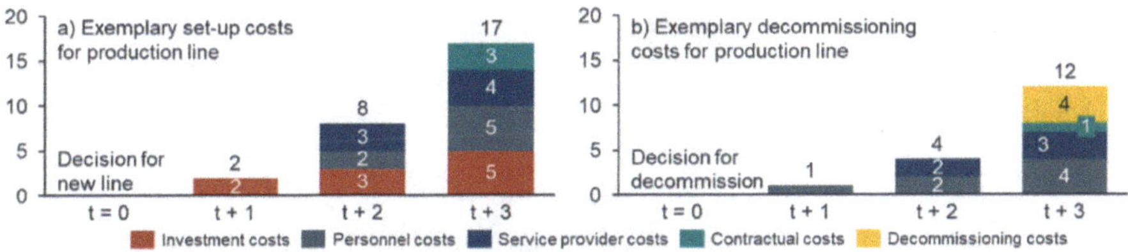

Fig. 1 Exemplary costs over time for the set-up costs (**a**) and the decommissioning costs (**b**) of production lines in a production network

3.2 Framework

This paper presents the possibility to "defreeze" decisions taken within a frozen horizon. For this, we propose a framework to the defreezing approach. For applicability, we present it with respect to production network configuration decisions, in which the optimal production line configuration for a given customer demand is calculated through optimization. However, the defreezing framework is suitable for all optimization models for planning decisions, which include a frozen horizon, and can be adopted accordingly. For the framework, we assume that the demand data is available for multiple periods and updated in each period, the production network configuration is adaptable and that short-term adjustments to the production network (planning) is possible at a cost. Our framework introduces an iterative defreezing mechanism, in which penalty terms have to be defined only if decisions in frozen horizons are to be changed. The concept of the defreezing mechanism is that for each period, the optimal production line configuration (based on the updated demand forecast) is identified. Afterwards, the new optimal configuration is compared to the optimal configuration from the previous period. Changes in decisions are penalized through sunk costs in an extension of the optimization problem.

The optimal configuration of the extended optimization problem thus includes the costs of plan adaptions. It is visualized in Fig. 2.

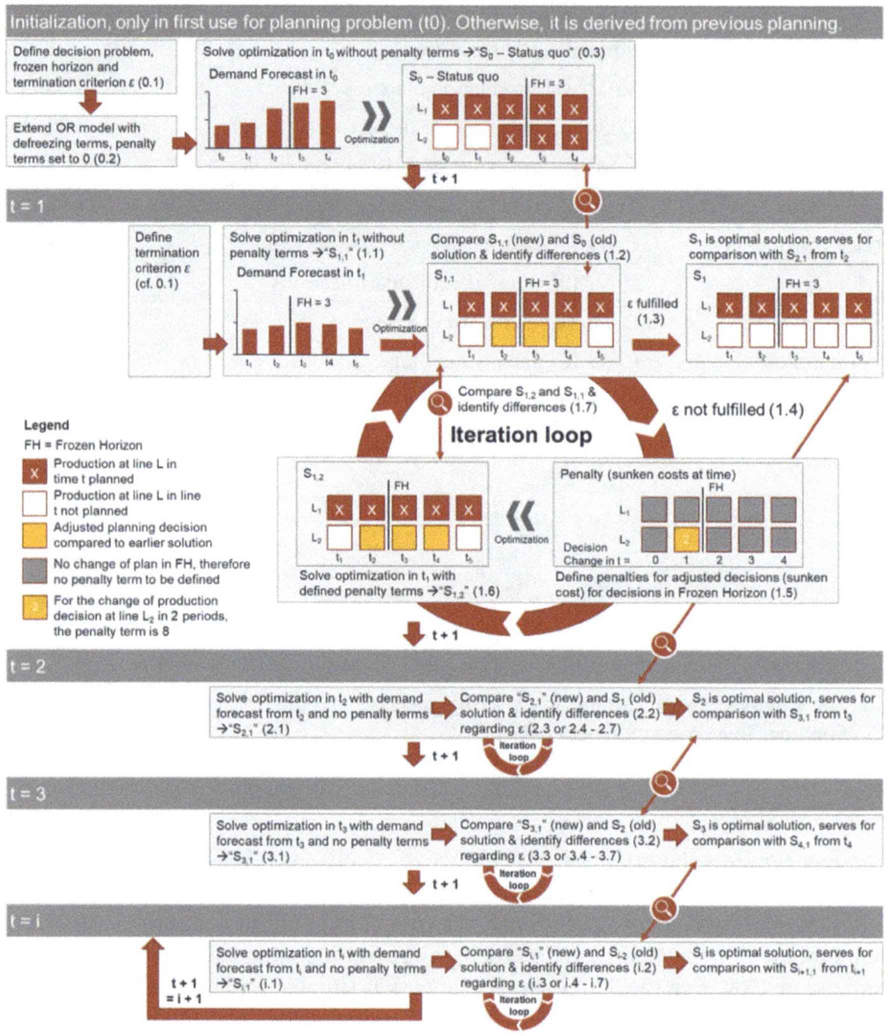

Fig. 2 The defreezing mechanism in the context of a network configuration decision problem

If the planning process must be initialized, first the problem and according decision variables must be identified. Given the decision problem, an appropriate frozen horizon must be chosen. Further, a termination criterion $\varepsilon \geq 0$ must be defined. This describes the number of allowed plan adjustments after an iteration loop. While ε can be set $= 0$ for planning problems with few decisions, ε can be set >0 for planning problems with many decisions, improving user-friendliness and adding flexibility to the approach (0.1). Subsequently, the optimization model is set up, including the terms which enable the defreezing mechanism. At the beginning, all penalty terms are set to zero (0.2). Then, the optimization model is solved given the input data (e.g., demand forecast) from the starting period t_0, generating the status quo solution S_0 for the planning problem (0.3). If the planning process is already in place, these steps can be skipped.

After the status quo solution S_0 is found through the optimization, the updated input data (e.g., demand forecast) from t_1 is used to find an updated optimal solution without penalty terms $S_{1,1}$ (1.1). The decision variables are then compared to S_0 (1.2). If the difference in adjustments is lower than ε, $S_{1,1}$ will be used as solution S_1 for the rolling horizon planning in t_1 (1.3). If the comparison between $S_{1,1}$ and S_0 does not satisfy ε, an iteration loop is started (1.4–1.7): For all decisions facing an adjustment within the frozen horizon, the respective penalty costs (sunk costs) are defined (1.5) (cf. Section 3.1). In the exemplary figure, the decision taken in t_0 to activate L_2 in t_2 is updated in t_1, therefore causing one period of sunk preparation costs. Afterwards, these penalty costs are included in the optimization model, and the optimal solution $S_{1,2}$ is computed (1.6). This solution is in turn compared to $S_{1,1}$ regarding ε (1.7), resulting in a final solution S_1 (ε satisfied) (1.3) or another iteration loop (ε not satisfied) (1.4–1.8), until ε is satisfied, showing a balance between the defreezing decision and occurring sunk cost. For the next time period, S_1 serves as baseline for the optimization problem given the new input data in t_2 and $S_{2,1}$ (analogous to S_0) (2.1). In t_2, the steps described for t_1 are repeated, finally resulting in the optimal solution S_2, which is in turn the baseline for the comparison of the solution $S_{3,1}$. It is also possible to define the penalty terms for each decision change before the optimization. While this might cause a significant amount of effort, the optimal solution can be derived in a single computation without iterating. Therefore, this approach is useful for decision problems, in which few decisions are taken, or when the penalty terms have little variability.

4 Discussion and Outlook

The presented defreezing method enables short-term decision changes within a frozen horizon to be evaluated and a holistic, optimized solution to be calculated. It allows plan changes within the frozen horizon by penalizing sunk costs incurred during planning. We expect the method to improve rolling horizon planning processes with high volatility and uncertainty through short-term shocks and comparably long frozen horizons, as the described framework increases the degrees of freedom for the optimization problem through softening hard constraints given by frozen horizon planning. Therefore, we assume that it will outperform traditional approaches in a numerical comparison. However, this comes at the price of complexity for the model and added efforts to iteratively identify sunk costs. Despite these drawbacks, it would be a suitable extension for production network configuration planning processes in a politically and economically uncertain global environment. Though the methodology has been described, it has yet to be validated in a real-world use case in future works. A suitable use case was identified at a European Tier 1 automotive supplier, which produces automotive components in a global production network with more than 30 assembly lines. Due to the high uncertainty and demand volatility in the automotive sector, as well as the fact that the planning processes use frozen horizons in their standard planning approach, we see high monetary potential in the validation of the method through this use case. When the use case is validated, it would be of great interest to compare the defreezing mechanism with standard rolling horizon planning. Further, a systematical edge case analysis would offer insights about the solution quality of the approach. Also, the approach to directly define all possible sunk cost should be investigated and compared with regard to the solution quality and required effort. Recommendations for either approach should be defined to help practitioners to choose the appropriate approach for their planning problem. Finally, if the approach poses an improvement to current planning process at the automotive supplier, it should be integrated into the planning tools. It should further automatically interact with other planning tools, forming an improved production planning suite. As the defreezing mechanism can be adapted for all decision problems which include a frozen horizon, further research could also implement it in other planning processes. This way, the mechanism could be included in interdependent planning processes with multiple decision variables and thus multiple frozen horizons. This would require to further refine the mechanism and to conduct research on the interactions between the planning processes, frozen horizons and their defreezing.

Acknowledgements. This research was funded by the German Federal Ministry for Education and Research (BMBF) through the research project 02J23C100-114 FENI-X.

Competing Interests The author(s) has no competing interests to declare that are relevant to the content of this manuscript.

References

1. Becker, A., Stolletz, R., Stäblein, T.: Strategic ramp-up planning in automotive production networks. Int. J. Prod. Res. **55**, 59–78 (2017). https://doi.org/10.1080/00207543.2016.1193252
2. Bruetzel, O., Hörger, M., Foran, A., Benfer, M., Nassehi, A., Lanza, G.: Robust planning of production networks at an automotive supplier. Int. J. Prod. Res. **63**, 3365–3383 (2025). https://doi.org/10.1080/00207543.2024.2436638
3. Lanza, G., Ferdows, K., Kara, S., Mourtzis, D., Schuh, G., Váncza, J., Wang, L., Wiendahl, H.-P.: Global production networks: design and operation. CIRP Ann. **68**, 823–841 (2019). https://doi.org/10.1016/j.cirp.2019.05.008
4. Norouzilame, F., Bruch, J., Bellgran, M.: Production plants within global production networks: synergies and redundancies. In: Operations Management in an Innovation Economy., Palermo, Sicily, University of Palermo's Campus (2014)
5. Benfer, M., Verhaelen, B., Peukert, S., Lanza, G.: Resilience measures in global production networks: a literature review and conceptual framework. DU. **75**, 491–520 (2021). https://doi.org/10.5771/0042-059X-2021-4-491
6. Terwiesch, C., Xu, Y.: The copy-exactly ramp-up strategy: trading-off learning with process change. IEEE Trans. Eng. Manag. **51**, 70–84 (2004). https://doi.org/10.1109/TEM.2003.822465
7. Schuh, G., Gützlaff, A., Schlosser, T.X., Rodemann, N., Haak, N.: Software-Based Identification of Adaptation Needs in Global Production Networks (2022). https://doi.org/10.15488/12169

8. Ahmed, O., Kamabe, H.: Proposing blockchain based framework to mitigate VUCA realm problems for automobile life cycle. Front. Blockchain. **8**, 1607271 (2025). https://doi.org/10.3389/fbloc.2025.1607271
9. Santo, M.: Drohender Handelskonflikt zwischen EU und USA. https://beschaffung-aktuell.industrie.de/news/drohender-handelskonflikt-zwischen-eu-und-usa/
10. Wilkes, W., Raymunt, M.: Mercedes, Porsche cut forecasts as US tariffs Hammer German Cars (2025). https://www.bloomberg.com/news/articles/2025-07-30/mercedes-sees-earnings-decline-over-tariffs-china-competition?embedded-checkout=true
11. Benfer, M., Hörger, M.: Bridging planning silos: a cross-functional decision support system for capacity, order, and supplier decisions in global production networks. CIRP Ann. S0007850625000757 (2025). https://doi.org/10.1016/j.cirp.2025.04.029
12. Sethi, S., Sorger, G.: A theory of rolling horizon decision making. Ann. Oper. Res. **29**, 387–415 (1991). https://doi.org/10.1007/BF02283607
13. Ahn, H., Olivar, S., Mehta, H., Song, Y.C.: Generative probabilistic planning for optimizing supply chain networks. https://arxiv.org/abs/2404.07511. (2024). https://doi.org/10.48550/ARXIV.2404.07511
14. Bhutta, M.K.S., Alhawari, O.I., Mohamed, Z.M.: An integrated production–distribution optimization model for multinational manufacturing corporations. Int. J. Manag. Sci. Eng. Manag. **19**, 29–45 (2024). https://doi.org/10.1080/17509653.2022.2157900
15. Azaron, A., Venkatadri, U., Farhang Doost, A.: Designing profitable and responsive supply chains under uncertainty. Int. J. Prod. Res. **59**, 213–225 (2021). https://doi.org/10.1080/00207543.2020.1785036
16. Sridharan, V., Berry, W.L., Udayabhanu, V.: Freezing the master production schedule under rolling planning horizons. Manag. Sci. **33**, 1137–1149 (1987). https://doi.org/10.1287/mnsc.33.9.1137
17. Glomb, L., Liers, F., Rösel, F.: A rolling-horizon approach for multi-period optimization. Eur. J. Oper. Res. **300**, 189–206 (2022). https://doi.org/10.1016/j.ejor.2021.07.043
18. Chand, S., Hsu, V.N., Sethi, S.: Forecast, solution, and rolling horizons in operations management problems: a classified bibliography. M&SOM. **4**, 25–43 (2002). https://doi.org/10.1287/msom.4.1.25.287
19. Dehli, M.: Lebenszykluskosten von Maschinen und Anlagen. In: Energieeffizienz in Industrie, Dienstleistung und Gewerbe, pp. 81–95. Springer Fachmedien Wiesbaden, Wiesbaden (2020). https://doi.org/10.1007/978-3-658-23204-7_5

A Framework Towards Automatic Manufacturability Assessment of Hairpin Technology Based on 3D CAD Models

Johannes Gerner(✉), Johannes Scholz, Florian Kößler, and Jürgen Fleischer

wbk Institute of Production Science, Karlsruhe Institute for Technology, Karlsruhe, Germany
johannes.gerner@kit.edu

Abstract. The growing demand for electric traction motors is characterized by a wide range of variants and short product lifecycles. Due to its high level of reproducibility and automation, hairpin technology has become the most relevant manufacturing technology for automotive stator windings. However, the production-oriented design of the hairpin coils poses a significant challenge in the development process. Although electrically optimized designs improve motor performance, they often create manufacturing challenges. In particular, forming the hairpin contour accurately while preserving electrical properties is difficult. Therefore, numerous iterations are required to adjust process parameters during the development of new hairpin stators. This increases the time required for the development of new stators, which contrasts with the desire for a shorter time to market. To address these challenges, this paper proposes a framework that uses a computer aided design (CAD) model of the hairpin stator as input to provide a digital process chain for the manufacturing of hairpin coils. Systematically evaluating the hairpin contour at an early stage, while respecting reasonable process parameters, has the potential to reduce manual iteration cycles and accelerate the overall development time. In addition, the selection of the correct process parameters, the production-oriented design of the hairpin coils and the interactions between the coils inside the winding head are also represented in the introduced process chain.

Keywords: Automated process planning · Electric traction motor · Hairpin technology · Bending process

1 Introduction

To achieve the decarbonization in the transportation sector, the transformation to electrified drives is essential [1]. Driven by this transformation, the production of performant and efficient electrified drive systems plays a key role to achieve these goals [2]. In order to meet the requirements of the automotive industry, a high demand for precise, robust and highly automated production technologies results. Therefore, hairpin technology has become the most relevant and reproducible manufacturing process for powerful stator windings [3]. The novelty of the technologies in the field of electric drive systems offers potential for the optimization of electric traction motors. This results in shorter product

L. Overmeyer and B.-A. Behrens (eds.), *Production at the Leading Edge of Technology*,
Lecture Notes in Production Engineering, https://doi.org/10.1007/978-3-032-19524-1_61

life cycles [4] and more customized electric traction motors for each application [5, 6]. Due to the tendency in the automotive industry to shorten life-cycles, an increasing variance in designs and declining sales [2], the result of new optimized electrical motor designs and the accompanying manufacturing restrictions is a large number of different hairpin coils within a single motor. Therefore, flexible and adaptable machines are required to cover the wide range of different hairpin geometries. Integrating information about the manufacturing process into the design of an electric traction motor offers two major advantages:

- Design of parts that are optimized holistically throughout their life cycle [7]
- Increase in cost, time and resource efficiency during the manufacturing ramp-up [8]

To enable a fast ramp-up of new hairpin geometries and to support the interaction between product design and production-oriented design during the product development process, this paper presents a framework for the automated assessment of the manufacturability of hairpin coils based on 3D CAD models. Section 2. describes the manufacturing process chain for hairpin coils, as well as the established bending processes. Furthermore, the state of the art in digital process planning and optimization for bending processes is presented. Section 3 outlines the requirements for the framework. Based on these requirements, the proposed framework is explained in Sect. 4. Finally, Sect. 5 concludes the paper with a summary and an outlook.

2 Related Work

2.1 Manufacturing of Hairpin Coils

The industrial manufacturing of hairpin stators is illustrated in Fig. 1 and consists of five process steps: forming, composing, inserting, twisting and contacting.

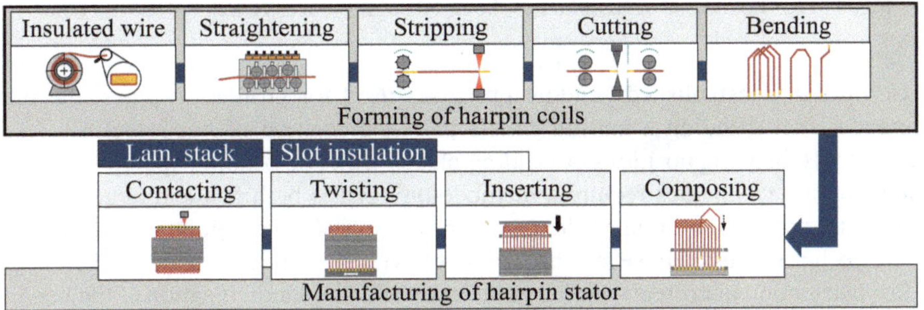

Fig. 1 Process chain for the production of hairpin stators

After straightening, stripping, cutting and bending the hairpin coils in the forming step, the hairpin coils are composed and inserted into the slots of the insulated lamination stack. Subsequently, the open coil ends are twisted in circumferential direction in order to complete the winding scheme. In the following process step the open coil ends are clamped and prepositioned before the coil pairs are contacted by laser welding. Due to the high variance of hairpin contours, the major challenges are the identification of the bending program and the derivation of the shape of the bending tool. In context of automotive series production, two principal bending techniques are established: tool-bound and sequential tool-bound bending. Tool-bound bending techniques consist of two main process steps: first, bending the planar hairpin coil by rotary draw bending, pipe bending or swing bending. Second, bending the spatial hairpin coil with a die bending process. However, with the sequential tool-bound bending, the hairpin coil is bent with discrete changes in the bending position and bending plane between the consecutive manufacturing operations. The bending techniques used in the sequential tool-bound bending are also rotary draw bending, pipe bending or swing bending (see Fig. 2) [9–11].

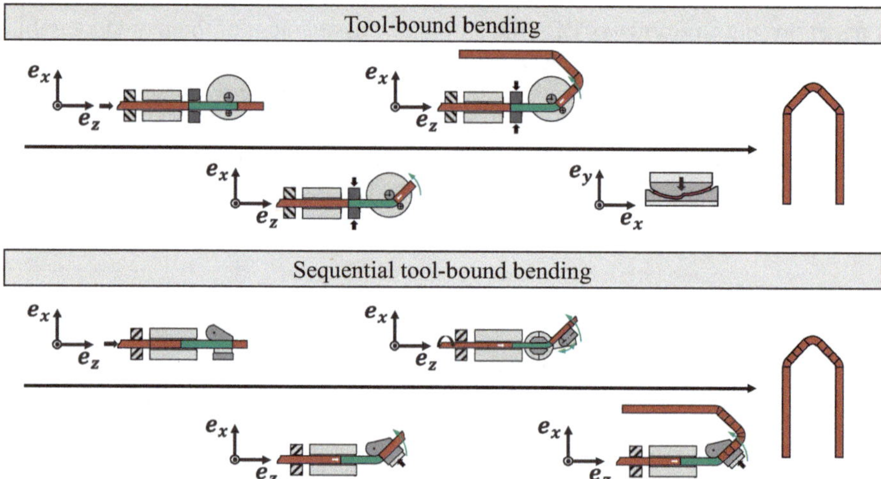

Fig. 2 Tool-bound and sequential tool-bound bending process chain for hairpin coils

Compared to established bending processes used for profiles or pipes, the main difference when bending a hairpin coil is the superposition of bends with different bending tools, bending positions or bending planes [12]. Die bending operations in tool-bound bending techniques require a methodology that is both fast and target-oriented for deriving production-oriented die geometry, in order to avoid iteration loops during the die design process. However, the challenge with sequential tool-bound bending is the fast calculation of correct bending parameters and the standardization of the bending radii to minimize the complexity of the bending machine and tools.

In order to maintain competitiveness, production systems for hairpin coils must be both cost-effective and highly flexible [6]. Achieving such flexibility in mass production often results in higher investment costs due a high number of bending axes required. Compared to tool-bound bending, sequential tool-bound bending requires a parallelization of bending operations at different stages to achieve a similar cycle time to that of tool-bound bending, in which the hairpin coil is manufactured in a single die bending operation. In context of hairpin coil bending, sequential tool-bound bending techniques are more flexible with regard to the programming of the spatial contour of the hairpin coil but generally require a higher number of bending axes to achieve the same cycle time compared to the tool-bound bending techniques [13–15]. Furthermore, a change of the wire dimensions always requires an adaptation of the wire guidance, independent of the bending technique. Due to the guidance of the planar hairpin coil, bending results of hairpin coils manufactured with tool-bound bending remain more stable compared to the sequential tool-bound bending process. Therefore, modular machines with a standardized interface for process components and a software-based adaptation of the bending tools enable a rapid and efficient setup of the machine. Compared to the state of the art, the adaptation of the bending program or the bending tool is done in iterative loops [10].

2.2 Approaches for Automatic Process Planning and Optimization

To optimize the design of a product with the trade-off between manufacturing costs and functionality during use, technology chain planning approaches are common. Based on product parameters like geometry features and quality requirements, suitable processes are selected according to their capabilities [16].

As described by Hausmann [8], it is necessary to check the winding head for collisions between the hairpin coils. This requires the spatial geometry of the hairpin coil. The 3D CAD model of the winding head provides this information. Furthermore, the 3D CAD model includes the geometric information which is required to select and analyze feasible manufacturing processes.

In context of bending processes, approaches for the derivation of manufacturing requirements based on 3D CAD models already exist. These requirements are used for an automated selection of bending techniques and the according tools [17, 18].

A major challenge in optimizing wire bending is defining the bending sequence. For N bends, there are N! possible sequences, but only some are collision-free. Identifying these feasible sequences is essential. [19, 20].

To determine feasible sequences, reachability checks and collision checks must be performed. Through a cost function the best feasible sequences can be identified [20].

The approaches mentioned are focused on bending processes in general and assume that the component design is finalized.

For the design of hairpin windings, optimization approaches have been developed with a focus on the electrical performance of the hairpin winding head. Bending restrictions and manufacturing costs are not considered in these approaches. However, restrictions regarding collisions in the winding head are considered [20, 21].

In Hausmann [8], an approach is provided which extends the optimized design of winding heads to include manufacturing restrictions. The geometry is optimized with respect to the possible bending angles for manufacturing. To check for collisions between the generated hairpins, a shell with the wire cross-section is created around the spline. This approach has great potential for the development of holistically optimized hairpin coils. However, in this approach, interactions between the bends and geometric changes of the hairpin are made independent of the bending process.

Taking into account the manufacturing technique used for manufacturing hairpin coils, Mogg-Walls [22] developed an algorithm for an automatic generation of hairpin winding heads by respecting intersections between hairpin coils. However, this approach focuses on classical design restrictions of bend hairpin coils and possibilities for new designs of winding heads enabled by additive manufacturing and does not take into account restrictions of highly productive bending processes used in industry.

With the goal of developing a closed-loop control system for the hairpin manufacturing, Wirth [23] extracts the required bending parameters for the bending process from the 3D CAD model of the hairpin coil and models the bending process with a numerical process model. This numerical process model also includes the springback behavior while bending the wire. It enables the prediction of the hairpin shape and leads to an increased time and cost efficiency during manufacturing ramp-up of new hairpin geometries and materials.

To sum up, the integration of product development and manufacturing development is an active field of research aimed at improving the resource and cost efficiency of components. Several approaches show the potential of using 3D CAD models to generate manufacturing information. However, the tools are used to optimize the manufacturing process of hairpin coils instead of using them as a feedback tool for the designer to optimize the design within the trade-off between manufacturing cost, manufacturability and functionality. Regarding the design of hairpin coils, existing approaches like the automated design or the use of numerical models to predict the bending shape show great potential. Nevertheless, a framework supporting the designer of a hairpin coil at the interface between product design and manufacturing is still missing. For this purpose, the paper addresses two major research questions (RQ):

- RQ 1: What are the requirements for a framework supporting the design of hairpin coils in the trade-off between product design and manufacturing?—Sect. 3.
- RQ 2: Which tools are required in such a framework and how it can work?—Sect. 4.

3 Requirements

To accelerate the manufacturing of hairpin coils, the product design in CAD environment has to provide information about the manufacturing process, such as the bending techniques, tools and machines used. Therefore, the core of the framework is 3D CAD model. The 3D CAD model includes an implicit description of bending angles, bending radii, bending position, and the wire dimensions shown in Fig. 3. Furthermore, the bending tools must be derived based on the spatial contour of the hairpin coil. In general, tools for sequential tool-bound bending techniques such as rotary draw bending, pipe bending and swing bending are defined by the bending radii on the die. Tool-bound bending typically consists of two subsequent process steps. First, the planar hairpin coil is manufactured by sequential tool-bound bending, and second, the spatial hairpin coil is manufactured by die bending. However, when manufacturing spatial hairpin coils, the geometry of the bending tool is derived from the inverse of the parametric hairpin contour. The manufacturing process, the mechanical properties of the wire, the geometrical tolerances [24], and the wire cross-section, and the contact between the bending tool and the wire influence the behavior of the wire during bending. To compensate springback effects, the process parameters must be calculated in accordance with the wire properties for both sequential tool-bound bending and die bending. In order to evaluate the capability of bending machines with respect to a specific hairpin contour, the bending tools, process parameters and collision space inside the machine must be defined. Finally, a cost model is required for each bending technique to assess the manufacturing process in terms of the cost of the bending tools and how the axes move according to the calculated bending parameters. Once the framework has been modeled, the loop between the challenges caused by the bending techniques and the parametric CAD model of the hairpin coil is closed. This provides information to optimize the spatial hairpin contour.

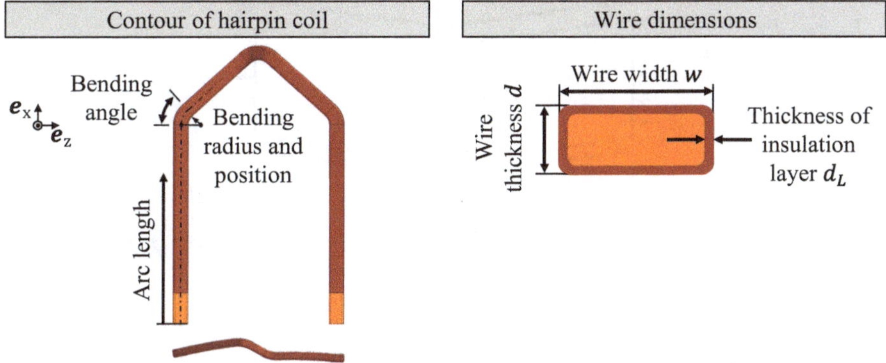

Fig. 3 Description of a hairpin coil with bending angles, bending radii and bending positions on the left and wire dimensions on the right-hand side

4 Framework

This section addresses the second research question and presents the methods and tools of the framework for manufacturing planning of hairpin coils based on a 3D CAD model. Evaluating the manufacturability of a part-based 3D CAD model requires a formalization of all information from the 3D CAD model with regard to the required manufacturing data. This description enables a continuous digital process chain for the manufacturing of hairpin coils, including the presented framework for automatic manufacturability evaluation. Fig. 4 shows the entire framework. The following discussion will focus on the workflow of the automatic manufacturability evaluation, located at the center of the Figure. Each of the described process steps requires input and output information.

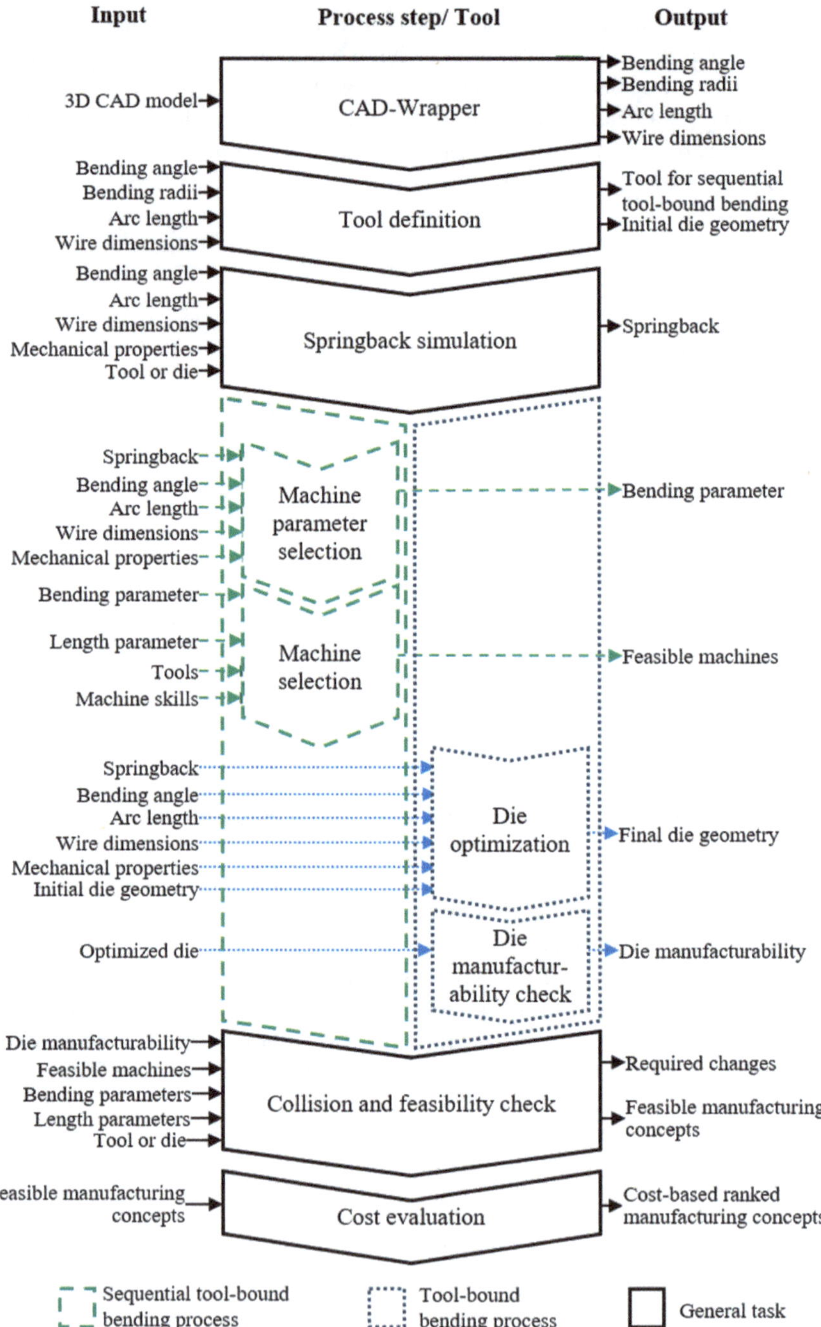

Fig. 4 Overview of the tasks including the in- and outputs along the framework

The first step in the framework is the CAD wrapper, which generates the required information from the 3D CAD model and is already described by Wirth [11]. In the context of hairpin coil manufacturing, the main parameters the bending angles, bending radii, wire dimensions and the arc length. The wire infeed is therefore defined by the arc length of the hairpin coil and, consequently, the bending positions. As mentioned in Sect. 2.1, the two established manufacturing processes are sequential tool-bound bending and tool-bound bending. Consequently, the framework provides tools for the manufacturing of hairpin coils with both processes. In the context of the sequential tool-bound bending, the pivotal parameter for the bending tool is the bending radius. The bending angle is defined by the movement of the machine's axis. However, the bending angles do not match the identified bending angles in the 3D CAD model due to springback effects and different contact points between the bending tool and wire [24]. To address this, a springback and kinematic simulation is required. This can be achieved by a finite element (FE) simulation that takes into account the mechanical properties of the wire and the bending tool used. Regarding the tool-bound bending, the die is first derived as the inverse of the parametric hairpin coil, but does not consider springback effects.

To consider these effects, a FE simulation must be conducted. Subsequently, the results of the simulation are processed in an optimization procedure to determine the optimal die. The geometry of the die defines the corresponding manufacturing requirements. These requirements can then be evaluated for feasibility using established technologies, such as CAD/CAM. After planning several manufacturing options, multibody simulations can be used to verify the manufacturing chains and detect collisions within the bending machine. Furthermore, cost models can be applied to the feasible manufacturing chains to identify the most cost-effective option. After the implementation, the digital framework reduces iterations during the manufacturing planning and enables an efficient execution to reach the start of operation faster.

5 Conclusion and Outlook

Hairpin technology is common in electric traction motors used in the mobility sector. However, manufacturers of both electric traction motors and hairpin bending machines are facing economic challenges and shortening product lifecycles. To overcome these challenges, the hairpin coil manufacturing must be considered early in the development process. A promising approach is to integrate an automated manufacturability check based on the 3D CAD model of the hairpin coil. This model describes the spatial hairpin contour, which is defined by the electrotechnical design. Integrating a manufacturability check and an economic evaluation reduces the number of iterations required and shortens time-to-market. Therefore, this paper presents a framework and the corresponding requirements for automatically assessing the manufacturability of hairpin coils. After the 3D CAD model is formalized to provide the relevant manufacturing information, the required bending tools are identified and springback effects are analyzed using a finite element simulation. Depending on the bending process used, the manufacturability of the hairpin coil is verified, and in case of tool-bound bending the die is optimized. Additionally, a collision analysis and an economic evaluation are conducted. In further research work, the implementation of the framework as a digital toolchain including the interfaces to the required simulation models is planned. With respect to a faster time-to-market, the extension of a CAD system by an assistant which provides manufacturing information shows great potential. Furthermore, an extension of the concept of the digital tool chain to the complete electric motor assembly enables the identification of collisions between the hairpin coils inside the stator assembly.

Competing Interests. The author(s) has no competing interests to declare that are relevant to the content of this manuscript.

References

1. Pariser Klimaschutzübereinkommen Consilium. https://www.consilium.europa.eu/de/policies/climate-change/paris-agreement. Last accessed 30 Aug 2024
2. Korne, T., Schmidt, K.-J. (eds.): Chancen und Risiken in der Automobilindustrie: Handlungsempfehlungen zur Transformation von Märkten, Produkten, Prozessen und Strukturen. Springer Fachmedien Wiesbaden, Wiesbaden (2025). https://doi.org/10.1007/978-3-658-48323-4
3. M. Grumbach, A. Locher, S. Kilian: Paradigm change in the driveline—electrification as the standard. MTZ Worldw 82(4) (2021). https://doi.org/10.1007/s38313-021-0625-3
4. Inc, J.: Innovation and development of electric vehicle powertrain technology: an Analysis. Jabil.com. Accessed Aug. **07** (2025) http://www.jabil.com/blog/automotive-industry-trends-point-to-shorter-product-development-cycles.html
5. Hemsen, J., Nowak, N., Eckstein, L.: Production cost modeling for permanent magnet synchronous machines for electric vehicles. Automot. Engine Technol. **8**(2), 109–126 (2023). https://doi.org/10.1007/s41104-023-00128-w
6. Fleischer, J., et al.: Agile Produktion elektrischer Traktionsmotoren als Antwort auf volatile Märkte und Technologien. Z. Für Wirtsch. Fabr. **116**(3), 128–132 (2021). https://doi.org/10.1515/zwf-2021-0025
7. Scholz, J., Klein, N., Kößler, F., Fleischer, J.: Systematic digital twin-based development approach for holistic sustainable electric traction motors. Sustainability. **17**(6), 2518 (2025). https://doi.org/10.3390/su17062518
8. Hausmann, L., Wirth, F., Fleischer, J.: Opportunities of model-based production-oriented design of stators with hairpin winding. In: 2020 10th International Electric Drives Production Conference (EDPC), pp. 1–8. IEEE, Ludwigsburg, Germany (2020). https://doi.org/10.1109/EDPC51184.2020.9388186
9. Gerner, J., Wirth, F., Fleischer, J.: Comparison of rotary draw bending and pipe bending processes in the context of hairpin technology. In: 2024 14th International Electric Drives Production Conference (EDPC), pp. 1–8, Regensburg, Germany (2024). https://doi.org/10.1109/EDPC63771.2024.10932856
10. Wirth, F., Gerner, J., Hausmann, L., Fleischer, J.: Closed-loop process control for sequential tool-bound bending of hairpin coils. In: 2023 13th International Electric Drives Production Conference (EDPC), pp. 1–8, Regensburg, Germany (2023). https://doi.org/10.1109/EDPC60603.2023.10372174
11. Wirth, F., Hausmann, L., Fleischer, J.: Model-based closed-loop process control for the manufacturing of hairpin coils. Prod. Eng. (2023). https://doi.org/10.1007/s11740-024-01271-5

12. Wirth, F.: Prozessgeregelte Formgebung von Hairpin-Steckspulen für elektrische Traktionsmotoren. Dissertation, Karlsruher Institut für Technologie, Karlsruhe, (2024). https://doi.org/10.5445/IR/1000177636
13. WAFIOS AG: CNC-Wickel-, Winde-und Biegemaschine, Modell FMU E für die Produktion von Hairpins, I-Pins und Statoranschlussbaugruppen aus Flachmaterial', Data sheet (2020). https://www.wafios.com/maschinen/maschinen-fuer-die-e-mobilitaet/hairpin-maschinen/FMU%20E. Last accessed 21 July 2025
14. WAFIOS AG: Transfer-Biegemaschine, Modell SpeedFormer für die Serienfertigung von Hairpins, I-Pins und Statoranschlussbaugruppen aus Flachmaterial. Data sheet (2021). https://www.wafios.com/maschinen/maschinen-fuer-die-e-mobilitaet/hairpin-maschinen/SpeedFormer. Last accessed 21 July 2025.
15. Otto Bihler Maschinenfabrik GmbH, Co. KG: Effiziente Automationslösungen für E-Bauteile. Whitepaper, Apr. (2024). https://meet.bihler.de/de/landing/e-mobility-whitepaper/?utm_campaign=10001-e-mobility-whitepaper&utm_source=bihler-homepage&utm_medium=homepage. Last accessed 21 July 2025
16. M. Fallböhmer, Generieren alternativer Technologieketten in frühen Phasen der Produktentwicklung. In: Berichte aus der Produktionstechnik, **2000**(23). Aachen, Shaker, (2000)
17. Nguyen, T.H.M., Duflou, J.R., Kruth, J.P.: A framework for automatic tool selection in integrated Capp for sheet metal bending. Adv. Mater. Res. **6–8**, 287–294 (2005). https://doi.org/10.4028/www.scientific.net/AMR.6-8.287
18. Salem, A.A., Abdelmaguid, T.F., Wifi, A.S., Elmokadem, A.: Towards an efficient process planning of the V-bending process: an enhanced automated feature recognition system. Int. J. Adv. Manuf. Technol. **91**(9–12), 4163–4181 (2017). https://doi.org/10.1007/s00170-017-0104-9
19. Bahubalendruni, M.R., Biswal, B.B.: A review on assembly sequence generation and its automation. Proc. Inst. Mech. Eng. Part C J. Mech. Eng. Sci. **230**(5), 824–838 (2016). https://doi.org/10.1177/0954406215584633
20. Baraldo, A., et al.: Automatic computation of bending sequences for wire bending machines. Int. J. Comput. Integr. Manuf. **35**(12), 1335–1351 (2022). https://doi.org/10.1080/0951192X.2022.2043563
21. England, M., Ponick, B.: Automatisierter Entwurf von Haarnadelwicklungen anhand von tabellarischen Belegungsplänen. E Elektrotechnik Informationstechnik. **136**(2), 159–167 (2019). https://doi.org/10.1007/s00502-019-0709-9
22. H. Mogg-Walls, A. Nassehi, M. Goudswaard, N. Simpson: Development of a computational design tool for the automatic routing of hairpin end-windings. In: IET Conference Proceedings, Nottingham UK pp. 539–546 (2024). https://doi.org/10.1049/icp.2024.2205.
23. Wirth, F., Hausmann, L., Fleischer, J.: Model-based closed-loop process control for the manufacturing of hairpin coils. Prod. Eng. (2024). https://doi.org/10.1007/s11740-024-01271-5
24. Wirth, F., Fleischer, J.: Influence of Wire Tolerances on Hairpin Shaping Processes. In: 2019 9th International Electric Drives Production Conference (EDPC), pp. 1–8, Esslingen, Germany (2019). https://doi.org/10.1109/EDPC48408.2019.9011999

BY

Spatial Correlation Analysis Between Punch and Scrap Web Sheared Surface Roughness for Indirect Punch Wear Assessment in Sheet-Metal Forming

Jiyoung Moon[1(✉)], Martin Unterberg[1], Philipp Niemietz[1], and Thomas Bergs[1,2]

[1] Manufacturing Technology Institute MTI of RWTH Aachen University, Aachen, Germany
j.moon@mti.rwth-aachen.de

[2] Fraunhofer Institute for Production Technology IPT, Aachen, Germany

Abstract. Sheet-metal forming operations would benefit from continuous punch condition monitoring to optimize maintenance scheduling and prevent unexpected failures, yet closed tool designs in high-precision processes prevent visual observation during production runs. While process signals such as force and acoustic emissions enable indirect tool monitoring, current unsupervised learning-based approaches using these process signals lack interpretability. Moreover, a recent supervised learning approach using a proxy for punch wear lacks quantitative validation of the relationship between punch wear characteristics and scrap web surface evolution, limiting reliable wear prediction model development. This study establishes a quantitative spatial correlation between punch and scrap web sheared surface roughness using high-resolution 3D profilometry. Results demonstrate that the wear of the punch surface induces surface roughness in the scrap web and, in turn, provide evidence that the scrap web shearing surface roughness can be used as an indicator for progressing punch wear in fine blanking.

Keywords: Sheet-metal forming · Punch wear · Indirect wear assessment

1 Introduction

In sheet-metal forming, punches represent some of the most critical and expensive components that significantly impact product quality and manufacturing costs [1]. Their condition directly affects dimensional accuracy, surface finish, and mechanical properties of the final product [2]. While traditional approaches such as those by Chumrum et al. [3] and Yoon et al. [4] assess punch wear via periodic tool removal and direct measurement, these methods require tool disassembly, which is impractical for industrial applications. Even recent in-line approaches, such as the image processing method by Schlegel et al. [5], are limited to processes where the tool edge remains visible. However, in high-precision processes such as fine blanking, the enclosed design of tools makes direct visual or tactile assessment during production impossible, creating a fundamental challenge for maintaining optimal punch condition.

L. Overmeyer and B.-A. Behrens (eds.), *Production at the Leading Edge of Technology*,
Lecture Notes in Production Engineering, https://doi.org/10.1007/978-3-032-19524-1_62

To address this limitation, indirect monitoring approaches using process signals have emerged as promising alternatives. Shanbhag et al. utilized acoustic emission (AE) signals for indirect tool wear monitoring in progressive die sheet-metal stamping, collecting AE data during accelerated wear tests and correlating extracted time-domain features with wear stages [6]. Niemietz et al. proposed an unsupervised approach using convolutional autoencoders that analyzed force and AE signals, tracking tool wear progression by detecting characteristic fluctuations through reconstruction errors [7]. Despite the advantage of eliminating labeling efforts, these unsupervised learning approaches are limited by interpretability issues and uncertainty in unlabeled predictions.

As an alternative approach, Unterberg et al. explored a supervised learning method using the sheared surface characteristics of scrap webs (the remaining part of the material after the blanking process) as indicators of punch wear. During the shearing and the stripping phases of the fine blanking process, the relative movement between punch and scrap web creates frictions that alter the surface characteristics of the scrap material. They hypothesized that adhesive wear on the punch primarily causes an increase in the scrap web sheared surface roughness, and therefore quantified punch wear by averaging six arithmetical mean roughness values measured on selected sheared surfaces [8]

However, the quantitative understanding of how specific punch wear characteristics translate to scrap web surface evolution remains insufficiently clarified, limiting the development of reliable wear prediction models. Furthermore, given the complexity of wear mechanisms, shearing processes, and varied geometrical characteristics of scrap webs, a single averaged roughness value lacked validation through examination of its correlation with punch wear state at any given stroke. To bridge this knowledge gap and establish scrap web sheared surface characteristics as a reliable proxy for the indirect monitoring approach, this research extends previous work through following key objectives:

1. Quantification of the punch wear and the surface roughness properties of the scrap web sheared surface through detailed segmentation analysis and surface roughness measurements
2. Discovery of the relationship between the punch wear and the scrap web sheared surface roughness by means of the spatial correlation analysis of the scrap web sheared surface at the last stroke of an experiment of 40,450 strokes and the final state of the punch wear.

2 Methodology

This chapter describes the methodology used to investigate the relationship between punch wear and scrap web sheared surface characteristics in the fine blanking process.

2.1 Experimental Setup

To investigate the relationship between punch wear and scrap web sheared surface characteristics, an industry-scale fine blanking experiment was conducted using a Feintool XFT 2500 speed machine. The tool was equipped with two punches, P1 and P2, enabling the production of two parts at the same time. A total of 40,450 strokes were completed over three days at a speed of 50 strokes/min. The material used was a high-strength sheet-metal of grade AISI 58CrV4 (DIN 1.8161) with a thickness of 5 mm.

2.2 Measurement of the Punch Wear

After the experiment, two worn punches were disassembled from the tool, and their 3D surface profiles were captured along their edges. Given the large number of measurement points required for comprehensive surface characterization, optical measurements were conducted using the Keyence VR-6000 Profilometer. The scanned 3D images were subjected to rotation and profile correction along the x- and y-axes using the analysis software of this device. For the correlation analysis with the corresponding scrap web sheared surfaces, the measured punch surfaces were divided into uniform segments of equal width for each individual outer edge, resulting in 30 edge segments per punch. Note that only flat surfaces were measured, resulting in different segment widths across each edge. Figure 1a illustrates the measured edges on the punch geometry. For each edge segment, an area with an immersion depth of up to 2 mm was used to measure areal surface roughness. One segment each on the left side of P1 and P2 showed irregular profiles due to punch production and were therefore excluded from the correlation analysis in Sect. 3 to avoid biases.

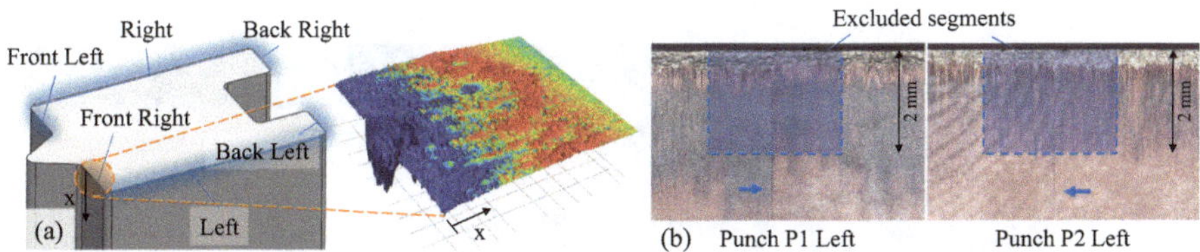

Fig. 1 **a** Measured edges on the punch and the scanned 3D image of the segment P2 Front Right (adapted from [9]). **b** Irregular profiles and excluded segments on P1 Left and P2 Left

Next, to investigate the distribution of adhesive wear on the punch surface, both convex (Fig. 2a) and concave (Fig. 2b) surface elements were identified using the unworn punch surface as the reference plane. The distribution of adhesive wear was determined by analyzing the chromium distribution on the punch surface. Energy-dispersive X-ray spectroscopy (EDX) analysis in Fig. 2c shows chromium distribution from the AISI 58CrV4 workpiece material in both convex and concave regions of punch P2 Front Right.

This material transfer indicates that adhesive bonding and subsequent material removal occurred during the forming process, with the transferred material potentially acting as abrasive particles during subsequent strokes.

2.3 Measurement of the Scrap Web Sheared Surface Roughness

Subsequently, areal surface roughness measurements were conducted on the collected scrap web from the last stroke to observe the correlation with the punch state. For comparison with the corresponding edge segments defined for the captured punch surfaces,

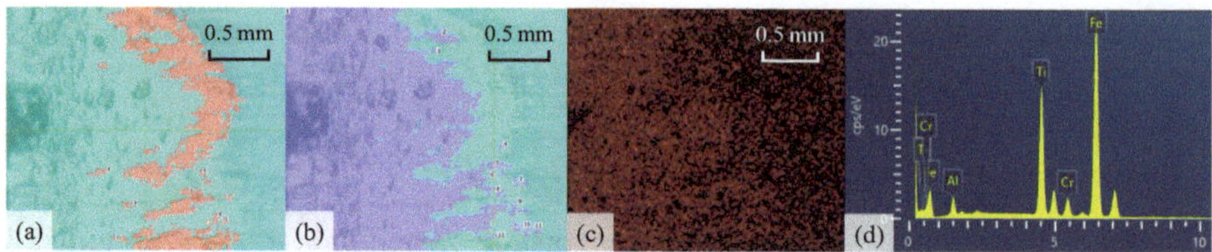

Fig. 2 **a** Convex elements (red area), **b** Concave elements (purple area), **c** Chromium elemental distribution map, **d** EDX spectrum of the segment P2 Front Right

the scrap web sheared surfaces were segmented using the same methodology described in Sect. 2.2. Within each edge segment, five sub-areas of equal height (1 mm) aligned with the shearing direction were sampled ($\text{Seg}_{i,1\ldots5}$), yielding a 5-point sequence per segment. Tear areas were excluded from the defined sub-areas where they intersected or overlapped. For each segment Seg_i, the 21 areal roughness parameters (Sa, Sq, Ssk, ...) of ISO 25178 standard were evaluated at the five sub-areas to construct 21 sequences. For each sequence, 9 statistical features (mean, median, standard deviation, variance, ...) were extracted using the Python TSFEL library, providing 189 sequence-derived statistics for each edge segment. Figure 3 illustrates the processes for extracting the roughness parameters for both punch and scrap web.

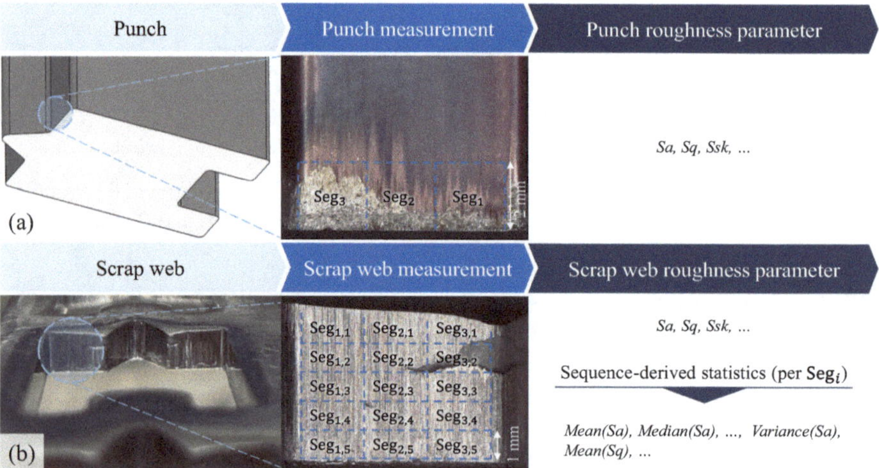

Fig. 3 Roughness parameter extraction workflow: **a** punch surface segmentation and measurement, **b** scrap web surface segmentation showing sequential sub-areas ($\text{Seg}_{i,1\ldots5}$) and derivation of sequence-based statistical parameters for correlation analysis

3 Results and Discussion

For the correlation analysis, each edge was scanned three times for the last scrap web to ensure measurement reliability. The average values from these scans were used for each scrap web sheared surface in the subsequent analysis. 60 segments were determined from 2 punches and the corresponding scrap webs, with each punch containing 30 segments and each segment providing roughness sequences along the shearing direction. Excluding two irregular segments, in total of 58 segments were used for the Pearson and Spearman correlation analysis.

The top results in Table 1 show mostly moderate correlations ($|r| \approx 0.6$). The key pattern is that negative and positive correlations coexist. High punch surface peak density (Spd) is associated with lower scrap web peakiness and roughness measures, such as Spd on the scrap web, Vvc and Sk. This negative relationship suggests that a higher density of small asperities on punch surface distributes contact forces during shearing among more micro-contacts, thereby reducing the stress per asperity. In practical terms, this corresponds to the scrap web sheared surface exhibiting suppressed maximum of highly peaked surface features (maximum of Spd) and roughness measures (maximum of Vvc, maximum of Sk).

Table 1 Top 10 results of the correlation analysis

Punch roughness parameter	Scrap web roughness parameter	Scrap web statistical feature	Correlation metric	Correlation coefficient
Spd	Vvc	Max	Spearman	−0.634
Spd	Spd	Max	Pearson	−0.620
Spd	Spd	Median	Pearson	−0.618
Spd	Spd	Root mean square	Pearson	−0.613
Sv	Sdr	Mean absolute deviation	Pearson	0.610
Spd	Spd	Mean	Pearson	−0.608
Spd	Sk	Max	Spearman	−0.600
Sv	Sdr	Mean absolute deviation	Spearman	0.593
Sku	Sdr	Mean absolute deviation	Pearson	0.590
Sz	Sdr	Mean absolute deviation	Pearson	0.589

The analysis also identifies positive relationship between deep and spiky punch surface features, such as Sv, Sku and Sz and the scrap web sheared surface variability of the developed interfacial-area ratio, captured by mean absolute deviation (MAD) of Sdr. This positive correlation suggests that deeper valleys and more spikey height distributions on punch surface can result in higher variability in scrap web sheared surface's interfacial area along the shearing direction.

Figure 4 visually depicts the relationship between punch and the scrap web sheared surface roughness parameters for each corresponding edge segment. For the general robustness against the outliers, only median of Spd and MAD of Sdr are considered among the scrap web sheared surface roughness parameters. For both punches P1 and P2, narrow edges such as back and front edges exhibit relatively lower Spd values than long edges such as left and right edges. From this observation, it can be inferred that narrow edges, which bear higher loads during the shearing process, tend to develop fewer asperities on their surfaces but with high stress per asperity, thereby intensifying the scratching conditions on the sheared surface of the scrap web. Additionally, corner segments demonstrate higher Sv values than segments located in the middle portions of the edges. This pattern suggests that geometric discontinuities at corners create more complex stress distributions and potentially enhanced debris accumulation, leading to alternating cutting along the shearing direction. As a result, the Sdr value on the scrap web sheared surface changes more from point to point.

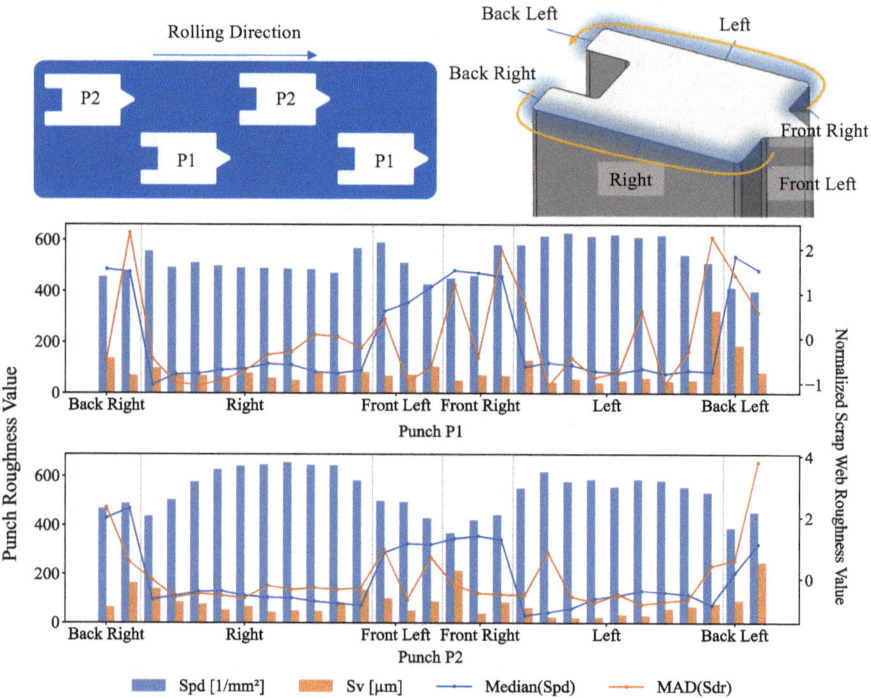

Fig. 4 Planar illustration of punch and scrap web sheared surface roughness of P1 and P2. The x-axis follows the yellow arrow sequence from Back Right to Back Left; the primary y axis corresponds to bar plots, while the secondary y-axis represents z-score normalized scrap web sheared surface roughness values. Inner dotted vertical lines indicate borders between edges

4 Summary and Outlook

In high-precision sheet-metal forming processes, in-situ direct punch condition assessment is impossible due to enclosed tool designs, making indirect monitoring approaches using process signals promising but requiring quantitative relationships between measurable parameters and actual tool wear states. While previous research hypothesized that adhesive wear mechanisms primarily influence scrap web surface roughness, no quantitative correlation has been established to validate this relationship for reliable indirect monitoring applications. This study established a quantitative relationship between punch wear and scrap web sheared surface roughness using comprehensive 3D profilometry and optical roughness measurements. The correlation analysis revealed moderate correlations ($|r| \approx 0.6$) between punch and scrap web sheared surface roughness parameters. Each individual parameter explains approximately 30–40% of the variability, leaving 60–70% of the variance unexplained. To accurately assess punch wear state from scrap web sheared surface characteristics, multiple complementary roughness parameters must be combined into a joint indicator rather than using individual correlations. Furthermore, the apparent asymmetry in some parameters between corresponding edges (e.g., left versus right) suggests that process-related factors such as material feed direction or tool alignment may introduce systematic variations that should be characterized. Investigation of the temporal evolution of these spatial patterns could also provide insights into wear progression mechanisms and their impact on product quality consistency over extended production runs. Future research should develop predictive models that integrate the established punch-scrap web correlations to enable proactive maintenance scheduling in industrial applications. Validation across different materials, punch geometries, and operating conditions is essential to establish the generalizability and reliability of indirect punch wear monitoring approach for broader manufacturing contexts.

Acknowledgements. This research was funded by the Deutsche Forschungsgemeinschaft (DFG, German Research Foundation) under Germany's Excellence Strategy—EXC-2023 Internet of Production—390621612. Furthermore, it is part of the Blockchain4DatenMarktplatz.NRW project, funded by the Federal Office for Economic Affairs and Export Control (BAFA) (46SK0233A) and managed by Project Management Jülich (PTJ).

Competing Interests The author(s) has no competing interests to declare that are relevant to the content of this manuscript.

References

1. Unterberg, M., Voigts, H., Weiser, I.F., Feuerhack, A., Trauth, D., Bergs, T.: Wear monitoring in fine blanking processes using feature based analysis of acoustic emission signals. Procedia CIRP **104**, 164–169 (2021). https://doi.org/10.1016/j.procir.2021.11.028
2. J. Reblitz, Evaluation of the friction and wear behavior of a-C:H coatings for lubricant-reduced sheet metal forming, in: Metal Forming 2024, Materials Research Forum LLC, pp. 663–673 (2024)
3. Chumrum, P., Koga, N., Premanond, V.: Experimental investigation of energy and punch wear in piercing of advanced high-strength steel sheet. Int. J. Adv. Manuf. Technol. **79**, 1035–1042 (2015). https://doi.org/10.1007/s00170-015-6902-z
4. Yoon, J., Won, C., Kim, H., Song, Y., Chung, G., Lee, S.: Abrasive wear in punching pin with cryogenic treatment for GPa-grade steels. Int. J. Precis. Eng. Manuf. **19**, 1179–1186 (2018). https://doi.org/10.1007/s12541-018-0139-3
5. Schlegel, C., Molitor, D.A., Kubik, C., Martin, D.M., Groche, P.: Tool wear segmentation in blanking processes with fully convolutional networks based digital image processing. J. Mater. Process. Technol. **324**, 118270 (2024). https://doi.org/10.1016/j.jmatprotec.2023.118270

6. Shanbhag, V.V., Pereira, P.M., Rolfe, F.B., Arunachalam, N.: Time series analysis of tool wear in sheet metal stamping using acoustic emission. J. Phys. Conf. Ser. **896**, 12030 (2017). https://doi.org/10.1088/1742-6596/896/1/012030
7. Niemietz, P., Unterberg, M., Trauth, D., Bergs, T.: Autoencoder based wear assessment in sheet metal forming (2020 IOP Conf. Ser.: Mater. Sci. Eng. 1157 012082), IOP Conf. Ser.: Mater. Sci. Eng. **1157**, 12098 (2021). https://doi.org/10.1088/1757-899X/1157/1/012098
8. Unterberg, M., Becker, M., Niemietz, P., Bergs, T.: Data-driven indirect punch wear monitoring in sheet-metal stamping processes. J. Intell. Manuf. **35**, 1721–1735 (2024). https://doi.org/10.1007/s10845-023-02129-w
9. Unterberg, M.: Data-driven indirect tool condition monitoring in sheet-metal stamping. Dissertation, 1st ed.

Increasing Resilience Using Standardization in Product-Network-CoDesign

Katharina Theuner[1(✉)], Moritz Hörger[1], Oliver Brützel[2], Thomas Troll[3], Martin Benfer[1], and Gisela Lanza[1]

[1] Wbk Institute of Production Science, Karlsruhe Institute of Technology (KIT), Karlsruhe, Germany
katharina.theuner@kit.edu
[2] WITTE Automotive GmbH, Velbert, Germany
[3] EFESO Management Consultants, Munich, Germany

Abstract. The increasing complexity and customization of products, combined with shorter innovation cycles and global crises, present significant challenges for companies and affect entire networks. To stay competitive, companies must ramp up production more frequently, pushing traditional sequential planning of products and production to its limits. Concepts such as Product-Production-CoDesign have emerged to enable a more parallelized development process across generations and life cycles. However, these approaches do not cover the entire global production network. Standardization enables global production networks, but leveraging it requires concurrent planning of the entire product portfolio and network design. This paper proposes a framework for Product-Network-CoDesign that systematically incorporates standardization into both product and production process development, taking into account its impact on relevant economic, strategic, and technical factors, as well as its influence on the resilience of the whole network. This systematic approach provides practitioners with decision-making guidance and enables increased resilience.

Keywords: Global production networks · Standardization · Resilience · Product development · Product-Network-CoDesign

1 Introduction

Disruptive events such as the re-election of Donald Trump and the subsequent tariff decisions have exposed vulnerabilities in global production networks and highlighted the critical importance of resilience. This term refers to a system's ability to maintain high performance despite environmental changes through active adaption [1]. To ensure long-term performance, companies must develop strategies to maintain the performance level, resulting in increased resilience [2]. This includes the efficient development and planning of agile production systems.

L. Overmeyer and B.-A. Behrens (eds.), *Production at the Leading Edge of Technology*, Lecture Notes in Production Engineering, https://doi.org/10.1007/978-3-032-19524-1_63

They are characterized by their changeability [3], and agility is seen as an enabler for dealing with current challenges [4]. Platforms and Standardization support this process [5] and are considered an enabler for global production networks [3].

At the same time, companies are increasingly affected by shorter product life cycles. On the one hand, this requires faster product development [6]. On the other hand, it also accelerates the need to design and implement corresponding production systems. As a result, both efficiency and speed in development and ramp-up must be significantly improved. However, a major challenge lies in the fact that, traditionally, product development and production planning have been carried out sequentially [5]. In light of the developments mentioned above, this sequential approach is no longer adequate [7]. Over the past decades, the concept of simultaneous engineering has therefore emerged [8], and is now commonly referred to as Product-Production-CoDesign (PPCD) [5]. This approach enables product development and production planning to be performed in parallel, allowing for faster and more integrated development processes. However, in practice, product development processes often lack a unified approach and consistent use of standards [9], which significantly hinders implementation.

Nevertheless, current developments remain insufficiently addressed, as PPCD focuses only on the interaction between product development and production planning. Influencing factors increasingly affect the entire global production network [3], and resilience cannot be assessed solely at the plant or product level, but must consider the network as a whole [10]. Since both network and portfolio planning are strategic tasks that must be addressed early, the CoDesign approach needs to be extended accordingly. While simultaneous product and production planning offers clear benefits, it does not meet the structural and strategic requirements at the network level. Integrated network and portfolio planning significantly contributes to enhancing long-term resilience and responsiveness to external disruptions.

Therefore, our paper proposes a framework for implementing CoDesign on a network level, reaching Product-Network-CoDesign (PNCD). It is based on the use of standardization and integrates an impact analysis and measurement of resilience.

The work is structured as follows: Sect. 2 provides an overview of related work, in Sect. 3 the framework is explained, and an evaluation approach is proposed, before a summary and outlook is given in Sect. 4.

2 Related Work

As described in [3], the development of a production network involves key strategic considerations, including the identification of target sales regions, the selection of suitable plant locations, and the choice and geographical placement of potential suppliers. In the work of Hochdörffer et al. [11], it becomes clear that, all of these elements require close coordination. In addition to network design, product portfolio development also constitutes a critical strategic task. To effectively meet customer demand, the production network must be aligned with and support the intended product portfolio [11]. At the same time, the production system itself must be efficiently planned, with a focus on operational execution to ensure reliable fulfillment of customer needs. In this context, decisions regarding cost structures and capacity planning play a decisive role [12].

Existing models address the interrelations between the different levels. For instance, some specifically focus on the production network as well as the planning and control tasks related to production. As described in [13], the network-related tasks particularly include network configuration, network sales planning, and network demand planning. Reinerth et al. [14] examine factory planning from the perspective of structure, process, and system. In this context, particular consideration is given to the production network, the production process, and the design of the production system.

As already mentioned in Sect. 1, product development and production planning are an initially sequential process [5]. Nevertheless, the idea of parallelizing processes has been developed over various other approaches, e.g., ease of manufacture (DFM) and ease of assembly (DFA), that enable a flexible manufacturing system [15]. More recent is the approach of Product-Production-CoDesign (PPCD), which refers to the collaborative and simultaneous planning and development of the product and production system. According to [5] their definition encompasses the application of efficient production operations, as well as the integration of related business models, and the consideration of several product generations.

The following approaches show how PPCD is used today. Albers et al. [5] present a case study in which information from earlier product development processes is used to enhance a design assistant, reducing development time. May et al. [16] link Design Thinking with CoDesign to support sustainable manufacturing, focusing on design for manufacturability and the early integration of user requirements before and during prototyping to ease future adaptations. Staehr et al. [17] propose a method for planning adaptable production systems by linking change drivers to specific design guidelines using Monte Carlo simulations.

These approaches show that, on the one hand, the production system can be made more flexible and, on the other hand, methods can be created to enable more efficient product development. It becomes evident that the presented approaches primarily concentrate on product development and production system planning. Their disadvantage is that they involve short- to medium-term decisions, meaning their scope is relatively limited, focusing only on the product and production planning, while the broader production network is largely neglected. Enhancing resilience, however, requires considering the entire network, as many influencing factors affect it [10]. To address this research deficit, we propose a concept for PNCD that contributes to enhancing resilience by building on established standards.

3 A Concept for Product-Network-CoDesign

Based on the presented models regarding the interaction of planning and development tasks across the network, product, and production system in Sect. 2, our approach determines four key components as central elements. An overview is given in Fig. 1. Building on the discussed approaches, our proposal focuses, in the medium-term, on product development and production system planning. The network is also considered as an additional strategic level, in line with the presented frameworks. However, these approaches often lack a long-term perspective on product planning, reflected in the portfolio. Therefore, we introduce portfolio development as an additional strategic layer.

From the previously outlined frameworks, it is evident that the connection between product and production already exists indicated by the solid line. Through CoDesign, this relationship is not only sequential but also simultaneous and reciprocal. Furthermore, established links between network planning and production system planning have already been described. There are also clear interrelations between portfolio development and product development [13,14].

What distinguishes our approach, however, is the systematic alignment between the product portfolio and the overall network. The concept of PNCD is particularly characterized by the parallel development and joint design of both dimensions. The dotted lines illustrate the relationship between PPCD and PNCD, with PPCD serving as the basis in the medium-term and evolving into PNCD in the long-term. These connections are not yet established but serve as a prerequisite for PNCD. This integrated perspective ensures strategic coherence across all planning levels and addresses a critical gap in current frameworks.

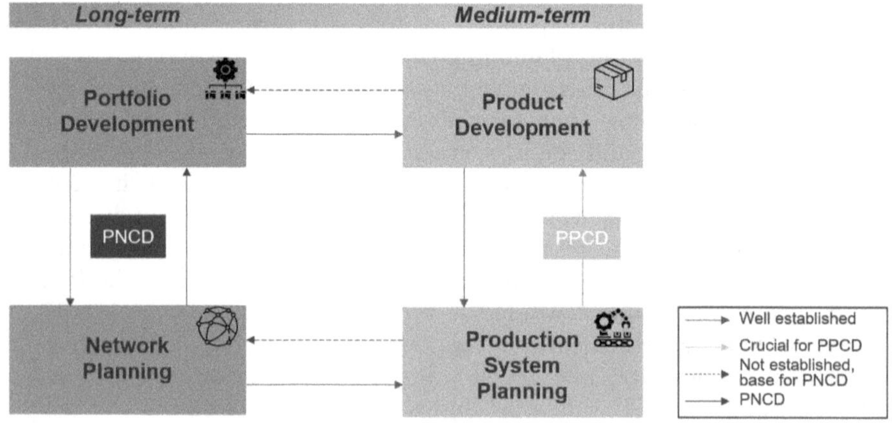

Fig. 1. A concept for Product-Network-CoDesign

3.1 A Framework for Implementing PNCD

Although standards can act as enablers, the trend toward highly customized products presents a conflicting development, often requiring product-specific production lines. A trade-off exists between highly customized products, which better meet individual customer needs but increase internal system complexity, and highly standardized products, which reduce complexity but limit the fulfillment of specific customer wishes [18]. Persistent external influences further complicate this balance, as the impact of standardization must be continuously evaluated. Standards can be defined on the one hand, by the product itself, for example through clearly specified product specifications and concrete design guidelines, and on the other hand, by the production system, such as through the use of defined methods and processes. To address this, the framework proposes a procedure for PNCD based on standards and includes the evaluation of resilience to ensure a comprehensive assessment.

The implementation is carried out in several stages, starting from an initial assessment of the current situation and extending to continuous adaptation, incorporating influencing factors and resilience measurement. A comprehensive overview is provided in Fig. 2 and is described below.

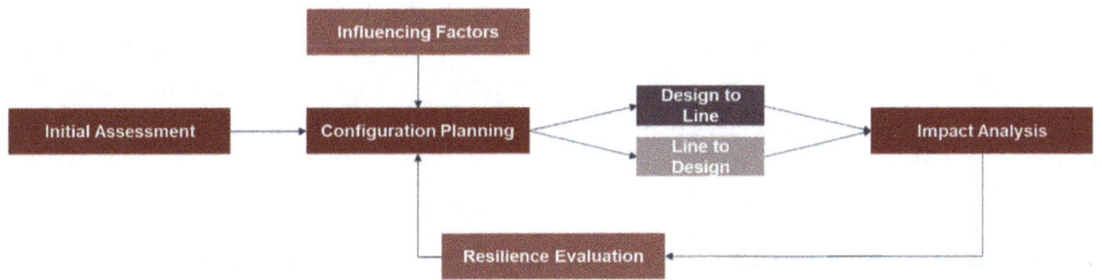

Fig. 2. A framework for implementing PNCD

In the first phase, **Initial Assessment**, an assessment of the current state is conducted. This includes identifying critical criteria related to the product, such as product characteristics and the Bill of Materials. The same applies to the production system, with a focus on process characteristics and an analysis of the existing production lines. The aim is to establish a structured understanding of the system's initial configuration.

Production networks are continuously affected by external **Influencing Factors**, as described in [3]. These have an impact on the resilience of production networks [10]. The collection and categorization of these factors are part of the second phase.

Within the next phase, **Configuration Planning**, a distinction between two main approaches can be made. This describes the implementation of PPCD with respect to the whole network. A starting point can be made either from the product or the production system.

Within the first approach, *Design-to-Line*, newly developed products can be adapted to existing production lines during product development. The objective is to ensure manufacturability within defined portfolios and customer requirements. Standardized development processes and modular production standards serve as enablers. This typically requires building on a platform or modular system.

In addition to the *Design-to-Line* approach, the *Line-to-Design* strategy can be applied. Here, production systems are adapted to accommodate newly developed products. The key challenge lies in enabling the system to manufacture both current and future product generations. This may involve the implementation of flexible, reconfigurable production lines. A standardized framework for production system design acts as an enabler by guiding the development of adaptable systems.

The configuration planning is influenced by a range of economic and strategic factors that form the phase **Impact Analysis**. These need to be identified and classified in order to determine the degree of standardization.

In order to assess the effects of configuration planning on resilience, resilience must first be made measurable. This is part of the last phase, **Resilience Evaluation**. This requires the definition of suitable resilience metrics that can be aligned with the identified influencing factors. These factors continuously impact the system and affect both product and process domains.

Insights gained from resilience measurement serve as a basis for reconfiguring the system. This creates a closed-loop process in which external influences are continuously analyzed, evaluated, and systematically integrated into future configuration decisions.

3.2 Conceptual Evaluation Framework in Case Studies

The proposed approach will be evaluated across multiple use cases, with the first application focusing on an automotive supplier. This use case evaluation has been partially completed. Influencing Factors were initially collected through expert interviews, followed by an initial assessment and configuration planning as described in Sect. 3. In this case, the *Design-to-Line* approach is applied so that the product can be produced on an existing manufacturing line. In several workshop sessions, a broad range of product and process factors potentially influenced by the configuration decision were identified and clustered into overarching categories.

Examples of major categories include process, product, and customer. On the process side, relevant aspects include performance, tolerances, and timing, reflected in key performance indicators such as Overall Equipment Effectiveness, cycle times and downtimes, as well as capacity measures like available area, number of shifts, or buffer capacity. Standardized procedures may improve performance and reduce development expenses due to reduced customization needs. On the product side, customer-specific requirements are critical, as standardized solutions can limit individualized features and potentially affect satisfaction. Expected sales volumes also influence whether a standardized or customized approach is more suitable. Risk considerations include acceptance of standardization by customers and developers, as well as the potential inhibition of innovation, since excessive standardization may restrict creativity and weaken long-term competitiveness.

4 Summary and Outlook

This paper proposes a framework for implementing Product-Network-CoDesign (PNCD) by leveraging standardization. Standards may be defined by the product, for example through clear specifications and design guidelines, or by the production system via established methods and processes. It provides a procedure for relevant steps and integrate the existing Product-Production-CoDesign (PPCD) into PNCD, aiming to contribute to increased resilience.

In a first use case, relevant influencing factors and KPIs were collected through expert interviews and workshops. These include performance indicators, customer requirements, and innovation risks, which shape configuration strategies and affect resilience evaluation.

The positive effect of standards on resilience can be seen, for example, in the use of alternative production sites when machines fail or in the availability of alternative suppliers through standardized components. Although a concrete resilience measurement method is not yet implemented, the framework provides a basis for such an approach.

Future work will focus on applying the framework to additional use cases, developing a method for resilience evaluation, and validating the concept through quantitative analysis. An optimization model is proposed to support strategic configuration decisions by integrating impact analysis with resilience considerations. This model will help determine the most economically viable production strategy for new products, and evaluate trade-offs between Design-to-Line, adapting products to existing production capabilities, and Line-to-Design, adapting production systems to product requirements through dedicated or modular flexible lines.

In summary, the framework provides a foundation for portfolio- and network-oriented CoDesign, aligned with standardization principles, with the objective of increasing the resilience of global production networks.

Acknowledgments. This research was funded by the German Federal Ministry for Education and Research (BMBF) through the research project 02J23C100-114 FENI-X.

Competing Interests. The author(s) has no competing interests to declare that are relevant to the content of this manuscript.

References

1. Tukamuhabwa, B.R., Stevenson, M., Busby, J., Zorzini, M.: Supply chain resilience: definition, review and theoretical foundations for further study. Int. J. Prod. Res. **53**(18), 5592–5623 (2015)

2. Janssen, B., Welsing, M., Schmitz, S., Schuh, G.: Enhancing the resilience of global production networks against unpredictable disruptions: a systematic literature review. In: Congress of the German Academic Association for Production Technology, pp. 502–513. Springer (2024)
3. Lanza, G., Ferdows, K., Kara, S., Mourtzis, D., Schuh, G., Váncza, J., Wang, L., Wiendahl, H.-P.: Global production networks: design and operation. CIRP Ann. **68**(2), 823–841 (2019)
4. Monauni, M.: Agility enablers in production networks-pooling and allying of manufacturing resources. Proc. CIRP **17**, 657–662 (2014)
5. Albers, A., Lanza, G., Klippert, M., Schäfer, L., Frey, A., Hellweg, F., Müller-Welt, P., Schöck, M., Krahe, C., Nowoseltschenko, K., et al.: Product-production-codesign: an approach on integrated product and production engineering across generations and life cycles. Proc. CIRP **109**, 167–172 (2022)
6. Gräßler, I., Pöhler, A., Hentze, J.: Decoupling of product and production development in flexible production environments. Proc. CIRP **60**, 548–553 (2017)
7. Jacob, A., Windhuber, K., Ranke, D., Lanza, G.: Planning, evaluation and optimization of product design and manufacturing technology chains for new product and production technologies on the example of additive manufacturing. Proc. CIRP **70**, 108–113 (2018)
8. Steimer, C., Cadet, M., Aurich, J.C., Stephan, N.: Approach for an integrated planning of manufacturing systems based on early phases of product development. Proc. CIRP **57**, 467–472 (2016)
9. Tortorella, G.L., Marodin, G.A., de Castro Fettermann, D., Fogliatto, F.S.: Relationships between lean product development enablers and problems. Int. J. Prod. Res. **54**(10), 2837–2855 (2016)
10. Benfer, M., Verhaelen, B., Peukert, S., Lanza, G.: Resilience measures in global production networks. Die Unternehm. **75**(4), 491–520 (2021)
11. Hochdörffer, J., Buergin, J., Vlachou, E., Zogopoulos, V., Lanza, G., Mourtzis, D.: Holistic approach for integrating customers in the design, planning, and control of global production networks. CIRP J. Manuf. Sci. Technol. **23**, 98–107 (2018)
12. Demartini, M., Tonelli, F., Pacella, M., Papadia, G.: A review of production planning models: emerging features and limitations compared to practical implementation. Proc. CIRP **104**, 588–593 (2021)
13. Schuh, G. (ed.): Produktionsplanung und-steuerung: Grundlagen, Gestaltung und Konzepte. VDI-Buch, vol. 3, Völlig neu bearbeitete auflage edn. Springer, Berlin (2006)
14. Reinerth, H., Lickefett, M.: Fabrikplanung. In: Fabrikbetriebslehre 1: Management in der Produktion, pp. 103–127 (2020)
15. ElMaraghy, H., Monostori, L., Schuh, G., ElMaraghy, W.: Evolution and future of manufacturing systems. CIRP Ann. **70**(2), 635–658 (2021)
16. May, M.C., Schäfer, L., Lachnit, T., Lanza, G.: Product-production-codesign thinking for sustainable manufacturing. In: Global Conference on Sustainable Manufacturing, pp. 185–193. Springer (2023)
17. Staehr, T., Stricker, N., Lanza, G.: Autonomous planning tool for changeable assembly systems. Proc. CIRP **88**, 104–109 (2020)
18. Kuhl, J., Krause, D.: Strategies for customer satisfaction and customer requirement fulfillment within the trend of individualization. Proc. CIRP **84**, 130–135 (2019)

Hierarchization of Coordination Efforts in Global Production Networks: A Systematic Literature Review

Kathrin Böttger[1(✉)], Günther Schuh[1,2], Seth Schmitz[1], and Benedict Janssen[1]

[1] Laboratory for Machine Tools and Production Engineering (WZL), RWTH Aachen University, Aachen, Germany
k.boettger@wzl.rwth-aachen.de

[2] Fraunhofer Institute for Production Technology IPT, RWTH Aachen University, Aachen, Germany

Abstract. In the context of increasingly interconnected and dynamic global markets, the complexity of global production networks (GPNs) continues to grow. This complexity requires increased coordination efforts to manage the value creation within GPNs effectively. Coordination efforts refer to the actions necessary to coordinate GPNs, focusing on resource sharing, information and knowledge exchange, determining the degree of centralization, and incentivizing. To optimize these coordination efforts, it is essential to first analyze and quantify them. This paper enumerates and categorizes the coordination efforts required within GPNs, thereby providing the transparency necessary to derive future optimization potentials. To achieve this, the paper presents the results of a systematic literature review in accordance with the PRISMA guidelines as described by Page et al. on the current state of literature regarding coordination efforts in GPNs and a subsequent hierarchization of these coordination efforts.

Keywords: Global production networks · Production management · Coordination

1 Motivation

Despite notable disruptions such as the COVID-19 pandemic and geopolitical tensions, global trade in goods and services has experienced strong growth over the past ten years. According to the World Trade Organization, forecasts indicate continued expansion in the coming years [1]. No company can afford to ignore the growth of world trade in manufactured goods, as businesses increasingly operate internationally in terms of both sales and production leading to a significant growth of complexity in Global Production Networks (GPNs). In order to remain competitive, managing and coordinating GPNs has become an increasingly important task for managers [2, 3].

As a result of the increasing interconnectivity of GPNs, more elements within GPNs require coordination, which in turn raises overall coordination efforts resulting in larger costs for e.g. indirect personnel, infrastructure, licenses as well as larger personnel costs

L. Overmeyer and B.-A. Behrens (eds.), *Production at the Leading Edge of Technology*, Lecture Notes in Production Engineering, https://doi.org/10.1007/978-3-032-19524-1_64

in general [2, 4]. This challenge is compounded by a lack of transparency around the resources needed to coordinate complex networks. Without clear transparency into coordination efforts, organizations cannot holistically reduce these efforts or identify the most effective measures such as digitalization initiatives for doing so [5]. The quantification of coordination efforts is especially important as a rise in efforts is directly linked to higher costs and weaker financial performance of a company [6].

To address the need for increased transparency over coordination efforts in GPNs, this paper presents a systematic literature review with the objective of compiling coordination efforts and sorting them based on their frequency of mention and subsequently organizing them in a hierarchical structure, enabling a breakdown of these efforts from main coordination efforts to micro-level efforts in GPNs. Specifically, it seeks to answer the following research question:

Which coordination efforts exist in GPNs and how can they be hierarchized based on the scope of effort?

The structure of the paper is as follows. First, the current state of research on coordination and coordination efforts in GPNs is presented. Next, the research methodology is elaborated. The following chapter then presents the results, directly answering the research question. Here the identified coordination efforts are systematically organized in a hierarchical structure, enabling a breakdown of these efforts from main coordination efforts to micro-level efforts in GPNs. The paper concludes by highlighting the need for further research to quantify and assess coordination efforts in GPNs.

2 Relevant Definitions and Prior Research on the Topic of Coordination Efforts in GPNs

In the following a brief introduction on the topic of both coordination and coordination efforts in GPNs is given and prior research on these topics is presented.

2.1 Relevant Definitions in the Context of This Paper

As defined by Lanza et al., a GPN consists of geographically dispersed production entities that are interconnected through material, information and financial flows [7].

Following the work by Remling and Friedli coordination in GPNs consists of four principal subjects: 1. *Centralization and Standardization*, which define organizational structures and allocate decision-making authority through formal competencies 2. *Information and Knowledge Exchange*, which describes information and knowledge flow and exchange mechanisms 3. *Resource Sharing*, which manages the distribution of scarce assets and 4. *Incentive Systems*, which motivate desired behaviors by rewarding collaboration or discouraging counterproductive actions [8].

Subsequently coordination efforts in GPNs refer to the necessary activities when coordinating in these networks which leads to a need for resources [9, 10].

2.2 Prior Research

Even though coordination of GPNs is a main driver for the successful value creation of GPNs it remains largely underinvestigated in the scientific literature [4, 11]. Many publications investigating coordination efforts are based on case studies. For example, both Norouzilame and Wiktorsson and Sayem et al. examine case studies and present coordination mechanisms employed by companies operating in different sectors and with varying company sizes [4, 12]. Specht et al. investigate decisions in the field of centralization in GPNs, which is one of the focus areas of coordination as highlighted in the previous subchapter [13]. Based on expert interviews, Wiech investigates barriers to inter-plant exchange and highlights coordination measures that can improve inter-plant exchange [14]. Badasjane presents a literature review on coordination mechanisms related to digital transformation GPNs highlighting coordination mechanisms in GPNs related to specific tasks [15]. A broader view is given by Rudberg and West who present a conceptual coordination model for GPNs, describing how coordination needs to be carried out in order to achieve a company's strategic goals [3]. While these authors discuss closely related coordination efforts, none of them provide a holistic overview over coordination efforts in GPNs. Additionally, the coordination efforts discussed in the literature vary greatly in granularity, making a summary of coordination efforts in GPNs and a subsequent hierarchization necessary.

3 Research Methodology

In accordance with the approach proposed by Moher et al. and guided by the PRISMA checklist, the present literature review is structured into four phases. Initially all relevant contributions from the selected databases are *identified*. In this step, authors, title, publication date, number of citations and abstract are transferred into a table. In the following phase the contributions are *screened* for duplication. Subsequently, the literature is *filtered* for publications in the field of engineering and economics and for publications that are accessible via the network of RWTH Aachen's university library. Additionally, only publications are considered that deal with coordination within a network rather than across a network's borders. Finally, the remaining articles are manually filtered for their *suitability* with respect to the research question [16, 17]. The investigation of the suitability of publications was carried out by two independently working researchers to ensure that no relevant publication was excluded.

The used databases are Science Direct, Scopus, Springer Link and Google Scholar. The selection of these databases was made based on their status as the most substantial scientific databases for English and German language literature. These databases collectively encompass several million articles [18].

The search strings utilized for the literature search are illustrated exemplary in the English language in Fig. 1. The search strings were formulated around the theme of "Coordination efforts in Global Production Networks" and comprise as many synonyms as possible to ensure a holistic search gathering all relevant publications. The search strings are formed in the same way in English and German and applied to titles, keywords and abstracts in the respective databases.

Global production network		Coordination effort
Global Production Network OR Global Manufacturing Network OR Supply Chain Network OR International Manufacturing Network OR Global Supply Chain OR Supply Chain	AND	Coordination effort OR Coordination efforts OR Coordination costs OR Organizational effort OR Organizational efforts OR Coordination tasks OR Coordination activities

Fig. 1 English search strings for the literature search

Figure 2 shows the results of the presented search process. It must be noted that from Google Scholar only the first 107 results were considered. This ensured feasibility and maintained quality as the relevance of publications notably decreased and additional entries became irrelevant to the research question. From 3,195 contributions found in the databases, 2,862 relevant contributions are analyzed in the literature using the method presented which led to a final 29 publications that remained for further analysis.

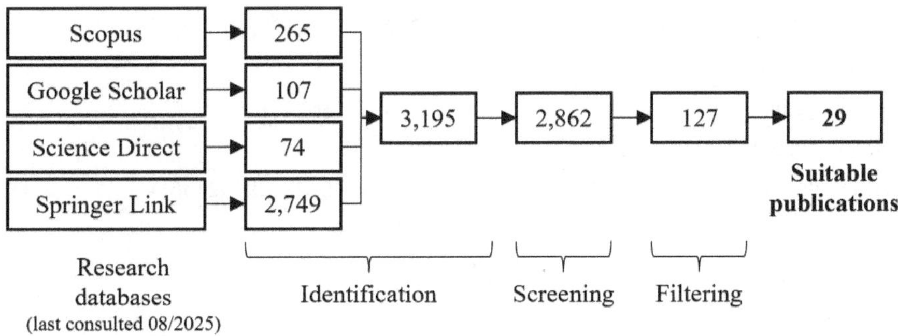

Fig. 2 Number of identified publications in the systematic literature review

4 Hierarchization of Coordination Efforts in GPNs

In the following sub-chapters first the overall categories of coordination efforts identified are presented. A hierarchization of identified coordination efforts is presented in Sect. 4.2.

4.1 Coordination Efforts in Global Production Networks

In total 130 coordination efforts are identified within the 29 publications relevant to this literature review. To maintain an overview of these coordination efforts they were categorized based on their theme and overall scope concerning coordination in GPNs in alignment with the thematic analysis proposed by Braun and Clarke. The methodology allows thematic clusters to be identified by first listing the results of the literature review and then analyzing the identified coordination efforts individually for recurring patterns which allows a clustering them into categories [19]. In this subchapter these main categories are presented (see Table 1) as well as the corresponding publications listed.

Table 1 Categorization and grouping of main coordination efforts in GPNs

Degree of centralization and standardization	
Determining the degree of centralization	[3, 4, 12, 13, 15, 21, 23, 24, 26–30]
Managing plant roles	[4, 15, 21, 24, 26, 30–33]
Coordination of standards and continuous improvement	[3, 4, 12, 14, 15, 20, 21, 23, 24, 26, 28, 29, 33–35]
Transferring culture and core values	[3, 4, 12, 26, 27]
Resource sharing	
Organizing shared initiatives	[3, 4, 12, 15, 20, 21, 24, 28–30]
Re-organization and re-allocation of resources	[4, 14, 30]
Information and Knowledge exchange	
Initiating knowledge exchange	[3, 4, 14, 15, 23, 26–28, 34, 36]
Information Technology implementation	[3, 4, 12, 14, 20, 22–24, 26, 28, 30, 32, 34, 36, 37]
Information Technology usage	[3, 4, 12, 20, 23, 24, 28, 32, 36, 37]
Incentive systems	
Implementing incentive systems	[14, 15, 23, 33, 36, 38]
Interaction among plants	
Fostering relationships between plants	[4, 14, 15, 25, 26, 28, 29, 32, 36]
Balancing cooperation and competition	[14, 15, 20, 23, 33, 36]
Decision management and conflict resolution	[22, 29, 32, 37, 39]

The main coordination categories in which the 130 coordination efforts could be summarized are *degree of centralization and standardization*, *resource sharing* and *information and knowledge exchange*. As shown in Table 1 the identified categories correspond to the main elements of coordination in GPNs proposed by Remling and Friedli that were presented in Sect. 2.1 [8]. Not previously mentioned in Chap. 2 is the *interaction among plants*, which cannot be accommodated within the framework proposed by Remling and Friedli and was therefore introduced here as an additional main category. In the following the main categories as well as corresponding coordination efforts are presented.

Degree of centralization and standardization. The subcategories in the category *degree of centralization and standardization* describe the efforts needed to maintain the adequate balance between centralized control over the network and decision autonomy of the plants. Specht et al. emphasize the importance of determining the degree of centralization within GPNs and that there is no optimal degree of centralization for a GPN but it depends on the desired rights, freedom and responsibilities of individual plants [13]. Additionally, Sayem et al. call for the implementation of organizational standards and the transfer of core values to the network, especially when GPNs are decentralized [12].

Resource sharing. *Resource sharing* describes the efforts needed to share resources across different plants and continuously adapting these resources. Two subcategories were identified within this context: *Organizing shared initiatives* and *Re-organization and re-allocation of resources*. Bendoly et al. highlight that sharing resources is an important coordination mechanism but also leads to an increase in organizational interdependencies that need to be managed, leading to higher coordination efforts [20]. Mundt proposes a model in which plants jointly finance central support functions by paying a fee which grants all sites access to services they could not afford alone [21].

Information and knowledge exchange. This category deals both with the general knowledge exchange between plants as well as the needed efforts to enable an efficient information exchange through information technology implementation and usage. Notably, quite a few publications focus on IT support for coordination in GPNs. Mittermayer and Rodriguez-Monroy investigate different forms of IT support in supply chains and the implementation of new software [22]. Additionally, Norouzilame and Wiktorsson emphasize the importance of IT channels for accessibility of databases while Wiech and Dreher both advocate IT tools for informal communication as a form of coordination tool [4, 23, 24].

Implementing incentive systems. *Implementing incentive systems* refer to the mechanisms and factors that guide the behaviour of individuals within GPNs to achieve a certain goal. Wiech points out that research regarding these mechanisms is characterized by contradicting findings and interpretations [23]. This may be the reason why this category was identified as having the lowest number of sub-coordination efforts.

Interaction among plants. To successfully coordinate a GPN the individual plants need to have established relationships that call for continuous efforts in sustaining them. Aligning goals in the network is identified as a key enabler for successful cooperation. Gunasekaran et al. and Kemppainen and Vepsäläinen both highlight the importance of fostering relationships among plants within GPNs which will allow for better network performance through stronger sharing of skills, competencies and knowledge [25, 26].

4.2 Hierarchization of Coordination Efforts

After having presented a broad overview of the clusters of coordination efforts, the related efforts and sub-efforts are discussed in this chapter. After an initial compilation of a long list of coordination efforts, they were grouped based on the approach by Braun and Clarke based on their overall scope and goal in GPNs [19]. Coordination efforts are described in literature with varying degrees of detail; accordingly, some could be hierarchized into three sub-levels, while others only into main and sub-categories. To maintain a manageable overview, only coordination efforts mentioned by at least three different authors are presented. If the overarching category was also mentioned in literature, it is cited in Table 2. It has to be noted that conceptually equivalent terms were grouped based on their contextual meaning, not on exact wording. The results of the hierarchization are visualized in Table 2 and are discussed in the following section.

Table 2 Clustering of coordination efforts

Degree of centralization and standardization	
Determining the degree of centralization	*Coordination of standards and continuous improvement*
o Balancing decision-making power [4, 13, 21, 26] o Set-up of reporting structures [12, 24, 26] o Central control over processes [12, 21, 24, 28, 30] **Managing plant roles** o Introducing responsibilities and plant roles [4, 21, 32, 33]	o Coordination of standards and rules [4, 14, 15, 21, 26, 28, 35] • Formalization [12, 14, 15, 21, 24, 28] • Homogenization [4, 15, 20, 25, 28, 29] • Enforcing [15, 21, 24, 34] o Continuous improvement in the network • Definition of KPIs [12, 15, 28, 33] • Definition and implementation of shared performance measurement [15, 20, 24, 28, 33] • Carrying out regular performance audits [15, 24, 34]
Resource sharing	
Organizing shared initiatives o Set-up of frame agreements for shared initiatives [12, 27, 28] o Organizing shared sourcing [3, 4, 12, 21, 29, 30]	o Organizing shared production planning [4, 21, 30, 35] o Organizing a shared inventory planning [4, 30, 35, 39]
Information and knowledge exchange	
Initiating knowledge exchange o Creating information transparency [4, 20, 23, 24, 26, 28] o Formation of cross-plant teams [3, 4, 14, 15, 20, 24] o Creating of competence groups [4, 15, 24] o Physical meetings [3, 4, 12, 14, 15] • For experts [3, 14, 15, 27] • For specific tasks [4, 12, 14, 15, 27, 28] o Virtual meetings [4, 12, 28]	**Information Technology implementation** o Software implementation [22, 26, 28] o Managing dependencies [4, 20, 22] o Integration of software between plants [4, 22, 28, 37] **Information Technology usage** o Making standards available and usable [3, 4, 20, 28] o Maintaining databases of knowledge in the network [3, 4, 20, 28]
Interaction among plants	
Fostering relationships between plants o Building relationships [14, 15, 25, 26] o Continuous communication between plants [3, 4, 14, 15, 24, 28, 35] o Defining communication frequencies [14, 20, 24, 25] **Balancing competition and cooperation** o Defining a network mission [14, 15, 33] o Mitigating inter-plant competition [14, 23, 33]	**Decision management and conflict resolution** o Iteration of decisions [8, 22, 29] o Resolving conflicts • Orchestrating dependencies [4, 20, 30] • Dealing with cultural differences [32, 36, 40]

Degree of centralization and standardization. Especially the *coordination of standards and continuous improvement* were highlighted by many authors. Sayem et al. focus on the standardisation of processes and procedures, such as standardised production processes and the exchange of standard documents [12]. Dreher specifies that coordination efforts include not only the standardisation of processes, but also the standardisation of process sequences. The author notes that this is an important aspect of coordination, as it improves the stability of GPN performance, since the order in which activities are carried out is no longer decided by individual sites [12, 24].

Resource sharing. Only one subcategory within *resource sharing* was mentioned several times by the literature investigated, namely *organizing shared initiatives.* The organization of shared initiatives includes not only the organization of the actual initiative such as *shared sourcing* proposed by Rudberg and West but also the *set-up of frame agreements* and governance for these shared initiatives [3, 27, 30].

Information and knowledge exchange. The largest coordination effort of *information and knowledge exchange* is *initiating knowledge exchange.* Within this category authors discuss the *formation of cross-plant teams and competence groups* to share information and knowledge in an agreed upon setting. This is particularly promoted by Badasjane and Dreher [15, 24]. IT systems as a method for *information and knowledge exchange* has already been discussed in the previous chapter. The sub-efforts both cover the *implementation of IT systems* as well as the *usage* of these systems.

Interaction among plants. A strong focus on the *interaction among plants* was laid by Wiech. The author highlights the importance of creating a shared focus on global goals and aligning goals so that the GPN becomes the focus of attention by plant leaders [23]. Additionally, the *resolution of conflict* as well as the *iteration of decisions* is needed in the context of coordinating interaction among plants. Mittermayer and Rodriguez-Monroy propose IT-systems as a tool to create rules and standards that mitigate conflicts between sites but also call for the collection of feedback from the sites to improve these rules [22].

5 Conclusion and Further Research

An increase in complexity of GPNs has a noticeable impact on the need for improved coordination and reduction of coordination efforts in GPNs. This study highlights the importance of coordination and reducing coordination efforts in GPNs: Through the conduction of the systematic literature review, a comprehensive overview over the literature on coordination efforts was created. Additionally, a hierarchization of coordination efforts was offered regarding the scope and granularity of efforts, enabling the pathway to quantification of coordination efforts. This study was able to expand on the current state of literature and define one new cluster of coordination elements that need to be considered namely *Interaction among plants.*

To complete a holistic picture of coordination efforts in GPNs the resources needed for the identified coordination efforts need to be investigated. Even though this might be company-specific, the enumeration of these resources will be necessary to then allow for the quantification of coordination efforts and the subsequent reduction of coordination efforts. Additionally, it is noteworthy that coordination efforts can be intertwined amongst each other. Therefore interdependencies among coordination efforts should be explored in the future [15].

Furthermore, a validation of the identified coordination efforts is required. Especially, as only English and German publications were considered, this step is important to ensure all relevant coordination efforts were considered. This subsequent phase involves the evaluation of the efficacy of the proposed framework through the utilization of authentic company case studies.

Acknowledgements. Funded by the Deutsche Forschungsgemeinschaft (DFG, German Research Foun-dation) under Germany's Excellence Strategy—EXC-2023 Internet of Production—390621612.

Competing Interests The author(s) has no competing interests to declare that are relevant to the content of this manuscript.

References

1. World Trade Organization: Global Trade Outlook and Statistics (2025)
2. Ferdows, K.: Made in the world: the global spread of production. Prod. Oper. Manag. **6**, 102–109 (1997)
3. Rudberg, M., West, B.M.: Global operations strategy: coordinating manufacturing networks. Int. J. Manag. Sci. **36**, 91–106 (2005)
4. Norouzilame, F., Wiktorsson, M.: Coordination within international manufacturing networks: a comparative study of three industrial practices. Am. J. Ind. Bus. Manag. **08**, 1603–1623 (2018). https://doi.org/10.4236/ajibm.2018.86107
5. Dabbish, L., Tsay, J., Herbsleb, J.: Transparency and Coordination in Peer Production (2014)
6. Moreno-Miranda, C., Dries, L.: The Role of Coordination Mechanisms and Transaction Costs Promoting Sustainability Performance in Agri-Food Supply Chains: Evidence From Ecuador (2024), https://doi.org/10.1002/agr.22003
7. Lanza, G., Ferdows, K., Kara, S., Mourtzis, D., Schuh, G., Váncza, J., Wang, L., Wiendahl, H.-P.: Global production networks: design and operation. CIRP Ann. **68**, 823–841 (2019). https://doi.org/10.1016/j.cirp.2019.05.008
8. Remling, D., Friedli, T.: The St. Gallen management model for international manufacturing networks. In: Friedli, T., Lanza, G., Remling, D. (eds.) Global Manufacturing Management. From Excellent Plants Toward Network Optimization. Management for Professionals, pp. 25–50. Springer International Publishing, Cham (2021)
9. Jansma, J.M., Ramsey, N.F., de Zwart, J.A., van Gelderen, P., Duyn, J.H.: FMRI study of effort and information processing in a working memory task. Hum. Brain Mapp. **28**, 431–440 (2007)
10. Steele, J.: What is (perception of) effort? Objective and subjective effort during attempted task performance (2020)
11. Netland, T.H.: Coordinating Production Improvement in International Production Networks: What's new? In: Johansen, J., Farooq, S., Cheng, Y. (eds.) International operations networks, pp. 119–132. Springer, London, London (2014)
12. Sayem, A., Feldmann, A., Ortega-Mier, M.: Coordination in international manufacturing: the role of competitive priorities and the focus of globally dispersed facilities. Sustainability **10**, 1314–1333 (2018)
13. Specht, F., Kaiser, J., Steier, G.L., Gleich, K., Friedli, T., Lanza, G.: Centralism or Autonomy? Insights into the decision allocation within International Manufacturing Networks. ZWF **119**, 23–29 (2024)
14. Wiech, M., Friedli, T.: Using plant leaders' perspectives to overcome barriers to inter-plant exchange. J. Manuf. Technol. Manag. **32**, 1167–1187 (2021)
15. Badasjane, V.: Coordinating digital transformation in International Manufacturing Networks (2023)

16. Moher, D., Liberati, A., Tetzlaff, J., Altman, D.G.: Preferred reporting items for systematic reviews and meta-analyses: the PRISMA statement. BMJ (Clinical research ed.), vol. 339 (2009). https://doi.org/10.1136/bmj.b2535
17. Page, M.J., McKenzie, J.E., Bossuyt, P.M., Boutron, I.: The PRISMA 2020 statement: an updated guideline for reporting systematic reviews. BMJ (Clinical research ed.), vol. 372 (2021)
18. Martín-Martín, A., Thelwall, M., Orduna-Malea, E., Delgado López-Cózar, E.: Google scholar, microsoft academic, scopus, dimensions, web of science, and OpenCitations' COCI: a multidisciplinary comparison of coverage via citations. Scientometrics **126**, 871–906 (2020). https://doi.org/10.1007/s11192-020-03690-4
19. Braun, V., Clarke, V.: Using thematic analysis in psychology. Qual. Res. Psychol. **3**, 77–101 (2006)
20. Bendoly, E., Bharadwaj, A., Bharadwaj, S.: Complementary drivers of new product development performance: cross-functional coordination, information system capability, and intelligence quality. Prod. Oper. Manag. **21**, 653–667 (2012)
21. Mundt, A.: The Architecture of Manufacturing Networks—Integrating the Coordination Perspective (2012)
22. Mittermayer, H., Rodríguez-Monroy, C.: Evaluating alternative industrial network organizations and information systems. Ind. Manag. Data Syst. **113**, 77–95 (2013). https://doi.org/10.1108/02635571311289674
23. Wiech, M.: Coordination of Inter-Plant Exchange—Make Plant Leaders Part of the Solution (2020)
24. Dreher, S.: Koordination des Produktionsanlaufs in globalen Produktionsnetzwerken. Springer Gabler (2023)
25. Gunasekaran, A., Irani, Z., Choy, K.-L., Filippi, L., Papadopoulos, T.: Performance measures and metrics in outsourcing decisions: a review for research and applications. Int. J. Prod. Econ. **161**, 153–166 (2015). https://doi.org/10.1016/j.ijpe.2014.12.021
26. Kemppainen, K., Vepsäläinen, A.P.: Trends in industrial supply chains and networks. Int. J. Phys. Distrib. Logist. Manag. **33**, 701–719 (2003). https://doi.org/10.1108/09600030310502885
27. de Waard, E., de Bock, P., Beeres, R.: The transaction costs of intra-organizational captive buying and selling relationships. J. Account. Organ. Chang. **15**, 257–277 (2019)
28. Kohli, R., Sherer, S.: Coordination begins at home: measuring coordination complementarities in supply chain environments. AMCIS 2004 Proc. **123**, 884–890 (2004)
29. Benfer, M., Hörger, M.: Bridging planning silos: a cross-functional decision support system for capacity, order, and supplier decisions in global production networks. CIRP Ann. **74**, 603–607 (2025). https://doi.org/10.1016/j.cirp.2025.04.029
30. Gao, X., Dong, S., Liu, C.: Overseas operations, global value chain position and technological innovation. Int. Rev. Econ. Financ. **94**, 103381 (2024). https://doi.org/10.1016/j.iref.2024.103381
31. Schober, F.: Negotiation of investment shares in interorganizational supply chains: a game-theoretic approach. ICIS 2005 Proc. **47**, 576–587 (2005)
32. Cheng, Y., Fredriksson, A., Fleury, A.: Rethinking international manufacturing in times of global turbulence. J. Manuf. Technol. Manag. **32**, 1113–1120 (2021). https://doi.org/10.1108/JMTM-10-2021-501
33. Kaiser, J., Friedli, T.: Company-Specific Plant Roles. A Reference Process for the Design and Deployment. ZWF, vol. 119, 30–35 (2024)
34. Lampón, J.F.: Efficiency in design and production to achieve sustainable development challenges in the automobile industry: modular electric vehicle platforms. Sustain. Dev. **31**, 26–38 (2023)

35. Hill, C.A., Scudder, G.D.: The use of electronic data interchange for supply chain coordination in the food industry. J. Oper. Manag. **20**, 375–387 (2002)
36. Powell, K.S., Lim, E., Ando, N.: Seeing the tree and the forest: Japanese auto firm multinational dispersion, cultural distance, and foreign manufacturing subsidiary ownership levels. Asian Bus. Manag. **20**, 163–187 (2021). https://doi.org/10.1057/s41291-019-00087-x
37. Aral, S., Bakos, Y., Brynjolfsson, E.: Information technology, repeated contracts, and the number of suppliers. Manage. Sci. **64**, 592–612 (2018). https://doi.org/10.1287/mnsc.2016.2631
38. Hsieh, C.-C., Wu, C.-H.: Capacity allocation, ordering, and pricing decisions in a supply chain with demand and supply uncertainties. Eur. J. Oper. Res. **184**, 667–684 (2006)
39. Bagul, A.D., Mukherjee, I.: Centralized vs decentralized sourcing strategy for multi-tier automotive supply network. Int. J. Product. Perform. Manag. **68**, 578–607 (2019)
40. Leseure, M.: Organisational design alternatives within international operations networks: a transaction cost perspective. Int. J. Serv. Oper. Manag. **7**, 419–439 (2010)

Mapping of Organizational Structures for Model-Based Systems Engineering Using an Organizational Graph

Gereon C. Bönsch(✉), Alexander Keuper, and Günther Schuh

Laboratory for Machine Tools and Production Engineering (WZL), RWTH Aachen University, Aachen, Germany
g.boensch@wzl.rwth-aachen.de

Abstract. Companies are increasingly using model-based systems engineering (MBSE) to deal with the growing complexity in product development and production and to increase efficiency and effectiveness. In particular, the organizational aspects of the transformation to model-based approaches pose challenges for companies. This paper presents a method for graph-based modeling of the interdependencies between activities, artifacts, roles and models in MBSE. Based on case studies of companies that have already successfully implemented MBSE in their organizations, an organizational graph is created. This resource description framework (RDF) graph contains the above-mentioned elements and their dependencies. It therefore provides an accessible orientation for companies looking for best practices as well as the basis for a future derivation of company-specific target organizational graphs.

Keywords: Model-based systems engineering · Product development · Organizational structure

1 Introduction

The push for digitalization and an emphasis on sustainability are currently posing significant challenges to companies within the mechanical and plant engineering sector. Both product design and production processes are further complicated by the need to integrate various disciplines, especially when producing cyber-physical systems (CPS), such as mechanical engineering, electronics, and software development [1–3]. This interdisciplinary nature makes coordination among specialized teams more difficult [4]. As complexity rises, traditional document-based approaches struggle to handle the increased constraints and scenarios, leading to longer development cycles, higher costs, and a greater potential for errors [5–7]. To tackle these challenges, model-based systems engineering (MBSE) has emerged as a promising solution. MBSE facilitates the creation of consistent system models and encourages an interdisciplinary approach to product development, covering areas of requirements engineering, design, simulation or production processes [8]. Transitioning from document-based methods to system models improves collaboration and minimizes the risk of costly mistakes [9].

L. Overmeyer and B.-A. Behrens (eds.), *Production at the Leading Edge of Technology*, Lecture Notes in Production Engineering, https://doi.org/10.1007/978-3-032-19524-1_65

However, despite the benefits of MBSE, there are hurdles to its implementation, particularly due to a lack of appropriate organizational structures. Implementing MBSE requires not only technical tools but also a thorough reevaluation of existing processes and roles [10]. Many companies struggle with this transition because of inadequate training and insufficient strategic guidance [4].

This paper presents a methodology for contributing to synchronizing technical system models with organizational structures for MBSE by mapping organizational structures via an organizational graph. Within the introduction, the potential of MBSE as well as current challenges are presented. This is followed by a presentation of related fundamentals in the state of the art. A detailed description of the methodology is then provided, concluding with a summary and outlook on further research.

2 State of the Art

This section describes relevant fundamentals regarding MBSE, organizational design and modeling as well as information representation via RDF.

MBSE supports development activities from requirements definition to validation in a model-based approach [11]. An overarching system model represents the architecture and behavior of the system and serves as the central information base, while domain models such as CAD objects are directly linked to it [12]. MBSE is closely associated with the Systems Modeling Language (SysML), which is based on the Unified Modeling Language (UML) [13]. While there is extensive research on the tools and methods landscape for MBSE, the organizational implementation is rarely addressed. Research by Henderson and Salado highlights organizational structure variables that can facilitate MBSE implementation. However, the results remain at a general level and offer neither company-specific advice nor support for specific design tasks [14, 15]. Huldt and Steinus look at the shift toward MBSE as a whole and begin by identifying the advantages of its introduction. However, this introduction is also hampered by various factors, ranging from cultural factors and management support to suitable processes and tooling. The authors see a particularly great need for improvement in organizational structures, but do not provide any concrete assistance in this regard [16]. Vasnev et al. present an approach for the gradual introduction of MBSE into a company's interface management and highlight the importance of organizational aspects. Although some of the steps in the process are transferable, it remains unclear to what extent the approach can be applied in other contexts. An overarching target vision for MBSE in the respective corporate context is not provided [17]. Based on their experiences at Boeing, Zhu and Wang show how common mistakes in process and organizational design can be avoided. However, the process and team structure described should be viewed in the context of large manufacturing enterprises. Transferability to other organizations or adaptation to their specific circumstances is not discussed [18].

Regarding organizations, a distinction is made between three concepts: instrumental, institutional, and functional concepts. The instrumental concept describes work processes as a sequence of interlinked activities, while the institutional concept represents the organizational structure that assigns activities to positions. The functional concept refers to planning and implementing organizational rules [19, 20]. Organizational design targets certain design areas such as task analysis, task synthesis or department formation [21].

A wide range of models is used in supporting these design tasks. Simple hierarchical trees often depict organizations as parent–child relationships but lack flexibility and focus solely on a functional view [22]. The Business Process Model and Notation (BPMN) is well suited to model workflows and activities but mostly neglects the functional view and provides low semantic depth limiting its interoperability [23]. UML Diagrams like activity diagrams for process representation or class diagrams for roles provide a richer, more formal language but are also more complex to create [24].

The Resource Description Framework (RDF) is a standard model for data interchange on the web that represents information using triples. An RDF triple consists of three components: subject, predicate, and object. This structure allows information to be represented in a machine-readable format. The subject identifies the resource or topic being described, typically through a Uniform Resource Identifier (URI). The predicate indicates a property or relationship of the subject and connects it to the object, which can either be another resource or a literal value.

These triples enable flexible descriptions of entities and their relationships, making them well-suited for representing complex connections [25]. Queries written in languages such as SPARQL are used to extract relevant information from an organizational graph [26]. These queries define the data domain, variables, and desired patterns, facilitating targeted retrieval of information from RDF-based models [27]. While an RDF is the most complex option for modeling organizational structures it is still chosen due to its expressiveness and the interoperability it provides.

3 Methodology

This paragraph outlines the methodology for creating an RDF based on a case study to provide detailed insights via queries. First, the case study is recorded in a formalized way. Then, the recorded elements are transferred into the RDF. Finally, queries are formulated to obtain specific information.

3.1 Record Case Study

The underlying case study presents an anonymized excerpt of the product development processes of two companies from the mechanical and plant engineering sector. The requirements specification procedure is described. The results were achieved in workshops with participants from product development, such as designers, process managers, and R&D management. A longlist of artifacts compiled in earlier work formed the basis for determining relevant artifacts for this case study, e.g. market segmentation or patent analysis [28]. These artifacts are the result of development activities, such as the recording of requirements or their classification. Activities are carried out by roles and supported by models. Figure 1 shows the connections and dependencies between these elements.

The first part of the matrix shows the relations between activities (A). The binary value $R_{intra,A} \in \{0,1\}$ indicates if an activity needs another activity as a prerequisite. For

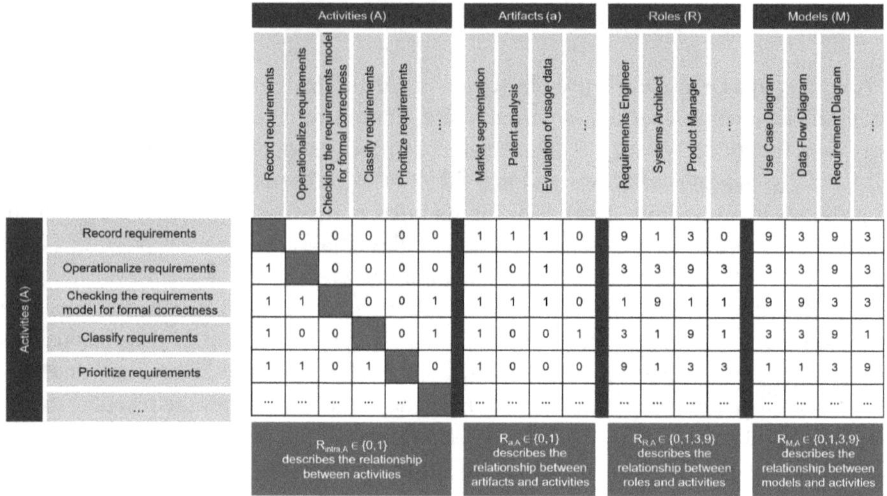

Activities (A)	Activities (A): Record requirements	Operationalize requirements	Checking the requirements model for formal correctness	Classify requirements	Prioritize requirements	...	Artifacts (a): Market segmentation	Patent analysis	Evaluation of usage data	...	Roles (R): Requirements Engineer	Systems Architect	Product Manager	...	Models (M): Use Case Diagram	Data Flow Diagram	Requirement Diagram	...
Record requirements		0	0	0	0	0	1	1	1	0	9	1	3	0	9	3	9	3
Operationalize requirements	1		0	0	0	0	1	0	1	0	3	3	9	3	3	3	9	3
Checking the requirements model for formal correctness	1	1		0	0	1	1	1	1	0	1	9	1	1	9	9	3	3
Classify requirements	1	0	0		0	1	1	0	0	1	3	1	9	1	3	3	9	1
Prioritize requirements	1	1	0	1		0	1	0	0	0	9	1	3	3	1	1	3	9
...	...	...	...	...	...		...	...	...	...	...	...	...	...	...	...	...	...

Fig. 1 Relationship matrix

example, to carry out the activity "operationalize requirements" the requirements need to be recorded first i.e. "record requirements".

In the next columns, the matrix also includes relationships between activities (A) and artifacts (a). The binary value $R_{a,A} \in \{0,1\}$ indicates if an artifact is associated with an activity. For example, if an artifact like "market segmentation" is linked to the activity "classify requirements", this suggests that market segmentation data may inform how requirements are classified.

Furthermore, the roles (R) involved in these activities are represented with their own set of relationships. The notation $R_{R,A} \in \{0,1,3,9\}$ describes varying degrees of connection between roles and activities. A higher numerical value might indicate a stronger suitability of a role to perform a specific activity. For example, if a product manager is well equipped to "prioritize requirements" and "classify requirements", this would be reflected by higher values in those corresponding cells.

Finally, models (M) are also incorporated into this relational framework through $R_{M,A} \in \{0,1,3,9\}$, which denotes how different models relate to activities. For instance, if "Use Case Diagram" provides a lot of value to the "check requirements model for formal correctness" activity, it may receive a higher value indicating a good fit for that context.

For $R_{R,A}$ and $R_{M,A}$ an exponential scale is chosen over a linear scale to emphasize the differences in relationships, facilitating the identification of particularly important or critical connections and giving them a higher evaluation.

3.2 Transfer into RDF

With the complete set of relations and dependencies charted in a matrix (Fig. 1) the information is translated into an RDF. The RDF allows to represent these relationships using subject-predicate-object triples in a machine-readable format. This translation ensures that the critical dependencies and associations from the matrix are maintained. First the four classes activities (A), artifacts (a), roles (R) and models (M) are created and disjointed. Then the elements from the matrix are added as individuals to their respective classes, leading to the partial graph shown in Fig. 2 depicting the model without relations and dependencies.

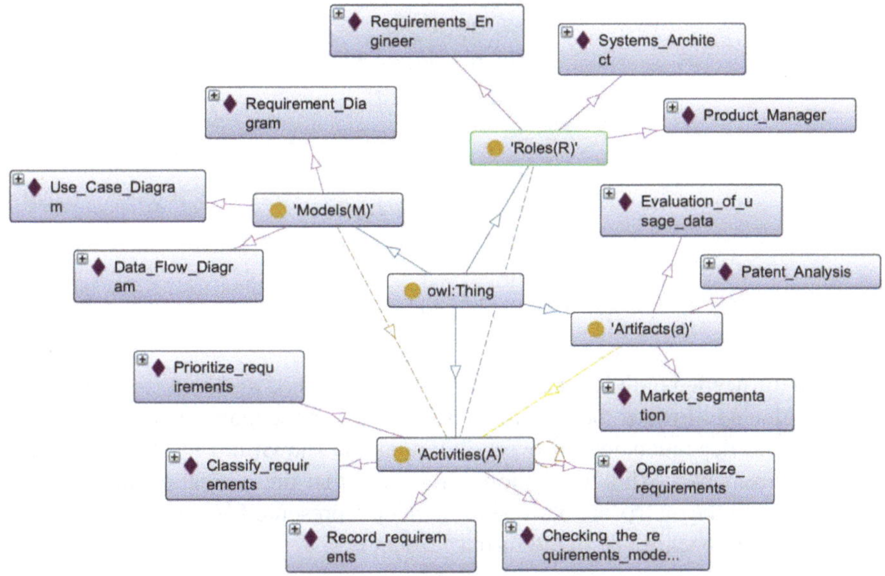

Fig. 2 Classes and individuals

After establishing the elements, the relationships and dependencies between them need to be implemented. To this end a set of object properties (Fig. 3) is implemented corresponding to the Ri,j values from the matrix.

$R_{intra,A} \in \{0,1\}$ is represented by the inverse properties "is_downstream_of" and "is_upstream_of". To use the same example this leads to the triple "operationalize_requirements is_downstream_of record_requirements" and the inverse "record_requirements is_upstream_of operationalize_requirements".

$R_{a,A} \in \{0,1\}$ is also represented by inverse properties being "requires" and "is_required_by", giving the example triple "market_segmentation requires classify_requirements" and its inverse "classify_requirements is_required_by market_segmentation".

$R_{R,A} \in \{0,1,3,9\}$ in contrast is not represented by two inverse properties but by the property "can_perform" for $R_{R,A} > 0$ and the subproperties "is_less_suitable", "is_suitable" and "is_very_suitable" outlining the varying degrees of suitability of the role in regard to the connected activity.

$R_{M,A} \in \{0,1,3,9\}$ is represented similar to $R_{R,A}$ with the property "is_applicable" for $R_{M,A} > 0$ and the subproperties "weak_fit", "medium_fit" and "good_fit" again outlining how good of a fit the model is for the connected activity.

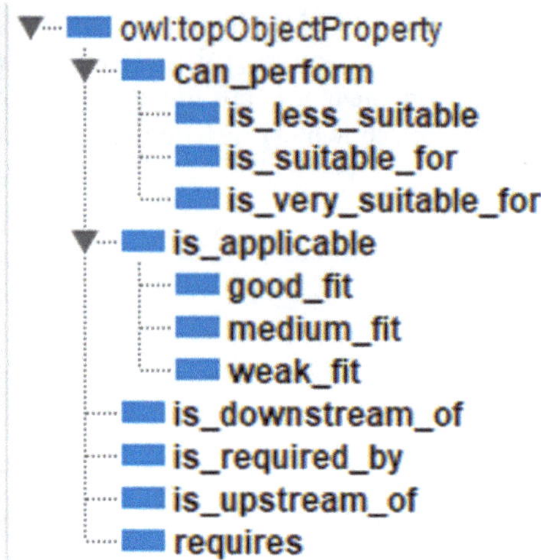

Fig. 3 Object properties

With all the needed individuals and properties established the properties are assigned to the individuals following the information contained in the matrix leading to a full set of relations and dependencies for every individual as shown in Fig. 4 for the activity "record_requirements". This translation will ensure that the critical dependencies and associations from the matrix are maintained and are represented using subject-predicate-object triples in a machine-readable format.

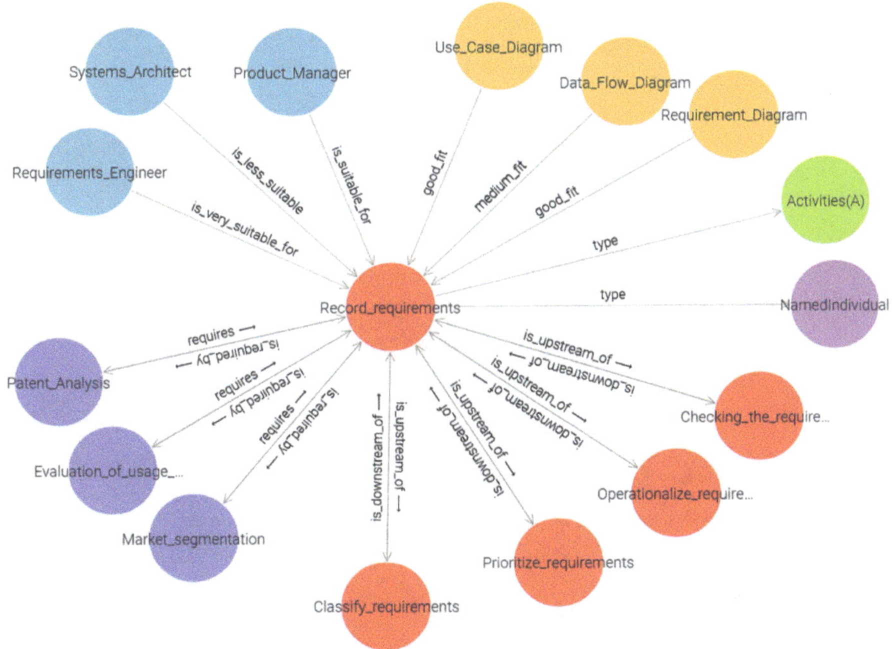

Fig. 4 Model snapshot

3.3 Obtain Information via Query

With all data contained in the matrix translated into a machine-readable format information can be obtained via queries. The following paragraphs will give examples of how this method can be applied.

Figures 5 and 6 show the input mask for queries. Classes, individuals, relationships, and values can be addressed and linked using Boolean operators. The query returns those instances that match the specified criteria. This makes it easy to formulate even complicated search queries without having to anticipate their structure or calculate them mathematically in advance, as would be the case e.g. with a matrix representation.

Question 1 Which role is best suited for recording the requirements?
Query 1 'Roles(R)' and is_very_suitable_for value Record_requirements.
Question 2 Which activities need to be completed before the requirements can be prioritized?
Query 2 'Activities(A)' and is_upstream_of value Prioritize_requirements.

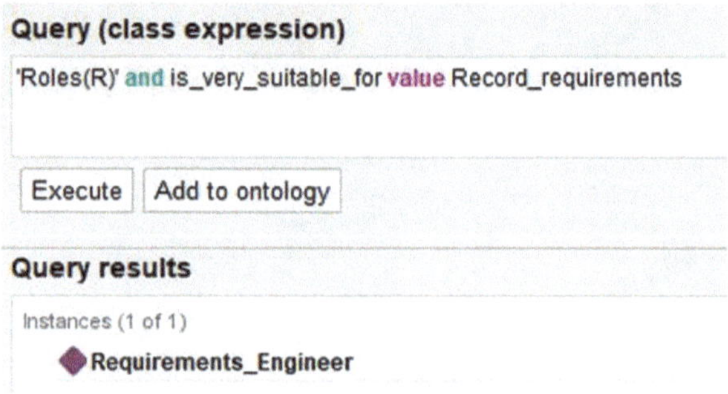

Query (class expression)
'Roles(R)' and is_very_suitable_for value Record_requirements
Execute
Add to ontology
Query results
Instances (1 of 1)
Requirements_Engineer

Fig. 5 Example Query 1

Query (class expression)
'Activities(A)' and is_upstream_of value Prioritize_requirements
Execute
Add to ontology
Query results
Instances (3 of 3)
Classify_requirements
Operationalize_requirements
Record_requirements

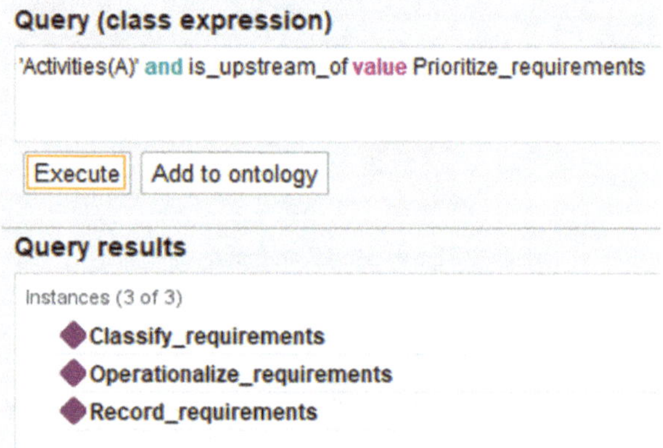

Fig. 6 Example Query 2

The example of extracting information via a query shows how easy it is to access. While the organizational graph grows with its observed area, the effort required to extract specific information remains stable. The resulting ontology enables organizations to build a dynamic knowledge base that reflects current practices while also supporting future developments in the MBSE domain. Moreover, the RDF ontology provides a flexible platform for analytical queries and visualizations, allowing decision-makers to gain deeper insights into their processes. This methodology offers significant potential for providing targeted decision support. Its ability to extract information through queries related to specific questions—regardless of graph size—enables organizations to make informed decisions in complex and evolving domains. The approach is highly flexible, allowing for seamless adaptation to changing requirements, while the integration of heterogeneous data sources facilitates cross-departmental information exchange.

4 Conclusion

Using the data gathered from case studies, an RDF-based organizational graph is constructed. This involves defining nodes representing activities, artifacts, roles, and models while edges indicate dependencies or relationships between them. RDF's flexibility allows for detailed representation of complex interconnections within MBSE frameworks. Information required for specific decision support is extracted via queries. Since the effort required to derive this information remains stable even in larger graphs, the approach is particularly suitable for larger and gradually growing areas of consideration. The approach is part of a larger undertaking to synchronize technical system models and organizational structures for MBSE in order to obtain company-specific organizational graphs to support organizational design [29].

The currently narrow scope of two case studies represents a significant limitation of this study. Future work should extend this scope by including further use cases. Targeted decision support should also be oriented towards specific questions of organizational development, once the basic feasibility has been demonstrated here.

Competing Interests. The author(s) has no competing interests to declare that are relevant to the content of this manuscript.

References

1. Riesener, et al.: Understand and Control Complexity. (2021)
2. Schuh, Sustainable Innovation. [S.l.]: Morgan Kaufmann. (2020)
3. Riesener, et al.: Complexity-oriented design for cyber-physical systems. Procedia CIRP **109** (2022)
4. Majumder: A Domain-Driven Model Generation Framework for CPS. (2023)
5. Hehenberger, et al.: Design, modelling, simulation and integration of cyber physical systems: methods and applications. Comput. Ind. **82** (2016)
6. Riesener, et al.: Requirements Engineering through Exploratory Analysis of Usage Data. (2021)
7. Di Carlo, et al.: Retrofitting a process plant in an Industry 4.0 perspective for improving safety and maintenance performance. Sustainability **13**(2) (2021)
8. Liu, et al.: A survey of model-driven techniques and tools for cyber-physical systems. Front. Inf. Technol. Electron. Eng. **21**(11) (2020)
9. Wu, et al.: Concept and engineering development of cyber physical production systems: a systematic literature review. Int. J. Adv. Manuf. Technol. **111**, 1–2 (2020)
10. Mueller, et al.: Challenges and requirements for the application of Industry 4.0: a special insight with the usage of cyber-physical system. Chin. J. Mech. Eng. **30**(5) (2017)
11. International Council on Systems Engineering: Systems Engineering Vision 2020
12. Zhang, et al.: Towards holistic system models including domain-specific simulation models based on SysML. Systems **9**(4) (2021)
13. Martin Eigner, et al.: Modellbasierte virtuelle Produktentwicklung. Berlin Heidelberg (2014)
14. Henderson, et al.: MBSE adoption experiences in organizations: lessons learned. Syst. Eng. **27**(1) (2024)
15. Henderson, et al.: The effects of organizational structure on MBSE adoption in industry: insights from practitioners. Eng. Manag. J. **36**(1) (2024)

16. Huldt, et al.: State-of-practice survey of model-based systems engineering. Syst. Eng. (2019). https://incose.onlinelibrary.wiley.com/doi/10.1002/sys.21466
17. Vasenev, et al.: Step-Wise MBSE Introduction into a Company: An Interface-Centric Case Study. (2023)
18. Zhu, et al.: Tackling Model-Based Development Process and Organizational Challenges. (2024)
19. Schulte-Zurhausen: Organisation. Franz Vahlen, München (2014)
20. Becker (ed.): Prozessmanagement. Springer Gabler, Berlin, Heidelberg (2012)
21. Fiedler: Organisation kompakt. Oldenbourg Verlag, Munich, Germany (2014)
22. Doherty, K.A.J., et al.: TreeGNG—Hierarchical Topological Clustering. Evere, Belgium: d-side (2005). https://www.esann.org/sites/default/files/proceedings/legacy/es2005-23.pdf
23. Di Martino, et al.: A tool for the semantic annotation, validation and optimization of business process models. Softw. Pract. Exp. (2023). https://onlinelibrary.wiley.com/doi/full/10.1002/spe.3184
24. Tong, et al.: Construction of RDF(S) from UML class diagrams. CIT **22**(4) (2015)
25. Allemang, et al. (eds.): Semantic Web for the Working Ontologist. Elsevier (2008)
26. Bilidas, et al.: In-memory parallelization of join queries over large ontological hierarchies
27. Prud'hommeaux, et al.: SPARQL query language for RDF
28. Bönsch, et al.: Artifact-oriented Product Development Process Structure for Implementing Model-Based Systems Engineering. (2024)
29. Bönsch, et al.: Concept for Synchronizing Technical System Models and Organizational Structures for Model-Based Systems Engineering. (2025)

Investigation of Key Parameters Influencing Mobile Robot Communication in 5G Networks

Henrik Lurz(✉), Sebastian Blankemeyer, and Annika Raatz

Institute of Assembly Technology and Robotics - Leibniz University Hannover, Garbsen, Germany
lurz@match.uni-hannover.de

Abstract. The collaborative manipulation or assembly of objects using the utilization of mobile robots is of significant importance for the agile production of large components, particularly in the context of flexible manufacturing. The integration of mobile robots that are synchronized to achieve a shared objective can eliminate the necessity for rigid automation systems, thereby substantially enhancing flexibility in production processes. In addition to the development of customized hardware, effective communication between the process participants is crucial to ensure minimal latency. This is particularly necessary for damage-free handling of the objects in order to reduce mechanical stresses in the component due to movement distortions. In this context, 5G technology has the potential to support the high bandwidth demands of video stream transmission while simultaneously providing the low-latency connectivity required for precise mobile robot control. However, specific adjustments are necessary to establish a network capable of achieving this, as there are currently no guidelines outlining how different settings will affect the performance. This paper examines the impact of various small cell configurations on the performance of an existing 5G campus network. Automated tests are developed to determine the optimal configuration for this network, which can subsequently be transferred to other 5G campus networks to identify their ideal settings. Finally, the optimal configuration for this 5G campus network will be determined by the measurement results to support the multi-mobile robot system.

Keywords: 5G communication · Mobile robotics · Control system

1 Introduction

In modern times the product variants are steadily increasing while product life time is decreasing which proposes a challenge for current production systems [1]. The production systems have to become more and more flexible due to frequent

L. Overmeyer and B.-A. Behrens (eds.), *Production at the Leading Edge of Technology*, Lecture Notes in Production Engineering, https://doi.org/10.1007/978-3-032-19524-1_66

adapting to new requirements or product lines [2]. One piece of the solution is the smart factory which is part of the fourth industrial revolution (Industry 4.0). A smart factory is an advanced manufacturing facility, leveraging cutting-edge technologies like artificial intelligence (AI), the Internet of Things (IoT), and robotics to enhance efficiency, boost productivity, and ensure high-quality outcomes through optimized operations [3]. Static robotics has been utilized in production for a long time, typically operating behind safety fences. However, in the context of smart factories, mobile robots present new opportunities by navigating shared pathways within production lines, enabling more flexible and dynamic integration into modern manufacturing processes [4]. This flexibility of mobile robots enables production lines to adapt to new requirements simply by reconfiguring their operations. As a result, mobile robots are exceptionally well-suited to meet the demands of a smart factory [5].

A mobile robot, while equipped with sensors and capable of navigating production lines, is not fully autonomous on its own. To function autonomously, it requires a central managing hub that either coordinates all mobile robots and assigns tasks or directly controls the individual robot to ensure efficient integration within the production process. However, a reliable wireless connection to fully enable such a system is a necessity. Currently, wireless communication in factories is typically established via WiFi. A significant drawback of WiFi is its reliance on unlicensed frequency bands, which can result in interference and increase susceptibility to jamming due to the lack of regulation [6]. A wireless technology that holds significant promise in Industry 4.0 and aims to address this challenge is 5G. However, there have been only a limited number of investigations into how different settings impact the performance of a 5G campus network, and there are currently no established guidelines for its setup for mobile robotics.

The objective of this paper is to examine the impact of the Time Division Duplex (TDD) pattern and the transmission power of the small cells within a 5G campus network. To facilitate this, we will describe the experimental setup featuring the MiR600 mobile robot, which was employed to validate the 5G system. Subsequently, a series of experiments will be conducted to assess the impact of these configurations on 5G connectivity performance. The analysis of the resulting data is expected to offer guidelines for parameterizing the small cells of a 5G system to better accommodate specialized applications. Finally, the measurement results can be used to set the optimal configurations for the 5G network with the existing multi-mobile robot system [7].

2 Related Work

5G technologies have become widely adopted, with applications spanning various sectors, including consumer markets, Agri-Food, and production machinery. Today, nearly all modern consumer devices are compatible with public 5G networks, and machinery is increasingly connected to cloud services for monitoring. However, the effectiveness of these applications depends on a well-adjusted 5G network that ensures optimal connectivity for user equipment. Several research studies have documented measurements from different 5G networks.

In their study, Ansari et al. conducted measurements in [8], focusing on the mid-band and high-band performance of 5G networks in the context of ultra-reliable low-latency communication. They established a testbed for 5G network assessment in a production-like environment. The paper notes that various small cell configuration parameters were utilized, though it does not specify which ones. Ultimately, they provided a comprehensive evaluation of latency for different message sizes in both downlink and uplink scenarios. Souto et al., in [9], presents an end-to-end performance evaluation of devices connected to a private 5G standalone network in indoor environments. Results show that setups with servers directly connected to the network core achieve superior performance across metrics including jitter, latency, throughput, and signal quality parameters compared to fully wireless setups. The hardware configurations are described in detail, but the settings for individual network components, particularly those related to 5G, are not elaborated upon. Rischke et al., in [10], investigates the actual one-way and round-trip delay characteristics of a private standalone campus network. Through extensive measurements, the study finds that round-trip times are discretely clustered between 12 ms and 40 ms, with delay distributions primarily determined by packet rates and inter-arrival times. Downlink and uplink delays are shown to be correlated, highlighting the influence of scheduling mechanisms on end-to-end latency. This research employs the default parameters and settings for a Nokia 5G system without delving into specifics. Stoian et al., in [11], evaluates the practical performance of service-differentiated 5G standalone (SA) networks in a laboratory setup, using a gNodeB and multiple user equipments in both indoor and Faraday cage environments. The study measures both transmission rate (throughput) and latency to analyze how enhanced mobile broadband and ultra reliable low latency communications slices can be separated and coexist on the same physical infrastructure. Results demonstrate through empirical measurement that differentiated network slices can achieve distinct performance profiles as configured. The configuration of the cells is explained in detail but missing a reasoning for selecting the specific options.

All of the aforementioned studies on measuring the performance of a 5G network provided detailed information about the hardware used, and some also mentioned the configuration of the cells. However, none of them explained the rationale behind the selection of specific settings. In this paper, we aim to address this gap by developing an automated testbed that enables us to assess the impact of various cell parameters on 5G network performance. Specifically, we will evaluate the effects of TDD and the transmission power of the cells. Ultimately, we intend to utilize the data collected to identify the ideal configuration for our cooperative mobile robot application.

3 Approach

Before outlining the network architecture of the 5G system, it is essential to explain the boundary conditions of the hardware setup. The architecture must support the cooperative mobile robot experiments referenced in Sect. 1. In our use case, this implies that up to four or more mobile robots are wirelessly connected to the 5G network, exchanging data with low latency requirements. Each mobile robot is equipped with multiple network-connected components that must be individually addressable. The software currently in use is the Robot Operating System (ROS), which depends on the name resolution of each component and, consequently, the effective routing of network traffic. The resulting network architecture is shown in Fig. 1 and shows the stationary network components on the left and the moving components on the mobile robot on the right.

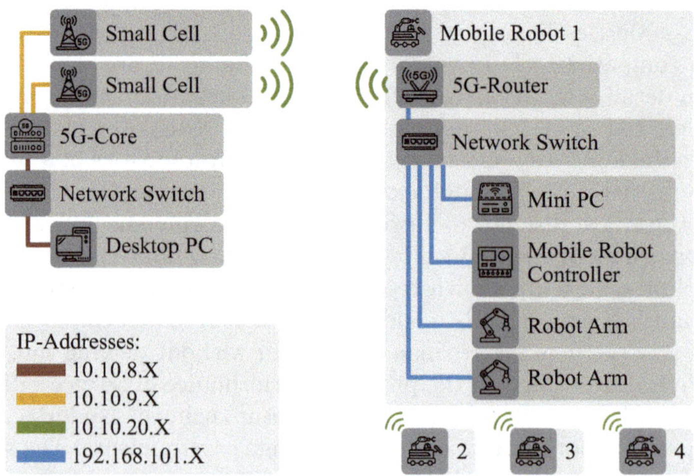

Fig. 1. The network architecture for the 5G system includes essential components such as the 5G core network, small cells for connectivity, and integrates multiple mobile robots

To establish a connectable 5G campus network, a 5G core is essential as it manages network resources and orchestrates communication. Two small cells are used to enable end devices to connect to the 5G network effectively. In our setup, a desktop computer symbolizes the control station of the mobile robots and connects to the 5G core via Ethernet. The mobile robot depicted on the right side of Fig. 1 consists of five primary components. The 5G router links to the 5G campus network and therefore establishes a wireless connection to the desktop computer. A Mini-PC in the mobile robot executes all necessary ROS programs for controlling the mobile robot and performing sensor evaluations. The mobile robot control unit receives and executes movement commands while publishing sensor data. Finally, two robotic arms mounted on each mobile robot are used to manipulate objects in the environment or during transportation. This setup is implemented for each of the four mobile robots and is illustrated in the bottom right corner of Fig. 1.

To enable communication among all components, four distinct networks are required. First, an Ethernet connection links the 5G core to static components, such as desktop computers functioning as the control station or external tracking systems. Second, an Ethernet connection is established between the 5G core and the two small cells. Third, a wireless connection establishes a link between the small cells and each 5G router on the mobile robots, thereby creating a dedicated network. Finally, an Ethernet connection provides a link between the 5G router and all network-capable hardware components within the mobile robot. This configuration requires specific routing settings for both the 5G core and the 5G router in the mobile robots. The 5G core forwards control messages from the desktop computer to the appropriate mobile robot, while the 5G router directs messages to the correct components.

4 Evaluation

The schematic network setup for the evaluation process is detailed in Sect. 3, while the specific components utilized are listed in Table 1. When assessing a 5G network, multiple performance metrics can be analyzed. In this study, we focus on downlink data transmission (from the control station to the mobile robot) and round trip time (RTT), as these metrics significantly impact our application performance when sending control commands to a mobile robot. Additionally, we have selected the Signal-to-Interference plus Noise Ratio (SINR), which measures the quality of a received signal by comparing the power of the desired signal to the total power of interference and background noise.

Table 1. Overview of the specific hardware components used in the evaluation process

Network component	Hardware	Software Version
5G Core	Dell EMC VEP1485	Athonet Core 1.24.1
Small Cell	T&W 5G Enterprise Cell	5.3.0.32983
5G Router	Teltonika RUTX50	RUTX_R_00.07.08.3

Two distinct experiments were conducted for the evaluation, with a distance of 10 m maintained between the small cell and the 5G router. The first experiment analyzes the impact of various TDD patterns on the selected performance metrics, while the second examines the effect of transmission (Tx) power per path on those same metrics. To eliminate outliers in both experiments, 100 static measurements were recorded for each setting and averaged during the analysis. The results of the two experiments are presented in Figs. 2 and 3, respectively. Figure 2 illustrates the impact of the five available TDD patterns. Each graph displays these patterns on the X-axis, with the first number indicating the number of downlink slots and the second number representing the uplink slots. The first graph illustrates the percentage change in round trip time (RTT), while the second graph displays the percentage change in data transmission speed compared to the initially selected TDD pattern. The measurements suggest that choosing the 3/1 or 7/2 TDD pattern can result in a 30 % reduction in RTT, which is attributable to the higher number of downlink slots relative to the number of uplink slots in these patterns. Additionally, data transmission speed experiences an even more significant improvement when switching to the 3/1 or 7/2 patterns, with results demonstrating an increase of approximately 250%. Although the SINR shows a slight increase with different TDD patterns, improving the quality of the received signal, the enhancement is minimal.

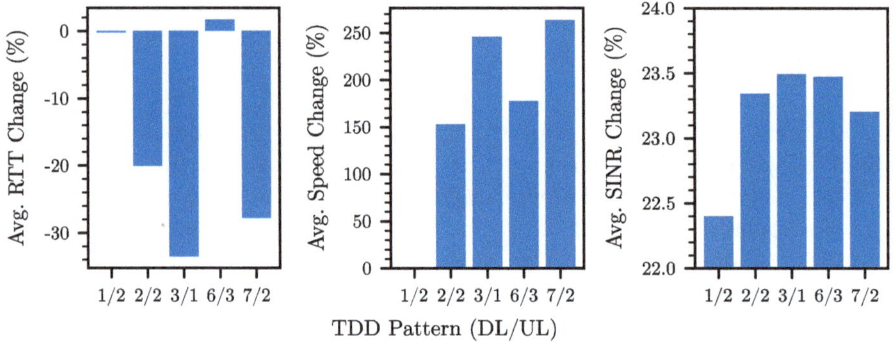

Fig. 2. Measurement results for different TDD patterns. Values on the X-axis represent the DL/UL configuration

Based on the results presented in Fig. 2, the 3/1 TDD pattern exhibits the lowest latency and is therefore the most suitable choice for the cooperative object transport using a multi-mobile robot system, which requires low-latency control messages. Consequently, we selected the 3/1 TDD pattern for the experiment involving changes to the transmission power per path, with the results displayed in Fig. 3.

The impact of varying Tx power per path is illustrated in Fig. 3, where the Tx power per path was increased from 0 dBm to 36 dBm in increments of 2 dBm. The first graph shows the percentage change in round trip time (RTT) compared to the initial measured average at 0 dBm. It indicates a significant decline in RTT, with reductions of over 30 % observed between 12 dBm and 18 dBm, reaching a maximal RTT reduction of 37% at 18 dBm compared to the initial value. The second graph displays the percentage change in data transmission speed relative to the initial average value. It reveals a sharp increase in data transmission when Tx power per path rises from 12 dBm to 16 dBm. The highest data transmission rate occurs at 16 dBm. The third graph illustrates that the quality of the received signal steadily improves up to 22 dBm, where it stabilizes around 30 dB. This suggests that increasing Tx power does not indefinitely enhance signal quality due to the onset of parasitic influences. Thus, higher Tx power does not always correlate with improved system performance.

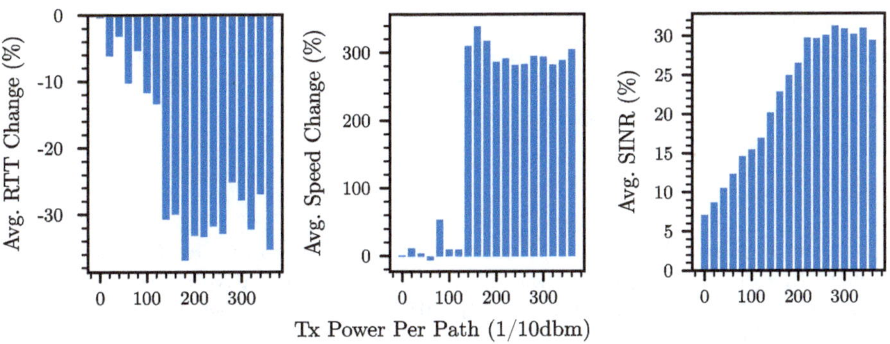

Fig. 3. Measurement results when increasing the Tx power per path from 0 to 36 dBm. A total of 100 measurements were conducted and averaged for each transmission level

5 Conclusion and Outlook

In this paper, we examined the impact of various configurations on the performance of a 5G campus network specifically tailored to facilitate cooperative mobile robot operations. A well-calibrated 5G network is essential for ensuring optimal performance in dynamic manufacturing environments where mobile robotics play a key role. While numerous studies have assessed the performance of 5G networks across different sectors, none have offered comprehensive guidelines regarding how specific settings, such as Time Division Duplex (TDD) patterns and transmission power levels, influence overall performance. For our specific setup, our investigation identified the 3/1 and 7/2 TDD configurations as the most effective for minimizing round trip times (RTT) and maximizing data transmission speeds. Furthermore, we explored the effects of transmission power per path, revealing that while increased power can enhance performance, it may also introduce interference that compromises signal quality. Notably, our measurements showed that a transmission power of 160 dBm resulted in the lowest RTT and nearly optimal data transmission performance for this specific 5G system.

Looking ahead, future research will focus on the implications of a multi-small cell setup, particularly how transitioning between cells affects performance and connectivity within 5G networks. Understanding the dynamics of handoffs between cells will be crucial for maintaining seamless communication, especially in environments with mobile robots in motion. Additionally, plans include investigating how obstructions from machinery influence overall network performance. Analyzing the effects of physical barriers on signal quality and connectivity will provide insights into the challenges encountered in real-world industrial settings.

Acknowledgment. The authors gratefully acknowledge the funding by the Bundesamt für Sicherheit in der Informationstechnik (BSI—Federal Office for Information Security)—Project no. 01MO23027B. The authors would like to thank the BSI for the support within the 5GProSec—IT security in the deployment of 5G in production ecosystems. Additionally, the authors gratefully acknowledge the funding by the Deutsche Forschungsgemeinschaft (DFG—German Research Foundation)—Project no. 541021498. The authors would like to thank the DFG for the support within the SPP 2433—Measurement technology on flying platforms.

References

1. Krüger, J., Wang, L., Verl, A., Bauernhansl, T., Carpanzano, E., Makris, S., Fleischer, J., Reinhart, G., Franke, J., Pellegrinelli, S.: Innovative control of assembly systems and lines. CIRP Annals 66(2), 707–730 (2017). https://doi.org/10.1016/j.cirp.2017.05.010
2. Blankemeyer, S., Wendorff, D., Raatz, A.: Robotic evaluation framework for 6D object pose estimation accuracy. Procedia CIRP, 134, 1113–1118 (2025). https://doi.org/10.1016/j.procir.2025.02.251
3. Soori, M., Arezoo, B., Dastres, R.: Internet of things for smart factories in industry 4.0, a review. Internet of Things and Cyber-Phys. Syst. **3**, 192–204 (2023). https://doi.org/10.1016/j.iotcps.2023.04.006
4. Hercik, R., Byrtus, R., Jaros, R., Koziorek, J.: Implementation of Autonomous Mobile Robot in SmartFactory. Applied Sciences 12(17), 8912 (2022). https://doi.org/10.3390/app12178912
5. Holfeld, B., Wieruch, D., Wirth, T., Thiele, L., Ashraf, S.A., Huschke, J., Aktas, I., Ansari, J.: Wireless Communication for Factory Automation: an opportunity for LTE and 5G systems. IEEE Communications Magazine 54(6), 36–43 (2016). https://doi.org/10.1109/MCOM.2016.7497764
6. Sossalla, P., Rischke, J., Baier, F., Itting, S., Nguyen, G.T., Fitzek, F.H.P.: Private 5G solutions for mobile industrial robots: a feasibility study. In: 2022 IEEE Symposium on Computers and Communications (ISCC), pp. 1–6. IEEE, Rhodes, Greece (2022). https://doi.org/10.1109/ISCC55528.2022.9912791
7. Lurz, H., Recker, T., Raatz, A.: Spline-based path planning and reconfiguration for rigid multi-robot formations. Proc. CIRP, **106**, 174–179 (2022). https://doi.org/10.1016/j.procir.2022.02.174
8. Ansari, J., Andersson, C., De Bruin, P., Farkas, J., Grosjean, L., Sachs, J., Torsner, J., Varga, B., Harutyunyan, D., König, N., Schmitt, R.H.: Performance of 5G Trials for Industrial Automation. Electronics 11(3), 412 (2022). https://doi.org/10.3390/electronics11030412
9. Souto, Á., Almeida, L., Fonseca, P., Meireles, M.A.C., Costa, G., Abreu, S.: Measurement of Communication Performance Between 5G Devices in a 5G SA Network. In: Anais Estendidos do XLII Simpósio Brasileiro de Redes de Computadores e Sistemas Distribuídos (SBRC 2024), pp. 265–270. Sociedade Brasileira de Computação - SBC, Brasil (2024). https://doi.org/10.5753/sbrc_estendido.2024.1791
10. Rischke, J., Vielhaus, C., Sossalla, P., Itting, S., Nguyen, G.T., Fitzek, F.H.P.: Empirical Study of 5G Downlink & Uplink Scheduling and its Effects on Latency. In: 2022 IEEE 23rd International Symposium on a World of Wireless, Mobile and Multimedia Networks (WoWMoM), pp. 11–19. IEEE, Belfast, United Kingdom (2022). https://doi.org/10.1109/WoWMoM54355.2022.00017
11. Stoian, T.-C., Marţian, A., Gheorghe, M.-G.: Service Differentiation on Network Slices within the 5G SA Mobile Communication System. In: 2023 31st Telecommunications Forum (TELFOR), pp. 1–4. IEEE, Belgrade, Serbia (2023). https://doi.org/10.1109/TELFOR59449.2023.10372752

Development of a Methodology to Increase the Adaptability of Manufacturing Companies Through Agile Resilience Management

Jonas Reinhold(✉), Björn Burzynska, Tanya Jahangirkhani, Dorit Schumann, Tabea Demke, Friederike Stefanowski, and Matthias Schmidt

Institute of Production Systems and Logistics (IFA), Leibniz University Hannover, Hannover, Germany

reinhold@ifa.uni-hannover.de

Abstract. Manufacturing facilities are becoming increasingly vulnerable to erratic shocks and uncertainties that have the potential to exert disruptive effects on operations. Global crises, such as the COVID-19 pandemic, the energy crisis, and the Suez Canal blockade, highlight the vulnerability of value networks that extend beyond company boundaries. In the absence of effective strategies to address these disruptions, manufacturing companies may face the risk of losing their competitive edge. While flexibility may suffice for minor risks, more substantial impacts necessitate a significant degree of adaptability and robustness. The promotion of organizational resilience, defined as the capacity to adapt to evolving circumstances swiftly, constitutes an ongoing strategic process that is pivotal to the effective creation of operational value. This publication delineates the methodological approach of the *PARMa* research project. Within the research project, the vulnerability of manufacturing companies to internal and external uncertainties is being addressed. Moreover, *PARMa* aims to strengthen the adaptability of production companies through proactive resilience management. In the event of sudden shocks, this should ensure rapid and targeted action, prevent operational disruptions and maintain competitiveness.

Keywords: Resilience · Management system · Risk management · Organizational agility

1 Introduction

Factories are exposed to various internal and external uncertainties that can lead to short-term and unpredictable shocks. These sudden shocks have the potential to cause significant disruption to operations due to existing vulnerabilities [1]. A prime example of this phenomenon is the global outbreak of the novel coronavirus, which has precipitated a severe disruption to global supply chains. It is evident that considerable challenges have been presented to companies by production losses, transport restrictions and fluctuations in demand. The energy crisis also poses a significant threat, as it

L. Overmeyer and B.-A. Behrens (eds.), *Production at the Leading Edge of Technology*, Lecture Notes in Production Engineering, https://doi.org/10.1007/978-3-032-19524-1_67

leads to drastically rising operating costs and instability in energy supplies. Moreover, the obstruction of the Suez Canal demonstrated the vulnerability of global trade routes when a single ship became stranded, thereby significantly disrupting the international flow of goods [2]. These crisis situations illustrate that the vulnerability of supply chains extends beyond company boundaries. As a consequence of the globalization of supply chains, the advent of unforeseen shocks and market threats poses a substantial risk to production companies in the imminent future. In order to manage potential risks, including wars, catastrophes, and other interruptions in supply, production or distribution proactively, companies must implement management tools and coping strategies. In the absence of adequate strategies, these events have the potential to exert a substantial detrimental effect on competitiveness. In extreme circumstances, they may even result in the forced exit of companies that are unable to adapt swiftly and efficiently to the market [3]. To prepare for sudden changes in conditions and swiftly reinstate operations in the event of disruptions, promoting organizational resilience is regarded as a promising concept [4].

1.1 Problem Definition and Research Gap

In the context of escalating uncertainty, it is evident that manufacturing companies are increasingly vulnerable to erratic shocks. In such cases, various approaches and concepts are available for maintaining the operational continuity of production systems. The concept of resilience describes one such approach. In accordance with the stipulations set out in ISO Standard 22316, the term *resilience* is defined as the capacity of an organization to accommodate alterations in its external environment and to demonstrate adaptability in response to such changes [5]. Consequently, resilience may also be described as adaptability. In addition to the concept of adaptability, other types of transformability are evident, including flexibility, changeability and robustness. Given the interdependence between adaptability and transformability, and the potential for synergies to emerge, it is inevitable to adopt a holistic approach to these capabilities. Investment in transformability, for instance, has been demonstrated to influence the resilience of a production system [4].

A survey conducted in the spring of 2023 revealed that 14% of respondents reported conducting risk analyses along the operational supply chain in response to uncertainties. Moreover, a significant proportion of the surveyed companies, amounting to 39%, expressed a desire to initiate the practice of conducting such analyses [6]. However, these models are deficient in their lack of a holistic perspective on the various types of adaptability and transformability, namely resilience, robustness, flexibility, and adaptability. Moreover, they often adopt a reactive stance rather than a proactive approach, and their approach is not user-oriented. For instance, in recent years, a significant number of companies have augmented their inventories in response to delivery issues. Nevertheless, it is becoming evident that these elevated inventories are not necessary for standard operations or during typical fluctuations. However, they are associated with considerable expenses [7].

In order to manage the effects of shocks in a changing and adaptable organization, it is necessary to implement a strategic control instrument. A Business Continuity Management System (BCMS) is an appropriate solution for this purpose, as it promotes the maintenance of operational capability in the event of a disruption. In the context of a BCMS, objectives for maintaining operational capability are set in line with organizational policy to achieve defined results [8, 12]. Thus, the capacity of manufacturing enterprises to adapt and evolve can be fostered through the implementation of a proactive and agile resilience management system, based on the structure of a BCMS. The proactive enhancement of organizational resilience can be regarded as a continuous strategic process that establishes the foundation for effective operational value creation.

2 Objectives of the Solution Concept

In the research project "Proactive Enhancement of Adaptability in Manufacturing Companies through Agile Resilience Management" (German acronym: PARMa), funded by the Federal Ministry of Research, Technology and Space (German acronym: BMFTR), the means by which companies can enhance their capacity for adaptation and transformation through the implementation of proactive resilient management strategies is to be investigated. Concrete measures will be derived from the findings and consolidated to enable companies to act quickly and purposefully in the event of sudden shocks and uncertainties. A systematic approach will be employed to address this issue. This approach entails identifying weaknesses, analyzing root causes in detail, and developing effective countermeasures. The focal point of this research project is to examine the capacity to respond effectively to sudden shocks in three distinct application scenarios (AS). These AS are predicated on the primary processes of procurement, production and distribution along the supply chain of manufacturing companies [9]. The term supply chain refers to the network of organizations, people, activities, information and resources that work together to move a product or service from raw material procurement to the end customer [10].

In the context of the AS **procurement**, disruptions have been shown to exert a substantial influence on manufacturing companies. The underlying causes of such disruptions may encompass a multitude of factors, including, but not limited to, supplier failures arising from financial instability or natural disasters, material shortages due to inadequate raw material procurement, and logistical challenges or inadequately designed warehousing processes. Such disruptions can result in several issues, including production delays, increased operating costs, and possible production downtime.

In the context of the AS **production**, disruptions may be attributed to a multitude of factors. Machine or facility failures, unreliable supply of spare parts, human error or software problems are all factors that can have a detrimental effect on production. These disruptions result in various undesirable consequences, including production downtime, increased scrap rates, and inefficient resource utilization. The resulting delays have been shown to increase operating costs, jeopardize delivery dates and significantly impair customer satisfaction.

In the context of the AS **distribution,** disruptions can also have far-reaching consequences. Such occurrences may be attributed to a multitude of factors, including transport delays, industrial actions such as strikes, capacity bottlenecks, IT failures, fluctuations in demand, inaccurate sales forecasts, or market changes. These disruptions have the effect of preventing the company from meeting its planned production volumes, which can result in either over- or underproduction. Overproduction leads to increased storage costs and capital commitments. Conversely, underproduction can lead to delivery bottlenecks and result in customer dissatisfaction.

The specific AS are applicable across various industries and serve within the framework of the research project to enable companies from the project-accompanying committee to increase their resilience through a continuous strategy process. The overarching objective of PARMa is to conceptualize resilience not solely as a capability, but also to proactively respond to disruptions and ensure production operability through the implementation of a holistic resilience management system.

To facilitate effective value creation processes within the production system, companies must identify and assess risks that have the potential to disrupt business operations and determine appropriate responses to these risks. To achieve this objective, it is imperative to employ a targeted approach in utilizing the types of changeability and adaptability, namely resilience, robustness, flexibility and transformability [11]. The novelty of the PARMa research project lies in its holistic approach, which considers all four types of adaptability and transformability simultaneously, enabling companies to achieve the best possible resilience. The existing methods have thus far provided only limited or isolated solutions [c.f. 13]. They are either purely qualitative in nature, one-dimensional in design, static and unable to learn, or require considerable effort to gather information. The evaluation methodology to be developed as part of the research project aims to describe the degree of resilience and investigate the effectiveness of measures to increase adaptability and transformability, thereby creating the basis for establishing agile organizational structures.

Unexpected shocks that lead to disruptions in the production system cannot usually be overcome by a company's existing ability to change alone. The ability to swiftly restore performance is dependent upon organizational resilience, which is characterized by proactive exploration of potential solutions, goal-oriented decision-making, and a shared awareness among all actors of the necessity for rapid adaptability [11]. Hence, the implementation of a strategic and data-driven resilience management ultimately provides the ultimate stage for a company to cope with unexpected shocks.

3 Workframe of the Research Project

To achieve the desired goals and requirements of the project, it is divided into six work packages, each with a distinct focal point. Figure 1 presents a visual representation of the overall work content, including subordinate tasks within the respective work packages.

The initial phase of the project involves establishing a shared understanding of the risks and potential disruptions to production, as well as developing a formal disruption model. Initially, internal weaknesses in the production system and specific vulnerabilities within the companies that could negatively impact operations are identified. These include, for example, essential machines, critical raw materials, and personnel with crucial expertise. A meticulously designed questionnaire helps the participating companies to identify their recognized deficiencies.

These findings are then supplemented by scientifically sound evidence, which is clustered by topic and presented in a feedback workshop. The evaluation of these results is

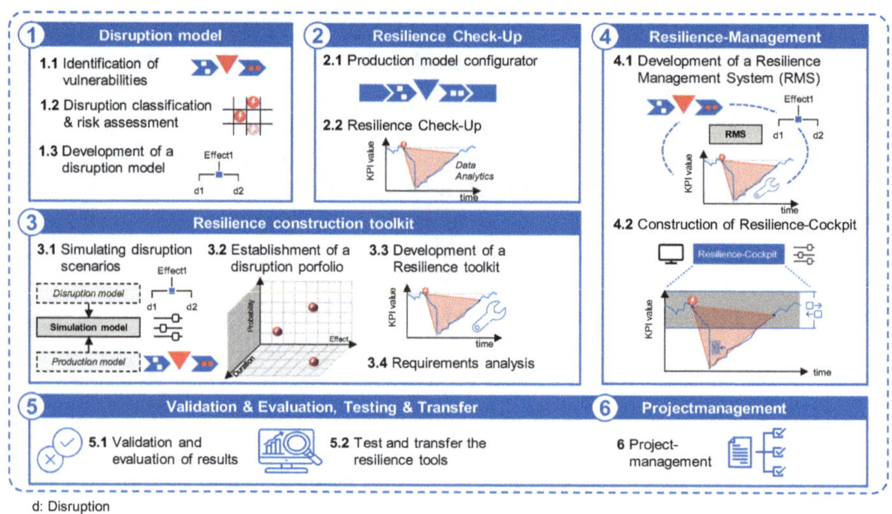

Fig. 1 The six work packages of the project

conducted in close cooperation with the respective companies. Subsequently, the events and developments that have the potential to activate the identified vulnerabilities and compromise the production process are described. The aforementioned disruptions are characterized by their probability of occurrence, duration, and intensity. These disruptions are further classified into distinct categories. Furthermore, it is acknowledged that disruptions frequently occur in conjunction, giving rise to intricate, interwoven scenarios. Such disruptions may encompass material shortages, delayed deliveries, staff shortages, or machine respectively production downtime. A selection of disruptions that pose a threat to the achievement of production targets are evaluated as business-relevant risks. This evaluation is conducted through two methods: a qualitative assessment through semi-structured expert interviews and a quantitative assessment based on data patterns. Furthermore, a formal disruption model is developed based on the identified and assessed disruptions. This model delineates and quantifies the characteristics of the disruptions and their effects on the production system. It is hypothesized that the shocks themselves, such as a fire at a supplier's premises or a global material shortage, do not need to be explicitly modeled. It is of greater significance how these shocks affect the previously identified vulnerabilities within the company. Consequently, the model serves as the foundational framework for simulating various disruption scenarios.

The next step in this research project is to develop a system for determining the current level of resilience among participating companies. To execute this task, it is necessary to methodically model the elements of a production system within the framework of a production model configurator. This enables companies to model and configure the various elements of their individual production systems (e.g., the number of workstations, the number and competence of available resources, procurement structures, etc.). In this case, it is necessary to examine the level of detail required to model the production process of the companies in order to quantify the effects of disruptions sufficiently. The resulting production model serves as the foundation for the simulation. The resilience check-up entails an examination of the past and current handling of uncertainties at the participating companies, where the existing risk management systems and measures are also compared. This involves the evaluation of two key factors: first, the capacity of the companies to adapt and evolve in response to external challenges; and second, their historical performance during periods of economic turbulence. To ascertain the present level of resilience, a standardized and transferable assessment logic based on ISO 22316 [5] is implemented. In this regard, the research project aims to assess the resilience demonstrated by the company in the face of past challenges and its existing capacity for adaptation. A series of measures will be implemented to enhance a company's capacity for adaptability and change. These measures are then allocated to the existing types of changeability and adaptability [11].

The third step involves developing a resilience toolkit that provides manufacturing companies with measures to enhance their adaptability and ability to change, taking into account their production structure, existing vulnerabilities, and potential shocks or disruptions that need to be mitigated. To achieve this overarching objective, four subsidiary objectives have been delineated. To this end, corresponding disruption scenarios are first simulated, and the resulting production system-specific effects are quantified. The impact of compensatory measures aimed at increasing adaptability and transformability is also evaluated. The procedure for creating a simulation model is based on the extant literature. The risk portfolio is developed as a secondary objective based on the quantified effects, the probability of occurrence of the disruptions under consideration, and their duration. In addition to the simulated effectiveness of measures to increase adaptability, these are consolidated into a resilience toolkit. This action successfully achieves the third sub-goal. The fourth sub-goal is to identify the prerequisites for an organization that is agile and adaptable, to obtain the necessary acceptance at all company levels (strategic, tactical, and operational).

The subsequent phase involves the implementation of an iterative resilience management system, which will enable manufacturing companies to establish a dynamic, adaptable, and ultimately resilient organizational structure in the long term. Therefore, the development of two distinct tools is planned: An application tool for an iterative resilience management system and a resilience cockpit for visualizing resilience levels and forecasting disruptions. The implementation of the resilience management system is to be carried out in an iterative manner, employing the steps that have been developed in previous phases. To this end, the existing simulation model is transferred to an application tool, where all findings and previous steps are combined. The resilience management system is supported by an application that utilizes a digital resilience cockpit where performance indicators are monitored. For validation, the resilience cockpit is integrated into the existing data infrastructure at the participating companies via suitable interfaces.

Throughout each research step, an evaluation, test, and validation of each intermediate result is conducted, as one of the sub-objectives entails the continuous validation and evaluation of project outcomes by the participating companies. The testing and validation processes can be executed by simulating disruption scenarios in the simulation model. This model comprises potential threats and the associated effects. In this manner, transparency is established, a continuous improvement process is initiated, the necessity for flexibility is derived, and ultimately, the quality of the results is ensured. Moreover, the outcomes are disseminated extensively to industry, among other domains, through the working groups of the associated partners, and are also published in scientific journals. This contributes to the effective dissemination of the results within the framework of transfer and public relations measures.

4 Outlook and Further Research Activities

Since the inception of the research project, work has been conducted concurrently on the development of the disruption model and the resilience check-up. The initial research results have been compiled. To develop the disruption model, participating companies were asked about typical disruptions in their processes in semi-structured expert interviews. In the subsequent phase, the responses will undergo meticulous evaluation through the implementation of a clustering method. Thereafter, the responses will be transferred to a disruption model. The aforementioned disruption model will subsequently be integrated into the resilience toolkit, in conjunction with the findings from the resilience check-up. The disruption scenarios will then be simulated within the simulation model. The value streams of the individual companies have already been recorded for the Resilience Check-Up, allowing the production and value-adding processes of these companies to be simulated as precisely and realistically as possible using the production model configurator later in the project. Subsequent phases of the research will involve using value stream analyses to determine the optimal structure for the production model configurator. This should reflect the most salient aspects of production as far as possible and represent the production of the participating companies in terms of the value stream for the resilience toolkit. The resilience check-up is administered via a comprehensive questionnaire that focuses on the subject of resilience in companies. By responding to the questions, an organization can obtain an objective resilience score. This score functions as an evaluation instrument, enabling companies to assess their present level of resilience. An initial version of the questionnaire has already been developed, which is based on ISO 22316 [5]. Upon completion of the questionnaire, validation will be initiated with the assistance of application partners, and a feasibility study will be conducted.

Acknowledgements. The joint project "Proactive Enhancement of Adaptability in Manufacturing Companies through Agile Resilience Management" (German acronym: PARMa), [funding code 02J23C021], is part of the funding guideline "Dynamische Wertschöpfungsnetzwerke im turbulenten Umfeld—Aufbau von Resilienz in produzierenden Unternehmen (Resipro)" within the programme "Zukunft der Wertschöpfung—Forschung zu Produktion, Dienstleistung und Arbeit", which is funded by the BMFTR, formerly known as Federal Ministry of Education and Research (German acronym: BMBF).

Competing Interests. The author(s) has no competing interests to declare that are relevant to the content of this manuscript.

References

1. Wiendahl, H.-P., ElMaraghy, H.A., Nyhuis, P., Zäh, M.F., Wiendahl, H.-H., Duffie, N., Brieke, M.: Changeable Manufacturing - Classification, Design and Operation. CIRP Annals 56 (2), 783–809 (2007).
2. Stich, V., Schröer, T., Linnartz, M., Marek, S., Herkenrath, C.: Wertschöpfungsnetzwerke in Zeiten von Infektionskrisen, acatech Studie, München (2021). Accessed 16 June 2025
3. Wiendahl, H.-H., Reichardt, J., Nyhuis, P.: Handbuch Fabrikplanung: Konzept, Gestaltung und Umsetzung wandlungsfähiger Produktionsstätten, 3. Edition ed. Carl Hanser Verlag GmbH Co KG, München (2024).
4. Hingst, L., Park, Y.-B., Nyhuis, P.: Life Cycle Oriented Planning of Changeability in Factory Planning Under Uncertainty. Publishing, Hannover (2021)
5. ISO 22316: Sicherheit und Resilienz: Resilienz von Organisationen—Grundsätze und Attribute, Internationale Organisation für Normung, Berlin (2017). https://www.beuth.de/de/norm/iso-22316/272586256. Accessed 9 June 2025
6. Neue Risiken für die Lieferkette und den Standort Deutschland: Maßnahmen gegen Lieferkettenprobleme (2023). Supply Chain Pulse Check. https://de.statista.com/statistik/daten/studie/1380801/umfrage/massnahmen-gegen-lieferkettenprobleme-von-unternehmen/. Accessed 15 June 2025
7. Fehr, M.: Volle Lager haben auch eine Schattenseite: Deutsche Industrie füllt ihre Lager auf, Frankfurter Allgemeine Zeitung (2022)
8. Bundesministerium für Bildung und Forschung (BMBF) Referat Zukunft von Arbeit und Wertschöpfung, Zukunft der Wertschöpfung: Forschung zu Produktion, Dienstleistung und Arbeit (2021)
9. Chen, I.J., Paulraj, A.: Towards a theory of supply chain management: the constructs and measurements. J of Ops Management 22 (2), 119–150 (2004).
10. Stevens, G.C.: Successful supply-chain management. Manag. Decis. **28**(8) (1990)
11. Nyhuis, P., Bleckmann, M., Demir, M., Jahangirkhani, T., Wenzel, A., Schmidt, M.: Risikoabhängige Auswahl der Veränderungsfähigkeitsarten/Risk-dependent selection of changeability types in factory systems. WT Werkstattstechnik **115**(01–02), 2–10, Düsseldorf (2025)
12. DIN EN ISO 22301: Sicherheit und Resilienz—Business Continuity Management System—Anleitung zur Verwendung von ISO 22301. Internationale Organisation für Normung, Berlin (2020)
13. DIN ISO 31000: Risikomanagement—Leitlinien. Internationale Organisation für Normung, Berlin (2018)

Software-Defined Value Network: An Industrial Testbed for Manufacturing-X

Johannes Clar(✉), David Dietrich, Nicolai Maisch, Colin Reiff, Georg Heiner Ziegler, Armin Lechler, Alexander Verl, and Oliver Riedel

ISW, University of Stuttgart, Stuttgart, Germany
johannes.clar@isw.uni-stuttgart.de

Abstract. Future Software-defined Value Networks enable data-driven innovation, diverse optimization opportunities, and resilient production environments. While significant progress has been made in standardizing communication protocols and data models, practical Industry 4.0 implementations require iterative prototyping and continuous refinement based on real-world experience. This paper presents a testbed designed to develop and explore industrial-scale software-defined manufacturing use cases, such as capability-based production. The testbed maps a full product life cycle within a circular economy context. Seven independent companies are interconnected within a data space, each contributing services like engineering, logistics, or production using robots or milling machines. Built on industry standards such as the Asset Administration Shell and OPC UA, the testbed demonstrates vertical and horizontal integration, enabling seamless cross-company automation. Following an overview of the conceptual setup, the testbed components are detailed.

Keywords: Software-defined value networks · Circular economy · Testbed · Manufacturing-X · Asset administration shell

1 Introduction

In the context of Industry 4.0, production systems are becoming increasingly adaptable through Software-defined manufacturing [1]. This shift enables the transition from standalone plants to integrated Software-defined Value Networks (SDVNs), aiming to create resilient, data-driven production environments [2]. A key use case is the ability of machines acting as both service providers and requesters, supporting flexible, service-oriented manufacturing [3]. To support such scenarios, the *Manufacturing-X* initiative was launched to build data ecosystems where diverse stakeholders collaborate to generate value across the network [4]. Implementing these use cases across company boundaries requires standardized, interoperable, and sovereign data exchange, supported by technologies such as (i) *OPC UA* for standardized communication in industrial automation, (ii) the *Asset Administration Shell (AAS)* as a standardized digital twin, and (iii) *Data Spaces*. In addition to standardization efforts, a survey

L. Overmeyer and B.-A. Behrens (eds.), *Production at the Leading Edge of Technology*, Lecture Notes in Production Engineering, https://doi.org/10.1007/978-3-032-19524-1_68

among industrial partners in [5] shows the need for application-specific solutions that translate complex, cross-company standards into tangible and actionable practices. This is taken as an opportunity to set up a testbed for the development and testing of SDVN.

The research questions addressed in this article are therefore: What components are required in an SDVN testbed to represent the diversity and complexity of cross-company processes in value chains? How can these components be used to research current technological enablers? To investigate these questions, Sect. 2 gives a brief introduction on current enabler technologies that pose requirements for an integrated testbed architecture. Section 3 introduces the testbed setup that uses these enabler technologies to digitally track a product's lifecycle from engineering to recycling. For this purpose, fictive companies are interconnected under diverse market conditions, creating monopolistic scenarios in some lifecycle phases and competitive environments where several providers compete for contracts in others. A summary is given and future research priorities to be developed and validated on the testbed are pointed out in Sect. 4.

2 State of the Art

Creating an SDVN requires vertical integration from the shop floor to the enterprise level as well as horizontal integration within layers and across company boundaries (Fig. 1). This integration aims to eliminate data silos and enable flexible connectivity across all levels of production. Therefore, some base technologies have emerged as industry conventions [6–9]:

- **OPC Unified Architecture (OPC UA)** is a vendor-independent standard that ensures data exchange in industrial automation [10]. Its standardization and domain-specific extensions enable interoperable communication at the operational technology (OT).
- The **AAS** is a standardized representation for physical or virtual assets, including relevant attributes, behaviors, and operational data across all phases of the lifecycle. Its modular submodel structure allows for scalable integration of diverse use cases and includes a standardized API for data access at the information technology (IT) level [11].
- The concept of **Data Spaces** enables cross-organizational collaboration under shared data governance [12], built on data sovereignty, fair access, and decentralized infrastructure [13]. Recent initiatives[1] have developed conventions and standardized software, with the aim of establishing data spaces for domain-specific operation.

Concepts such as [6] illustrate how the different base technologies can be combined to build resilient and flexible manufacturing data spaces with both horizontal and vertical integration (Fig. 1). To successfully implement such concepts, adopting a design thinking approach can be highly beneficial [14]. A core

[1] Examples include Gaia-X, Catena-X, and Factory-X, among others.

principle of this approach is rapid prototyping, which allows iterative testing and continuous system optimization [15]. To prototypically implement and iteratively refine data-based manufacturing use cases, an industry-scale testbed is essential, as technological and organizational barriers still hinder the adoption of diverse technologies in practical implementation. Therefore, prototyping the integrated Manufacturing-X architecture is essential to gain practical insights into collecting, transmitting, and using data across SDVNs.

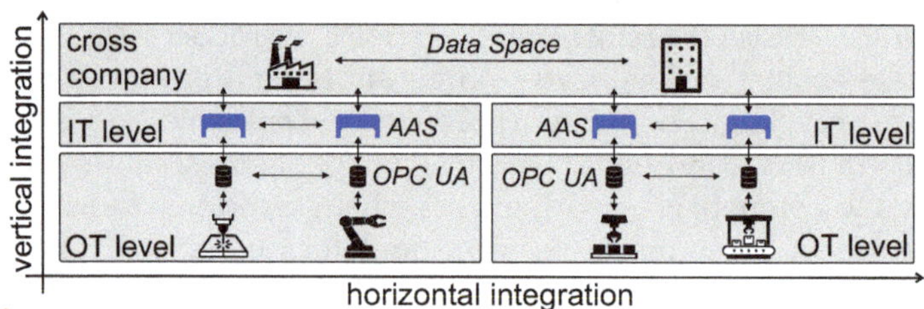

Fig. 1. Architecture for integrating multiple levels of automation

3 Setup of the SDVN Testbed

To prototype a Manufacturing-X ecosystem in an industrial testbed, various fictive companies including real machines are integrated into an adaptive value network. This is showcased by producing a customizable Multi Key Tool (MKT) (Fig. 2, center) with optional features such as a bottle opener, and configurable administrative options. Starting with product configuration via a graphical user interface that generates the product's AAS containing all relevant design features and the submodel of required production capabilities.[2] Through different orchestration approaches, the AAS of available production resources can be matched with required capabilities [16], and production requests are sent to the appropriate companies. Once triggered, the companies retrieve design data from the product's AAS and execute the production steps accordingly. After execution, key production data is written back into the AAS, enabling traceability across the supply chain. Once all production steps are completed, the product's AAS is assigned a Digital Nameplate[2] and handed over to the customer (Fig. 2). These data-driven processes are described in detail below.

3.1 Configurator

The configurator allows customers to select their desired product customization. It offers a variety of options, including: (i) **milling features** such as a bottle

[2] https://industrialdigitaltwin.org/en/content-hub/submodels.

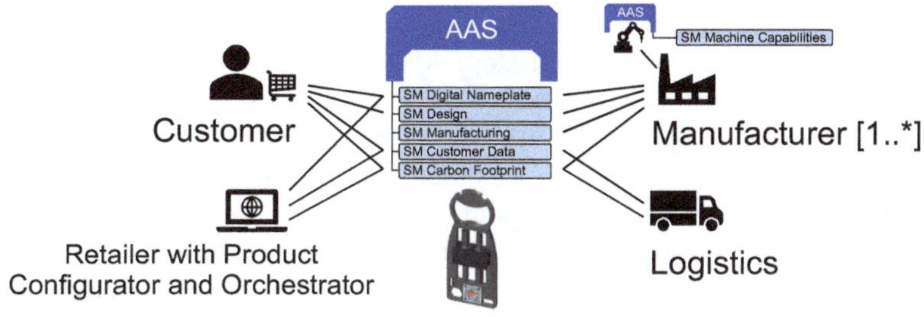

Fig. 2. Collaboration across different value chain partners

opener and spoke wrench; (ii) **additional components** like screwdriver bits, a bit holder, and custom packaging; (iii) **engraving options** including a QR code, ruler, personalized text, and a logo; (iv) **color selection** for the final product. In addition, customers can choose a preferred production mode: (i) **cost-effective**: optimized for low manufacturing costs; (ii) **eco-friendly**: focused on sustainability using recycled materials and avoiding air freight; (iii) **express**: prioritizing speed with fast-track production and logistics.

Finally, a contact person is selected, which determines which contact card is included in the product packaging. In the context of software-defined value networks, a key objective is to enable the autonomous and flexible manufacturing of products across organizational boundaries. This requires the ability to dynamically identify, select, and coordinate resources and services from multiple independent companies. Two conceptual coordination approaches can be considered to achieve this: centralized orchestration and decentralized negotiation.

In the **centralized orchestration approach**, a central entity, referred to as the orchestrator, takes responsibility for the production process's end-to-end planning and coordination. The orchestrator has global visibility into the services and resources offered by participating companies. Based on product requirements, the orchestrator evaluates the offerings' capabilities, capacities, and constraints and generates an optimized production plan. This plan specifies which company will perform which task, in what sequence, and when. While this model facilitates global optimization and control, it introduces a potential single point of failure and requires extensive trust and data sharing among all parties.

In contrast, the **decentralized negotiation approach** is inspired by market-based mechanisms. In this model, the initiating company issues a digital product tender or request that includes detailed specifications of the required product features, quality parameters, and constraints. Participating companies can autonomously evaluate the tender and submit proposals indicating their ability to fulfill specific parts of the production process. The initiating company selects from the received offers based on criteria such as cost, lead time, quality, or sustainability. This method fosters scalability, autonomy, and flexibility while reducing the need for centralized control. However, it may lead to suboptimal global outcomes due to limited coordination and lack of global optimization.

Both approaches represent different models of dynamic collaboration in SDVNs. The choice between them depends on factors such as product complexity, digital infrastructure maturity, trust relationships among companies, and performance requirements. While the decentralized approach is well-suited for cross-company collaboration, the centralized variant could prove useful for company-internal orchestration of manufacturing resources where trust and data availability are less constrained. Both variants are implemented in the testbed to enable a realistic comparison of their scalability and performance in future experiments. In such experiments, identical production scenarios would be executed using both coordination approaches in the testbed, measuring metrics like planning time, resource utilization, fault tolerance, and communication overhead under varying network sizes and complexity levels.

3.2 Production Companies

Once planned by the orchestrator or negotiated through the decentralized approach, configured orders are distributed to the production companies. Although these companies themselves are fictive, each one comprises a real machine and an associated IT infrastructure. The following describes these companies by their function within the product lifecycle.

(i) Supply: The first phase after planning in any configured MKT's lifecycle is the supply of unmachined parts. This is realized by only one supply company, creating a monopoly for this essential process. All parts are stored in an internal storage system. When an order is placed, a warehouse management system processes the incoming AAS and generates a job formatted according to the OPC UA job specification. This job is then sent via OPC UA to the storage control system, where it is executed by a Cable-Driven Parallel Robot (Fig. 3a). Data produced during job execution are sent to the data management system via OPC UA, where relevant data is then written into the AAS. The AAS is then provided to other SDVN members through a data space (see Fig. 1). The communication architecture is structured similarly in all the companies presented here.

(ii) Transport: Deliveries within the SDVN are required between the various lifecycle phases. A fictive logistics company fulfills these processes by physically connecting all companies with a magnetic planar motion system (Fig. 3b). This system allows multiple transportation units to be moved simultaneously and completely independently. The resulting flexibility allows for the simulation of various transportation methods such as road, sea or air freight, allowing the logistics provider to offer different options.

(iii) Manufacturing: The variety of configuration options allows optional features in the MKT. The production process has to adapt based on the chosen configuration. To represent this, two manufacturing companies with different sets of capabilities are included in this SDVN. One of these companies offers the capabilities of a 3-axis milling machine. For a fully automated process, a small industrial robot is installed to feed the mill (Fig. 3c). In addition, the other company offers surface finishing using an ultrashortly pulsed laser (Fig. 3d).

(iv) Assembly: In this phase of the lifecycle, a comprehensive scenario is represented in this SDVN. Two companies with a similar set of capabilities offer product assembly. At this point, a decision must be made based on the selected production mode regarding which company will receive the order. To perform the complex assembly task, one company uses a collaborative 6-axis robot along with auxiliary devices and an automated tool changer (Fig. 3e). The competitor relies on a 2-armed collaborative robot with different tools at each arm (Fig. 3f). After phase (iv), the product is ready to be handed over to the customer for **(v) Use**. This next phase is not modeled in this SDVN.

(vi) Recycling: In recycling, the MKT is disassembled, individual parts are assessed, and returned to the supplier's warehouse. Since this process resembles assembly, the same companies as presented in (iv) can be employed.

This modular network architecture allows for the flexible substitution of individual companies. Moreover, interfaces for automated guided vehicles are integrated within the logistics company, allowing network extensions.

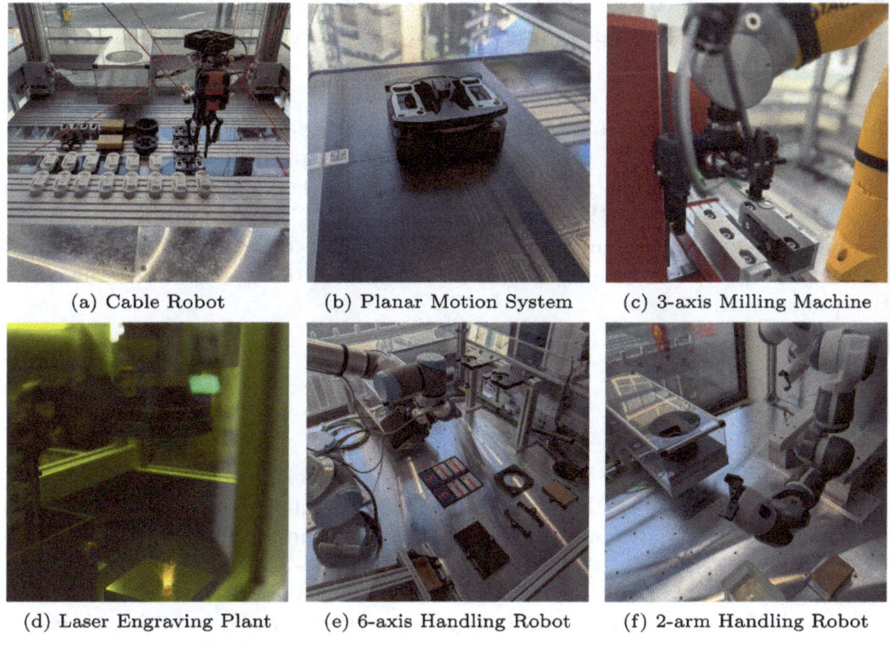

(a) Cable Robot (b) Planar Motion System (c) 3-axis Milling Machine (d) Laser Engraving Plant (e) 6-axis Handling Robot (f) 2-arm Handling Robot

Fig. 3. Overview of the machines integrated in the SDVN

4 Summary and Future Research Priorities

The transformation from conventional value chains to cross-company SDVNs is supported by an adaptive testbed introduced in this work to identify and overcome challenges at an industrial scale. The lifecycle of an exemplary product is modeled here, beginning with the product configuration and the generation of its corresponding AAS. A centralized or decentralized orchestration process distributes the AAS to the fictive producing companies where the design parameters are read and the manufacturing data are aggregated. Vertical communication between the enterprise level of these companies and their real machines is realized via OPC UA. However, due to its limited size, this SDVN cannot replicate the complexity of real value networks. Nevertheless, this SDVN implements a reference architecture for fully integrated Manufacturing-X ecosystems.

Future work based on this testbed includes developing methods to collaboratively manage a growing digital twin using the AAS and integrating new submodel specifications, aiming to standardize the architecture for efficient integration of new participants at machine and company levels. Additionally, comparing the centralized and decentralized orchestration variants under realistic conditions will help evaluate trade-offs in scalability and performance. Incorporating (partly) autonomous and decentralized software components can enhance production process flexibility through the use of type 3 AAS [17]. Therefore, asset and service discovery mechanisms also require further research. Another focus will be the adaptation to already existing data space specifications such as Catena-X or Tractus-X.[3]

Acknowledgment. Partly funded by the Federal Ministry for Economic Affairs and Energy (BMWE) through the projects growING (grant no. 13IPC036G) and Factory-X (grant no. 13MX001L). Partly funded by the German Federal Ministry of Research, Technology and Space (BMFTR) within the "Research Campus Public-Private Partnership for Innovation" funding initiative (02P23Q820) and managed by the Project Management Agency Karlsruhe (PTKA). The authors are responsible for the content of this publication.

References

1. Neubauer, M., Frick, F., Ellwein, C., Lechler, A., Verl, A.: Ein Paradigmenwechsel für die industrielle Produktion/A paradigm shift for industrial production–Software-defined manufacturing. wt Werkstattstechnik online **112**(06), 383–389 (2022). https://doi.org/10.37544/1436-4980-2022-06-33
2. Dietrich, D., Zürn, M., Reiff, C., Neubauer, M., Lechler, A., Verl, A.: Software-defined value networks: motivation, approaches, and research activities. In: Bauernhansl, T., Verl, A., Liewald, M., Möhring, H.C. (eds.) Production at the Leading Edge of Technology, Lecture Notes in Production Engineering, pp. 514–524. Springer Nature Switzerland, Cham (2024). https://doi.org/10.1007/978-3-031-47394-4_50
3. Wu, D., Greer, M.J., Rosen, D.W., Schaefer, D.: Cloud manufacturing: strategic vision and state-of-the-art. J. Manuf. Syst. **32**(4), 564–579 (2013). https://doi.org/10.1016/j.jmsy.2013.04.008

[3] https://catena-x.net/ and https://eclipse-tractusx.github.io/.

4. Plattform Industrie 40: White paper on manufacturing-x (2022). https://www.plattform-i40.de/IP/Redaktion/EN/Downloads/Publikation/Manufacturing-X_long.html
5. Dietrich, D., Zürn, M., Briem, A.K., Koch, D., Lober, W., Lind, J., Lechler, A., Verl, A.: Software-defined value networks: industrial requirements and research gap. In: Advances in Automotive Production Technology–Digital Product Development and Manufacturing–Stuttgart Conference on Automotive Production (SCAP2024) (2025)
6. Neubauer, M., Steinle, L., Reiff, C., Ajdinović, S., Klingel, L., Lechler, A., Verl, A.: Architecture for manufacturing-X: bringing asset administration shell, eclipse dataspace connector and OPC UA together. Manuf. Lett. **37**, 1–6 (2023). https://doi.org/10.1016/j.mfglet.2023.05.002
7. Justmann, K., Leurs, L., Ruskowski, M.: Mapping of OPC UA FX to the asset administration shell and capability, skill and service model. In: 2024 IEEE 29th International Conference on Emerging Technologies and Factory Automation (ETFA), pp. 1–4 (2024). https://doi.org/10.1109/ETFA61755.2024.10710837
8. Moreno, T., Almeida, A., Toscano, C., Ferreira, F., Azevedo, A.: Scalable digital twins for industry 4.0 digital services: a dataspaces approach. Prod. Manuf. Res. **11**(1), (2023)
9. Drath, R., Mosch, C., Hoppe, S., Faath, A., Barnstedt, E., Fiebiger, B., Schlögl, W.: Diskussionspapier–Interoperabilität mit der Verwaltungsschale, OPC UA und AutomationML. Technical Report AutomationML eV and Industrial Digital Twin Association (IDTA) and OPC Foundation and VDMA (2023)
10. OPC Foundation: Opc 10000-1: Ua part 1: Overview and concepts (2006) . https://opcfoundation.org/developer-tools/documents/view/158
11. Industrial Digital Twin Association eV: Specification of the asset administration shell part 1: metamodel–IDTA number: 01001-3-0-2. https://industrialdigitaltwin.org/wp-content/uploads/2025/03/IDTA-01001-3-0-2_SpecificationAssetAdministrationShell_Part1_Metamodel.pdf
12. Möller, F., Jussen, I., Springer, V., Gieß, A., Schweihoff, J.C., Gelhaar, J., Guggenberger, T., Otto, B.: Industrial data ecosystems and data spaces. Electron Mark **34**(1), (2024)
13. Nagel, L., Lycklama, D.: Design principles for data spaces (2021). https://doi.org/10.5281/zenodo.5244997
14. Mesa, D., Renda, G., Gorkin III, R., Kuys, B., Cook, S.M.: Implementing a design thinking approach to de-risk the digitalisation of manufacturing SMEs. Sustainability **14**(21), (2022). https://doi.org/10.3390/su142114358
15. Camburn, B., Viswanathan, V., Linsey, J., Anderson, D., Jensen, D., Crawford, R., Otto, K., Wood, K.: Design prototyping methods: state of the art in strategies, techniques, and guidelines. Des. Sci. **3**, (2017). https://doi.org/10.1017/dsj.2017.10
16. Pfrommer, J., Schleipen, M., Beyerer, J.: PPRS: production skills and their relation to product, process, and resource. In: 2013 IEEE 18th Conference on Emerging Technologies & Factory Automation (ETFA), pp. 1–4. IEEE (2013). https://doi.org/10.1109/ETFA.2013.6648114
17. Sakurada, L., La Prieta, F.D., Leitao, P.: A methodology for integrating asset administration shells and multi-agent systems. In: 2023 IEEE 32nd International Symposium on Industrial Electronics (ISIE) (2023). https://doi.org/10.1109/ISIE51358.2023.10227964

Stakeholder Identification and Classification for Digital Twins of Production Machines

Fynn Hendrik Dierksen(✉), Marietta Riebeling, and Klaus Dröder

Institute of Machine Tools and Production Technology, Technische Universität Braunschweig, Braunschweig, Germany
f.dierksen@tu-braunschweig.de

Abstract. Digital twins are increasingly an integral part of engineering for special machinery and mechanical components in general, as they enable insights into machine operation, behavior and condition and improve system reliability, providing substantial benefit and information to the various stakeholders. However, the success and further utilization of digital twins depends on the perceived added value for these stakeholders within the manufacturing company, the customer or regulation authorities. A lack of integration of stakeholders can lead to a rejection of the digital twin at the user level. Therefore, this study aims to develop a methodology for the machine-specific identification and classification of stakeholders within the digital twin framework, ensuring an alignment of diverse interests and requirements of all relevant parties. Based on the developed method, the approach is applied to a production machine as a sample use case. A comprehensive network with 39 stakeholders is identified. These stakeholders can be categorized into four primary groups: system users, system developers, regulatory and legislative bodies, and decision-makers. For each stakeholder, the fundamental requirements are identified and a classification is carried out both thematically and in relation to the observation period. Consequently, the proposed stakeholder classification method enables a tailored implementation process of digital twins according to the unique requirements of each asset and its associated stakeholders connected to the production machine. This targeted approach enhances stakeholder engagement, maximizes the utility and effectiveness of digital twins, and contributes to the overall success of projects in the mechanical and special machine engineering sectors.

Keywords: Digital twin · Stakeholder identification · Stakeholder classification

1 Introduction

In digitalized and competitive markets, manufacturing companies must efficiently adjust production processes to fluctuating market conditions and diverse customer requirements, while ensuring full operational functionality. This requires various process data and information to be accessible to all relevant parties of interest. Nowadays, digital twins are a valuable tool to provide meaningful data, with market projections forecasting over 39.8% annual growth for digital twins, highlighting their economic importance

L. Overmeyer and B.-A. Behrens (eds.), *Production at the Leading Edge of Technology*,
Lecture Notes in Production Engineering, https://doi.org/10.1007/978-3-032-19524-1_69

and strategic value. Studies have shown increasingly numbers of adaption, with merely 29% of companies have implemented them, and 65% are planning adoption strategies [1].

For the adoption and integration strategy for digital twins, effective stakeholder management is crucial: analysing and integrating the interests of various stakeholders ensures project success and helps demonstrate the technology's benefits for each stakeholder [2]. For a successful adoption and integration for digital twins, the individual use-case has to be at the centre of a project. The acceptance of the technology and therefore the success of the entire digital twin is dependent on the effective integration of stakeholder interests and requirements. As a result, an effective stakeholder management is crucial: analysing and integrating the interests of various stakeholders has to be the basis for new digital twins. However, as this technology is currently not wide-spread and the variance in the designs and applications is highly diverse. As a result, the knowledge of the various stakeholders is also highly diverse, requiring structures and methodologies for an effective integration of the individual demands. This paper presents a flexible methodology to identify, analyse, and classify key stakeholders for individual project environments, forming a basis for a structured and transparent project structure for the integration of effective digital twins. Chapter 2 reviews stakeholder analysis research; Chap. 3 develops a methodology for the identification of stakeholder needs; Chap. 4 gives a short overview for a demonstrative use case it to a picker-cell use case.

2 State of the Art

Stakeholder involvement is crucial for managing unexpected, potentially project-threatening events, and should be prioritized at the project's inception [3]. The conventional stakeholder management approach, originating in project management, comprises three domains: stakeholder identification, needs analysis, and stakeholder classification [4].

Stakeholder identification and needs analysis share methodological overlap and include approaches such as the Delphi method, brainstorming, the 635 method, design thinking, and interviews. The Delphi method and interviews are survey-based [5]. Experts or project partners are often involved due to their environmental familiarity [6]. However, these survey-based methods have a risk of overlooking hidden stakeholders, which are stakeholders which are affected parties of interest affected by the project without direct involvement in the decision-making process [7]. Creative methods such as brainstorming or the 635 method support identifying such stakeholders by encouraging uninhibited group development [8].

When these stakeholders are identified, a classification into influential categories is useful to analyze the different impact and prioritization about addressing the stakeholder needs [9]. The commonly used matrix method adds a two-dimensional view, classifying stakeholders based on how strongly two criteria are met [6]. Alternatively, role-based classifications group stakeholders by their project roles, allowing for differentiated views of individuals and groups [10].

The approach of analysing and involving stakeholders and deriving requirements has already been implemented in previous research. The utilisation of methods such as interviews or the classification of stakeholders is also evident, as demonstrated in [5]. The application of the analysis approach is centred on the utilisation of the battery product passport, thus rendering it highly product-specific. Reference [11] provides a detailed examination of the stakeholder requirements for maintenance processes in mechanical engineering. From the outset, the focus is placed on a selected group of stakeholders. Consequently, there is a paucity of consideration given to the involvement of other stakeholders. Moreover, the emphasis is confined to the domain of maintenance, which imposes a particular constraint.

So far, there are no other known use cases that take a holistic view of stakeholder integration in the digital twin in the area of machine construction. However, the authors of [12] also emphasise that such a consideration is necessary to enable stakeholders to understand the goal, benefits and cost-efficient modelling of a digital twin.

The objective of this paper is thus to develop a method that systematically enables the stakeholders of a digital twin to be identified and their requirements for the digital twin to be derived.

3 Methodology for Stakeholder Identification and Classification of Production Machines

In order to achieve the goal of a requirements profile for digital twins based on stakeholder needs, the methodology consists of three parts: stakeholder identification, needs analysis, and classification. Each part builds on established research methods (see Chap. 2). The needs identified should form the basis for determining the technical requirements for the digital twin.

Stakeholder Identification

The identification of stakeholders follows a four-stage process (see Fig. 1). It begins with a systematic literature search. These findings are expanded through brainstorming to determine baseline stakeholders specific to the context [13]. Within the second step the network is extended to include additional actors such as satellite stakeholders, aiming to form a comprehensive stakeholder network for the digital twin. Finally, this network is reviewed with a technical expert. The resulting stakeholder network forms the basis for the subsequent needs analysis during the development and implementation of the digital twin.

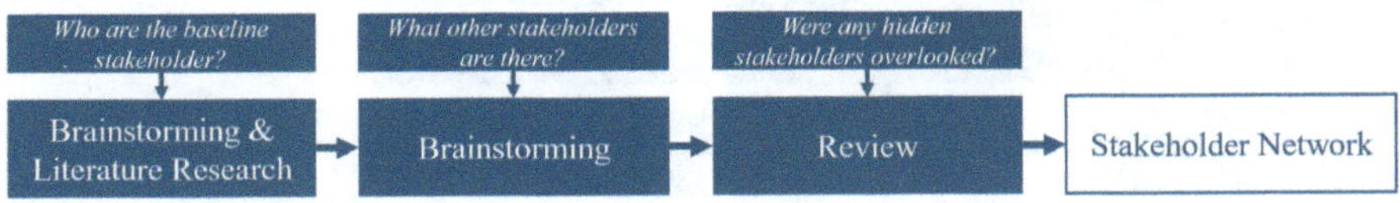

Fig. 1 Method for stakeholder identification

Analysis of Stakeholder Needs

The stakeholder needs of the digital twin are derived from the stakeholder network. The process is divided into two parts. First, stakeholder roles are described based on the previously developed network. These role descriptions are linked to the individual project and used to derive requirements [14]. The network and role descriptions are then supplemented by analysing existing implementation approaches, gathered through a literature search similar to Brunn et al. [14]. The goal is to understand the involvement of the stakeholders in alternative implementations and which objectives have been achieved within the previous approaches. The procedure is illustrated in Fig. 2.

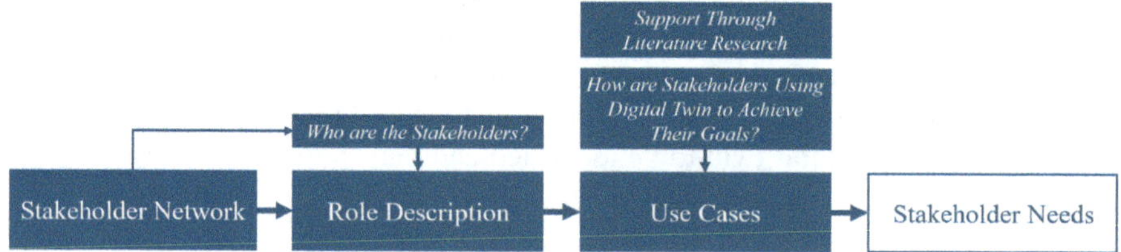

Fig. 2 Method for Identifying stakeholder needs

Stakeholder Classification

The classification of stakeholders is essential for enabling a targeted, collaborative development process that addresses stakeholder needs.

Role-based classifications offer consistent and comprehensible results through clear criteria and are flexible across different contexts. The role-based classification is carried out using the product, process and resource criteria. This classification facilitates a preliminary grouping process, thereby enabling stakeholders to be combined into consideration groups. This facilitates the subsequent analysis and incorporation of requirements. In order to establish a hierarchy of data requirements in the event of a conflict, stakeholders are then prioritised according to the frequency of the data requirements in the categories real-time, weekly, monthly.

The developed approach combines this role-based pre-classification with a matrix method, whose dimensions (e.g. data requirements, role in the value creation process) can be adapted to individual production facilities and data-driven systems (see Fig. 3).

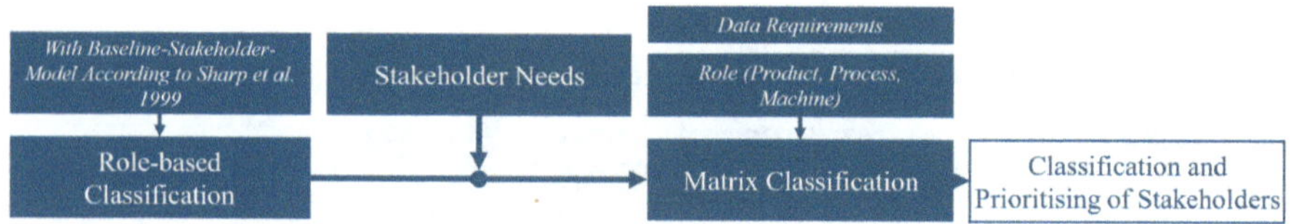

Fig. 3 Method for stakeholder classification

4 Demo-Application of the Developed Method

To make the presented three-step approach of Chap. 3 more transparent, it is demonstrated for a sample use case. In the following, a picker cell is investigated, which has the primary function of handling objects in wide-spread pick-and-place operations. The objective of this process is twofold: firstly, to separate the items, and secondly, to place them into a separate tray or packaging. Process efficiency and speed are of key relevance, often achieved through specifically designed kinematics such as parallel robots. This process is typically part of a larger manufacturing process chain and therefore defined by a large number of contact points with other production processes as well as a possibly large overlap of competencies and interested parties.

4.1 Stakeholder Identification

The stakeholder identification procedure (see Fig. 1) begins with defining a basic stakeholder network through brainstorming and literature review. Seven relevant publications on stakeholder roles in mechanical engineering were identified as the basis for the network [3, 15–18]. Stakeholders are categorised using the baseline stakeholder approach from [13], with system developers, regulators, legislators, system users, and decision-makers as core categories. For increased clarity, the following illustrations will focus on the system user. Satellite stakeholders, who primarily interact with baseline stakeholders, serve to expand the network. The stakeholders that are connected to the 'system user' are the derived baseline stakeholders. The surrounding stakeholders that have no connecting line are the satellite stakeholders (see Fig. 4).

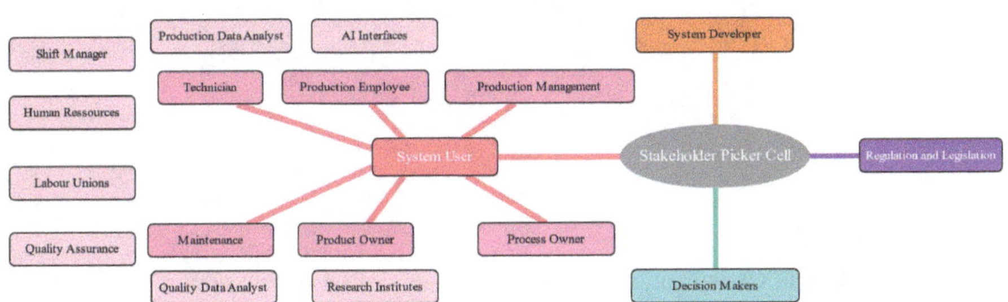

Fig. 4 Stakeholder network of the picker cell with focus on the system user

4.2 Derivation of Stakeholder Needs

The method shown in Fig. 2 is used to derive stakeholder requirements, including data needs. First, role descriptions are created for each baseline stakeholder, outlining their activities related to the picker cell. Data requirements are then derived using literature-based use cases. The derived needs from literature are displayed in Fig. 5. These identified needs and frequencies form the basis for stakeholder classification and the derivation of technical requirements of the stakeholders.

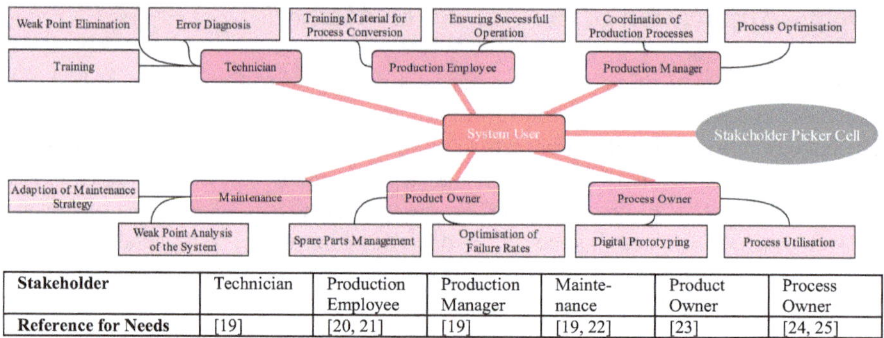

Stakeholder	Technician	Production Employee	Production Manager	Mainte-nance	Product Owner	Process Owner
Reference for Needs	[19]	[20, 21]	[19]	[19, 22]	[23]	[24, 25]

Fig. 5 Identified stakeholder needs—focus on system users

4.3 Classification of Stakeholders

The classification step, shown in Fig. 3, uses a matrix-based framework with two criteria. The first dimension distinguishes stakeholder focus: product, process, or machinery. The second categorises the urgency of data needs as 'Real-time', 'Weekly' or 'Monthly'. This method reflects the role of each stakeholder in the digital twin development process. Results are illustrated in Fig. 6.

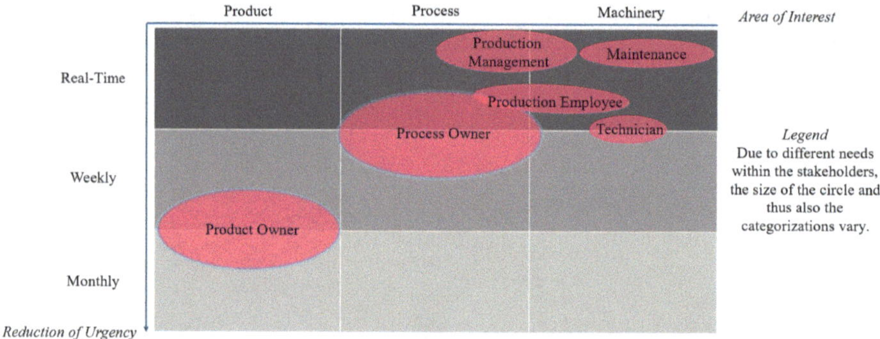

Fig. 6 Classification of stakeholders—focus on system users

An analysis of stakeholders using a digital twin for the picker cell shows a clear link between data requirements and both planning level and decision-making power. Operators and technicians at the operational level rely on real-time data to control processes and reduce downtime. The production manager, combining operational, tactical, and strategic responsibilities, also benefits from up-to-date data. Without it, decision quality and system optimisations suffer.

Effective integration of the digital twin requires coordination between all stakeholder groups. Effective integration of the digital twin requires coordination between all stakeholder groups. This allows the prioritisation of needs.

5 Summary and Future Outlook

As a core Industry 4.0 technology, digital twins digitally represent industrial facilities and enable bidirectional data exchanges for data-driven decisions. The structure and complexity of a digital twin depends on system requirements, making system-specific stakeholder analysis essential. Early stakeholder involvement promotes greater user acceptance. Due to the lack of standard procedures, this paper developed a methodology combining theoretical stakeholder concepts and qualitative methods.

The methodology has three stages: stakeholder identification, needs analysis via role descriptions and a two-part classification. The first classification follows the baseline model from the literature, the second is based on data needs and value creation. The resulting stakeholder network includes system users, developers, decision-makers, and regulators with varying informational needs.

The study highlights the diverse interests and data requirements involved. As the inclusion of individual information needs is a core aspect of modern process design, a well thought-through stakeholder analysis can be considered a key factor for successful digital twin development. With the research and approach presented, a baseline for a machinery-oriented approach to stakeholder analysis was created, enabling a stronger foundation for future digital twin implementations. Consequently, the derived requirements provide a foundation for the subsequent development of technical requirements for the stakeholders, necessitating further consideration and therefor a tailored implementation of the digital twin for production machines. This also enables the identification and targeted combination of requirements more quickly. Future research should validate the method further in practice, assess scalability, and develop criteria for cost-effective integration of stakeholder needs across company sizes and production models.

Competing Interests. The author(s) has no competing interests to declare that are relevant to the content of this manuscript.

References

1. Hexagon, A.B.: Statistiken zu digitalen Zwillingen 2025. https://hexagon.com/de/resources/insights/digital-twin/statistics. Accessed 09 May 2025
2. Kober, C. et al.: Challenges of Digital Twin Application in Manufacturing, pp. 162–168. IEEE (2022). https://doi.org/10.1109/IEEM55944.2022.9989654
3. Karlsen, J.T.: Project stakeholder management. Eng. Manag. J. **14**, 19–24 (2002). https://doi.org/10.1080/10429247.2002.11415180
4. El-Aboodi, S.: Definitionsphase. Springer Fachmedien Wiesbaden, Wiesbaden, pp. 81–138 (2024). https://doi.org/10.1007/978-3-658-43561-5_6
5. Merkle, L. et al.: Stakeholder Analysis of Digital Twins for Battery Systems. IEEE, pp. 1–8 (2020). https://doi.org/10.1109/EVER48776.2020.9242960
6. Krips, D.: Inhalt Einer Stakeholderanalyse. DVP Projektmanagement. Springer Berlin Heidelberg, Berlin, Heidelberg, pp. 11–34 (2017). https://doi.org/10.1007/978-3-662-55634-4_3
7. Teng, Y., et al.: Analysis of stakeholder relationships in the industry chain of industrialized building in China. J. Clean. Prod. **152**, 387–398 (2017). https://doi.org/10.1016/J.JCLEPRO.2017.03.094

8. Furnham, A.: The brainstorming myth. Bus. Strateg. Rev. **11**, 21–28 (2000). https://doi.org/10.1111/1467-8616.00154
9. Freeman, R.E.: Strategic Management. Cambridge University Press (2015)
10. Suhanda, R.D.P., et al.: RACI matrix design for managing stakeholders in project case study of PT. XYZ. IJIES **5**, 122–133 (2021). https://doi.org/10.25124/ijies.v5i02.134
11. Chen, S. et al.: Understanding Stakeholder Requirements for Digital Twins In Manufacturing Maintenance, pp. 2008–2019. IEEE (2023). https://doi.org/10.1109/WSC60868.2023.10408657
12. Kober, C. et al.: Digital Twins: A Critical Perspective and Research Trends, pp. 749–754. IEEE (2024). https://doi.org/10.1109/IEEM62345.2024.10857032
13. Sharp, H. et al.: Stakeholder identification in the requirements engineering process, pp. 387–391. IEEE (1999). https://doi.org/10.1109/DEXA.1999.795198
14. Bunn, M.D., et al.: Stakeholder analysis for multi-sector innovations. J. Bus. Ind. Market. **17**, 181–203 (2002). https://doi.org/10.1108/08858620210419808
15. Liyanage, R., et al.: Digital Twin Ecosystems: Potential Stakeholders and Their Requirements. Lecture Notes in Business Information Processing, vol 463, pp. 19–34. Springer International Publishing, Cham (2022). https://doi.org/10.1007/978-3-031-20706-8_2
16. Perno, M., et al.: Developing a framework for scoping digital twins in the process manufacturing industry. Adv. Transdisc. Eng. (2020). https://doi.org/10.3233/ATDE200185
17. Kiran Sankar, M.S., et al.: The emergence of digitalization to the manufacturing sector in the sustainability context: a multi-stakeholder perspective analysis. J. Clean. Prod. **468**, 142983 (2024). https://doi.org/10.1016/j.jclepro.2024.142983
18. Onaji, I., et al.: Digital twin in manufacturing: conceptual framework and case studies. Int. J. Comput. Integr. Manuf. **35**, 831–858 (2022). https://doi.org/10.1080/0951192X.2022.2027014
19. Karaarslan, E. et al.: Digital Twin Driven Intelligent Systems and Emerging Metaverse. Springer Nature Singapore, Singapore (2023)
20. Friederich, J., et al.: A framework for data-driven digital twins of smart manufacturing systems. Comput. Ind. **136**, 103586 (2022). https://doi.org/10.1016/J.COMPIND.2021.103586
21. Lensing, K.: KI-basierte Assistenzsysteme für die Industrie 4.0. Springer Reference Technik. Springer Berlin Heidelberg, Berlin, Heidelberg, pp. 1–29 (2020). https://doi.org/10.1007/978-3-662-45537-1_163-1
22. Herrmann, C. et al.: An Integrated Approach for the Evaluation of Maintenance Strategies to Foster Sustainability in Manufacturing (2007)
23. Stray, V., et al.: Agile Processes in Software Engineering and Extreme Programming, vol. 383. Springer International Publishing, Cham (2020)
24. Arnold, D., et al.: Materialfluss in Logistiksystemen. Springer, Berlin Heidelberg, Berlin, Heidelberg (2019)
25. Jones, D., et al.: Characterising the digital twin: a systematic literature review. CIRP J. Manuf. Sci. Technol. **29**, 36–52 (2020). https://doi.org/10.1016/J.CIRPJ.2020.02.002

Efficient Dynamic Bottleneck Detection Using the Active Period Method: Challenges and Solutions from Industrial Case Studies

Jörg Drees[1(✉)], Marco Bernreuther[2], Martin Jestädt[1], Steffi Stambera[1], and Tim Teriete[2]

[1] iFAKT GmbH, Stuttgart, Germany
j.drees@ifakt.de

[2] Fraunhofer Institute for Manufacturing Engineering and Automation, Stuttgart, Germany

Abstract. Bottlenecks are individual resources that limit the performance of an entire production system. Detecting, analyzing and eliminating them is critical to optimizing production performance. Traditional static bottleneck detection is inadequate in today's complex manufacturing environment, where bottleneck resources frequently shift due to the diversity of products and their varying workloads. Instead, existing digital production data can be used to dynamically detect bottlenecks. This paper describes practical challenges and solutions from two industrial case studies where the Active Period Method (APM) was used for dynamic bottleneck detection. Numerical experiments show the efficiency of this approach and how the challenges from the two industrial use cases are handled. Finally, a main goal of this work is to provide this new and efficient implementation of the APM to the whole scientific community, therefore a Gitlab repository will be provided.

Keywords: Dynamic bottleneck · Bottleneck detection · Active period method

1 Introduction

Manufacturing companies need to continuously optimize their production processes in order to improve or at least maintain their competitive position. However, optimizing the performance of individual resources does not necessarily mean that the overall production system will perform better. Instead, the performance of an entire production system is limited by one or a few resources, called bottlenecks [5]. Detecting, analyzing and eliminating these bottlenecks is critical to optimizing production performance.

Traditional bottleneck detection methods (e.g., the operator balance chart) are inadequate in today's increasingly complex manufacturing environment.

L. Overmeyer and B.-A. Behrens (eds.), *Production at the Leading Edge of Technology*, Lecture Notes in Production Engineering, https://doi.org/10.1007/978-3-032-19524-1_70

These methods have only a static view of bottlenecks. However, in increasingly flexible production systems, the occurrence of dynamic bottlenecks is prevalent [17]. Methods such as the Active Period Method (APM, see [13]) use production data to dynamically detect bottlenecks. Although the APM is extensively discussed in the literature, its application in industrial practice can be challenging [3].

This paper describes practical challenges and solutions from two industrial case studies where the APM was used for dynamic bottleneck detection. The objective of this paper is to help practitioners apply the APM by presenting the lessons learned from the case studies and providing a new and efficient implementation of the APM. The rest of the paper is outlined as follows: Sect. 2 categorizes existing data-driven bottleneck detection methods. Section 3 discusses the case studies conducted and the challenges they faced. Section 4 presents our solutions and provides an efficient implementation of the APM. Section 5 concludes this paper with a brief summary of the findings and an outlook for further research.

2 Data-Driven Bottleneck Detection Methods

A proven illustration of bottlenecks is the funnel model shown in Fig. 1. The process step or executing resource is illustrated as a funnel. Incoming orders with the corresponding products fall into this funnel, visualized as balls whose volume depends on the amount of work involved. The diameter of the lower opening of the funnel corresponds to the current performance of the resource, limited by its maximum capacity [10]. Now there are several definitions in the literature that define the characteristics of a bottleneck [15] and corresponding bottleneck detection methods [18]. For the following overview, data-driven bottleneck detection methods are categorized based on West et al. below according to their data requirements [21].

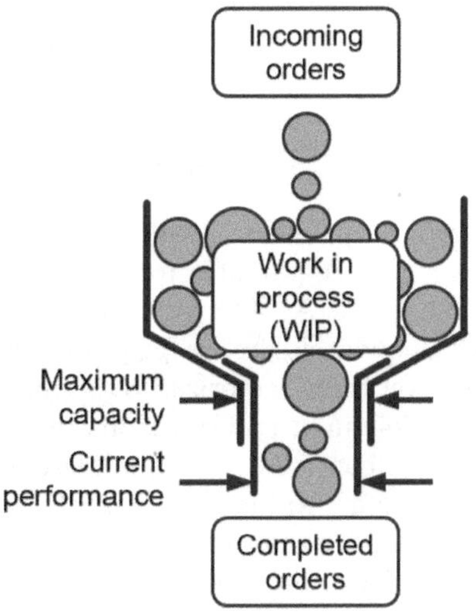

Fig. 1. The funnel model [10]

The first category of bottleneck detection methods uses machine states. Hopp and Spearman present the Utilization Method, which detects the average bottleneck as the resource with the highest utilization rate in the observation period, considering both processing and downtime [6]. Roser et al. group all machine states into either active or inactive states. They define the bottleneck as the resource with the longest average active period [12]. One year later, Roser et al. present the APM as a further development of this approach. In addition to the average bottleneck for the period under consideration, the dynamic bottleneck at a given point in time is detected as the resource with the currently longest uninterrupted active period [13].

The second category of bottleneck detection methods uses buffer levels. Kuo et al. present a systems-theoretic approach for detecting bottlenecks in serial production lines. If the blocking frequency is larger than the starving frequency of the next process step, the bottleneck is downstream. If the starving frequency is larger than blocking frequency of the previous process step, the bottleneck is upstream [8]. Li et al. further develop the Arrow Method with the ability to detect bottlenecks in production systems in real time [9]. Instead of blocking and starvation frequencies, Klenner et al. use a comparison of currently measured stocks with maximum stocks [7]. Other methods use the queue before a resource, as evaluated by Thürner et al., such as the Maximum Workload Method, the Queue Length Method, or the corrected workload method [20].

The third category of bottleneck detection methods uses process times. Betterton and Silver present a method that detects the bottleneck as the resource with the smallest work in process interdeparture time variance [2]. However, this method only detects average bottlenecks over a given period of time.

The selection of an appropriate data-driven bottleneck detection method depends on the use case and, in particular, on the data available. Due to the usually good availability and data quality of machine states and the independence from the material flow, the APM considered in this paper is generally widely applicable.

3 Use Cases and Challenges

Two companies with markedly divergent production processes have been selected as reference partners for the application examples presented herein.

Use Case A: The automotive supplier operates a component production with subsequent assembly for power trains. The manufacturing process under consideration is characterized by a multi-stage, mechanized component production for double-clutch transmissions in batch production (manufacturing-to-order) with call-off control by the final assembly. The area under consideration includes the upstream production lines of three individual components (gears/clutch bodies). Modern production stations use programmable logic controllers (PLCs) to control the processes, which can provide a wide range of status information if configured accordingly, including whether the station was active or passive. It is assumed that the workstation only reports its status to a higher-level system in the event of a status change (status event). The database used contains historical MES machine data for a period of one year, derived from status events per workstation from more than 15 workstations, resulting in a total of 2.4 million status events. The APM requires the distinction between active and inactive periods. A preliminary analysis of the data is performed by evaluating each possible station status. The challenge with this data set is the large number of resources with parallelism. Due to the high variability of orders and machine availability, the data must be collected over a long period of time to obtain representative results.

Use Case B: The manufacturing process in question is used in the medical technology sector, in particular in the production of high-performance ventilators. The value chain is designed for customer-specific variants in make-to-order production. The area under investigation is a final assembly line in which a product family is integrated in a one-piece flow with upstream component production and subsequent detailed testing processes. In this particular use case, manual assembly processes are considered. It is imperative to translate the order confirmations of the individual work plan operations by the workers into station status events. This goal can be achieved by systematically recognizing the initiation and completion of a processing step at a station by the workers in the Enterprise Resource Planning (ERP). Consequently, it is possible to draw conclusions about the active periods. The database used contains ERP production order confirmation data for a quarter with a complex mix of variants with more than 70,000 events on more than 10 workstations. The challenges associated with this use case are that manual operations require a workaround for order confirmations for status events. In addition, intensive preparation and cleansing of the event data is required for good analysis results.

Challenges enumerated in the use cases render the implementation of inventory-based methods, such as blockage/starvation, impractical due to the absence of inventory tracking and the frequently intricate material flow [4]. In contrast, the APM does not require this differentiation of station status and it's dynamic detection allows for a more nuanced analysis of bottlenecks. A main hurdle with using the APM is that it is computationally intensive due to its iterative nature. Various approaches try to mitigate this by introducing matrices as a period state tracker for computation [16,19]. This methodology should be viewed critically, since the size of the computation matrix increases squared to the number of observed stations and states in the observation period. The authors point out that the observation granularity can be reduced without affecting the result to much. However, this approach is prone to user error when selecting a suitable aggregation level because the effect of the aggregation cannot be estimated in advance. Furthermore, machine state data is prone to errors. This is even more apparent when dealing with manual data entries, where batch wise data entry is common. This can create contradicting states at a singular timestamp which the user might be unaware of. Because of the APM's high susceptibility to data errors, labor intensive data cleansing is of great importance for accurate results. This calls for a simpler to employ and performance-oriented implementation of the APM for real-world usage.

4 Solutions and Efficient Implementation

As described in the previous section the main hindrance when trying to employ the APM for bottleneck detection is its relative complex computation, which greatly limits its use for practitioners without employing proprietary solutions. Our approach tries to overcome this hurdle by integrating the APM into an easy-to-use package for open-source development. Rather than earlier approaches, which used proprietary or niche software [1,14] we developed the tooling for bottleneck detection in the de facto standard for data science, Python [11]. This also increases compatibility of the method with other widely used data pipeline, analysis, and visualization tools. To enable practitioners to detect bottlenecks via the APM, we also simplify its usage by making the method robust against a number of real-life edge cases, see Fig. 2. Some of these have already been discussed in theory [1,16] before but are not available for real-world usage in a simple way.

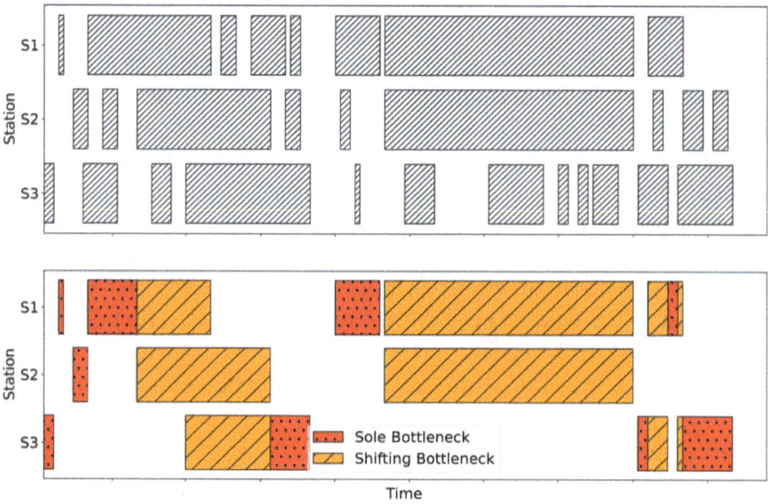

Fig. 2. Visualization of momentary bottleneck edge cases (bottom) and the corresponding station activity Gantt chart (top)

Since the required data is prone to errors our implementation needed to be robust against the most common quality issues of event data. The first of which is the duplication of states at the same timestamp. This case can be further differentiated into states that are compatible with each other, such as any number of states that can be interpreted as the same state and states that give contradictory information about the station's current active state (the data indicates active and inactive periods at a station, at the same timestamp). This case can occur when the resolution of the observed timestamp changes at any stage in the data pipeline or the internal clocks of systems are not perfectly synchronized. The latter is especially problematic, since the station's state cannot be inferred correctly by the bottleneck detection algorithm, which leads to deviations from the correct result. Our implementation checks for these cases beforehand and informs the user if the quality of the data is insufficient.

The second edge case is that a station is never the sole bottleneck in a time period even if it has the longest active period in a time frame. This contradicts the original detection algorithm's description in which the longest uninterrupted active duration is evaluated to a part as the sole bottleneck. This case is depicted in the left half of Fig. 2 and occurs when three or more stations' longest active periods overlap, so the middle stations are never the sole bottleneck, even if they have the highest utilization in the observed period. Therefore, the detection method needed to check whether any active period is fully overlapping with other periods and should never be a sole bottleneck.

The last edge case discussed here is the total simultaneous active period of multiple stations which should always be interpreted as shifting bottlenecks, as depicted in the right half of Fig. 2, since no single station is the sole bottleneck in the time period.

A main hurdle to employing the APM in a production system to detect bottlenecks is its complicated theory. Our own observations from the research project showed that the confidence of practitioners in the method without fully understanding it is low. This is especially true when it contradicts other classical measures for bottleneck detection such as mean utilization or the operator balance chart. We try to mitigate this problem by providing a variety of visualization options to the user, since we found that the graphical representation of a station Gantt chart increased understanding of the method's functionality.

In the remainder of this section, we will describe our approach in more detail and evaluate it numerically. Our implementation of the APM uses object-oriented programming (OOP) and introduces two classes: *Machine* and *ActivePeriodMethod.* The *Machine* class servers as an abstraction of physical stations and is easy to interpret compared to abstract data structures. It holds the station's name and it's states history sorted by their start times with their duration and type. Furthermore, it provides methods for updating states, calculating its usage period as well as checking and returning it's next active period. Here, the last two methods are implemented in a way that allows for efficient calculations of $\mathcal{O}(1)$, i.e. constant, complexity, as long as the provided current time increases for consecutive calls to the methods, as will be the case for the APM.

Similarly, the *ActivePeriodMethod* class holds all stations (*Machines*) and data on momentary and average bottlenecks. Furthermore, it provides methods for calculating momentary and average bottlenecks as well as visualizing them. To simplify usability, we also provide a wrapper function *detect_bottlenecks*, which makes directly working with the classes optional. Let n denote the number of events, then the performed steps and their computational loads are as follows:

1. Raw bottleneck calculation:
   - Calculation of the raw bottlenecks, i.e. without sole/shifting distinction, that is based on the *Machine* classes efficient methods;
   - Computational load of $\mathcal{O}(n)$, i.e. linear, complexity;
2. Momentary bottleneck calculation
   - Calculation of the momentary: bottlenecks, i.e. sole/shifting distinction, based on the raw bottlenecks;
   - Computational load of $\mathcal{O}(l)$, i.e. linear, complexity, where l denotes the number of raw bottlenecks with $l \ll n$;

3. Average bottleneck calculation:
   - Aggregation of momentary sole/shifting bottlenecks to average data;
   - Insignificant computational load;
4. Result visualization:
   - Visualization of momentary/average bottlenecks as Gantt/bar chart;
   - Insignificant computational load;

Algorithm 1 summarized the momentary bottleneck calculation. The computational complexity shown in Fig. 3 has also been studied in numerical experiments. A random data generator that is also part of the provided package has been used to generate the problem instances. The empirical complexity of the algorithm is approximated by the function $f(n) = 10^{-3.9} \cdot n^{1.10}$, which aligns well with the linear complexity in n and l that is to be expected. Furthermore, it can be seen, that the computations are efficient for the industrial use cases. Processing the 2.4 million events of use case A takes less than 1000 s and processing the 70,000 events of use case B takes less than 10 s.

Algorithm 1: Detect Bottlenecks

```
Input: raw_bottlenecks
Output: momentary_bottlenecks
Initialize shifting_bottlenecks and sole_bottlenecks;
for current_bottleneck, next_bottleneck in raw_bottlenecks do
    Fetch last_shifting_bottleneck from shifting_bottlenecks if it is on the same
    machine and with an overlap w.r.t. current_bottleneck;
    Calculate overlap of current_bottleneck and next_bottleneck;
    if overlap exists then
        if overlap overlaps with last_shifting_bottleneck then
            Combine both to one shifting bottleneck and update in
            shifting_bottlenecks;
        else
            Add overlap as shifting bottleneck to shifting_bottlenecks;
        Add overlap of next_bottleneck to shifting_bottlenecks;
    Add sole part of current_bottleneck to sole_bottlenecks;
    if next_bottleneck is last bottleneck then
        Add sole part of next_bottleneck to sole_bottlenecks;
```

5 Conclusion and Outlook

Our implementation of the APM focuses on real-life employment by practitioners to remove hurdles when trying to employ this valuable method in manufacturing environments. In order to save further computing power in line with green IT principles, the algorithm could be made online-efficient by using the APM to update bottlenecks with only partially new data. This would replace the method of always running the algorithm on the entire data set.

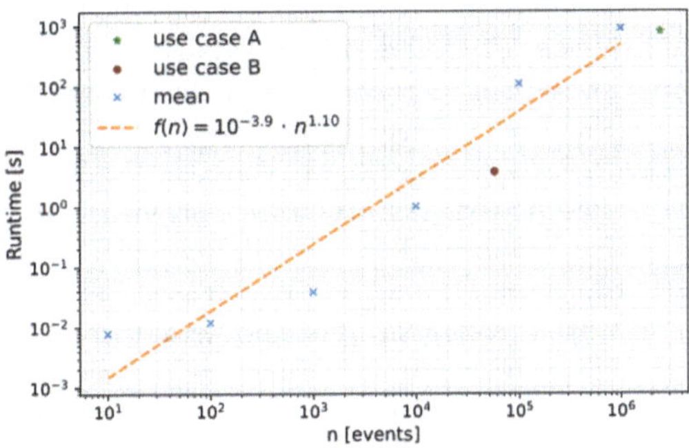

Fig. 3. Computational complexity analysis of APM, with industrial use cases

Code Reference: https://gitlab.cc-asp.fraunhofer.de/mcb/active-period-method

Notes and Comments. Shown results are part of DAWOPT, a project funded by the Ministry of Economic Affairs, Labour and Tourism Baden-Württemberg.

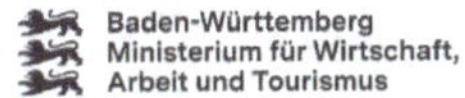

References

1. Arff, B. et al.: Analysis and Visualization of Production Bottlenecks as Part of a Digital Twin in Industrial IoT. Appl. Sci. **13**(6), 3525 (2023).
2. Betterton, C.E., Silver, S.J.: Detecting bottlenecks in serial production lines – a focus on interdeparture time variance. IJPR **50**(15), 4158–4174 (2011).
3. Brochado, A.F. et al.: A data-driven model with minimal information for bottleneck detection - application at Bosch thermotechnology. IJMSEM **18**(4), 318–331 (2023).
4. Biller, S. et al.: Bottlenecks in Bernoulli serial lines with rework. IEEE Trans. Autom. Sci. Eng. **7**, 208–217 (2010).
5. Goldratt, E.M., Cox, J.: The Goal: A Process of Ongoing Improvement. 3rd edn. North River Press, Great Barrington (2004)
6. Hopp, W.J., Spearman, M.L.: Factory Physics: Foundations of Manufacturing Management. 2nd edn. Irwin, Chicago (2001)
7. Klenner, F. et al.: Smart Data Analytics zur Identifikation dynamischer Engpässe in flexiblen Fertigungssystemen. at **64**(7), 540–554 (2016)
8. Kuo, C.-T., Lim, J.-T., Meerkov, S.M.: Bottlenecks in serial production lines: A system-theoretic approach. MPE **2**, 233–276 (1996)
9. Li, L., Chang, Q., Ni, J.: Data driven bottleneck detection of manufacturing systems. IJPR **47**(18), 5019–5036 (2009)
10. Nyhuis, P., Wiendahl, H.-P.: Logistische Kennlinien: Grundlagen, Werkzeuge und Anwendungen. 3rd edn. Springer Vieweg, Berlin (2012)
11. Python Software Foundation: The Python Language Reference, version 3.11, https://docs.python.org/3.11/reference/index.html, Accessed 2025/07/21

12. Roser, C., Nakano, M., Tanaka, M.: A practical bottleneck detection method. In: Peters, B.A., Smith, S.J., Medeiros, D.J. (eds.) Proceedings of the 2001 Winter Simulation Conference, pp. 949–953. IEEE, Arlington (2001)
13. Roser, C., Nakano, M., Tanaka, M.: Shifting bottleneck detection. In: Yücesan, E. et al. (eds.) Proceedings of the 2002 Winter Simulation Conference, pp. 1079–1086. IEEE, San Diego (2002)
14. Roser, C., Nakano, M., Tanaka, M.: Time shifting bottlenecks in manufacturing. In: The Abstracts of the International Conference on Advanced Mechatronics: Toward Evolutionary Fusion of IT and Mechatronics: ICAM (2004)
15. Roser, C., Lorentzen, K., Deuse, J.: Reliable shop floor bottleneck detection for flow lines through process and inventory observations. Proc. CIRP **19**, 63–68 (2015)
16. Roser, C. et al.: An enhanced data-driven algorithm for shifting bottleneck detection. In: Dolgui, A. et al. (eds.) Advances in Production Management Systems. Artificial Intelligence for Sustainable and Resilient Production Systems. APMS 2021, pp. 683–689. Springer, Cham (2021)
17. Schwenken, J. et al.: Identifikation und Prognose dynamischer Engpässe. Konzeption und Anwenderanforderungen für eine praxistaugliche Verfahrensauswahl und -weiterentwicklung. ZWF **117**(5), 294–299 (2022)
18. Skoogh, A. et al.: Throughput bottleneck detection in manufacturing: a systematic review of the literature on methods and operationalization modes. Prod. Manuf. Res. **11**(1), 2283031 (2023)
19. Subramaniyan, M. et al.: An algorithm for data-driven shifting bottleneck. Cogent Eng. **3**(1), 1239516 (2016)
20. Thürer, M. et al.: Bottleneck detection in high-variety make-to-Order shops with complex routings: an assessment by simulation. Prod. Planning Control **33**(15), 1481–1492 (2021)
21. West, N., Syberg, M., Deuse, J.: A Holistic Methodology for Successive Bottleneck Analysis in Dynamic Value Streams of Manufacturing Companies. In: Andersen, A.-L. et al. (eds.) Towards Sustainable Customization: Bridging Smart Products and Manufacturing Systems. CARV 2021, MCPC 2021. Lecture Notes in Mechanical Engineering. Springer, Cham (2022)

Demand-Based Information Exchange for Hybrid Processes

Jonas Recker(✉), Timm Schulz-Isenbeck, Sven Schümmelfeder, Alexander Keuper, and Günther Schuh

Laboratory for Machine Tools and Production Engineering (WZL) of RWTH Aachen University, Aachen, Germany
j.recker@wzl.rwth-aachen.de

Abstract. The manufacturing industry is currently facing a highly volatile, uncertain and complex competitive environment, which companies must address. This challenge has resulted in an increasing number of companies adopting agile processes, enabling them to react more quickly to changing market conditions. Existing plan-driven processes both in production as well as development are often supplemented by agile components to create so-called hybrid processes. However, the development and production of mechatronic products is becoming increasingly challenging due to the need of synchronization of parallel plan-oriented processes and the agile way of working. To facilitate the synchronization of hybrid projects, standardizing the required information flow for agile and plan-driven processes is essential. This ensures transparency in a hybrid environment and enables the traceability of decisions. Moreover, standardized information exchange provides an opportunity to establish systematic knowledge management based on the respective role, thus supporting efficient reuse of information. This paper presents a concept for systematic identification of information demands in hybrid project organizations. First, relevant activities, artifacts and roles in a hybrid project organization are described to derive superordinate project organization types. Following, interdependencies between these organization types and their related artifacts are identified. Thus, relevant information flows and requirements for each role in the hybrid environment are specified. Finally, measures tailored to specific information demands of hybrid project organizations are derived. As a result, the approach enables the systematic, demand-oriented exchange of information between different entities in hybrid production and development processes.

Keywords: Hybrid project organization · Agile knowledge management · Information exchange

1 Introduction

In the contemporary landscape of the manufacturing industry, organizations are increasingly confronted with a highly volatile, uncertain, and complex competitive environment. This dynamic context necessitates a paradigm shift in how companies approach

L. Overmeyer and B.-A. Behrens (eds.), *Production at the Leading Edge of Technology*, Lecture Notes in Production Engineering, https://doi.org/10.1007/978-3-032-19524-1_71

their processes and operations [1]. In the context of intensified competition, optimizing the company's competitive positioning is imperative. Increasing the responsiveness and adaptability of development and production processes is crucial for reducing time-to-market [2]. To remain competitive, many organizations have begun to adopt agile methodologies that empower them to respond swiftly to changing market conditions and customer demands [3].

The integration of agile components into established production and development processes presents both opportunities and challenges [4]. While agility fosters innovation and responsiveness, it also requires the synchronization with parallel plan-oriented processes [5]. Especially the development and production of mechatronic products still relies partially on plan-oriented processes due to the high level of product complexity and heavily regulated markets [6]. The software development of mechatronic products is often agile, resulting in hybrid development projects that combine the advantages of agile and plan-driven (sub) projects. However, a key challenge lies in synchronizing information. A study shows that 25% of managers' and teams' working week is spent searching for information, with 74% of managers stating that inadequate communication negatively impacts the speed and quality of work [7]. Thus, establishing a cohesive information exchange framework is critical for ensuring transparency in hybrid environments, enhancing traceability of decisions made throughout the project lifecycle [8]. By promoting efficient reuse of information, organizations can leverage existing knowledge assets to drive innovation and improve operational efficiency [5].

Therefore, there is a need for a systematic approach to identify information demands for hybrid project organizations. This paper aims to derive a conceptual demand-based information exchange framework for hybrid development and production projects.

2 Fundamentals Hybrid Product Development

The following section briefly describes the fundamentals of product development with focus on project management and information exchange.

In mechanical engineering, there are various process models for the product development process, containing methods that provide a logical structure for the development process [9]. By using such models, product developers can determine the status of projects and reflect on their approach [10]. The two most popular process models are VDI 2221 and VDI 2206, also known as the V-model. The VDI 2221 procedure is divided into seven steps, each of which leads to a technical artifact [11].

A product development process consists of the interaction of activities, roles, and technical artifacts that together enable the structured development of a product. The phases of product development include activities that transform a given input into an output, often focusing on intellectual and organizational activities [12]. An activity in the development process is defined as a coherent, goal-oriented unit of action that might include individual steps of the problem-solving process [13]. Activities can be sequential, parallel or partially iterative, and are designed to facilitate the efficient processing of information [14]. Technical artifacts are defined as elements that are created, modified, and utilized during the development process to store engineering knowledge. The documentation, implementation and verification of various aspects of product design, functionality and behavior is a key element, as is the support of interdisciplinary collaboration [15]. A role is an organizational unit that is assigned specific tasks, competencies, and responsibilities within the organizational structure to ensure structured activities and clear lines of command and communication within a company [16].

Conventional plan-driven project management is distinguished by a structured and rigid framework, and it can be described using sequential and parallel process models. In a sequential process, such as the stage-gate-process, all project steps are carried out one after the other in a defined order [9]. The parallelization of individual process steps can result in a significant reduction in process time [17]. The structured nature of conventional methodologies, coupled with adequate process documentation and effective role assignments, fosters effective project organization. However, stringent planning often results in limited flexibility, thereby constraining the adaptability to unexpected issues during the development process [18]. Contrary to conventional project management methodologies, agile project management employs flexible iterations and sprints, facilitating the rapid identification of solutions [5]. Agile methods such as Scrum or Design Thinking are commonly used in the industry [19].

To be able to synchronize development projects with different management styles efficiently, a systematic information exchange system in line with corporate objectives is required [20]. A structured information exchange system comprises the acquisition, processing and output of information and enables the linking of information to gather knowledge [21]. Information demands in product development refer to the specific need for relevant data and knowledge within activities, depending on the artifacts under development and the respective roles involved [22].

3 Related Work

The subsequent section provides a brief overview of relevant approaches in the fields of agile projects, product development and manufacturing industry. The approaches presented are examined from a variety of perspectives within the target area to emphasize any shortcomings.

A range of approaches is analyzed in the context of the classification of diverse project organizations. Reference [14] presents an approach that facilitates the identification of design-related requirements, which are instrumental to generate a development phase-specific configuration method. A comparable approach is proposed in [23], wherein the different methods, such as agile development, are made assessable with the aid of six criteria. Reference [24] identifies contextual factors, derives fundamental types, establishes causal relationships, and consolidates results into a practical framework for selecting appropriate hybrid development process types for specific projects. Existing approaches to structure project organizations have been found to neglect the key distinction between activities, artifacts, and roles, which is crucial for the accurate categorization of organizational forms. However, there is a paucity of models that systematically take these aspects into account and thus enable more accurate classification.

The classification of either activities, artifacts or roles is proposed in a variety of approaches. As demonstrated in [25, 13], there is a range of methodologies and instruments available for the systematic documentation of activities, their subsequent relationship to one another, and their classification. A set of features for the classification of projects is presented, based on the specification of activities. An analysis of artifacts is conducted by [13], who methodically aggregate information exchange demands through five key questions. The analysis is then documented using prescribed templates. The framework delineated by [25] facilitates the conceptualization of information flows between disparate roles and activities. Reference [14] undertake an analysis of the necessity for different roles based on the requirements that have been depicted. Reference [13] provide a comprehensive overview of the various domains and their interactions involved in the development process and demonstrate the potential for diverse roles and the information exchange to be structured. Existing approaches to the classification of activities, artifacts and roles are predicated on isolated, feature-oriented considerations. These often focus exclusively on individual dimensions and do not offer a comprehensive, integrated classification that considers all relevant features of each dimension.

The approaches analyzed illustrate that the exchange of information in hybrid project organizations is often complex and not standardized. The in-depth analysis of approaches presented among others reveals that structured communication is particularly crucial in hybrid contexts with changing information needs in specific activities.

4 Methodology

To address the deficits a concept for the systematic identification of information demands and measures in hybrid project organizations is proposed as follows. The first solution module consists of the feature-based description of activities, artifacts and roles, to derive the information demands. The second step presents a model that enables the identification of project organization types based on the information demands. Finally, the third solution module presents specific measures to support the exchange of information in the individual project organization types as required.

4.1 Identification of Information Demands

The first step in the methodology is to identify information demands to derive superordinate project organization types in the further course of the methodology. To identify the organizational types, a feature-based description of the roles, activities, and artifacts of agile and hybrid product development is carried out. The features and characteristics are based on the work of AYS and have been extended in accordance with the underlying object area of the methodology to be developed [24].

Figure 1 shows an exemplary overview of the features and characteristics that describe activities, artifacts, and roles of the product development process.

The identification of information demands based on the characterization of activities, roles, and artifacts thus allows the mapping of project organization types to the information demands. During this, eight different project organization types of hybrid development were identified. The following section describes the derivation of the project type based on the characteristics of the activities, roles, and artifacts.

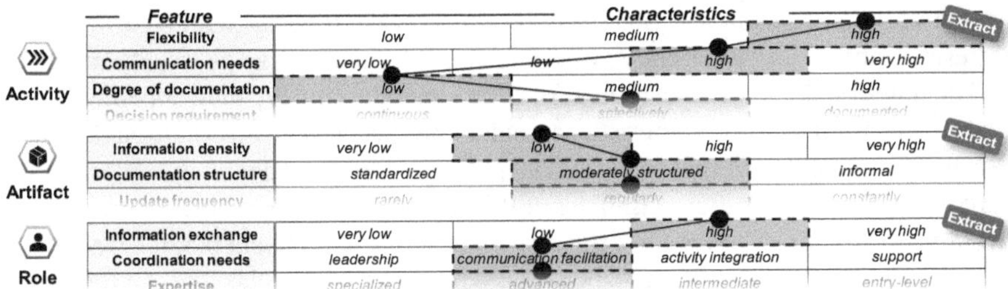

Fig. 1 Identification of the information demand based on activities, artifacts and roles

4.2 Identification of Project Organization Types

The second step involves linking activities, artifacts, and roles to derive project types for the respective combination. The linking is conducted within a conceptualized project organization cube (see Fig. 2). The dimensions of activity, artifact, and role are plotted on the axes, forming a three-dimensional evaluation space. Each axis is divided into two areas, so that the evaluation space consists of eight proportional cubes.

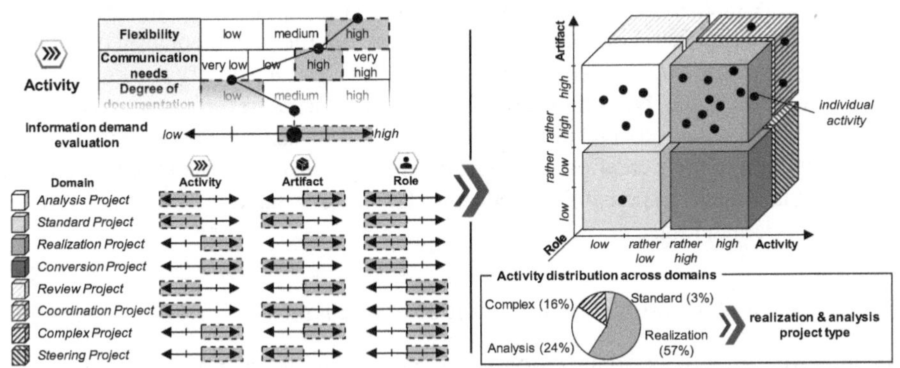

Fig. 2 The project organization cube assigns the project organization types

Each activity is evaluated according to its characteristics from the first step. Therefore, the evaluation indicates, to which project type domain each activity is assigned. The evaluation of all activities within the project then allows to derive the overall project organization type, as depicted in Fig. 2.

Prior to the evaluation of each activity, the artifacts and roles associated with an activity are prioritized. This step is necessary because it is possible that several artifacts or roles are connected to one activity. Individual artifacts are broken down into several, or several artifacts are aggregated into one. To prioritize the artifacts, a pairwise comparison between the artifacts is carried out. The most important artifact is considered decisive for further evaluation. Accordingly, prioritization of the connected roles needs to be conducted. In the second step, the information demand in each dimension is evaluated according to the different characteristics of the first solution element. To do this, values between 0 and 1 are assigned to individual characteristics, with 0 for the lowest characteristic and 1 for the highest, dividing the intermediate steps evenly across the scale. The numerical values determined for each characteristic are then added up and are divided by the number of characteristics to calculate the average information demand in the given dimension. In the third and final step, the information demands in each dimension are linked with each other within the multidimensional project organization cube. If the average information demand in a dimension is between 0.00 and 0.5, the (rather) low range is selected on the axis. Results between 0.51 and 1.00 are classified in the (rather) high range. Linking the information demands in each dimension allows classification in the evaluation space and the project organization type can be derived. Different organizational types can be derived through analyzing interactions between defined roles, conducted activities, and utilized artifacts. Firstly, a *standard project* is characterized by low requirements in all three dimensions and comprises routine, clearly structured tasks with known processes. A *conversion project* requires a high level of operational activity to carry out specific tasks without a defined solution space and a low level of responsibility and standardized results. The focus of a *coordination project* lays on coordination between participants and the organization, while content and operational activities remain minimal. The aim to create complex content with high information density is met in an *analysis project*, while at the same time the need for coordination is minimized. A *steering project* requires a high level of responsibility and coordination combined with a high level of operational execution, but the documentation of results is kept simple. In a *realization project* a high level of operational activity is combined with a defined solution space, but with little coordination effort. A *review project* is used for the professional evaluation of results and the making of strategic decisions without being actively involved in their creation. Lastly, a *complex project* combines high operational activity, management responsibility and complex results, making it interdisciplinary, dynamic and demanding.

4.3 Derivation of Measures for Information Exchange

Depending on the project organization types represented in the project, measures can be applied to provide targeted support for the development process. In their approach, SCHLATTMANN AND SEIBEL develop a systematic procedure for increasing the efficiency of teamwork, focusing on the distribution of roles and the structure and organization of meetings. The fundamental principle underpinning this approach is the activity-independent structuring of meetings, from which guidelines for various dimensions (e.g. the preparation of meetings) are derived [13]. This concept is expanded in Table 1 and applied to two topic-specific dimensions. Within the framework of this application, both the frequency of meetings and the type of information flow are specified for the eight developed forms of project organization types.

Table 1 Overview of example activities and specific measures for the identified project types

Project type	Exemplary activity	Measures
Standard project	Monthly creation of a sales report	*Meeting frequency*: Opening and closing meeting *Information flow*: Retrieve information on demand
Conversion project	Series production of a product after the release	*Meeting frequency*: Meetings in the event of disruptions *Information flow*: Direct flow of information to the relevant responsibilities, recording of production information
Coordination project	Coordination between the design team and the development team on a specific product feature	*Meeting frequency*: Several times a week (3–4 times) *Information flow*: Regular status queries and review of questions and tasks
Analysis project	Market research on new customer needs	*Meeting frequency*: Approx. 1 × per week *Information flow*: Update of artifacts
Steering project	Creation of a project schedule and capacity allocation	*Meeting frequency*: Several times a week to daily *Flow of information*: Direct flow of information in the event of changes
Realization project	Development of a functional prototype	*Meeting frequency*: Several times a week (2–3 times) *Flow of information*: Continuous flow of information on progress
Review project	Quality check of reports	*Meeting frequency*: If required, weekly *Information flow*: In-depth, documentation in the form of protocols
Complex project	Development of a new product in interdisciplinary teams	*Meeting frequency*: Daily *Information flow*: High level of detail, direct communication

5 Conclusion

In summary, this paper highlights the critical role of information exchange that arises from integrating agile and plan-driven processes in the manufacturing industry. The methodology is structured around three solution models. Initially, information demands are identified through a feature-based analysis of activities, artifacts, and roles. Subsequently, project organization types are identified accordingly. Finally, targeted measures are derived to facilitate effective information exchange tailored to specific project organization types. The results indicate that a clear typification of project activities and roles significantly enhances communication efficiency and decision-making in hybrid environments. By defining standardized information demands for various organizational types, the approach fosters transparency and traceability throughout the project lifecycle. A data-based tool has been developed that enables feature-based evaluation of activities, artifacts, and roles. The tool's functionality encompasses the capture and analysis of the relevant characteristics of each dimension, followed by systematic organization to create a detailed classification. Suitable project organization forms and the associated measures are proactively provided. This enables a precise and efficient selection of measures that are optimally adapted to the specific requirements and characteristics of a project's activity.

Future research should focus on empirical validation of the proposed methodology and the tool across diverse industries and projects to refine its applicability. Additionally, the integration of emerging technologies, such as artificial intelligence, could further streamline the information management processes within hybrid organizations.

Acknowledgements. Funded by the Deutsche Forschungsgemeinschaft (DFG, German Research Foundation) under Germany´s Excellence Strategy—EXC-2023 Internet of Production—390621612.

Competing Interests. The author(s) has no competing interests to declare that are relevant to the content of this manuscript.

References

1. Brandl, F.J., Roider, N., Hehl, M., Reinhart, G.: Selecting practices in complex technical planning projects: a pathway for tailoring agile project management into the manufacturing industry **33**, 293 (2021)
2. Gunasekaran, A.: Agile manufacturing: enablers and an implementation framework **36**, 1223 (1998)
3. Amajuoyi, P., Benjamin, L.B., Adeusi, K.B.: Agile methodologies: adapting product management to rapidly changing market conditions **19**, 249 (2024)
4. Schuh, G., Dölle, C.: Sustainable Innovation. Springer, Berlin Heidelberg, Berlin, Heidelberg (2021)
5. Riesener, M., Dolle, C., Ays, J., Ays, J.L.: Hybridization of development projects through process-related combination of agile and plan-driven approaches. In: 2018 IEEE International Conference on Industrial Engineering and Engineering Management (IEEM), p. 602. IEEE.
6. Bender, B., Gericke, K.: Pahl/Beitz Konstruktionslehre. Springer, Berlin Heidelberg, Berlin, Heidelberg (2021)
7. Atlassian: The State of Teams: The Future of Work is Here (2025)
8. Alzoubi, H.M., Yanamandra, R.: Investigating the mediating role of information sharing strategy on agile supply chain, p. 273 (2020)
9. Schuh, G.: Innovationsmanagement. Springer, Berlin Heidelberg, Berlin, Heidelberg (2012)

10. Ponn, J., Lindemann, U.: Konzeptentwicklung und Gestaltung technischer Produkte: Systematisch von Anforderungen zu Konzepten und Gestaltlösungen. Springer, Berlin Heidelberg, Berlin, Heidelberg (2011)
11. VDI e.V.: VDI 2221: Design of technical products and systems (2019)
12. Ulrich, K.: Product Design and Development
13. Schlattmann, J., Seibel, A.: Produktentwicklungsprojekte - Aufbau, Ablauf und Organisation, 2nd edn. Springer Vieweg, Berlin, Heidelberg (2024)
14. Kuhn, M. Gestaltung agiler Entwicklungsnetzwerke, 1st edn
15. Bönsch, G.C., Keuper, A., Schuh, G.: Artifact-oriented Product Development Process Structure for Implementing Model-Based Systems Engineering
16. Bullinger, H.-J.: Handbuch Unternehmensorganisation: Strategien, Planung, Umsetzung, 3rd edn. Springer (2009)
17. Bochtler, W., Laufenberg, L., Eversheim, W.: Simultaneous Engineering. Springer, Berlin Heidelberg, Berlin, Heidelberg (1995)
18. Jakoby, W.: Projektmanagement für Ingenieure. Springer Fachmedien Wiesbaden, Wiesbaden (2021)
19. Accenture, New Work SE, Statista. Digitale Trends und die neue Arbeitswelt. https://de.statista.com/statistik/daten/studie/987155/umfrage/umfrage-zur-relevanz-neuer-arbeitsmethoden-fuer-die-taegliche-arbeit/. Accessed 28 April 2025.
20. Krcmar, H.: Einführung in das Informationsmanagement, 2nd edn. Springer, Berlin Heidelberg, Berlin, Heidelberg (2015)
21. VDI e.V.: VDI 5610: Knowledge management for engineering Fundamentals, concepts, approach (2009)
22. Yassine, A.A., Sreenivas, R.S., Zhu, J.: Managing the exchange of information in product development **184**, 311 (2008)
23. Stewart, S., Giambalvo, J., Vance, J., Faludi, J., et al.: A Product development approach advisor for navigating common design methods. Process. Environ. **4**, 4 (2020)
24. Ays, J.: Kontextbasierte Integration agiler und plangetriebener Entwicklungsprozesse, 1st edn
25. Wang, R., Nellippallil, A.B., Wang, G., Yan, Y., et al.: A process knowledge representation approach for decision support in design of complex engineered systems **48**, 1 (2021)

Enhancing Systematic Problem-Solving with Large Language Models

Jan Chytraeus(✉), Yuxi Wang, and Joachim Metternich

Institute of Production Management, Technology and Machine Tools, Technical University of Darmstadt, Darmstadt, Germany
j.chytraeus@ptw.tu-darmstadt.de

Abstract. Sustainable problem-solving is a key success factor in production. The increasing complexity of production processes leads to more complex problems, which emphasizes the importance of a systematic problem-solving process (SPSP). The recurrence of previously solved problems indicates that the root cause remains unidentified due to various factors, and that problems are not solved sustainably. Recent advances in artificial intelligence (AI) and especially in the area of Large Language Models (LLMs) offer a wide range of possibilities for improving the SPSP. However, there are uncertainties regarding the current obstacles in systematic problem-solving (SPS). This article analyzes SPS through a literature review that identifies existing obstacles. It then examines how LLMs can contribute to overcoming them. Based on the results, opportunities for more efficient problem-solving are derived and a framework for an LLM-based SPSP assistant system is proposed.

Keywords: Systematic problem-solving process · Large language models

1 Introduction

The complexity of modern production processes increases the number of potential problem causes, thereby complicating problem-solving for employees [1, 2]. Furthermore, the loss of expert knowledge and methodological competence due to demographic change and skilled worker shortages also negatively impacts the efficiency of problem-solving [3, 4]. This knowledge drain particularly affects less-experienced operators and technicians, who often lack access to structured guidance, as well as engineers and managers who depend on expert input for decision-making [5]. Studies show that 60% of problems reoccur [6], indicating that often only symptoms are addressed rather than the actual root causes [7]. Sustainable problem-solving, however, requires the identification and elimination of root causes [8]. LLMs offer potential support in this context by structuring unstructured knowledge, retrieving relevant information, and providing context-aware assistance in natural language, thereby helping to mitigate knowledge drain and enhance SPS [9, 10]. Recent studies demonstrate their industrial use in tasks such as prescriptive and adaptive maintenance [11] and automated analysis of production documentation [9]. This paper refers to the widely used SPSP according to Toyota, which is a key success

L. Overmeyer and B.-A. Behrens (eds.), *Production at the Leading Edge of Technology*, Lecture Notes in Production Engineering, https://doi.org/10.1007/978-3-032-19524-1_72

factor in companies. The process comprises seven steps: *initial problem perception, clarify the problem, locate area/point of cause, investigation of root cause, countermeasure, evaluation* and *standardization* [12]. The recurring occurrence of problems highlights that SPS faces various obstacles. To identify obstacles in SPS and analyze how LLMs can improve SPS, the following research questions (RQ) are defined:

RQ 1: What obstacles exist in SPS in production?

RQ 2: How can the capabilities of LLMs be utilized to overcome the existing obstacles of SPS?

RQ 3: How can an LLM-based SPSP assistant system be designed to address the identified obstacles?

This paper contributes to current research by identifying obstacles in SPS, mapping the capabilities of LLMs to these obstacles, and proposing a conceptual framework for an LLM-based SPSP assistant system. After reviewing related work (Sect. 2) and outlining the methodology (Sect. 3), the paper presents the results, including obstacles, the mapping to LLM capabilities and the proposed framework (Sect. 4), before concluding with future research directions (Sect. 5).

2 Related Work

Various digitalization approaches exist to support employees in SPS, particularly through data analytics. Manufacturing Analytics (MA) increases process transparency through raw data analysis and accelerates the response to problems [2]. Another approach combines Case-Based Reasoning (CBR) and Bayesian Networks (BN), enabling the reuse of solutions from similar past cases and the representation of generalized knowledge. Integrating data from methods such as FMEA and 8D further enhances the system's ability to handle uncertainties and incomplete data [13]. Caglar et al. extend this approach by providing suitable problem-solving methods and a digital learning platform that offers theoretical support for method selection and application. In addition, they integrate dialogue-based elements [14]. However, these approaches primarily focus on data provision or its augmentation through methodological support and dialogue-based elements, while the design of human-centered assistant systems that actively guide employees through the entire SPSP remains largely unaddressed. An LLM-based assistant system could address this gap by providing situational and methodological support through natural language interaction [15, 16]. LLMs have already proven to be effective assistant systems in diverse domains [17]. This potential is further illustrated by recent research on Root Cause Analysis (RCA). Ahmed et al. demonstrate that fine-tuned LLMs can identify root causes and mitigation steps for cloud incidents [18]. Chen et al. present *RCACopilot*, an LLM-based system that automatically collects diagnostic data, categorizes causes, and provides explanations [19]. Freire et al. develop an LLM-based knowledge sharing system that integrates 5-Why analysis to enhance RCA [9]. These studies illustrate that LLMs can enhance RCA, suggesting that SPS in production may likewise benefit from their application. Such an integration could contribute to improving the efficiency of SPS. However, despite their potential, LLMs still exhibit limitations, including reasoning and generalization errors, hallucinations, biases, and security risks [20].

3 Methodology

3.1 Literature Review

For the present work, a PRISMA-oriented literature review was conducted [21]. Initially, an appropriate search strategy was developed. Based on the components of *RQ 1*, relevant key and search terms were defined, and corresponding search strings were created. The research was carried out in the databases ScienceDirect, Web of Science Core Collection, and IEEE Xplore, with the search strings adapted to the specific requirements. The following search string is provided as an example: (manufacturing OR production OR "shopfloor management") AND ("problem solving" OR "root cause analysis") AND (challenge OR obstacle OR barrier OR limitation). The selection of the literature was based on clearly defined inclusion and exclusion criteria. Only publications addressing obstacles in the context of problem-solving processes in production were considered. Publications published before 2015, not written in English, or without full-text access were excluded. The initial search yielded 375 publications. After all selection steps and snowballing, in which the publication year criterion was partially relaxed to also include relevant works published before 2015, eleven of these sources as well as one further identified work were used to answer *RQ 1*, resulting in a final corpus comprising twelve scientific contributions (see Fig. 1). The identified statements were thematically clustered into obstacles. The result of the literature review is presented in Sect. 4.1.

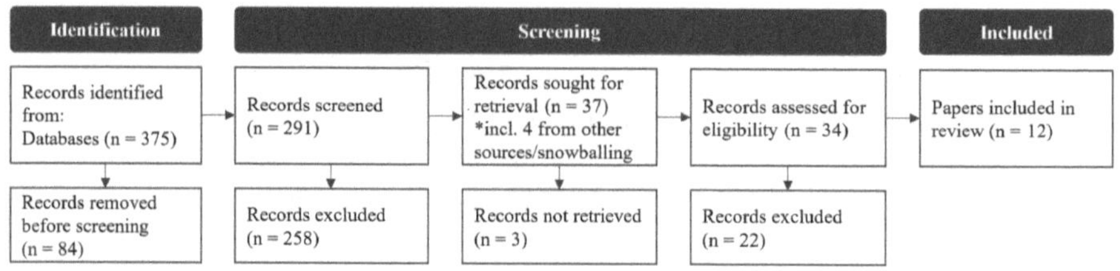

Fig. 1 PRISMA flow diagram (based on [21])

3.2 Framework Design

The framework was developed using the Design Science Research Methodology (DSRM) as proposed by Peffers et al. [22]. This paper addresses the initial steps of the methodology, namely problem identification and motivation as well as the definition of objectives for a solution, which together provide the foundation for further requirement definition in collaboration with stakeholders. In the problem analysis, obstacles in SPS were identified through a systematic literature review in order to specify the problems to be addressed and determine the areas of focus. In the objective definition, these obstacles were logically mapped to the capabilities of LLMs to illustrate how such models can help address the identified obstacles. Based on this, a conceptual system framework was developed to show how the integration of LLM capabilities could support SPS. The framework is structured in multiple layers, with each layer addressing specific obstacles by leveraging LLM capabilities and illustrating their potential contribution. This framework is introduced in Sect. 4.3.

4 Results

4.1 Identified Obstacles

The results of the literature review are presented in Table 1. The most frequently identified obstacle was "lack of methodological competence & application". Other commonly cited obstacles were "lack of time and resources" and "lack of motivation". The findings highlight the urgent need to methodologically support existing SPS and to enhance the efficiency of their execution.

Table 1 Obstacles in SPS in production

No	Obstacle	Author	Frequency
O1	Lack of methodological competence and application	[23–29]	7
O2	Lack of time and resources	[23, 25, 26, 29, 30]	5
O3	Lack of motivation	[23, 26, 29]	3
O4	Focus on symptoms instead of root cause	[23, 29, 31]	3
O5	Lack of leadership/role model function	[25, 26, 32]	3
O6	Lack of consistent and complete data	[13, 26]	2
O7	Lack of knowledge/expert knowledge	[29, 33]	2
O8	Lack of communication	[26, 29]	2

4.2 Mapping Capabilities to Obstacles

The mapping of capabilities was conducted through a logically deduced process and remains theoretical. The fundamental capabilities of LLMs in domain-specific applications can be summarized in four categories: memorization (C1), reasoning (C2), generalization (C3) and diversification (C4) [34]. Memorization (C1) is the ability to recall and reproduce specific facts, instructions, or patterns learned from training data or context (e.g., past failures, standards, manuals) [34]. In production environments, this can enhance the application of domain knowledge across various scenarios [35]. Thus, memorization can address obstacles such as a lack of methodological competence (O1), time constraints (O2), and deficits in expert knowledge (O7). Reasoning (C2) is the ability to think logically, step-by-step, draw inferences, evaluate alternatives and support decisions [34]. In deviation management, for example, the ability to reason based on workers' descriptions and contextual information can help identify possible causes, even in the absence of historical data [35]. Reasoning thus supports root cause analysis (O4) and enhances the methodological approach of employees (O1). Generalization (C3) is the ability to apply what has been learned to new and unseen situations by recognizing patterns and making analogies rather than exact matches [34]. For example, workers can use LLMs to generate or modify NC code using natural language. In addition, LLMs can help improve inputs such as problem descriptions, thereby supporting the reuse and documentation of knowledge [36, 37]. Generalization bridges data gaps (O6), improves communication (O8), assists in root cause analysis (O4), and helps to compensate for a lack of expert knowledge (O7). Diversification (C4) is the ability to produce many different outputs or ideas that support creativity, exploration, and contingency planning. For example, LLMs incorporate the function of the worker into their solution proposals and adapt the level of detail as well as the length of the response to the worker's preferences [38]. Obstacles that can be addressed through diversification are insufficient communication (O8) and a lack of motivation (O3). Overall, it can be concluded that the capabilities of LLMs address the key obstacles in SPS, except for the lack of leadership (O5) (see Table 2), thereby addressing *RQ 2*.

Table 2 Capabilities of LLMs addressing obstacles in SPS in production

No	Obstacle	C1	C2	C3	C4
O1	Lack of methodological competence & application	x	x		
O2	Lack of time and resources	x			
O3	Lack of motivation				x
O4	Focus on symptoms instead of root cause		x	x	
O5	Lack of leadership/role model function				
O6	Lack of consistent and complete data			x	
O7	Lack of knowledge/expert knowledge	x		x	
O8	Lack of communication			x	x

4.3 Conceptual Framework

To answer *RQ 3*, a conceptual framework for an LLM-based SPSP assistant system is proposed below (see Fig. 2). Current research indicates that LLM applications are often based on multi-layered architectures, which combine functionalities such as input processing, reasoning/validation, and knowledge integration [19, 39]. This design is particularly relevant for LLM-based assistants, as it enables complex interactions to remain transparent for users. Moreover, it ensures that individual components can be adapted without destabilizing the overall system [40]. Consequently, it can be concluded that a multi-layered approach is also appropriate for the proposed framework, where different functions interact closely. Building on these insights, the framework is vertically structured into three layers (Input Layer, Validation Layer, Data Layer) and horizontally aligned with the steps of the SPSP. The system logic reflects the individual steps of the SPSP and systematically guides users through each step to ensure a standardized execution of the process. The input layer provides a user interface for natural language interaction and supports multimodal inputs and outputs (e.g., image, speech, text). The validation layer employs evaluation logic to verify whether the criteria for the specific SPSP step are met. If the criteria are fulfilled, the system transitions to the next step. If not, it requests additional information until all criteria are satisfied. The criteria are derived from the requirements of the SPSP and best practices from industry. The data layer integrates a central database with knowledge sources such as a methodology library, the evaluation criteria, and a case library of solved problems. These support the validation of user input and the selection of solutions.

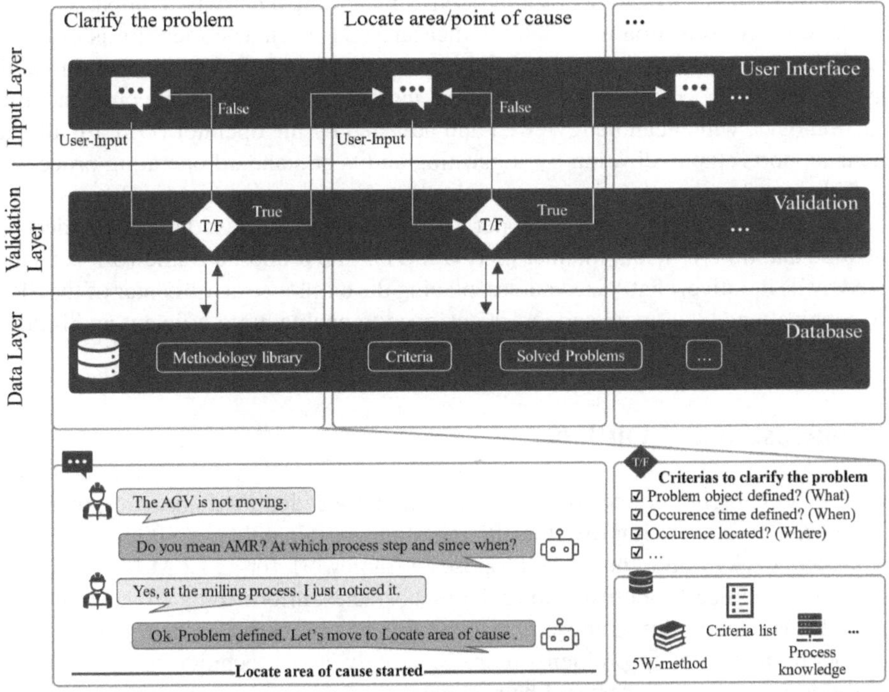

Fig. 2 Framework for an LLM-based SPSP assistant system

The following user scenario illustrates the functionality of the system in a practical context and serves as an illustrative evaluation. On a production line, a machine operator notices that no material is being delivered because an autonomous mobile robot (AMR) has stopped. The operator uses the assistant system and documents the problem (e.g., image, speech, text). The system checks the entries for completeness of the W-questions (C3). The provision of additional process knowledge (C1) and targeted queries (C2) ensure that all problems and data are recorded in a standardized, error-free, and efficient manner (O2, O6, O7) (Fig. 2). To localize the cause, the system conducts a dialogue-based is/is-not analysis with the operator. Based on process knowledge, it checks whether other AMRs are in operation and suggests verifying whether they are also affected (C1, C2). This ensures the correct methodological application, independent of the operator's competence level (O1). In the subsequent root cause analysis, the operator is shown an Ishikawa diagram pre-filled with historical AMR failures and current production data (C1, C3). The operator can expand the diagram together with the system and evaluate suggestions. This supports the distinction between symptoms and causes (O4). The application of the 5-Why method is guided (C2). The system assists in formulating questions and offers possible answers (O1). A comment function involves relevant stakeholders such as logistics staff or team leaders (C2) to facilitate cross-departmental communication (O8). For measure identification, the system suggests options (C1) such as visual markings or automatic warnings. These can be provided in different formats depending on employee preferences (C4), thereby supporting motivation (O3). For prioritization, the system offers methodological support, e.g., by guiding a pairwise comparison (C2), enabling efficient selection of a suitable measure (O1, O2). The measure is evaluated using defined criteria. For example, the system compares AMR downtime and material delivery performance with key performance indicators and derives recommendations for action (C2), which can be reviewed and adjusted by the operator (O1). Finally, the system supports standardization by suggesting audits or standard operating procedures (e.g., rules for keeping charging areas clear) based on recorded and similar cases (C1, C2). These can be extended by the operator. In this way, knowledge is provided in a structured and time-efficient manner (O1, O2, O7). The proposed framework is characterized by its iterative, chat-based nature, placing the employee at the center of the SPSP. The technical and situational support is designed to enable more efficient and intuitive problem-solving.

5 Conclusion and Outlook

Given the growing importance of SPS and its numerous identified obstacles, there is an urgent need for action to improve the efficiency and sustainability of SPS in production. Against this background, this work proposes a framework for an LLM-based assistant system that specifically addresses the existing obstacles, informed by a systematic mapping of LLM capabilities to those obstacles. Practically, such a system could support operators in documenting problems more accurately, assist technicians and experts in identifying root causes faster, and help managers standardize countermeasures across production lines, thereby reducing downtime and retaining organizational knowledge.

As a limitation, the study remains conceptual and has not yet been validated in practice. Empirical evaluation in industrial settings will be essential to assess the usability, acceptance, and effectiveness of such a system. Therefore, further empirical studies are required. For future research, there is a need to systematically determine and define the technical, functional, and user-centered requirements for the framework. In future iterations, the framework could incorporate approaches such as knowledge graphs to complement the capabilities of LLMs. This may help to address known limitations of LLMs, such as hallucinations or incorrect reasoning, and enable more transparent and domain-specific support for SPS, due to the incomplete and heterogeneous nature of domain-specific industrial data. Such approaches can organize scattered production documents, connect data from machines and processes, and provide explicit relationships that make LLM reasoning more reliable in manufacturing contexts. In particular, the criteria required for validation of user inputs at each step of the LLM-based SPSP assistant system should be defined in order to ensure the quality of LLM outputs and the overall result quality of the SPSP.

Acknowledgements. The authors gratefully acknowledge the financial support of the research project "PrePAIR" (project number 13IK037I) funded by the European Union and supported by the German Federal Ministry for Economic Affairs and Energy (BMWE) on the basis of a decision by the German Bundestag.

Competing Interests. The author(s) has no competing interests to declare that are relevant to the content of this manuscript.

References

1. Reinhart, G., Bengler, K., Dollinger, C., Intra, C. et al.: Der mensch in der produktion von morgen. In: Handbuch Industrie 4.0, Carl Hanser Verlag GmbH & Co. KG, München, p. 51 (2017)
2. Meister, M., Beßle, J., Cviko, A., Böing, T., et al.: Manufacturing Analytics for problem-solving processes in production **81**, 1 (2019)
3. Kletti, J., Rieger, J., (eds).: Die perfekte Produktion. Springer Fachmedien Wiesbaden, (2022)
4. Abele, E., Reinhart, G.: Zukunft der Produktion. Carl Hanser Verlag GmbH & Co, KG, München (2011)
5. Massingham, P.R.: Measuring the impact of knowledge loss: a longitudinal study **22**, 721 (2018)
6. Klamma, R.: Vernetztes Verbesserungsmanagement mit einem unternehmensgedächtnis-repository, (2000)
7. Jung, B., Schweißer, S., Wappis, J., (eds).: 8D—Systematisch probleme lösen, 4th edn. Hanser, (2020)
8. Maximilian Meister.: Methodik zur adaptiven Planung digital unterstützter Problemlösungsprozesse im Produktionsanlauf, (2021)
9. Freire, S.K., Wang, C., Foosherian, M., Wellsandt, S., Ruiz-Arenas, S., Niforatos, E.: Knowledge sharing in manufacturing using large language models: User evaluation and model benchmarking, (2024)
10. Lewis, P., Perez, E., Piktus, A., Petroni, F., Karpukhin, V., Goyal, N., Küttler, H., Lewis, M., Yih, W., Rocktäschel, T., Riedel, S., Kiela, D.: Retrieval-augmented generation for knowledge-intensive NLP tasks, (2020)
11. Tao, L., Huang, Q., Wu, X., Zhang, W., Wu, Y., Li, B., Lu, C., Hai, X.: LLM-R: A framework for domain-adaptive maintenance scheme generation combining hierarchical agents and RAG, (2024)
12. Liker, J.K.: Toyota way: 14 management principles from the world's greatest manufacturer. McGraw-Hill Education; McGraw Hill, New York (2013)
13. Meister, F., Khanal, P., Daub, R.: Digital-supported problem solving for shopfloor steering using case-based reasoning and Bayesian networks **119**, 140 (2023)
14. Caglar, T.R., Andrushchenko, E., Müller-Stein, L., Jochem, R.: Development and validation of a smart failure management system for SMEs in manufacturing. In: 2024 IEEE international conference on Industrial Engineering and Engineering Management (IEEM). IEEE, p. 1295 (2024)

15. Khan, M.S.U., Afzal, M.Z., Stricker, D.: SituationalLLM: Proactive language models with scene awareness for dynamic, contextual task guidance. **5,** 61 (2025)
16. Baek, J., Jauhar, S.K., Cucerzan, S., Hwang, S.J.: Research agent: Iterative research idea generation over scientific literature with large language models, (2024)
17. Naveed, H., Khan, A.U., Qiu, S., Saqib, M., Anwar, S., Usman, M., Akhtar, N., Barnes, N., Mian, A.: A comprehensive overview of large language models, (2023)
18. Ahmed, T., Ghosh, S., Bansal, C., Zimmermann, T., Zhang, X., Rajmohan, S.: Recommending root-cause and mitigation steps for cloud incidents using large language models, (2023)
19. Chen, Y., Xie, H., Ma, M., Kang, Y., Gao, X., Shi, L., Cao, Y., Gao, X., Fan, H., Wen, M., Zeng, J., Ghosh, S., Zhang, X., Zhang, C., Lin, Q., Rajmohan, S., Zhang, D., Xu, T.: Automatic root cause analysis via large language models for cloud incidents, (2023)
20. Kostikova, A., Wang, Z., Bajri, D., Pütz, O., Paaßen, B., Eger, S.: LLLMs: A data-driven survey of evolving research on limitations of large language models. (2025)
21. Page, M.J., McKenzie, J.E., Bossuyt, P.M., Boutron, I., et al.: The PRISMA 2020 statement: An updated guideline for reporting systematic reviews. Int. J. Surg. **88**, 105906 (2021)
22. Peffers, K., Tuunanen, T., Rothenberger, M.A., Chatterjee, S.: A design science research methodology for information systems research **24**, 45 (2007)
23. Alexander, D.: Getting root cause analysis to work for you, (2004)
24. Ilevbare, I.M., Probert, D., Phaal, R.: A review of TRIZ, and its benefits and challenges in practice, vol. 33, p. 30 (2013)
25. Prelovskaya, O., Iashchenko, V.V.: Problem-solving process in manufacturing of electrical and electronic components for the automotive industry: implementation and evaluation concerns. In: 2021 IEEE Conference of Russian Young Researchers in Electrical and Electronic Engineering (ElConRus), p. 1912. IEEE, (2021)
26. Antony, J., McDermott, O., Sony, M.: Revisiting Ishikawa's original seven basic tools of quality control: A global study and some new insights, vol. 70, p. 4005 (2023)
27. Pietsch, D., Matthes, M., Wieland, U., Ihlenfeldt, S., et al.: Root cause analysis in industrial manufacturing: A Scoping review of current research. Chall Promises AI-Driven Approaches **8**, 277 (2024)
28. Yin Kwok, K., Rao Tummala, V.M.: A quality control and improvement system based on the total control methodology (TCM), vol. 15, p. 13 (1998)
29. Fraunhofer-Institut für Produktionsanlagen und Konstruktionstechnik. Wie setzen Unternehmen systematische Problemlösung um und welche Herausforderungen ergeben sich dabei? https://www.ipk.fraunhofer.de/de/kompetenzen-und-loesungen/unternehmens-und-produktionsmanagement/resiliente-prozesse/systematische-problemloesung/systematische-problemloesung-in-unternehmen.html. Accessed 23 May 2025
30. Hong, W.-J., Shen, C.-Y., Wu, P.-Y.: Multi-source wafer map retrieval based on contrastive learning for root cause analysis in semiconductor manufacturing, vol. 36, p. 259 (2025)
31. Lokrantz, A., Gustavsson, E., Jirstrand, M.: Root cause analysis of failures and quality deviations in manufacturing using machine learning, vol. 72, p. 1057 (2018)
32. Gangidi, P.: A systematic approach to root cause analysis using 3 × 5 why's technique, vol. 10, p. 295 (2019)
33. Wehner, C., Kertel, M., Wewerka, J.: Interactive and intelligent root cause analysis in manufacturing with causal Bayesian networks and knowledge graphs, (2024)
34. Li, J., Yang, Y., Bai, Y., Zhou, X. et al.: Fundamental capabilities of large language models and their applications in domain scenarios: A survey. In: Proceedings of the 62nd annual meeting of the association for computational linguistics (Volume 1: Long Papers). Association for Computational Linguistics, Stroudsburg, PA, USA, p. 11116 (2024)
35. Wang, Y., Cai, Y., Chytraeus, J., Zhao, K. et al.: Retrieval-augmented generation-based decision support system for deviation analysis on the shop floor, (2025)

36. Abdelaal, M., Lokadjaja, S., Engert, G.: GLLM: Self-corrective G-code generation using large language models with user feedback, (2025)
37. Cicek, E., Wang, Y., Metternich, J.: Enhancing deviation management on the shop floor using knowledge graph based retrieval-augmented generation system 2025, accepted for publication
38. Wang, Y., Chytraeus, J., Metternich, J.: Large language models for deviation management. In: Production at the leading edge of technology, Springer Cham (2024)
39. Roy, D., Zhang, X., Bhave, R., Bansal, C., Las-Casas, P., Fonseca, R., Rajmohan, S.: Exploring LLM-based agents for root cause analysis, (2024)
40. Buschmann, F.: Pattern-oriented software architecture: A system of patterns. John Wiley & Sons, Chichester (2008)

Analysis of Asset Administration Shell-Based Negotiation Processes for Scaling Applications

David Dietrich(✉), Armin Lechler, and Alexander Verl

ISW, University of Stuttgart, Stuttgart, Germany
david.dietrich@isw.uni-stuttgart.de

Abstract. The proactive Asset Administration Shell (AAS) enables bidirectional communication between assets. It uses the Language for I4.0 Components in VDI/VDE 2193 to facilitate negotiations, such as allocating products to available production resources. This paper investigates the efficiency of the negotiation, based on criteria, such as message load, for applications with a scaling number of assets. Currently, the focus of AAS standardization is on submodels and their security to enable interoperable data access. Their proactive behavior remains conceptual and is still a subject of scientific research. Existing studies examine proactive AAS architecture examples with a limited number of assets, raising questions about their scalability in industrial environments. To analyze proactive AAS for scaling applications, a scenario and evaluation criteria are introduced. A scalable implementation is developed using current architectures for proactive AAS, upon which experiments are conducted with a varying number of assets. The results reveal the performance limitations, communication overhead, and adaptability of the AAS-based negotiation mechanism scaling. This information can improve the further development and standardization of the AAS.

Keywords: Asset administration shell · Industry 4.0 language · Scaling application

1 Introduction

The planning of production processes typically relies on central software systems, such as manufacturing execution systems (MES). However, decentralized systems are becoming increasingly popular due to factors such as scalability, flexibility, and data sovereignty. Research on agent systems in the 2000s has advanced to the internet of things and digital twins with distributed intelligence and peer interaction. The negotiation process as one possible peer interaction consists of service requester SR and service provider SP establishing a flexible horizontal cooperation for distributed tasks in a bidding procedure. Corresponding use

L. Overmeyer and B.-A. Behrens (eds.), *Production at the Leading Edge of Technology*, Lecture Notes in Production Engineering, https://doi.org/10.1007/978-3-032-19524-1_73

cases in industrial manufacturing that require decentral negotiation include plug-and-produce scenarios [1] or flexible, capability-based planning [2]. Enablers for Software-defined Value Networks with data spaces such as Catena-X[1] also form the basis for interoperable negotiations along value chains and for use cases such as cross-company production planning [3]. For a successful industrial use, current technologies to realize digital twins need scalable solutions.

1.1 Foundations

One standard, which attempts to realize digital twins is the Asset Administration Shell (AAS[2]) [4], which is specified by the Industrial Digital Twin Association [5]. The AAS is a structured collection of an asset's data throughout its entire lifecycle and serves as a single point of access to this data. Therefore, the specification contains a metamodel of the AAS, the definition of a REST-API, a package file format called AASX, and a security concept. A distinction is made between the integration of the AAS, which can be passive in a file or active on a server with the possibility to communicate with its asset, or other AAS, and services. For the communication with other AAS and services the classification of communication and presentation (CP) is used which distinguishes between a passive, active or industry 4.0 compliant communication. The term proactive AAS is commonly used to describe an AAS with CP44 compliant industry 4.0 communication with control sovereignty over its asset or other AAS.

The CP44-compliant I4.0 communication between a service requester SR and service provider SP is specified in [6,7] and displayed in Fig. 1. The specification

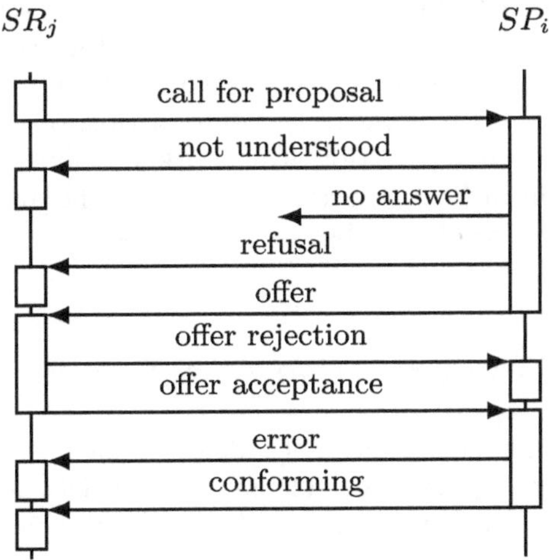

Fig. 1. Single-stage bidding procedure [7]

[1] https://catena-x.net/.

[2] For the sake of consistency, no distinction is made between singular and plural forms; the abbreviation *"AAS"* is used uniformly throughout.

is subdivided into vocabulary, the structure of message, and the interaction protocols, which, for instance, enable negotiation processes like bidding procedures. However, it lacks a formal definition for use in software tools or communication protocols.

1.2 Related Work

The concept of proactive AAS is increasingly mentioned alongside the specifications in current research. Although numerous applications use the AAS as a data model to implement business logic and interaction capabilities, only few works implement these as proactive AAS according to the standards. Belyaev and Diedrich [8] presents an architecture for an AAS that extends the passive part with a component manager, algorithms, and an interaction manager to realize the communication capabilities and logic. A model of proactive behavior with a corresponding state machine is presented in [9]. In [10], a division of the functionalities between the AAS server and a separate application, as well as the infrastructure required for communication, is discussed. Different approaches to realize the architecture, for example, monolithic or agent-based, as well as interfaces for the interaction aligned with the specified REST interface are described in [11]. A bidding process for a production planning scenario is introduced in [12] using business process models.

It is remarkable that, throughout the described evaluation use cases in the literature, only prototypical implementations with a handful of AAS are considered. Even though the theoretical approaches of the architectures concern scaling, an evaluation of these, as well as a scaling of the interaction approaches to a large number of assets and thus the suitability for an industrial scale has not yet been considered. This arises the research question: How can current approaches of proactive AAS be scaled and what performance do they have regarding their interaction—for example, in terms of negotiation time or message load?

To answer this research question, this work introduces a modeling of the proactive AAS system in Sect. 2. Section 3 presents the analyzed scenario with its implementation and results. A discussion of these is carried out in Sect. 4 and concluded in Sect. 5.

2 Modeling of the Proactive AAS System

This section addresses metrics and methodologies for describing and analyzing a scalable system of proactive AAS. The system consists of m service providers (SP_i) and n service requester (SR_j) which define roles in the negotiation process. For the negotiation of the service a description, boundary conditions, and optimization criteria are extracted from the introduced literature. Regarding the multiple interactions after another its execution needs to be included in the modeling. For simplicity, each SP refers to a set of offered services described by $\Sigma_i^{SP} \subseteq \Sigma$ and each SR refers to a single requested service $\sigma_j^{SR} \in \Sigma$ for a set of semantically described services $\Sigma = \sigma_1, \sigma_2, \ldots, \sigma_K$. This approach neglects a hierarchical composition and extensive service modeling. Other attributes of a SP are the costs c_i, the process duration d_i, and the delivered quality level q_i^{out}. For a SR, the attributes are the budget b_j, and the quality requirement q_j^{req}. Regarding the interaction of messages, each SP and SR has an expected response time $t_i^{SP,exp}$ or $t_j^{SR,exp}$ specified in [6].

A service provider SP_i is considered capable $\kappa_{ij}^c = \sigma_j^{SR} \in \Sigma_i^{SP} \wedge (q_j^{req} \leq q_i^{out})$ of processing a service request j if the descriptions and quality levels match. In addition, it is feasible $\kappa_{ij}^f = \kappa_{ij}^c \wedge (c_i \leq b_j)$ when budget and quality are met. The interaction is considered actionable $\kappa_{ij}^a = \kappa_{ij}^c \wedge (t_{now} - t_i^{last} > t^{exp}) \wedge (t_{now} - t_{iu}^{agr} > d_u)$ if no other offer is provided within the expected response time and if the work of the last agreement with preceding requester, u, is performed. Due to the condition in [6], that an offer is binding, only an actionable service provider may respond an offer to a call for proposals (CFP). The service requester in turn collects offers within t^{exp} and chooses the best according to costs c_i.

In the modeled system, a service provider can fulfill multiple requests after another. To analyze the negotiation process, SP and SR are initialized at t_{start} and simulated simultaneously. Each SR counts its CFP χ_j^{cfp}, and its success with resulting costs c_{ν_k}. The relation of the SR_k to the fulfilling SP_m is described by the mapping ν_k with $k \in 1, 2, \ldots, m$. Furthermore, each SR and SP counts its messages $\chi_{SP_i}^{mes}$ and $\chi_{SR_j}^{mes}$. The system is balanced, so the simulation ends at t_{end}, with a simulation time $\Delta t = t_{end} - t_{start}$, as soon as CFP of all SR have been fulfilled or when no feasible providers are found.

2.1 Criteria

A generic metric to measure the scalability Ψ of a distributed system is proposed by [13] as:

$$\Psi = F(\lambda_2, QoS_2, C_2)/(F(\lambda_1, QoS_1, C_1)). \tag{1}$$

F evaluates the system on two scales 1 and 2 regarding revenue λ, quality of service QoS and costs C, where F is proportional to λ/C. To obtain a scaling behavior, [13] introduces a scaling strategy based on the scaling factor k.

Applied to the AAS negotiation process, we consider revenue $\lambda = n$ as the number of service requests and costs $C = m$ as the number of service providers.

The quality of service of the system $QoS = \Delta t$ is defined as the time required to achieve a balanced system. As the evaluation of the system shall increase on a decreasing QoS the evaluation $F = \lambda/(C * QoS)$ is used. Thus from Eq. 1 results:

$$\Psi = \frac{F_2}{F_1} = \frac{\lambda_2 C_1 QoS_1}{\lambda_1 C_2 QoS_2} = \frac{n_2 m_1 \Delta t(k_1)}{n_1 m_2 \Delta t(k_2)} \tag{2}$$

To get further insights in the system, the number of average negotiations per call for proposal $\tau^{cfp} = \sum_{j=1}^{n} \chi_j^{cfp}/n$ and the average messages $\tau^{mes} = \sum_{i=1}^{m} \chi_{SP_i}^{mes}/m + \sum_{j=1}^{n} \chi_{SR_j}^{mes}/n$ are also considered as metrics. Regarding the outcome of the negotiations, we also propose the average successful requests $\tau^s = s/n$ for $s = (\sum_{k \in n | s_k = 1} 1)$ and the total costs per successful request $\tau^c = (\sum_{k \in n | s_k = 1} c_{\nu_k})/\tau^s$.

2.2 Methodology

Regarding the technology-agnostic methodology of the interaction in Fig. 1, two questions arise when systems of proactive AAS are scaled:

- How can service requesters initially find out which AAS they can request?
- According to what principles do service providers create offers?

Regarding the first question, the three network communication methods unicast, multicast, and broadcast are considered. It can be determined that current evaluation scenarios assume one or more known providers [12] or publish a broadcast CFP [10]. For an optimized multicast behavior, the common semantic service description σ can be used either for a discovery service which determines capable SP, or as a topic in the call.

Addressing the second question, current implementations process single calls without respect to others, which resembles a first-in first-out principle regarding scalability. Together, they enable the AAS scalability of the research question.

3 Implementation and Results

For the evaluation of the scalability, the following scenario, is assumed:

- Factor k is scaled binary logarithmic from 2^0 to 2^8 with twice as many service requesters $n = 2k$ as service providers $m = k$, resembling different manufacturing systems. By that Eq. 2 simplifies to $\Psi = \Delta t(k_1)/\Delta t(k_2)$.
- Out of different production capabilities $\Sigma = A, B, C, D, E$, each SR_j is described by a random service request σ_j^{SR}, while SP_i receives uniformly distributed one to three capabilities.
- Costs, work duration, and budget $c_i, d_i, b_i = \mathcal{N}(\mu_{c,d,b}, \sigma_{c,d,b})$ are normally distributed with $\mu_c = 10$, $\mu_b = 1,1 * \mu_c = 11$, $\sigma_c = \sigma_b = 1$, $\mu_d = 20s$, and $\sigma_d = 2s$.
- The required quality level $q_j^{req} \in \{0, 1, 2, 3\}$ reaches uniformly distributed from zero to three with provided quality level likewise from two to five.
- The response times $t_i^{SP,exp} = t_j^{SR,exp} = t^{exp} = 5s$ are set as equal.

Hence a multicast request based on the semantic description σ should not affect ψ but only reduce the message load τ^{mes}, only a multicast scenario with first-in first-out offer creation is analyzed regading scalability. For each factor k ten simulation runs will be conducted as sample size. The generated parameters of SP and SR in each simulation run are modeled in a passive properties of an AAS with a unique ID.

Regarding the implementation of the negotiation process in a proactive AAS, no common standard or open-source implementation was found, but are realized individually by tools like Node-red,[3] Flowable[4] or custom programs. As d_i and t^{exp} are expected to play a major role compared to runtime performance, the negotiation process is implemented in Python with MQTT for communication between the assets. The message structure is compliant with [6] and the respective interactive elements needed from the above model as topics "cfp/σ_j" for a multicast CFP and "message/AAS-ID" for responses to an existing conversation based on the ID of the target AAS. Each SR and SP with its logic, the interaction manager, an MQTT-client as messenger, and the passive AAS compliant to [5], are encapsulated in a respective docker container building a proactive AAS. An orchestration service conducts the scaling strategy using replicas in a docker swarm and collects the results.

The scenario was conducted on a virtual machine in Proxmox with Ubuntu 24.04, 90GB RAM, 32×86-64-v4 processors on a single docker swarm manager node. Its results regarding the criteria in Sect. 2.1 are displayed in Fig. 2. Figure 2a shows the mean Ψ and its standard deviation over the ten simulation runs per k on a binary logarithmic scale with $\bar{\Delta}t(k_1) = 21.7s$. According to [13], it resembles a scaling behavior which is unscalable. Regarding the other criteria in Fig. 2b, an exponential behavior in the average message load $\bar{\tau}^{mes}$ can be observed. The average success rate $\bar{\tau}^{s}$ starts at almost zero and approaches nearly one at a scale of 2^3 SP. The average CFP $\bar{\tau}^{cfp}$ increases from about one between $k = 1$ and $k = 2^3$ to four at $k = 2^7$. After a peak over 9.5 at $k = 1$ the average costs $\bar{\tau}^{c}$ decrease to eight.

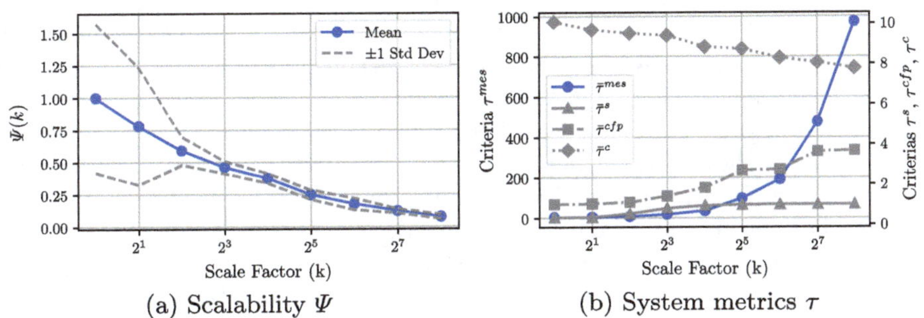

(a) Scalability Ψ

(b) System metrics τ

Fig. 2. Results the negotiation process for scaling strategy with factors k

[3] https://nodered.org/.
[4] https://www.flowable.com/.

4 Discussion

This section gives a brief discussion of the previously presented results with respect to threats to their validity and possible improvements. $\bar{\tau}^s$ shows a low success rate for scale factors lower than 2^3 and results from the random distribution of capabilities and their quality leading to a partial mismatch between SR and SP. The average CFP supports this thesis, as with around one CFP only few requests are scheduled after a first one where a SP was capable but not actionable. The increasing average CFP also shows, that for higher k a SR needs more attempts to find actionable providers. This may be due to the fact, that a SR chooses the cost-best proposal leading to a queue for other SR with less budget. This cost reduction can also be observed at $\bar{\tau}^c$. The increasing message load $\bar{\tau}^{mes}$ resembles the increasing communication overhead of even the multicast communication with plenty of answers for each CFP and may lead to network limitations for even a leightweight MQTT communication at scale. These metrics are also reflected in the unscalable behavior of Ψ with simulation times up to $360s$. In comparison, the duration of in average two requests per provider including duration and response time would lead to about $30s$.

A major influence of the performance may be the orchestration, especially when at the maximum $n_{k=8} = 2 \cdot 2^8 = 512$ service providers are started simultaneously leading to a delayed behavior. Nevertheless, all 512 SR are started within a timeout of $90s$, making a significant influence on Ψ but keeping its trend clearly intact. Further performance limitations and thus threats to validity may lie in the custom implementation like the cost optimal selection or message processing, but are expected as minor role due to the observed response time. Compared to industrial use, simplification in the service description are made. The methods to scale systems of proactive AAS from Sect. 2.2 based on multicast CFP and first-in first-out offer provision have been demonstrated successfully.

The results show various fields of action to further improve scalability of proactive AAS. On the one hand the negotiation behavior of the proactive AAS needs further respect. This includes the questions raised in Sect. 2.2 representing the common standard in the analysis. Besides the cost optimal selection of SR, also the provision of binding offers of SP needs further attention. For example, ingoing CFP within the response time may also be cached and the offer is provided for a best fit. On the other hand the scalability can be improved by architecture patterns known of decentralized software systems, such as supernodes, zones or service meshes. They could enable a subdivision of SR and SP at scale and support message flow and decision logic by mapping cached CFP to best-fit offers across zones at a system level.

5 Conclusion

This work presented an analysis of the negotiation process, as part of the industry 4.0 language in VDI/VDE 2193, at scale. It shows, that performance limitations are to be expected, when current methodologies tested with few proactive

AAS are scaled to industrial manner with hundreds of assets. With regard to the research question of how proactive AAS approaches can be scaled, a scalable behavior of SR and SP for CFP and offers was designed. Furthermore, a modeling for the proactive AAS system and target criteria to measure scalability are developed. The implementation based on a scalable docker swarm with corresponding results are introduced and discussed.

However, several aspects are identified to future work improving scalability. First, regarding the negotiation process, the behavior of proactive AAS and architectural patterns of the system need to be improved for a better negotiation an scheduling. Second, the implementation of the business logic and algorithms in a proactive AAS, which in this case was done in a custom python implementation, needs to be addressed in the further standardization. Last, the interaction elements of I4.0 Messages and their semantics need further attention for interoperable bidding procedures.

Acknowledgment. This research and development project is funded by the German Federal Ministry of Research, Technology and Space (BMFTR) within the "Research Campus Public-Private Partnership for Innovation" funding initiative (02P23Q820 ff) and managed by the Project Management Agency Karlsruhe (PTKA). The authors are responsible for the content of this publication.

References

1. Bennulf, M., Danielsson, F., Svensson, B.: A conceptual model for multi-agent communication applied on a Plug & Produce system. Procedia CIRP **93**, 347–352 (2020). https://doi.org/10.1016/j.procir.2020.04.004
2. Dietrich, D., Neubauer, M., Lechler, A., Verl, A.: Iterative Planning as a Holistic Framework for Production System-Wide Optimization Control Loops. In: Silva, F.J.G., Pereira, A.B., Campilho, R.D.S.G. (eds.) Flexible Automation and Intelligent Manufacturing: Establishing Bridges for More Sustainable Manufacturing Systems. Lecture Notes in Mechanical Engineering, pp. 611–621. Springer Nature Switzerland, Cham (2024a). https://doi.org/10.1007/978-3-031-38241-3_69
3. Dietrich, D., Zürn, M., Reiff, C., Neubauer, M., Lechler, A., Verl, A.: Software-defined value networks: motivation, approaches, and research activities. In: Bauernhansl, T., Verl, A., Liewald, M., Möhring, H.C. (eds.) Production at the Leading Edge of Technology. Lecture Notes in Production Engineering, pp. 514–524. Springer Nature Switzerland, Cham (2024b). https://doi.org/10.1007/978-3-031-47394-4_50
4. Zhang, J., Ellwein, C., Heithoff, M., Michael, J., Wortmann, A.: Digital twin and the asset administration shell. Softw. Syst. Model. (2025). https://doi.org/10.1007/s10270-024-01255-0
5. Industrial Digital Twin Association eV: Specification of the asset administration shell part 1: metamodel–IDTA number: 01001-3-0-2 (2025)
6. Verein Deutscher Ingenieure, Verband der Elektrotechnik Elektronik Informationstechnik (2020a) VDI/VDE 2193 Blatt 1: Sprache für I4.0-Komponenten: Struktur von Nachrichten

7. Verein Deutscher Ingenieure, Verband der Elektrotechnik Elektronik Informationstechnik (2020b) VDI/VDE 2193 Blatt 2: Sprache für I4.0-Komponenten: Interaktionsprotokoll und Ausschreibungsverfahren
8. Belyaev, A., Diedrich, C.: Aktive Verwaltungsschale von I4.0-Komponenten: Erscheinungsformen von Verwaltungsschalen. Automation 2019 (2019)
9. Schröder, T., Diedrich, C.: System architecture for the implementation of proactive administration shells based on cognitive science concepts; [Systemarchitektur für die Implementierung proaktiver Verwaltungsschalen auf Basis kognitionswissenschaftlicher Konzepte]. At-Automatisierungstechnik **67**(7), 599–616 (2019). https://doi.org/10.1515/auto-2019-0030
10. Grunau, S., Redeker, M., Göllner, D., Wisniewski, L.: The implementation of proactive asset administration shells: evaluation of possibilities and realization in an order driven production. In: Jasperneite, J., Lohweg, V. (eds.) Kommunikation und Bildverarbeitung in der Automation. Technologien für die intelligente Automation, vol. 14, pp. 131–144. Springer Berlin Heidelberg, Berlin, Heidelberg (2022). https://doi.org/10.1007/978-3-662-64283-2_10
11. Stolze, M., Cainelli, G., Kosel, C., Belyaev, A., Diedrich, C.: Konzepte zur Realisierung proaktiver Verwaltungsschalen und deren Kommunikation (2024a). https://doi.org/10.25673/116045
12. Stolze, M., Belyaev, A., Kosel, C., Diedrich, C., Barnard, A.: Realizing automated production planning via proactive AAS and business process models. In: IEEE International Conference on Emerging Technologies and Factory Automation, ETFA (2024). https://doi.org/10.1109/ETFA61755.2024.10710731
13. Jogalekar, P., Woodside, M.: Evaluating the scalability of distributed systems. In: Proceedings of the Thirty-First Hawaii International Conference on System Sciences, pp. 524–531. IEEE Computer Society (1998). https://doi.org/10.1109/HICSS.1998.649248

Author Index

L. Overmeyer and B.-A. Behrens (eds.), *Production at the Leading Edge of Technology*,
Lecture Notes in Production Engineering, https://doi.org/10.1007/978-3-032-19524-1

Zeitfracht Medien GmbH
Ferdinand-Jühlke-Straße 7
99095 Erfurt, Deutschland
produktsicherheit@kolibri360.de